# Physical constants and data

| | |
|---|---|
| Speed of light | $c = 2.997925 \times 10^8$ m/s |
| Gravitational constant | $G = 6.67 \times 10^{-11}$ N · m²/kg² |
| Avogadro's number | $N_A = 6.022 \times 10^{26}$ particles/kmol |
| Boltzmann's constant | $k = 1.38066 \times 10^{-23}$ J/K |
| Gas constant | $R = 8314$ J/kmol · K<br>$= 1.9872$ kcal/kmol · K |
| Planck's constant | $h = 6.6261 \times 10^{-34}$ J · s |
| Electron charge | $e = 1.60218 \times 10^{-19}$ C |
| Electron rest mass | $m_e = 9.1094 \times 10^{-31}$ kg<br>$= 5.486 \times 10^{-4}$ u |
| Proton rest mass | $m_p = 1.6726 \times 10^{-27}$ kg<br>$= 1.007276$ u |
| Neutron rest mass | $m_n = 1.6749 \times 10^{-27}$ kg<br>$= 1.008665$ u |
| Permittivity constant | $\epsilon_0 = 8.85419 \times 10^{-12}$ C²/N · m² |
| Permeability constant | $\mu_0 = 4\pi \times 10^{-7}$ N/A² |
| Standard gravitational acceleration | $g = 9.80665$ m/s² $= 32.17$ ft/s² |
| Mass of earth | $5.98 \times 10^{24}$ kg |
| Average radius of earth | $6.37 \times 10^6$ m |
| Average density of earth | 5,570 kg/m³ |
| Average earth-moon distance | $3.84 \times 10^8$ m |
| Average earth-sun distance | $1.496 \times 10^{11}$ m |
| Mass of sun | $1.99 \times 10^{30}$ kg |
| Radius of sun | $7 \times 10^8$ m |
| Sun's radiation intensity at the earth | 0.032 cal/cm² · s $= 0.134$ J/cm² · s |

inciples of Physics

**McGRAW-HILL
BOOK COMPANY**

New York
St. Louis
San Francisco
Auckland
Bogotá
Hamburg
London
Madrid
Mexico
Milan
Montreal
New Delhi
Panama
Paris
São Paulo
Singapore
Sydney
Tokyo
Toronto

FIFTH EDITION

# Principles of Physics

## F. BUECHE

Professor of Physics
University of Dayton

**Principles of Physics**

Copyright © 1988, 1982, 1977, 1972, 1965 by McGraw-Hill, Inc. All rights reserved.
Printed in the United States of America. Except as permitted under the United States
Copyright Act of 1976, no part of this publication may be reproduced or distributed in any
form or by any means, or stored in a data base or retrieval system, without the prior
written permission of the publisher.

1234567890 VNHVNH 89210987


ISBN 0-07-008892-6

This book was set in Century Old Style by Progressive Typographers, Inc. The editors
were Irene M. Nunes, Karen S. Misler, and James W. Bradley; the designer was Betty
Binns Graphics/Karen Kowles; the production supervisor was Leroy A. Young. The
drawings were done by Oxford Illustrators, Ltd. Von Hoffman Press, Inc. was printer and
binder.

Library of Congress Cataloging-in-Publication Data

Bueche, F. (Frederick) (date).
    Principles of physics.

    Includes bibliographical references and index.
    1. Physics.     I. Title.
QC23.B8496     1988          530          87-3564
ISBN   0-07-008892-6

# Contents in brief

# Contents

## 28   The physics of the very large and very small   691

## Appendixes

# Preface

**T**HIS text is designed for the noncalculus physics course taken by those who intend to pursue careers in science and its application. It can also be used profitably by intelligent nonscientists who desire a better understanding of the world in which we live. Although mathematics through simple algebra and trigonometry is used, the trigonometry required is taught in the text and the algebra as presented is not difficult. I believe firmly — and experience confirms — that the average college student is capable of mastering the material in this book without undue difficulty.

One of the major guidelines I have followed in preparing this text is that, because time is so limited, the principles of physics must receive overriding emphasis. Although relatively few in number, these principles form the framework on which we base our understanding of nature. It is important that students understand how each facet of their study is related to these principles. In that way, they see that many seemingly unrelated facts are simply different glimpses of the same fundamental concept. Physics thus becomes an understandable, related whole rather than a collection of individual facts and formulas to be memorized.

Because the principles of physics permeate the world around us, we can elicit many of them from careful observation of familiar phenomena. I try to present each fundamental principle by drawing upon the wealth of observational data each student already possesses. When students are led in this way, it is easy for them to achieve first a qualitative and then a quantitative appreciation of the principle. Although many physical concepts are quite complex until they are well understood, the skillful teacher — and that is what this textbook tries to be — can lead the student to understanding while bypassing the pitfalls that seem to obscure the concept.

No concept is completely understood until one can use it to solve problems that require both qualitative and quantitative reasoning. This text provides a multitude of **worked examples** to illustrate quantitative applications. In addition, the **questions and guesstimates** at the end of each chapter test the students' ability to reason out the meaning of challenging situations, and the end-of-chapter **problems** provide ample opportunity for quantitative practice.

You will notice that the difficulty level of each problem is indicated, with the more difficult problems being preceded by a single or a double solid square. However, the particular section to which the problem applies is not listed, although this is done in the instructor's manual. I do this for two reasons: (1) many problems require ideas from several sections, and (2) a major part of problem solving is to determine which concept to apply. I wish my students to be able to furnish this aspect of problem analysis, and I assume you wish the same for your students.

The solution methods for all problems are given in a solutions manual prepared by Joseph J. Kepes. In addition, he has prepared a workbook that gives the student additional practice.

One of the major difficulties students have in learning physics is the result of our trying to cover too much in too little time. By emphasizing principles, I have been able to divide the text into only 28 chapters, including four on modern physics. This amounts to about one chapter a week for the traditional two-semester course. If even this proves to be too formidable a menu, further abbreviations that can be instituted are pointed out in the instructor's manual.

Despite my emphasis on principles, you will find many **applications** and interesting **sidelights** scattered throughout the text. The applications are carefully screened for relevance to the principles being taught; care is taken that the principles stand out clearly and do not become lost among the extraneous material. The **special notes** interjected throughout the text are intended to extend the students' horizons, provide interesting sidelights to the material being studied, and encourage an appreciation of the historical and cultural aspects of physics. I have noticed that students often flip through the text to read these notes, which are conveniently set apart from the text proper. Students seem to relish these interesting, often self-contained excursions into optional material.

### New to the Fifth Edition

In response to extensive surveys and reviews, numerous changes in organization, coverage, and presentation have been made in this new edition of *Principles of Physics.* If you used the fourth edition and wanted some changes, look for them here; they probably have been made. A cursory examination will show that the figures and the use of color have been improved greatly. Notice how the most important statements and equations are enhanced by color and by their positioning on the page. Further emphasis is achieved by the judicious use of boldface and italic type. Many of the end-of-chapter problems are new, and most of the others use new numbers. **Minimum learning goals** serve as a summary for each chapter. They have the advantage of requiring the active participation of the student for their use, thus serving as a self-test.

One change deserves special comment: statics now precedes linear motion. Many experienced teachers have long recognized the great difficulty students experience with the topic of motion. Recent educational research confirms their observations. We have therefore returned to the "old-fashioned" format so dear to the hearts of many experienced teachers: simple statics is studied first and then linear motion is introduced, with particular care being taken to avoid misunderstandings and to remove mistaken concepts. On a more practical level, motion problems tend to encourage memorization, something we wish to deemphasize. By starting with statics, we can easily show the students that success in physics is *not* achieved through memorization.

## Acknowledgments

Many people have helped me in the preparation of this edition through comments on this and previous editions. Among them are Edward Adelson, Ohio State University; Alex Azima, Lansing Community College; Walter Benenson, Michigan State University; Edward S. Chang, University of Massachusetts at Amherst; Richard D. Haracz, Drexel University; H. James Harmon, Oklahoma State University; Edwin Kaiser, Ocean County College; Sanford Kern, Colorado State University; Willem Kloet, Rutgers University; Robert V. Lange, Brandeis University; Richard V. E. Lovelace, Cornell University; Roger D. McLeod, University of Lowell; Richard Mowat, North Carolina State University; Lewis J. Oakland, University of Minnesota; Norman Pearlman, Purdue University; Neal Peek, University of California at Davis; Frederick M. Phelps, Central Michigan University; Donald Presel, Southeastern Massachusetts University; Richard Prior, University of Arkansas at Little Rock; James Purcell, Georgia State University; P. Venugopalo Rao, Emory University; Russell Rickert, West Chester State College; Nanjudiah Sadanand, Central Connecticut State University; Phyllis Salmons, Embry-Riddle Aeronautical University; Leo Sartori, University of Nebraska at Lincoln; Sam Scales, Southern College of Technology; J. C. Sprott, University of Wisconsin at Madison; and Thomas Weber, Iowa State University.

I thank them, as well as others whose names I have inadvertently omitted, for their thoughtful suggestions. Joseph J. Kepes, who prepared the workbook and instructor's manual, has as usual been extremely helpful. I also extend my gratitude to the McGraw-Hill editorial staff, particularly to Irene Nunes, whose influence is seen throughout the text, and to James W. Bradley, who shepherded the text through the production process.

*F. Buech*

# Principles of Physics

# Introduction

S you begin your study of physics, there are probably several questions in your mind concerning the nature of this course and the field of physics in general. Let us therefore take time at the outset to make a few general comments you may find valuable.

## What is physics?

We cannot give a one-sentence answer that does justice to this question. Some typical short answers are:

**1** Physics is the study of the laws of nature and their application to nonliving things.

**2** Physics is the science of matter and energy and of the relations between them.

**3** Physics is the most basic of all sciences. It deals with the structure and behavior of matter.

**4** Physics is the body of knowledge gained from the study of natural phenomena.

**5** Physics is what physicists do.

All of these describe the field of physics, but the last answer is perhaps more informative than the others. So let us examine what physicists do.

Until the late 1800s, the field of science was termed *natural philosophy*. That is why, even today, your physics professor probably has a Ph.D. degree, a doctor of philosophy degree. Back in Aristotle's time (about 350 B.C.), no distinction was made between science and philosophy. Philosophers reasoned about all things, natural as well as supernatural. By 1600, however, when Galileo lived, philosophy had begun to divide into two channels of thought. Natural philosophers concerned themselves primarily with questions that could be answered by present or possible future experiments — the way falling objects behave, for example. The other channel became present-day philosophy and theology, with an emphasis on ethics and logic.

As time went on and as knowledge about our universe became more abundant, natural philosophy began to fragment. Those who were most interested in the way atoms react to form molecules were classified as chemists. Others devoted their time to the study of living things and were the pioneers in the field of biology. Still others studied the nature of the earth's crust, the area of study we recognize as geology. In the same way, such specialties as archaeology, architecture, medicine, and psychology branched off from the mainstream of natural philosophy.

Despite this fragmentation, it is still true that physics is basic to all other sciences. If you analyze any scientific problem to its fundamentals, you will find that its solution is founded on physical principles. This is obvious in engineering, chemistry, structural design, and geology. Because the behavior of living cells is governed by the laws of physics, biology and physiology are also connected with physics. As our understanding of living things progresses, this connection will become increasingly important. Because physics underlies all the other sciences, serious students of science are encouraged to learn the laws of physics.

Today physicists work in all branches of science and engineering. Their knowledge of physics is invaluable as they investigate the fundamental aspects of these various fields. Moreover, working with specialists in such diverse fields as medicine, geology, and enginering, physicists are able to make important contributions in the applied sciences.

 **Is physics difficult?**

It can be if you do not understand how to study it. Physics is not a set of facts and rules to be memorized. Instead, you should learn a few simple rules — the laws of physics — and try to understand their full meaning by applying them to the universe around you. Therefore, the first fact you must understand is that **last-minute cramming and memorization are fruitless ways to try to learn physics.**

Doing physics is solving the problems and riddles with which nature confronts us. At the end of each chapter in this text, you will find a number of "Questions and Guesstimates" that require you to reason out answers with little or no mathematics. These questions test your understanding of the material of the chapter and give you practice in the mental processes that are so necessary for solving problems.

Learning physics is much like learning to play a game. You must first learn the few rules of the game, and then apply them to the situations that arise while you are playing it. Of course, the more you play the game, the better you become at it. Don't expect to become a physics expert by reading a physics book. As with chess or poker or any other game, you can learn the rules by reading the book, but hands-on experience is necessary if you are to become a proficient player.

That is why we caution you to study physics in the following way. Read the text first. Think about what you are reading. (Slow readers have an advantage here.) You are reading the rules of the game. Then practice playing the game by answering the questions and solving the problems at the end of the chapter. Obtaining the answer is no more important than understanding the way it is obtained because you probably won't see this exact question or problem again. Instead, you will be tested on how well you understand the method used to solve the problem, which applies to a multitude of similar problems. Memory doesn't help much. The understanding achieved by thoughtful practice is all-important.

If you sometimes have difficulty understanding a physical concept, remember that your teachers are available to help you. Other students are also often a good source of aid. Do not use them as a crutch, however; they will not be there to help you at test time. The Study Guide that accompanies this text includes many solved examples for each chapter, and you may find it helpful in your studies.

## Many people enjoy physics. Why?

There are several reasons physicists and many of those who study physics find it enjoyable. First, it is a joy to find out how the world behaves. Knowledge of the laws of nature allows us to look on the world with a fuller appreciation of its beauty and wonder. Second, we all enjoy discovering something new. Scientists take great satisfaction in exposing a facet of nature that was previously not seen or perhaps not understood. Imagine how Columbus must have felt when he sighted America. Scientists share a similar excitement when their work results in the discovery of a new aspect of nature. Fortunately, it seems that the more we discover about nature, the more there is to discover. The excitement of discovery drives science forward. Third, most of us enjoy the successful completion of a demanding task. That is why people of all ages work puzzles. Each question or problem in science is a new puzzle to be solved. We enjoy the satisfaction of success. Fourth, science benefits humanity. A substantial fraction of those who embark on scientific work do so because they wish to contribute to the progress of civilization. Call it idealistic, perhaps, but ask yourself what medical tools we would have today without the work of countless scientists in physics, chemistry, biology, and the related sciences. Our present civilization is heavily indebted not only to those in science but also to those in the general populace who know enough about science to support its progress.

With these preliminary remarks in mind, let us begin our study of physics. Remember what we have said: don't try to learn physics by memorizing or cramming. To succeed in physics, you must become a problem solver, and you will find this acquired skill valuable in many aspects of your later life. And, if you heed these cautionary words about how to study physics, you too will enjoy the satisfaction of puzzles solved and jobs well done.

# 1 Vectors and their use

**E**ACH profession has both tools and vocabulary that are peculiar to it. For example, physicians use a stethoscope, carpenters use a level, astronomers use a telescope, and biologists use a microscope. Physicists are no exception; they use tools and words that may not be familiar to you. For that reason, we begin our study of physics by learning about some of the words and tools we shall need as we describe the world of physics.

## 1.1 Vector and scalar quantities

Whenever you measure a quantity, you express your result in terms of a number. For example, your height might be 165 centimeters (cm),* a quantity having both a numerical value, 165 (called its *magnitude*), and a *unit of measure,* centimeters in this case. Equivalently, your height could be expressed as 65 inches (in) or 5.4 feet (ft). In each case, the quantity has a magnitude and a unit of measure. Like such other quantities as the volume of a box or the number of candies in a jar, height has no direction associated with it. Quantities that have no direction associated with them are called *scalar quantities.*

Other quantities do have a direction associated with them. For example, a police officer is interested in more than just the magnitude of your car's motion on a

---

* A centimeter is about 0.394 inch.

4

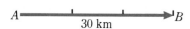

A ——————————→ B
30 km

**FIGURE 1.1**

The vector arrow indicates a displacement of 30 km east.

one-way street; the officer becomes very upset if the direction of the motion is not correct. Motion is a quantity that involves direction as well as magnitude. We might give its magnitude as 40 kilometers per hour (km/h) and its direction as eastward; the direction is needed if the motion is to be described fully. There are many other quantities that also involve direction. Among them are forces (pushes and pulls) and the movement you must undergo to travel from one city to another. These quantities are called vector quantities.

Vector quantities have direction as well as magnitude.

Let us now learn how to handle vector quantities.

One convenient feature of a vector quantity is that we can represent it by an arrow on a diagram. For example, suppose a car travels 30 km east. We say that the car has undergone a *displacement* of 30 km east. Obviously the displacement is a vector quantity. Not only does it have magnitude, 30 km, but it also has direction, east. We can represent the displacement by a vector arrow, as shown in Fig. 1.1. The arrow is made three units long to represent its 30-km magnitude and is directed eastward to show the direction of the displacement.

## 1.2 Vector addition

Everyone knows that when you add two apples and three apples, you have a total of five apples. This is an example of how scalar quantities are added. The sum of two scalars is simply the sum of their two magnitudes. Adding 40 cubic centimeters (cm³) of water to 20 cm³ of water gives 60 cm³. Again, scalar quantities add numerically.

Vector quantities do not add this way, however. Let us first illustrate this point using displacements. Remember,

The displacement from a point $A$ to a point $B$ is a vector quantity. Its magnitude is the straight-line distance from $A$ to $B$; its direction is that of an arrow that points from $A$ to $B$.

Let us consider what happens as you undergo a displacement of 30 km eastward and then 10 km northward, as shown in Fig. 1.2. We are interested in the total displacement resulting from these two displacements, namely, the displacement from $A$ to $C$. This displacement, represented by the arrow labeled $\mathbf{R}$, is called the *resultant displacement*. It is the sum of the two displacement vectors.

Obviously the resultant displacement from $A$ to $C$ is a vector and has a direction different from that of either of the original displacements. Moreover, its magnitude is certainly not 30 km + 10 km = 40 km. Instead, the pythagorean theorem for a right triangle* gives the magnitude of the resultant displacement as

$$\text{Magnitude of } \mathbf{R} = \sqrt{(10 \text{ km})^2 + (30 \text{ km})^2} = \sqrt{1000 \text{ km}^2} = 31.6 \text{ km}$$

As we see, vector addition is quite different from scalar addition.

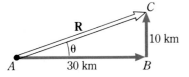

**FIGURE 1.2**

The vector diagram of a trip in which the traveler went 30 km east and then 10 km north.

---

* The pythagorean theorem for a right triangle is (hypotenuse)² = (side 1)² + (side 2)².

Often the direction of the resultant vector is as important as its magnitude. One way to find its direction is to measure the angle $\theta$ in Fig. 1.2 with a protractor. Then, if the drawing is exactly to scale, the measured angle $\theta = 18.4°$. We are able to state that the resultant displacement is 31.6 km at 18.4° north of east.

Before proceeding, we must mention how we indicate the vector nature of a quantity when a symbol is used for it. Suppose we are concerned with a displacement of 40 m directed north. Let us represent the displacement by the letter $D$. If we want to consider only its magnitude, we designate the displacement as $D$. Thus we would write in this case $D = 40$ m. If we are concerned with the displacement's direction as well as its magnitude, however, we emphasize this point by representing the displacement in boldface type: $\mathbf{D}$ (or, when writing by hand, $\vec{D}$ or $\underset{\sim}{D}$). Be careful, then. When we write a symbol in boldface type, we are telling you that it represents a vector quantity and that we wish you to pay attention to direction.

## 1.3 Graphical addition of vectors

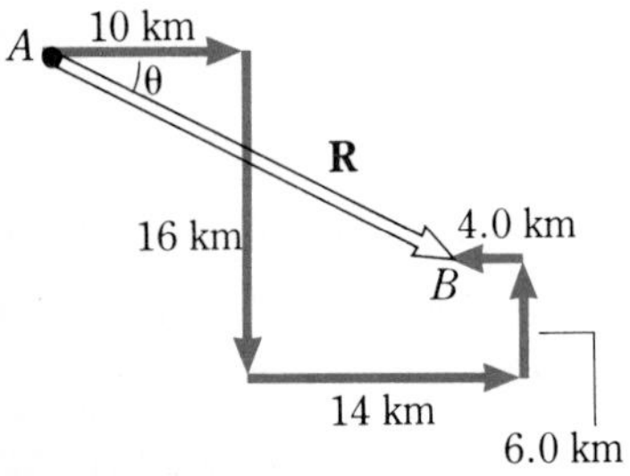

We can always find the resultant displacement of several successive displacements by using a scale diagram. This was done in Fig. 1.2 for two such displacements. Notice that the method consists of laying out the vectors to scale at appropriate angles. The tail of the second vector must be placed at the tip of the first, and, of course, the resultant points from the tail of the first vector to the tip of the second.

This method for finding the resultant, called the *graphical method,* is easily extended to more than two displacements. For example, suppose we add the following successive displacements: 10 km east, 16 km south, 14 km east, 6.0 km north, and 4.0 km west. Drawing the vectors to scale, we add them tip to tail and obtain the diagram shown in Fig. 1.3. The resultant displacement $\mathbf{R}$ extends from the tail of the first vector to the tip of the last. Be sure you understand this diagram. Measurement on the scale diagram shows $\mathbf{R}$ to have a magnitude of 22.4 km and a direction of 26.5° south of east.

Figure 1.4 shows how we use the graphical method to add two displacements that are not at right angles to each other: 10 km directed 45° east of north and 5 km directed south. As before, we lay out the vectors to scale at their appropriate angles. The resultant points from the tail of the first vector to the tip of the second.

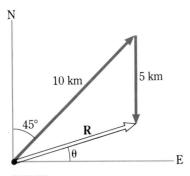

**FIGURE 1.3**

The graphical addition of five successive displacements.

**FIGURE 1.4**

A vector diagram of a trip of 10 km northeast followed by one of 5 km south.

### Example 1.1

Add the following displacements graphically:

| Displacement (cm) | 25 | 10 | 30 |
|---|---|---|---|
| Angle (degrees) | 30 | 90 | 120 |

The angles are measured relative to the east, as indicated in Fig. 1.5. It is customary to measure angles in this way.

*Reasoning*   We lay out the vector diagram as in Fig. 1.6. (It would be well for you to sketch the diagram yourself from the data given and compare your sketch with Fig. 1.6.) Measurements on the diagram show $R = 49$ cm and $\theta = 82°$.   ∎

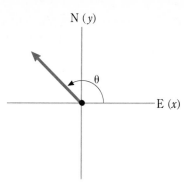

FIGURE 1.5
It is customary to measure angles relative to the east (or $x$) axis, as shown.

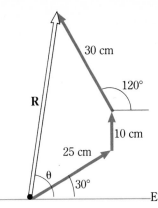

FIGURE 1.6
Addition of the displacements given in Example 1.1.

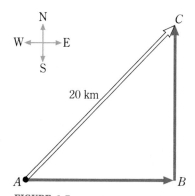

FIGURE 1.7
A displacement of 20 km northeast is resolved into component displacements **AB** east and **BC** north. Both **AB** and **BC** are components of vector **AC**.

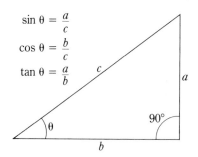

$\sin \theta = \dfrac{a}{c}$

$\cos \theta = \dfrac{b}{c}$

$\tan \theta = \dfrac{a}{b}$

FIGURE 1.8
The trigonometric functions of a right triangle.

## 1.4 Rectangular components of vectors

Although the graphical method for adding vectors is simple and straightforward, it is cumbersome and only as accurate as our scale drawings. We therefore need another method that does not have these drawbacks. Such a method exists and, as we shall see, is applicable to all vector quantities. It is called the *rectangular component method* for adding vectors. Before we describe the method, however, we must learn how to find rectangular components.

Suppose a person travels from point $A$ to a point $C$ that is 20 km northeast of $A$. The appropriate vector arrow representing this displacement is the arrow from $A$ to $C$ in Fig. 1.7. It is clear, however, that you can also go from $A$ to $C$ by the path $ABC$. In other words, you can undergo first the displacement from $A$ to $B$ and then the displacement from $B$ to $C$. The net effect is the same either way: you have undergone a displacement from $A$ to $C$. Thus we see that the displacement vector from $A$ to $C$ can be replaced by the two vectors **AB** and **BC**. We call these two vectors the **rectangular components** of the original vector. In the next section, we shall see that vectors are easily added through their rectangular components, but we must first learn how to use trigonometry to find these rectangular components.

Let us review the simple trigonometric functions of a right triangle. (See also Appendix 3.) In terms of the labeled sides of the right triangle in Fig. 1.8, these functions are defined as follows:

$$\sin \theta = \frac{a}{c} \qquad \cos \theta = \frac{b}{c} \qquad \tan \theta = \frac{a}{b} \qquad (1.1)$$

Trigonometric tables (Appendix 4) and most calculators give the values of these functions for various angles. If we know the angle $\theta$ in Fig. 1.8 and the length of the hypotenuse, we can easily find all the sides of the triangle as follows.

In Fig. 1.8, suppose $\theta = 30°$ and side $c = 30$ cm. Then we can use the definition of the sine to write

$$a = c \sin \theta = (30 \text{ cm})(\sin 30°)$$

$c_x = (20 \text{ cm}) (0.80) = 16 \text{ cm}$
$c_y = (20 \text{ cm}) (0.60) = 12 \text{ cm}$

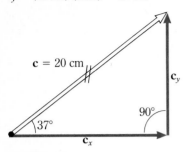

**FIGURE 1.9**
The two slash marks on the vector **c** indicate that it has been replaced by its components.

The tables of Appendix 4 tell us that $\sin 30° = 0.500$, and so

$$a = (30 \text{ cm})(0.500) = 15.0 \text{ cm}$$

Similarly, we have that

$$b = c \cos \theta = (30 \text{ cm})(0.866) = 26.0 \text{ cm}$$

Thus finding the sides of a right triangle when its hypotenuse and acute angle are known is a simple matter. Let us now apply this knowledge to finding the components of a vector.

Figure 1.9 shows a 20-cm displacement vector that makes an angle of 37° with the $x$ axis. (We shall now consider $x$ and $y$ directions instead of east and north. If you wish, you may associate $x$ with east and $y$ with north.) The original vector **c** is equivalent to the vector sum of its two rectangular components $\mathbf{c}_x$ and $\mathbf{c}_y$. We can find the magnitude of these components by using the method just presented:

$$c_x = c \cos 37° = (20 \text{ cm})(0.80) = 16 \text{ cm}$$
$$c_y = c \sin 37° = (20 \text{ cm})(0.60) = 12 \text{ cm}$$

The original 20-cm displacement at an angle of 37° to the $x$ axis is equivalent to the sum of two rectangular component vectors: $\mathbf{c}_x = 16$ cm in the positive $x$ direction and $\mathbf{c}_y = 12$ cm in the positive $y$ direction.

It is possible in this way to replace any vector by its rectangular components. Once you have learned how to do this, it is a simple matter to add (or subtract) vectors of all types. Before proceeding, however, be sure you can find the $x$ and $y$ components of the vectors shown in Fig. 1.10. Notice that the direction of each component is indicated by an algebraic sign. When we write $\mathbf{c}_x = -15$ mm, we mean that the component points in the negative $x$ direction. Similarly, $\mathbf{c}_y = 30$ mm means that the component points in the positive $y$ direction. Thus the direction of a component vector is given by the algebraic sign appended to its numerical value.

**FIGURE 1.10**
Can you show that the components of these vectors are as stated?

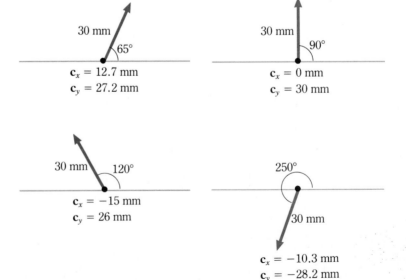

## 1.5 Trigonometric addition of vectors

Now that we know how to find vector components, it is easy to add displacements. Suppose, for example, that a bug crawling along a tabletop undergoes the displacements shown in Fig. 1.10:

30 mm at 65° to the positive $x$ axis (east)

30 mm at 90°

30 mm at 120°

30 mm at 250°

where we continue to measure angles as in Fig. 1.5.

We could find the bug's resultant displacement by using a scale diagram, as is done in Fig. 1.11, but the method is quite cumbersome in this case. A better approach is to use the components of the various vectors to find the components of the resultant. Let us call the $x$ component of the resultant $\mathbf{R}_x$. To obtain it, we simply add the $x$ components of the individual vectors, components we already found in Fig. 1.10:

$$\mathbf{R}_x = 12.7 + 0 + (-15.0) + (-10.3) \text{ mm}$$
$$= 12.7 + 0 - 15.0 - 10.3 = -12.6 \text{ mm}$$
$$\mathbf{R}_y = 27.2 + 30.0 + 26.0 - 28.2 = 55.0 \text{ mm}$$

These are the rectangular components of the resultant displacement. Notice that $\mathbf{R}_x$ is negative and must therefore point in the negative $x$ direction.

If you now refer to Fig. 1.12, you can see what $\mathbf{R}$ looks like. Its components are $\mathbf{R}_x = -12.6$ mm and $\mathbf{R}_y = 55$ mm. To find the magnitude of $\mathbf{R}$, we use the pythagorean theorem:

$$R = \sqrt{(55 \text{ mm})^2 + (12.6 \text{ mm})^2} = \sqrt{3184 \text{ mm}^2} = 56.4 \text{ mm}$$

To find the angle $\theta$ the resultant makes with the $x$ axis, we first find angle $\phi$ in Fig. 1.12. Notice that

$$\tan \phi = \frac{\text{opposite side}}{\text{adjacent side}} = \frac{R_y}{R_x} = \frac{55.0}{12.6} = 4.37$$

The tables or a hand calculator yields $\phi = 77°$. Since $\theta + \phi = 180°$ (from Fig. 1.12), we have

$$\theta = 180° - \phi = 180° - 77° = 103°$$

Figures 1.11 and 1.12 both confirm these values for $R$ and $\theta$. When you use the trigonometric method, it is wise to sketch the situation graphically to see whether your end result is realistic.

Before proceeding, we should point out an important feature of vector addition that is easily understood in terms of components. Since the only thing that matters in finding $\mathbf{R}$ is how large $\mathbf{R}_x$ and $\mathbf{R}_y$ are, it does not matter which vector we consider first. As long as we add the effects of all the vectors, the order in which we take them

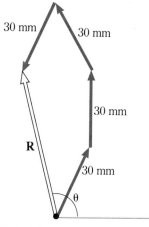

FIGURE 1.11

The resultant of the displacements in Fig. 1.10 is 56 mm at 103°.

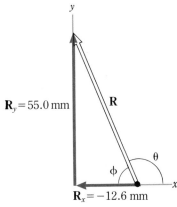

FIGURE 1.12

Compute $\mathbf{R}$ from its components.

is unimportant. This is also true for graphical addition. The diagram may be drawn using the vectors in any sequence; the resultant vector will always be the same.

---

### Example 1.2

Add the vector displacements (given in meters) in part *a* of Fig. 1.13.*

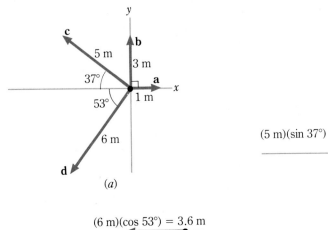

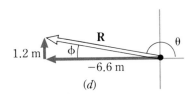

**FIGURE 1.13**

With the component method, the vectors in (*a*) can be added to give the resultant in (*d*).

**Reasoning**  We have labeled the vectors **a**, **b**, **c**, and **d**. The *x* and *y* components of **a** and **b** are obvious. We find the components of the other two vectors in parts *b* and *c* of the figure. Be sure you agree with the magnitudes given there. Notice in particular how the components are found.

Let us now tabulate the data in Fig. 1.13 in order to find $\mathbf{R}_x$ and $\mathbf{R}_y$:

|  | a | b | c | d |
|---|---|---|---|---|
| $\mathbf{R}_x$ | +1.00 | 0 | −4.00 | −3.60 |
| $\mathbf{R}_y$ | 0 | +3.00 | +3.00 | −4.80 |

We therefore have that

$$\mathbf{R}_x = 1.00 + 0 - 4.00 - 3.60 = 1.00 - 7.60 = -6.60 \text{ m}$$

$$\mathbf{R}_y = 0 + 3.00 + 3.00 - 4.80 = +1.20 \text{ m}$$

---

*·In stating data for examples and problems, we shall sometimes give numbers such as 2 m and 40 cm. You should assume that all these numbers are accurate to at least three significant figures unless they are readings from a graph, in which case more uncertainty may exist. When these numbers are used in working the examples, we carry at least three significant figures. To conserve space, however, we sometimes do not print all the digits until the answer. Where feasible, answers are given to three significant figures. If you are using a calculator, it is wise to carry along one nonsignificant figure and round off the answer.

Be sure you understand how these values are obtained.

We now use these components to sketch **R** in Fig. 1.13*d*. From the sketch,

$$R = \sqrt{(6.6 \text{ m})^2 + (1.2 \text{ m})^2} = 6.71 \text{ m}$$

Also from Fig. 1.13*d*,

$$\tan \phi = \frac{1.20}{6.60} = 0.182$$

from which $\phi = 10°$. Then, from Fig. 1.13*d*,

$$\theta = 180° - 10° = 170° \quad \blacksquare$$

---

*Exercise*   Add the two vectors in Fig. 1.13*b* and *c*.   *Answer: 7.81 m at 193°*   ■

---

**1.6** **Forces as vector quantities**

When you push or pull on an object, you *exert a force* on it. Although the concept of force is not a simple one when examined in detail, for now we can think of a force as simply a push or a pull. As our studies continue, your understanding of the concept and of the various types of forces will grow. In any case, a push or a pull—a force—has direction associated with it. In other words, **force is a vector quantity.**

The magnitude of a force must be expressed in a unit of measure. You are probably most familiar with the unit called the pound. When you stand on the floor, your feet exert a downward force on the floor that has a magnitude equal to your weight, perhaps 120 pounds (lb). In scientific work, however, we use a different unit for force. It is called the newton (N), after one of the greatest physicists of all time, Isaac Newton:

$$1 \text{ N} = 0.225 \text{ lb} \qquad 1 \text{ lb} = 4.45 \text{ N}$$

We shall learn in Chap. 4 precisely how the newton is defined. For now, it is sufficient if you can relate forces in newtons to forces in pounds. As an example, when a 30-lb child sits on your lap, you should realize that the child pushes down on you with a force of $30 \times 4.45 = 133$ N.

Since forces are vector quantities, we can represent them by vector arrows and add them, using the methods we have just learned. For example, consider the force **F** with which the man in Fig. 1.14 pulls on the wagon handle. As we see in part *b*, the force has *x* and *y* components that we can easily find. You should be able to show that $F_x = F \cos \theta$ and $F_y = F \sin \theta$.

**FIGURE 1.14**

The man pulls on the wagon handle with a force **F** that has *x* and *y* components.

(*a*)

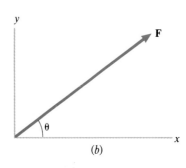

(*b*)

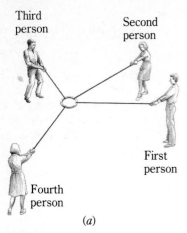

(a)

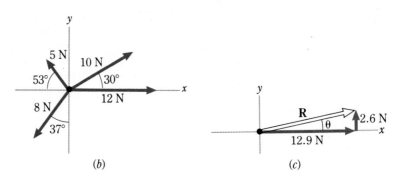

(b)                              (c)

**FIGURE 1.15**
By adding the vectors in (b), the resultant illustrated in (c) is obtained.

Often we wish to find the resultant effect of several forces pulling on the same object. To see how to do this, consider the four persons in Fig. 1.15a, who are pulling on strings attached to a ring. The forces these strings exert on the ring are shown in b. We wish to find the resultant force **R** exerted on the ring. To do this, find the x and y components and add them. We can tabulate the force components as follows:

| | Force (N) | | | |
|---|---|---|---|---|
| | 12.0 | 10.0 | 5.0 | 8.0 |
| $\mathbf{F}_x$ | 12.0 | 8.7 | −3.0 | −4.8 |
| $\mathbf{F}_y$ | 0 | 5.0 | 4.0 | −6.4 |

Be sure that you can calculate these force components from the data in Fig. 1.15b. We can now write the components of the resultant force:

$$\mathbf{R}_x = 12.0 + 8.7 - 3.0 - 4.8 = 12.9 \text{ N}$$
$$\mathbf{R}_y = 0 + 5.0 + 4.0 - 6.4 = 2.6 \text{ N}$$

The resultant force vector **R** must therefore be as shown in Fig. 1.15c.
Using the pythagorean theorem, we find

$$R = \sqrt{(2.6 \text{ N})^2 + (12.9 \text{ N})^2} = 13.2 \text{ N}$$

and the angle $\theta$ is found from Fig. 1.15c:

$$\tan \theta = \frac{2.6}{12.9}$$
$$\theta = 11.5°$$

## 1.7 Subtraction of vector quantities

Many physical situations lend themselves to analysis by vector subtraction. For example, if you walk 10 blocks east and then retrace your path by going 4 blocks west, you subtracted a 4-block displacement from a 10-block displacement. If you wish, you could say you *added* a 10-block eastward displacement and a $-4$-block eastward displacement. The resultant displacement is 6 blocks eastward in either case (Fig. 1.16).

**FIGURE 1.16**
Two equivalent ways of describing a trip consisting of a 10-block eastward displacement and a 4-block westward displacement.

With this equivalency of the two descriptions in mind, we see that the subtraction of a vector is equivalent to the addition of the same vector with its direction reversed. The following rule applies to vector subtraction:

To subtract a vector, reverse its direction and add it.

In mathematical symbols,

$$\mathbf{A} - \mathbf{B} = \mathbf{A} + (-\mathbf{B})$$

where $-\mathbf{B}$ is simply vector $\mathbf{B}$ with its direction reversed.

### Example 1.3
Subtract vector $\mathbf{B}$ from vector $\mathbf{A}$ in Fig. 1.17$a$.

*Reasoning* To subtract a vector, we reverse its direction and add it. You should be able to show that the components of these vectors are

$$\mathbf{A}_x = 8.7 \text{ m} \qquad \mathbf{A}_y = 5.0 \text{ m}$$
$$\mathbf{B}_x = -6.0 \text{ m} \qquad \mathbf{B}_y = 0 \text{ m}$$

We wish to compute $\mathbf{R}$, where $\mathbf{R} = \mathbf{A} + (-\mathbf{B}) = \mathbf{A} - \mathbf{B}$.

$$\mathbf{R}_x = \mathbf{A}_x - \mathbf{B}_x = 8.7 \text{ m} - (-6.0 \text{ m}) = 14.7 \text{ m}$$
$$\mathbf{R}_y = \mathbf{A}_y - \mathbf{B}_y = 5.0 \text{ m} - 0 \text{ m} = 5.0 \text{ m}$$

we have

$$R = \sqrt{R_x^2 + R_y^2} = \sqrt{(14.7 \text{ m})^2 + (5.0 \text{ m})^2} = 15.5 \text{ m}$$

Can you show that the angle $\mathbf{R}$ makes with the $+x$ axis is given by

$$\tan \theta = \frac{5.0}{14.7} = 0.34$$

from which $\theta = 19°$? We check our answer graphically using the diagram in Fig. 1.17$b$. Notice that we reverse the direction of $\mathbf{B}$ and add it. ∎

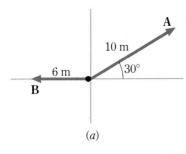

(a)

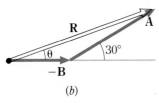

(b)

**FIGURE 1.17**
To find $\mathbf{A} - \mathbf{B}$, reverse the direction of $\mathbf{B}$ and add it to $\mathbf{A}$.

# *Sailing into the wind*

When most of us think about the motion of a sailboat, we assume that the boat moves in the direction of the wind, as shown in (*a*) of the accompanying illustration. A sailboat can sail against the wind, however, by following a zigzag motion called tacking. Now that we know how to resolve force vectors into components, we can understand this puzzling type of motion.

As an approximation, let us assume the sail to be a flat surface. The wind blowing against the sail exerts a force F on it if it is perpendicular to the wind, as in (*a*). If the sail is not perpendicular to the wind, only the component of F perpendicular to the sail exerts a force on it. Thus we have the force $F_s$ pushing on the sail in (*b*) and (*c*). This is the force that propels the boat.

If you examine a sailboat closely, you will notice a finlike plane, the keel, extending from its bottom. This keel prevents the boat from moving sideways through the water—at least to a first approximation—and causes the boat to always move along its lengthwise direction. Because of this, only the component of $F_s$ that is lengthwise to the boat causes motion. The sideways component of $F_s$ is rendered ineffective by the keel.

Look now at (*b*) and (*c*). In each case, the boat moves in the direction of $F_m$, the lengthwise component of $F_s$. As a result, it moves perpendicular to the wind in (*b*) and into the wind in (*c*). Although the direction of movement can never be exactly opposite the wind direction, a sailboat can, by following a zigzag path—tacking—achieve a resultant motion that is straight backward. These same principles apply to sail-driven iceboats and to roller skaters carrying sails.

No boat going in the direction of the wind can exceed the wind speed, but it is possible for a boat going perpendicular to the wind, or even against it, to exceed the wind speed. Can you explain why?

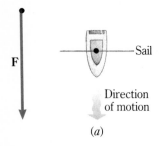

WIND

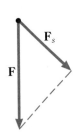

F — Sail

Direction of motion

(*a*)

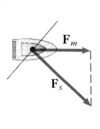

$F_s$

F

$F_m$

$F_s$

(*b*)

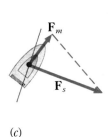

$F_s$

F

$F_m$

$F_s$

(*c*)

---

## 1.8 Metric prefixes and scientific notation

Those of you who have done any cooking know that recipes are often written in terms of teaspoons, tablespoons, ounces, cups, and so forth. This becomes very frustrating because a cook must remember all sorts of conversion factors: how many teaspoons in a tablespoon, how many ounces in a cup, and on and on. The system of measurement used by scientists avoids this confusing situation. It is based on the metric system, in which the units are related to each other by powers of 10. The complete system of units is called (in French) the *Système International,* or SI for short. We shall discuss this system in detail in Chap. 4. For now, we are interested only in how it relates units of the same type, for example, units of length.

Instead of using many different names, such as foot, mile, and inch, to represent lengths, the international system uses *prefixes* (Table 1.1). Although the SI standard unit of length is the meter, many other length units can be obtained by using these prefixes. From the table, we see that the prefix *kilo-* stands for 1000. Therefore a kilometer (km) is 1000 m. Similarly, a millimeter (mm) is $\frac{1}{1000}$ m. These and the other prefixes carry over to all types of measure. A kilogram is 1000 grams, a kilobuck is 1000 dollars, and a kilowatt is 1000 watts. A nanosecond is $\frac{1}{1,000,000,000}$ second, and a megawatt is 1,000,000 watts. Since it is clear from the prefixes how the various units are related, you need learn only the prefixes to know all the units used to measure a quantity; you know at once that a milligram is $\frac{1}{1000}$ gram, for example.

The numbers used in science are often very small or very large. For example, the diameter of an atom is only about 0.0000000002 m, and the age of the universe is about 15,000,000,000 years. Rather than use such cumbersome numbers, scientists make use of a shorthand called *scientific notation.* To see its basis, notice that

$$10 \times 10 \times 10 = 10^3$$

$$\frac{1}{10} \times \frac{1}{10} \times \frac{1}{10} = \frac{1}{10^3} = 10^{-3}$$

Therefore we can write 1000 as $1 \times 10^3$ and 0.001 as $1 \times 10^{-3}$. Similarly, since $6 \times 10^7$ means to multiply 6 seven times by 10, we have that

---

■ **TABLE 1.1** **Prefixes and scientific notation**

| *Prefix* | *Symbol* | *Value* | *Notation* |
|---|---|---:|---:|
| tera | T | 1,000,000,000,000 | $10^{12}$ |
| giga | G | 1,000,000,000 | $10^{9}$ |
| mega | M | 1,000,000 | $10^{6}$ |
| kilo | k | 1,000 | $10^{3}$ |
| hecto | h | 100 | $10^{2}$ |
| deka | da | 10 | $10^{1}$ |
|  |  | 1.0 | $10^{0}$ |
| deci | d | 0.1 | $10^{-1}$ |
| centi | c | 0.01 | $10^{-2}$ |
| milli | m | 0.001 | $10^{-3}$ |
| micro | $\mu$ | 0.000001 | $10^{-6}$ |
| nano | n | 0.000000001 | $10^{-9}$ |
| pico | p | 0.000000000001 | $10^{-12}$ |
| femto | f | 0.000000000000001 | $10^{-15}$ |
| atto | a | 0.000000000000000001 | $10^{-18}$ |

$6 \times 10^7 = 60{,}000{,}000$. As other examples, $3.1 \times 10^5 = 310{,}000$ and $0.0563 \times 10^4 = 563$. Similarly, since $6 \times 10^{-7}$ means to divide 6 seven times by 10, $6 \times 10^{-4} = 0.0006$, $373 \times 10^{-5} = 0.00373$, and $5.6 \times 10^{-3} = 0.0056$. Be sure you understand how each of these equivalent forms is found. Test your understanding by solving the following examples. In doing so, recall that $10^0 = 1$.

*Exercise* Fill in the blanks: (*a*) $56{,}321{,}000 = 5.6321 \times \underline{\hspace{1cm}} = 56{,}321 \times \underline{\hspace{1cm}}$; (*b*) $43.7 \times 10^5 = \underline{\hspace{1cm}} \times 10^0$; (*c*) $0.0000516 = 5.16 \times \underline{\hspace{1cm}}$; (*d*) $6.15 \times 10^{-2} = \underline{\hspace{1cm}} \times 10^0$ *Answer: (a) $10^7$, $10^3$; (b) 4,370,000; (c) $10^{-5}$; (d) 0.0615* ∎

---

## 1.9 Algebra involving scientific notation

These are a few simple rules that apply to algebraic manipulations in scientific notation.

### Addition and subtraction

As you know very well, when you add or subtract numbers such as 271.05 and 0.17, you align the decimal points:

$$
\begin{array}{r}
271.05 \\
+\quad 0.17 \\
\hline
271.22
\end{array}
\qquad
\begin{array}{r}
271.05 \\
-\quad 0.17 \\
\hline
270.88
\end{array}
$$

This same requirement carries over to numbers in scientific notation, where it means that the numbers to be added or subtracted must be expressed as the same power of 10. Therefore, to add $3.5 \times 10^3$ and $4.80 \times 10^4$, we must express them as the same power of 10:

$$
\begin{array}{r}
3.5 \times 10^3 \\
+48.0 \times 10^3 \\
\hline
51.5 \times 10^3
\end{array}
\qquad \text{or} \qquad
\begin{array}{r}
0.35 \times 10^4 \\
4.80 \times 10^4 \\
\hline
5.15 \times 10^4
\end{array}
$$

Similarly, $3.5 \times 10^3 - 4.80 \times 10^4 = -4.45 \times 10^4$.

*For addition and subtraction, the powers of 10 must be the same.*

*Exercise* Evaluate $857 + 2.050 \times 10^3$, $0.0375 + 65.0 \times 10^{-3}$, $0.717 \times 10^4 - 2.8 \times 10^2$. *Answer: $2.907 \times 10^3$, $102.5 \times 10^{-3}$, $0.689 \times 10^4$* ∎

---

### Multiplication

To multiply numbers expressed in scientific notation, we make use of the algebraic rule for exponents: $10^a \times 10^b = 10^{a+b}$. Thus

$$(3 \times 10^5) \times (4 \times 10^8) = 3 \times 4 \times 10^5 \times 10^8 = 12 \times 10^{13}$$

$$(1.60 \times 10^3) \times (0.40) \times (5.0 \times 10^{-2}) = (1.60) \times (0.40) \times (5.0) \times 10^3 \times 10^{-2}$$
$$= 3.2 \times 10^1$$

Notice that, unlike in addition and subtraction, the powers of 10 in multiplication can be different.

### Division

We make use of the fact that $1/10^a = 10^{-a}$ to obtain

$$\frac{8 \times 10^4}{2 \times 10^3} = \frac{8}{2} \times \frac{10^4}{10^3} = \frac{8}{2} \times 10^4 \times 10^{-3} = 4 \times 10^1$$

$$\frac{16 \times 10^5}{2 \times 10^{-3}} = 8 \times 10^5 \times 10^3 = 8 \times 10^8$$

*Exercise*  Evaluate the following (recall that $10^0 = 1$): $(0.91 \times 10^{-3}) \times (6.0 \times 10^5)$,  $(0.91 \times 10^{-3}) \div (6.0 \times 10^5)$,   $820 \div (4.0 \times 10^{-3})$,   $(7.0 \times 10^4) \div (3.5 \times 10^{-6})$.  *Answer: $5.46 \times 10^2$, $0.152 \times 10^{-8}$, $205 \times 10^3$, $2.0 \times 10^{10}$*  ∎

---

### Example 1.4
A strand of hair is 20 cm long and has a radius of 0.030 mm. What is its volume?

*Reasoning*  The hair is a cylinder of length $L = 0.20$ m and radius $r = 0.030 \times 10^{-3}$ m. For a cylinder, $V = \pi r^2 L$, and so

$$V = (3.1416)(3.0 \times 10^{-5} \text{ m})^2 (0.20 \text{ m})$$

$$= (3.1416)(9.0 \times 10^{-10} \text{ m}^2)(0.20 \text{ m}) = (3.1416)(1.80 \times 10^{-10} \text{ m}^3)$$

$$= 5.65 \times 10^{-10} \text{ m}^3 \quad ∎$$

---

*Exercise*  We shall see in Chap. 10 that the mass, in kilograms, of an iron atom can be found from the computation

$$\text{Mass} = \frac{7.86 \times 10^3 \text{ kg}}{55.8 \times 6.02 \times 10^{26}}$$

Evaluate this expression.  *Answer: $2.34 \times 10^{-25}$ kg*  ∎

---

## Minimum learning goals*

When you finish this chapter, you should be able to

1  Define (*a*) scalar quantity, (*b*) vector quantity, (*c*) displacement vector, (*d*) rectangular component.

2  Find the resultant of several displacement vectors by the graphical method.

3  When a displacement and its angle are given, find its $x$ and $y$ components.

4  Find the magnitude and angle of a vector from its $x$ and $y$ components.

5  Use the trigonometric method to add several vectors.

6  Subtract one vector from another.

7  State the power of 10 that corresponds to a given metric prefix.

8  Change a force in pounds to its equivalent in newtons, and vice versa.

9  Transform numbers in scientific notation to their common form, and vice versa.

10  Add, subtract, multiply, and divide numbers expressed in scientific notation.

---

* Notice the word *minimum. All* students should achieve these goals.

## Questions and guesstimates

**1** When you lie down on your bed tonight, what will be the resultant displacement your body has undergone since you rose this morning?

**2** Two helicopter landing pads are a few kilometers apart. A woman takes off from one in a helicopter and eventually lands at the other. In the meantime, her husband walks from one to the other. Compare the wife's total displacement with that of the husband.

**3** The sum of two vectors is zero. What can you conclude about their rectangular components?

**4** Two displacements **A** and **B** are added. What must be the relation of **A** to **B** if the magnitude of their sum is (*a*) *A + B*; (*b*) zero?

**5** A force **A** acts at 130° while another, **B**, acts at 50°. What must be the relation between their components if their resultant is to be at 90°?

**6** What must be the components of a force **A** if it is to balance simultaneously a force **B** in the *x* direction and a force **C** in the *y* direction?

**7** Estimate the total resultant displacement you have undergone in the last (*a*) 1.5 h; (*b*) 24 h.

**8** What are some physical situations in which vector quantities are subtracted? Can these quantities be thought of as being added rather than subtracted?

**9** Represent each person in a city of 200,000 by a vector extending from toe to nose. Estimate the resultant of these vectors at (*a*) noon; (*b*) midnight.

**10** Vector **A** lies in the *xy* plane. What range can its angle $\theta$ take on if (*a*) its *x* component is negative? (*b*) its *x* and *y* components are both negative? (*c*) its *x* and *y* components have opposite signs?

## Problems

*The problems at the end of every chapter are divided into three levels of difficulty: standard, somewhat difficult (indicated by one square ■), and most difficult (two squares ■). Problems labeled (G) should be solved graphically. All others should be solved analytically. All angles are measured as in Fig. P1.1.*

**1** To go from my house to a certain store, I must go four blocks north and six blocks west. What is my resultant displacement (magnitude and angle) as I make this trip? (G)

**2** Find the resultant displacement of a car that goes 20 km west and then 16 km south. (G)

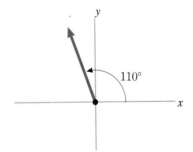

**FIGURE P1.1**

**3** In being positioned on a milling machine table, a device is given the following displacements: 5.0 cm at 0°, 12.0 cm at 80°, 7.0 cm at 110°, 9.0 cm at 210°. Find the magnitude and angle of the resultant displacement. (G)

**4** A treasure map says, "Start at the large tree. Go 80 paces straight east, then 50 paces at 70° north of east, then 60 paces at 30° west of north, then 18 paces straight south to the treasure." How far from the tree and at what direction is the treasure? (G)

**5** Point *B* is 300 km from point *A* and at an angle of 50° north of east. A road goes 40 km east from *A* and then ends. Starting at the end of the road, how far and at what angle to the east must one travel in order to reach *B*? (G)

■ **6** To go from St. Louis to Miami, a plane must fly 1670 km at 47° south of east. To go from Ottawa to Miami, it must fly straight south for 2060 km. How far and in what direction must the plane fly to go from St. Louis to Ottawa? (G)

**7** To go from Miami to Chicago, a plane must fly about 1780 km at 30° west of north. How far north of Miami is Chicago? How far west?

**8** A force of 150 N pulls at an angle of 110° measured counterclockwise to the positive *x* axis. What is the *x* component of this force? The *y* component?

**9** A force of 50 N pulls in the *xy* plane at an angle of 40°. Find its *x* and *y* components. Repeat for angles of 200° and 310°.

**10** A displacement of 20 m is made in the *xy* plane at an angle of 70°. Find its *x* and *y* components. Repeat for angles of 120° and 250°.

**11** A certain force has a magnitude of 400 N, and its $x$ component is $-160$ N. Find its $y$ component and its direction. Two possible answers exist. Find both.

**12** In an $xy$ coordinate system, point $P$ is 50 cm from the origin and its $y$ coordinate is $-18$ cm. Find $P$'s $x$ coordinate and its direction from the origin. There are two possible answers. Find both.

**13** An object undergoes the following successive displacements in the $xy$ plane: 25 cm at $\theta = 0°$ and 45 cm at $\theta = 110°$, with all angles measured counterclockwise from the positive $x$ axis. Find its resultant displacement.

**14** Suppose you start at point $A$ and walk first 350 m south and then 230 m at 30° west of north to point $B$. Find the displacement from $A$ to $B$.

**15** Find the resultant of the forces in Fig. P1.2. Give the counterclockwise angle relative to the positive $x$ axis.

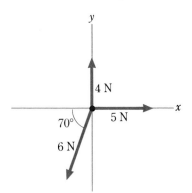

FIGURE P1.2

**16** Find the resultant of the forces in Fig. P1.3. Give the counterclockwise angle relative to the positive $x$ axis.

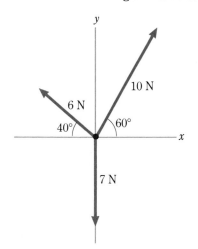

FIGURE P1.3

**17** Solve Prob. 4 using trigonometry.

**18** Solve Prob. 3 using trigonometry.

**19** Solve Prob. 6 using trigonometry.

**20** Solve Prob. 5 using trigonometry.

**21** What third force must be added to the following two forces to give a zero resultant: 30 N at 70° and 80 N at 135°?

**22** What third force must be added to the following two forces to give a zero resultant: 30 N at 120° and 20 N at 200°?

**23** A mine shaft goes straight down for 60 m. From its lower end, a horizontal tunnel goes 50 m west and then 30 m north to its end. How far is the end of the tunnel from the top of the shaft? What angle does a straight line from the top of the shaft to the end of the tunnel make with the vertical?

**24** A room has a ceiling 2.5 m high, and the floor is 3 m × 5 m. Find the length of the diagonal line from a ceiling corner to the opposite floor corner. What angle does the line make with the floor?

**25** A force **A** is added to a force that has $x$ and $y$ components 3 N and $-5$ N. The resultant of the two forces is in the negative $x$ direction and has a magnitude of 4 N. Find the $x$ and $y$ components of **A**.

**26** We wish to add vector **B** to vector **A** so as to produce a resultant vector in the $y$ direction that has a magnitude of 15 m. If **A** is 30 m at 180°, what should be the components of **B**?

**27** Two displacements **A** and **B** lie in the $xy$ plane. Displacement **A** is 25 cm at 130° to the positive $x$ axis, and **B** is 15 cm in the negative $x$ direction. Find the displacement that results from subtracting **A** from **B**.

**28** Two forces **A** and **B** act in the $xy$ plane. Force **A** is 50 N and directed along the positive $y$ axis. Force **B** is 40 N and directed at 30° below the positive $x$ axis. If **B** is subtracted from **A**, what is the resultant force?

**29** Using scientific notation, write the following lengths in meters with one digit to the left of the decimal, as in $3.65 \times 10^4$: (*a*) 604 km; (*b*) 0.163 mm; (*c*) 47 nanometers (nm); (*d*) 0.00295 micrometer ($\mu$m); (*e*) 73 megameters (Mm).

**30** Using scientific notation, write the following masses in grams with one digit to the left of the decimal, as in $6.21 \times 10^{-3}$: (*a*) 241 kg; (*b*) 20.6 nanograms (ng); (*c*) 0.0051 microgram ($\mu$g); (*d*) 17.3 gigagrams (Gg); (*e*) 0.135 picogram (pg); (*f*) 891,000 milligrams (mg).

**31** Carry out the computation $(6.21 \times 10^{-3}) \times (3.74 \times 10^4) \div (5.40 \times 10^6)$. Write your answer in scientific notation with one digit to the left of the decimal.

**32** Carry out the computation $(8.60 \times 10^5) \times (6.17 \times 10^{-2}) \div (1.79 \times 10^{-4})$. Write your answer in scientific notation with one digit to the left of the decimal.

**33** Carry out the computation $(2.1 \times 10^4) \div 0.006543$, giving your answer in scientific notation.

**34** Carry out the computation $0.026 \div 4724$, giving your answer in scientific notation.

**35** Evaluate the expressions (*a*) $(6.6 \times 10^{-34})(3 \times 10^8)/(438 \times 10^{-9})$; (*b*) $(6.025 \times 10^4)^{1/2}$; (*c*) $(2 \times 10^{-3})^4/(5 \times 10^6)$; (*d*) $\sqrt{4 \times 10^5}$.

**36** Evaluate the expressions (*a*) $(6.20 \times 10^3)^{-5}$; (*b*) $\sqrt{3 \times 10^5}$; (*c*) $(8 \times 10^{-3})^{1/3}$; (*d*) $(8 \times 10^4)^{1/3}$.

**37** When displacement **B** is added to displacement **A**, it gives a displacement **C** that has $\mathbf{C}_x = -3$ m, $\mathbf{C}_y = +6$ m, and $\mathbf{C}_z = +4$ m. Displacements **A** and **B** are identical in direction, but the magnitude of **A** is only half that of **B**. Find the components of **A**.

**38** A 500-N force and a 700-N force are perpendicular to each other and pull on a single object. Find the third force that must act on the object if the resultant is to be zero. Show the relative directions of the forces on a diagram and evaluate the pertinent angles.

**39** The cubical box in Fig. P1.4 is subject to a diagonal force of 50 N. Three forces of equal magnitude, each parallel to a side of the cube, are to be applied at the coordinate origin. What are the directions and magnitudes of the smallest forces that will balance the 50-N force?

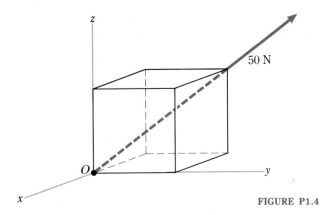

FIGURE P1.4

# 2 Static equilibrium

$\mathbf{A}$N important part of physics has to do with objects and systems that are at rest and remain at rest. This portion of physics is called **statics.** Not only is this branch of mechanics important to us throughout our study of physics, but it also affords us an opportunity to consolidate our knowledge of vectors. Thus we study statics in this chapter and discover that two basic conditions must be satisfied if an object is to remain at rest. We learn how to use these conditions and then become acquainted with their chief consequences.

## 2.1 Objects in equilibrium

What conditions are necessary for an object to remain at rest? This is the question we discuss in this chapter. An object that is at rest and remains at rest is said to be in **static equilibrium.** Once we know the conditions for equilibrium — and there are only two of them — the analysis of situations involving objects at rest is quite straightforward. It consists of applying these two conditions to the object in question.

To begin our discussion, consider the situation in Fig. 2.1$a$: an object supported by a hand. Everyone knows why the object does not fall, but let us examine this very simple situation in detail so that we may easily understand more complicated situations.

What forces act on the object? We know that if it is released, the object will fall. This tells us that a force is pulling downward on it. The object falls because the earth

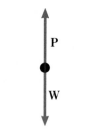

(a) Actual situation

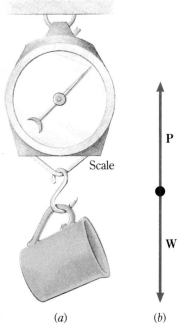

P

W

(b) Free-body diagram

**FIGURE 2.1**
The object does not move because
the resultant force on it is zero.

Scale

P

W

(a)                    (b)

**FIGURE 2.2**
The free-body diagram for the
object hanging from the scale is
shown in (b).

pulls downward on it with a gravitational force, a force we discuss in detail in Chap. 7. This gravitational pull that the earth exerts on an object is called the **weight W** of the object:

Each object near the earth is pulled toward the center of the earth by a force we call the weight of the object.

It is the weight of the object in Fig. 2.1a that would cause it to fall if the hand were not supporting it.

To support the object, the hand must push upward on it. Let us represent this upward push by **P**. Experiment tells us that the object will remain motionless (that is, in static equilibrium) if **W** and **P** are equal in magnitude. In other words, the vertical forces acting on the object must balance if equilibrium is to be achieved.

In the analysis of objects at equilibrium, it is helpful to sketch a *free-body diagram* for the system. This type of diagram shows the object and the forces acting on it in the simplest possible way. The free-body diagram for our object-hand system is shown in Fig. 2.1b. The object is represented by the dot; the forces acting on the dot, **P** and **W** in this case, are the forces that act directly on the object. At equilibrium, these vertical forces must cancel; in other words, their vector sum must be zero.

As another example of static equilibrium, consider the situation shown in Fig. 2.2a: an object hanging at the end of a scale. Since it is the object we are interested in, we discuss only those forces acting on it, which are shown in the free-body diagram of Fig. 2.2b. Notice that, for the purpose of this discussion, we have isolated the object hanging from the hook. There are many forces we are not concerned with here—the force that supports the scale, for example. We are interested only in those forces that act directly on the object, which are **W** (the downward pull of gravity) and the upward pull **P** of the scale. If the object is to be in equilibrium, these two forces must balance. For equilibrium, the vector sum of the vertical forces that act on the object must be zero.

Before we leave the situation shown in Fig. 2.2, let us discuss what the scale reads. Most scales read the force they must exert on the object they are supporting. The scale in Fig. 2.2 exerts a force of magnitude $P$ on the object; hence the scale reads $P$. Since $P = W$, however, the scale gives the weight of the object. Similarly, when you stand on a bathroom scale, the scale pushes up on you to support your weight. This supporting force has the same magnitude as your weight, and so the scale reads your weight. Later we shall see that sometimes a scale does not give the magnitude of your weight (the earth's pull on you); for example, a scale reads zero when an astronaut in an orbiting spaceship stands on it.

Let us consider still another object in static equilibrium, the block shown in Fig. 2.3. Four forces act on the block if we consider the force of friction to be negligible. The earth pulls downward on it with the force **W**. If the block is not to crash down through the supporting surface, the surface must push upward on the block with a force **P** such that $P = W$. For the block to remain in equilibrium, the vector sum of the vertical forces acting on it must be zero. Similarly, if the block is not to move sideways, the forces **F**$_1$ and **F**$_2$ must balance; the vector sum of the horizontal forces acting on the block must be zero.

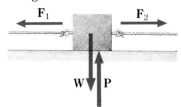

$F_1$          $F_2$

W   P

**FIGURE 2.3**
For this object to remain at rest,
what must be the relation between
**W** and **P**? Between **F**$_1$ and **F**$_2$?

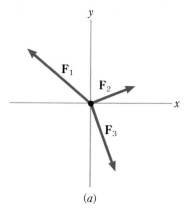

(a)

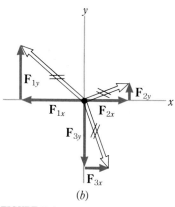

(b)

**FIGURE 2.4**

At equilibrium, what equations must the force components satisfy? The cross-hatch marks are placed on the vectors in (b) to show that they have been replaced by their components.

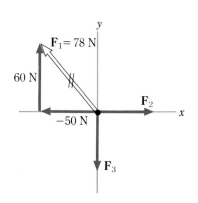

The situations we just discussed lead us to the following conclusion. If an object is to be in equilibrium, the vector sum of the horizontal forces acting on it must be zero and the vector sum of the vertical forces must be zero. Extensive experiments confirm this conclusion. Therefore, it can be stated that

For an object to be in equilibrium, the vector sum of the forces acting on it must be zero.

We now wish to state this conclusion in mathematical form. To do so, let us consider the object (a dot) shown in Fig. 2.4$a$. It is subjected to three forces, $\mathbf{F}_1$, $\mathbf{F}_2$, and $\mathbf{F}_3$. It proves convenient to divide these forces into their rectangular components, as shown in Fig. 2.4$b$. What can we say about these force components if the object is to be in equilibrium? (Notice that in $b$ we place cross-hatch marks on the $\mathbf{F}_1$, $\mathbf{F}_2$, and $\mathbf{F}_3$ arrows to remind ourselves that they have been replaced by their components.)

For an object to be in equilibrium, the horizontal (or $x$-directed) forces that act on it must have a zero resultant. Therefore, in Fig. 2.4$b$,

$$\mathbf{F}_{1x} + \mathbf{F}_{2x} + \mathbf{F}_{3x} = 0$$

Similarly, the $y$-directed forces must have a zero resultant, and so

$$\mathbf{F}_{1y} + \mathbf{F}_{2y} + \mathbf{F}_{3y} = 0$$

The forces acting on the object in Fig. 2.4 must obey these equations if the object is to be in equilibrium.

Let us illustrate this result by the example shown in Fig. 2.5. The object is subjected to the force $F_1 = 78$ N, which has a 60-N $y$ component and a $-50$-N $x$ component. (The $x$ component points in the negative $x$ direction and is therefore negative.) We want to know the magnitudes of $\mathbf{F}_2$ and $\mathbf{F}_3$ if the object is to be in equilibrium. For equilibrium to exist,

$$\mathbf{F}_{1x} + \mathbf{F}_{2x} + \mathbf{F}_{3x} = 0$$

which in the present case becomes

$$-50 \text{ N} + F_2 + 0 = 0$$

from which we find at once that $F_2 = 50$ N. Also, for equilibrium,

$$\mathbf{F}_{1y} + \mathbf{F}_{2y} + \mathbf{F}_{3y} = 0$$

**FIGURE 2.5**

What are the magnitudes of $\mathbf{F}_2$ and $\mathbf{F}_3$ if the object is in equilibrium? Notice that cross-hatch marks have been placed on the $\mathbf{F}_1$ vector to show that it has been replaced by its components.

which in the present case becomes

$$60 \text{ N} + 0 - F_3 = 0$$

from which we find $F_3 = 60$ N. We find, therefore, that the object in Fig. 2.5 is in equilibrium if $F_2 = 50$ N and $F_3 = 60$ N. As you see, we have made good use of the fact that the resultant force on an object must be zero if the object is to be in equilibrium.

We can extend this result to an object subjected to many forces. Call them $\mathbf{F}_1$, $\mathbf{F}_2$, $\mathbf{F}_3$, . . . , $\mathbf{F}_N$, where the three dots in this expression represent all the other forces, such as $\mathbf{F}_4$, $\mathbf{F}_5$, up to $\mathbf{F}_{N-1}$. Then, for the object to be in equilibrium, the components of the forces must sum to zero. Therefore, at equilibrium,

$$\mathbf{F}_{1x} + \mathbf{F}_{2x} + \mathbf{F}_{3x} + \cdots + \mathbf{F}_{Nx} = 0$$
$$\mathbf{F}_{1y} + \mathbf{F}_{2y} + \mathbf{F}_{3y} + \cdots + \mathbf{F}_{Ny} = 0$$

and if some of the forces are in the $z$ direction,

$$\mathbf{F}_{1z} + \mathbf{F}_{2z} + \mathbf{F}_{3z} + \cdots + \mathbf{F}_{Nz} = 0$$

These cumbersome equations can be written in compact form by using the mathematical symbol $\Sigma \mathbf{F}_x$ for the words "the sum of all the $x$ components of the forces," $\Sigma \mathbf{F}_y$ for the $y$ components, and $\Sigma \mathbf{F}_z$ for the $z$ components. Then the first condition for equilibrium can be written

$$\begin{aligned} \Sigma \mathbf{F}_x &= 0 \\ \Sigma \mathbf{F}_y &= 0 \qquad \text{at equilibrium} \\ \Sigma \mathbf{F}_z &= 0 \end{aligned} \qquad (2.1a)$$

These three equations can be simplified even further if we notice that they simply state that the vector sum of all the forces acting on the object must be zero:

$$\Sigma \mathbf{F} = 0 \qquad \text{at equilibrium} \qquad (2.1b)$$

Equations 2.1 are the mathematical statement of the first condition for equilibrium. You should always remember that they are to be applied to only one object at a time. They refer to the forces *on* that object, and you must always be prepared to explain by what means each force is exerted on the object that has been isolated for consideration. It must be the push of a table, the pull of a string, or any other very real physical means of force exertion.

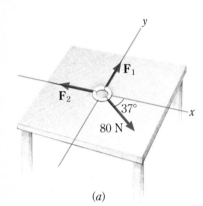

(a)

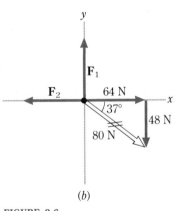

(b)

FIGURE 2.6
Find $\mathbf{F}_1$ and $\mathbf{F}_2$ if the ring is to be at equilibrium.

### Example 2.1

The ring in Fig. 2.6a is lying on a table and being pulled by three strings that hold it at rest. One of the strings has a tension of 80 N. Find the tension in the two others. *By the **tension** in a string, we mean the force with which it pulls on the object to which it is attached.*

*Reasoning*   The ring is at equilibrium, and so the forces acting on it must obey Eq. 2.1. There are $z$-directed forces acting on the ring — the downward pull of gravity and the upward push of the table — but these forces cancel and are of no present concern to us. The $x$- and $y$-directed forces acting on the ring are shown in Fig. 2.6b. Only the 80-N force does not lie along an axis. Why are its components 64 N and 48 N? At equilibrium,

$$\Sigma F_x = 0 \quad \text{or} \quad 64\ \text{N} - F_2 = 0$$

$$\Sigma F_y = 0 \quad \text{or} \quad F_1 - 48\ \text{N} = 0$$

We solve these two equations and find $F_2 = 64$ N and $F_1 = 48$ N. These are the tensions in the two strings. ∎

**Exercise** Replace the 80-N force in Fig. 2.6a by an unknown force $F_3$. If $F_1 = 42$ N, find $F_2$ and $F_3$. *Answer: $F_2 = 56$ N, $F_3 = 70$ N* ∎

## 2.3 Solving problems in statics

With a little practice, you will be able to use Eq. 2.1 to solve many problems in statics. You should, however, follow a few simple rules so that you do not become confused:

**1** Isolate a body. Which object are you going to talk about? The forces acting on this object are the *only* ones you need in writing Eq. 2.1.

**2** Draw the forces acting on the body you have isolated and label them in a free-body diagram. (Use symbols such as $\mathbf{F}_1$, $\mathbf{P}$, and $\mathbf{Q}$ for any forces whose values are not yet known.)

**3** Split each force into its *x, y,* and *z* components and label the components in terms of the symbols given in rule 2 and the proper sines and cosines.

**4** Write down Eq. 2.1.

**5** Solve the equations for the unknowns.

**Example 2.2**
The object in Fig. 2.7a weighs 400 N and hangs at rest. Find the tensions in the three cords that hold it. (Remember, tension is the force with which a cord pulls on an object.)

**FIGURE 2.7**
Since the junction point of the cords in (a) is at equilibrium, the *y* forces must cancel each other in (c). The same holds true for the *x* forces.

**Reasoning** Because the object is at equilibrium, the vector sum of the forces acting *directly* on it must be zero. There are only two such forces, the tension in the lower cord and the pull of gravity, 400 N. Therefore, the tension in the lower cord must be 400 N. It is this tension that supports the object.

Following our rules, we must isolate an object for discussion and make it the point

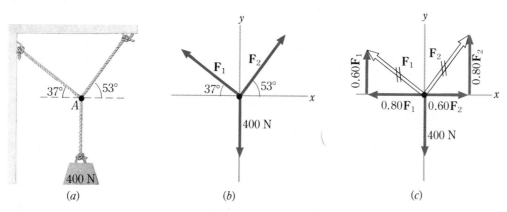

(a)     (b)     (c)

of our attention. Choose the junction of the three cords at $A$. This choice is reasonable because one of the forces acting here is known (400 N) and we seek the other two. The free-body diagram is shown in Fig. 2.7*b*, and Fig. 2.7*c* shows the vectors resolved into components.

Writing Eqs. 2.1, we have, from Fig. 2.7*c*,

$$\Sigma F_x = 0 \qquad 0.60F_2 - 0.80F_1 = 0 \tag{a}$$

$$\Sigma F_y = 0 \qquad 0.60F_1 + 0.80F_2 - 400 \text{ N} = 0 \tag{b}$$

To carry out step 5, we now solve these equations for $F_1$ and $F_2$. Recall from algebra that two general methods are often used.

*Method 1:* Multiply Eq. *a* by 0.6 and Eq. *b* by 0.8. Thus

$$0.36F_2 - 0.48F_1 = 0 \tag{c}$$

$$0.64F_2 + 0.48F_1 - 320 \text{ N} = 0 \tag{d}$$

Addition of Eqs. *c* and *d* gives

$$1.00F_2 - 320 \text{ N} = 0$$

$$F_2 = 320 \text{ N}$$

Substituting this value for $F_2$ in Eq. *c* gives

$$0.48F_1 = (0.36)(320 \text{ N})$$

$$F_1 = 240 \text{ N}$$

*Method 2:* Solve Eq. *a* for $F_1$ in terms of $F_2$. This gives $F_1 = 0.75F_2$. Substitute this in Eq. *b* to obtain

$$0.80F_2 + (0.60)(0.75F_2) - 400 \text{ N} = 0$$

from which $F_2 = 320$ N. Substituting this value for $F_2$ in Eq. *a* gives $F_1 = 240$ N. ∎

---

### Example 2.3

The system shown in Fig. 2.8*a* is in equilibrium. Find the weight $W$ of the block and the tensions in the cords. Notice that the tension in the horizontal cord is 100 N.

*Reasoning* The tension in the lower cord is equal to $W$. Why? We isolate the junction of the cords as our object and sketch Fig. 2.8*b* and *c*. Using Fig. 2.8*c*, we can write Eq. 2.1 to give

$$\Sigma F_x = 0 \qquad 0.60F - 100 \text{ N} = 0$$

$$\Sigma F_y = 0 \qquad 0.80F - W = 0$$

The first of these equations yields $F = 167$ N. Substituting this value in the second equation gives $W = 133$ N.

In both this and the previous example, the general procedure is the same. The physics involved is concerned with identifying and drawing the forces, finding their components, and applying the condition for equilibrium. Most of the work is algebraic and occurs when the simultaneous equations are solved. Do not allow the algebra to confuse you. Problems like this are quite straightforward and should present no difficulty. ∎

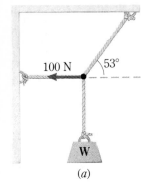

(a)

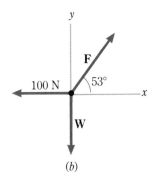

(b)

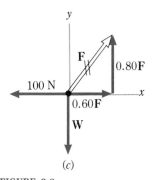

(c)

**FIGURE 2.8**

Appropriate force diagrams for the junction point in (*a*) are shown in (*b*) and (*c*).

(a)

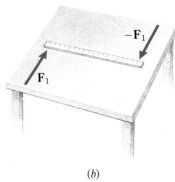

(b)

**FIGURE 2.9**
Even though $\Sigma \mathbf{F} = 0$ for it, the stick in (b) is not in equilibrium.

An object may not remain at rest even if the first condition for equilibrium is satisfied. There is a second condition that must be satisfied if an object is to be in static equilibrium. It is easy to show this by referring to Fig. 2.9. We see there a meter stick supported by a tabletop. The stick is at equilibrium in part *a* because the pull of gravity on it is balanced by the upward push of the table and we have $\Sigma \mathbf{F} = 0$.

Now consider what happens when you push near its two ends with equal but oppositely directed forces $\mathbf{F}_1$ and $-\mathbf{F}_1$: the meter stick does not remain at rest. Even though $\mathbf{F}_1$ balances $-\mathbf{F}_1$ and therefore the condition $\Sigma \mathbf{F} = 0$ is satisfied, the stick begins to rotate. There must be another condition, one involving rotation, that must be satisfied if the object is to be in equilibrium. We discuss that second (and final) condition for equilibrium in the next section. First, however, we must discuss how forces cause rotation.

To learn how forces and rotations are related, we can perform the experiment in Fig. 2.10. We see there a wheel that consists of two disks cemented together. It is free to rotate on a stationary axle that we call the *axis*, or *pivot*, of rotation. By hanging objects from the two cords, we can determine the turning effect of a force. The force $\mathbf{F}_2$ tries to turn the wheel clockwise (in the same direction the hands on a clock turn) while $\mathbf{F}_1$ tries to turn the wheel counterclockwise. By experimenting with different radii $r_1$ and $r_2$ for the two disks, we find that the two turning effects balance whenever

$$r_1 F_1 = r_2 F_2$$

Therefore it is clear that the turning effect depends both on the magnitude of the force and, in some way, on its distance from the pivot.

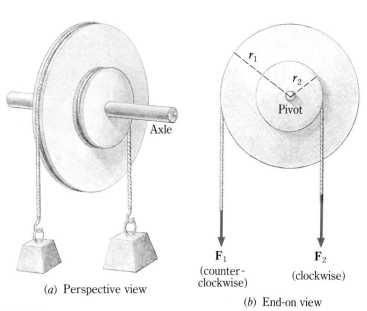

(a) Perspective view

(b) End-on view

**FIGURE 2.10**
How must $\mathbf{F}_1$ and $\mathbf{F}_2$ be related if the wheel is not to turn?

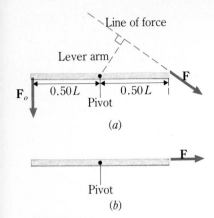

**FIGURE 2.11**

The turning effect of **F** is dependent on the product of **F** and the lever arm. What is the turning effect of the force in (b)?

We can learn more about turning effects from Fig. 2.11. A meter stick pivoted at its center is subjected to two forces $F_o$ and $F$, as shown. Experiment shows that the system balances whenever **F** has a magnitude that satisfies the condition

$$(0.5L)(F_o) = \text{lever arm} \times F$$

where the meaning of the lever arm is shown in Fig. 2.11. In terms of the line labeled "line of force" (which is the endless line along which the force vector lies), we have that

The **lever arm** of a force is the length of the perpendicular dropped from the pivot to the line of force.

From this, we conclude that the turning effect of a force about a pivot is proportional to the product of the force and its lever arm.

We call the turning effect of a force about a pivot the **torque*** $\tau$ due to the force about the pivot:

The turning effect of a force, the torque, is equal to the product of the force and its lever arm:

$$\tau = \text{lever arm} \times \text{force}$$

It is of particular interest to notice what the torque is when the line of force passes through the pivot, as in Fig. 2.11b. Then the lever arm is zero, and so

$$\tau = 0 \times F = 0$$

*When the line of force goes through the pivot, the torque due to the force about that pivot is zero.*

---

### Example 2.4

Find the lever arms and torques for the forces in Fig. 2.12.

***Reasoning*** From its definition, the lever arm is zero for $F_1$, $a$ for $F_2$ and $F_3$, and $b$ for $F_4$. The torques are

| | |
|---|---|
| $F_1$ | 0 |
| $F_2$ | $aF_2$ counterclockwise |
| $F_3$ | $aF_3$ clockwise |
| $F_4$ | $bF_4$ counterclockwise |

Be sure you understand how we obtain these results. ∎

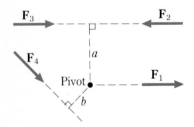

**FIGURE 2.12**

Find the lever arms and torques for the forces with respect to the pivot.

---

\* Torque is pronounced "tork," as in fork, and is represented by the greek letter tau ($\tau$).

**The second condition for equilibrium**

Now that we know how to express the turning effect of a force in terms of torque, it is a simple matter to state the second, and final, condition for static equilibrium. Careful experiments show that, for an object to remain motionless, the clockwise torques acting on it must be balanced by the counterclockwise torques. If we call *counterclockwise torques positive and clockwise torques negative,* we can state that

For an object to be in equilibrium, the sum of the torques acting on it must be zero.

This is the *second condition for equilibrium.*

We can write this condition in mathematical form by using the symbolism $\Sigma\tau$ to represent "the sum of all the torques." Then the second condition for equilibrium becomes $\Sigma\tau = 0$. Throughout this discussion, we have tacitly assumed that the motion of the object under consideration is restricted to a plane, that is, to two dimensions. A large share of the cases of interest are of this type.

All the requirements for a body to be in equilibrium are now known. To summarize, in two dimensions,

$$\Sigma F_x = 0 \qquad \Sigma F_y = 0 \qquad \Sigma\tau = 0 \tag{2.2}$$

Before leaving this section, we should mention that the terms **moment** and **moment of force** are sometimes used instead of torque. In that case, the lever arm is frequently referred to as the **moment arm**. The concepts are the same, of course.

---

### Example 2.5

In Fig. 2.13, we see a beam of length $L$ pivoted at one end and supporting a 2000-N object at the other. Find the tension $T$ in the supporting cable that runs upward to the ceiling. Assume that the weight of the beam is negligible.

*Reasoning*   Notice at the outset that the lower (vertical) cord supports the 2000-N object. Thus the tension in it is 2000 N. We follow the procedure outlined in Sec. 2.3. First we isolate the beam as the object for discussion. Its free-body diagram is shown in Fig. 2.13*b*. (Notice that we ignore the weight of the beam because we were told we could.) Because we know very little about the force on the beam at its hinged end, we represent this force in a general way by giving its $x$ and $y$ components, labeled **H** and **V**. We next divide the tension **T** into its $x$ and $y$ components.

Now we are prepared to write Eqs. 2.2:

$$\Sigma F_x = 0 \qquad 0.50\,T - H = 0$$
$$\Sigma F_y = 0 \qquad 0.87\,T + V - 2000\text{ N} = 0$$

To write the torque equation, we must choose a pivot. It seems natural to take the pivot at point $P$, since this is where the rod is pivoted. (We shall soon see, however, that we have a much wider choice.) With a pivot at $P$, the forces **H** and **V** have zero-length lever arms and so cause no torque about $P$. The torque due to the

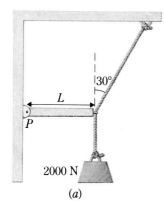

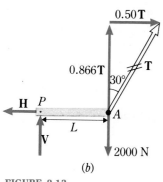

2000 N

(*a*)

0.50**T**

0.866**T**    **T**

30°

**H**  *P*

**V**

*L*

*A*

2000 N

(*b*)

**FIGURE 2.13**

The beam is isolated as the object for discussion, and its free-body diagram is shown in (*b*). We assume the beam's weight to be negligible.

2000-N object is $-L \times 2000$ N since it has a lever arm of length $L$. This torque is clockwise about $P$ and is therefore negative.

We choose to find the torque due to **T** by using its components. The component 0.866**T** causes a counterclockwise (positive) torque whose value is $(L)(0.866\,T)$. Notice that the other component, 0.50**T**, really acts at the end of the rod, at $A$. Its line therefore goes through the pivot, and thus its lever arm and torque are zero.

With these considerations in mind, we have that

$$\Sigma \tau = 0$$

$$-(L)(2000 \text{ N}) + (L)(0.866\,T) = 0$$

$$T = 2310 \text{ N}$$

If we wished, we could find the forces **H** and **V** by substituting in the equations $\Sigma \mathbf{F}_x = 0$ and $\Sigma \mathbf{F}_y = 0$. ■

---

***Exercise*** Find $H$ and $V$ in Fig. 2.13. *Answer: H = 1150 N, V ≈ 0* ■

---

## 2.6 The center of gravity

Two complications were deftly avoided in Example 2.5. One involved the choice of pivot, a matter we discuss in the next section. The other was avoided by assuming the beam to have negligible weight. Because this is not a generally valid assumption, we must now consider what effect the weight of an object has when we write a torque equation. In particular, where can we consider the pull of gravity to be applied to an object so that we can state its lever arm?

Of course, gravity pulls on all parts of any object. It turns out, however, that, for torque purposes, the pull of gravity (the object's weight) appears to act at one point. We call this point the **center of gravity** (c.g.) of the object. Let us now see how this point is located experimentally.

Suppose we wish to locate the center of gravity of the object in Fig. 2.14. We first

**FIGURE 2.14**

An experimental method for determining the center of gravity of an object.

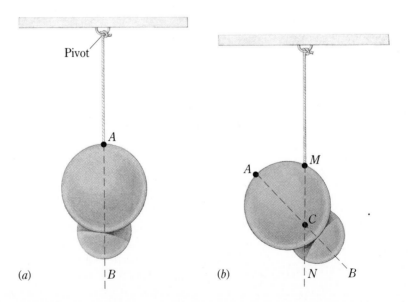

support it from a string and pivot, as in part *a*. The weight of the object causes it to take the equilibrium position shown. At equilibrium, the torques about the pivot add to zero. Only two forces act on the object—the upward pull of the string and the downward pull of gravity. The torque due to the string is zero because its force line passes through the pivot. Since the sum of the torques must be zero, the torque due to the pull of gravity must also be zero. This is true only if the lever arm for the weight is zero, and this is true only if the force of the weight acts along the line *AB* so that its line passes through the pivot. Hence, we learn from Fig. 2.14*a* that the pull of gravity acts at a point somewhere along line *AB*.

Let us now repeat this experiment for another suspension point, as in Fig. 2.14*b*. Using the same reasoning, we see that the pull of gravity acts at a point along the line *MN*. We see, then, that the point at which the weight acts is on both lines *MN* and *AB*. It must therefore act at their point of intersection, point *C*. A check may be made by using a third suspension point. Its vertical line will also intersect the others at *C*. We therefore conclude that *C* is the center of gravity for this object.

The **center of gravity** of an object is the point at which the weight of an object can be considered to act for the purpose of computing torques.

## 2.7 The position of the pivot is arbitrary

Often an object at equilibrium has an obvious pivot, and it is common to use that point as the pivot when computing torques. In other situations, no obvious pivot exists. We see in this section that, when applying the torque equation to an object at equilibrium, we are free to choose as pivot any convenient point.

Consider the situation in Fig. 2.15, where a sign painter of weight $\mathbf{W}_p$ stands in equilibrium on a board of weight $\mathbf{W}_b$ and length $L$. The board's center of gravity is at its geometric center, and so $\mathbf{W}_b$ is shown to act there in Fig. 2.15*b*. We now show that the final form of the torque equation we write for this equilibrium situation does not depend on the position we choose for the pivot.

Let us choose point $A$ as the pivot. You should verify that the torque equation $\Sigma \tau = 0$ becomes

$$-T_1(a) - W_b(0.50L - a) - W_p(0.50L - a + b) + T_2(L - a) = 0$$

Let us group the terms involving the arbitrary length $a$:

$$-a(T_1 - W_b - W_p + T_2) - 0.50 W_b L - W_p(0.50L + b) + T_2 L = 0$$

We can easily show, however, that the factor multiplying $a$ is zero *provided the system is at equilibrium*. At equilibrium, $\Sigma \mathbf{F}_y = 0$, and so

$$T_1 + T_2 - W_b - W_p = 0$$

Because this is the factor that multiplies $a$ in the equation, the term in question is zero. Therefore, the torque equation becomes

$$-0.50 W_b L - W_p(0.50L + b) + T_2 L = 0$$

which is independent of $a$ and the position chosen for the pivot. In this case at least, the position chosen for the pivot is arbitrary.

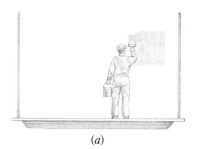

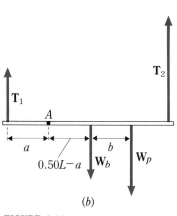

**FIGURE 2.15**

The position of the pivot is arbitrary.

# Mobiles

The mobile shown in part *a* of the figure is an interesting example of the principles of static equilibrium. For the mobile to hang as it does, each of its elements must satisfy the conditions for equilibrium. The designer of the mobile has to balance each portion in such a way that the torques acting on each horizontal bar cancel. Let us illustrate this by considering a specific situation.

Suppose each bird in *a* has a weight of magnitude $W$ and each horizontal bar is uniform, with length $L$ and weight $1.50W$. The vertical cords have negligible weight. At what point must the lower bar be suspended? This can be answered by using the free-body diagram in *b*. We wish to find the distance $x$ from one end to the suspension point. Taking torques about the suspension point, we can write

$$\Sigma \tau = 0$$

$$(2W)(x) - (1.5W)(0.50L - x) \\ - (W)(L - x) = 0$$

This equation can be solved to give $x$ in terms of $L$: $x = 0.39L$. Does this seem reasonable? Can you show that the suspension point for the upper bar must be at a distance $0.25L$ from one end?

We have shown the mobile hanging in a single plane, the plane of the page. Most mobiles, however, twist out of a plane and float from position to position in a slight breeze. Does our calculation still apply?

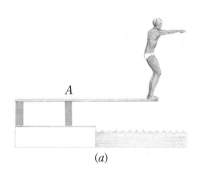

*(a)*

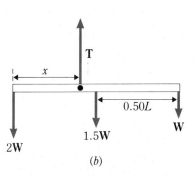

*(b)*

---

Although we have obtained this result for a specialized situation, it is possible to give a more general proof. We therefore have the following general result:

In writing the torque equation for a body at equilibrium, the position chosen for the pivot is arbitrary.

*In practice, we usually choose the pivot so that the line of an unknown force passes through it.* Then the torque due to that force is zero and does not appear in the torque equation.

---

### Example 2.6

The 900-N man in Fig. 2.16 is about to dive from a diving board. Find the forces exerted by the pedestals on the board. Assume that the board has negligible weight.

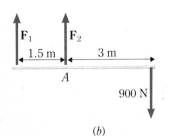

*(a)*

*(b)*

**FIGURE 2.16**

A 900-N man stands at the end of a diving board. We guess, incorrectly, that the forces exerted by the two pedestals on the board are as shown.

*Reasoning* We isolate the board and draw the forces on it, as shown in part *b*. A little reflection should convince you that we have drawn the force $\mathbf{F}_1$ in the wrong direction.* This is a deliberate error, and we shall see how it affects the result. Choosing point $A$ as pivot, we have

$$\Sigma\tau = 0 \qquad -(900\ \text{N})(3\ \text{m}) - (F_1)(1.5\ \text{m}) = 0$$

$$\Sigma\mathbf{F}_y = 0 \qquad\qquad F_1 + F_2 - 900\ \text{N} = 0$$

Solving these gives $F_1 = -1800$ N and $F_2 = 2700$ N. Notice that the negative sign on $F_1$ tells us that our force was drawn in the wrong direction. Its magnitude is 1800 N. ∎

---

*Exercise* Use the left end of the board as pivot to find the magnitude of $\mathbf{F}_2$. *Answer: 2700 N.* ∎

---

### Example 2.7
For the uniform 50-N beam in Fig. 2.17, how large is the tension in the supporting cable and what are the components of the force exerted by the hinge on the beam?

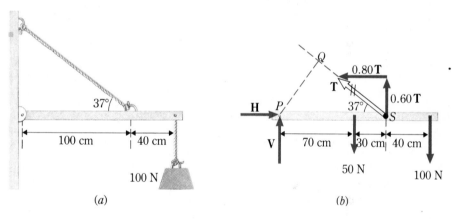

**FIGURE 2.17**
The forces acting on the beam in (*a*) are shown in detail in (*b*). Notice that the force component of $0.6\,T$ pulls on the beam at point $S$, and so its lever arm is 100 cm.

*Reasoning* If we isolate the beam, the forces acting on it are shown in Fig. 2.17*b*. Notice that the weight of the beam, 50 N, is taken as acting at the beam's center of gravity. Further, notice how the tension in the cable has been replaced by its components. We can eliminate the force components at the wall, $\mathbf{H}$ and $\mathbf{V}$, by taking point $P$ as the pivot. Then

$$\Sigma\tau = 0 \qquad (0.60\,T)(1.0\ \text{m}) - (50\ \text{N})(0.70\ \text{m}) - (100\ \text{N})(1.40\ \text{m}) = 0$$

Why doesn't the component $0.80\,T$ appear in this equation?

$$\Sigma\mathbf{F}_x = 0 \qquad\qquad H - 0.80\,T = 0$$

$$\Sigma\mathbf{F}_y = 0 \qquad V + 0.60\,T - 50\ \text{N} - 100\ \text{N} = 0$$

---

* The pedestal at the right supports the seesaw-like board. The left pedestal must pull downward on the board to keep it from rotating clockwise.

First solving for $T$ in the torque equation and then substituting in the other two equations, we get

$$T = 292 \text{ N} \qquad H = 234 \text{ N} \qquad V = -25.2 \text{ N}$$

What does the minus sign on $V$ tell us? ▪

---

### Example 2.8
A person holds a 20-N weight, as shown in Fig. 2.18a. Find the tension in the supporting muscle and the component forces at the elbow.

*Reasoning* The system can be replaced by the simplified model shown in *b*. We assume the lower arm to weigh 65 N, and the dimensions given are typical. Notice that the situation is very similar to that of Example 2.7, a beam supported by a cable. We use the free-body diagram in *c* to write the equilibrium conditions:

$$\Sigma F_x = 0 \qquad\qquad\qquad\qquad H - T_m \sin 20° = 0$$

$$\Sigma F_y = 0 \qquad\qquad\qquad V + T_m \cos 20° - 65 \text{ N} - 20 \text{ N} = 0$$

$$\Sigma \tau = 0 \qquad (T_m \cos 20°)(0.035 \text{ m}) - (65 \text{ N})(0.10 \text{ m}) - (20 \text{ N})(0.35 \text{ m}) = 0$$

where the left end of the lower arm (the elbow) is taken as the pivot. Notice that the force $T_m \sin 20°$ actually acts through the pivot, and so its torque is zero.
Solving these equations, we find

$$T_m = 410 \text{ N} \qquad H = 140 \text{ N} \qquad V = -300 \text{ N}$$

**FIGURE 2.18**

We can analyze the forces in the human arm by use of the models in (*b*) and (*c*).

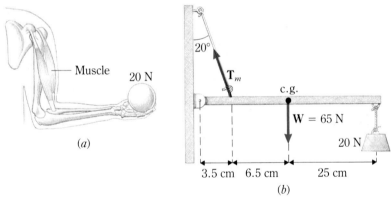

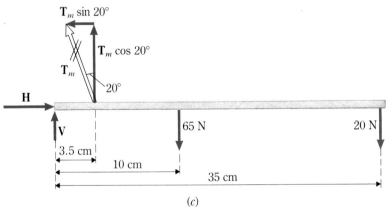

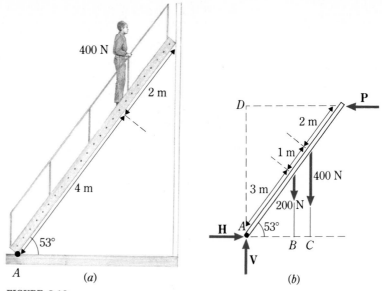

FIGURE 2.19
A 400-N person stands on a 200-N ramp. On the assumption that the wall is smooth, the forces acting on the ramp are as shown in (b).

The negative sign on $V$ tells us that its direction is opposite what we assumed it to be.

All these forces are much larger than the weight being held. Can you show that $T_m$ becomes very large as the arm is outstretched? Why is it very tiring to hold a weight in your outstretched hand? ∎

---

### Example 2.9
The uniform 200-N ramp in Fig. 2.19 leans against a smooth wall. (By the term *smooth*, we mean that the force at the wall is perpendicular to the wall surface. No friction exists.) If a 400-N person stands on the ramp as shown, how large are the forces at the wall and at the ground?

**Reasoning** Isolate the ramp for use as our object. The forces acting on it are shown in *b*. We then have (taking point $A$ as the pivot)

$$\Sigma F_x = 0 \qquad\qquad\qquad\qquad\qquad\qquad H - P = 0$$

$$\Sigma F_y = 0 \qquad\qquad\qquad\qquad V - 200 \text{ N} - 400 \text{ N} = 0$$

$$V = 600 \text{ N}$$

$$\Sigma\tau = 0 \qquad (0.80)(6 \text{ m})(P) - (0.60)(3 \text{ m})(200 \text{ N}) - (0.60)(4 \text{ m})(400 \text{ N}) = 0$$

In this problem, notice particularly the lever arms. By definition, the lever arm is the perpendicular dropped from the pivot $A$ to the line of force. The lever arms are $AB$ for the 200-N force, $AC$ for the 400-N force, and $AD$ for **P**. Solving the equations simultaneously gives

$$V = 600 \text{ N}$$

$$P = H = 275 \text{ N} \quad\blacksquare$$

# Architectural arches

Arches are used for many purposes, from simple doorways to the magnificent arches of the Gothic cathedral shown in illustration (a). All such spans involve the forces and torques we have been discussing. Because they involve rather complex physics, arches were beyond the construction abilities of many civilizations. Even the Greek temples lacked true arches; their doorways consisted of a single stone slab or wooden beam held up by pillars. It was not until Roman times that people were first able to construct arched spans. Let us see what physics is involved in their construction.

A typical stone archway is shown in (b). Its two sides act as pillars to support the curved portion. Look first at the wedgelike stone at the top of the arch. It is kept from falling by the stones at each side. These stones, in turn, are wedged between their neighbors, and a similar situation applies to all the other stones that form the curve. Friction forces between the stones keep them from sliding from position. Even more important, however, is that the stones outside the arch must press inward on the wedged stones to keep them from moving to the right or left. We say that the arch is *buttressed* by the walls to its left and right, as in (c).

We can determine the forces a buttress must supply to any simple arch. Represent by $L$ the side length

of the arch in (d), and suppose that each length $L$ supports an effective load W at its center. The buttress must supply the forces H and V, and the topmost point of the right-hand side of the arch exerts an essentially horizontal force P on the left-hand side. Isolate the left side of the arch and write the equilibrium equations for it:

$$\Sigma F_x = 0 \qquad\qquad H = P$$
$$\Sigma F_y = 0 \qquad\qquad V = W$$
$$\Sigma \tau = 0 \qquad W(\tfrac{1}{2}L \cos \theta) = P(L \sin \theta)$$

Simultaneous solution yields

$$H = \frac{W}{2 \tan \theta}$$
$$V = W$$

These are the forces the buttress must supply. Clearly, the steeper the arch (that is, the larger $\theta$ is), the less horizontal force the buttress must exert. Even so, the steep arches of Gothic cathedrals require extensive buttressing. The history of the architectural designs of cathedrals over the centuries is a fascinating tale of our understanding of arches. You may be interested in pursuing this topic further.*

---

* See, for example, "Gothic Structural Experimentation," R. Mark and W. W. Clark, *Scientific American 251*:176 (1984).

S. Seitz/Woodfin Camp

(a)

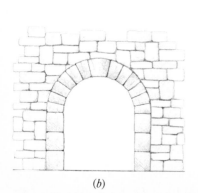

(b)

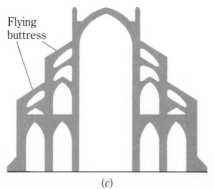

Flying buttress

(c)

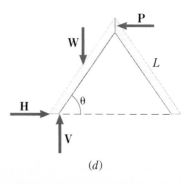

P

W

L

H

θ

V

(d)

## 2.8 Back injury from lifting

You probably have been warned that there is a right and a wrong way to lift a heavy object. Let us apply what we have learned to see why this is true. As a typical situation, look at the man lifting a 60-N bowling ball in Fig. 2.20a. Back strain is likely to occur if the tension in the back muscle becomes too large or if the compression of the spine on the hip is excessive. We can easily calculate these forces by idealizing the situation, as shown in b. This model replaces the spine by a horizontal beam pivoted at the hip. Call the tension in the back muscle $T_m$ and let the components of the force at the hip be $H$ and $V$. Take the weight of the upper portion of the man's body to be 250 N with the dimensions shown.

As the man holds the ball at equilibrium, the appropriate equations become

$\Sigma \mathbf{F}_x = 0$ $\qquad\qquad\qquad\qquad\qquad\qquad H - T_m \cos 12° = 0$

$\Sigma \mathbf{F}_y = 0$ $\qquad\qquad\qquad\qquad\qquad T_m \sin 12° + V - 60 - 250 = 0$

$\Sigma \tau = 0$ $\qquad (250)(0.50L) + (60)(L) - (T_m \sin 12°)(0.67L) = 0$

where all forces are in newtons. (Be sure you understand how the torque equation arises.) Divide through the last equation by $L$ and solve to find $T_m = 1330$ N. Substitution in the other equations yields $V = 32$ N and $H = 1300$ N.

Notice how very large these forces are. Although the bowling ball weighs only 60 N, the tension in the back muscle is 1330 N and the force on the spine is about that large. Obviously, when you bend over to lift an object, you place tremendous strain on your back. If you squat down to lift an object and hold your back fairly erect, however, the forces are much smaller. You should be able to show this using the model in Fig. 2.20b.

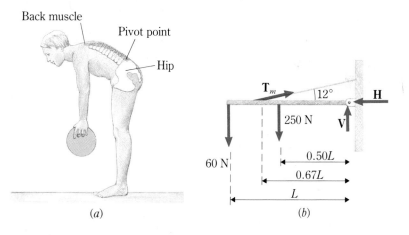

**FIGURE 2.20**
The forces in the man's back can be found using the model in (b).

## 2.9 Why objects tip over

Did you ever notice what happens when an erect soft drink can is tilted? If you tilt it too far, as in Fig. 2.21a, it falls over when released. If released from the position shown in b, however, the can returns to a fully upright position. This general type of behavior occurs in many other situations. Let us see what is involved.

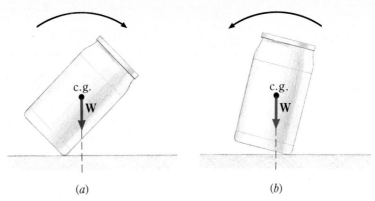

FIGURE 2.21
Which soft drink can will topple over?

Only two forces act on the can when it is released, the pull of gravity **W** and the upward push of the table. If we take the contact point at the table as the pivot, only **W** causes a nonzero torque. In *b*, the torque due to **W** is counterclockwise and causes the can to return to an upright position. In *a*, however, **W** causes a clockwise torque that causes the can to topple over. Thus it is clear why the two situations are so different.

A more important application of this idea can be seen in Fig. 2.22. Why will the car in *a* topple over while the one in *b* will not? You can see at once one of the chief advantages of an auto that has a wide track and a low center of gravity. Those who ride tractors or lawnmowers on hilly terrain are also concerned by the two possible types of behavior shown in Fig. 2.22.

There is a simple way to tell whether or not an object will topple over in a given type of motion. Suppose the object were to undergo a small motion of the type in question. If the motion lowers the center of gravity of the object, the object will topple; otherwise, it will not. Check Figs. 2.21 and 2.22 to see that this statement is true. We shall learn in later chapters that this behavior follows from an important law of nature, the law of energy conservation.

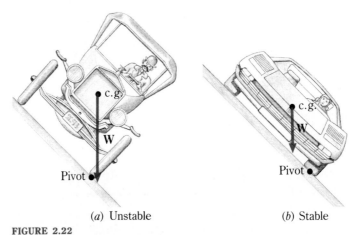

(a) Unstable          (b) Stable

FIGURE 2.22
What determines whether or not a vehicle sitting at rest on an incline is stable?

## Minimum learning goals

When you finish this chapter, you should be able to

1 Define (*a*) static equilibrium, (*b*) free-body diagram, (*c*) weight, (*d*) tension, (*e*) lever arm, (*f*) torque, (*g*) clockwise versus counterclockwise, (*h*) line of force, (*i*) center of gravity.

2 Find the torque due to a given force with reference to a specified pivot point.

3 State the two conditions for equilibrium in words and in equation form.

4 Give the position of the center of gravity of simple objects and determine the position of the center of gravity of more complex objects.

5 Solve simple problems for objects at equilibrium.

## Questions and guesstimates

1 A traffic light hung from a cable that stretches across a street invariably causes the cable to sag. Why don't the workers remove the sag when they hang the cable?

2 Draw the free-body diagrams for a 300-N girl in the following equilibrium situations: (*a*) she stands on one foot; (*b*) she hangs from a bar with one hand; (*c*) she stands on her head; (*d*) she does a handstand with a single hand resting on a stool.

3 Refer to Fig. 2.8. Will the tension in the upper cable increase or decrease as its angle with the vertical is decreased? What will be the tension in it when the cable is vertical?

4 The center of gravity of a hollow spherical shell is inside the shell. Name a few other objects for which the center of gravity is not on the object. Where, approximately, is the center of gravity of a mixing bowl? A clothes hanger?

5 We are told that slender people are less apt to have back trouble than obese people. Why should this be true?

6 A child watches a parade by sitting on its father's shoulders with its legs straddling his neck. Discuss the various ways the father could lower the child to the ground. Which ways could lead to serious injury to the father's back?

7 A strong horizontal wind topples a tree. Why is it not really correct to say that the wind pulled the tree out of the ground? Explain what actually happens.

8 A boy stands in a large garbage pail. Fastened to the handle of the pail is a rope that passes over a pulley suspended from the ceiling. The boy pulls on the free end of the rope and tries to lift the pail and himself. What happens to the tension in the rope and the force the boy exerts on the bottom of the pail as he slowly pulls harder and harder on the rope? Can the boy lift himself and the pail off the floor?

9 A woman is unable to loosen the nut that holds a lawnmower blade in place because she cannot apply enough force to the wrench she has available. She slips the handle of the wrench into a pipe 80 cm long and then removes the nut easily. Explain.

10 Each of the following tools makes use of torque. Describe the torques that exist in each case: wire cutters, wheelbarrow, crescent wrench, bottle opener, claw hammer, nutcracker.

## Problems

*The problems at the end of every chapter are divided into three levels of difficulty: standard, somewhat difficult (indicated by one square ■), and most difficult (two squares ▮).*

1 An object that weighs 20 N is fastened by a cord to the bottom of an object that weighs 30 N. The 30-N object is supported by a cord fastened to the ceiling. Find the tension in (*a*) the upper cord and (*b*) the lower cord.

2 A physics book (weight = 10.5 N) sits at equilibrium on top of a 31-N dictionary that sits on a tabletop. Find the force with which (*a*) the table pushes up on the dictionary and (*b*) the dictionary pushes up on the physics book.

**3** Three cords pull on an object. Their forces are in the *xy* plane and are 200 N at 37°, 300 N at 90°, and a force **F**. (We specify angles in the usual way in the *xy* plane.) Find **F** if the object is to remain at equilibrium.

**4** A point object is subjected to the following forces, all in the *xy* plane: 150 N at 120°, 80 N at 270°, and a force **F**. Find **F** if the object is to be at equilibrium.

**5** Two children are fighting over a wagon. One child pulls outward on the wagon handle with a force of 500 N, with the handle positioned 30° above the horizontal. With what force must the other child pull straight back on the wagon if it is not to move?

**6** In Fig. P2.1, two objects of equal weight, 80 N, hang from a cord that passes over a frictionless pulley. What are the tensions in the three cords if (*a*) the weight of the pulley is negligible and (*b*) the pulley weighs 30 N?

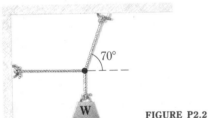

80 N    80 N    **FIGURE P2.1**

■ **7** In Fig. P2.2, the weight $W = 2000$ N. What is the tension in (*a*) the horizontal cord and (*b*) the cord running to the ceiling?

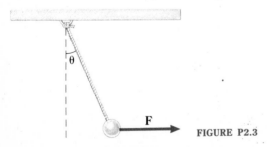

70°

**W**    **FIGURE P2.2**

■ **8** The tension in the horizontal cord in Fig. P2.2 is 400 N. What is the weight of the object?

■ **9** It is found that the system in Fig. P2.3 comes to

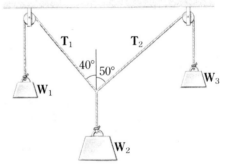

θ

**F**

**FIGURE P2.3**

equilibrium with $\theta = 40°$ when $F = 200$ N. How much does the ball at the end of the cord weigh?

■ **10** If the ball in Fig. P2.3 weighs 400 N and $F = 300$ N, what is the angle $\theta$ when the system is at equilibrium?

■ **11** The tension in the cord attached to the vertical wall in Fig. P2.4 is 60 N. Find (*a*) *W* and (*b*) the tension in the cord attached to the ceiling.

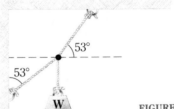

53°

53°

**W**    **FIGURE P2.4**

■ **12** In Fig. P2.4, $W = 240$ N. Find the tension in (*a*) the cord running to the ceiling and (*b*) the cord running to the wall at the left.

■ **13** For the equilibrium situation shown in Fig. P2.5, find $W_2$ and $W_3$. Assume that $W_1 = 800$ N and that the pulleys are frictionless, so that they do not alter the tensions in the cords.

$T_1$    $T_2$

40° 50°

$W_1$    $W_3$

$W_2$

**FIGURE P2.5**

■ **14** Find $W_1$ and $W_3$ in Fig. P2.5 if the system is at equilibrium with $W_2 = 250$ N. Assume the pulleys to be frictionless so that they do not alter the tensions in the cords.

■ **15** For the equilibrium situation shown in Fig. P2.6, find $W_1$, $W_2$, $T_1$, and $T_2$ if $W_3 = 800$ N. Assume the pulleys to be frictionless so that they do not alter the tensions in the cords.

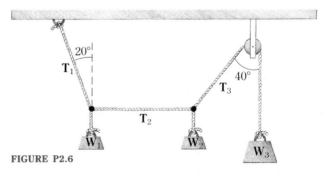

20°

$T_1$

40°

$T_3$

$T_2$

$W_1$    $W_2$    $W_3$

**FIGURE P2.6**

**16** In Fig. P2.6, the system is at equilibrium when $W_1 = 300$ N. Find the values of $W_2$ and $W_3$. Assume the pulleys to be frictionless so that they do not alter the tensions in the cords.

**17** After breaking a leg, a student finds herself in a hospital in traction. Assume that the pulleys in Fig. P2.7 are frictionless; then the tension in the cord is everywhere the same, namely, 140 N. How large is the force stretching the leg? How large an upward force does the device exert on the foot and leg together?

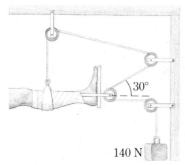

140 N

FIGURE P2.7

**18** The pulleys in Fig. P2.8 are frictionless and have negligible weight. At equilibrium, how large is $W_1$ if $W_2 = 280$ N? Also find $T_1$, $T_2$, $T_3$, and $T_4$.

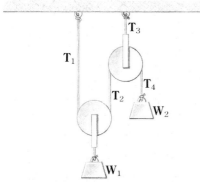

FIGURE P2.8

**19** For the situation in Fig. P2.9, with what force must the 500-N man pull downward on the rope to support himself free from the floor? Assume that the pulleys have negligible friction and weight.

**20** Find the torques about pivot $B$ provided by the forces in Fig. P2.10 if $L = 4.0$ m.

**21** Find the torques about pivot $A$ provided by the forces in Fig. P2.10 if $L = 6.0$ m.

**22** For each force in Fig. P2.11, what is the (a) lever arm and (b) torque about point $P$ as pivot? The square has sides of length $L$.

**23** How large a force $F$ applied to the wheelbarrow handles in Fig. P2.12 will be able to lift a 700-N load at the center of gravity indicated? Give your answer in terms of $a$ and $b$.

FIGURE P2.9

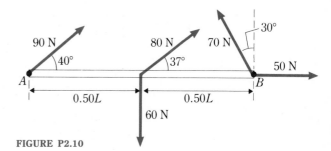

FIGURE P2.10

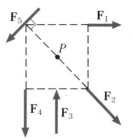

FIGURE P2.11

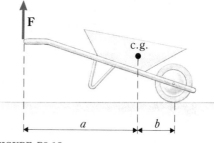

FIGURE P2.12

**24** For the nail puller in Fig. P2.13, how large a force is applied to the nail when the force on the handle is 250 N? Express your answer in terms of $a$ and $b$. Assume the force applied to the nail to be vertical.

FIGURE P2.13

**■ 25** The vertical ropes with tensions $T_1$ and $T_2$ in Fig. P2.14 support a uniform 80-N plank and two weights. If $T_1$ is 210 N and $W_2$ is 245 N, find $W_1$ and $T_2$.

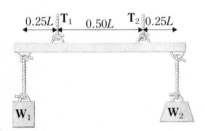

FIGURE P2.14

**■ 26** In Fig. P2.14, the 80-N plank is uniform, $W_1 = 390$ N, and $W_2 = 300$ N. Find $T_1$ and $T_2$.

**■ 27** The uniform 245-N plank in Fig. P2.14 is supported by two ropes. If each rope can withstand a tension of only 1000 N and if $W_2$ is to be twice as heavy as $W_1$, what is the greatest value $W_1$ can have? Assume the ropes holding the weights are very strong.

**■ 28** To locate a person's center of gravity, he is placed on two scales as in Fig. P2.15. The scale on the left reads 260 N, and that on the right reads 200 N. Find the distance $x$ to the center of gravity in terms of $L$. Assume that the scale readings have been corrected by subtracting the readings when the person was not in place.

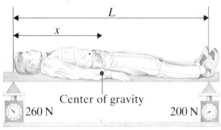

FIGURE P2.15

**■ 29** In Fig. P2.16, the beam is uniform and weighs 300 N. Find (a) the tension in the upper rope and (b) the $H$ and $V$ component forces exerted by the pin if $W = 800$ N.

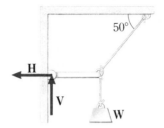

FIGURE P2.16

**■ 30** The uniform 500-N beam in Fig. P2.17 supports a load as shown. (a) How large can the load be if the horizontal rope is able to hold 3000 N? (b) What are the components of the force at the base of the beam?

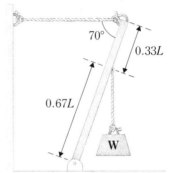

FIGURE P2.17

**■ 31** When you stand on tiptoe, the situation is much like that shown in Fig. P2.18. The magnitude of **F**, the push of the floor, will be equal to the person's weight if the person is standing on one foot. Find (a) the tension in the Achilles tendon and (b) $H$ and $V$ at the ankle in terms of $F$.

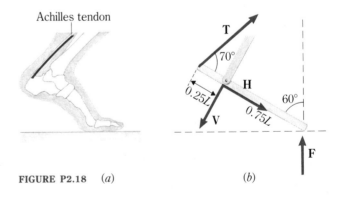

FIGURE P2.18    (a)    (b)

**■ 32** In Fig. P2.19, the beam weighs 900 N and $T_3 = 870$ N. Find $T_1$, $T_2$, $W$, and the force with which the beam pushes down on the frictionless pin at its base.

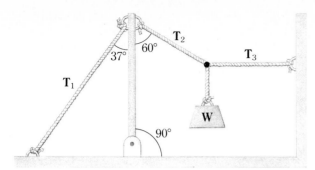

FIGURE P2.19

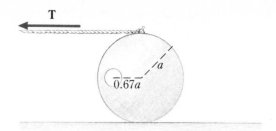

FIGURE P2.22

**33** As you discovered in Sec. 2.9, it is advantageous for a car to have a low center of gravity and a wide track. Suppose a car's center of gravity is a distance $h$ above the roadway and the width of the car where the wheels touch the road is $d$. If you try to tip the car over sideways, through how large an angle will you have to tilt it if $d = 3h$?

**34** The uniform block in Fig. P2.20 is twice as tall as it is wide. It is prevented from sliding by a small ridge. If $\theta$ is slowly increased, at what value will the block topple over?

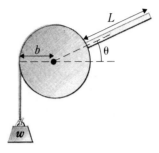

FIGURE P2.23

an axle. What weight $w$ must an object hanging from the wheel's rim have if the system is to be at equilibrium in the position shown?

**38** The stiff bar of negligible weight in Fig. P2.24 has a length $L$. At its ends are very small frictionless wheels. Two weights, $w$ and $W$, hang from the bar, each at a distance $0.25L$ from one of the ends. When the bar is placed across a vee-shaped trough, it comes to equilibrium at an angle $\theta = 10°$ to the horizontal. Find the ratio $w/W$.

FIGURE P2.20

**35** The bowling ball in Fig. P2.21 weighs 70 N and rests against the walls of a frictionless groove. How large are the forces that the walls of the groove exert on the ball? Consider the ball to be a uniform sphere.

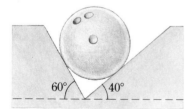

FIGURE P2.21

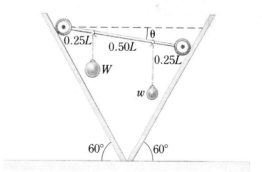

FIGURE P2.24

**36** In Fig. P2.22, the uniform cylinder of radius $a$ originally had a weight of 80 N. After an off-axis cylindrical hole was drilled through it, it weighed 65 N. The axis of the cylinder is parallel to that of the hole. If we assume the cylinder does not slip, what tension in the cord will keep the cylinder from moving?

**37** In Fig. P2.23, a uniform bar of length $L$ and weight $W$ protrudes from a wheel of radius $b$ that can rotate freely on

**39** A covered cylindrical can (height $h$, diameter $d$) full of water sits upright on the floor. (a) When the can is tipped slowly on one edge of its rim, at what angle to the vertical does it topple over? (b) How large a horizontal force on its highest point is required to tip the can this far? The weight of the can plus contents is $W$.

# 3 Uniformly accelerated motion

$O$BJECTS that remain at rest are the subject of the previous chapter. We learned there what conditions must hold if an object is to be in equilibrium. Now we turn our attention to objects in motion. First, we learn in this chapter how to describe an object's motion in terms of displacement, velocity, and acceleration. These quantities have precise meanings in physics and are interrelated through their defining equations. Once we have learned these interrelations, we can describe the motion of such varied objects as molecules, baseballs, and rockets. To fully understand these motions, however, we need to find out how forces and torques are related to acceleration. These topics are discussed in several of the chapters that follow.

## 3.1 Speed

When you say that a car is moving at a speed of 80 kilometers per hour (80 km/h), everyone knows what you mean; the car will go 80 km in 1 h provided it maintains this speed. In 0.5 h the car will go $0.5 \times 80 = 40$ km, and in 2 h it will go $2 \times 80 = 160$ km. In general, the distance a car travels if its speed is unchanging is

Distance traveled = (speed)(time taken)

Solving for speed, we find that the defining equation for this quantity is

$$\text{Speed} = \frac{\text{distance traveled}}{\text{time taken}} \tag{3.1}$$

We use this same equation to define the average speed of a car whose speed is not constant. If the car goes 200 km in 4.0 h, then its average speed is

$$\text{Average speed} = \frac{200 \text{ km}}{4.0 \text{ h}} = 50 \text{ km/h}$$

As you see, the units of speed are a distance unit divided by a time unit. For example, the speed of a snail might be 2.0 centimeters per hour (cm/h) and the speed of a glacier might be 1.5 m/year. Average speed is always the distance traveled divided by the time taken.

Notice that speed has no direction; it is a scalar quantity. A car's speedometer measures how fast the car is going, but it tells us nothing about the direction of travel. The car could be traveling on a straight road across the prairie, or it could be circling a racetrack; its average speed is still 100 km/h if it travels 200 km in 2 h.

---

## 3.2 Average velocity

In everyday conversation, we use the terms *speed* and *velocity* interchangeably. In science, however, these two quantities have different meanings. As we shall see, velocity is a vector quantity (unlike speed, which is a scalar). Moreover, an object's average speed is often not equal to its average velocity, even in magnitude. Let us now give the scientific definition of velocity.

Suppose an object takes a time $t$ to move from point $A$ to point $B$ by the path shown in Fig. 3.1. The displacement from $A$ to $B$ is shown as the vector **s** between the two points. As indicated, the displacement is 800 cm eastward and is a vector quantity. We define the *average velocity* of the object as it travels from $A$ to $B$ by

$$\text{Average velocity} = \frac{\text{displacement vector}}{\text{time taken}}$$

In symbols,

$$\bar{\mathbf{v}} = \frac{\mathbf{s}}{t} \qquad \mathbf{s} = \bar{\mathbf{v}}t \tag{3.2}$$

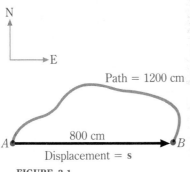

FIGURE 3.1
The displacement **s** from $A$ to $B$ is 800 cm eastward.

The bar above the **v** is used to indicate that this is an average velocity. Notice that, since $\bar{\mathbf{v}}$ is proportional to **s**, velocity is a vector quantity; it has direction, the same direction as the displacement vector. In Fig. 3.1, the velocity is eastward because the displacement **s** is eastward.

Let us take a numerical example. Suppose that the straight-line distance from $A$ to $B$ in Fig. 3.1 is 800 cm and that it takes the object 20 s to go from $A$ to $B$ along the 1200-cm path. Then, since **s** = 800 cm eastward and $t = 20$ s,

$$\bar{\mathbf{v}} = \frac{800 \text{ cm eastward}}{20 \text{ s}} = 40 \text{ cm/s eastward}$$

Notice that the velocity, being a vector quantity, has both magnitude (40 cm/s) and direction (eastward). The average *speed* of the object, however, is

$$\text{Average speed} = \frac{\text{path length}}{\text{time taken}} = \frac{1200 \text{ cm}}{20 \text{ s}} = 60 \text{ cm/s}$$

which is not equal to the magnitude of the average velocity. As we see, an object's average speed need not equal the magnitude of its average velocity.

## 3.3 Instantaneous velocity

Let us consider the motion of a falling ball; a photograph of this type of motion is shown in Fig. 3.2. The position of such a ball at uniformly spaced times is captured by illumination from a flashing strobe light with equal time intervals between flashes. Call the time interval between successive flashes $\Delta t$ (read "delta tee"). Notice that the ball speeds up as it falls. This is obvious from the fact that it falls a larger distance during each successive time interval. Let us now discuss how we can determine the velocity of the ball as it passes point $C$. Such a velocity at a single point is called the *instantaneous velocity*.

The direction of the velocity is clear. It is the same as the direction of motion, downward. To find a rough magnitude of the ball's velocity at $C$, we can compute its average velocity from $A$ to $B$. We call the coordinate that measures the ball's position $r$. As the ball goes from $A$ to $B$, its displacement is $\Delta\mathbf{r}$. If we call the time between strobe flashes $\Delta t$, the time the ball takes to move from $A$ to $B$ is $\Delta t$. Therefore, for the region from $A$ to $B$, the ball's average velocity is

$$\overline{\mathbf{v}} = \frac{\text{displacement}}{\text{time taken}} = \frac{\Delta\mathbf{r}}{\Delta t}$$

This is not the exact velocity of the ball at $C$, however, because the velocity is continuously increasing. If we speed up the strobe light (in other words, make $\Delta t$ smaller), the images of the ball will be closer to each other and points $A$ and $B$ will be much closer to $C$. If we perform our calculation for these new points $A$ and $B$, the average velocity we obtain should be closer to the ball's actual velocity at $C$ than the first value we calculated.

We can imagine a case in which the strobe light is flashing so rapidly that the time interval between flashes approaches zero, which we represent by $\Delta t \rightarrow 0$. Then $A$ and $B$ will be so close to $C$ that the average velocity we compute is almost exactly equal to the velocity at $C$. We call the velocity at $C$ the instantaneous velocity at that point and represent it by $\mathbf{v}$. It is defined mathematically in terms of the experimental procedure we have just outlined and is therefore

$$\text{Instantaneous velocity} = \mathbf{v} = \lim_{\Delta t \to 0} \frac{\Delta\mathbf{r}}{\Delta t} \tag{3.3}$$

The symbol $\lim_{\Delta t \to 0}$ is read "in the limiting case where $\Delta t$ approaches zero." It represents mathematically the experimental procedure in which $\Delta t$ is made so

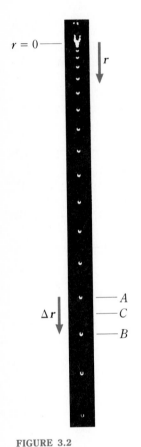

**FIGURE 3.2**

The strobe photograph shows the ball's location at successive times. The ball falls from $A$ to $B$ in a time $\Delta t$. *(Education Development Center)*

small that the average velocity between $A$ and $B$ becomes essentially the instantaneous velocity* at $C$.

There is an interesting relation between the magnitudes of the instantaneous velocity at a point such as $C$ and the speed at $C$. When we make $\Delta t$ very small, the object cannot change its direction of motion appreciably during the time it takes to go from $A$ to $B$. As a result, the straight-line distance from $A$ to $B$ equals the path length covered by the object as it goes from $A$ to $B$. Therefore, because the path length and the displacement have the same magnitude, the instantaneous velocity and the speed at $C$ also have the same magnitude. Thus we see that *the magnitude of the instantaneous velocity at a point is equal to the speed at that point.*

## 3.4 One-dimensional motion

For most of the remainder of this chapter, we restrict our discussion to motion along a straight line. We shall learn in later chapters how to generalize our results to two-dimensional motion. As an example of one-dimensional motion, consider the car in Fig. 3.3$a$. At the instant shown, its motion is in the positive $x$ direction. Therefore the vector representing its velocity is also in the positive $x$ direction. If the car were going in the reverse direction, however, its velocity vector would point in the negative $x$ direction. Hence, in one-dimensional motion, it is possible to show direction by means of plus and minus signs. A positive velocity indicates a velocity vector pointing in the positive $x$ direction, and a negative velocity indicates that the velocity vector points in the negative $x$ direction. Plus and minus signs can also be used to indicate the direction of displacement, force, and acceleration.

Let us discuss the motion of the car in Fig. 3.3$a$. At a time $t$, its displacement from the coordinate origin is $x$. Suppose that the car was at $x = 0$ when $t = 0$ and that it is moving at 20 m/s. Then its position as a function of time yields the data

| $t$ (s) $\rightarrow$ 0 | 1 | 2 | 3 | 4 | 5 | 6 |
|---|---|---|---|---|---|---|
| $x$ (m) $\rightarrow$ 0 | 20 | 40 | 60 | 80 | 100 | 120 |

---

* Those of you who know calculus will recognize that Eq. 3.3 can be written in terms of the derivative. The right-hand side is what we define as the derivative of $\mathbf{r}$ with respect to $t$, written $d\mathbf{r}/dt$. Therefore Eq. 3.3 can also be written $\mathbf{v} = d\mathbf{r}/dt$.

**FIGURE 3.3**

Motion along a straight line can be shown by a graph. In this case, the car's speed is constant at 20 m/s.

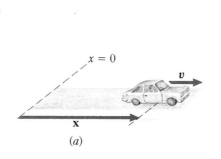

$x = 0$

$v$

x

($a$)

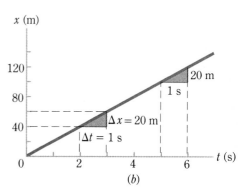

$x$ (m)

120

80

40

$\Delta x = 20$ m

$\Delta t = 1$ s

20 m

1 s

0    2    4    6    $t$ (s)

($b$)

In other words, the car's displacement increases by 20 m each second. We can plot these data to give the graph of x as a function of t shown in Fig. 3.3b. Check the car's position at $t = 5$ s by plotting the point $x = 100$ m at $t = 5$ s.

We can attach important meaning to the two small triangles in b. Notice that the vertical length of each is 20 m and the horizontal length is 1 s. The triangles therefore tell us that the car goes 20 m in the positive x direction every second. Think of the vertical ($\Delta x$) and horizontal ($\Delta t$) sides of the triangles in the following way. The vertical side, of length $\Delta x$, is the displacement the car undergoes during a time interval $\Delta t$. Thus the average velocity of the car is

$$\bar{\mathbf{v}} = \frac{\text{displacement}}{\text{time taken}} = \frac{\Delta x}{\Delta t}$$

where $\Delta x$ is the vector displacement in the x direction. If $\Delta x$ is positive, the velocity is in the positive x direction; if it is negative, the velocity is in the negative x direction. Thus the triangles in Fig. 3.3b can be used to find the velocity of the car.

As another example of straight-line motion, let us return to the falling ball of Fig. 3.2. In this case, the velocity is continuously increasing instead of being constant. We can use the photograph to measure the position r of the falling ball as a function of time. Typical data obtained from such a photograph are

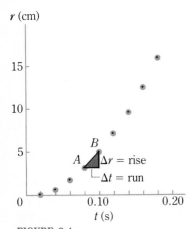

r (cm)

FIGURE 3.4

A plot of the data for an experiment such as the one in Fig. 3.2.

| $t$(s) | $\to 0$ | 0.02 | 0.04 | 0.06 | 0.08 | 0.10 | 0.12 | 0.14 | 0.16 |
|---|---|---|---|---|---|---|---|---|---|
| $r$ (cm) | $\to 0$ | 0.20 | 0.78 | 1.76 | 3.14 | 4.90 | 7.06 | 9.60 | 12.5 |

Notice that the displacement r is a vector whose positive direction is downward. These data are plotted in Fig. 3.4.

To find the ball's average velocity between points A and B on the graph, we must compute $\Delta r/\Delta t$. From the graph, $t_B - t_A = \Delta t = 0.100 - 0.080 = 0.020$ s. From the table or graph, $r_B - r_A = \Delta r = 4.90 - 3.14 = +1.76$ cm, so that

$$\bar{\mathbf{v}}_{AB} = \frac{\Delta \mathbf{r}}{\Delta t} = \frac{\mathbf{r}_B - \mathbf{r}_A}{t_B - t_A} = \frac{+1.76 \text{ cm}}{0.020 \text{ s}} = +88 \text{ cm/s}$$

This, within experimental error, is the average velocity between A and B. Since it is positive, $\mathbf{v}_{AB}$ is in the positive direction, downward.

The vertical side of the triangle in Fig. 3.4, called the *rise* of the graph, divided by the horizontal side, called the *run* of the graph, gives the average velocity. You may remember from your math classes that this ratio, rise divided by run, is the *slope* of the line that forms the third side of the triangle. In Fig. 3.4, therefore, $\Delta r/\Delta t$ is the slope of the line joining A and B. We thus arrive at the following conclusion: *the average velocity between points A and B on a graph of displacement versus time is the slope of the line joining these two points.*

If we pass to the limiting case of A and B being very close together, the line connecting these two points becomes tangent to the graph. Hence

The slope of the graph of displacement versus time at any point is equal to the instantaneous velocity at that point.

We therefore see that the slope of the graph of displacement versus time has great importance. It tells us the instantaneous velocity of the moving object.

## Example 3.1

Figure 3.5a shows a ball after it has been tossed straight up, and Fig. 3.5b shows the ball's y coordinate as a function of time. Find the ball's instantaneous velocity at (a) P, (b) Q, and (c) N. Also, find the average velocity (d) between A and Q and (e) between A and M.

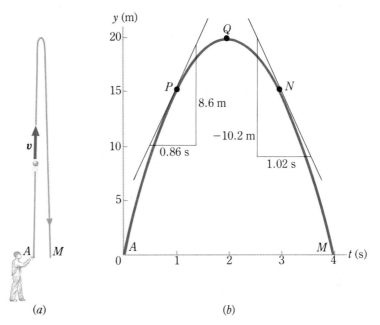

FIGURE 3.5
The essentially straight-line motion in (a) is shown in the graph in (b).

(a)                    (b)

**Reasoning**   The graph shows that the ball rises to a height of 20 m and then falls back down. We use the fact that **v** at any point is given by the slope of the tangent line at that point.

(a) Draw the tangent to the curve at P:

$$\mathbf{v}_P = \text{slope at } P = \frac{8.6 \text{ m}}{0.86 \text{ s}} = 10.0 \text{ m/s}$$

(b) $\mathbf{v}_Q = $ slope at $Q = 0$

The ball stops at Q and then begins to fall.

$$(c) \ \mathbf{v}_N = \text{slope at } N = \frac{-10.2 \text{ m}}{1.02 \text{ s}} = -10.0 \text{ m/s}$$

The sign here is negative because the slope is negative at N. The ball is now moving in the negative y direction; in other words, it is falling. The slope of the curve gives both the magnitude and the direction of the velocity. A negative slope means the velocity is in the negative y direction.

(d) Draw a straight line (a chord) from A to Q (not shown). It rises 20 m in 2.0 s. Then, because $\bar{v} = $ rise/run,

$$\bar{\mathbf{v}}_{AQ} = \text{slope of chord from } A \text{ to } Q = \frac{20 \text{ m}}{2.0 \text{ s}} = 10.0 \text{ m/s}$$

3.4 One-dimensional motion      49

(e) $\bar{v}_{AM}$ = slope of chord from $A$ to $M = \dfrac{0 \text{ m}}{4.0 \text{ s}} = 0$ m/s

We can see that this result is correct because the ball is at the same position for $A$ and $M$. Thus its total displacement is zero. Then

$$\bar{v}_{AM} = \frac{\text{displacement}}{\text{time taken}} = \frac{0 \text{ m}}{4.0 \text{ s}} = 0 \text{ m/s}$$

As we have said, the scientific definition of average velocity differs from the common meaning of average speed. ■

## 3.5 Acceleration

When a car speeds up, we say it has accelerated. In science, however, the term *acceleration* has a more precise meaning than this. We define it in the following way.

Suppose that at a certain instant an object has a velocity (*not* speed) $v_o$ and that its velocity at a later time $t$ is $v_f$. (The subscripts $o$ and $f$ stand for "original" and "final.") The average acceleration $\bar{a}$ of the object during this time interval is

$$\bar{a} = \frac{\text{change in velocity}}{\text{time taken}} = \frac{v_f - v_o}{t} \tag{3.4}$$

In other words, acceleration is the change in velocity (*not* speed) per unit time. The units of acceleration are a velocity unit (for example, meters per second or kilometers per hour) divided by a time unit. Therefore acceleration has units of length divided by time squared, typically meters per second squared.

To see what this definition means in a practical situation, consider a car that starts from rest and attains a speed of 20 m/s in 12 s as it travels in the positive $x$ direction. We have $v_o = 0$ and $v_f = 20$ m/s, both in the positive $x$ direction, and the time taken $t = 12$ s. Therefore

$$\bar{a} = \frac{v_f - v_o}{t} = \frac{(20 - 0) \text{ m/s}}{12 \text{ s}} = 1.67 \text{ m/s}^2$$

in the positive $x$ direction.

Suppose now, instead, that the car is going in the positive $x$ direction and slows from 20 m/s to 0 m/s in 12 s. Then

$$\bar{a} = \frac{v_f - v_o}{t} = \frac{(0 - 20) \text{ m/s}}{12 \text{ s}} = -1.67 \text{ m/s}^2$$

in the positive direction. The car's acceleration is negative in this case because its velocity in the positive $x$ direction decreases; the change in velocity, $v_f - v_o$, is negative, and so $a$ is negative. This means that $a$ is in the negative $x$ direction. In this particular case, we say that the car has decelerated, but *deceleration* is a colloquial term and it is more precise to use the term *acceleration* with the appropriate sign to show direction.

## Example 3.2

Figure 3.5*b* shows the time variation of the vertical (*y*) position of a ball thrown straight upward. Graph the ball's velocity and find its acceleration.

*Reasoning*  As we pointed out, the slope of this graph at any point gives the instantaneous velocity at that point. For example, **v** = 10 m/s at *P* and − 10 m/s at *N*. The velocity is negative at *N* because we are taking the upward direction as positive and the velocity at *N* is downward. If we take slopes at various points along the curve of Fig. 3.5*b*, we find the following velocity data (check one point and see):

| Time (s) | → | 0 | 0.5 | 1.0 | 1.5 | 2.0 | 2.5 | 3.0 | 3.5 | 4.0 |
|---|---|---|---|---|---|---|---|---|---|---|
| Velocity (m/s) | → | 20 | 15 | 10 | 5 | 0 | −5 | −10 | −15 | −20 |

These data are graphed in Fig. 3.6.

Notice from the tabulated data that the velocity in the positive *y* direction decreases 5 m/s each 0.5 s. This tells us at once that the acceleration of the ball is

$$\bar{a} = \frac{\text{change in velocity}}{\text{time taken}} = \frac{-5 \text{ m/s}}{0.5 \text{ s}} = -10 \text{ m/s}^2$$

Thus it appears that an object that is thrown upward has, during its free flight, a downward-directed acceleration of magnitude 10 m/s².

We can also find the ball's acceleration from Fig. 3.6. Notice that, in the typical triangle shown, the rise is the change in velocity and the run is the time taken. Thus, from the triangle, we have

$$\bar{a} = \frac{\text{change in velocity}}{\text{time taken}} = \frac{-14 \text{ m/s}}{1.4 \text{ s}} = -10 \text{ m/s}^2$$

which agrees with our previous result. During its entire trip (both up and down), the ball's acceleration is about 10 m/s² directed downward. When it is going up, the ball slows down 10 m/s each second, and when it is moving down, its speed increases 10 m/s each second. As we shall see in Sec. 3.9, more exact measurements give the ball's acceleration as 9.8 m/s². ■

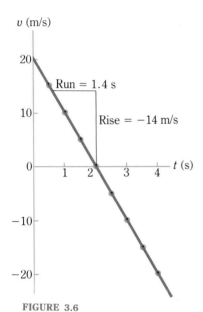

**FIGURE 3.6**

The velocity of the ball in Fig. 3.5*a*. What is the ball's acceleration?

## 3.6 Uniformly accelerated linear motion

Situations in which acceleration varies are often difficult to analyze mathematically. For that reason, we shall restrict ourselves to situations in which acceleration is constant. (We say that an object in such a situation is *uniformly* accelerated.) Although this may seem like a drastic simplification, many physical systems approximate this condition. For example, objects falling freely under the action of gravity near the earth's surface have constant acceleration. Let us now see how to describe the linear motion of objects that undergo uniform (constant) acceleration.

Because the motion is along a straight line, we can simplify our discussion by using plus and minus signs to show direction. Moreover, we shall represent the vector displacement by *x*, the *x*-directed velocity by *v*, and the *x*-directed acceleration by *a*.

**FIGURE 3.7**

It takes a time interval $t$ for the ball to move from $A$ to $B$.

For example, the object in Fig. 3.7 is moving with a constant acceleration $a$ in the $x$ direction. It passes point $A$ with velocity $v_o$ and passes point $B$ with velocity $v_f$ at a time $t$ later. The displacement from $A$ to $B$ is $x$.

For the trip from $A$ to $B$, we can state the following results.

(*a*) The average velocity $\bar{v}$ for the trip is

$$\bar{v} = \frac{\text{displacement}}{\text{time}} = \frac{x}{t}$$

from which

$$x = \bar{v}t \qquad (3.2)$$

(*b*) Because the acceleration is constant, the average and instantaneous accelerations are the same. The definition of acceleration becomes

$$a = \frac{v_f - v_o}{t} \qquad v_f = v_o + at \qquad (3.4)$$

Since the object is undergoing uniform acceleration, its velocity changes linearly with time from $v_o$ to $v_f$. As a result, the average velocity between $A$ and $B$ is simply the average of these two values:

$$\bar{v} = \frac{v_o + v_f}{2} \qquad (3.5)$$

This equation for $\bar{v}$ allows us to find the average velocity for the trip from $A$ to $B$ if the initial and final velocities are known.

We now have three equations that apply to uniformly accelerated motion, Eqs. 3.2, 3.4, and 3.5. They are sufficient to deal with any ordinary situation in which acceleration is uniform.

---

**Example 3.3**

Suppose a car starts from rest and accelerates uniformly to 5.0 m/s in 10 s as it travels along the $x$ axis. Find its acceleration and the distance it travels in this time.

**Reasoning** To begin, let us write down what is known and what is unknown for the 10-s trip. Associating points $A$ and $B$ with the end points of the trip, we have

$$v_o = 0 \qquad v_f = 5.00 \text{ m/s} \qquad t = 10.0 \text{ s} \qquad a = ? \qquad x = ?$$

From Eq. 3.4, we have

$$a = \frac{v_f - v_o}{t} = \frac{(5.00 - 0) \text{ m/s}}{10.0 \text{ s}} = 0.500 \text{ m/s}^2$$

From Eq. 3.5,

$$\bar{v} = \frac{v_f + v_o}{2} = \frac{(5.00 + 0) \text{ m/s}}{2} = 2.50 \text{ m/s}$$

and finally, from Eq. 3.2,

$$x = \bar{v}t = \left(2.50\ \frac{m}{s}\right)(10.0\ s) = 25.0\ m \quad \blacksquare$$

## Example 3.4

Suppose a car traveling at 5.00 m/s is brought to rest in a distance of 20.0 m. Find its acceleration and the time taken to stop. Assume that motion is along the $x$ axis and that acceleration is constant.

*Reasoning*  Following the same procedure as before, we write down the knowns and unknowns for the 20-m trip:

$$v_o = 5.00\ \text{m/s} \qquad v_f = 0 \qquad x = 20.0\ \text{m} \qquad a = ? \qquad t = ?$$

We can first find the average velocity:

$$\bar{v} = \frac{v_o + v_f}{2} = 2.50\ \text{m/s}$$

Now we can use $x = \bar{v}t$ to obtain

$$t = \frac{x}{\bar{v}} = \frac{20.0\ \text{m}}{2.50\ \text{m/s}} = 8.00\ \text{s}$$

The acceleration is obtained from Eq. 3.4:

$$a = \frac{v_f - v_o}{t} = \frac{(0 - 5.00)\ \text{m/s}}{8.00\ \text{s}} = -0.625\ \text{m/s}^2$$

where the negative sign signifies that the car's velocity in the positive $x$ direction is decreasing. In other words, the car is slowing down.  $\blacksquare$

These two illustrations show clearly that following a systematic approach to these problems is advantageous. First write down what is known and what is unknown. Then use the appropriate equation (3.2, 3.4, and/or 3.5) to find the unknowns. In some cases, such as the following, this is not as easily done.

## Example 3.5

A car starts from rest and accelerates at 4.00 m/s² through a distance of 20.0 m. How fast is it then going? How long did it take to cover the 20.0 m?

*Reasoning*  We write down the knowns and unknowns:

$$v_o = 0 \qquad a = 4.00\ \text{m/s}^2 \qquad x = 20.0\ \text{m} \qquad v_f = ? \qquad t = ?$$

Our three equations are

$$x = \bar{v}t \tag{3.2}$$

$$\bar{v} = \frac{v_o + v_f}{2} \tag{3.5}$$

$$a = \frac{v_f - v_o}{t} \tag{3.4}$$

None of these equations can be used directly to obtain any of the unknowns. Each equation contains at least two unknowns, and we have to solve simultaneous equations to obtain any unknowns. In the next section, we discuss two additional equations that are useful in such situations. ∎

## 3.7 Two derived equations for uniformly accelerated motion

Example 3.5 can be solved easily if we obtain two more equations to use along with Eqs. 3.2, 3.4, and 3.5. To do this, we must solve the three known equations simultaneously. Once we have done this, we shall not have to repeat the process, but shall simply make use of the results.

If we substitute the value for $\bar{v}$ given by Eq. 3.5 for $v$ in Eq. 3.2, we obtain

$$x = \tfrac{1}{2}(v_o + v_f)t \tag{3.6}$$

Now we replace $t$ by its value from Eq. 3.4 to get

$$x = \tfrac{1}{2}(v_f + v_o)\left(\frac{v_f - v_o}{a}\right)$$

After clearing fractions and rearranging, we have

$$v_f^2 - v_o^2 = 2ax \qquad\qquad v_f^2 = v_o^2 + 2ax \tag{3.7}$$

(Notice, in passing, that this is just the equation we need to solve the previous example.)

The second equation is found by making a different substitution in Eq. 3.6. Replace $v_f$ in that equation by its value obtained from Eq. 3.4:

$$x = \tfrac{1}{2}v_o t + \tfrac{1}{2}(v_o + at)t$$

which simplifies to

$$x = v_o t + \tfrac{1}{2}at^2 \tag{3.8}$$

We now have five equations to use in the solution of problems involving uniformly accelerated motion along a straight line:

$$x = \bar{v}t \tag{3.9a}$$

$$\bar{v} = \tfrac{1}{2}(v_f + v_o) \tag{3.9b}$$

$$v_f = v_o + at \tag{3.9c}$$

$$v_f^2 = v_o^2 + 2ax \tag{3.9d}$$

$$x = v_o t + \tfrac{1}{2}at^2 \tag{3.9e}$$

Returning now to the problem posed at the end of the preceding section, we had there that

$$v_o = 0 \qquad a = 4.0 \text{ m/s}^2 \qquad x = 20 \text{ m} \qquad v_f = ? \qquad t = ?$$

Using Eq. 3.9$d$, we have at once

$$v_f^2 = v_o^2 + 2ax = 0 + (2)(4.0 \text{ m/s}^2)(20 \text{ m}) = 160 \text{ m}^2/\text{s}^2$$
$$v_f = \pm\sqrt{160 \text{ m}^2/\text{s}^2} = \pm\sqrt{160} \text{ m/s} = \pm 12.6 \text{ m/s}$$

Only the positive sign concerns us here because we consider the motion to be in the positive direction. Therefore $v_f = 12.6$ m/s. (What is the meaning of the answer $-12.6$ m/s?) Now we can use Eq. 3.9$c$ to find $t$:

$$at = v_f - v_o$$
$$(4.0 \text{ m/s}^2)t = 12.6 \text{ m/s} - 0$$
$$t = 3.15 \text{ s}$$

---

**Example 3.6**
Find the time a car takes to travel 98 m if it starts from rest and accelerates at 4.0 m/s$^2$.

*Reasoning*   The knowns and unknown are

$$v_o = 0 \qquad a = 4.0 \text{ m/s}^2 \qquad x = 98 \text{ m} \qquad t = ?$$

The appropriate equation is

$$x = v_o t + \tfrac{1}{2}at^2$$
$$98 \text{ m} = (0 \text{ m/s})(t) + \tfrac{1}{2}(4.0 \text{ m/s}^2)(t^2)$$
$$t = \sqrt{49} = 7.0 \text{ s}$$

You should check to see that the units of $t$ are as stated.   ∎

---

**Example 3.7**
A car is moving at 60 km/h (16.7 m/s) when it begins to slow down with a deceleration of 1.50 m/s$^2$. How long does it take to travel 70 m as it slows down?

*Reasoning*   The knowns and unknown are

$$v_o = 60 \text{ km/h} = 16.7 \text{ m/s} \qquad x = 70 \text{ m} \qquad a = -1.50 \text{ m/s}^2 \qquad t = ?$$

The acceleration is negative because the velocity is decreasing in the direction of positive motion. Notice that we have changed the units of velocity to meters per second so that we won't be using two sets of distance and time units in the same problem. We must always make this kind of change. A method for changing from one set of units to another is explained in the next section.
   If we now choose to solve this problem using Eq. 3.9$e$, we run into trouble with algebra:

$$x = v_o t + \tfrac{1}{2}at^2$$
$$70 = 16.7t + \tfrac{1}{2}(-1.50)t^2 = 16.7t - 0.75t^2$$

where you should supply the units. We can solve this quadratic equation with the quadratic formula, but it is usually simpler to find one of the other unknowns first and avoid the complicated algebra. For example, using Eq. 3.9$d$, we have

$$v_f^2 = v_o^2 + 2ax = (16.7 \text{ m/s})^2 + (2)(-1.50 \text{ m/s}^2)(70 \text{ m})$$
$$= 279 \text{ m}^2/\text{s}^2 - 210 \text{ m}^2/\text{s}^2$$

from which

$$v_f = \pm 8.3 \text{ m/s} = +8.3 \text{ m/s}$$

Why do we choose the positive sign? Now, using Eq. 3.9$c$,

$$v_f = v_o + at$$
$$8.3 \text{ m/s} = 16.7 \text{ m/s} - (1.50 \text{ m/s}^2)t$$

from which $t = 5.6$ s. ∎

*Exercise* Show that the quadratic equation for $t$ also gives $t = 5.6$ s. Why must we discard the alternative value, $t = 16.7$ s? ∎

## 3.8 Conversion of units

Because speed, acceleration, and most other physical quantities can be measured in various units, we often need to convert from one unit to another. In the preceding example, we had to convert from kilometers per hour to meters per second. In this section, we learn how to make such conversions.

We first must know how units are related. The relations 60 s = 1 min and 1 mm = 0.001 m are typical of many you already know. A tabulation of many others appears on the inside cover of this text. Using these relations in converting from one unit to another is a fairly simple process.

The conversion method we use is based on the following fact: if you multiply a quantity by unity, the quantity remains unchanged. For example, suppose we wish to change 6600 s to its equivalent in minutes. We know that 60 s = 1 min. To obtain unity from this equation, we can divide it by either 1 min or 60 s to obtain

$$\frac{60 \text{ s}}{1 \text{ min}} = 1 \qquad \text{or} \qquad 1 = \frac{1 \text{ min}}{60 \text{ s}}$$

Since each of these *conversion factors* is unity, multiplying or dividing another quantity by either one does not change that quantity. Therefore we can write

$$6600 \text{ s} = (6600 \text{ s}) \left( \frac{1 \text{ min}}{60 \text{ s}} \right) = 110 \text{ min}$$

Notice that we carry units along and cancel them just as though they were algebraic symbols.

As you see, the conversion factor 1 min/60 s changed time in seconds to time in minutes. We did not use the conversion factor 60 s/1 min because it gives

$$6600 \text{ s} = (6600 \text{ s}) \left( \frac{60 \text{ s}}{1 \text{ min}} \right) = \frac{396,000 \text{ s}^2}{1 \text{ min}}$$

Although this is a correct result, it does not tell us what we wish to know, the time in minutes. We must select the conversion factor that is suitable for the change we wish to accomplish.

Sometimes we wish to make a conversion that requires more than one conversion factor. For example, to convert 5.0 m/s to kilometers per hour, we proceed as follows. Since we know that 1 km = 1000 m and 1 h = 3600 s, we can write four conversion factors:

$$\frac{1 \text{ km}}{1000 \text{ m}} = 1 \qquad \frac{1000 \text{ m}}{1 \text{ km}} = 1 \qquad \frac{1 \text{ h}}{3600 \text{ s}} = 1 \qquad \frac{3600 \text{ s}}{1 \text{ h}} = 1$$

We therefore have

$$5.0 \text{ m/s} = \left(5.0 \frac{\text{m}}{\text{s}}\right)\left(\frac{1 \text{ km}}{1000 \text{ m}}\right) = \frac{5.0}{1000} \frac{\text{km}}{\text{s}}$$

Continuing, we have

$$5.0 \frac{\text{m}}{\text{s}} = \frac{5.0}{1000} \frac{\text{km}}{\text{s}} = \left(\frac{5.0}{1000} \frac{\text{km}}{\text{s}}\right)\left(\frac{3600 \text{ s}}{1 \text{ h}}\right) = 18.0 \frac{\text{km}}{\text{h}}$$

Thus 5 m/s = 18 km/h. As you see, converting units requires a judicious use of conversion factors.

---

### Example 3.8

A car is capable of accelerating from rest to 30 m/s in 9.0 s. Find the acceleration of the car in meters per second squared and kilometers per hour squared.

***Reasoning*** We know that $t = 9.0$ s, $v_o = 0$, and $v_f = 30$ m/s. Therefore

$$a = \frac{v_f - v_o}{t} = \frac{30 \text{ m/s}}{9.0 \text{ s}} = 3.33 \text{ m/s}^2$$

Making use of the relations 1 h = 3600 s and 1000 m = 1 km, we have

$$3.33 \text{ m/s}^2 = \left(3.33 \frac{\text{m}}{\text{s}^2}\right)\left(\frac{1 \text{ km}}{1000 \text{ m}}\right)\left(\frac{3600 \text{ s}}{1 \text{ h}}\right)\left(\frac{3600 \text{ s}}{1 \text{ h}}\right)$$

$$= 43{,}200 \text{ km/h}^2 \quad \blacksquare$$

---

***Exercise*** Express 60 cm³ in cubic meters. *Answer: $6 \times 10^{-5}$ m³* ■

---

## 3.9 Freely falling bodies

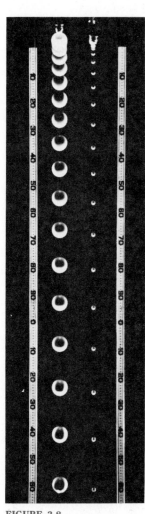

FIGURE 3.8
The falling objects are made visible at equal time intervals by means of a strobe light. Although the objects are of different size and weight, they fall in unison. (*Education Development Center*)

Consider the experiment shown in Fig. 3.8: two objects falling freely under the pull of gravity. A strobe light captures the objects' positions at equally spaced time intervals. Notice that, despite differences in size and weight, the objects fall with the same acceleration. This fact was first pointed out by Galileo (1564–1642). Measurements show that, on earth, a *freely falling* object accelerates downward with an acceleration of 9.8 m/s². After it is dropped, a freely falling object has the following speeds after successive 1-s intervals: 9.8 m/s, 19.6 m/s, 29.4 m/s, and so on. The

# Theories of free fall

The study of the behavior of falling objects has a long and interesting history. It is a prime example of the difference between good science and poor science. We begin the story at the time of the famous philosopher Aristotle (384–322 B.C.).

In Aristotle's time, it was known that a light object falls through air more slowly than a heavy object. Noting this fact, Aristotle based his theory of falling objects on the idea that all objects are composed of four elements: earth, air, fire, and water. Objects composed primarily of earth and water try to reach their natural resting place, the earth; hence, when they are allowed to do so, they fall to the earth. Objects composed of air try to rise to *their* natural resting place, the sky. According to Aristotle, a stone falls fast because it is composed primarily of earth and thus strongly seeks its natural resting place. A feather, however, is mostly air and hence seeks the earth less strongly; it thus falls more slowly than a stone. Moreover, Aristotle concluded that the speed of a falling object is constant. If you let a feather (or a sheet of facial tissue) fall through the air, you will see how he came to this conclusion. The fact that a stone falls with ever-increasing speed eluded him, however, because he had no way to measure the flight of such swiftly falling objects. Aristotle was a highly respected philosopher, and so few people were inclined to question his theories and conclusions. For this reason, little progress was made in understanding how falling objects behave until the time of Galileo, nearly 2000 years later.

By the year 1250, science as we know it began to appear. Roger Bacon (1214–1294) was among the first to espouse the idea that experience (that is, experiment) is necessary for the development of reliable theories about natural behavior. However, it appears that even he did not recognize the importance of controlling the variables that influence the results of an experiment. As late as 1605, in his noted treatise *The Advancement of Learning,* Francis Bacon (1561–1626) insisted—contrary to then existing practice—that theories be based upon experimentally determined facts.

It was Galileo Galilei (1564–1642) who finally led the way to the development of true science, performing important experiments in astronomy, optics, and mechanics. The most important aspect of his work was his recognition that meaningful experiments must be *controlled.* By this we mean that, insofar as possible, only one variable at a time should be changed during an experiment. Thus Galileo recognized that comparing the way a feather and a stone fall is a nearly uninterpretable experiment because there are so many differences between the two objects. He designed ingenious experiments to time accurately the way similar objects of different weights fall and was able to establish that an object's weight does not influence its acceleration, provided the effects of air friction are negligible. Furthermore, he found that freely falling objects do not fall with constant speed, as Aristotle believed, but instead undergo constant acceleration.

Over the years, the methods of science have been further refined, but experiment remains at the heart of all good science. Without carefully controlled experiments to provide us with unambiguous results, we can only guess about the behavior of the world about us. To be of value, scientific theories must be based upon experimental fact.

downward-directed velocity increases by 9.8 m/s each second; acceleration is 9.8 m/s$^2$ downward.

Despite this assertion, we know that a marble, a feather, and a piece of facial tissue all fall in different ways. These are not *freely* falling bodies, however. As the feather falls, the friction of the air against it tends to stop its fall. This friction force nearly balances the pull of gravity on the feather and hence it is decidedly not free to fall. Similarly, a facial tissue falls slowly because of air effects. A marble, on the other hand, weighs much more than a feather or a facial tissue. The pull of gravity on it is far larger than the air friction retarding its fall. We can therefore consider the marble to be falling freely under the action of gravity unless its speed becomes so high that the force of air friction becomes very large.

The fall of objects subject to no appreciable force except the pull of gravity is exceptionally simple, however. Experiment shows that they fall with a downward acceleration (on earth) of 9.80 m/s$^2$. We call this the **acceleration due to gravity** and represent it by the symbol $g$. The value of $g$ varies slightly from place to place on the earth. Typical values are given in Table 3.1.

Look back to the situation in Fig. 3.5, which shows the motion of a ball subject only to the force of gravity. We analyzed this motion in Example 3.2 and in Fig. 3.6. You will recall that we found the ball's acceleration to be about 10 m/s$^2$ downward, whether the ball was rising or falling. This is another example of the fact that a freely falling body has an acceleration of 9.8 m/s$^2$ downward. It does not matter whether the ball is rising or falling; its acceleration is still $g$ downward.

We shall soon analyze the free-fall motion of objects in several examples, but before we do so, notice the following facts. First, whether the object is rising or falling, its acceleration is 9.8 m/s$^2$ downward. If we choose to consider up as positive, then the acceleration due to gravity, being directed downward, is $-9.8$ m/s$^2$. It is very important that we keep track of the signs of displacement, velocity, and acceleration because the signs tell us the direction of these quantities. Second, because the acceleration is constant (9.8 m/s$^2$ downward), motion under the action of gravity is uniformly accelerated motion and our five motion equations apply. However, we shall replace $x$ by $y$ in the equations to emphasize the vertical nature of the motion.

You should be very careful in applications that involve up-and-down motion. It is absolutely necessary to decide at the beginning which direction is positive. The choice is arbitrary, but once you have made it in a particular problem, you must retain that choice throughout the problem.

■ **TABLE 3.1**
**Acceleration due to gravity $g$**

| Place | $g$ ($m/s^2$) |
|---|---|
| Beaufort, N.C. | 9.7973 |
| New Orleans | 9.7932 |
| Galveston | 9.7927 |
| Seattle | 9.8073 |
| San Francisco | 9.7997 |
| St. Louis | 9.8000 |
| Cleveland | 9.8024 |
| Denver | 9.7961 |
| Pikes Peak | 9.7895 |

*Example 3.9*

A stone is dropped from a bridge. If it takes 3.0 s for the stone to hit the water beneath the bridge, how high above the water is the bridge? Ignore air friction. (Notice here that the problem ends the instant before the stone hits the water. It is only during this interval that the stone is a freely falling body.)

*Reasoning*    Taking up as positive, we have these knowns and unknown:

$$a = -9.8 \text{ m/s}^2 \qquad t = 3.0 \text{ s} \qquad v_o = 0 \qquad y = ?$$

Use Eq. 3.9$e$, replacing $x$ with $y$:

$$y = v_o t + \tfrac{1}{2} a t^2$$

$$= 0 + \tfrac{1}{2}(-9.8 \text{ m/s}^2)(9.0 \text{ s}^2) = -44 \text{ m}$$

Why is $y$ negative?    ■

## Galileo Galilei (1564 – 1642)

AIP Niels Bohr Library

Galileo, the son of a musician, began his career as a medical student, but soon abandoned medicine for mathematics and the physical sciences. When he was 25 years old, he was appointed professor at the University of Pisa, Italy, and 3 years later he accepted a professorship at the University of Padua. His investigations into the behavior of falling objects were carried out early in his career. When, in 1609, Galileo learned of the newly invented telescope, he constructed one for his own use. Pointing it at the heavens, he discovered many previously unknown luminous objects in the sky. These newly found stars, as well as his closer glimpses of the moon and planets, raised doubts in his mind concerning the traditional view of the earth as the center of a perfect universe. His widely publicized views were ridiculed by many and confirmed and praised by others. In 1611 he was invited to Rome to receive many honors, including an audience with the Pope, but by 1616 his attacks on the church-sanctioned concept of an earth-centered universe drew the wrath of the clergy and he found it wise to mute his discussion of the subject. After nearly 15 years of silence and work, he defiantly published his observations and conclusions, which still contradicted the church's doctrines. The result was trial and house arrest, until Galileo, by now an old man broken in health and spirit, denied his discoveries. Even so, he used the remaining decade of his life to complete his studies on motion. These are described in his greatest work, the *Dialogues Concerning Two New Sciences,* which was published in Holland after being smuggled out of Italy.

---

***Exercise***   What is the stone's velocity just before it strikes the water?   *Answer: − 29.4 m/s* ■

---

### Example 3.10
A ball is thrown upward with a speed of 15.0 m/s. How high does it go? What is its speed just before the thrower catches it? How long is it in the air? Ignore air friction.

***Reasoning***   *Method 1:* It is important that we define the trip we are interested in. Consider the motion from $A$ to $B$ in Fig. 3.9. Taking up as positive, we have the following knowns and unknown:

$$a = -9.8 \text{ m/s}^2 \qquad v_o = 15.0 \text{ m/s} \qquad v_f = 0 \qquad y = ?$$

We take $v_f = 0$ because the ball's speed decreases as the ball rises. At point $B$, the ball is just ready to fall and its speed is zero. Use Eq. 3.9d, replacing $x$ with $y$:

$$2ay = v_f^2 - v_o^2$$
$$-19.6y = 0 - 225$$

Supply the units in this equation and show that $y = 11.5$ m from $A$ to $B$.

B

A  C

The ball is thrown upward at point
$A$ with a speed of 15 m/s. Since it
stops at point $B$, its velocity there
is zero.

For the second part of the problem, we are concerned with the trip from $B$ to $C$. This is a new trip, and so we have new knowns. Taking up as positive, we have

$$a = -9.8 \text{ m/s}^2 \qquad y = -11.5 \text{ m} \qquad v_o = 0 \qquad v_f = ?$$

Why is $y$ negative for this trip? Use

$$2ay = v_f^2 - v_o^2$$
$$(-19.6 \text{ m/s}^2)(-11.5 \text{ m}) = v_f^2 - 0$$
$$v_f = \pm\sqrt{225 \text{ m}^2/\text{s}^2} = -15.0 \text{ m/s}$$

We use the negative sign because the final velocity is downward. Notice that the final speed is the same as the speed with which the ball was thrown. This is an example of a famous basic law of physics, the *law of conservation of energy*, which we shall examine in considerable detail in Chap. 5.

*Method 2:* There is another way of solving this problem. Consider the full trip from $A$ to $B$ to $C$. The starting point is $A$, and the end point is $C$. Taking up as positive, we know

$$a = -9.8 \text{ m/s}^2 \qquad v_o = 15.0 \text{ m/s} \qquad y = 0$$

We write that $y$, the vector displacement from the starting point $A$ to the end point $C$, is zero, since the ball returned to the thrower. Use

$$2ay = v_f^2 - v_o^2$$
$$0 = v_f^2 - (15.0 \text{ m/s})^2$$
$$v_f = \pm 15.0 \text{ m/s} \qquad \text{as before}$$

For our trip, the minus sign is correct. Before we investigate the reason for choosing the plus or minus answer, let us compute the time for the trip. Use

$$y = v_o t + \tfrac{1}{2} at^2$$
$$0 = 15t - 4.9t^2$$

Supply the units to this equation, and show that $t = 0$ and $t = 3.06$ s. Here again we have a choice of answers.

The reason for the two answers in each case is not hard to determine. Our knowns for this way of solving the problem are true for two trips, the one from $A$ to $B$ to $C$ and the trip from $A$ that stops immediately after it starts. For this latter trip, all the known conditions are true, and the time taken for it is zero. In addition, the end velocity is still the same as the starting velocity. ∎

### Example 3.11
How fast must a ball be thrown straight upward if it is to return to the thrower in 3.0 s? See Fig. 3.9 and neglect air friction.

*Reasoning*  We take up as positive and notice that the displacement vector from beginning to end has zero vertical length. Then

$$a = -9.8 \text{ m/s}^2 \qquad t = 3.0 \text{ s} \qquad y = 0 \qquad v_o = ?$$

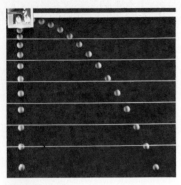

**FIGURE 3.10**
A flash photograph of two golf balls, one falling from rest and the other projected horizontally. The interval between light flashes is $\frac{1}{30}$ s, and the horizontal lines are 15 cm apart. (*Educational Development Center*)

We can use Eq. 3.9e, $y = v_o t + \frac{1}{2}at^2$, to give

$$0 = (v_o)(3.0 \text{ s}) - (4.9 \text{ m/s}^2)(9.0 \text{ s}^2)$$

We then find

$$v_o = 14.7 \text{ m/s} \quad \blacksquare$$

*Exercise* How high did the ball go? *Answer: 11.0 m* ∎

### 3.10 Projectile motion

Very seldom does a baseball, a bullet, or a golf ball follow a straight-line path. These objects are typical projectiles, and we call their motion *projectile motion*. A typical projectile path is shown in Fig. 3.10. We see there a strobe photograph of two balls. Ball 1 is dropped straight downward and falls with an acceleration of 9.8 m/s², as we have already discussed. Ball 2 is projected horizontally at the same time ball 1 is dropped. Notice, though, that it falls just as fast as ball 1. Apparently, ball 2 falls vertically with an acceleration of 9.8 m/s² even though it is moving horizontally at the same time. Indeed, measurements on the photograph will convince you of this fact:

A projectile moves horizontally at constant speed as it falls vertically with acceleration *g*.

We shall justify this behavior in terms of Newton's laws in Chap. 4. For now, however, we simply accept as experimental fact that, when air friction effects are negligible, a projectile undergoes two perpendicular motions simultaneously: it moves vertically with downward acceleration *g* and horizontally with constant horizontal velocity.

We show this motion in detail in Fig. 3.11. A baseball is projected horizontally at *A* with velocity $v_h$. If air friction is negligible, the baseball maintains this same horizontal velocity until it strikes something. Simultaneously, its downward vertical velocity increases 9.8 m/s each second as it undergoes free fall. Let us now analyze this type of motion.

**FIGURE 3.11**
The projected ball undergoes two independent motions at right angles to each other.

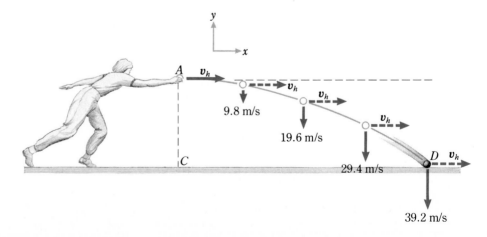

Because the two perpendicular motions are independent of each other, we can analyze them separately. Consider first the horizontal motion, which is extremely simple because it is motion with constant velocity $v_h$. Thus in the horizontal part of the motion, the ball's acceleration is zero and we have

$$v_o = v_f = \bar{v} = v_h$$
$$x = \bar{v}t = v_h t$$

The vertical motion is not much more complicated because the ball simply moves in the $y$ direction under the acceleration due to gravity. Thus the $y$-direction motion is that of a freely falling body. Therefore the vertical motion is not new to us, and we should be able to analyze it with no difficulty.

Our procedure, then, is to recognize that the free motion of a ball, bullet, or any other projectile contains within it two separate problems. The horizontal problem is simply motion with constant velocity, and the vertical motion is the motion of a free body in a vertical line. We compute each portion of the projectile-motion problem separately and then combine the solutions to obtain the answer.

## Example 3.12

Consider the situation in Fig. 3.11. Suppose the ball leaves the thrower's hand at $A$ and travels horizontally 2.0 m above the ground with a velocity of 15 m/s. Where will it hit the ground? (That is, how far from $C$ is $D$?)

**Reasoning**  We start by splitting the problem into two parts:

| Horizontal | Vertical (down positive) |
|---|---|
| $v_o = v = \bar{v} = 15$ m/s | $v_o = 0$ |
| $x = \bar{v}t = 15t$ | $a = 9.8$ m/s$^2$ |
| To find $t$, we solve the vertical problem. | $y = 2.0$ m |
| | To find $t$, we use |
| | $y = v_o t + \frac{1}{2}at^2$ |
| | $2.0 = 4.9t^2$ |
| | $t = 0.639$ s |

Once the time of flight, that is, the time it takes for the ball to drop to the ground, is found from the vertical problem, the result can be used in the horizontal problem:

$$x = (15 \text{ m/s})(0.639 \text{ s}) = 9.58 \text{ m}$$

In other words, the ball travels only 9.58 m horizontally before gravity causes it to fall to the ground. Actually, one usually throws a ball somewhat upward if one wants it to travel any great distance. If the ball has an initial upward component to its velocity, it will take longer to fall to the ground and so will have more time to travel in a horizontal direction.  ■

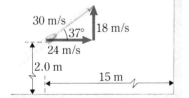

**FIGURE 3.12**

Where will the projectile hit the wall? Is it going up when it hits, or is it already on the way down?

## Example 3.13

An arrow is shot with a velocity of 30 m/s at an angle of 37° above the horizontal. It is initially 2.0 m above the ground and 15 m from a wall, as shown in Fig. 3.12. (*a*) At what height above the ground does it hit the wall? (*b*) Is it still going up just before it hits, or is it already on its way down? Neglect air friction.

Assume that the arrow acts as a projectile not influenced by the air. Consider the trip from the starting point to the wall. For part *a*, because the horizontal component of the velocity is 30 cos 37° and the vertical component is 30 sin 37° m/s, we have

| Horizontal | Vertical (up positive) |
|---|---|
| $v_o = v = \bar{v} = 24$ m/s | $v_o = 18$ m/s |
| $x = 15$ m | $t = \frac{5}{8}$ s |
| $= \bar{v}t$ | $a = -9.8$ m/s² |
| $15 = 24t$ | $y = ? = $ distance above launch point |
| $t = \frac{5}{8}$ s | $y = v_o t + \frac{1}{2}at^2$ |
| We now use this value in the vertical problem. | $= (18)(\frac{5}{8}) - (4.9)(\frac{5}{8})^2$ |
| | $= 11.2 - 1.9$ |
| | $= 9.3$ m above the starting point |

Hence, the arrow strikes the wall 9.3 m above the starting point. Since this point is 2.0 m above the ground, the arrow strikes the wall 11.3 m above the ground.

For part *b*, let us find the vertical component of the velocity just before the arrow hits the wall. We have, in the vertical problem,

$$v = v_o + at = 18 \text{ m/s} - (9.8 \text{ m/s}^2)(\tfrac{5}{8} \text{ s}) = +11.9 \text{ m/s}$$

Since up is positive, a positive velocity means an upward velocity. Hence, the arrow is on its way up when it hits.

The arrow's total velocity just before it strikes has an upward component of 11.9 m/s and a horizontal component of 24 m/s. These can be added in the usual way to find the magnitude of the velocity at that point:

$$v_{\text{total}} = \sqrt{(11.9)^2 + (24)^2} \text{ m/s} = 26.8 \text{ m/s} \quad \blacksquare$$

*Exercise* How far away must the wall be if the arrow is to hit at this same height but on the way down? *Answer: 73.2 m* ▪

## Minimum learning goals

When you finish this chapter, you should be able to

1 Define (*a*) speed, (*b*) velocity, (*c*) acceleration, (*d*) gravitational acceleration, (*e*) free fall, (*f*) conversion factor.

2 Describe how you measure (*a*) the average velocity of an object as it moves from *A* to *B* and (*b*) the instantaneous velocity at any point.

3 Describe an object's velocity when you are given the graph of its motion showing *x* as a function of *t*.

4 State the five uniform-motion equations, explain the symbols in them, and state the restrictions on their use.

5 Solve simple problems involving uniformly accelerated motion.

6 Convert from one set of units to another.

7 Find the distance traveled over level ground by a projectile shot at a known angle and speed from a given height.

## Questions and guesstimates

**1** Give an example of a case in which an object's velocity is zero but its acceleration is not.

**2** Can an object's velocity ever be in a direction other than the direction of its acceleration? Explain.

**3** Sketch graphs of velocity and acceleration as a function of time for a car as it strikes a telephone pole. Repeat for a billiard ball in a head-on collision with the edge of the billiard table.

**4** Are any of the following true? (*a*) An object can have a constant velocity even though its speed is changing. (*b*) An object can have a constant speed even though its velocity is changing. (*c*) An object can have zero velocity even though its acceleration is not zero. (*d*) An object subjected to a constant acceleration can reverse its velocity.

**5** A rabbit enters the end of a drainpipe of length *L*. Its motion from that instant is shown in Fig. P3.1. Describe the motion in words.

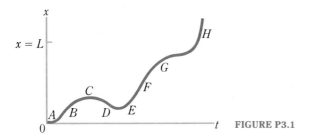

FIGURE P3.1

**6** A high-school girl who is an average runner completes a 100-m dash by running twice around an indoor circular track that is 50 m in circumference. Estimate her average speed and average velocity.

**7** A stone is thrown straight up in the air. It rises to a height *h* and then returns to the thrower. For the time the stone is in the air, sketch the following graphs: *y* as a function of *t*; *v* as a function of *t*; *a* as a function of *t*.

**8** Under what condition is it wrong to say that an object's acceleration is negative when the object is thrown upward? Does the sign of the acceleration depend at all on the direction of motion? Can an object's acceleration be positive when the object is slowing down?

**9** The acceleration due to gravity on the moon is only about one-sixth that on earth. Estimate the ratio of the height to which you could throw a baseball on the moon to the corresponding height on earth.

**10** How could you best analyze Fig. 3.8 to obtain the value of *g*? Assume the time interval between strobe flashes is known.

**11** Airplane enthusiasts hold meets at which they show off their skills. One event is to drop a sack of sand exactly in the center of a circle on the ground while flying at a predetermined height and speed. What is so difficult about that? Don't they just drop the sack when they are directly above the circle?

**12** Parents should be prepared for nearly anything. Suppose your child wants to find out how fast a slingshot can shoot a stone. Devise a method for finding out. Assume the only tool you have is a meterstick.

**13** If you want to hit a distant stationary object with a rifle, you do not aim by lining up the object with the hole in the barrel of the gun. How do you aim?

## Problems

*The problems at the end of every chapter are divided into three levels of difficulty: standard, somewhat difficult (indicated by one square ▪), and most difficult (two squares▪). Unless otherwise stated, assume uniform acceleration.*

**1** A particle shooting eastward from a nuclear reactor along a straight line travels 10.0 m in $6.3 \times 10^{-4}$ s. What is its average velocity?

**2** An atomic accelerator shoots out particles moving at $2.9 \times 10^8$ m/s. How long does it take such particles to travel 3.0 mm?

**3** In a television tube, electrons are shot from a gun at one end of the tube and strike the picture screen at the other end, where the light we see is emitted. Suppose electrons shoot from the gun with a velocity of $8.0 \times 10^7$ m/s toward the screen 20 cm away. How long does it take the electrons to travel from gun to screen?

**4** Sonar-type depth gauges measure the length of time it takes a sound pulse to go from the water surface to the lake bottom and back. If a lake is 12.0 m deep and the speed of sound in water is 1450 m/s, how long does it take a pulse sent down from the lake surface to return?

**5** During a videogame, a point moves 8.0 cm in the positive $x$ direction and then 6.0 cm in the positive $y$ direction, all in 2.5 s. (*a*) What was the point's average velocity during this time interval? (*b*) What was its average speed?

**6** To get to school in the morning, a boy goes three blocks east and four blocks north in 20 min. Find his average (*a*) speed and (*b*) velocity.

**7** Figure P3.2 shows the motion of an ant along a straight line. Find the ant's average velocity from (*a*) $A$ to $E$; (*b*) $B$ to $E$; (*c*) $C$ to $E$; (*d*) $D$ to $E$; (*e*) $C$ to $D$.

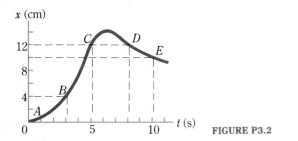

FIGURE P3.2

**8** The motion of a bug along a wire strung along the $x$ axis is shown in Fig. P3.2. Find the bug's average velocity from (*a*) $B$ to $D$; (*b*) $D$ to $E$; (*c*) $A$ to $D$; (*d*) $A$ to $B$.

**9** Person $A$ can run at a top speed of 5.0 m/s, whereas person $B$ can run at only 3.0 m/s. They are to race 300 m. To make the race more even, $A$ is required to start $t$ seconds later than $B$. How large should $t$ be if the race is to end in a tie?

**10** For the situation in Prob. 9, the handicap is to be made this way: person $B$ is to be given a head-start distance $s$, and $A$ must run the full 300 m. Both start at the same time. How large should $s$ be if the race is to end in a tie?

**11** A girl walks eastward along a street; Fig. P3.3 is a graph of her displacement from home. Find her average velocity for the whole time interval shown and her instantaneous velocity at $A$, $B$, and $C$.

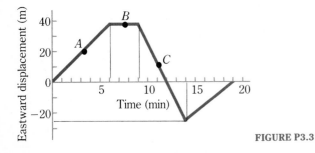

FIGURE P3.3

**12** Referring to Fig. P3.3 for the motion of a dog along an eastwest street, (*a*) find the average velocity for the time interval $t = 7$ min to $t = 14$ min. Also find the instantaneous velocity at (*b*) $t = 13.5$ min and (*c*) $t = 15$ min.

**13** The graph of a particle's motion along the $x$ axis is given in Fig. P3.4. Find the (*a*) average velocity for the interval from $A$ to $C$. Also find the instantaneous velocity at (*b*) $D$ and (*c*) $A$.

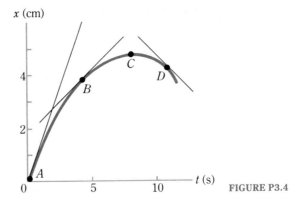

FIGURE P3.4

**14** For the motion along the $x$ axis graphed in Fig. P3.4, (*a*) find the average velocity from $C$ to $D$. Also find the instantaneous velocity at (*b*) $B$ and (*c*) $C$.

**15** Two girls start running straight toward each other from two points 200 m apart. One runs with a speed of 5.0 m/s, and the other moves at 7.0 m/s. How close are they to the slower one's starting point when they meet?

**16** A truck traveling 15 m/s due west is on a collision course with a car going due east at 30 m/s. They are 800 m apart. How long does it take them to hit? How far are they from the truck's original position when they collide? Assume both speeds remain constant.

**17** A car moving on a straight road accelerates from 3.1 m/s to 6.9 m/s in 5.0 s. What is its average acceleration?

**18** An airplane in straight-line flight slows from 345 km/h to 217 km/h in 40 s. What is its average acceleration in meters per second squared?

**19** A car moving at 25 m/s skids to a stop in 14.0 s. Find the average acceleration and the distance the car travels while stopping.

**20** A car manufacturer claims that its car can accelerate from rest to 28 m/s in 14 s. Find the average acceleration and the distance covered in this time.

**21** A bullet moving at 150 m/s strikes a tree and penetrates 3.5 cm before stopping. Find its acceleration and the time taken to stop.

**22** In a television, electrons shoot down the picture tube and strike fluorescent material on the screen, causing light to be given off and producing the picture. Suppose that in a certain set, the electrons are accelerated from rest to $2.0 \times 10^8$ m/s in about 1.20 cm. What is the acceleration of the electrons; how long does the acceleration take?

**23** A rough value for the maximum deceleration of a skidding automobile is 6.5 m/s². Using this value, determine how long it takes a car going 30 m/s to stop. How far does the car go in this time?

■ **24** A truck initially traveling at 30 m/s decelerates at 1.50 m/s². Find (*a*) how long it takes the truck to stop, (*b*) how far it moves in that time, and (*c*) how far it moves in the third second after the brakes are applied.

■ **25** A proton moving with a speed of $1.0 \times 10^7$ m/s passes through a sheet of paper 0.020 cm thick and emerges with a speed of $3.00 \times 10^6$ m/s. Assuming uniform deceleration, find the deceleration and the time taken to pass through the paper.

■ **26** The driver of a car going 25 m/s suddenly notices a corner at which she wishes to turn. At the instant she applies the brakes, the corner is 60 m ahead. The car decelerates uniformly and reaches the corner 3 s later. (*a*) How fast is the car moving as it passes the corner? (*b*) What is the magnitude of its acceleration during the 3 s?

■ **27** A locomotive is pulling a train 137 m long (including the locomotive). It accelerates uniformly from rest and reaches a crossing 1.80 km from its starting point after 900 s. (*a*) How long after the locomotive reaches the crossing does the train clear the crossing? Assume that the acceleration remains constant and ignore the width of the crossing. (*b*) How fast is the train moving as it clears the crossing?

■ **28** The first car of a stationary train is blocking a crossing. Just as the train begins to move, a waiting motorist notices that it takes 20 s for one railcar to pass through a distance equal to its length. Find the train's acceleration in terms of the length *L* of the railcar. Assuming constant acceleration, how long after the train starts will the following 40 railcars have passed the motorist?

■ **29** A car is traveling at 30 m/s along a road parallel to a railroad track. How long does it take the car to pass an 800-m-long train traveling at 20 m/s (*a*) in the same direction and (*b*) in the opposite direction?

■ **30** Just as a car starts to accelerate from rest at 1.40 m/s², a bus moving with a constant speed of 15 m/s passes it in a parallel lane. (*a*) How long before the car overtakes the bus? (*b*) How fast is the car going then? (*c*) How far has the car gone at that point?

■ **31** Two trains, both traveling at 20 m/s, are headed toward each other on the same track. When they are 3.0 km apart, they see each other and begin to decelerate. (*a*) If their accelerations are uniform and equal, what must the magnitudes of these accelerations be if the trains are to barely avoid collision? (*b*) If only one train slows with this acceleration, how far will it go before collision occurs?

■ **32** A police car is at rest alongside a road when a car passes at a constant speed of 32 m/s. At 2 s later, the police car accelerates from rest at 3.60 m/s². (*a*) If it maintains this acceleration, how far does the police car move before catching the other car? (*b*) What is the police car's speed then?

**33** Change to centimeters: (*a*) 6.0 in; (*b*) 3.0 ft; (*c*) 2.5 yd. (1 in = 2.54 cm, 1 ft = 12 in, and 1 yd = 3 ft.)

**34** Change to gallons: (*a*) 12.0 liters; (*b*) 4800 cm³; (*c*) 360 in³. (1 liter = 1000 cm³, 1 gal = 3.785 liters, 1 m³ = 35.3 ft³, and 1 liter = 61 in³.)

**35** Change to meters per second: (*a*) 5.0 km/h; (*b*) 37 in/h; (*c*) 25 mi/h. (1.609 km = 1 mi)

**36** Change to miles per hour: (*a*) 60 km/h; (*b*) 260 m/s; (*c*) 73 ft/ms. (5280 ft = 1 mi)

**37** A flower pot slips off a window ledge that is 4.0 m above a bug sitting below. (*a*) How fast is the pot moving when it hits the bug? (*b*) How much time does the bug have to move after the pot starts on its journey?

**38** A frightened diver hangs by his fingers from a diving board with his feet 8.0 m above the water. (*a*) How long after his fingers give out does he strike the water? (*b*) How fast is he going then?

■ **39** A stone is thrown straight upward with a speed of 20 m/s. How high does it go? How long does it take to rise to maximum height?

■ **40** A girl throws a stone straight down from the top of a 15-m-high building with a speed of 6.0 m/s. How long does it take the stone to reach the ground? How fast is it moving just before it hits?

■ **41** A stone is thrown straight upward from the ground and goes as high as a nearby building. The stone returns to the ground 3.0 s after it is thrown. How high (in meters) is the building?

■ **42** A girl standing on the top edge of an 18-m-high building tosses a coin upward with a speed of 12 m/s. How long does it take the coin to hit the ground? How fast is the coin going just before it strikes the ground?

■ **43** A physics student who uses a calculator even to evaluate $2 \times 2$ comes up with the following scheme to measure a building height. A timing mechanism is set up. It measures the time an object dropped from the top of the building takes to fall the last 2.0 m before it hits the ground. By experiment, the object is found to take 0.125 s to move this last 2.0 m. How high is the building?

■ **44** A ball is thrown straight upward with a speed *v* from a point *h* meters above the ground. Show that the time the ball takes to strike the ground is

$$\frac{v}{g}\left(1 + \sqrt{1 + \frac{2hg}{v^2}}\right)$$

**45** A monkey perched 20 m above the ground in a tree drops a coconut directly above your head as you run beneath the tree with a speed of 1.5 m/s. (*a*) How far behind you does the coconut hit the ground? (*b*) If the monkey really wants to hit you, how long before you pass beneath the perch should the coconut be dropped?

**46** Two balls are dropped from different heights. One is dropped 1.5 s after the other, but they both strike the ground at the same time, 5.0 s after the first is dropped. (*a*) What is the difference in the heights from which they were dropped? (*b*) From what height was the first ball dropped?

**47** A boy wants to throw a can straight up and then hit it with a second can. He wants the collision to occur 4.0 m above the throwing point. In addition, he knows that the time he needs between throws is 3.0 s. Assuming he throws both cans with the same speed, what must the initial speed be?

**48** An elevator in which a woman is standing moves upward at 4.0 m/s. If the woman drops a coin from a height 1.4 m above the elevator floor, how long does it take the coin to strike the floor? What is the speed of the coin relative to the floor just before impact?

**49** A marble rolls off the edge of a table 1.20 m high (Fig. P3.5). Find the distance $x$ if the marble is rolling at 2.00 m/s. What are the vertical and horizontal components of its velocity just before it hits the floor?

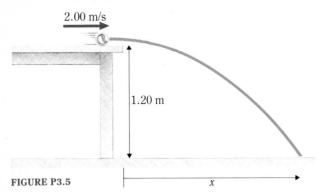

2.00 m/s

1.20 m

**FIGURE P3.5**

$x$

**50** A fire hose shoots water horizontally from the top of a building toward the wall of a building 20 m away. If water leaves the hose at 5.0 m/s, how far below the hose level does it strike the wall? *Hint:* Consider the water as consisting of a series of particles shooting along the stream.

**51** An electron traveling horizontally leaves an electron gun at the end of a television tube with a speed of $10^8$ cm/s. If the fluorescent screen at the opposite end of the tube is 40 cm away, how far below its original level has the electron fallen by the time it hits the screen?

**52** At a circus, the "human cannonball" is shot out of a cannon with a speed of 20 m/s. The cannon barrel is pointed 40° above the horizontal. How far from the end of the cannon should the net used to catch the person be placed? Assume that the net is at the same level as the end of the barrel.

**53** A ball is thrown at 20 m/s at an angle of 30° below the horizontal from a bridge 30 m above the water. (*a*) Where, relative to a point on the water directly below the throwing point, does the ball hit the water? (*b*) For how long is the ball in the air?

**54** Repeat Prob. 53 for a ball thrown at an angle of 30° above the horizontal.

**55** The stunt driver in Fig. P3.6 wishes to shoot off the incline and land on the platform. How fast must the motorcycle be moving if the stunt is to succeed?

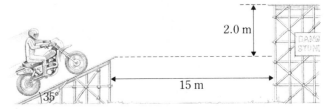

2.0 m

15 m

35°

**FIGURE P3.6**

**56** A projectile is shot from the ground with a velocity $v_o$ at an angle $\theta$ above the horizontal, level ground. It returns to the ground at a distance $R$ from the shooting point. Show that the range $R$ of the projectile is given by

$$R = \frac{2v_o^2 \sin \theta \cos \theta}{g}$$

provided that friction forces are negligible. Using the trigonometric formula $2 \sin \theta \cos \theta = \sin 2\theta$, show that the range is maximum when $\theta = 45°$.

**57** A railcar is moving horizontally with a speed of 20 m/s and decelerating at 5.0 m/s² when a light bulb 3.0 m above the floor comes loose and drops. Where, relative to a point on the floor directly below its original position, will the bulb strike the floor?

**58** Two trucks are parked back to back facing in opposite directions on a straight, horizontal road. The trucks quickly accelerate simultaneously to 3.0 m/s in opposite directions and maintain these velocities. When the backs of the trucks are 20 m apart, a boy in the back of one truck throws a stone at an angle of 40° above the horizontal at the other truck. How fast must he throw, relative to the truck, if the stone is to land in the back of the other truck?

**59** A flatcar moving along straight, horizontal tracks is decelerating at 0.50 m/s². A girl standing on the car throws a ball upward at an angle $\theta$ to the vertical. What value should $\theta$ have if the ball is to return to her hands?

# 4 Newton's laws

IN Chap. 3, we discussed velocity and acceleration without specifically considering what caused an object to move: forces. We now investigate how forces cause accelerations. In carrying through this study, we state and discuss Newton's three laws of motion, laws of primary importance in physics.

## 4.1 The discovery of physical laws

Physical laws are statements of the way matter behaves. These are laws over which we have no control; they have existed and will exist forever. The purpose of all basic research in the physical sciences is to discover these laws. Understanding in science is knowing the laws of nature and their consequences.

Incorrect or incomplete observations can lead to incorrect statements of physical laws. For example, Aristotle believed that he had discovered a law of nature when he stated that heavier bodies accelerate toward the earth faster than lighter bodies, but in fact he had not. There is no such physical law. The law of nature that applies to this situation was discovered many centuries later by Galileo, as we saw in Chap. 3.

Even the law that Galileo discovered for the acceleration of freely falling bodies is now known to be far from complete and general. We know that objects in a spaceship behave very differently from Galileo's falling bodies. To the occupants of a spaceship orbiting the earth, objects do not seem to fall at all; they appear to be completely weightless. Of course, Galileo had no way of knowing this, and so it is only natural that his proposed law of nature was incomplete. Nor were his measurements accurate enough to show that the same body accelerates differently under gravity at different places on earth.

It is a general principle in science that no law of nature is ever fully and completely

## Isaac Newton (1642–1727)

AIP Niels Bohr Library

Before the age of 30, Newton had invented the mathematical methods of calculus, demonstrated that white light contained all the colors of the rainbow, and discovered the law of gravitation.

His was a lonely and solitary life. His father died before he was born, and after his mother remarried, he was raised by an aged grandmother. In 1661 he was admitted to Cambridge University, where he worked for the next eight years except for one year at home to escape the plague. During those years, he made his major discoveries, although none were published at that time. His genius was nonetheless recognized, and in 1669 he was appointed Lucasion Professor of Mathematics at Cambridge University, a position he retained until 1695.

His major scientific work was completed prior to 1692, when he suffered a nervous breakdown. After his recovery, he determined to lead a more public life and soon became Master of the Mint in London. He was elected president of the Royal Society in 1703 and held that position until his death.

known. Scientists do their best to state nature's laws as they know them. However, it would be presumptuous to say that this or that natural law as now understood will not have to be modified as we learn more about our universe. There is an excellent chance that nearly all the natural laws we use in this text are correct in all essentials or that their limitations are known. We know this because these laws have been tested against experience in all manner of ways. There is always the chance, however, that someday one of these laws as presently stated and understood will disagree with the results of some new and ingenious experiment. Our statement of the law must then be changed to conform with that new experience, as well as with the results we already know. When we state a physical law, therefore, it can be considered correct only in the light of present knowledge.

As you progress in your study of physics, you will notice that only rarely is an accepted idea concerning a physical law found to be wrong. It is, however, quite common to find that the accepted idea must be extended, amplified, and modified as our knowledge of the universe becomes wider. The reputation of the scientist who proposes a law sometimes does not ensure its correctness. Even such a great scientist as Isaac Newton had some badly mistaken ideas about the way in which nature behaves. For this reason, the wise scientist does not accept a physical law only because of the reputation of its promulgator. A law discovered by a young "unknown" through careful experiments is likely to be more reliable than the philosophic opinion of a "great" scientist.

There is a hierarchy of physical laws, and it is the ultimate aim of scientists to reduce to a minimum the number of laws needed to describe the universe. We could have a different law for each different physical situation, but this would be intolerable because no one could remember such a large number of laws. However, many of these individual laws would conform to a more general law that encompasses them all. For example, the individual laws

1 A 10-N object in free fall near the surface of the earth accelerates at 9.8 m/s$^2$.
2 A 12-N object in free fall near the surface of the earth accelerates at 9.8 m/s$^2$.

were encompassed in a single law by Galileo:

All bodies in free fall near the surface of the earth accelerate at approximately 9.80 m/s².

We shall see that this statement of Galileo's is only one part of a much more general law first stated by Newton some years later. A statement of a natural law that contains in it the substance of many lesser laws is preferable to any of the lesser laws.

## 4.2 Newton's first law of motion

Isaac Newton was one of the foremost physicists of all time. We begin a study of his work by examining his three laws of motion, first published in 1687 in a classic compendium entitled *Principia Mathematica Philosophiae Naturalis*.

Newton's first law of motion pertains to objects that have zero resultant force acting on them. We are familiar with many such situations. The resultant (net) force acting on a book lying on a table is zero; the downward pull of gravity on the book is balanced by the upward push of the table. You can list many similar situations, and, of course, you can state what happens to the object in each case: it remains at rest. We can summarize this behavior as follows:

An object at rest remains at rest if there is zero resultant force acting on it.

This statement is one part of Newton's first law of motion.

You may wonder why we hold Newton in such esteem for such an obvious conclusion. We do so because the law has a second part, a far less obvious finding. It concerns the behavior of a moving object that has zero net force acting on it.

Prior to Newton's time, nearly everyone assumed that a force was needed to keep an object moving. If you give a push to a book lying on a table, it slides for a while but soon stops. To keep it moving, you must continue to push. Indeed, it is nearly obvious that moving objects left to themselves soon stop. Newton discovered that this observation, although correct, does not apply to objects that have zero resultant force acting on them.

He reasoned as follows. Suppose a block is sliding along a table as in Fig. 4.1. It is true that the weight **W** and the upward push of the table **P** balance each other, and so $\Sigma F_y = 0$ for the block. There is a third force acting on the block, however: the horizontal friction force that the table exerts on it. Thus the resultant force acting on the block is not zero. Newton recognized that the block stops moving because of this unbalanced friction force. Indeed, the smaller the friction force, the farther the block slides. Newton concluded that, in the absence of friction, the block does not slow at all.*

If you apply this line of reasoning to other objects, you will see that it leads to a similar conclusion. Although it is impossible in everyday life to eliminate friction completely, Newton proposed the idea that, in the absence of friction, sliding objects would not slow down. Moreover, for a moving object to be deflected, Newton concluded, an unbalanced deflecting force must act on it. These conclusions are summarized in the second part of Newton's first law:

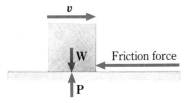

**FIGURE 4.1**

What is the net force acting on the sliding block?

---

* Galileo had stated this same conclusion, as Newton pointed out.

If the resultant force acting on a moving object is zero, the object continues its motion with constant velocity.

Notice that the word used is *velocity*, not *speed*. This law states that neither the magnitude *nor the direction* of the object's velocity will change. In other words, the object will continue to move along a straight line.

We can summarize Newton's first law of motion as follows:

If the vector sum of the external forces acting on an object is zero, the velocity of the object remains constant.

Of course, this statement holds for $v = 0$ as well as for any other value of $v$.

## 4.3 Inertia and mass

Closely linked with the first law is the concept of inertia. A common definition of this term is

Inertia is the tendency of an object at rest to remain at rest and of an object in motion to remain in motion with its original velocity.

We have much common experience concerning inertia. We know, for example, that a loaded truck has much more inertia than a child's wagon, since the wagon is much easier to set in motion than the truck. Moreover, the wagon is much easier to stop than the truck when they are both moving at the same velocity. It is difficult to alter the state of motion of an object that has large inertia.

To make the concept of inertia quantitative, we define a quantity called *mass*. In the SI system of units, the system used in science, mass is defined in the following way. A certain metal cylinder carefully preserved near Paris, France, is defined to have a mass of exactly one kilogram (1 kg). (Figure 4.2 shows a duplicate cylinder kept at the U.S. National Bureau of Standards in Washington, D.C.) By definition, any object that has the same inertia as this standard kilogram object has a mass of 1 kg. An object having three times as much inertia is defined as having a mass of 3 kg, and so on. We shall soon see how we can carry out such a comparison by means of an experiment involving Newton's second law of motion. Before we do that, however, let us jump ahead to Newton's third law because we need to know about it before we discuss applications of the second law.

**FIGURE 4.2**
The platinum-iridium cylinder shown here is a copy of the standard kilogram mass. It is kept at the U.S. National Bureau of Standards, whose responsibility it is to preserve this secondary standard of mass. *(National Bureau of Standards)*

## 4.4 Action and reaction: the third law

You may know that the earth orbits the sun because of the gravitational force exerted on the earth by the sun. Newton was able to treat this type of motion successfully after he discovered the law of gravitation, a topic we discuss in Chap. 7. Did you ever wonder, however, about the gravitational force the earth exerts on the sun? To measure this force directly, we would have to carry out measurements on the sun — a seemingly impossible task. Fortunately, it is often possible to state the value of such inaccessible forces by using another law discovered by Newton, the action-reaction law.

Push your finger against a wall and the wall pushes back on your finger. As

**■ TABLE 4.1    Situations involving Newton's third law**

| Situation | Action force | Reaction force |
|-----------|-------------|----------------|
| Club hits golf ball | Club's force on ball | Ball's force on club |
| Kitten slaps ball of yarn | Paw's force on yarn | Yarn's force on paw |
| Satellite orbits earth | Earth's pull on satellite | Satellite's pull on earth |
| Magnet attracts nail | Magnet's pull on nail | Nail's pull on magnet |
| Diver pushes off from diving board | Foot's push on board | Board's push on foot |

another example, consider what happens when you kick a football. Your foot exerts a force on the ball, but you can also feel an oppositely directed force exerted by the ball on your foot. An object sitting on a table pushes down on the table, but the table pushes up on the object.

Newton examined a multitude of such situations and arrived at a quantitative conclusion, **Newton's third law:**

If object $A$ exerts a force $\mathbf{F}$ on object $B$, then object $B$ exerts a force $-\mathbf{F}$ on object $A$, equal in magnitude but opposite in direction to $\mathbf{F}$.

One of these forces (either one) is called the *action force;* the other is called the *reaction force.* The third law states that the reaction force is exactly equal in magnitude to the action force and opposite in direction. It states more than that, however, because it tells us that the forces act on two different objects. The action force is exerted on one body, and that body exerts a reaction force back on the other body.

Because of the third law, we can say that the action and reaction forces are equal in magnitude and opposite in direction in each of the examples in Table 4.1. Remember, *the action and reaction forces act on different objects.* From time to time we shall use this law to tell us the force on one body when the force on another is known.

## 4.5 The SI system of units

In the next section, we discuss Newton's second law, which relates three quantities: force, mass, and acceleration. Before we state the law, however, it is important that you understand the units used in measuring these quantities. Scientists throughout the world have agreed to use what is called the SI system of units, where SI stands for *Système International.*

In the SI, seven quantities are assigned basic units (Table 4.2), and the units for all other quantities are defined in relation to these seven basic units in ways that we shall point out as our studies progress. For now, we are interested only in the units for mass, length, and time.

We explained in Sec. 4.3 how the kilogram, the basic unit of mass, is defined. It is the mass of the standard kilogram object kept in France. As we explain in Sec. 4.7, we can determine the mass of another object either by comparing its inertia with that of the standard kilogram or by weighing.

The basic unit of time is the second. An atomic process, a comparatively easily

**■ TABLE 4.2
Basic SI units**

| Quantity | Name | Symbol |
|----------|------|--------|
| Length | Meter | m |
| Mass | Kilogram | kg |
| Time | Second | s |
| Temperature | Kelvin | K |
| Electric current | Ampere | A |
| Number of particles | Mole | mol |
| Luminous intensity | Candela | cd |

**FIGURE 4.3**
The heart of the U.S. National Bureau of Standards atomic clock system. *(National Bureau of Standards)*

measured frequency of vibration associated with cesium atoms, is used in its definition. One second is defined to be the time taken for exactly 9,192,631,770 of these vibrations. A portion of a very accurate cesium clock maintained by the National Bureau of Standards is shown in Fig. 4.3. It maintains time to an accuracy of better than 1 part in $1 \times 10^{13}$. Two such clocks synchronized at a certain instant are expected to differ by less than 1 s after the passage of 300,000 years.

The basic unit of length is the meter, which has been defined in various ways. In 1983 it was redefined in terms of the speed of light: one meter is the distance light travels through vacuum in 1/299,792,458 s. Because speed = distance/time, this definition *decrees* the speed of light to be exactly 299,792,458 m/s in vacuum.

The basic units of time (the second) and length (the meter) are defined in terms of precise experiments. Scientists anywhere in the universe, when told about these definitions, can duplicate our units for length and time without actually transporting our rulers or clocks to their own laboratories. However, the kilogram unit is based on a metal cylinder kept on earth, and so the unit of mass lacks universality. It is likely that, within the foreseeable future, the mass standard will be redefined in terms of an atom's mass. Then scientists everywhere will be able to obtain copies of it easily.

Now that we know the basic units for length, mass, and time, we can design experiments to define units for other quantities in terms of them. For example, we need to define a unit for force; in the SI, this unit is the newton (N). To define it, we can use the experiment shown in Fig. 4.4, in which a 1-kg mass is subjected to a resultant force $\mathbf{F}_{net}$. (The mass might be the 1-kg standard mass.) We know, of course, that because it is unbalanced, $\mathbf{F}_{net}$ causes the mass to accelerate in the direction of the force, and we can measure this acceleration. We define the force unit in relation to this experiment:

The SI unit of force, the newton, is that unbalanced force that gives a 1-kg mass an acceleration of 1 m/s².

$$1 \text{ N} = 1 \text{ kg} \cdot \text{m/s}^2$$

Other forces can be measured by comparing them with the defined 1-N force. Can

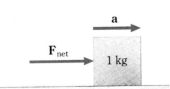

**FIGURE 4.4**
A net force of 1 N will give a 1-kg mass an acceleration of 1 m/s².

you devise an experiment involving balancing forces that could be used to measure an unknown force?

Although the newton is the preferred force unit, two other force units are sometimes used: the dyne, where 1 dyne equals exactly $10^{-5}$ N, and the pound (lb), where 1 lb $= 4.4482$ N. These are not SI units, however, and so we avoid using them in this text.

## 4.6 Newton's second law

Now that we have precise definitions for force, mass, and acceleration, we can investigate how these quantities are related. Common experience tells us that an unbalanced force acting on an object causes the object to accelerate. For example, the force exerted by a bat on a ball causes a change in the ball's velocity. Newton recognized that unbalanced forces cause accelerations, and his second law relates the resultant force on an object to the object's acceleration.

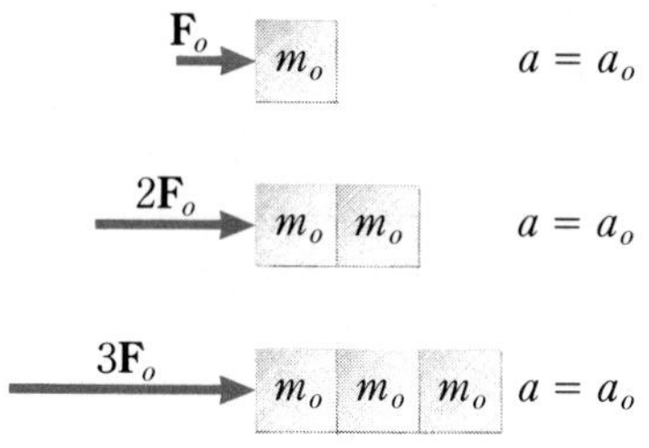

We know from experience that it is more difficult to accelerate a massive object than a less massive object. To quantify this behavior, we can carry out the experiment shown schematically in Fig. 4.5. The individual objects, each of mass $m_o$, might be 1-kg masses floating without friction on an air table. As indicated, the net force must be increased in proportion to the increased mass if the acceleration $a_o$ is to be preserved. Experiment therefore tells us that in a situation such as this,

$$\mathbf{F}_{net} \sim m \qquad \text{for acceleration constant}$$

where the symbol $\sim$ is read "is proportional to."

The variation of this experiment shown in Fig. 4.6 can be used to show that the net force required to give a mass $m_o$ an acceleration changes in proportion to the acceleration. As we see in the figure,

$$\mathbf{F}_{net} \sim \mathbf{a} \qquad \text{for mass constant}$$

Notice that $\mathbf{a}$ is in the same direction as $\mathbf{F}_{net}$.

We can combine these two proportions into an equation if we use a proportionality constant:

$$\mathbf{F}_{net} = (\text{constant})m\mathbf{a}$$

As a check, we notice that if $\mathbf{a}$ is held constant, then $\mathbf{F}_{net} \sim m$, but if $m$ is held constant, then $\mathbf{F}_{net} \sim \mathbf{a}$.

To evaluate the proportionality constant, we recall that a force of 1 N gives a 1-kg mass an acceleration of 1 m/s². Thus the experimental equation becomes

$$1 \text{ N} = (\text{constant})(1 \text{ kg})(1 \text{ m/s}^2)$$

from which we see that the constant is unity and the units are related through

$$1 \text{ N} = 1 \text{ kg} \cdot \text{m/s}^2$$

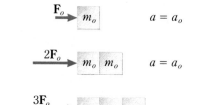

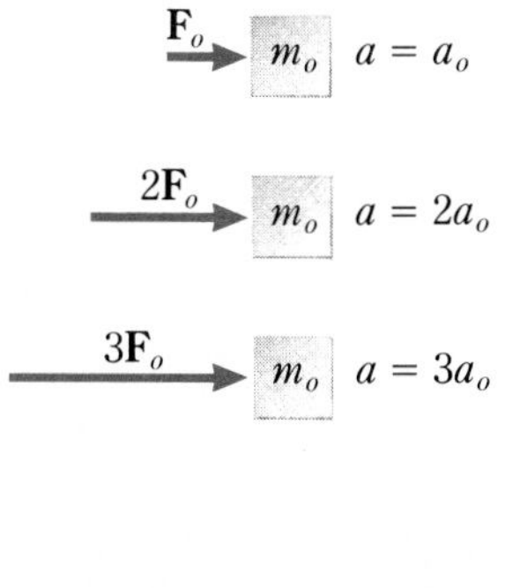

FIGURE 4.5

For constant acceleration, **F** is proportional to $m$.

**FIGURE 4.6**
For constant mass, **F** is proportional to $a$.

## The methods of science

You have no doubt heard people refer to the "scientific method." Just what is this method that scientists follow in their work? In truth, there is no single procedure to which successful scientists adhere. Instead, there are certain features common to all good scientific work. Let us look at these features.

All good scientific work requires precise experimental data. Only by careful measurement can we find out how the universe works. Experimental scientists devote much of their time to measuring how nature behaves. In designing their experiments, they must often use great ingenuity to make intelligible the effects they observe. For example, can we conclude that heavier objects always fall faster than lighter ones because a stone falls faster than a feather? No. There are too many differences in these two experiments to decide what causes the different rates of fall. Scientific experiments must be well controlled so that the effect of each variable can be seen unmistakably.

Once experimental data are available, scientists try to discern unifying trends in them, to make sense out of the many results by showing that they represent a *law of nature,* a concise statement of how physical quantities behave. For example, Newton was able to interpret data that show how forces affect the motion of objects. He summarized the data available to him in the unifying statement $F = ma$.

Theoretical scientists devote much of their time to devising theories of physical behavior. They try to explain natural phenomena by postulating general ways in which nature behaves. For example, Aristotle noticed that a feather falls more slowly than a stone and that fuzz falls more slowly than a coin. To explain this phenomenon, he postulated a theory that we now know to be wrong. We can prove it wrong by showing that many experiments contradict its predictions. In disagreement with Aristotle's theory, we can show that a coin and a feather fall at the same rate in a vacuum. This, then, is the way we sort out theories. We test their predictions by experiment. Only one prediction need be proved incorrect to disprove a theory.

To be satisfactory, useful, and of lasting value, a theory must be testable. You can explain the

---

We arrive then at the following experimental result:

To give a mass $m$ an acceleration $\mathbf{a}$, a net force given by

$$\mathbf{F}_{net} = m\mathbf{a} \tag{4.1a}$$

is required.

This relation, first proposed by Newton, is called Newton's second law. You should note two important points: (1) $\mathbf{F}_{net}$ is the resultant force on the object whose mass is $m$ and (2) the acceleration $\mathbf{a}$ is in the same direction as $\mathbf{F}_{net}$.

Because a force component in the $x$ direction causes an acceleration in the $x$ direction, we can divide the equation $\mathbf{F}_{net} = m\mathbf{a}$ into three equivalent equations, one for each coordinate direction:

# Finding on Radioactivity May Upset Physics Law

**By WALTER SULLIVAN**

Experiments conducted deep in a South Dakota gold mine have reportedly produced evidence for an extremely rare form of radioactive decay whose existence, if proved, would overthrow one of the basic laws of physics.

Known as the law of lepton conservation, it says the total number of lightweight or weightless particles — leptons — emerging from an atomic reaction must equal the number entering into it.

According to the experimenters, the observations recorded a form of radiation known as double beta decay and also imply the existence of the majoron, a hitherto undetected particle some theorists say may be a primary constituent of the universe.

The gold-mine findings are reported in a paper being presented today to the Division of Particles and Fields of the American Physical Society. The experimenters were Dr. Frank T. Avignone and Harry S. Miley of the University of South Carolina, and Ronald L.

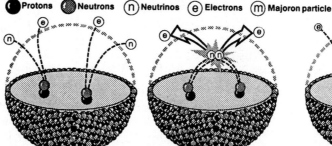

**Possible Double Beta Decays**

Efforts to record a rare form of radioactivity in which two neutrons of an atomic nucleus decay simultaneously focus on three possible modes.

● Protons   ◉ Neutrons   (n) Neutrinos   (e) Electrons   (m) Majoron particle

**Two-Neutrino Emission:** Neutrinos and electrons from each decaying neutron are ejected from nucleus.

**Neutrino Annihilation:** "Lefthanded" and "right-handed" neutrinos meet and annihilate each other. Resulting energy helps eject electrons.

**Majoron Emission:** Neutrinos derived from each neutron interact, producing majoron particle which is ejected.

---

movements of the sun, moon, and stars by assigning a guiding god to each, as the ancients did. Such a theory is not only untestable, it has no value in making further predictions. We therefore discard it as unsatisfactory.

As you see, the methods used by scientists involve four major features: gathering precise data, making the data coherent by unifying them in terms of concise laws, devising theories to further interpret and unify the laws and data, and testing these proposed theories by new experiments. Scientists have no set prescription that guides their work in detail. All scientists agree, however, that the above-mentioned features are part of their work.

$$(\mathbf{F}_{net})_x = \Sigma\mathbf{F}_x = m\mathbf{a}_x$$

$$(\mathbf{F}_{net})_y = \Sigma\mathbf{F}_y = m\mathbf{a}_y \qquad\qquad (4.1b)$$

$$(\mathbf{F}_{net})_z = \Sigma\mathbf{F}_z = m\mathbf{a}_z$$

This is the form of Newton's second law we use most often.

---

### Example 4.1

A 900-kg car is to accelerate from rest to 12.0 m/s in 8.0 s along a straight road. How large a force is required?

***Reasoning***   Take the $x$ axis to be along the roadway in the direction of motion. Because we need to solve the equation $(\mathbf{F}_{net})_x = m\mathbf{a}_x$ to find the required force, we must first find $a_x$. We note that, for the car's trip, $v_o = 0$, $v_f = 12.0$ m/s, and $t = 8.0$ s. Using Eq. 3.9c, $v_f = v_o + at$, we find that

$$a_x = \frac{v_f - v_o}{t} = \frac{(12.0 - 0)\text{ m/s}}{8.0\text{ s}} = 1.50\text{ m/s}^2$$

Then, from Newton's second law,

$$(F_{net})_x = ma_x = (900\text{ kg})(1.50\text{ m/s}^2) = 1350\text{ N}$$

How is this force supplied to the car?  ■

---

*Exercise*  How far does the car go in the 8.0 s?  *Answer: 48 m*  ■

---

## 4.7 Mass and its relation to weight

We have defined mass in terms of the 1-kg standard mass, and other masses are defined by comparison with this standard. Suppose a given force is applied first to the standard 1-kg mass and then to an unknown object. If the force gives the same acceleration to the two objects and if we assume the objects to be free of all other unbalanced forces, then the masses of the two objects are the same. This follows directly from $F_{net} = ma$, for if the two $F$'s and the two $a$'s are the same, then the masses must also be the same. Similarly, an object of mass $n$ kilograms will acquire an acceleration of only $1/n$th that imparted to the standard kilogram by the same force. We see, then, that the unknown mass of an object can be determined by comparing its acceleration with that of the standard kilogram when both are subject to the same force.

In practice, however, we usually determine masses in more convenient ways based on weighing. Of course, mass is a measure of an object's inertia whereas weight is the force with which gravity pulls on the object. Like apples and eggs, mass and weight are entirely different quantities, but they are obviously related in some way because massive objects are heavy.

There is a simple experiment that tells us the relation between mass and weight. It involves an object in free fall, as in Fig. 4.7. We know that an object in free fall accelerates downward with the free-fall acceleration $g$, the acceleration due to gravity. Moreover, the force that causes the object to accelerate downward is the pull of gravity on it, in other words, its weight $W$. And of course the object must obey Newton's second law, $F_{net} = ma$. For the free-fall situation in Fig. 4.7, $F_{net}$ is simply $W$, the weight of the object. Also, the acceleration $a$ in this case is $g$, the free-fall acceleration. Hence, in this experiment, $F_{net} = ma$ becomes

$$W = mg \tag{4.2}$$

This is an extremely important relation because it tells us how the weight $W$ of an object, any object, is related to its mass. Even though weight is a force and mass is a measure of inertia, these two quite different quantities are proportional, with the proportionality constant being $g$.

Because mass and weight are proportional, we can compare the masses of objects by comparing their weights. (For example, an object that weighs 3.7 times as much as the standard kilogram has a mass of 3.7 kg.) However, weights must be com-

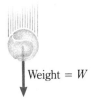

Weight = $W$

**FIGURE 4.7**

The unbalanced force on the object is $W$, giving it a free-fall acceleration $g$.

pared at the same location so that the $g$ is the same for the two masses. For example, we cannot weigh one mass on earth and the other on the moon. A 1-kg mass weighs 9.8 N on earth but only about 1.67 N on the moon, where $g$ is about 1.67 m/s$^2$. Therefore, if one object is weighed on the moon and another is weighed on earth and their weights are equal, their masses are unequal. Of course, in all ordinary weighing experiments, the value for $g$ is constant, and so masses are easily determined by comparing weights.

## 4.8 | Friction forces

Before we illustrate the use of Newton's second law, let us discuss friction because friction forces play an important part in nearly all applications of Newton's laws.

There are three major categories of friction forces:

**1** Viscous friction forces occur when objects move through gases and liquids. Typical of these is the friction force the air exerts on a moving car. The air exerts a retarding force on the car as the car slides through the air. We discuss this type of force in Chap. 9.

**2** Rolling friction forces arise as, for example, a rubber tire rolls on pavement, primarily because the tire deforms as the wheel rolls. The sliding of molecules against each other within the rubber causes energy to be lost. We shall see in Chap. 5 that energy losses and friction are closely related.

**3** Sliding friction forces occur when the two surfaces in contact with each other oppose the sliding of one surface over the other.

It is this third type of friction that is our main concern in this chapter.

Try the experiment shown in Fig. 4.8. Push lightly against your textbook with a horizontal force. It does not move. Apparently, the tabletop also pushes horizontally on the book with an equal and opposite force. This equal force, supplied by friction, is designated by **f** in the figure. Notice the following about the friction force **f** exerted on the book by the tabletop: *it opposes the sliding motion of the book, and it is directed parallel to the sliding surfaces.*

Slowly increase the force with which you push on the book, as shown in the graph of Fig. 4.9. When the magnitude of the pushing force reaches a certain critical value $f_c$, the book suddenly begins to move. Then, to keep it moving, a smaller force of magnitude $f_k$ is sufficient. (The subscript $k$ stands for *kinetic*, which means "moving.") We see from this simple experiment that two friction forces are important. The maximum static friction force $f_c$ must be overcome to start the object moving, and then a smaller friction force $f_k$ opposes the motion of the sliding object.

The major reasons for this behavior can be seen in Fig. 4.10. As you see, the surfaces in contact are far from smooth. Even highly polished surfaces look like this when observed at high magnification. The jagged points from one surface penetrate those of the other surface, and this causes the surfaces to resist sliding. Once sliding has begun, however, the surfaces do not have time to "settle down" onto each other completely. As a result, less force is required to keep them moving than to start the motion.

As you might expect from this model of its origin, the friction force depends on how forcefully two surfaces are pushed together. We describe this feature of the situation by what is called the *normal force* $F_N$ (*normal* meaning *perpendicular*

**FIGURE 4.8**
The friction force **f** opposes the sliding of the book.

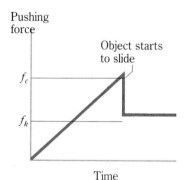

**FIGURE 4.9**
The book in Fig. 4.8 begins to slide when the pushing force equals or exceeds $f_c$.

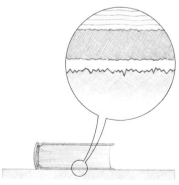

**FIGURE 4.10**
When they are magnified, the surfaces are seen to be rough.

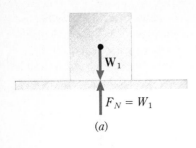

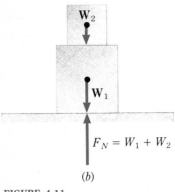

**FIGURE 4.11**
The normal force $F_N$ is the perpendicular force exerted by the supporting surface on the object it supports.

■ **TABLE 4.3  Coefficients of friction**

| Materials in contact | $\mu_s$ | $\mu_k$ |
|---|---|---|
| Rubber on concrete | ~0.9 | ~0.7 |
| Wood on snow | 0.08 | 0.06 |
| Steel on Teflon | 0.04 | 0.04 |
| Steel on steel | 0.75 | 0.57 |
| Wood on wood | 0.7 | 0.4 |
| Metal on metal (lubricated) | 0.10 | 0.07 |
| Glass on glass | 0.9 | 0.4 |

in this context). The normal force is the perpendicular force a supporting surface exerts on any surface resting on it. In Fig. 4.11a, the block pushes down on the supporting surface with a force equal to the block's weight. The supporting surface pushes back with an equal, opposite force, and so $F_N = W_1$ in this case. In Fig. 4.11b, the force pushing down on the supporting surface is the sum of the weights of the two blocks, and so the supporting force is $F_N = W_1 + W_2$ in this case.

Experiment shows that $f_c$ and $f_k$ are often directly proportional to $F_N$. In equation form,

$$f_c = \mu_s F_N \qquad f_k = \mu_k F_N \qquad (4.3)$$

where $\mu$ is the Greek letter mu. The factors $\mu_s$ and $\mu_k$ are called the *static* and *kinetic coefficients of friction,* respectively. They vary widely depending on what the surfaces are made of and how clean and dry they are. Typical values are given in Table 4.3.

Although friction forces are seriously dependent on the smoothness and cleanliness of the surfaces, the following approximate statements can be made: (1) at low speeds, $f_k$ does not change much with speed as one surface slides over the other, and (2) for a given value of $F_N$, $f_c$ and $f_k$ are fairly insensitive to the size of the area of contact between the surfaces.

### Example 4.2
Find the friction force that opposes the motion of the 500-N box in Fig. 4.12 if the kinetic coefficient of friction is 0.60. The woman pushes with a force of 800 N, and the box is already moving.

**FIGURE 4.12**
Notice that the normal force equals 500 N + (800 N) sin 30°.

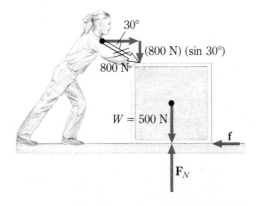

*Reasoning* We must first find the normal force. The floor supports not only the weight of the box (500 N) but also the vertical component of the woman's push. To provide this support, the floor must push upward with an equal force. Therefore

$$F_N = 500 \text{ N} + (800 \text{ N})(\sin 30°) = 900 \text{ N}$$

Knowing that $f = \mu F_N$, we can write

$$f = (0.60)(900 \text{ N}) = 540 \text{ N}$$

Notice the direction of **f** in the diagram. The friction force is always parallel to the contact surface and is in such a direction as to tend to stop the sliding of one surface over the other. ∎

*Exercise* Find $f$ if the box remains motionless. *Answer: 693 N* ∎

## 4.9 Application of Newton's second law

We now have the necessary background for applying Newton's second law to a variety of situations. Before we exhibit its use through examples, let us point out the general procedure we shall follow.

**1** Sketch a picture of the system.

**2** Isolate the object for which we wish to write $\mathbf{F} = m\mathbf{a}$.

**3** Draw a free-body diagram for the isolated object, showing all forces acting on it. Do not include forces that do not act directly on the object.

**4** Choose a coordinate system for the free-body diagram and find the components of the forces.

**5** Write the $\mathbf{F} = m\mathbf{a}$ equations in component form for the forces in the free-body diagram. When we do so, **F** should be in newtons, $m$ in kilograms, and **a** in m/s². Remember that $m = W/g$.

**6** Solve the component equations for the unknowns.

**7** Check the reasonableness of the results.

Sometimes, when more than one object is moving, you may have to repeat steps 2 to 5 for objects other than the isolated one. Although, in the interest of brevity, we do not show each step in the following examples, this omission does not diminish their importance.

---

*Example 4.3*

The wagon in Fig. 4.13a weighs 90 N and is being pulled with a force of 100 N. (a) At what rate does the wagon accelerate? (b) With how large a force is the ground pushing up on the wagon? Assume negligible friction forces.

*Reasoning* If we isolate the wagon as the object for discussion, we can draw its free-body diagram as in Fig. 4.13b. The force **P** is the push of the ground upward on the wagon, the normal force. Notice that the pull of the rope has both a vertical and a horizontal component. After this force has been split into its $x$ and $y$ components, the force diagram is as shown in part $c$.

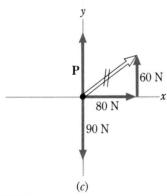

(a)

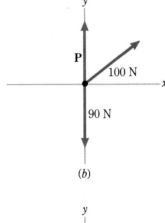

(b)

(c)

**FIGURE 4.13**

The rope pulling on the wagon helps $P$ balance the weight of the wagon and also causes the wagon to accelerate to the right.

We expect the wagon to move neither upward nor downward. Hence both its vertical acceleration and the net force in the vertical direction must be zero. Therefore

$$\Sigma F_y = 0$$

becomes

$$P - 90\ \text{N} + 60\ \text{N} = 0$$

from which $P = 30$ N. This is the answer to part $b$.

The wagon accelerates horizontally because of the net force in that direction. From the diagram,

$$\Sigma F_x = ma_x$$

becomes

$$80\ \text{N} = \left(\frac{90\ \text{N}}{9.8\ \text{m/s}^2}\right)\mathbf{a}_x$$

where $m$ has been replaced by $W/g$. Solving for $\mathbf{a}_x$ gives $\mathbf{a}_x = 8.7$ m/s². ■

---

**Exercise**   Suppose a friction force of 30 N impedes the motion of the wagon. What is its acceleration?   *Answer: 5.4 m/s²* ■

---

**Example 4.4**

A 3300-lb car is traveling at 38 mi/h when its brakes are applied and it skids to rest. The skidding tires experience a friction force about 0.70 times the weight of the car. How far does the car go before stopping? Take the motion to be along the $x$ axis.

**Reasoning**   First we change all data to SI units:

$$\text{Weight} = (3300\ \text{lb})\left(\frac{4.45\ \text{N}}{1\ \text{lb}}\right) = 14{,}700\ \text{N}$$

$$\text{Mass} = \frac{W}{g} = \frac{14{,}700\ \text{N}}{9.8\ \text{m/s}^2} = 1500\ \text{kg}$$

$$\text{Speed} = \left(38\ \frac{\text{mi}}{\text{h}}\right)\left(1.61\ \frac{\text{km}}{\text{mi}}\right)\left(\frac{1\ \text{h}}{3600\ \text{s}}\right) = 0.0170\ \text{km/s}$$

This problem is a combination of an $F = ma$ problem and a motion problem. We can use $\mathbf{F}_{\text{net}} = m\mathbf{a}$ to find $\mathbf{a}$ and then the five uniform-motion equations to find the stopping distance. We know the car weighs 14,700 N, and we are told that

$$\text{Stopping force} = (0.70)(\text{weight}) = 10{,}300\ \text{N}$$

This is the unbalanced force acting on the car. Take the direction of motion as positive. Then the stopping force is negative and $\Sigma F_x = ma_x$ gives

$$-10{,}300\ \text{N} = (1500\ \text{kg})(a)$$

$$a = -6.9\ \text{m/s}^2$$

Why is the acceleration negative?

Now we can solve the motion problem for the skidding auto. Our knowns and unknowns are

$$v_o = 17.0 \text{ m/s} \qquad v_f = 0 \qquad a = -6.9 \text{ m/s}^2$$

$$x = ?$$

We can use $v_f^2 - v_o^2 = 2ax$ to find $x = 20.9$ m.

It is instructive to notice that the stopping force we have used is a reasonable one. We therefore conclude that a car moving at 17 m/s (about 60 km/h) requires a distance of about 20 m to stop. Can you show that the stopping distance increases as the *square* of the initial speed of the car? ∎

---

*Exercise* What is the coefficient of friction between tires and roadway? *Answer:* 0.70 ∎

---

## 4.10 Weight and weightlessness

A fascinating physical phenomenon called weightlessness is sometimes observed when objects are accelerating. Although we postpone discussing weightlessness in orbiting spaceships until we have studied motion in a circle, we can discuss other examples of weightlessness here. A great deal of insight into this phenomenon can be obtained by considering an object suspended from the ceiling of an elevator, as in Fig. 4.14. The reading on the spring scale is the quantity commonly called the object's weight. However, since we have defined weight as the force of gravity on an object, we call the spring-scale reading the *apparent weight* of the object.

The free-body diagram in Fig. 4.14b shows the forces acting on the object. There are only two: the pull of gravity (the weight of the object) $W$ and the upward force $\mathbf{T}$ with which the hook pulls on the pail.* The upward pull of the hook is equal to the reading on the scale and therefore to the object's apparent weight.

### Case 1: Elevator at rest

Since $\mathbf{a}_y = 0$ in this case, $\Sigma \mathbf{F}_y = m\mathbf{a}_y$ becomes

$$\Sigma \mathbf{F}_y = 0 \qquad \text{or} \qquad T - W = 0$$

So $T = W$ in this case, the scale reads $W$, and the apparent weight of the object equals the gravitational force on it.

### Case 2: Elevator with constant velocity

Because the velocity is constant, the acceleration is zero. Thus the analysis used in Case 1 applies here also, and the scale reading is $W$. The apparent weight equals the actual weight.

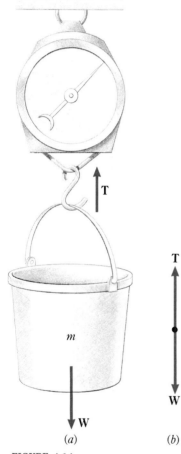

**FIGURE 4.14**
The spring scale reads the force $T$ with which the hook pulls on the pail, and this is equal to the apparent weight of the object.

---

* We neglect the small weight of the hook.

## Case 3: Elevator accelerating upward

Let us call the acceleration $\mathbf{a}_y$. If up is taken as the positive direction, $\Sigma \mathbf{F}_y = m\mathbf{a}_y$ becomes

$$T - W = ma_y$$

from which

$$\text{Apparent weight} = T = W + ma_y$$

The apparent weight of the object is more than its rest value. The support, in this case the hook, must not only balance the gravitational force but also provide an unbalanced force $T - W$ that causes the upward acceleration. Notice how important it is to define a positive direction for the forces and acceleration.

## Case 4: Elevator accelerating downward

If up is still taken as the positive direction, the acceleration now becomes negative. We have from $\Sigma \mathbf{F}_y = m\mathbf{a}_y$,

$$T - W = -ma_y$$

from which

$$\text{Apparent weight} = T = W - ma_y$$

Clearly, the apparent weight of the object is less than the pull of gravity on it.

An interesting case occurs when $a_y = g$, the acceleration due to gravity. In that case, since $W = mg$, we find

$$T = mg - mg = 0$$

and the *object appears to have no weight; it appears to be weightless.* This is not surprising when we consider that an object can acquire a downward acceleration $g$ only when it is in free fall. When the object is in free fall, the hook attaching it to the scale cannot be pulling upward on it and so exerts no tension force to hold the object in place. The reading on the scale goes to zero, and the object appears to be weightless.

Although this situation is artificial, it does show that the apparent weight of an object depends critically on its acceleration. We shall see in Chap. 7 that these ideas can be carried over to the motion of an object in a circular path. It will then become clear why objects in orbit appear to be weightless. Indeed, a spaceship coasting anywhere in space is like the freely falling elevator; all objects within it appear to be weightless.

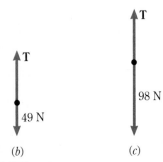

**FIGURE 4.15**

Since the 10-kg mass will fall as the pulley turns clockwise, the tension in the rope must be less than the weight of the 10-kg mass but more than the weight of the 5-kg mass.

## Example 4.5

The two masses in Fig. 4.15 are tied to opposite ends of a massless rope, and the rope is hung over a massless and frictionless pulley.* Find the acceleration of the masses. (This device is called an *Atwood's machine.*)

---

* We specify that the rope and pulley be massless so that we can neglect their inertias. Because the pulley is both massless and frictionless, the tension in the rope is the same on both sides of the pulley.

**Reasoning** As you may have guessed, the 10.0-kg mass will fall and the 5.0-kg mass will rise. The frictionless pulley does nothing but provide a support for the rope. The tension $T$ in the rope is the same throughout its length.

To solve a problem involving two or more bodies, we write $F = ma$ for each body separately, but first we must decide which direction of motion we want to designate as positive. In this case, we know that the pulley will turn clockwise, and so it is convenient to take the motion as positive when it is in that direction.

The free-body diagrams for the masses are shown in Fig. 4.15$b$ and $c$. Notice that the 10-kg mass weighs $(10)(9.8)$ N and that its mass is 10 kg. Applying $\mathbf{F}_{net} = m\mathbf{a}$ to each mass and taking the clockwise direction as positive, we get (from $c$ and $b$)

$$98 \text{ N} - T = (10.0 \text{ kg})(a)$$

$$T - 49 \text{ N} = (5.0 \text{ kg})(a)$$

Adding the lower equation to the upper yields

$$49 \text{ N} = (15 \text{ kg})(a)$$

$$a = 3.3 \text{ m/s}^2$$

where we have used the fact that $1 \text{ N} = 1 \text{ kg} \cdot \text{m/s}^2$. Substituting this value for $a$ in either of the two force equations, we find

$$T = 65 \text{ N}$$

To check the answer, we see that the tension in the rope is larger than the 49-N weight of the lighter mass, and so this mass rises. In addition, the tension is smaller than the 98-N weight of the larger mass, and so this mass falls. ∎

---

### Example 4.6

The two objects in Fig. 4.16 are connected by a string, with one object hanging from a frictionless pulley and the other sitting on a table. The friction force retarding the motion of the object on the table is 0.098 N. Find the acceleration of the objects.

**Reasoning** Some students presented with this problem state that there will be no movement because the 200-g object is much lighter than the 400-g object. This is a fallacy, however. The weight of the 400-g object is downward and is balanced by the upward push of the table. The friction force is the only force pulling backward on the 400-g object, as is easily seen from the free-body diagram in $b$.

Isolating each body and writing $F = ma$ for each, after changing all quantities to SI units, we have (Fig. 4.16$b$ and $c$)

$$T - 0.098 = 0.40a$$

$$(0.20)(9.8) - T = 0.20a$$

where the forces are in newtons and the masses in kilograms. In writing these equations, we have taken as positive the direction of motion that occurs as the 200-g object falls downward. Notice also that, since $W = mg$, the weight of the 0.20-kg object is $(0.20)(9.8)$ N.

Adding the two equations and solving for $a$, we get

$$a = 3.10 \text{ m/s}^2$$

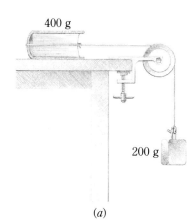

400 g

200 g

(a)

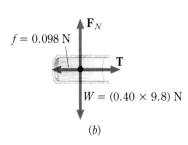

$f = 0.098$ N

$\mathbf{F}_N$

$T$

$W = (0.40 \times 9.8)$ N

(b)

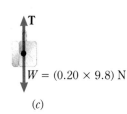

$T$

$W = (0.20 \times 9.8)$ N

(c)

**FIGURE 4.16**
Although the 0.098-N friction force retards the motion of the 400-g object, the weight of the 200-g object is large enough to cause the objects to move. The weight of the 400-g object is balanced by the push of the table.

Using the fact that $1 \text{ N} = 1 \text{ kg} \cdot \text{m/s}^2$, you should show that $a$ really is in the units stated.

Substitution in either of the equations yields

$$T = 1.34 \text{ N} \quad \blacksquare$$

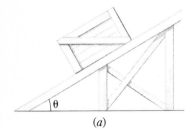

(a)

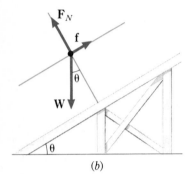

(b)

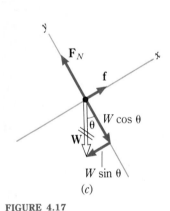

(c)

**FIGURE 4.17**

In dealing with an object on an incline, it is convenient to take the $x$ and $y$ axes parallel and perpendicular to the incline, as shown. The forces are then split into components along these axes.

### Example 4.7

A police officer investigating an accident notices that a car made skid marks 7.0 m long on the dry, level pavement. Estimate a lower limit for the speed of the car before it started to skid. Assume braking occurs at all four wheels.

**Reasoning**   Let us assume that the car decelerated uniformly to rest. If it has a mass $m$, its weight is $mg$. The normal force must balance this, and so $F_N = mg$. From $f = \mu F_N$, we then know that the friction force stopping the car is

$$f = \mu mg$$

where $\mu$ is the coefficient of friction between tires and pavement. This is the unbalanced force on the car, and it causes the car to decelerate.

To find the acceleration, we write $F = ma$ for the car, with $F = \mu F_N = f$. Then $F_{\text{net}} = ma$ becomes, taking the direction of motion as positive,

$$-\mu mg = ma$$

The mass of the car cancels, and we obtain $a = -\mu g$. Why is the acceleration negative?

We can now solve the motion problem involving the skidding car. The knowns (or assumed knowns) are

$$v_f = 0 \qquad x = 7.0 \text{ m} \qquad a = -\mu g$$

We wish to find $v_o$, the initial velocity. To do so, we can use $v_f^2 = v_o^2 + 2ax$:

$$0 = v_o^2 - 2(\mu g)(7.0 \text{ m})$$

$$v_o = 11.7 \sqrt{\mu} \text{ m/s}$$

In a practical situation, $\mu$ would be at least 0.50, and so the smallest possible value of $v_o$ is $11.7 \sqrt{0.50}$ m/s $= 8.3$ m/s. Why is this the lower limit on the speed?   $\blacksquare$

### Example 4.8

The box of mass $m$ in Fig. 4.17 slides down the incline with acceleration $a$. A friction force $f$ impedes its motion. Find the acceleration in terms of $m$, $f$, and $\theta$, which is the angle of the incline.

**Reasoning**   We show the free-body diagram for the box in Fig. 4.17$b$. As usual, $W$ is the pull of gravity on the box and $F_N$ is the normal force with which the incline supports the box. Because the box is sliding downward, the friction force is directed up the incline so as to stop the sliding.

For convenience, we take the $x$ and $y$ axes as being parallel and perpendicular to the incline, as shown in Fig. 4.17$c$. Because the direction of motion is along the incline, we are interested in the force components in the $x$ direction. Therefore it is necessary to split the weight vector into its $x$ and $y$ components. Notice how this is done because it is a technique you must master.

Let us now write $F_{\text{net}} = ma$ for each coordinate. Because we know the box will not crash through the incline or float above it, no acceleration occurs in our $y$ direction. Hence $F_{net} = ma$ gives $F_{net} = 0$ in the $y$ direction, and so, from Fig. 4.17c,

$$\Sigma F_y = 0$$
$$F_N - W \cos \theta = 0$$
$$F_N = W \cos \theta$$

We can also write, from Fig. 4.17c, that

$$\Sigma F_x = ma_x$$
$$f - W \sin \theta = ma_x$$

Solving for $a_x$ and remembering that $W = mg$, we find

$$a_x = (f/m) - g \sin \theta$$

In the special case when there is no friction, so that $f = 0$,

$$a_x = -g \sin \theta \qquad \text{no friction} \tag{4.4}$$

Why is the acceleration negative? We see that when there is no retarding friction, the acceleration of a body on an incline does not depend on the nature of the body. This means that a child's wagon will move down a hill with the same acceleration as a coasting automobile if retarding friction forces are negligible.

In the limiting case, when $\theta = 0$ and the ground is flat, Eq. 4.4 says that the acceleration is zero, since $\sin \theta = 0$. This is true, of course. On the other hand, if $\theta = 90°$, that is, if the incline is straight up and down, the box falls straight down. If there is no friction, the box should fall with its free-fall acceleration. Equation 4.4 is also true in that limit, since $\sin 90° = 1$, and Eq. 4.4 becomes $a = -g$. ∎

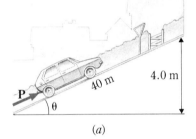

(a)

(b)

**FIGURE 4.18**
The force **P** is partly balanced by the component of the weight acting down the incline. Acceleration up the incline results from the unbalanced portion of **P**.

### Example 4.9
A particular 1200-kg automobile is to accelerate at 0.50 m/s² up an incline that rises 4.0 m in each 40 m. How large a force must push on the car to accelerate it in this way? Ignore friction.

**Reasoning**  Refer to Fig. 4.18. The force **P** pushing the car up the incline is the force desired. The unbalanced force in the $x$ direction is $P - W \sin \theta$. From Newton's second law, we have

$$\Sigma F_x = ma_x$$
$$P - W \sin \theta = ma_x$$

Since $m = 1200$ kg, $W = mg$, and, from the figure, $\sin \theta = 4.0/40$,

$$P - (1200 \text{ kg} \times 9.8 \text{ m/s}^2)(4.0/40) = (1200 \text{ kg})(0.50 \text{ m/s}^2)$$

$$P = 1780 \text{ N}$$

Can you describe how the wheels furnish this force? ∎

**Exercise**  Find $P$ if a 300-N friction force opposes the motion.  *Answer: 2080 N* ∎

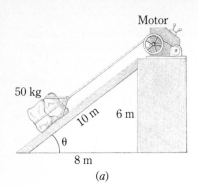

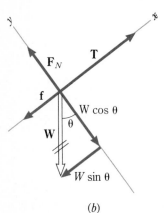

FIGURE 4.19
Since the block is moving up the incline with constant speed, the pull resulting from the motor must exactly balance the sum of the friction force and the component of the weight down the incline.

FIGURE 4.20
(a) The 7-kg block falls downward, pulling the 5-kg block up the plane. (b) The free-body diagram for the 7-kg block. (c) The free-body diagram for the 5-kg block.

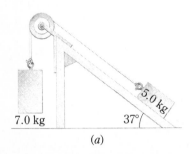

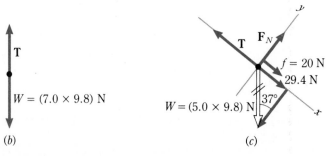

### Example 4.10
A motor is to pull a 50-kg block up the incline in Fig. 4.19. The coefficient of friction between block and incline is 0.70. What is the tension in the rope if the block is moving at constant speed?

***Reasoning*** Notice here that the acceleration is zero. Hence there is no unbalanced force on the block. Since the sum of the $x$ forces is zero, we see from Fig. 4.19b that $\Sigma F_x = 0$ becomes

$$T - f - W \sin \theta = 0$$

or, since $W = mg$ and $\sin \theta = \frac{6}{10}$,

$$T = f + (50 \times 9.8 \text{ N})(\tfrac{6}{10})$$

But $f = \mu F_N$, and $F_N = W \cos \theta$ because the $y$ forces must balance. Hence

$$f = \mu F_N = (0.70)(50 \times 9.8 \text{ N})(\tfrac{8}{10})$$

Substituting this value in the expression for $T$ gives $T = 568$ N. ∎

### Example 4.11
Find the acceleration for the system shown in Fig. 4.20a. The friction force on the 5.0-kg block is 20 N.

***Reasoning*** We isolate each body and write $F = ma$ for each. The free-body diagrams are shown in Fig. 4.20b and c. When friction forces are present, it is necessary to know the direction of motion, since $f$ will oppose the motion. In this case, we surmise that the 7.0-kg block falls. Taking the direction of motion as positive and noting that $W \sin \theta$ is $(5.0 \text{ kg} \times 9.8 \text{ m/s}^2)(0.60) = 29.4$ N, we have the following $F = ma$ equations:

$$(7.0 \text{ kg} \times 9.8 \text{ m/s}^2) - T = (7.0 \text{ kg})(a)$$
$$T - 29.4 \text{ N} - 20 \text{ N} = (5.0 \text{ kg})(a)$$

Adding the equations gives

$$19.2 \text{ N} = (12.0 \text{ kg})(a)$$

from which $a = 1.60 \text{ m/s}^2$. The tension is found by substituting in either of the two starting equations. The result is $T = 57.4$ N. ∎

***Exercise*** What is the friction coefficient between the block and the plane? *Answer: 0.51* ∎

## Minimum learning goals

When you finish this chapter, you should be able to

1 Define (*a*) physical law, (*b*) inertia, (*c*) mass, (*d*) net force, (*e*) newton, (*f*) normal force, (*g*) friction force, (*h*) coefficient of friction.

2 State Newton's first law and give several examples to illustrate each part of it.

3 State Newton's third law and point out the action-reaction pair of forces in any simple situation.

4 State Newton's second law in words and in equation form. Identify what is meant by $F_{net}$, *m*, and *a*. Explain why it is so important to isolate a body when using this law.

5 Explain how an object's mass can be determined by applying a single force in succession to it and to a known mass.

6 Give the relation between mass and weight. Explain when it is allowable to determine mass by weighing.

7 Given two of the following for a given situation, find the third: $F_N$, $\mu$, *f*. Compute $F_N$ for situations similar to those shown in Figs. 4.11 and 4.17.

8 Compute how far an object of known mass will slide in a given time when its initial speed and the forces acting on it are given.

9 Resolve the forces acting on an object supported by an incline into components parallel and perpendicular to the incline. Point out which component forces cause the object to slide along the incline.

10 Compute the acceleration of an object up or down an incline when the forces acting on the object are given.

11 Distinguish between weight and apparent weight. Explain when the two can differ.

## Questions and guesstimates

1 Why does a passenger tend to slide across the seat as a car quickly turns a corner? Why does a carton of eggs fall off the seat if the car stops too quickly?

2 Distinguish clearly between mass, weight, and inertia.

3 Clearly identify the action-reaction forces in each of the following: a child kicks a can; the sun holds the earth in orbit; a ball breaks a window; a parent spanks a child; a ball bounces on a table; a boat tows a water skier.

4 For objects on the moon, the acceleration due to gravity is about 1.67 m/s². How much does an object that has a measured mass of 2 kg on earth weigh on the moon? On earth? What is its mass on the moon?

5 Because objects on the moon weigh only about one-sixth as much as they do on earth, you would almost certainly be able to lift a heavy football player if the two of you were on the moon. Could you easily stop him if he was running at a fair rate across the moon's surface?

6 Is it possible for an object on earth to accelerate downward at a rate greater than *g*?

7 Suppose a brick is dropped from a height of several centimeters onto your open hand when it is lying flat on a tabletop. Why will your hand probably be injured in this situation even though you could easily catch the brick in your free hand without injury?

8 It is generally believed that, on the average, an intoxicated person receives slighter injuries from a fall than a sober person. Why might this belief be valid?

9 Consider the large mops used to sweep the halls in a school. It is easy to slide the mop along the floor if the handle makes only a small angle with the floor. If the angle between handle and floor is too large, however, you cannot move the mop along no matter how hard you push. Explain. Can you find a relation between the critical angle for sliding and the coefficient of friction between floor and mop?

10 An object is being weighed in an elevator. If the elevator suddenly begins to accelerate upward, explain what happens if the weighing device is (*a*) a spring balance, (*b*) an analytical two-pan beam balance, (*c*) a triple-beam balance (an unequal-arm balance).

11 A car at rest is struck from the rear by a second car. The injuries (if any) incurred by the two drivers are of distinctly different character. Explain what happens to each driver.

12 Estimate the minimum distance in which a car can be accelerated from rest to 10 m/s if its motor is extremely powerful.

13 When a high jumper leaves the ground, where does the force that accelerates her or him upward come from? Estimate the force that must be applied to the jumper in a 2-m-high jump.

14 Estimate the force your ankles must exert as you strike the floor after jumping from a height of 2.0 m. Why should you flex your legs in such a situation?

# Problems

**1** How large a horizontal force must be exerted on a 7.0-g bullet to give it an acceleration of 20,000 m/s²? With this acceleration, how fast will the bullet be going after it has moved 1.50 cm from rest?

**2** A horizontal unbalanced force of 5000 N accelerates a 1000-kg car from rest along a horizontal straight roadway. (*a*) What is the acceleration of the car? (*b*) How long does it take the car to reach a speed of 14.0 m/s?

**3** A certain 1500-kg car can accelerate from rest to 20 m/s in 8.0 s. (*a*) What is its acceleration? (*b*) How large a force is needed to produce this acceleration?

**4** A horizontal force of 20 N is required to slide a box along a level floor at a constant speed of 0.30 m/s. How large is the friction force opposing the motion?

**5** If a tow rope is pulled upward at an angle of 20° to the horizontal with a force of 300 N, it can slide a 40-kg box along the floor at a constant speed of 25 cm/s. How large a friction force opposes the box's motion?

**6** A water skier is being pulled by a boat at a constant speed of 12.0 m/s. The tension in the cable pulling the skier is 140 N. How large is the retarding force exerted on the skier by the water and the air?

**7** A 50-kg parachutist is gliding to earth with a constant speed of 8.0 m/s. The parachute has a mass of 6.0 kg. (*a*) How much does the parachutist weigh? (*b*) How large a force upward does the air rushing past the parachutist and chute exert on them?

**8** A horizontal force of 5000 N is required to give a 1000-kg car an acceleration of 0.20 m/s² on a level road. How large a retarding force opposes the motion?

■ **9** An advertisement claims that a certain 950-kg car can be accelerated from rest to 60 km/h in 8.0 s. How large a net force must act on the car to give it this acceleration?

**10** A 1500-kg car is to be towed by another car. If the towed car is to be accelerated uniformly from rest to 2.0 m/s in 9.0 s, how large a force must the tow rope hold?

**11** A 1300-kg car moving at 20 m/s is to be stopped in 80 m. How large must the stopping force be? Assume uniform deceleration.

**12** Each block in Fig. P4.1 weighs 50 N, and $T = 20$ N. Find the normal force in each case.

FIGURE P4.1

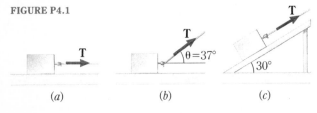

(*a*)          (*b*)          (*c*)

**13** Each block in Fig. P4.2 weighs 35 N, and $P = 20$ N. Find the normal force in each case.

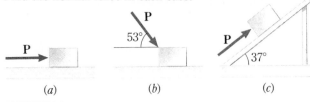

(*a*)          (*b*)          (*c*)

FIGURE P4.2

■ **14** In Fig. P4.2, suppose the block weighs 40 N, $P = 50$ N, and the appropriate coefficient of friction is 0.30. (*a*) What is the friction force in each case? (*b*) What is the block's acceleration?

■ **15** In Fig. P4.1, the weight of the block is 70 N, $T = 60$ N, and the appropriate coefficient of friction is 0.35. (*a*) What is the friction force in each case? (*b*) What is the block's acceleration?

**16** A 5-kg box slides down a 30° incline under the action of gravity. If the box slides with constant speed, how large is the friction force that impedes its motion?

■ **17** A 300-g block is sitting on an adjustable incline. The angle of the incline is slowly increased, and the block begins to slide when the angle is 41°. What is the coefficient of friction between block and incline? Is this the static or the kinetic coefficient?

■ **18** In Fig. P4.1*b*, $T$ is large enough to accelerate the block but not large enough to lift it. The coefficient of friction between block and surface is $\mu$. Show that the block undergoes maximum acceleration when $\theta$ is chosen such that $\cos \theta + \mu \sin \theta$ is a maximum. (If you know calculus, you will be able to show that this condition reduces to $\tan \theta = \mu$.)

■ **19** In Fig. P4.2*b*, the static coefficient of friction is 0.40. If the block weighs 200 N, at what value of $P$ will it begin to move?

■ **20** According to a rule of thumb, the friction force between dry concrete and skidding tires is about nine-tenths of the car's weight. If the skid marks left by a car coming to rest are 20 m long, about how fast was the car going just before the brakes were applied? Justify the rule of thumb.

■ **21** If the coefficient of friction between a car's tires and a roadway is 0.70, what is the least distance in which the car can accelerate from rest to 15 m/s?

■ **22** An electron ($m = 9.11 \times 10^{-31}$ kg) in a television tube is accelerated from rest to $5 \times 10^7$ m/s in 0.80 cm. Find the average accelerating force on the electron. How many times larger than $mg$ is it?

■ 23 A boy is running across a slippery floor at 4.0 m/s when he decides to slide. If the coefficient of friction between his shoes and the floor is 0.250, how far will he slide before stopping?

■ 24 What is the shortest distance in which a car going 30 m/s can stop on a level roadway if the maximum friction coefficient (the static coefficient) between its tires and the pavement is 0.75?

■ 25 A 900-kg automobile traveling at 20 m/s collides with a tree and goes 1.60 m before stopping. How large is the average retarding force exerted by the tree on the automobile?

■ 26 The friction force retarding the motion of a 150-kg box across a level floor is 400 N. (*a*) What is the coefficient of friction between box and floor? (*b*) Assuming that the friction coefficient does not change as the speed increases, how large an acceleration can you give the box by pulling it with a force of 750 N that is inclined at an angle of 53° above the horizontal?

■ 27 (*a*) Find the acceleration of the 4.0-kg block in Fig. P4.3 if the coefficient of friction between block and surface is 0.60. (*b*) Repeat if the 50-N force is pushing *down* on the block at an angle of 30° below the horizontal (that is, if the force shown in the figure is reversed in direction).

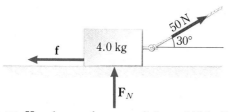

FIGURE P4.3

■ 28 How large a force parallel to a 30° incline is needed to give a 5.0-kg box an acceleration of 2.0 m/s² up the incline (*a*) if friction is negligible, (*b*) if the coefficient of friction is 0.30?

■ 29 An 8.0-kg box is released on a 30° incline and accelerates down the incline at 0.30 m/s². Find the friction force impeding its motion. How large is the coefficient of friction?

■ 30 An 8.0-g bullet enters a 2.0-cm-thick piece of plastic with a speed of 140 m/s. It passes through the plastic and emerges with a speed of 70 m/s. How large is the average force retarding the bullet's passage through the plastic?

■ 31 If you pull straight up on a 5.0-kg mass with a cord that is just capable of holding a 20.0-kg mass at rest, what is the maximum upward acceleration you can impart to the 5.0-kg mass?

■ 32 A 60-kg prisoner wishes to escape from a third-story window by going down a rope made of bedsheets tied together. Unfortunately, the rope can hold only 500 N. How fast must the prisoner accelerate down the rope if it is not to break?

■ 33 A 50-kg woman stands on a spring scale inside an elevator. (The scale reads the force with which it pushes upward on the woman.) What does the scale read when the elevator is accelerating (*a*) upward at 3.0 m/s², (*b*) downward at 3.0 m/s²?

■ 34 A 200-g mass is hung from a thread; from the bottom of the 200-g mass, a 300-g mass is hung by a second thread. Find the tensions in the two threads if the masses are (*a*) standing still, (*b*) accelerating upward at 20 m/s², (*c*) moving downward at a constant acceleration of 5.0 m/s², (*d*) falling freely under the action of gravity.

■ 35 A book sits on the roof of a car as the car accelerates horizontally from rest. If the static coefficient of friction between car and book is 0.45, what is the maximum acceleration the car can have if the book is not to slip?

■ 36 A carton of eggs rests on the seat of a car moving at 20 m/s. What is the least distance in which the car can be uniformly slowed to a stop if the eggs are not to slide? The value of $\mu$ between carton and seat is 0.15.

■ 37 A cement block sits on the floor of a station wagon descending a 20° incline while decelerating at 1.5 m/s². How large must the static coefficient of friction between floor and block be if the block is not to slide?

■ 38 Prove that the acceleration of a car moving on a horizontal road cannot exceed $\mu g$, where $\mu$ is the coefficient of friction between tires and road. What is the similar expression for the acceleration of a car going up an incline of angle $\theta$? Why is it counterproductive to cause the car to "burn rubber" when it is "scratching off"? Does it matter whether the car has two-wheel or four-wheel drive?

■ 39 A passenger in a large ship sailing in a quiet sea hangs a ball from the ceiling of her cabin by means of a long thread. She notes that whenever the ship accelerates, the pendulum ball lags behind the point of suspension, and the pendulum no longer hangs vertically. How large is the ship's acceleration when the pendulum stands at an angle of 5° to the vertical?

■ 40 A uniform ladder of mass $m$ and length $L$ leans against a smooth wall and makes an angle $\theta$ with the floor. What must the value of $\mu_s$ between floor and ladder be if the ladder is not to slide?

■ 41 In Fig. P4.4, the tension in the rope pulling the two blocks is 40 N. Find the acceleration of the blocks and the tension in the connecting cord if the friction force on the blocks is negligible. Repeat if the coefficient of friction between blocks and surface is 0.20.

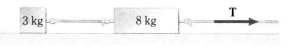

FIGURE P4.4

**42** In Fig. P4.4, how large must $T$ be to give the blocks an acceleration of 0.40 m/s² (*a*) if friction forces are negligible, (*b*) if the coefficient of friction between blocks and surface is 0.30? Also find the tension in the connecting cord in each case.

**43** In Fig. P4.5, block 1 has a mass of 2.50 kg and block 2 has a mass of 1.60 kg. (*a*) Ignoring friction, what are the acceleration of the blocks and the tension in the connecting cord? (*b*) Repeat for a friction force of 8.0 N retarding block 1.

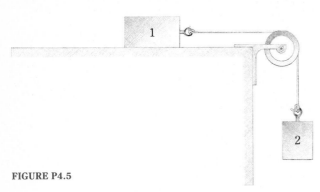

**FIGURE P4.5**

**44** In Fig. P4.5, object 1 is a 3000-g mass and object 2 is a 2000-g mass. When the system is released, object 2 falls 80 cm in 1.50 s. How large a friction force opposes the motion of object 1? Assume that no friction forces exist in the rest of the system.

**45** In Fig. P4.6, find the tension in the cord and the time needed for the masses to move 150 cm starting from rest. Assume the pulley to be frictionless and massless.

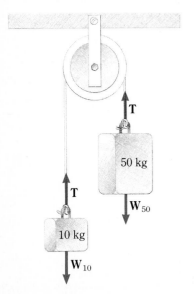

**FIGURE P4.6**

**46** Two blocks with masses $m_1 = 3$ kg and $m_2 = 4$ kg are touching each other on a frictionless table, as in Fig. P4.7. If the force shown acting on $m_1$ is 5 N, (*a*) what is the acceleration of the two blocks and (*b*) how hard does $m_1$ push against $m_2$? (*c*) Repeat *a* and *b* if **F** is in the reverse direction and pushes on $m_2$ rather than on $m_1$.

**FIGURE P4.7**

**47** The pulley in Fig. P4.8 is massless and frictionless. Find the acceleration of the mass in terms of $F$ if there is no friction between the surface and the mass. Repeat for a friction force of $f$.

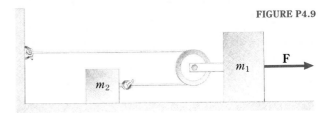

**FIGURE P4.8**

**48** There is negligible friction between the blocks and table of Fig. P4.9. Compute the tension in the cord and the acceleration of $m_2$ if $m_1 = 300$ g, $m_2 = 200$ g, and $F = 0.60$ N. *Hint:* Note that $a_2 = 2a_1$.

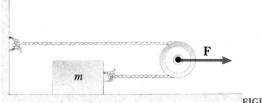

**FIGURE P4.9**

**49** In Fig. P4.10, the coefficient of friction is the same at the top and bottom of the 700-g block. If $a = 70$ cm/s² when $F = 1.30$ N, how large is the coefficient of friction?

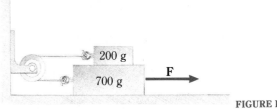

**FIGURE P4.10**

**50** A 2.0-kg block starts to slide from rest down a 37° incline. How far does it slide in the first 3.0 s (*a*) if friction is negligible, (*b*) if $\mu$ between block and surface is 0.40?

**51** Find the tensions in the two cords and the accelerations of the blocks in Fig. P4.11 if friction is negligible. The pulleys are massless and frictionless, $m_1 = 200$ g, $m_2 = 500$ g, and $m_3 = 400$ g.

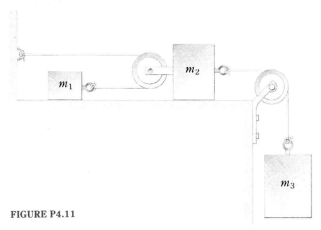

FIGURE P4.11

**52** Find the acceleration of the blocks in Fig. P4.12 and the tension in the cord (*a*) if there is negligible friction, (*b*) if $\mu = 0.15$.

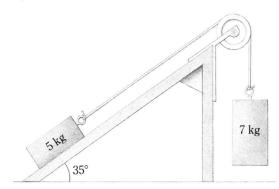

FIGURE P4.12

**53** A 700-kg car is at rest on a hill that is inclined 12° to the horizontal. How far does the car move in the first 10.0 s after the brakes are released (*a*) if the car rolls freely down the hill, (*b*) if a 1000-N friction force retards its motion?

**54** Two frictionless carts of masses $M_1$ and $M_2$ sit at rest on a straight horizontal track. They are a distance $D$ apart, and a rope stretches between them. The occupants of cart 1 pull on the rope in such a way that the tension in it is constant, and the carts move toward each other. (*a*) Where, relative to the original position of cart 2, do the two carts collide? (*b*) What is the ratio of their speeds just before collision?

**55** The force **F** in Fig. P4.13 pushes a block of mass *M*, which in turn pushes a block of mass *m*. There is no friction between *M* and the supporting surface. If the friction coefficient between the two blocks is $\mu$, how large must *F* be if the block of mass *m* is not to slip?

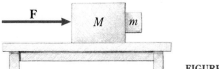

FIGURE P4.13

# 5 Work and energy

**O**NE of the goals of scientists is to discover ways of unifying and simplifying the various facts and concepts in their field of study. In previous chapters we discussed forces and their effect in causing motion. In principle, we can describe all motions in terms of the forces that cause them. However, a concept introduced in this chapter, the conservation of energy, greatly unifies and simplifies the description of motion in many instances. The principle of conservation of energy is a unifying concept that is important not only in mechanics but also in other branches of physics. In order to understand the principle, we first discuss the concept of work and how it leads to the concept of energy.

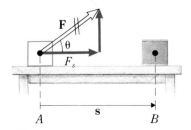

**FIGURE 5.1**

The work done by **F** in the displacement of the object from $A$ to $B$ is $F_s s$, which is ($F \cos \theta$)($s$).

## 5.1 The definition of work

When you sit at your desk studying this book, you are not doing work. This does not mean that you are lazy or that learning physics is an effortless process. It is simply stating a fact that arises from the definition of work the scientist uses. The word *work* is used in so many colloquial ways that giving it a precise meaning when we use it in physics becomes particularly important.

Physicists and all other scientists define the work done by a force in the following way. Suppose, as shown in Fig. 5.1, a force **F** pulls an object from $A$ to $B$ through a displacement **s**. We represent the component of **F** in the direction of **s** by $F_s$. Then the work done by **F** during the displacement **s** is

Work done by $\mathbf{F} = F_s s$           (5.1*a*)

Work is a scalar quantity; it has no direction associated with it.

In the SI, forces are measured in newtons and distances are measured in meters, and so the unit for work is the newton-meter (N · m). This unit is given a special name, the *joule* (J).

A joule is the work done by a force of one newton as it acts through a distance of one meter along the line of the force.

Other units sometimes used to measure work are the foot-pound (ft · lb), the erg, and the electronvolt (eV), where

$$1 \text{ ft} \cdot \text{lb} = 1.356 \text{ J}$$
$$1 \text{ erg} = 1 \times 10^{-7} \text{ J} \quad \text{exactly}$$
$$1 \text{ eV} = 1.602 \times 10^{-19} \text{ J}$$

Quantities expressed in these units must always be changed to joules before they can be used in our fundamental SI-based equations.

We can express the defining equation for work in a form different from Eq. 5.1a by noticing in Fig. 5.1 that

$$F_s = F \cos \theta$$

where $\theta$ is the angle between **F** and **s**. Substituting this value for $F_s$ in Eq. 5.1a, we obtain

$$\text{Work done by } \mathbf{F} = Fs \cos \theta \tag{5.1b}$$

In summary,

The work $W$ done by a force **F** acting on an object during a displacement **s** is

$$F_s s \quad \text{or} \quad Fs \cos \theta$$

In these equivalent expressions, $F_s$ is the component of **F** in the direction of the displacement **s** and the angle $\theta$ is the angle between **F** and **s**.

---

### Example 5.1

In Fig. 5.2, a person exerts a vertical force **F** on a pail as the pail is carried a horizontal distance of 8.0 m at constant speed. How much work does **F** do?

***Reasoning*** The definition of work is $W = Fs \cos \theta$. In Fig. 5.2, **F** is vertical and **s** is horizontal. Hence $\theta = 90°$ and

$$W = Fs \cos 90° = 0$$

No work is done by the vertical force because it has no component in the direction of motion. ∎

---

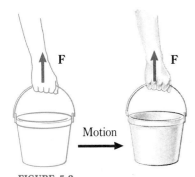

**FIGURE 5.2**

No work is done on the pail by **F**, since it has no component in the direction of the displacement.

### Example 5.2

How much work do you do on an object of weight $mg$ as (*a*) you lift it a distance $h$ straight up and (*b*) you lower it through this same distance?

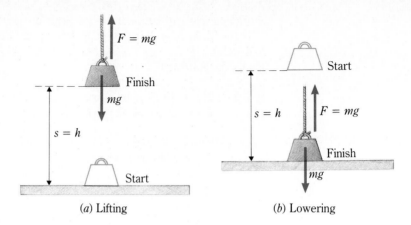

FIGURE 5.3

The work done by the lifting force is *mgh* in (*a*) and − *mgh* in (*b*).

(*a*) Lifting      (*b*) Lowering

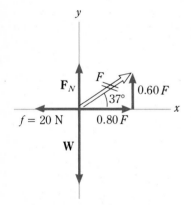

(*a*)

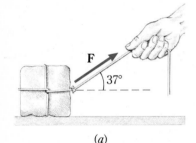

(*b*)

FIGURE 5.4

The horizontal component of **F** does work on the object, while the work done by the vertical component is zero.

*Reasoning*   The situation is shown in Fig. 5.3*a*. To lift the object, you must pull up on it with a force equal to its weight *mg*.* The displacement is *h* upward, and the lifting force is in the same direction. Therefore, from the definition of work,

$$W = Fs \cos 0° = (mg)(h)(1) = mgh$$

This is the work you do as you lift the mass a distance *h*.

   (*b*) Figure 5.3*b* shows what happens when you lower the mass. Now **F** and **s** are in opposite directions. Therefore $F = mg$ and $\theta = 180°$. We then find from $W = Fs \cos \theta$ that

$$W = (mg)(h)(\cos 180°) = mgh(-1) = -mgh$$

The work you do is negative in this case because **F** and **s** have opposite directions.  ▪

─────────

*Exercise*   How much work does the pull of gravity do on the object of Example 5.2 as (*a*) it is lifted and (*b*) it is lowered?   *Answer: (a) − mgh; (b) mgh*  ▪

─────────

## Example 5.3

The box in Fig. 5.4 is being pulled along the floor at constant speed by a force **F**. Let us say that the friction force opposing the motion is 20 N and that the box has a mass of 30 kg. Find the work done by the pulling force as the box is moved 5.0 m.

*Reasoning*   We know that the net force on the box is zero because it is moving with constant speed. Hence the *x*-directed forces must cancel, and so, from Fig. 5.4*b*, we can write

$$0.80F = 20 \text{ N}$$

─────────

* A force slightly larger than *mg* is needed to give the object an initial acceleration upward, but once the object is moving, a force *mg* upward balances the pull of gravity and the object keeps moving with constant velocity.

To find the work done by **F**, we can use

$$W = F_s s$$

Since $F_s$ is $0.8F = 20$ N, we have

$$W = (20 \text{ N})(5.0 \text{ m}) = 100 \text{ N} \cdot \text{m} = 100 \text{ J}$$

Notice that the $y$ component of **F** does no work. No displacement occurs in the $y$ direction. Is it necessary to know the mass of the box to solve this problem? ▪

---

## 5.2 Power

When you purchase an automobile, you are interested in the horsepower of its engine. You know that usually an engine with large horsepower is most effective in accelerating the automobile. Let us now learn the precise meaning of power.

Power $P$ measures the rate at which work is being done. Its defining equation is

$$\text{Power} = \frac{\text{work done}}{\text{time taken to do the work}}$$

or, in symbols,

$$P = \frac{W}{t} \tag{5.2}$$

When work is measured in joules and $t$ is in seconds, the unit for power is the joule per second, which is called the watt (W), after the Scottish inventor of the steam engine, James Watt. For motors and engines, power is usually measured in horsepower (hp), where

$$1 \text{ hp} = 746 \text{ W}$$

Of course, since the watt is the SI unit for power, we must use it in our equations. For example, an electric motor that has an output of $\frac{1}{4}$ hp is capable of producing

$$\left(\frac{1}{4} \text{ hp}\right)\left(746 \frac{\text{W}}{\text{hp}}\right) = 186 \text{ W}$$

of power. That means that the motor can perform 186 J of work each second.

We obtain another convenient relation for power when we notice that the work done on an object by a force $F_x$ as it displaces the object a distance $x$ is $F_x x$. Using this expression in Eq. 5.2, we find

$$P = \frac{W}{t} = \frac{F_x x}{t} = F_x \left(\frac{x}{t}\right)$$

Now, since $x/t$ is the velocity $v_x$ with which the object is moving in the $x$ direction, we have that

$$P = F_x v_x \tag{5.3}$$

Of course, both Eqs. 5.2 and 5.3 assume that the power output is constant. If $F_x$ and/or $v_x$ varies with time, then Eq. 5.2 gives the average power during the time interval $t$ and Eq. 5.3 gives the instantaneous power at the instant at which $F_x$ and $v_x$ have the values being used.

Equation 5.2 is used to define a frequently encountered unit used for work and energy. Noting that

Work = power × time

we measure power in kilowatts and time in hours. Then the work done by the source of the power has the units of kilowatts × hours, and this unit of work is called the kilowatthour (kWh). It can be related to the joule through

$$1 \text{ kWh} = (1 \text{ kWh}) \left(1000 \frac{\text{W}}{\text{kW}}\right)\left(3600 \frac{\text{s}}{\text{h}}\right) = 3.60 \times 10^6 \text{ W} \cdot \text{s} = 3.60 \times 10^6 \text{ J}$$

### Example 5.4

The motor in Fig. 5.5 is lifting a 200-kg object at a constant speed of 3.00 cm/s. What power, in horsepower, is being developed by the motor?

*Reasoning*  Because the object is moving with constant speed, the upward pull due to the motor equals the weight of the object. Therefore the force $F_y$ exerted on the object by the motor is

$$F_y = mg = (200 \text{ kg})(9.8 \text{ m/s}^2) = 1960 \text{ N}$$

Then, because $v_y = 0.0300$ m/s, Eq. 5.3 gives

$$P = F_y v_y = (1960 \text{ N})(0.0300 \text{ m/s}) = 58.8 \text{ W}$$

This gives

$$P = (58.8 \text{ W}) \left(\frac{1 \text{ hp}}{746 \text{ W}}\right) = 0.0788 \text{ hp}$$

for the power output of the motor. ∎

*Exercise*  What would the motor's power output be if the load was being lowered at a speed of 3.0 cm/s?  *Answer:* − 58.8 W  ∎

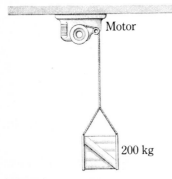

**FIGURE 5.5**
We wish to find the power output of the motor as it lifts the object with a speed of 3.00 cm/s.

Motor

200 kg

## 5.3  Kinetic energy

If an object can do work, we say that the object possesses energy. For that reason, *energy is often said to be the ability to do work*. Although, as we shall see, the concept of energy is too complex to be described fully in such a brief statement, associating energy and work still proves useful. There are many kinds of energy, and we begin our study of them by considering kinetic energy.

A moving baseball can break a window, a moving hammer can drive a nail, and a stone moving upward can actually lift itself against the force of gravity. Obviously, moving objects have the ability to do work; in other words, they possess energy. We

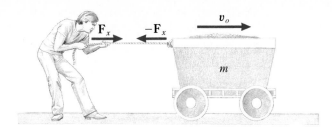

call the energy possessed by an object because of its motion *kinetic energy* (KE). Let us now determine how much work a moving object can do because of its motion; this is the object's kinetic energy.

To take a concrete situation, suppose that a loaded wagon of mass $m$ is coasting with velocity $v_o$, as in Fig. 5.6. As indicated, a person is pulling backward on the wagon with a constant force $-F_x$, trying to stop its motion. According to the action-reaction law, the wagon exerts an equal forward force on the person. As the wagon and the person move through a distance $x$, the work done on the person is

Work on person $= F_x x$

Let us now relate this work to the change in the wagon's motion.

Because of the retarding force $-F_x$ being exerted on it, the wagon is decelerating, and, from $F = ma$,

$$a_x = \frac{-F_x}{m}$$

Using the motion equation $v_f^2 - v_o^2 = 2a_x x$ to replace $a_x$, we find that

$$F_x = -m\left(\frac{1}{2x}\right)(v_f^2 - v_o^2)$$

If we place this expression for $F_x$ in the equation for the work done on the person, we find that

Work on person $= \frac{1}{2}mv_o^2 - \frac{1}{2}mv_f^2$           (5.4)

This expression tells us how much work a moving object can do as it slows from a velocity $v_o$ to a velocity $v_f$. Indeed, if the wagon is brought to rest, then $v_f = 0$ and the work done is $\frac{1}{2}mv_o^2$. We therefore conclude that

An object of mass $m$ traveling with velocity $v$ is capable of doing an amount of work equal to $\frac{1}{2}mv^2$ as it is brought to rest.

Using this fact as a rationale, we define the energy an object has because of its motion as $\frac{1}{2}mv^2$. We call this type of energy *kinetic\* energy of translation*, and so, for an object of mass $m$ moving with velocity $v$,

$$KE = \frac{1}{2}mv^2$$           (5.5)

---

\* From the Greek word *kinetikos*, to move.

Energy has the same units as work, namely joules in the SI. Once again, we emphasize that the kinetic energy of an object represents the work the object can do because of its motion.

## 5.4 The work-energy theorem

As we have just seen, work and energy are closely related. In this section, we obtain a relationship between the work done on an object and the change in the object's kinetic energy. We could use the situation depicted in Fig. 5.6 for this purpose and compute the work done on the wagon by the rope pulling backward on it. Instead, let us take the more general situation shown in Fig. 5.7: a wagon of mass $m$ moving in the positive $x$ direction under the influence of several forces. Call the resultant force acting on the wagon $\mathbf{F}_{net}$. Then, because the motion is in the $x$ direction, $\mathbf{F}_{net} = m\mathbf{a}$ becomes

$$F_{net} = ma_x$$

As we did in the preceding section, we replace $a_x$ by the initial and final velocities of the object together with the distance moved ($x$) to obtain

$$F_{net}x = \tfrac{1}{2}mv_f^2 - \tfrac{1}{2}mv_o^2$$

However, $F_{net}x$ is simply the work done on the wagon by the resultant force acting on it. Therefore our result can be summarized as

Work done on object by $F_{net}$ = change in object's KE

$$\text{Work done by } F_{net} = \tfrac{1}{2}mv_f^2 - \tfrac{1}{2}mv_o^2 \qquad (5.6)$$

This is called the *work-energy theorem*.

In applying this relationship, you should recognize that forces in the direction of motion speed up the motion; they increase the kinetic energy. Retarding forces such as friction, however, do negative work on the object. This follows from the fact that the direction of the retarding force is opposite the direction of the displacement, and so $F_x x \cos \theta$ becomes $F_x x \cos 180°$, which is $-F_x x$. Hence retarding forces decrease the kinetic energy.

Forces in the direction of motion increase the kinetic energy of an object, whereas stopping forces decrease it.

The work-energy theorem is an extremely powerful relationship, and we shall be using it throughout our study of physics.

**FIGURE 5.7**

The resultant force acting on the wagon causes its kinetic energy to change.

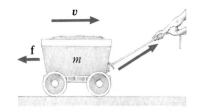

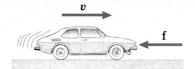

**FIGURE 5.8**

The net force acting on the coasting car is $f$.

**Example 5.5**

A 2000-kg car traveling at 20 m/s coasts to rest on level ground in a distance of 100 m. How large is the average friction force tending to stop it? See Fig. 5.8.

*Reasoning*  The average friction force $f$, which is the only unbalanced force acting on the car, causes it to lose kinetic energy. Therefore, from the work-energy theorem,

$$\text{Change in KE} = \text{work done by friction force}$$

$$\tfrac{1}{2}mv_f^2 - \tfrac{1}{2}mv_o^2 = fs \cos \theta$$

$$0 - \tfrac{1}{2}(2000 \text{ kg})(20 \text{ m/s})^2 = (f)(100 \text{ m})(-1)$$

$$f = 4000 \text{ kg} \cdot \text{m/s}^2 = 4000 \text{ N}$$

where we have used the fact that $1 \text{ N} = 1 \text{ kg} \cdot \text{m/s}^2$.  ∎

**Example 5.6**

If the friction force on the car of Example 5.5 is constant at 4000 N, how fast is the car going after 50 m?

*Reasoning*  The loss in kinetic energy by the time the car reaches the 50-m mark is caused by friction. Therefore, from the work-energy theorem,

$$\text{Change in KE} = \text{work done by friction force}$$

$$\tfrac{1}{2}mv_f^2 - \tfrac{1}{2}mv_o^2 = fs \cos \theta$$

$$\tfrac{1}{2}(2000 \text{ kg})(v_f^2) - \tfrac{1}{2}(2000 \text{ kg})(20 \text{ m/s})^2 = (4000 \text{ N})(50 \text{ m})(-1)$$

Solving for $v_f$ gives $v_f = 14.1$ m/s.  ∎

## 5.5 Gravitational potential energy

As we have seen, some objects are able to do work by virtue of their motion. They have kinetic energy. Other objects can do work because of either their position or their configuration. Such objects are said to have *potential energy.* Let us begin our study of potential energy by discussing the energy an object has because of gravitational forces.

Consider the system in Fig. 5.9, where the pulleys are assumed frictionless. Because both objects have the same weight $mg$, when object $B$ is given a slight downward push to start it moving, it will fall slowly toward the floor with constant speed. At the same time, object $A$ will rise slowly. By the time $B$ has fallen the distance $h$ to the floor, $A$ has been lifted a distance $h$.

We now ask: How much work is done on object $A$ by the rope as the object is raised from the floor with constant speed? Because the tension in the rope is equal to $A$'s weight, $mg$, the lifting work done by the rope is, from work $= F_y y$,

$$\text{Lifting work done} = (\text{tension})(\text{distance}) = mgh$$

Who or what external agent does this work? The weight of object $B$ pulls object $A$ up, and so it does the work. We must therefore conclude that object $B$ possessed the ability to do work when it hung in its original position above the floor. This work is

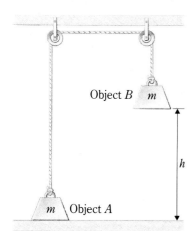

**FIGURE 5.9**

As object $B$ falls, it does work by lifting object $A$.

Object $B$  $m$

$m$  Object $A$

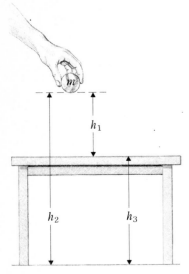

**FIGURE 5.10**

The gravitational potential energy of the ball can be either $mgh_1$ or $mgh_2$. Notice that the two values differ by the additive constant $mgh_3$.

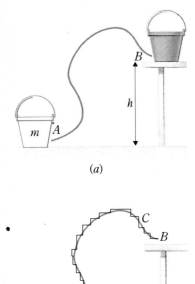

(a)

(b)

**FIGURE 5.11**

The path in (a) can be approximated by the series of horizontal and vertical steps in (b).

$mgh$, where $h$ is the distance through which the object of mass $m$ can fall. With this fact in mind, we define

Gravitational potential energy (GPE) = $mgh$        (5.7)

The SI unit for all forms of energy is the same as the unit for work, namely, the joule.

It should be pointed out that the gravitational potential energy of an object is not an absolute quantity. Different people may assign to it values that differ by an additive constant. For example, consider the ball in Fig. 5.10. If one considers the height of the ball above the table, its potential energy could be written $mgh_1$. With respect to the floor, however, its potential energy is $mgh_2$. We shall see later why we do not exclude either of these values but simply consider both correct. The reason is associated with the fact that we are always interested in differences in heights (or potential energies), and so additive constants such as $mgh_3$ in

$$mgh_1 = mgh_2 - mgh_3$$

always cancel out.

Furthermore, it is possible to obtain negative potential energies. For example, suppose one chooses to measure distances from the top of the table in Fig. 5.10. When the ball is at a distance $y$ above the table, its potential energy is $mgy$. As it is lowered to the tabletop, its gravitational potential energy decreases to zero. When it is lowered still farther, its $y$ coordinate becomes negative, and so its gravitational potential energy becomes negative. This simply means that the ball has less potential energy below the table than it has at the tabletop, the position arbitrarily taken as the zero level. In order to restore the ball to the zero level, it must be lifted.

## 5.6 The gravitational force is conservative

We have already seen several situations in which work was done in lifting an object. To lift an object straight up at constant speed, a force equal to the weight of the object $mg$ is required. As a result, the work done in lifting an object straight up through a distance $h$ is $mgh$. We now show that even if the object is not lifted straight up, this same result is true.

Suppose we wish to lift the pail in Fig. 5.11a from the floor to the tabletop. How much work must be done? Let us lift it along the path shown by the line from $A$ to $B$, with the lifting force being directed vertically upward throughout the motion.

To compute the work done in lifting the pail from $A$ to $B$, we approximate the actual path by the jagged path shown in $b$. If we make the jag lengths very small, we can make the two paths identical for all practical purposes. We know that the lifting force is vertical. Therefore, it does no work in the tiny horizontal movements of the jagged path. Work is done by the lifting force only in the vertical movements. When the pail is raised, positive work is done, but when it is lowered (as near point $C$), negative work is done. As a result, the downward movements cancel the work done on the equivalent upward movements. We therefore conclude that the work done depends only on the net effect of all the vertical movements. In going from $A$ to $B$, the pail of mass $m$ is lifted a net distance $h$. As a result, the work done is $mgh$, which is the same as the work done in lifting the pail straight up from $A$ through a distance $h$ and then moving it sideways to $B$. In fact, since the path shown from $A$ to $B$ is perfectly arbitrary, we conclude that

If point $A$ is a distance $h$ below point $B$, the work done against the force of gravity in lifting a mass $m$ from $A$ to $B$ is $mgh$. This result holds for any path taken between $A$ and $B$.*

Of course, if the mass is lowered from $B$ to $A$, the work done against the gravitational force is $-mgh$.

The gravitational force is an example of a **conservative force.** *A force is conservative if the work done in moving an object from point A to point B against the force is not dependent on the path taken for the movement.* We shall see later that the electrostatic and nuclear forces are also conservative. Friction forces, on the other hand, are not conservative. You can easily verify this by sliding your textbook from one point to another across a table. You obviously have to do more work when you slide it by way of a complicated long path than when you follow a straight-line path. This would not be the case if the friction force were conservative.

---

### Example 5.7

A rope pulling upward on a 40-kg object is to accelerate the object from rest to a speed of 0.30 m/s in a distance of 50 cm. What is the required tension in the rope?

*Reasoning*    Let us first solve this using the work-energy theorem. The situation is sketched in Fig. 5.12. We know that the net force acting on the object is $T - (40)(9.8)$ N, where $T$ is the tension in the rope. Then

Change in KE of object = work done by net force

$$\tfrac{1}{2}mv_f^2 - \tfrac{1}{2}mv_o^2 = F_{net}\, s \cos\theta$$

$$\tfrac{1}{2}(40 \text{ kg})(0.30 \text{ m/s})^2 - 0 = [T - (40 \text{ kg})(9.8 \text{ m/s}^2)](0.50 \text{ m})(1)$$

This yields $T = 396$ N.

Alternatively, we could recognize that the work done on the object is accomplished by two forces, the upward pull $T$ of the rope and the downward pull $mg$ of gravity. Then the work-energy theorem can be written as

Change in KE of object = work done by $T$ + work done by $mg$

This becomes

$$\tfrac{1}{2}mv_f^2 - \tfrac{1}{2}mv_o^2 = Ts - mgs$$

Where the work done on the object by the gravitational force is negative because the force is downward but the displacement is upward. This equation gives the same numerical equation we had previously and yields $T = 396$ N.  ∎

---

*Exercise*    If the rope will break when the tension in it exceeds 600 N, what maximum speed can be given to the object in the 50 cm through which it is lifted? *Answer: 2.28 m/s*  ∎

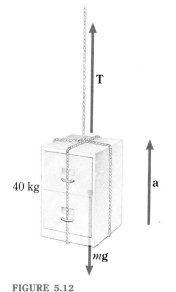

**FIGURE 5.12**
To accelerate the object upward, $T$ must be larger than $mg$.

---

* This assumes that $h$ is not so large that $g$ changes appreciably from $A$ to $B$.

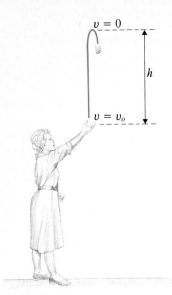

**FIGURE 5.13**

The kinetic energy of the coin is changed to gravitational potential energy as the coin rises. As it falls, the potential energy is changed back to kinetic energy.

**The interconversion of kinetic and potential energy**

Each time you toss an object into the air or drop it, you see an example of the interchange of kinetic energy and gravitational potential energy. For example, when you toss a coin upward, as in Fig. 5.13, its kinetic energy changes to gravitational potential energy. This is quantitatively correct, as we shall now prove.

In Fig. 5.13, a coin of mass $m$ is thrown upward with an initial velocity $v_o$. For the trip to the top of its path, $v_f = 0$ and $a = -g$. Thus, from the motion equation, $v_f^2 - v_o^2 = 2ay$, we have

$$0 - v_o^2 = -2gh$$

where $h$ is the distance the coin rises. Solving this equation for $h$ gives $h = v_o^2/2g$. We can use this value for $h$ in the expression for the gravitational potential energy of the coin at the top of its path:

$$GPE = mgh = mg\left(\frac{v_o^2}{2g}\right) = \tfrac{1}{2}mv_o^2$$

Thus we have shown that *the gravitational potential energy of an object at the top of its path equals the kinetic energy the object had at the bottom.*

As we see, the coin's original kinetic energy changes to gravitational potential energy as the coin rises. A similar situation exists as the coin falls freely: it loses gravitational potential energy but gains kinetic energy. Again, the magnitudes of the energy changes are the same. This is an example of a sweeping generalization we can make concerning energy. It is embodied in the law of conservation of energy, which we shall discuss soon.

**Other forms of energy**

You will recall that energy is the ability to do work. When you keep that definition in mind, it is obvious that there are many forms of energy. Coal, oil, gasoline, and all other fuels possess energy because they can be burned (a chemical reaction) to do work. The energy that can be liberated from materials by chemical reaction is called *chemical energy.* Another example is that atomic nuclei can be used to do work in nuclear reactors, and so they possess energy, which we call *nuclear energy.* Still another type of energy results from the fact that electric charges can do work; they possess *electric energy.*

In mechanical systems, energy is frequently stored in elastic devices. The stretched rubber band a child uses to propel a paper wad has *elastic potential energy* stored in it. Similarly, the compressed spring in a popgun possesses potential energy because it can be used to shoot a pellet from the gun. As we see, there are many forms of energy.

Perhaps one of the most important forms of energy is *thermal energy,* or, in colloquial parlance, *heat energy.* This energy allows the hot steam in a steam engine to do work. It is also the energy in the hot gas that drives the pistons in an automobile engine. Thermal energy is present in all objects, and we study it in detail in later chapters.

Heat can be produced by rubbing, that is, by friction. For example, primitive peoples started their fires by rubbing pieces of wood together. As another example, your skin suffers a floor burn when you accidentally slide across the gym floor. These are but a few of the many ways in which friction work leads to heating. We can trace this heating effect of friction forces to a very complicated mechanical action: the friction force acts on the molecules and atoms of the sliding surfaces in such a way as to set them into violent motion. The energy associated with this motion within a substance is what we call thermal energy. For now, it is sufficient if we realize that *work done by friction forces results in the production of thermal energy.*

## 5.9 The law of conservation of energy

Despite the fact that there are numerous forms of energy, it has become evident from the results of many experiments that the following law of nature exists:

Energy can be neither created nor destroyed. When a loss occurs in one form of energy, an equal increase occurs in other forms of energy.

This statement is called the **law of conservation of energy.** Notice that it does *not* state that kinetic energy is conserved or that gravitational potential energy is conserved or even that the sum of kinetic and potential energy is conserved. Instead, it tells us that a loss in one form of energy is always accompanied by an equal total increase in other forms of energy.

The law of conservation of energy offers us a very powerful tool for dealing with many physical situations. To use it, we notice that it may be restated as follows:

For any process, the sum of all the energy changes (gains positive, losses negative) is zero.

In using the law, we must always remember that any friction forces that slow the motion of an object generate thermal energy (TE).

Friction-related energy losses appear as an equal gain of thermal energy.

Let us now examine a few examples that show how we use this law.

### Example 5.8
A 900-kg car is moving horizontally at 20 m/s when its brakes are applied and it skids to a stop in 30 m. How large is the friction force between the car's wheels and the road?

*Reasoning*  Only two energy changes take place: the car loses kinetic energy and thermal energy is generated at the skidding wheels. From the law of conservation of energy,

$$\text{Change in KE} + \text{change in TE} = 0$$

$$\text{Change in KE} + \text{work done against friction} = 0$$

$$(\tfrac{1}{2}mv_f^2 - \tfrac{1}{2}mv_o^2) + fs = 0$$

where $f$ is the friction force and $s$ is the distance the car skids before stopping.

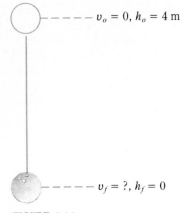

$v_o = 0$, $h_o = 4$ m

$v_f = ?$, $h_f = 0$

FIGURE 5.14

Find the final speed of the ball.

Notice that friction work always leads to a *gain* in thermal energy. Hence the friction work term $fs$ is always positive. Because $v_f = 0$ and $v_o = 20$ m/s, we have

$$0 - \tfrac{1}{2}(900 \text{ kg})(20 \text{ m/s})^2 + f(30 \text{ m}) = 0$$

from which $f = 6000$ kg $\cdot$ m/s$^2$ = 6000 N. Can you show that the friction coefficient between the tires and the road is 0.68 in this instance? ■

**Exercise** How much thermal energy is generated as the car skids to a stop? *Answer: 180 kJ* ■

---

**Example 5.9**

A 3.0-kg ball falls 4.0 m. How fast is it going just before it hits the ground?

**Reasoning** The situation is shown in Fig. 5.14. If we assume negligible air friction, the only energy changes involve kinetic and potential energies. From the law of energy conservation,

$$\text{Change in KE} + \text{change in GPE} = 0$$

$$(\tfrac{1}{2}mv_f^2 - \tfrac{1}{2}mv_o^2) + (mgh_f - mgh_o) = 0$$

$$\tfrac{1}{2}m(v_f^2 - v_o^2) + mg(h_f - h_o) = 0$$

Because $h_f = 0$ and $h_o = 4$ m, $h_f - h_o = -4.0$ m. Notice that, as we said previously, only the difference in heights matters. Further, because $h_f - h_o$ is negative, the potential energy term is negative; the ball loses gravitational potential energy. The amount of kinetic energy it gains is exactly equal to the amount of potential energy it loses.

Because $v_o = 0$, the energy equation becomes

$$\tfrac{1}{2}(3.0 \text{ kg})(v_f^2 - 0) + (3.0 \text{ kg})(9.8 \text{ m/s}^2)(-4.0 \text{ m}) = 0$$

which yields $v_f = 8.85$ m/s. Notice that the ball's mass is not needed because it cancels from the equation. ■

**Exercise** Use the motion equations to solve this problem. ■

---

**Example 5.10**

Suppose the ball of Example 5.9 is moving at 6.0 m/s just before it hit. How large is the average friction force acting on it?

**Reasoning** First notice that, unlike Example 5.9, this example cannot be easily solved using the motion equations. Because of the friction force, the acceleration of the falling ball is not 9.8 m/s$^2$. However, the energy method can still be easily applied. The law of conservation of energy tells us that

$$\text{Change in KE} + \text{change in GPE} + \text{change in TE} = 0$$

The ball loses gravitational potential energy and gains kinetic and thermal energies.

The thermal energy term is simply $fs$, where $s = 4.0$ m in this case. We therefore have

$$\tfrac{1}{2}m(36 \text{ m}^2/\text{s}^2 - 0) + m(9.8 \text{ m/s}^2)(-4.0 \text{ m}) + f(4.0 \text{ m}) = 0$$

Checking our signs, we see that the object gains kinetic energy and loses potential energy and that the change in thermal energy is positive, as it must always be when the change is caused by friction losses. Notice that the mass does not cancel in this situation. Replacing it by 3.0 kg, we find that $f = 15.9$ N. ▪

### Example 5.11

A 300-kg roller coaster car starts from rest at point $A$ in Fig. 5.15 and begins to coast down the track. If the retarding friction force is 20 N, how fast is the car going at point $B$? At point $C$?

*Reasoning*  Qualitatively, the car loses potential energy and gains kinetic and thermal energies. From the law of energy conservation, for the trip from $A$ to $B$,

Change in GPE + change in KE + change in TE = 0

$$mg(h_B - h_A) + \tfrac{1}{2}m(v_f^2 - 0) + fs = 0$$

where $s$ is the distance from $A$ to $B$ along the track. Using $m = 300$ kg, $h_B - h_A = -10.0$ m, $f = 20$ N, and $s = 40$ m, we can write this as

$$-29{,}400 \text{ J} + \tfrac{1}{2}(300 \text{ kg})v_f^2 + (20 \text{ N})(40 \text{ m}) = 0$$

Because potential energy is lost, its term should be negative, as we find it to be. Solving gives $v_f = 13.8$ m/s.
   For the trip from $A$ to $C$, we have

$$mg(h_C - h_A) + \tfrac{1}{2}m(v_f^2 - 0) + (20 \text{ N})(60 \text{ m}) = 0$$

with $h_C - h_A = (8.0 - 10.0)$ m $= -2.0$ m. Substituting and solving give $v_f = 5.59$ m/s. ▪

---

*Exercise*  How fast will the car be moving at $C$ if it has a speed of 5.0 m/s at $A$ and friction forces are negligible?  *Answer: 14.9 m/s* ▪

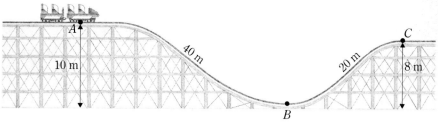

**FIGURE 5.15**
The gravitational potential energy that the cart has at $A$ is changed to kinetic energy and thermal energy generated by friction as it moves to points $B$ and $C$.

FIGURE 5.16

The kinetic energy of the car when it is at $A$ is partly lost to potential energy and work against friction forces as it moves to $B$.

## Example 5.12

The 2000-kg car in Fig. 5.16 is at point $A$ and moving at 20 m/s when it begins to coast. As it passes point $B$, its speed is 5.0 m/s. (*a*) How large is the average friction force that retards the car's motion? (*b*) Assuming the same friction force, how far beyond $B$ does the car go before stopping?

**Reasoning** (*a*) In going from $A$ to $B$, the car loses kinetic energy to potential energy and to thermal energy generated by friction work. From the conservation law,

$$\text{Change in KE} + \text{change in GPE} + \text{change in TE} = 0$$

$$\tfrac{1}{2}m(v_B^2 - v_A^2) + mg(h_B - h_A) + fs_{AB} = 0$$

In this case, $m = 2000$ kg, $v_B = 5.0$ m/s, $v_A = 20$ m/s, $h_B - h_A = 8.0$ m, and $s_{AB} = 100$ m. Solving gives $f = 2180$ N.

(*b*) As the car goes from $B$ to the point at which it stops, its kinetic energy at $B$ is changed to thermal energy because of the friction work done. From the law of conservation of energy, we have

$$\text{Change in KE} + \text{change in TE} = 0$$

$$\tfrac{1}{2}m(0 - v_B^2) + fs_{BE} = 0$$

where $s_{BE}$ is the distance from $B$ to the point at which the car stops. Using $f = 2180$ N, we find that $s_{BE} = 11.5$ m. ▪

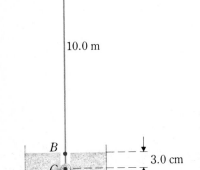

FIGURE 5.17

The gravitational potential energy of the ball at $A$ has been lost doing friction work by the time it comes to rest at $C$.

## Example 5.13

A 2-kg ball falls into a box of sand from a height 10.0 m above the sand, as shown in Fig. 5.17. It comes to rest 3.0 cm beneath the surface of the sand. How large is the average force exerted on it by the sand?

**Reasoning** We notice that all the potential energy at $A$ is changed to kinetic energy at $B$. This kinetic energy is then lost to friction work (thermal energy) as the ball penetrates to $C$. Let us consider the trip from $A$ to $C$. At both these points, $KE = 0$. The law of conservation of energy becomes

$$\text{Change in GPE} + \text{change in TE} = 0$$

$$mg(h_C - h_A) + fs_{BC} = 0$$

We know that $m = 2.0$ kg, $h_C - h_A = -10.03$ m, and $s_{BC} = 0.030$ m. Substituting and solving yield $f = 6550$ N. ▪

## Example 5.14

Consider the pendulum (a ball at the end of a string) in Fig. 5.18*a*. It is released from rest at point $A$. How fast is the ball moving (*a*) at $B$, (*b*) at $C$?

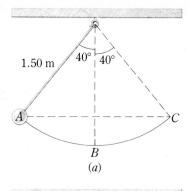

1.50 m  40° 40°

A - - - - - - - - - ▷C

B

(a)

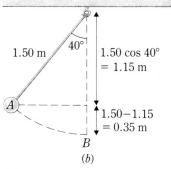

1.50 m  40°

1.50 cos 40°
= 1.15 m

A

1.50−1.15
= 0.35 m

B

(b)

**FIGURE 5.18**

As the pendulum swings back and forth, its kinetic energy and potential energy keep interchanging.

**Reasoning**  We should notice at the outset that the pendulum cord does no work on the ball because the cord always pulls in a direction perpendicular to the ball's direction of motion. Hence there is no component of the tension in the direction of the motion, and so the tension in the cord does no work.

As the ball swings back and forth, its potential and kinetic energies keep varying. From the law of conservation of energy,

Change in GPE + change in KE = 0

When the ball loses potential energy, it must gain a like amount of kinetic energy. (*a*) For the trip from *A* to *B*,

$$mg(h_B - h_A) + \tfrac{1}{2}m(v_B^2 - v_A^2) = 0$$

Since $h_B - h_A = -0.35$ m, $v_A = 0$, and $m$ cancels from the equation, we have $v_B = 2.62$ m/s.
    (*b*) For the trip from *A* to *C*,

$$mg(h_C - h_A) + \tfrac{1}{2}m(v_C^2 - v_A^2) = 0$$

but $h_C = h_A$, and so $v_C = v_A = 0$.
    As we see, the pendulum is an interesting example of interconversion between potential and kinetic energies. Over and over, the potential and kinetic energies interchange as the pendulum swings back and forth. At any point in the motion, the potential energy lost in the fall from point *A* has been changed to kinetic energy. Of course, all this assumes friction forces to be negligible.  ■

*Exercise*  How fast is the ball moving after it has fallen 2.0 cm from its original position?  *Answer: 0.626 m/s*  ■

**Example 5.15**
How large a force is needed to accelerate a 2000-kg car from rest to 15.0 m/s in 80 m if an average friction force of 500 N opposes the motion? See Fig. 5.19.

*Reasoning*  We assume the car is moving along level ground, so that its potential energy change is zero. Solving this problem with the work-energy theorem, we have

Change in KE of car = work done by $F_{net}$ on car

$$\tfrac{1}{2}m(v_f^2 - v_o^2) = F_{net}s \cos \theta$$

In this case, two forces act on the car: the accelerating force *F* and the 500-N friction force. Therefore the net force in the direction of motion is $F - 500$ N and $\theta$ is zero. The above equation therefore becomes

$$\tfrac{1}{2}(2000 \text{ kg})(225 \text{ m}^2/\text{s}^2 - 0) = (F - 500 \text{ N})(80 \text{ m})$$

Solving gives $F = 3300$ N.  ■

Start                              Finish
$v_o = 0$                        $v_f = 15$ m/s

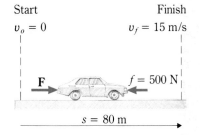

F          $f = 500$ N

$s = 80$ m

**FIGURE 5.19**

How large is the force responsible for the acceleration?

*Exercise*  How much thermal energy is generated as the car travels the 80 m? *Answer: 40,000 J*  ■

5.9 The law of conservation of energy  109

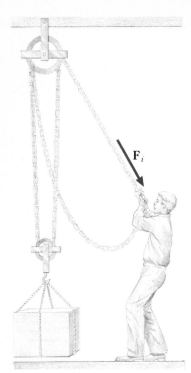

**FIGURE 5.20**

When the worker pulls out a length of chain $s_i$, the load rises a distance $s_o$.

## 5.10 Machines

Machines are devices we use to help us do work. They are of many kinds, and new types are being devised every day. Despite their diversity, there are certain principles that apply to all. These are the principles we examine in this section.

One of the most important facts concerning machines is that *machines cannot create energy;* when they operate continuously, they can do no more work than the amount of energy supplied to them from an external source. This is simply a statement of the law of conservation of energy as it applies to machines. As we show in the special note on page 112, this fact precludes the existence of perpetual motion machines.

There are other statements we can make that apply to all machines. Consider the pulley system (a chain hoist) shown in Fig. 5.20. A man supplies an input force $F_i$ to the machine by pulling on a chain. As he pulls the chain a distance $s_i$, the amount of work he does is

Input work (or energy) $= F_i s_i$

Meanwhile, the machine exerts a force $F_o$ on the load as it lifts the load a distance $s_o$. Hence the output work done by the machine is

Output work $= F_o s_o$

If there are no energy losses inside the machine (such as those due to friction), then the law of energy conservation tells us that

Input work = output work

$$F_i s_i = F_o s_o \tag{5.8}$$

This relation applies only to an ideal machine, which is defined as one in which all the input energy is delivered to the output end.

Most machines, however, depart seriously from ideality. Because of friction and other mechanisms that reduce the effectiveness of the input energy, the output work is less than the input energy. To describe this feature of machines, we define

$$\text{Percent efficiency} = \frac{\text{output work}}{\text{input energy}} \times 100 \tag{5.9a}$$

which can be written as

$$\text{Percent efficiency} = \frac{F_o s_o}{F_i s_i} \times 100 \tag{5.9b}$$

No machine can have an efficiency as high as 100 percent, although some come close. Electric motors, for example, have efficiencies of about 95 percent.

Because we often use machines to help us lift heavy loads, we characterize their usefulness in this respect by their *mechanical advantage.* If a machine requires an input force $F_i$ to produce an output force $F_o$ we define

$$\text{Actual mechanical advantage (AMA)} = \frac{F_o}{F_i} \tag{5.10}$$

An automobile jack might typically require an input force of 100 N to lift a 5000-N load. Thus the actual mechanical advantage of the jack is

$$\text{AMA} = \frac{F_o}{F_i} = \frac{5000 \text{ N}}{100 \text{ N}} = 50$$

How could you determine the efficiency of a car jack?

We can easily determine the mechanical advantage of an ideal machine because Eq. 5.8 applies to it. From that equation, $F_o/F_i = s_i/s_o$, and so we have (for an ideal machine only)

$$\text{Ideal mechanical advantage (IMA)} = \frac{s_i}{s_o} \tag{5.11}$$

Also from Eq. 5.8 we notice that

$$F_o = F_i \frac{s_i}{s_o}$$

Therefore, to obtain a large output force, we must have an input force that works through a much larger distance than the output force does. This is simply a reflection of the fact that, in an ideal machine, the input energy $F_i s_i$ must equal the output energy $F_o s_o$; therefore large $F_o/F_i$ requires large $s_i/s_o$.

Another important relation for machines follows directly from Eqs. 5.9, 5.10, and 5.11:

$$\text{Percent efficiency} = \frac{\text{AMA}}{\text{IMA}} \times 100 \tag{5.12}$$

You can prove this easily by dividing Eq. 5.10 by Eq. 5.11 and comparing the result with Eq. 5.9.

Let us illustrate the use of these equations by referring to the simple machine in Fig. 5.21. It is called a wheel and axle and is used to enable a small input force $F$ to lift a heavy load $W$. We can compute the machine's IMA by noting that, when the wheel and axle turn through one revolution, the two cords wind a length equal to the circumferences of their respective circles. Thus $s_i = 2\pi b$ and $s_o = 2\pi a$. Then

$$\text{IMA} = \frac{s_i}{s_o} = \frac{2\pi b}{2\pi a} = \frac{b}{a}$$

If the machine were 100 percent efficient, a force $F$ could lift a weight

$$W = \frac{b}{a} F$$

By making the radius of the wheel much larger than the radius of the axle, we obtain a very effective lifting device.

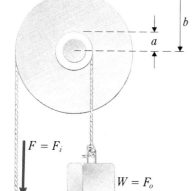

**FIGURE 5.21**

The wheel and axle have an IMA given by the ratio of the radius of the wheel to the radius of the axle.

# Perpetual motion machines

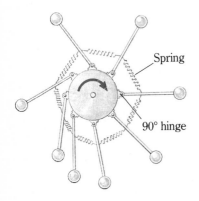

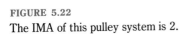

Throughout the ages, inventors have sought to design a machine that could do work in excess of the energy supplied to it. If it could be constructed, a device of this type could both supply the work needed to operate itself and perform useful work. In other words, it could do work without an external source of energy. Such a machine creates energy, and the law of conservation of energy tells us that no such machine can exist. Despite that fact, people are still inventing purported perpetual motion machines, investors are contributing money to their development, and reputable newspapers occasionally report what they think is a viable candidate. Nevertheless, such machines are impossible to construct.

One early proposal for a perpetual motion machine is shown in the figure. The rigid rods and the massive balls attached to them are connected to the wheel by 90° hinges. These hinges allow the rods to move from tangential to radial positions relative to the wheel. Because a net clockwise torque is applied to the wheel by the weights of the balls, the wheel can do continuous work as it turns under the force of gravity. Hence, without a source of energy, this perpetual motion machine is capable of doing useful work until its hinges and bearing wear out. The law of energy conservation tells us this machine cannot possibly work, but why not?

Experience with all such machines indicates that their inventors either have been unable to construct a working model or have used subterfuge to conceal a hidden energy source. Those who know physics and believe in its laws reject the idea that a perpetual motion machine can be built. Even so, it is fun to examine proposals for such machines and find the underlying flaw that destroys their feasibility.

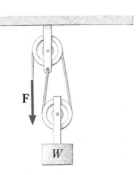

**FIGURE 5.22**
The IMA of this pulley system is 2.

Pulley systems are also interesting machines. The one in Fig. 5.22 lifts an object of weight $W$ when the force $F$ pulls the cord out from the top pulley. This pulley is fastened to the ceiling, while the lower pulley moves upward as the force $F$ pulls the cord out. Notice that the lower pulley moves upward a distance $0.5s_i$ when the cord is pulled out a distance $s_i$ from the top pulley. (Each of the two cords supporting the lower pulley shortens $0.5s_i$, giving a total shortening $s_i$ for the cord within the system.) Therefore we have

$$\text{IMA} = \frac{s_i}{s_o} = \frac{s_i}{0.5s_i} = 2.00$$

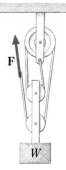

**FIGURE 5.23**

This system has an IMA of 4.

This pulley system has an IMA of 2. You should be able to show that the ideal mechanical advantage of the pulley system in Fig. 5.23 is 4.

The actual mechanical advantage of these pulley systems is considerably less than their ideal mechanical advantage. Not only is there friction in the pulleys, but the system must also lift the useless weight of the movable pulley. Even so, pulley systems are widely used to lift heavy objects.

---

**Example 5.16**

To lift a 2000-N object with the pulley system in Fig. 5.24, an input force of 800 N is required. Find the IMA, AMA, and efficiency for the system.

*Reasoning*   Inspection of the diagram indicates that $s_i = 3s_o$. Hence

$$\text{IMA} = \frac{s_i}{s_o} = \frac{3s_o}{s_o} = 3.00$$

By definition, however,

$$\text{AMA} = \frac{F_o}{F_i} = \frac{2000 \text{ N}}{800 \text{ N}} = 2.50$$

This gives

$$\text{Percent efficiency} = \frac{\text{AMA}}{\text{IMA}} \times 100 = \frac{2.5}{3.0} \times 100 = 83\% \quad \blacksquare$$

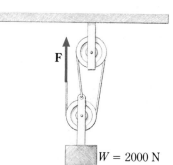

**FIGURE 5.24**

What is the IMA of this system?

---

## Minimum learning goals

When you finish this chapter, you should be able to

1 Define (*a*) work, (*b*) joule, (*c*) power, (*d*) watt, (*e*) kilowatthour, (*f*) kinetic energy, (*g*) work-energy theorem, (*h*) gravitational potential energy, (*i*) law of conservation of energy, (*j*) efficiency of a machine, (*k*) IMA and AMA of a machine.

2 Compute the work done on an object by a specified force when the object is moved through a given distance.

3 Compute power in simple situations. Change from watts to horsepower and vice versa.

4 Compute the change in kinetic energy of an object subjected to a known net force acting through a known distance.

5 Compute the change in gravitational potential energy of an object as it is moved from one place to another.

6 Distinguish between conservative and nonconservative forces.

7 Give several examples in which potential and kinetic energies are interchanged. Repeat for interchange of kinetic and thermal energies.

8 State what happens to the energy lost when work is done against friction forces.

9 Use the law of conservation of energy in simple situations in which potential, kinetic, and thermal energies are interchanged.

10 Compute the IMA, AMA, and efficiency of a simple machine when appropriate data are given.

## Questions and guesstimates

1 A conscientious hobo in a boxcar traveling from Chicago to Peoria pushes on the front of the boxcar all the way. Having at one time been a physics student, he thinks his pushing does a great deal of work since both $F_s$ and $s$ are large. Where is the flaw in his reasoning?

2 A person holds a bag of groceries while standing still talking to a friend. A car sits stationary with its motor running. From the standpoint of work and energy, how are these two situations similar?

3 As a rocket reenters the atmosphere, its nose cone becomes very hot. Where does this heat energy come from?

4 Reasoning from the interchange of kinetic and potential energies, explain why the speed of a satellite in a noncircular orbit about the earth keeps changing. Is its speed largest at apogee (farthest point from earth) or at perigee (closest point)?

5 Describe a situation in which the gravitational potential energy of an object is negative. Will everyone agree that it is negative? Can an object have negative kinetic energy?

6 A car cannot accelerate along an extremely slippery roadway. Suppose a car of mass $m$ accelerates from rest to a speed $v$ along a horizontal roadway. Assume that its wheels do not slip. How much work does the friction force between wheels and pavement do in the process?

7 Is energy a vector or a scalar quantity?

8 The coefficient of sliding friction for a block on an incline is large enough so that the block does not move by itself. The block is pulled up the incline at constant speed by a force parallel to the incline. Compare the work done by (a) the pulling force, (b) the friction force, and (c) gravity. Repeat for the block moving down the incline.

9 Automobiles, bicycles, and many other devices have gear systems that can be changed by shifting. Considering these to be ideal machines, discuss why shifting is used.

10 About what horsepower is a human being capable of producing for a short period, as in climbing a flight of stairs?

11 Estimate the force a driver experiences when her car hits another car head on. Assume both cars to be similar and traveling at 25 m/s. Discuss the effect of seat belts and other safety factors.

12 A human heart consumes about 1 J of energy per heartbeat. About how many joules of energy must food furnish to a person each day to supply this energy? For comparison purposes, one nutritionist's calorie of food energy is equivalent to 4184 J.

## Problems

*Many of these problems are solved most easily by energy methods. Use energy methods when appropriate.*

1 A horizontal force of 25 N pulls a box along a table. How much work is done in pulling the box 80 cm?

2 To pull a certain child in a wagon, a force of 200 N at an angle of 30° above the horizontal is required. How much work is done by this force in pulling the wagon 8.0 m?

3 A woman pushes a push-type lawnmower with a force of 200 N at an angle of 25° below the horizontal. How much work does she do as she pushes the mower 40 m?

4 A horizontal force $F$ pulls a 20-kg box across the floor at constant speed. If the coefficient of friction between box and floor is 0.60, how much work does $F$ do in moving the box 3.0 m?

■ 5 A 1200-kg car skids to a stop in 30 m. How large a

friction force acts between its four skidding tires and the pavement if the coefficient of friction is 0.70? How much work does the friction force do on the car?

**6** The coefficient of friction between a 20-kg box and the floor is 0.40. How much work does a pulling force directed 37° above the horizontal do on the box in pulling it 8.0 m across the floor at constant speed?

**7** A 5.0-kg bucket is lowered into a vertical well shaft by a rope-and-winch system. How much work does the rope do on the bucket as it is lowered 9.0 m?

**8** How much lifting work is done by a 50-kg woman as she climbs a flight of stairs 4 m high?

**9** A friction force of 25 N opposes the sliding of a 5-kg box along the floor. What is the power supplied to the box as it is dragged along the floor at a speed of 40 cm/s?

**10** A 2000-kg loaded coal cart is being pulled up a 10° incline at constant speed by a cable parallel to the incline. The friction force impeding the motion is 800 N. How much work is done on the cart by the cable as the cart is pulled 30 m?

**11** By changing the angle of a variable-angle incline, an assembly-line worker finds that a 500-g block slides down the incline with constant speed when the incline angle is 40°. How much work does the friction force do on the block as it slides 80 cm?

**12** A body of weight $w$ is pulled up a frictionless incline of angle $\theta$ with speed $v$. Show that the work done on the body in time $t$ is $wvt \sin \theta$.

**13** A certain tractor can pull with a steady force of 14,000 N while moving at 3.0 m/s. How much power in watts and in horsepower is the tractor developing under these conditions?

**14** A small electric motor is needed to lift a 200-g mass at a steady rate of 5.0 cm/s. Give the power in watts and horsepower that the motor must be capable of producing.

**15** It takes 8.0 s for a 30-kg girl to climb a 9-m-high flight of stairs. How much horsepower is she developing?

**16** The friction force tending to stop a car traveling at a low speed is due mainly to energy loss in the rolling tires. For a typical compact car moving at 9.0 m/s, the total force resisting motion is about 200 N. (*a*) How much useful power in horsepower must the motor provide to maintain this speed? (*b*) At 36 m/s, the combined rolling and air friction is 1100 N for the same car. What horsepower is needed at this speed?

**17** A pump is needed to lift water through a height of 2.5 m at a rate of 500 g/min. What must the minimum horsepower of the pump be?

**18** How large a force is needed to accelerate a proton ($m = 1.67 \times 10^{-27}$ kg) from rest to $2 \times 10^7$ m/s in

0.50 cm? (A proton is a hydrogen atom that has lost its electron.)

**19** How large a force is needed to stop an 800-kg car in 50 m if the car is coasting with a speed of 25 m/s?

**20** Assuming a coefficient of friction $\mu$ between a car and the roadway, use energy methods to show that the stopping distance for a car whose original speed is $v$ is $v^2/2\mu g$ if the car brakes equally on all four wheels. What result applies if only two wheels brake and the weight of the car is equally distributed to the four wheels?

**21** Suppose that $10^{17}$ electrons strike the screen of a television tube each second. If each electron is accelerated to a speed of $10^9$ cm/s, starting from rest, how many watts are expended in maintaining this beam of electrons? ($m_e = 9.1 \times 10^{-31}$ kg)

**22** An atom-smashing machine known as a Van de Graaff generator can accelerate a beam of protons ($m_p = 1.67 \times 10^{-27}$ kg) from rest to $10^9$ cm/s. If the machine accelerates $5 \times 10^{16}$ protons per second, how many watts is it producing?

**23** An 80-kg hiker climbs a 500-m-high hill. (*a*) How much work does the hiker do against gravity? (*b*) Does this amount of work depend on the path the hiker takes? (*c*) If the hike takes 90 min, what average horsepower is expended?

**24** A 25,000-kg truck takes 40 min to climb from an elevation of 1800 m to 2900 m along a mountain road. (*a*) How much work does the truck do against gravity? (*b*) What average horsepower does the truck expend against gravity?

**25** When an object is thrown upward, it rises to a height $h$. How high is the object, in terms of $h$, when it has lost one-third of its original kinetic energy?

**26** A 2-kg object dropped from a height of 12 m is moving at 8 m/s just before it hits the ground. How large is the average friction force retarding the motion?

**27** A motor is to lift a 900-kg elevator from rest at ground level in such a way that it has a speed of 4.0 m/s at a height of 20 m. (*a*) How much work does the motor do? (*b*) What fraction of the total work goes into kinetic energy?

**28** A 3-kg mass starts from rest at the top of a 37° incline that is 5.0 m long. Its speed as it reaches the bottom is 2.0 m/s. Use energy methods to find the average friction force that retards its sliding motion.

**29** A 3-kg mass has a speed of 5.0 m/s at the bottom of a 37° incline. How far up the incline does the mass slide before it stops if the friction force retarding its motion is 20 N?

**30** Starting from rest, a locomotive pulls a series of boxcars up a 3° incline. After the train has moved 2000 m, its speed is 10.0 m/s. Assume that the entire train has a mass of $7.0 \times 10^5$ kg. (*a*) How much work must the locomotive do? (*b*) What fraction of this work is done against gravity? (*c*) Assuming the acceleration to be uniform, how long does the process take? (*d*) What average horsepower does the locomotive expend?

■ **31** An electric motor is to power a pump that lifts 500 g of water originally in a tank through a height of 2.0 m in 90 s. Assume that the water is moving at 1.60 m/s when it exits at the top. What minimum output in horsepower should the motor have? The speed of the water in the tank is negligible.

■ **32** Repeat Prob. 31 if the motor is operating on the moon, where the acceleration due to gravity is only one-sixth that on earth.

**33** A 1.8-g bullet with a speed of 420 m/s strikes a block of wood and comes to rest at a depth of 6 cm. (*a*) How large is the average decelerating force? (*b*) How long does it take to stop the bullet?

**34** A 100-kg ballplayer is running at 8.0 m/s when he is stopped in a distance of 2.00 m by an opposing player. (*a*) How large is the average force the opposing player exerts? (*b*) How long does it take to stop the runner?

■ **35** A 200-g ball is thrown straight upward with a speed of 12 m/s. (*a*) How high does it rise if friction forces are negligible? (*b*) If it rises only 6.0 m, how large an average friction force impedes its motion? (*c*) Under the average friction force of *b*, how fast is the ball moving when it returns to the thrower?

■ **36** A 500-g block is released from the top of a 30° incline and slides 140 cm down the incline to the bottom. What is the speed of the block when it reaches the bottom (*a*) if friction is negligible and (*b*) if the friction force is 0.90 N? (*c*) In *b*, how large is the coefficient of friction?

■ **37** A bead of mass *m* starts from rest at point *A* in Fig. P5.1 and slides along the frictionless wire. Find its speed (*a*) at *B* and (*b*) at *C*.

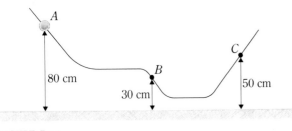

**FIGURE P5.1**

■ **38** The bead of mass *m* sliding on the wire in Fig. P5.1 has a speed of 2.0 m/s up the wire (to the left) as it passes *A*.

The wire is frictionless, and the bead eventually passes *B* and *C*. Find its speed at each of these points.

■ **39** In Fig. P5.1, the distance along the wire from *A* to *C* is 400 cm. If the 0.70-g bead starts from rest at *A* and finally stops just as it reaches *C*, how large is the average friction force retarding its motion?

■ **40** In Fig. P5.1, the 0.70-g bead starts from rest at *A* and has a speed of 0.50 m/s when it reaches *B*, 2.0 m from *A* along the wire. How large is the average friction force retarding its motion?

■ **41** At high speeds, the friction forces acting on a car increase in proportion to $v^2$, where *v* is the car's speed. If this is considered the major factor involved and if a car is rated at 32 km per gallon of gas at 80 km/h, what is its distance per gallon rating at 110 km/h?

■ **42** A car of mass *m* rolls from rest down a hill of height *h* and length *L*. Show that when the car reaches the bottom of the hill, its speed is

$$v = \sqrt{2gh - \frac{2Lf}{m}}$$

where *f* is the average friction force retarding the motion.

**43** A ball of mass *M* is suspended as a pendulum bob from a cord 4.0 m long. A force on the mass pulls the cord to one side until it makes an angle of 50° with the vertical, and then the system is released. With what speed is the mass moving as it passes directly underneath the point of suspension? (Ignore air friction.)

**44** For the pendulum described in Prob. 43, how fast is the ball moving when the cord makes an angle of 20° with the vertical?

■ **45** The pendulum of length *L* in Fig. P5.2 is released from point *A*. As it swings down, the string strikes the peg at *B*, and the ball swings through *C*. (*a*) How fast is the ball moving as it passes through *C*? (*b*) If we neglect friction, the ball approaches a limiting speed as the string winds on the pin. What is that speed? Assume the very thin string does no work on the very tiny ball.

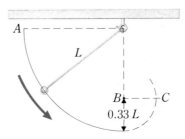

**FIGURE P5.2**

■ **46** If the masses in Fig. P5.3 are released from the position shown, (*a*) find an expression for the speed of either mass just before $m_1$ strikes the floor. Ignore the mass and friction of the pulley. (*b*) Repeat if $m_1$ has a downward velocity $v_o$ at the instant shown in the figure.

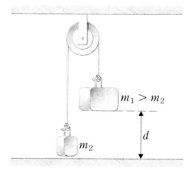

FIGURE P5.3

**47** A 2-kg block starts to slide up a 20° incline with an initial speed of 200 cm/s. It stops after sliding 37 cm and slides back down. Assuming the friction force impeding its motion to be constant, (*a*) how large is the friction force and (*b*) what is the block's speed as it reaches the bottom?

**48** A man finds that, on level ground, his 800-kg car accelerates from rest to 15.0 m/s in 10 s and then coasts to rest from 15.0 m/s in 500 m. Compute the average horsepower delivered by the car. (Note that the friction work done in stopping is not the same as that done in starting. Why? Assume instead that the average friction force is the same in the two cases.)

**49** A 600-kg object is to be lifted by a pulley system using a force of 400 N. The machine found suitable for this purpose lifts the load 0.50 m when the applied force moves 10 m. Find the (*a*) IMA, (*b*) AMA, and (*c*) efficiency of the machine.

**50** A pulley system lifts a 200-kg load when a force of 150 N is applied to it. If the efficiency of the device is 87 percent, find (*a*) AMA, (*b*) IMA, (*c*) $s_i/s_o$.

**51** What must be the ratio of the radii of a wheel-and-axle device if the device is to lift 25 kg with an applied force of 30 N? Assume the efficiency of the device is 90 percent.

**52** For a particular type of car jack, the operator moves her hand (the input force) through 60 cm for every 1.50 cm the load is lifted. (*a*) What is the IMA of the jack? (*b*) Assuming 20 percent efficiency, how large an applied force is needed to lift 4000 N?

**53** An electric motor is labeled 0.55 kW. On the assumption that it is 90 percent efficient, how many horsepower can it deliver?

**54** A $\frac{1}{4}$-hp motor has attached to its shaft a pulley of diameter 8.0 cm. If the shaft rotates at 1800 revolutions per minute (rev/min), how large a load is the belt running on the pulley capable of pulling? Assume that the motor is 90 percent efficient and that the power input to it is $\frac{1}{4}$ hp.

**55** A certain 50-W motor runs with a shaft speed of 1600 rev/min. Because of reducing gears, the final (output) shaft rotates at 18 rev/min. (*a*) If the machine is 30 percent efficient, with what force can it pull the belt on a 3.0-cm-radius pulley at the output shaft? (*b*) If the gear system is reversed so that the output shaft rotated at 180,000 rev/min, what force is available to pull the belt on the same pulley? Assume the power input to the motor is 50 W.

**56** A boxcar is moving with a constant speed of 3.0 m/s along a straight horizontal track when it strikes a barrier and stops almost immediately. To what maximum angle with the vertical does a 90-cm pendulum hanging from the boxcar ceiling swing as a result of the impact?

**57** A child's 200-g toy car is driven by an electric motor that has a constant output power. The car can climb a 20° incline at 20 cm/s and can travel on a horizontal table at 40 cm/s. The friction force retarding it is $kv$, where $k$ is a constant and $v$ is its speed. How steep an incline can it climb with a speed of 30 cm/s?

**58** The bead in Fig. P5.4 is sliding along a wire with a speed of 4.0 m/s at the point indicated. It slides to point $Q$ before stopping. What is the coefficient of friction between bead and wire?

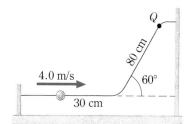

FIGURE P5.4

**59** The frictionless system in Fig. P5.5 is released from rest. After the right-hand mass has risen 80 cm, the object of mass $0.50m$ falls loose from the system. What is the speed of the right-hand mass when it returns to its original position?

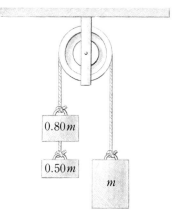

FIGURE P5.5

# 6 Linear momentum

THE law of conservation of energy, discussed in the previous chapter, is not the only conservation law obeyed by nature. A second example, the law of conservation of linear momentum, is the subject of this chapter. Like the law of energy conservation, the linear momentum conservation law has profound consequences. We examine the law and some of its implications in the following discussion. In addition, we use the law to derive the relationship between the pressure of an ideal gas and the motion of its molecules. We shall use momentum and its conservation law frequently as we continue our studies of the laws of physics.

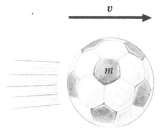

**FIGURE 6.1**

The linear momentum of this object is $m\mathbf{v}$, a vector quantity.

## 6.1 The concept of linear momentum

An experience common to us all is that a moving object possesses a quality that causes it to exert a force on anyone or anything trying to stop it. The faster the object is moving, the harder it is to stop. In addition, the more massive the object is, the more difficulty we have in stopping it. For example, an automobile moving at 2 m/s is stopped much less easily than is a tricycle traveling at the same speed. Newton called this quality of a moving object its *motion*. Today we term it the **linear momentum** of the moving object.

We define linear momentum in the following way. Consider the object of mass $m$ and velocity $\mathbf{v}$ shown in Fig. 6.1. For the object,

$$\text{Linear momentum} = \mathbf{p} = m\mathbf{v} \tag{6.1}$$

where $\mathbf{p}$ is the symbol used for linear momentum. In the SI, the units for linear momentum are kg · m/s. Notice that the momentum of an object is large if the object has large mass and large velocity. From its defining equation, linear momentum is a vector quantity. Its direction is the same as that for $\mathbf{v}$.

## 6.2 Newton's second law restated

There is an important relationship between the net force applied to an object and the change in linear momentum that the force causes. When a net force $\mathbf{F}_{net}$ accelerates an object, it obviously causes the velocity, and thus the momentum, of the object to change. Let us now investigate this relationship by seeing how Newton's second law looks when it is written in terms of linear momentum.

Consider the object of mass $m$ in Fig. 6.2. Because of the net force $\mathbf{F}$ applied to it, it experiences an acceleration, and the mass and force are related through $\mathbf{F} = m\mathbf{a}$. From the definition of acceleration as $\mathbf{a} = (\mathbf{v}_f - \mathbf{v}_o)/t$, however, $\mathbf{F} = m\mathbf{a}$ becomes

$$\mathbf{F} = \frac{m(\mathbf{v}_f - \mathbf{v}_o)}{t}$$

This can be rewritten as

$$\mathbf{F} = \frac{m\mathbf{v}_f - m\mathbf{v}_o}{t} \qquad \text{or} \qquad \mathbf{F} = \frac{\Delta \mathbf{p}}{t} \tag{6.2}$$

where $\Delta \mathbf{p}$ is the change in linear momentum that occurs in time $t$. Thus we have related the net force acting on the object to its change in linear momentum.

Equation 6.2 is the form in which Newton stated his second law, rather than as $\mathbf{F} = m\mathbf{a}$. In words, Eq. 6.2 says that the net force that acts on an object equals the time rate of change of the object's linear momentum. In some situations, Eq. 6.2 is preferable to $\mathbf{F} = m\mathbf{a}$ because this latter form applies only if the mass of the object is constant. Nowadays, for example, we often accelerate atomic particles to such high speeds that their mass increases greatly with a further increase in speed. (This effect was first predicted by Einstein in his theory of relativity, discussed in Chap. 25.) In such a situation, Eq. 6.2 is valid but $\mathbf{F} = m\mathbf{a}$ is not. Whenever the mass of the object being accelerated is changing, Eq. 6.2 is the form of Newton's second law that one should use.

Often we wish to apply the concept of momentum change to situations in which the applied force is not constant. For example, suppose a bat strikes a ball of mass $m$ to change its velocity from $\mathbf{v}_o$ to $\mathbf{v}_f$. Then we use Eq. 6.2 to define the average force $\overline{\mathbf{F}}$ exerted by the bat on the ball. We have, after multiplying through by $t$,

$$\overline{\mathbf{F}}t = \Delta \mathbf{p} \tag{6.3}$$

In the case of the bat hitting the ball, this becomes

$$\overline{\mathbf{F}}t = m\mathbf{v}_f - m\mathbf{v}_o$$

The product $\overline{\mathbf{F}}t$ is called the **impulse** of the force. Because change in momentum is fairly easy to measure, the impulse can be evaluated even though the average force and the time of contact may be extremely difficult to determine.

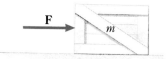

**FIGURE 6.2**

The net applied force $\mathbf{F}$ increases the momentum of the mass. Momentum has direction, and the increase will be in the direction of $\mathbf{F}$.

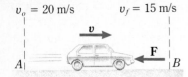

$v_o = 20$ m/s    $v_f = 15$ m/s

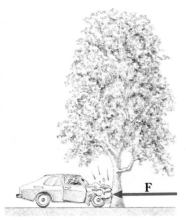

**FIGURE 6.3**
The car takes 3 s to travel from $A$ to $B$. Determine **F**.

## Example 6.1

In Fig. 6.3, a 1500-kg car traveling along a straight line reduces its speed from 20 m/s at $A$ to 15 m/s at $B$ in 3.0 s. How large is the average force retarding its motion?

*Reasoning*    From Eq. 6.2, the momentum form of Newton's second law, we can write

$$\overline{\mathbf{F}} = \frac{m\mathbf{v}_f - m\mathbf{v}_o}{t}$$

Take the direction of motion as the positive direction. Then $\mathbf{v}_o = +20$ m/s, $\mathbf{v}_f = +15$ m/s, and $t = 3.0$ s. After making these substitutions, we find that $\overline{\mathbf{F}} = -2500$ N. (Notice that we are using plus and minus signs to show direction.) The fact that $\overline{\mathbf{F}}$ is negative tells us that it is in the negative direction, a fact we show in Fig. 6.3. ▪

*Exercise*    How far is it from $A$ to $B$? How could you determine $F$ using energy considerations?    *Answer: 52.5 m* ▪

## Example 6.2

The 1200-kg car in Fig. 6.4, initially moving at 20 m/s, strikes a tree and comes to rest in a distance $s = 1.5$ m. Estimate the average stopping force the tree exerts on the car.

*Reasoning*    We can apply the momentum form of Newton's second law (Eq. 6.2) to the car:

$$\overline{\mathbf{F}} = \frac{\Delta \mathbf{p}}{t} = \frac{m\mathbf{v}_f - m\mathbf{v}_o}{t} = \frac{0 - (1200 \text{ kg})(20 \text{ m/s})}{t}$$

To proceed further, we must estimate the time $t$ the car takes to stop. As a rough approximation, we can assume that the car decelerates uniformly during the collision, so that its average velocity is

$$\bar{v} = \tfrac{1}{2}(v_f + v_o) = \tfrac{1}{2}(0 + 20 \text{ m/s}) = 10 \text{ m/s}$$

Then we can find $t$ from $s = \bar{v}t$:

$$t = \frac{s}{\bar{v}} = \frac{1.5 \text{ m}}{10 \text{ m/s}} = 0.150 \text{ s}$$

This can be substituted in the equation for $\overline{\mathbf{F}}$ to yield

$$\overline{\mathbf{F}} = \frac{-24,000 \text{ kg} \cdot \text{m/s}}{0.15 \text{ s}} = -1.60 \times 10^5 \text{ N}$$

which is a force of about 18 tons.

Notice how very large this force is. Notice also that it depends strongly on the distance the car travels in coming to rest; the force decreases as the distance increases. It is for this reason that bumpers on modern cars are designed to "give" during collisions. ▪

## 6.3 Conservation of linear momentum

We saw in Chap. 5 that energy is conserved and that knowing this is important in understanding the world about us. Linear momentum also obeys a conservation law, as we now show.

Consider the collision of two particles shown in Fig. 6.5a. The particles might be balls, molecules, or any other two objects. Newton's action-reaction law tells us that the particles exert forces on each other that are of equal magnitude but oppositely directed. Let us compute the change in momentum of the particle on the left in Fig. 6.5 as a result of the collision. From Eq. 6.2, Newton's law in momentum form, we have for the *average* force

$$\overline{\mathbf{F}}_1 t = m_1\mathbf{v}_{1f} - m_1\mathbf{v}_{1o}$$

Similarly, for the particle on the right

$$\overline{\mathbf{F}}_2 t = m_2\mathbf{v}_{2f} - m_2\mathbf{v}_{2o}$$

The same time interval $t$ appears in both equations because that is the interval during which the balls are touching each other. If we add these two expressions, we obtain

$$(\overline{\mathbf{F}}_1 + \overline{\mathbf{F}}_2)(t) = (m_1\mathbf{v}_{1f} - m_1\mathbf{v}_{1o}) + (m_2\mathbf{v}_{2f} - m_2\mathbf{v}_{2o})$$

Since the vector force $\mathbf{F}_1$, the action force, is equal to the reaction force $\mathbf{F}_2$ but directed oppositely, we have $\mathbf{F}_1 = -\mathbf{F}_2$ and the left side of this equation is zero. Hence

$$0 = (m_1\mathbf{v}_{1f} - m_1\mathbf{v}_{1o}) + (m_2\mathbf{v}_{2f} - m_2\mathbf{v}_{2o})$$

In words, this says that

$$0 = \text{change in momentum of particle } 1 + \text{change in momentum of particle } 2$$

Consequently, the total change in momentum of the isolated two-particle system is zero.

We can extend this line of reasoning to much more complicated systems. To do so, we define what is called an **isolated system: a group of objects on which the net resultant force from outside is zero.** For such a group (or system) of objects, if one object in the group experiences a force, there must exist an equal but opposite reaction force on some other object in the group. As a result, the change in momentum of the group of objects as a whole is always zero.

These considerations are summarized in the **law of conservation of linear momentum:**

The total linear momentum of an isolated system is a constant.

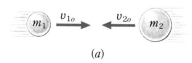

(a)

$\mathbf{F}_1 = -\mathbf{F}_2$

(b)

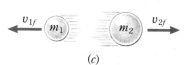

(c)

**FIGURE 6.5**

When the two bodies in (*a*) collide, they exert oppositely directed forces on each other, as in (*b*). Because the forces are of equal magnitude, what can we say about the momenta in (*c*) as compared to (*a*)?

$x$

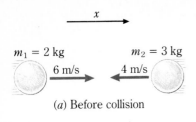

$m_1 = 2$ kg      $m_2 = 3$ kg

6 m/s      4 m/s

(a) Before collision

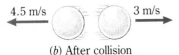

4.5 m/s      3 m/s

(b) After collision

$v = 0$

(c) After collision

**FIGURE 6.6**

The situations in (b) and (c) are physically possible results of the collision of the bodies in (a). In both instances, the momentum is the same as before collision, namely zero. Hence momentum is conserved, although kinetic energy is not.

$m_1$      $m_2$

(a) Before

$v_f$

(b) After

**FIGURE 6.7**

Momentum is conserved in this collision, even though kinetic energy is not. Where do you think most of the lost kinetic energy goes?

Even if the objects are not completely isolated, the law is often useful. For example, in a collision of two cars, the skidding of the wheels along the pavement causes unbalanced external forces to act on the two-car system. Even so, the force of one car on the other at the instant of collision is often much larger than the skidding forces on the road. Therefore, the large changes in momentum that occur at the instant of collision are almost all the result of the force of one car on the other. As a result, the law of conservation of momentum can be applied to the two-car system at the instant of collision even though the system is not isolated.

In applying the conservation law, we must remember that momentum is a vector. To illustrate the importance of this, refer to Fig. 6.6. If we take the $x$ direction to be positive, the total momentum of the system before collision (Fig. 6.6a) is

$$\text{Momentum before} = m_1\mathbf{v}_{1o} + m_2\mathbf{v}_{2o}$$
$$= (2 \text{ kg})(6 \text{ m/s}) + (3 \text{ kg})(-4 \text{ m/s})$$
$$= 12 - 12 = 0$$

where the value for $\mathbf{v}_{2o}$ is negative because $\mathbf{v}_{2o}$ is in the negative $x$ direction. Even though each object has momentum, their combined momentum is zero! This is, of course, a very special case, chosen because it emphasizes so dramatically the fact that momentum is a vector. However, this particular case of zero total momentum is interesting in several other respects.

What must be true after the collision? The law of conservation of momentum tells us that the momentum of this isolated system is not changed by the collision. Hence, in this case, the momentum after collision must still be zero. One possible way in which this could be achieved is shown in Fig. 6.6b. Notice that the momentum of each body has a magnitude of 9 kg · m/s, one being positive, the other negative. This is definitely a possible solution for the problem, since momentum is conserved. We have the right, though, to ask whether this is the only solution to the problem.

It is a simple matter to show that the situation shown in Fig. 6.6b is not what happens in one certain case. Suppose that one of the bodies has a wad of chewing gum stuck on the side where the collision occurs. If the gum is sticky enough, the two bodies remain stuck together after the collision. (If you don't like this way of fastening the bodies together, we can use magnets on the two bodies to hold them together after the collision. You can perhaps think of other ways this could be accomplished.) What can these bodies do if they stick together?

The law of conservation of momentum allows only one answer in this case. Since the momentum of the system is zero before collision, it must still be zero after collision. Now, however, since the bodies are stuck together, they must move as a unit, and their velocities are in the same direction. Unless their final velocity is zero, the momentum after collision will not have the required zero value. Therefore, in this case, the moving objects collide, stick together, and come to a complete stop. In this case, the kinetic energy of the colliding objects is lost during the collision; most of the lost kinetic energy appears as thermal energy in the chewing gum.

---

**Example 6.3**

As shown in Fig. 6.7, a 30,000-kg truck traveling at 10.0 m/s collides with a 1200-kg car traveling at 25 m/s in the opposite direction. If they stick together after the collision, how fast and in what direction are they moving?

**Reasoning** Calling the truck's direction of motion positive, we apply the law of conservation of momentum. Let $v_f$ be their combined speed after collision. From the conservation law,

$$\text{Momentum before} = \text{momentum after}$$

$$(m_1 v_{1o})_\text{truck} + (m_2 v_{2o})_\text{car} = (m_1 + m_2) v_f$$

$$(30{,}000 \text{ kg})(10.0 \text{ m/s}) + (1200 \text{ kg})(-25 \text{ m/s}) = (31{,}200 \text{ kg})(v_f)$$

Notice that the car's original velocity is negative. Solving for $v_f$ gives $v_f = 8.65$ m/s. The positive sign for $v_f$ indicates that the final motion is in the positive direction, that is, in the direction in which the truck was moving before collision. ∎

---

### Example 6.4

Figure 6.8$a$ shows an x-ray photograph of a pistol just after a bullet has been fired. (You can see the bullet in the gun barrel if you look carefully.) The hot combustion gases from the exploded gunpowder are accelerating the projectile part of the bullet down the barrel. If the masses of the gun and bullet are $M$ and $m$, respectively, and the exit velocity of the bullet is $v_{bf}$, find the recoil velocity of the gun $v_{gf}$.

**Reasoning**  If you look carefully at the figure, you can see a hand holding the gun. The external force the hand exerts on the gun is small relative to the internal forces exerted by the exploding powder, however, and so, for the instant of the explosion, we can assume that the gun is isolated and that momentum is conserved. The situations are shown in Fig. 6.8$b$. From the law of momentum conservation,

$$\text{Momentum before} = \text{momentum after}$$

$$mv_{bo} + Mv_{go} = mv_{bf} + Mv_{gf}$$

Since $v_{bo} = v_{go} = 0$, this becomes

$$0 = mv_{bf} + Mv_{gf}$$

Therefore the recoil velocity of the gun is

$$v_{gf} = -\frac{m}{M} v_{bf}$$

Notice that the more massive the gun, the less the recoil velocity will be. ∎

**FIGURE 6.8**

Before the gun was fired, its momentum was zero. Hence the sum of the momenta must still be zero after it is fired. *(Hewlett-Packard)*

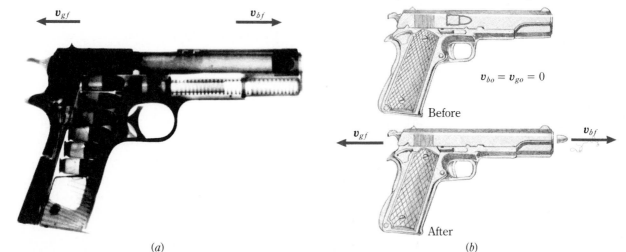

$(a)$

$(b)$

# Momentum in particle physics

The conservation law for linear momentum has been used to learn much about the basic particles in our universe. Two of the earliest such uses of the law resulted in the discovery of two subatomic particles, the neutron and the neutrino, both uncharged particles. The neutron has a mass that is nearly equal to the mass of the proton ($1.67 \times 10^{-27}$ kg), but the neutrino has such a small mass that we are not sure of its value; it may even be zero. (The name *neutrino* means "tiny neutron" in Italian; it was named by the Italian physicist Enrico Fermi.)

In 1930, Walter Bothe discovered that beryllium atoms emit a very penetrating radiation when bombarded by high-energy particles. The nature of this radiation was first determined 2 years later by James Chadwick. Chadwick was not able to observe directly the particles that we now know to make up this radiation because they are uncharged and thus difficult to capture or even detect. Instead, he let the particles collide with hydrogen and nitrogen atoms. The motions of these atoms can be measured, as we shall see in later chapters. Whenever a particle collides with an atom, the atom is given energy and momentum. Such collisions are perfectly elastic, and so the energy before collision can be equated to the energy after collision. A second equation describing the collision can be obtained by equating the momenta before and after collision. Because he could measure the energy and momentum of the atoms, Chadwick had enough data to solve the energy and momentum equations for the mass of the incoming particle, the neutron. It was in this way that he found the neutron mass to be $1.67 \times 10^{-27}$ kg.

The neutrino is far more elusive than the neutron because of its lack of both charge and mass. A neutrino can pass through the earth with very little chance of being stopped. Its presence became known through examination of beta decay, a process in which a radioactive atom emits an electron. Like a gun recoiling as it discharges a bullet, the atom should recoil as it throws out an electron. Careful measurements showed, however, that when the atom and electron alone are considered, energy and momentum cannot be conserved in the process. Hence it was proposed that another particle, one then undetected, was emitted along with the electron. By using the conservation laws, researchers surmised that the particle must be uncharged, be nearly massless, and travel at a speed close to the speed of light. It was not until the 1950s that the existence of this elusive particle, the neutrino, was confirmed. However, scientists throughout the world accepted its existence much earlier because of their faith in the laws of momentum and energy conservation.

## 6.4 Elastic and inelastic collisions

Kinetic energy is lost in many collisions. For example, when the two objects in Fig. 6.6c collide, they remain motionless after collision. All their original kinetic energy is changed to other forms of energy upon collision. Similarly, when two cars collide, a portion of their original kinetic energy is lost as work is done to mangle the cars. We call any collision during which kinetic energy is lost an *inelastic collision*.

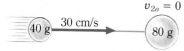

$v_{2o} = 0$

40 g — 30 cm/s → 80 g

**FIGURE 6.9**

If the head-on collision is perfectly elastic, what are the velocities after collision?

Under certain special conditions, scarcely any kinetic energy is lost in a collision. In the ideal case, when no kinetic energy is lost, the collision is said to be a *perfectly elastic collision*. Collisions between hard balls, such as billiard balls, are nearly perfectly elastic. Collisions between individual molecules, atoms, and subatomic particles often result in no loss of kinetic energy; these are perfectly elastic collisions.

---

### Example 6.5

In Fig. 6.9, a 40-g ball traveling to the right at 30 cm/s collides head on with an 80-g ball that is at rest. If the collision is perfectly elastic, what is the velocity of each ball after collision? (By "head on," we mean that all motion takes place on a straight line.)

*Reasoning* During the collision, momentum is conserved. Hence, letting the velocities of the 40- and 80-g balls after collision be $v_{1f}$ and $v_{2f}$, respectively, we have

$$\text{Momentum before} = \text{momentum after}$$

$$(0.040 \text{ kg})(0.30 \text{ m/s}) + 0 = (0.040 \text{ kg})(v_{1f}) + (0.080 \text{ kg})(v_{2f})$$

$$2v_{2f} + v_{1f} = 0.30 \text{ m/s}$$

Notice that this equation has two unknowns, $v_{1f}$ and $v_{2f}$. To proceed further, we need another independent equation. This is available in the present case because we are told that the collision is perfectly elastic, and so kinetic energy is conserved. Therefore,

$$\text{KE before} = \text{KE after}$$

$$\tfrac{1}{2}(0.040 \text{ kg})(0.30 \text{ m/s})^2 + 0 = \tfrac{1}{2}(0.040 \text{ kg})(v_{1f}^2) + \tfrac{1}{2}(0.080 \text{ kg})(v_{2f}^2)$$

or $\qquad\qquad 2v_{2f}^2 + v_{1f}^2 = 0.090 \text{ m}^2/\text{s}^2$

Solving for $v_{1f}$ in the momentum equation and substituting in the kinetic energy equation, we have

$$6v_{2f}^2 - (1.20 \text{ m/s})(v_{2f}) = 0$$

which gives two alternate values for $v_{2f}$, zero and 0.200 m/s. Substitution of these values in the first equation gives 0.300 m/s and $-0.100$ m/s for $v_{1f}$. The velocity of 0.300 m/s implies that the 40-g ball goes right through the 80-g one. Since this is a physical impossibility, we discard this answer. We therefore find that, after the collision, the 40-g ball is going to the left at 10.0 cm/s and the 80-g ball is going to the right at 20.0 cm/s. ∎

---

*Exercise* What happens if the balls have equal masses $m$? *Answer: They interchange velocities* ∎

---

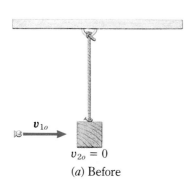

$v_{1o}$

$v_{2o} = 0$

(*a*) Before

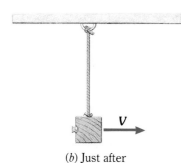

$V$

(*b*) Just after

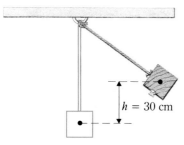

$h = 30$ cm

(*c*) Highest position

**FIGURE 6.10**

The momentum is the same in (*a*) and (*b*) but not in (*c*). Kinetic energy is changed to gravitational potential energy in going from (*b*) to (*c*).

### Example 6.6

In Fig. 6.10, a 10-g pellet of unknown speed is shot into a 2.000-kg block of wood suspended from the ceiling by a cord. The pellet hits the block and becomes lodged in it. After the collision, the block and pellet swing to a height 30 cm above the original position. What is the speed of the pellet before collision? (This device is called a **ballistic pendulum**.)

***Reasoning*** It is important to notice the following facts: (*a*) momentum is conserved during the collision but is lost as the pendulum rises under the unbalanced force of gravity; (*b*) many energy changes occur during the inelastic collision; but after the collision, as the pendulum swings up, its kinetic energy changes to potential energy.

After the collision, the kinetic energy of the bullet-pendulum combination changes to gravitational potential energy as the pendulum swings to a stop at the height *h*. Therefore, applying the law of energy conservation to the situation *after* the collision, we have for parts *b* and *c* of Fig. 6.10

$$(KE + GPE)_{\text{bottom}} = (KE + GPE)_{\text{top}}$$
$$\tfrac{1}{2}(m_1 + m_2)(V^2) + 0 = 0 + (m_1 + m_2)(gh)$$

which gives the velocity *V* of the pendulum just after the collision to be

$$V = \sqrt{2gh} = \sqrt{2(9.8 \text{ m/s}^2)(0.30 \text{ m})} = 2.42 \text{ m/s}$$

Looking now at the collision itself, we recall that, during it, linear momentum is conserved but kinetic energy is not. Therefore

Momentum before = momentum after

$$(0.010 \text{ kg})(v_{1o}) + 0 = (2.000 + 0.010 \text{ kg})(2.42 \text{ m/s})$$

from which we find $v_{1o} = 486$ m/s as the original speed of the pellet. ∎

## 6.5 Rockets and jet propulsion

**FIGURE 6.11**
A jet-propelled cart.

Although we think of rockets and jet engines as being relatively new devices, Newton understood the principle of their operation. He devised a jet propulsion system like the one shown in Fig. 6.11 and explained how the law of conservation of momentum applies to it. The steam generated by the boiling water shoots out of the rear of the engine and has momentum in the rearward direction. Because the water and engine initially have zero momentum, however, the cart and engine must now move (that is, recoil) in the forward direction with a momentum equal in magnitude to that of the escaping steam. Thus the cart acquires forward momentum.

In modern rockets and jet engines of all types, fuel is burned and very hot gases are formed. The gas molecules shoot out of the rear of the engine much as a stream of bullets would shoot from a fantastically fast-repeating gun. Like the gun recoiling, the rocket or the jet plane recoils in the direction opposite the motion of the ejected gas. Since the momentum acquired by the gas molecules is directed toward the rear, the rocket must acquire an equal momentum in the opposite direction (forward) because momentum is conserved.

A close examination of this kind of propulsion system shows that the interior of the engine restricts the hot gas molecules in such a way that they are shot preferentially rearward. In the process, however, according to Newton's law of action and reaction, the molecules exert a forward force on the engine, thereby thrusting the rocket forward. Both these forces occur within the engine. No force is exerted on the craft from outside. The craft is not propelled by interaction of the expelled hot gases with the atmosphere, and, in fact, a rocket operates best in outer space, where there is no air. Air exerts a friction force that retards the motion of the rocket and is therefore undesirable.

## Example 6.7

A Centaur rocket shoots hot gas from its engine at a rate of 1300 kg/s. The speed of the molecules is 50,000 m/s relative to the rocket. How large a forward push (**thrust**) is given to the rocket by the exiting gases?

**Reasoning** The impulse exerted on the expelled gas each second is

$$\overline{F}t = mv_f - mv_o$$

If we take $t = 1$ s, $m$ is the mass of gas expelled in 1 s (1300 kg), $v_f = 50,000$ m/s, and $v_o = 0$. Putting in the values gives $\overline{F} = 65,000,000$ N, or about 7000 tons. This is the action force needed to expel the hot gases. An equal and opposite reaction force exerts a forward thrust on the rocket. ∎

## 6.6 Momentum components

Because momentum is a vector, we can speak of its components. For example, the object in Fig. 6.12 has a momentum **p**. This momentum vector is composed of the three components $\mathbf{p}_x$, $\mathbf{p}_y$, and $\mathbf{p}_z$, and the vector sum of these components is equivalent to the original vector **p**. Therefore we can use the components of **p** in place of **p** itself for discussion purposes.

Similarly, the total momentum **p** of a system of objects can be replaced by its components. If the system is isolated, the law of conservation of momentum tells us that there is no change in the total momentum or in its components. We can therefore state that *the components of the total momentum are conserved for an isolated system.* Let us see how we can make use of this fact.

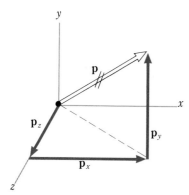

**FIGURE 6.12**
The momentum vector can be replaced by its components.

## Example 6.8

In nuclear reactors, uranium nuclei undergo fission (explode) and thereby produce energy. Suppose a nucleus of mass $m$ explodes into three equal-mass pieces, as shown in Fig. 6.13. The original nucleus was at rest, and two of the pieces have the velocities shown. Determine the velocity components for the third piece.

**Reasoning** The speed of two of the fragments is $v_o$, and we wish to find the velocity of the third piece. To do so, we call the unknown velocity **v**, and its components are $v_x$, $v_y$, and $v_z$. Applying the law of momentum conservation in component form and recognizing that the original nucleus had zero momentum, we have

(Momentum before)$_x$ = (momentum after)$_x$

$$0 = (0.33m)v_o + (0.33m)(0.8v_o) + (0.33m)(v_x)$$

from which we find $v_x = -1.8v_o$,

(Momentum before)$_y$ = (momentum after)$_y$

$$0 = 0 + (0.33m)(0.6v_o) + (0.33m)(v_y)$$

from which $v_y = -0.6v_o$, and

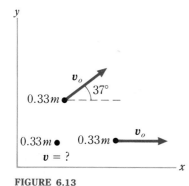

**FIGURE 6.13**
The original nucleus, of mass $m$, was at rest. What is the velocity of the third fragment?

# The long-ball hitter

New York Mets

Baseball players and golfers often wish to hit a ball as far as possible. Even in football and tennis, the player sometimes wishes to impart the maximum possible velocity to the ball. Let us examine the physics of such situations.

When a ball is hit, its behavior is governed by the momentum given to it. We know that the change in the ball's momentum is related to the applied force through the impulse equation:

Change in momentum
   = average force × contact time

Thus we see that two factors influence the way the ball responds: (1) the average force applied and (2) the time the bat (or club) is in contact with it. Let us examine both.

It is obvious that the speed of the bat influences the behavior of the ball. To bunt a baseball, the batter holds the bat essentially stationary, or even lets it retract, as the ball strikes it. As a result, only a small forward velocity is imparted to the ball. To send the ball flying away at high speed, however, the bat should be swung with high velocity. Because a baseball bat must be quickly accelerated to its highest speed, the batter's arms and wrists must be very strong. Only then can the required large torque be furnished to the bat. Indeed, even the shoulder and back muscles of a golfer or baseball player are important in supplying a large acceleration to the club or bat.

Suppose now that a bat of mass $M$ and speed $V_o$ strikes a ball of mass $m$ coming toward it with speed $v_o$. If we ignore the effect of the person

holding the bat and simply consider the bat to collide squarely and elastically with the ball, the conservation laws of kinetic energy and momentum give the final velocity of the ball $v$ to be

$$v = \frac{2MV_o + (M - m)(v_o)}{M + m}$$

If $M$ is much larger than $m$, the maximum speed of the ball is

$$v \cong 2V_o + v_o$$

Hence, to this approximation, the ball will have a speed equal to the sum of its original speed and twice the speed of the bat. Clearly, both the pitched speed and the speed of the bat are important. (For golf, $v_o = 0$.)

In practice, however, the bat does not act freely, of course. When a batter bunts, $V_o$ is close to zero or might even be negative. Alternatively, to hit the ball far, the batter exerts a continuous force on the bat during collision. Then we must replace the momentum and energy conservation equations by the impulse equation stated at the start of this note. Both the average force of the bat on the ball and the time of contact depend on the way the batter swings the bat. To keep the bat in contact with the ball for a maximum amount of time, the batter must follow through on the swing. That is why the slugger in the photograph swings far around: to maintain contact with the ball for as long as possible. If the ball is to achieve maximum range, the bat must be moving at high speed and must remain in contact with the ball during a large part of the swing.

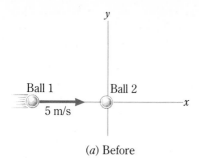

(a) Before

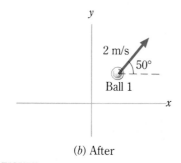

(b) After

**FIGURE 6.14**

What was the second ball doing after collision?

$$(\text{Momentum before})_z = (\text{momentum after})_z$$

$$0 = 0 + 0 + (0.33\ m)(v_z)$$

from which $v_z = 0$.

We therefore conclude that the third piece had a velocity with components $v_x = -1.8v_o$, $v_y = -0.6v_o$, and $v_z = 0$. ∎

**Exercise** Determine the velocity components if the angle in Fig. 6.13 is 90° instead of 37°. *Answer: $v_x = -v_o$, $v_y = -v_o$, $v_z = 0$* ∎

### Example 6.9

Ball 1, moving at 5.0 m/s, collides with a ball of equal mass (ball 2) that is at rest, as shown in Fig. 6.14a. After collision, ball 1 has the velocity shown in Fig. 6.14b. What is the velocity of ball 2 after collision?

**Reasoning** We know that both balls are moving in the $xy$ plane after the collision if ball 1 is moving in that plane. Why?

From the conservation of momentum for the $x$ and $y$ coordinates,

$x$ coordinate: $m(5.0\ \text{m/s}) + 0 = m(2.0 \cos 50°\ \text{m/s}) + mv_x$

$y$ coordinate: $0 + 0 = m(2.0 \sin 50°\ \text{m/s}) + mv_y$

The first equation gives $v_x = 3.71$ m/s, and the second gives $v_y = -1.53$ m/s. From this, we see that ball 2 moves at an angle $\theta$ below the positive $x$ axis, where $\tan \theta = 1.53/3.71$, or $\theta = 22°$. The speed of ball 2 is $\sqrt{(1.53)^2 + (3.71)^2}$ m/s $= 4.01$ m/s. ∎

**Exercise** Is this collision perfectly elastic? *Answer: No, about 20 percent of the kinetic energy is lost* ∎

## 6.7 The pressure of an ideal gas

The law of conservation of linear momentum has been and continues to be a very important tool in our quest to learn about atoms and about the even tinier particles found in nature. Even before all scientists agreed that atoms exist, some had used the concepts of this chapter to show how atoms in a gas give rise to the measured pressure of the gas.* These early workers pictured a gas as being composed of tiny ball-like entities shooting in random directions at high speeds, a view we accept today. The earliest calculations were extremely crude, but over the years more sophisticated models have confirmed the results. Let us see how we can use a very simple model to predict the pressure of a gas.

Consider what happens when a ball-like molecule strikes the wall of the boxlike container of Fig. 6.15. During impact, the ball exerts a momentary force on the wall. There are perhaps $10^{15}$ such collisions taking place each second due to the multitude of other molecules striking the wall. Hence their combined effect is to exert what appears to be a steady force on the wall. **We define the pressure $P$ on the wall to**

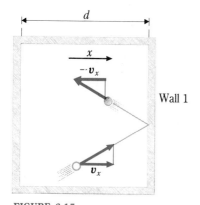

**FIGURE 6.15**

The gas molecule reverses the $x$ component of its velocity when it collides elastically with the wall.

---

* The atomic theory of matter became almost universally accepted only after about 1890.

**be the force per unit area exerted by the gas on the wall.**\* Its SI unit is the newton per square meter, which is called the pascal (Pa).

Now that we know the physical basis for gas pressure, it becomes clear how we must approach the computation of pressure values. We must find the average force on the wall due to the impact of a single molecule and then multiply this force by the total number of molecules striking the wall.

Suppose there are $N$ molecules in the container in Fig. 6.15. The $x$ dimension of the container is $d$, and the area of the wall struck by the molecule, wall 1, is $A$. Thus the volume of the container is $V = Ad$.

Consider the single molecule shown in Fig. 6.15. Assume that it can move unhindered throughout the container. Further, we picture it as undergoing elastic collision with the wall, so that a collision with wall 1 simply reverses the direction of $v_x$, as indicated. Because it is moving in the $x$ direction with speed $v_x$, the molecule travels an $x$ distance $x = v_x t$ in time $t$. The distance in the $x$ direction traveled between collisions with wall 1 is $2d$, however. Hence the number of collisions the molecule makes with this wall in $t$ seconds is

$$\text{Collisions in time } t = \frac{v_x t}{2d}$$

Now let us compute the momentum change the molecule undergoes during one collision. From Fig. 6.15,

$$\Delta p_x \text{ per collision} = (\text{final momentum})_x - (\text{original momentum})_x$$
$$= m(-v_x) - m(v_x) = -2mv_x$$

Since there are $v_x t/2d$ collisions with wall 1 in time $t$, the momentum change imparted by the wall in time $t$ is

$$\Delta p_x \text{ in time } t = (-2mv_x)\left(\frac{v_x t}{2d}\right) = -\frac{mv_x^2 t}{d}$$

We can now use the momentum form of Newton's second law to tell us the average force exerted on the molecule by the wall; $\overline{F} = \Delta p/t$ gives

$$\text{Average force on molecule} = \frac{\Delta p_x}{t} = -\frac{mv_x^2}{d}$$

By the action-reaction law, the molecule exerts a force of equal magnitude on the wall. Hence

$$\text{Average force on wall per molecule} = \frac{mv_x^2}{d}$$

Each of the $N$ molecules in the container exerts a similar force on the wall. We can thus write that the average force $\overline{F}$ on the wall due to all the molecules in the container is

---

\* A more complete discussion of pressure appears in Chap. 9.

$$\bar{F} = N\left(\frac{mv_x^2}{d}\right)$$

where $v_x^2$ is now the average value for all the molecules in the container. To find the pressure $P$ of the gas, we recall that

$$P = \text{force per unit area} = \frac{\bar{F}}{A} = \frac{Nmv_x^2}{Ad}$$

Since $Ad$ is simply the volume of the container, the pressure is

$$P = \left(\frac{N}{V}\right) mv_x^2$$

where $N/V$ is the number of gas molecules per unit volume.

We can put this in a better form if we recognize that there is nothing special about the $x$ direction. Therefore

$$v_x^2 = v_y^2 = v_z^2$$

Moreover, if we recall a little solid geometry, we can see from Fig. 6.16 that

$$v^2 = v_x^2 + v_y^2 + v_z^2$$

$$v^2 = 3v_x^2 \quad \text{and} \quad v_x^2 = \frac{v^2}{3}$$

Replacing $v_x^2$ with $v^2/3$ in the pressure equation gives

$$P = \left(\frac{N}{V}\right)\left(\frac{mv^2}{3}\right)$$

In terms of the kinetic energy of the molecule, this is

$$P = \left(\frac{2}{3}\right)\left(\frac{N}{V}\right)\left(\frac{mv^2}{2}\right) \tag{6.4}$$

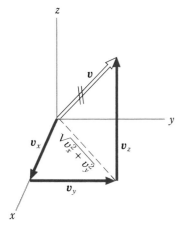

**FIGURE 6.16**

The pythagorean theorem in three dimensions gives $v^2 = v_x^2 + v_y^2 + v_z^2$.

Notice that the pressure of a gas is equal to two-thirds of the kinetic energy per unit volume, since $N/V$ is the number of molecules in unit volume. We shall see in Chap. 10 that gas pressure can also be used as a measure of the temperature of a gas. Hence it appears that we shall find temperature and kinetic energy to be intimately related. In fact, we shall learn that a hot object is one whose constituent molecules possess a great deal of kinetic energy.

Although Eq. 6.4 was obtained by using a greatly oversimplified model, the result is general. As long as a gas is far removed from conditions under which it will liquefy, Eq. 6.4 is applicable, to good approximation. A gas that obeys this equation exactly is an ideal gas, and Eq. 6.4 is one form of the ideal-gas law. More is said about this in Chap. 10. It is surprising that the crude method of computation we have used (and that was used by the early workers in physics) leads to a result that agrees so well with experiment. The assumption that the molecules do not collide with one another is far from true. In air at ordinary pressures, a molecule collides with another after traveling a distance of only about $10^{-5}$ cm. Nonetheless, our result is valid.

## Example 6.10

Standard atmospheric pressure is $1.01 \times 10^5$ N/m². In an ideal gas at standard pressure and temperature (0°C), there are $2.69 \times 10^{25}$ molecules in each cubic meter of volume. Find the average speed of the nitrogen molecules in air under standard conditions. One molecule of nitrogen has a mass of $4.65 \times 10^{-26}$ kg.

*Reasoning*   We use Eq. 6.4. From the data we have, $N/V = 2.69 \times 10^{25}$ m⁻³, $m = 4.65 \times 10^{-26}$ kg, and $P = 1.01 \times 10^5$ N/m², all in the proper SI units. Hence, from Eq. 6.4,

$$1.01 \times 10^5 \text{ N/m}^2 = \tfrac{1}{3}(2.69 \times 10^{25} \text{ m}^{-3})(4.65 \times 10^{-26} \text{ kg})(v^2)$$

from which, remembering that $1 \text{ N} = 1 \text{ kg} \cdot \text{m/s}^2$, we have

$$v = 492 \text{ m/s}$$

Notice how very fast the molecules are moving.   ■

## Minimum learning goals

When you finish this chapter, you should be able to

**1** Define (*a*) linear momentum, (*b*) impulse, (*c*) isolated system, (*d*) elastic versus inelastic collision, (*e*) recoil, (*f*) ballistic pendulum, (*g*) pressure, (*h*) ideal gas.

**2** State Newton's second law in terms of momentum.

**3** Find the change in momentum of an object due to a given impulse, or vice versa.

**4** State the law of conservation of linear momentum and use it in simple situations.

**5** Analyze the collision of two objects that stick together on impact.

**6** Analyze situations in which an object originally at rest explodes into multiple pieces.

**7** Analyze situations in which two objects move along a straight line, undergo a perfectly elastic collision, and then continue to move along the same straight line.

**8** Give plausible reasons for the fact that kinetic energy is not conserved in most collisions.

**9** Explain the operating principle of rockets, jet engines, and similar devices propelled by a jet.

**10** Explain why a gas in a container exerts pressure on the container walls.

**11** Be able to use the relation among $P$, $v$, $m$, and $N/V$ to solve simple problems, such as that typified by Example 6.10.

## Questions and guesstimates

**1** When a cannon is fired, it recoils for some distance against a cushioning device. Why is it necessary to make the support so that it "gives" in this way?

**2** A wad of gum is shot at a block of wood. In which case does the gum exert the larger impulse on the block, when it sticks or when it rebounds?

**3** When a balloon filled with air is released so that the air escapes, the balloon shoots off into the air. Explain. Would the same thing happen if the balloon were released in a vacuum?

**4** Explain why a rocket can accelerate even in outer space, where there is no air against which it can push.

**5** An inventor constructs a sailboat with a large electric fan mounted on it. He directs the fan at the sail and blows air at it, expecting thereby to move in the direction of this artificial wind. To his surprise, the boat moves slowly in the opposite direction. Can you tell him why it does so?

**6** A ball dropped onto a hard floor has a downward momentum, and after it rebounds, its momentum is upward. The ball's momentum is not conserved in the collision, even though it may rebound to the height from which it was dropped. Does this contradict the law of momentum conservation?

**7** Reasoning from the impulse equation, explain why it is unwise to hold your legs rigidly straight when you jump to the ground from a wall or table. How is this related to the commonly held belief that a drunken person has less chance of being injured in a fall than one who is sober?

**8** Explain, in terms of the impulse equation, the operating principle of impact-absorbing car bumpers and similar impact-absorbing devices.

**9** A baseball player has the following nightmare. He is accidentally locked in a railroad boxcar. Fortunately, he has his ball and bat along. To start the car moving, he stands at one end and bats the ball toward the other. The impulse exerted by the ball as it hits the end wall gives the car a forward motion. Since the ball always rebounds and rolls along the floor back to him, the player repeats this process over and over. Eventually the car attains a very high speed, and the player is killed as the boxcar collides with another car sitting at rest on the track. Analyze this dream from a physics standpoint.

**10** Explain how a Mexican jumping bean jumps.

**11** Two blocks of unequal mass are connected by a spring, with the whole system lying on an essentially frictionless table. The blocks are pushed together and tied with a string so that the spring is compressed. Describe the motion of the blocks, when the string is cut.

**12** A 70-kg woman jumps from a roof 10 m above the ground. (*a*) What is her approximate speed just before she strikes the ground? (*b*) She lands on her feet but allows her legs to "give." About how long does it take her to come to rest? (*c*) About how large an average force does the ground exert on her?

**13** Suppose you lay your hand flat on a tabletop and then drop a 1.0-kg mass squarely on it from a height of 0.50 m. Estimate the average force exerted on your hand by the mass. Why is injury very likely in this case even though you can *catch* the mass easily when it is dropped from this height?

## Problems

*Use momentum methods when practicable.*

**1** What is the linear momentum of (*a*) a 900-kg car moving eastward at 80 km/h? (*b*) A 9.0-g bullet moving vertically upward at 300 m/s?

**2** What is the linear momentum of a 500-g ball after it has fallen freely from rest through a distance of 4.0 m?

**3** A mass *m* undergoes free fall. What is its linear momentum after it has fallen a distance *h*?

**4** Show that the linear momentum and kinetic energy of a mass *m* are related through $KE = p^2/2m$.

**5** How large a force is needed to stop a coasting 900-kg car in 3.0 s if the car's original speed is 25 m/s?

**6** Determine the average force required to accelerate a 1200-kg car from rest to 18 m/s in 6.0 s.

**7** A coasting 1300-kg car moving at 25 m/s slows to 5.0 m/s in 8.0 s. What average stopping force acted on the car?

**8** A 7.0-g bullet moving at 200 m/s shoots through a sheet of plastic 2.0 cm thick and emerges with a speed of 140 m/s. The time the bullet takes to pass through the sheet is $1.18 \times 10^{-4}$ s. Find the average stopping force on the bullet.

**9** A 900-kg car going 20 m/s strikes a wall and stops in 3.0 m. (*a*) How long does it take the car to stop if uniform deceleration is assumed? (*b*) What is the stopping force?

■ **10** A 120-g ball moving at 18 m/s strikes a wall perpendicularly and rebounds straight back at 12 m/s. After the initial contact, the center of the ball moves 0.27 cm closer to the wall. Assuming uniform deceleration, show that the time of contact is 0.00075 s. How large an average force does the ball exert on the wall?

■ **11** A proton ($m = 1.67 \times 10^{-27}$ kg) moving at $4.0 \times 10^7$ m/s shoots through a sheet of foam plastic 0.25 cm thick and emerges with a speed of $1.5 \times 10^7$ m/s. (*a*) How long does it take to pass through the plastic? (Assume uniform deceleration.) (*b*) What average force retards the proton's motion through the plastic?

■ **12** A 50-g arrow moving at 20 m/s strikes a pumpkin and drills a 35-cm hole straight through it. When it emerges, the arrow has a speed of 8.0 m/s. (*a*) Assuming uniform deceleration, find the time it takes the arrow to pass through the pumpkin. (*b*) What average force opposes the motion?

■ **13** An unlucky bystander happens to be in the center of a shootout, and a 5.0-g bullet moving at 80 m/s lodges in her shoulder. The bullet undergoes uniform deceleration and stops in 6.0 cm. Find (*a*) the time it takes for the bullet to stop, (*b*) the impulse on the shoulder, and (*c*) the average force experienced by the bystander.

■ **14** A stream of water from a hose is hitting a window. The window is vertical, the stream is horizontal, and the water stops when it hits. About 20 cm³ (that is, 20 g) of water with speed 1.50 m/s strikes the window each second. Find (*a*) the impulse on the window exerted in time *t* and (*b*) the force exerted on the window.

**15** During a switching operation, a train car of mass $M_1$ coasting along a straight track with speed *v* strikes and couples to a car of mass $M_2$ that is sitting at rest. Find their speed after coupling.

**16** While engaging in target practice, a woman shoots a 3.0-g bullet with horizontal velocity 150 m/s into a 5.0-kg watermelon sitting on top of a post. The bullet lodges in the watermelon. With what speed does the watermelon fly off the post?

**17** Two identical balls collide. Ball 1 is traveling to the right at 20 m/s, and ball 2 is standing still. Find the direction and magnitude of their velocity if they stick together after collision.

■ **18** The 20-g bullet in Fig. P6.1 moves with speed 4000 cm/s, strikes an 8000-g block resting on a table, and embeds in the block. Find (*a*) the speed of the block after collision and (*b*) the friction force between table and block if the block moves 80 cm before stopping.

**FIGURE P6.1**

■ **19** While coasting along a street at 0.50 m/s, a 15-kg girl in a 5-kg wagon sees a vicious dog in front of her. She has with her only a 3.0-kg bag of sugar that she is bringing from the grocery store, and she throws it at the dog with a forward velocity of 4.0 m/s relative to her original motion. How fast is she moving after she throws the bag of sugar? What is her final direction of motion?

■ **20** Near the Fourth of July, a very foolish boy places a firecracker (of negligible mass) in an empty can (mass, 50 g), ignites the firecracker and then quickly plugs the end of the can with a wooden block (mass, 200 g). He throws the can straight up, and it explodes at the top of its path. If the block shoots out with a speed of 3.0 m/s, how fast is the can going?

■ **21** A 500-g pistol lies at rest on an essentially frictionless table. It accidentally discharges and shoots a 10-g bullet parallel to the table. How far has the bullet moved by the time the gun has recoiled 2.00 mm?

■ **22** Find the speed of the 20-g bullet in Fig. P6.1 if the bullet embeds in the 3000-g block and causes it to slide 1.50 m before coming to rest. The coefficient of friction between block and table is 0.40.

■ **23** A 2.0-kg block rests over a small hole in a table. A woman beneath the table shoots a 15.0-g bullet through the hole into the block, where it lodges. How fast was the bullet going if the block rises 75 cm above the table?

■ **24** A 60-kg astronaut becomes separated in space from her spaceship. She is 15.0 m away from it and at rest relative to it. In an effort to get back, she throws a 700-g wrench in a direction away from the ship with a speed of 8.0 m/s. How long does it take her to get back to the ship?

■ **25** A 3.0-kg melon is balanced on a bald man's head. His wife shoots a 50-g arrow at it with speed 25 m/s. The arrow passes through the melon and emerges at 10 m/s. Find the speed of the melon as it flies off the man's head.

■ **26** A ball is falling freely under the action of gravity. When its downward speed is 8.0 m/s, it explodes into two equal parts. One part goes straight up to a height of 12.0 m above the explosion point. What is the velocity of the other part just after the explosion?

■ **27** In Fig. P6.2, both balls have the same mass. The ball on the left is displaced to the outlined position and released. It collides with the stationary ball and sticks to it. (*a*) How fast is the combination moving just after the collision? (*b*) What fraction of the original kinetic energy is lost in the collision?

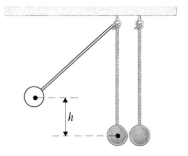

**FIGURE P6.2**

■ **28** The two pendulum balls in Fig. P6.2 have different masses; the ball on the left has a mass $m_1$. When it is let swing from the height shown, it strikes the ball on the right and sticks to it. The combination then swings to a height $h/4$. Find the mass $m_2$ of the ball on the right in terms of $m_1$.

■ **29** Many fast neutrons are produced in nuclear reactors. So that they will slow down, they are allowed to collide with other particles of comparable mass. Suppose a neutron with speed *v* collides head on with a proton at rest. The masses of these two particles are nearly identical. If the collision is perfectly elastic, what is the speed of the neutron after the collision?

■ **30** In Fig. P6.2, the two identical masses are displaced to a height *h*, one to the left and the other to the right. They are released simultaneously and undergo a perfectly elastic collision at the bottom. How high does each swing after collision?

■ **31** The mass on the left in Fig. P6.2 is pulled aside and released. Its speed at the bottom is $v_o$. It then collides with the mass on the right in a perfectly elastic collision. Find the velocities of the two masses just after collision if the mass on the left is 3 times as large as the mass on the right.

■ **32** A neutron ($m = 1.67 \times 10^{-27}$ kg) moving with speed $v_o$ strikes a stationary particle of unknown mass and rebounds perfectly elastically straight back along its original path with speed $0.70v_o$. What is the mass of the struck particle?

■ **33** A neutron (mass $= m_o$) moving with speed $v_o$ strikes the stationary nucleus of a copper atom ($m = 63m_o$) and rebounds straight back in a perfectly elastic collision. Find the velocity of the copper nucleus after the collision. Assume that it is able to move freely.

■ **34** An electron ($m = 9.1 \times 10^{-31}$ kg) traveling at $2.0 \times 10^7$ m/s undergoes a head-on collision with a hydrogen atom ($m = 1.67 \times 10^{-27}$ kg) at rest. Assuming the collision to be perfectly elastic and the motion to be along a straight line, find the final velocity of the hydrogen atom.

■ **35** According to a police report, car 1 was sitting at rest waiting for a stoplight when it was hit from the rear by an identical vehicle, car 2. Both cars had their four-wheel brakes on, and from their skid marks it is surmised that they skidded together about 6.0 m in the original direction of travel before coming to rest. Assuming a stopping force of about 0.7 times the combined weights of the cars (that is, $\mu = 0.7$), what must have been the approximate speed of car 2 just before the collision?

■ **36** Ball $A$ in Fig. P6.3 is released from point $A$. It slides along the frictionless wire and collides with ball $B$. If the collision is perfectly elastic, find how high ball $B$ rises after the collision ($m_A = m_B/2$).

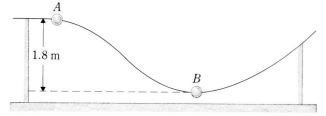

**FIGURE P6.3**

■ **37** A particle of mass $m_o$ is at rest when it suddenly explodes into three pieces of equal mass $\frac{1}{3}m_o$. One piece flies out along the positive $x$ axis with a speed of 30 m/s, while another goes in the negative $y$ direction with a speed of 20 m/s. (*a*) Find the components of the velocity of the third piece. (*b*) Repeat if the third piece has a mass $\frac{1}{2}m_o$ and the other two each have mass $\frac{1}{4}m_o$.

■ **38** Two equal-mass cars are moving along perpendicular streets with equal speed $v_o$. At the corner made by the intersection of the two streets, the cars collide and stick together. What are the direction and speed of the motion of the wreckage just after the collision?

■ **39** Two protons are moving along the $x$ axis, one with velocity $v_o$ and the other with velocity $-v_o$. They undergo a perfectly elastic collision. After collision, one goes off at an angle of 37° to the positive $x$ axis in the $xy$ plane. What happens to the other? What are their velocities after collision?

■ **40** A particle has $x, y, z$ velocity components $(-v_o, 0, 0)$, and a second particle of the same mass has velocity components $(0, v_o, 0)$. After the particles collide, one has a velocity $(-\frac{1}{2}v_o, 0, 0)$. Find the velocity components for the other. Is the collision perfectly elastic?

■ **41** A particle of mass $m$ moving in along the positive $x$ axis with velocity $v_o$ strikes an identical stationary particle at the origin. After collision, the particles move in the $xy$ plane away from the origin, making angles of 37° and 53° with the negative $x$ axis. What are the velocities of the two particles after collision?

■ **42** A ball of mass $m_o$ is moving with speed $v_o$ to the left along the positive $x$ axis, toward the origin. It strikes a glancing blow on a ball of mass $m_o/4$ at rest at the origin. After collision, the incoming ball is moving to the left with speed $v_o/2$ at an angle 37° above the negative $x$ axis. Find the speed and direction of motion of the other ball.

■ **43** Repeat Prob. 42 if the incoming ball is reflected back to the right at an angle of 37° to the positive $x$ axis with a speed of $v_o/5$.

**44** Suppose that half the molecules of a gas are removed from a container without changing the average kinetic energy per molecule within the container. By what factor does the gas pressure change?

**45** A number of molecules are contained at atmospheric pressure in a chamber closed by a piston device, as in Fig. P6.4. If the piston is pushed down in such a way that the kinetic energy of the molecules is not changed appreciably, what is the pressure in the container when the volume is only one-third its original size?

**FIGURE P6.4**

**46** Air in a certain room has a pressure of $0.98 \times 10^5$ N/m², and the mass of 1 m³ of the air is 1.28 kg/m³. Find the average speed of the air molecules.

**47** Show that the total translational kinetic energy of the molecules in a volume $V$ of an ideal gas at pressure $P$ is $3PV/2$.

**48** Two boys, 15 m apart, are playing catch while standing in a boxcar at its two ends. The boxcar can move freely on its horizontal tracks. Together, the boys and the boxcar have a mass $M$; the mass of the ball is $m$. At time $t = 0$, boy 1 throws the ball to boy 2, who catches it. Boy 2 waits 8.0 s and then throws the ball back to boy 1, who catches it. Both throwing speeds are 25 m/s. How far is the boxcar from its original position at (a) $t = 4.0$ s and (b) $t = 15.0$ s?

**49** The Atwood machine in Fig. P6.5 has a third mass attached to it by a limp string. After being released, the $2m$ mass falls a distance $D$ before the limp string becomes taut. Thereafter, both masses on the left rise at the same speed. What is this final speed? Assume an ideal, frictionless pulley.

**50** The $2m$ mass in Fig. P6.5 is supported so that it cannot fall. When the support for the lower mass on the left is removed, this mass falls freely a distance $L$ before the limp string connecting it to the other mass becomes taut. Thereafter, the three masses move in unison. What is their final speed?

**51** A vertical, uniform chain of total mass $M$ and length $L$ is being lowered onto a table at a constant speed $V$. At time $t = 0$, the lower end of the chain touches the table. Find the force exerted by the chain on the table (as a function of $t$) as the chain is deposited on the table.

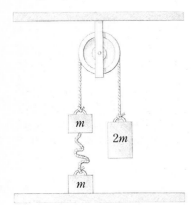

**FIGURE P6.5**

# 7 Motion in a circle

**A** spaceship orbiting the earth and the earth circling the sun are familiar examples of motion in a near-circular path. Objects that spin on an axis and rotating wheels are also well known to us. In this chapter, we learn how to describe motions such as these.

## 7.1 Angular displacement $\theta$

To describe the motion of an object along a line, we need a coordinate along the line, and we often take it to be the $x$ coordinate. To describe the motion of an object on a circular path or the rotation of a wheel on an axle (or axis), we need a coordinate to measure angles, the rotational counterpart of linear displacement. You are probably familiar with the usual ways for doing this, but let us summarize them in review.

Consider the two positions of the wheel in Fig. 7.1. In going from position $a$ to position $b$, the wheel has turned through the angle $\theta$. There are three common ways in which $\theta$ is measured. We can measure it in degrees (deg), and we know that one full circle is equivalent to $360°$. Alternatively, we can measure it in revolutions (rev). One full circle is one revolution, and so we know that

$$1 \text{ rev} = 360°$$

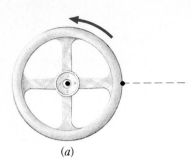

*(a)*

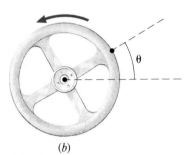

*(b)*

**FIGURE 7.1**

The angle $\theta$ describes the angular distance through which the wheel has turned.

The third method for measuring $\theta$ is less familiar to the average person, but it is widely used in science. This way of measuring angles, called *radian measure,* is defined in the following way. Notice that the angle $\theta$ in Fig. 7.2 subtends an arc of length $s$ on the circle of radius $r$. The *radian* unit for measuring angles is defined through the equation

$$\theta \text{ in radians} = \frac{\text{arc length}}{\text{radius}}$$

$$\theta = \frac{s}{r} \qquad (\theta \text{ in radians}) \tag{7.1}$$

For $\theta = 360°$, one full circle, $s = 2\pi r$, and so $\theta = s/r$ becomes

$$\theta = 360° = \frac{2\pi r}{r} \text{ radians} = 2\pi \text{ rad}$$

where *rad* is the abbreviation for radian. Thus we have the following relationship between the units we use to measure angles:

$$1 \text{ rev} = 360° = 2\pi \text{ rad}$$

It is sometimes convenient to remember that $1 \text{ rad} = 57.3°$.

Notice that Eq. 7.1 defines an angle in radians as the ratio of two lengths, and so the dimensions cancel. Therefore the radian is a dimensionless unit. We use the terms *radian, degree,* and *revolution* to specify angles; because they are not units in the usual sense, however, we cannot expect them to follow through equations the way a dimension-specifying unit does.

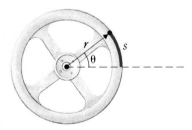

**FIGURE 7.2**

In radian measure, $\theta = s/r$.

---

**Example 7.1**

Convert a $70°$ angle to radians and revolutions.

**Reasoning**   Using the conversion factors $2\pi$ rad/360° and 1 rev/360°, we have

$$70° = (70 \text{ deg})\left(\frac{2\pi \text{ rad}}{360 \text{ deg}}\right) = 1.22 \text{ rad}$$

$$70° = (70 \text{ deg})\left(\frac{1 \text{ rev}}{360 \text{ deg}}\right) = 0.194 \text{ rev} \quad \blacksquare$$

---

**Exercise**   Change 0.21 rad to degrees and revolutions.   *Answer: 12.0°, 0.0334 rev.*   ■

---

## 7.2  Angular velocity $\omega$

When we state that a phonograph record is turning at 33 rev/min, we are giving its angular velocity. We are describing how fast it rotates. Analogously to linear motion, where average velocity is defined to be displacement divided by time, we define

$$\text{Average angular velocity} = \frac{\text{angle turned}}{\text{time taken}}$$

$$\overline{\omega} = \frac{\theta}{t} \qquad\qquad (7.2)$$

where $\omega$ (Greek omega) is angular velocity. Typical units for $\omega$ are radians per second, degrees per second, and revolutions per minute.

In advanced work, direction is assigned to rotational quantities, and so they are vector quantities. For example, the direction of the angular velocity for the rotating wheels in Figs. 7.1 and 7.2 is taken perpendicular to the page, out of the page. If the wheels had been rotating in the opposite direction, then the angular-velocity vector would point into the page. Because we are concerned only with rotations in a plane, we can express the vector nature of rotational quantities by plus and minus signs to indicate whether the vector is to be directed out of or into the plane. For the present, the vector nature of rotational quantities is unimportant for us.

Just as we did in linear motion, we make a distinction between average and instantaneous angular velocity. You will recall that instantaneous linear velocity is obtained by measuring the linear distance moved in a time so small that the velocity does not change appreciably. Doing the same in rotation, we define the *instantaneous angular velocity* as

$$\omega = \lim_{\Delta t \to 0} \frac{\Delta\theta}{\Delta t} \qquad\qquad (7.3)$$

In this expression, $\Delta\theta$ is the small angular distance moved by a rotating wheel in the small time $\Delta t$, and the limit notation tells us to take the value of the ratio as the time interval $\Delta t$ approaches zero, as discussed in Chap. 3.

---

***Example 7.2***
The wheel in Fig. 7.2 turns through 1800 rev in 1 min. Find its average angular velocity in rad/s.

***Reasoning*** From the defining equation,

$$\overline{\omega} = \frac{\theta}{t} = \frac{1800 \text{ rev}}{60 \text{ s}} = 30 \text{ rev/s}$$

Then

$$30 \text{ rev/s} = \left(30 \, \frac{\text{rev}}{\text{s}}\right)\left(\frac{2\pi \text{ rad}}{\text{rev}}\right) = 60\pi \text{ rad/s} = 188 \text{ rad/s} \quad \blacksquare$$

---

***Exercise*** Through how many radians does the wheel turn in 15 s? *Answer: 47.1 rad* ∎

---

## 7.3 Angular acceleration $\alpha$

Average linear acceleration, defined in Chap. 3 by the equation

$$a = \frac{v_f - v_o}{t}$$

measures the rate at which the velocity of an object is changing. The quantity $v_f - v_o$ is the change in velocity during time $t$. You will recall that typical units for acceleration are meters per second squared and feet per second squared.

In the case of rotating objects, we are often interested in how they speed up or slow down. Hence we are concerned with angular acceleration, that is, the rate of change of angular velocity. We define the average angular acceleration $\alpha$ (alpha) of a rotating wheel or any other object by the relationship

$$\text{Average angular acceleration} = \frac{\text{change in angular velocity}}{\text{time taken}}$$

$$\alpha = \frac{\omega_f - \omega_o}{t} \qquad (7.4)$$

Its units are those of angular velocity divided by time. For example, if $t$ is measured in seconds and $\omega$ in radians per second, the angular acceleration is expressed in radians per second per second. Although it is not wrong to measure $\omega$ in radians per second when $t$ is in minutes, giving units of radians per second per minute, it is generally preferable to use the same unit for $t$ in both places.

If the angular acceleration is uniform, we know that, as with linear motion, the average angular velocity is

$$\overline{\omega} = \tfrac{1}{2}(\omega_f + \omega_o)$$

---

*Example 7.3*
A wheel starts from rest and attains a rotational velocity of 240 rev/s in 2.0 min. What is its average angular acceleration?

*Reasoning*  We know that

$$\omega_o = 0 \qquad \omega_f = 240 \text{ rev/s} \qquad t = 2.0 \text{ min} = 120 \text{ s}$$

From the definition of angular acceleration,

$$\alpha = \frac{\omega_f - \omega_o}{t} = \frac{(240 - 0) \text{ rev/s}}{120 \text{ s}} = 2.00 \text{ rev/s}^2 \quad \blacksquare$$

---

*Exercise*  What is the wheel's angular speed (in radians per second) 130 s after starting from rest?  *Answer: 1630 rad/s*  ▪

---

**7.4** **Angular motion equations**

As you have probably recognized by now, there is a great deal of similarity between the linear and angular motion equations. The $\theta$ in angular motion corresponds to $x$ in linear motion, $\omega$ corresponds to $v$, and $\alpha$ corresponds to $a$. Moreover, we have defined $\omega$ and $\alpha$ by equations that are identical to those for $v$ and $a$ except for the interchange of symbols. This leads us to the conclusion that all the motion equations we learned for linear uniformly accelerated motion carry over to uniformly accelerated angular motion as follows:

| Linear | Angular | |
|---|---|---|
| $s = \bar{v}t$ | $\theta = \bar{\omega}t$ | (7.5a) |
| $v_f = v_o + at$ | $\omega_f = \omega_o + \alpha t$ | (7.5b) |
| $\bar{v} = \frac{1}{2}(v_f + v_o)$ | $\bar{\omega} = \frac{1}{2}(\omega_f + \omega_o)$ | (7.5c) |
| $2as = v_f^2 - v_o^2$ | $2\alpha\theta = \omega_f^2 - \omega_o^2$ | (7.5d) |
| $s = v_o t + \frac{1}{2}at^2$ | $\theta = \omega_o t + \frac{1}{2}\alpha t^2$ | (7.5e) |

There is no need to learn new equations for angular motion. Simply replace the linear-motion variables with their angular counterparts. We shall see in the next chapter that even the equations for kinetic energy and momentum have easily guessed angular analogs. Let us now see how the problem solution methods we used for linear motion carry over to angular motion.

### Example 7.4

A roulette wheel turning at 3.0 rev/s coasts to rest uniformly in 18.0 s. What is its deceleration? How many revolutions does it turn through while coming to rest?

**Reasoning**  This is a typical angular-motion problem. The knowns and unknowns are

$$\omega_o = 3.0 \text{ rev/s} \qquad \omega_f = 0 \qquad t = 18.0 \text{ s} \qquad \alpha = ? \qquad \theta = ?$$

From the defining equation for $\alpha$, Eq. 7.5b,

$$\alpha = \frac{\omega_f - \omega_o}{t} = \frac{(0 - 3.0) \text{ rev/s}}{18.0 \text{ s}} = -0.167 \text{ rev/s}^2$$

What is the meaning of the negative sign on this answer?
To find $\theta$, let us use Eq. 7.5e:

$$\theta = \omega_o t + \frac{1}{2}\alpha t^2$$
$$= (3.0 \text{ rev/s})(18.0 \text{ s}) + \frac{1}{2}(-0.167 \text{ rev/s}^2)(18.0 \text{ s})^2 = 27 \text{ rev}$$

Notice how important it is to retain the proper sign on $\alpha$.  ∎

## 7.5 Tangential quantities

When a spool unwinds a string or a wheel rolls along the ground, both rotational and linear motions occur. We wish now to find out how these two types of motion are related. The relation between linear and angular distances, $s$ and $\theta$, is inherent in Eq. 7.1, the definition of angular measure. To see this, let us look at Fig. 7.3a.
A point on the rim of a wheel traces an arc length $s$ as the wheel turns through the angle $\theta$. From Eq. 7.1, the definition of radian measure, we have that

$$s = r\theta$$

provided $\theta$ is measured in radians. We call the distance $s$ a **tangential distance** because it is measured tangential to the wheel's rim.

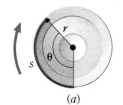

(a)

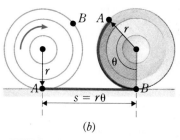

(b)

**FIGURE 7.3**
As the wheel turns through an angle $\theta$, it lays out a tangential distance $s = r\theta$.

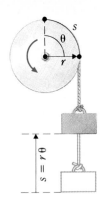

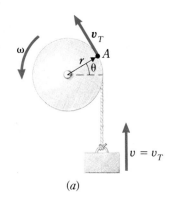

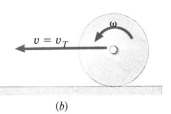

**FIGURE 7.4**

How much string does the spool wind as it turns through one revolution?

**FIGURE 7.5**

The angular velocity $\omega$ is related to the tangential velocity $v_T$ by $v_T = \omega r$. In this relation, $\omega$ must be in radian measure.

If we look at Fig. 7.3b, we see that the linear distance the wheel rolls, $s = r\theta$, equals the tangential distance turned by the rim. This fact allows us to relate linear motion to angular motion for a rolling wheel. Further, if we look at the spool in Fig. 7.4, we see that there is a similar relationship for the way in which string is wound on its rim. As a point on the rim turns through a tangential distance $s$, a length $s$ of string is wound on the rim. In all cases such as these, we have, from $s = r\theta$,

$$\text{Tangential distance} = r\theta \qquad (\theta \text{ in radians}) \tag{7.6}$$

and this equals the linear distance rolled or the length of string wound. We emphasize that $\theta$ must be measured in radians because Eq. 7.6 is based on the definition of radian measure, Eq. 7.1.

As the spool in Fig. 7.4 turns, the mass at the end of the string rises with a certain velocity. Similarly, as the wheel in Fig. 7.3 rotates, it rolls without slipping along the ground and its center moves with a definite linear velocity. The magnitude of the linear velocity in each of these cases is the same as the velocity of a point on the rim of the spool or wheel. We call this velocity the **tangential velocity** $v_T$. Let us now relate $v_T$ to the angular velocity $\omega$ of the wheel.

As the spool in Fig. 7.5a turns with constant speed through an angle $\theta$ in a time $t$, its rotational velocity is

$$\omega = \frac{\theta}{t}$$

Because $\theta = s/r$, where $r$ is the radius of the spool, we can substitute this value in the equation for $\omega$ and obtain

$$\omega = \frac{s/r}{t} = \frac{s}{t}\frac{1}{r}$$

However, $s/t$ is simply the speed with which the mass in Fig. 7.5a is lifted, and that speed is equal to the tangential speed $v_T$ of point $A$. Therefore the equation for $\omega$ yields $\omega = v_T/r$, or

$$\text{Tangential velocity} = v_T = \omega r \tag{7.7}$$

Here, too, radian measure must be used. In a similar way, we can show that the center of the wheel in Fig. 7.5b moves with a speed $v_T = \omega r$, provided the wheel does not slip. Thus we see that Eq. 7.7 is an important relationship between the rotational motion of an object and the linear motions that result from the rotation.

Another quantity of interest is what we call the tangential acceleration. If $\omega$ is increasing for a rotating wheel, then $v_T$ must also be increasing. The angular acceleration $\alpha$ is

$$\alpha = \frac{\omega_f - \omega_o}{t}$$

where $\omega_f - \omega_o$ is the change in the angular velocity during the time interval $t$. Because $\omega = v_T/r$, we can write this as

$$\alpha = \frac{v_{Tf} - v_{To}}{rt} \qquad \text{or} \qquad \frac{v_{Tf} - v_{To}}{t} = \alpha r$$

This, however, is simply the rate of change of tangential velocity, or the *tangential acceleration* $a_T$. Therefore

Tangential acceleration $= a_T = \alpha r$ (7.8)

This is also the linear acceleration of the center of a rolling wheel or of a given point on an unwinding string. Can you show this from a consideration of the fact that acceleration is the rate of change of velocity — tangential velocity in this case?

---

### Example 7.5

A car with 80-cm-diameter wheels starts from rest and accelerates uniformly to 20 m/s in 9.0 s. Find the angular acceleration and final angular velocity of one wheel.

*Reasoning*   We know that the linear velocity and acceleration of the wheel's center are given by the tangential relations to be

$$v_T = \omega r \quad \text{and} \quad a_T = \alpha r$$

In our case, the final value for $v_T$ is 20 m/s. Therefore

$$\omega_f = \frac{v_{Tf}}{r} = \frac{20 \text{ m/s}}{0.40 \text{ m}} = 50 \text{ rad/s}$$

Notice that we must insert the proper angular measure since angular units do not carry through the equations. Why is the answer in radians per second and not revolutions per second?

To find the acceleration, we first solve the linear-motion problem. Our knowns and unknowns are

$$v_o = 0 \quad v_f = 20 \text{ m/s} \quad t = 9.0 \text{ s} \quad a = ?$$

We have

$$a = \frac{v_f - v_o}{t} = \frac{20 - 0}{9.0} \text{ m/s}^2 = 2.22 \text{ m/s}^2$$

Then, since this value of $a$ is really $a_T$, we can write

$$\alpha = \frac{a_T}{r} = \frac{2.22 \text{ m/s}^2}{0.40 \text{ m}} = 5.55 \text{ rad/s}^2$$

Notice that, once again, we must furnish the proper angular units (radians) for our result. ∎

---

*Exercise*   Through how many revolutions does each wheel turn during the 9.0 s?
*Answer: 35.8 rev* ∎

---

### Example 7.6

In an experiment such as that shown in Fig. 7.5a, suppose that the mass starts from rest and accelerates downward at 8.6 m/s². If the radius of the spool is 20 cm, what is its rotation rate after 3.0 s?

*Reasoning*  Let us work this as a rotational-motion problem. We have that $\omega_o = 0$, $t = 3.0$ s, and

$$\alpha = \frac{a_T}{r} = \frac{8.6 \text{ m/s}^2}{0.20 \text{ m}} = 43 \text{ rad/s}^2$$

To find $\omega_f$, we can use the definition of angular acceleration:

$$\alpha = \frac{\omega_f - \omega_o}{t}$$

After rearrangement, this becomes

$$\omega_f = \omega_o + \alpha t = 0 + (43 \text{ rad/s})(3 \text{ s}) = 129 \text{ rad/s}$$

Notice that here, too, we have to supply the designation (radian) for the angle. ∎

## 7.6 Centripetal acceleration

A very interesting situation arises when an object travels along a circular path with constant speed. For example, in Fig. 7.6 we see a car traveling with constant speed $v$ around a circular track. Let us say that its speed is 20 m/s. Although the speed is 20 m/s at positions 1 and 2 and at all other points along the track, *the car is undergoing acceleration.* To understand this statement, we must remember two facts: (1) speed and velocity are not the same thing and (2) acceleration is defined to be the time rate of change of velocity (a vector), not of speed (a scalar). Because the *direction* of the velocity at position 1 is not the same as that at position 2, the velocity changes as the car moves along the track. From the definition of average acceleration, we have for the average acceleration of the car between 1 and 2

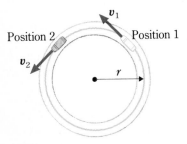

**FIGURE 7.6**

Even though the *speed* of the car is constant around the track, its *velocity* is continually changing because the direction of the velocity vector is not constant.

$$\bar{a} = \frac{\text{change in velocity}}{\text{time taken}}$$

Let us now compute the acceleration of the car.

The situation is redrawn in Fig. 7.7. The $y$ component of the car's velocity is $v_y$ at position 1 and $-v_y$ at position 2. The $x$ component of the velocity at 1 is the same as at 2. Hence we find that in going from 1 to 2,

$$(\text{Change in velocity})_y = v_{yf} - v_{yo} = -v_y - v_y = -2v_y$$

**FIGURE 7.7**

Notice that, in going from 1 to 2, the particle's velocity changes by $-2v_y$.

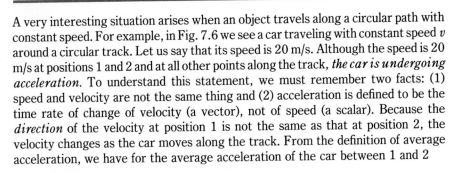

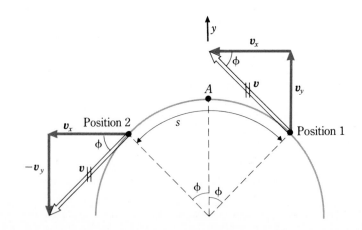

The time the car takes to go from 1 to 2 is, from $s = \bar{v}t$,

$$t = \frac{s}{v}$$

where $v$ is the tangential velocity of the car along the track and $s$ is the arc length from 1 to 2. Moreover, from the definition of radian measure, $\theta = s/r$, we have that

$$2\phi = \frac{s}{r} \quad \text{or} \quad s = 2r\phi$$

since $s$ subtends an angle $2\phi$ in this case. Thus we find that

$$t = \frac{s}{v} = \frac{2r\phi}{v}$$

We now know the change in velocity, $-2v_y$, and the time taken, $2r\phi/v$. Hence we can compute

$$\bar{a} = \frac{\text{change in velocity}}{\text{time taken}} = \frac{-2v_y}{2r\phi/v} = -\frac{vv_y}{r\phi}$$

From Fig. 7.7, though, we see that $v_y = v \sin \phi$. Therefore

$$\bar{a} = -\frac{v^2 \sin \phi}{r\phi}$$

This is the *average* acceleration of the car as it moves from position 1 to position 2. However, we are interested in the *instantaneous* acceleration at a point such as $A$. To obtain it, we simply let $\phi$ shrink to a very small value. Then, because $\sin \phi \cong \phi$ in radians when $\phi$ is small (use your calculator to check that this is true), the instantaneous acceleration is

$$a = -\frac{v^2 \sin \phi}{r\phi} \cong -\frac{v^2\phi}{r\phi} = -\frac{v^2}{r}$$

This is the acceleration of the car as it goes by point $A$. Since the speed is constant, all points on the circle are equivalent, and so $a = -v^2/r$ no matter where we choose point $A$.

Let us now find the direction of this acceleration. Figure 7.7 shows that $v_y$ changes from the positive $y$ direction to the negative $y$ direction. That is, the change in the velocity vector, $\mathbf{v}_2 - \mathbf{v}_1$, is a vector in the negative $y$ direction. Because the acceleration has the same direction as $\mathbf{v}_2 - \mathbf{v}_1$, it too is in the negative $y$ direction. At point $A$, however, this is the direction toward the center of the circle. We therefore find that *the car's acceleration is directed radially inward toward the center of the circle.* We call this the **radial acceleration** or the **centripetal acceleration** of an object moving in a circle. To summarize:

An object moving with constant speed $v$ along a circular path of radius $r$ experiences a centripetal (radial) acceleration $a_c$ toward the center of the circle. The magnitude of the acceleration is

$$a_c = \frac{v^2}{r} = \omega^2 r \tag{7.9}$$

where we have made use of the fact that $v = \omega r$.

## 7.7 Centripetal force

Newton's first law states that a net force must act on an object if the object is to be deflected from straight-line motion. Therefore an object traveling on a circular path must have a net force deflecting it from its straight-line path. For example, the car in Fig. 7.6 will slip from the circular track if the track is too slippery to provide the required friction force at the wheels. Similarly, the ball in Fig. 7.8 being twirled in a circular path is compelled to follow this path by the centerward pull of the string. If the string breaks when the ball is at point $B$, the ball will follow the straight-line path indicated by the broken line tangential to the circle.

Now that we know about centripetal acceleration, computing the force needed to hold an object of mass $m$ in a circular path is a simple task. When traveling on a circle, the object experiences an acceleration toward the center of the circle, the centripetal acceleration

$$a_c = \frac{v^2}{r}$$

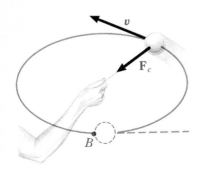

**FIGURE 7.8**

If the string breaks when the ball is at $B$, the ball will follow the tangential broken line.

where $r$ is the circle's radius and $v$ is the tangential speed of the object on the circular path. A force in the same direction, toward the center of the circle, must pull on the object to furnish this acceleration. It is the force $\mathbf{F}_c$ in Fig. 7.8, for example. From $F_{net} = ma$, we find this required force to be

$$F_c = ma_c = \frac{mv^2}{r} \tag{7.10}$$

It is called the **centripetal force.**

The force required to hold an object of mass $m$ and speed $v$ in a circular path of radius $r$ is called the centripetal force and is given by $mv^2/r$. It is directed toward the center of the circle.

All objects that travel in a circle (or an arc of a circle) require a centripetal force. The earth is pulled toward the sun by gravitational attraction. This pull furnishes the needed centripetal force and causes the earth to circle the sun. Similarly, the moon and other satellites circle the earth because of the earth's gravitational pull. This pull furnishes the needed centripetal force. We shall see more examples of centripetal forces later.

It is of interest to notice that no work is done by the centripetal force. To do work, a force must have a component in the direction of motion. The centripetal force is directed along the radius of a circle, however, whereas the motion occurs tangent to the circle. Because the tangent is perpendicular to the radius, the centripetal force has no component in the direction of motion. It therefore does no work. It simply changes the direction of the motion.

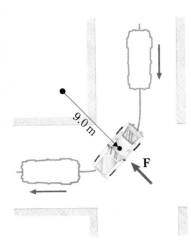

**FIGURE 7.9**

For the car to turn the corner as shown, the friction force $\mathbf{F}$ between the tires and the pavement must furnish the centripetal force needed to hold the car in a circular path.

*Example 7.7*

A 1200-kg car is turning a corner at 8.0 m/s, and it travels along an arc of a circle in the process (Fig. 7.9). If the radius of the circle is 9.0 m, how large a horizontal force must the pavement exert on the tires to hold the car in the circular path?

*Reasoning*   The force required is the centripetal force:

$$F = m\,\frac{v^2}{r} = (1200\text{ kg})\left(\frac{64\text{ m}^2/\text{s}^2}{9.0\text{ m}}\right) = 8530\text{ N}$$

This force must be supplied by the friction force of the pavement on the wheels. If the pavement is wet, so that there is little friction between tires and road, the friction force on the tires may not be large enough. In that event, the car will skid out of a circular path (into a more nearly straight line) and may not be able to make the curve.  ∎

---

*Exercise*   What minimum coefficient of friction must exist for the car not to slip? *Answer:* 0.73  ∎

---

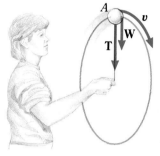

**FIGURE 7.10**

When the ball is in the position shown, its weight provides part of the necessary centripetal force.

## Example 7.8

A ball tied to the end of a string is swung in a vertical circle of radius *r*, as in Fig. 7.10. What is the tension in the string when the ball is at point *A* if the ball's speed is *v* at that point? Do not neglect the force of gravity.

*Reasoning*   For the ball to travel in a circle, the forces acting on it must furnish the required centripetal force. In this case, two forces act on the ball at point *A* — the pull **T** of the string and the pull of the earth (the weight **W** of the ball). Both forces are radial at *A*, and so their vector sum must provide the required centripetal force. We therefore have

$$T + W = \frac{mv^2}{r}$$

which gives the tension in the string as

$$T = \frac{mv^2}{r} - W = m\left(\frac{v^2}{r} - g\right)$$

Notice that if $v^2/r = g$, the tension in the string is zero. Then the centripetal force is just equal to the weight. This is the case in which the ball just makes it around the circle. If the speed of the ball is less than the value given by this relation, that is, if $v < \sqrt{rg}$, the required centripetal force is less than the ball's weight. In that case, the ball falls down out of the circular path.  ∎

---

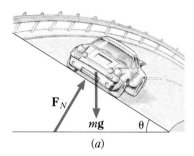

(a)

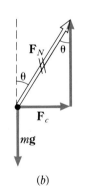

(b)

**FIGURE 7.11**

For proper banking, the normal force balances *mg* and supplies the centripetal force as well.

## Example 7.9

*Banking curves.* A curve in a road has a 60-m radius. It is to be banked so that no friction force is required for a car going at 25 m/s to safely make the curve. At what angle should it be banked?

*Reasoning*   The situation is shown in Fig. 7.11. Because the friction force between tires and road is to be zero, we show only two forces acting on the car: its weight *mg* and the normal force $\mathbf{F}_N$ exerted by the roadway. The horizontal component of $\mathbf{F}_N$, labeled $\mathbf{F}_c$ in the figure, must supply the required centripetal force.
From the free-body diagram,

$$F_c = F_N \sin\theta$$

where $\theta$ is the angle at which the curve is banked. Furthermore, because the car does not move vertically, the vertical forces must balance. Hence

$$F_N \cos \theta = mg$$

$$F_N = \frac{mg}{\cos \theta}$$

We can substitute this value for $F_N$ in the equation for $F_c$ to give

$$F_c = \left(\frac{mg}{\cos \theta}\right) \sin \theta = mg \tan \theta$$

Because the required centripetal force is $mv^2/r$, we equate this to the value we found for $F_c$ and obtain

$$\frac{mv^2}{r} = mg \tan \theta$$

$$\tan \theta = \frac{v^2}{rg}$$

In our case, $v = 25$ m/s and $r = 60$ m, and so we find

$$\tan \theta = \frac{(25 \text{ m/s})^2}{(60 \text{ m})(9.8 \text{ m/s}^2)} = 1.063$$

The angle that has this value for its tangent is $46.7°$, and so the curve must be banked at an angle of about $47°$. Notice that the mass of the car, and hence its weight, are not involved. When banked at this angle, even a slippery road guides a 25-m/s car around the curve safely. ∎

---

*Exercise*   Which way will the car tend to slide if its speed is 20 m/s?   *Answer: Down the incline* ∎

---

## 7.8  A common misconception

People sometimes jump to completely erroneous conclusions when interpreting their experiences. For example, a man seated in the center of a car seat sometimes thinks that he has been pushed to the side of the car as it rounds a corner. He might even assert that the force pushing him was so great that it threw him against the side hard enough to injure him. This is nonsense, of course. There was no mysterious ghost pushing him toward the side of the car. Certainly no material object was pushing him in that direction. He must therefore be mistaken.

The same man would not claim that a mysterious force suddenly threw him violently against the dashboard as the car stopped suddenly. He knows that his forward momentum can be lost only if some force retards his motion. Hence, when the car stopped suddenly, he continued going forward until the dashboard began to exert a force on him to stop him from moving forward. This is merely an example of

Newton's idea that things continue in motion until a force acts on them to stop them.

Similarly with the car turning the corner: the friction between pavement and tires pushed horizontally on the car and altered its straight-line motion. It is too bad about the man sitting in the middle of the nearly frictionless seat. The friction force between the seat of his pants and the car seat was too small to alter his straight-line motion. Hence he slid along in a straight line until he hit the side of the car, which then exerted a force on him so that he could travel in the same curved path as the car.

## 7.9 Newton's law of gravitation

One of the most interesting examples of nearly circular motion, planetary motion, was the subject of intensive study by astronomers many centuries ago. Astronomy was a highly developed but largely empirical science long before the time of Newton. When Newton began his study of the action of forces, he had available highly precise data concerning the orbits of the planets in the solar system. Astronomers of the time could accurately predict the future motions of the planets, but they were unable to justify the rules that they had found applicable over the centuries. Two keys were needed before the planetary motions could be explained.

The first was the law relating force and acceleration. This key was found when Newton discovered his three laws of motion. However, no progress could be made until the forces acting on the planets had been discovered. This discovery was also made by Newton, who set out to explain the motion of a typical planet about the sun.

As shown in Fig. 7.12, each planet travels in a nearly circular path with the sun at its center. If the planet is to be forced out of its normal straight-line motion, a net force must act on it. Obviously, this force must be similar to the radial force **F** pulling the planet toward the sun in Fig. 7.12. Newton analyzed the motion of the planets mathematically and showed that the force exerted on the planet by the sun must have the form

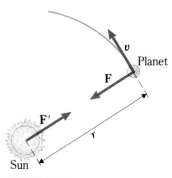

**FIGURE 7.12**

The sun and the planet attract each other with equal-magnitude forces.

$$F = G \frac{m_s m_p}{r^2} \tag{7.11}$$

where $m_s$ and $m_p$ are the masses of the sun and of the planet, respectively, and $r$ is the distance between their centers. The quantity $G$ is a constant whose value was not determined until much later.

If the sun attracts the planet, however, then the law of action and reaction tells us that the planet must attract the sun. This reaction force is shown as **F'** in Fig. 7.12. Because the action and reaction forces are equal and opposite, $\mathbf{F'} = -\mathbf{F}$.

Looking at other astronomical bodies, Newton saw that the moon orbiting the earth also obeys an equation similar to Eq. 7.11. Other planets have moons, and these also obey Eq. 7.11. Generalizing from these observations, Newton stated his *law of universal gravitation:*

Two spheres with masses $m_1$ and $m_2$ that have a distance $r$ between centers attract each other with a radial force of magnitude

$$F = G \frac{m_1 m_2}{r^2} \tag{7.12}$$

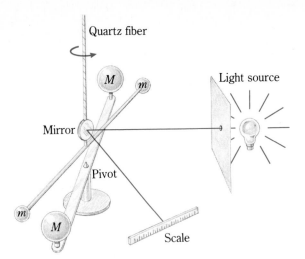

A schematic diagram of the Cavendish balance. Observe how the light beam is used to detect the twist of the fiber.

If the objects are not spherical, the attraction still exists but the value to be used for $r$ in Eq. 7.12 is more complicated. However, if the objects are small compared to the distance of separation (that is, if they can be considered to be pointlike), then $r$ can be taken to be equal to the distance between any two points on the objects, to a fair approximation. In any case, Newton concluded that each object attracts every other object with a gravitational force.

The value of the gravitational constant $G$ in Eq. 7.12 is not predicted by theory and can be determined only by experiment. Its value was first found by Henry Cavendish in 1798 with a device called the Cavendish balance. One form of this device is shown schematically in Fig. 7.13. Two identical masses $m$ are suspended from an extremely thin and delicate quartz fiber. The two large masses $M$ can be moved close to the small masses $m$, and the attraction between $M$ and $m$ causes the fiber to twist. After the system is calibrated so that the amount of force needed to produce a given twist is known, the attraction force between $m$ and $M$ can be computed directly from the observed twist of the fiber. Then, since $m, M, r,$ and $F$ are all known, these values can be used in Eq. 7.12 to solve for $G$.

In practice, this is an extremely delicate experiment since the attractive force is so very small. To measure the very small twist of the fiber, a beam of light is reflected from a mirror attached to the fiber system. By this means, the slight twist of the fiber results in a more easily measured deflection of the light beam. (The use of a light beam in this way is referred to as an *optical lever.*) Because of the delicacy of the fiber, even the slightest movement of the air close to it will disrupt the measurements. Therefore great care is needed to eliminate air movement and vibration if reliable results are to be obtained. The value currently accepted for $G$ is

$$G = 6.672 \times 10^{-11} \text{ N} \cdot \text{m}^2/\text{kg}^2$$

---

### Example 7.10

Two coins, each of mass 8.0 g, are 200 cm apart. Find the ratio $F/W$, where $F$ is the attractive force one coin exerts on the other and $W$ is the weight of each coin on earth.

***Reasoning*** The coins are far enough apart and small enough that the gravitation law can be applied with $r = 200$ cm. The attractive force is

$$F = G\frac{m_1 m_2}{r^2} = 6.67 \times 10^{-11} \text{ N} \cdot \text{m}^2/\text{kg}^2 \frac{(0.0080 \text{ kg})^2}{(2.0 \text{ m})^2}$$

$$= 1.07 \times 10^{-15} \text{ N}$$

The weight of each coin is

$$W = mg = (0.0080 \text{ kg})(9.8 \text{ m/s}^2) = 0.0784 \text{ N}$$

Therefore the ratio is

$$\frac{F}{W} = 1.36 \times 10^{-14}$$

Clearly, the gravitational force between ordinary objects is negligible. ■

---

## 7.10 The gravitational force and weight

Every object on the earth is attracted by the mass of the earth. Since the earth's mass is so large, this attractive force is many orders of magnitude larger than the attractive force between any two objects on the earth's surface. Let us see what we can conclude from a knowledge of the gravitational force.

It can be shown that the attractive force that a uniform spherical mass exerts on an object outside the sphere can be computed by assuming that all the mass of the sphere is concentrated at its center. If we call the earth's mass $m_e$, an object of mass $m$ on the earth's surface experiences a force

$$F = G\frac{mm_e}{R_e^2} \tag{7.13}$$

where $R_e$ is the radius of the earth. This attractive force that the earth exerts on an object is simply the weight of the object, $mg$. If we replace $F$ in Eq. 7.13 by $mg$, we find the following value for $g$, the acceleration due to gravity:

$$g = \frac{Gm_e}{R_e^2} \tag{7.14}$$

Notice that we could use this equation to find the mass of the earth if its radius were known.

As pointed out in Sec. 3.9, the weight of an object of mass $m$ depends upon its location on the earth's surface. We notice from Eq. 7.14 that $g$, and therefore the weight, vary with distance from the center of the earth. Since the earth bulges somewhat at the equator, there are slight variations in $g$ and in weights from place to place on the earth. (In addition, the rotation of the earth causes an object's apparent weight to be less at the equator than at the poles.)

Now that people have journeyed to the moon and may someday journey to distant planets, we extend the definition of weight to include these cases. We define the weight of an object on the moon (or other large body) to be the gravitational attractive force exerted on the object by that large body. Since the mass and radius of the moon were known long ago, it was possible to calculate that objects weigh only about one-sixth as much on the moon as on earth.

# 7.11 Orbital motion

Perhaps the most majestic examples of circular motion are found in the heavens. The earth and other planets travel around the sun in nearly circular paths; the earth's moon follows a nearly circular path around the earth, and the moons of other planets do much the same around their planets. In addition, planets created by people — in other words, artificial satellites — trace nearly circular paths around the earth. Let us now examine this type of motion, *orbital motion.*

In Fig. 7.14, we see the moon or some other satellite following a circular path around the earth. The satellite's mass is $m_s$, and its speed is $v$. We represent the orbit radius by $r$ and the mass of the earth by $m_e$. A centripetal force of magnitude $m_s v^2 / r$ is required to hold the satellite in orbit. This force is furnished by the gravitational attraction of the earth for the satellite:

$$\text{Gravitational force} = G \frac{m_s m_e}{r^2}$$

Equating the gravitational force to the required centripetal force gives

$$G \frac{m_s m_e}{r^2} = \frac{m_s v^2}{r} \tag{7.15}$$

It is of interest to notice that $m_s$, the satellite mass, cancels from this equation. We therefore conclude that *the mass of a satellite is unimportant in describing the satellite's orbit.* At a given orbital speed $v$ and radius $r$, the moon and a baseball will orbit in the same way. Thus any satellite orbiting at radius $r$ must have the following speed, given by Eq. 7.15:

$$v = \sqrt{\frac{Gm_e}{r}} \tag{7.16}$$

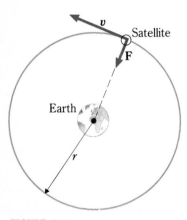

**FIGURE 7.14**

The centripetal force on the satellite is supplied by the gravitational attraction of the earth.

---

***Example 7.11***

Assuming the orbit of the earth about the sun to be circular (it is actually slightly elliptical) with radius $1.5 \times 10^{11}$ m, find the mass of the sun.

***Reasoning***   The centripetal force needed to hold the earth in an orbit of radius $R$ is furnished by the gravitational attraction of the sun. We therefore have (using $m_e$ and $m_s$ for the masses of the earth and sun, respectively)

Centripetal force = gravitational force

$$\frac{m_e v^2}{R} = G \frac{m_e m_s}{R^2}$$

$$m_s = \frac{v^2 R}{G}$$

where $v$ is the speed of the earth in its orbit around the sun. Since the earth travels around its orbit once each year ($3.15 \times 10^7$ s), we have

$$v = \frac{2\pi(1.5 \times 10^{11} \text{ m})}{3.15 \times 10^7 \text{ s}} = 2.99 \times 10^4 \text{ m/s}$$

from which

$$m_s = 2.01 \times 10^{30} \text{ kg} \quad \blacksquare$$

## Example 7.12

Radio and television signals are sent from continent to continent by "bouncing" them from *synchronous satellites.* These satellites circle the earth once each 24 h, and so if the satellite circles eastward above the equator, it always stays over the same spot on the earth because the earth is rotating at this same rate. Weather satellites are also designed to hover in this way. (*a*) What is the orbital radius for a synchronous satellite? (*b*) What is its speed?

*Reasoning* The satellite circles the earth once each 24 h, and so its speed is

$$v = \frac{\text{distance gone}}{\text{time taken}} = \frac{2\pi r}{t}$$

where $r$ is the orbital radius and $t = 24$ h. Substituting for $v$ in Eq. 7.15, we have

$$\frac{4\pi^2 r^2}{t^2} = \frac{Gm_e}{r}$$

from which

$$r^3 = \frac{Gm_e t^2}{4\pi^2}$$

Because $G = 6.67 \times 10^{-11}$ N · m²/kg², $m_e = 5.98 \times 10^{24}$ kg, and $t = 24$ h $=$ 86,400 s, substituting these values yields $r = 4.22 \times 10^7$ m. We can now substitute this value for $r$ in the expression for $v$ to give

$$v = \frac{2\pi r}{t} = \frac{2\pi(4.22 \times 10^7 \text{ m})}{86,400 \text{ s}} = 3070 \text{ m/s}$$

Thus we see that if an object is to be in a synchronous orbit around the earth, it must have a speed of 3070 m/s and be a distance $4.2 \times 10^7$ m $- R_e = 3.6 \times 10^7$ m above the surface of the earth. $\blacksquare$

*Exercise* What is the angular velocity (in radians per second) of a synchronous satellite? *Answer: $7.27 \times 10^{-5}$ rad/s* $\blacksquare$

## 7.12 Weightlessness

We often hear that objects appear to be weightless in a spaceship circling the earth or on its way to a distant point in space. Let us examine this effect in detail. First, we should restate our definition of weight: the pull of gravity on an object. On the earth, the weight of an object is the earth's gravitational pull on the object. Similarly, an object's weight on the moon is taken to be the moon's gravitational pull on the object.

Ordinarily we measure the weight of an object by placing it on a scale. If only a rough measure is needed, however, we can simply notice the force the object exerts

$$a = 0$$
$$T = W$$

(a)

a downward
$$W - T = ma$$
$$T = W - ma$$

(b)

**FIGURE 7.15**

The weight of an object in an elevator seems to a person in the elevator to vary, depending on the motion of the elevator.

on our hand when we hold it fixed. Usually the force read by the scale, that is, the force exerted by the object on the scale, and the force on our hand are equal to the pull of gravity on the object, that is, to its the weight. This is not always true, however, as we shall now see, and so *we reserve the phrase* **apparent weight** *for the reading of the scale and the force on our hand together with other common nonbasic ways for judging an object's weight.*

To illustrate this point, let us consider the apparent weight of an object of mass $m$ in an elevator. In Fig. 7.15a, if the elevator is at rest, Newton's second law tells us that, since the acceleration is zero, the resultant force on the object is zero. Calling the gravitational force on the object (its weight) $W$ and the tension in the string holding it $T$, we have

$$T - W = 0 \quad \text{or} \quad T = W$$

when $a = 0$. In this instance, the tension in the string is $W$ and the apparent weight of the object (the scale reading) is equal to its actual weight $W$.

This situation prevails as long as $a = 0$. Under that condition, $T = W$ and the apparent weight is equal to the actual weight. Even if the elevator is moving up or down at constant speed, the acceleration is still zero and the apparent weight still equals the actual weight.

Let us now examine the situation in Fig. 7.15b, the elevator is accelerating downward. If we apply Newton's second law as before, we find

$$W - T = ma$$

which gives

$$T = W - ma$$

Notice that the tension in the string (and therefore the scale reading) are less than $W$ by the amount $ma$. To the person in the accelerating elevator, the object appears to weigh less than $W$. Its apparent weight is $W - ma$.

The most spectacular observation occurs when the elevator is falling freely so that $a = g$, the acceleration in free fall. Then, since $W = mg$ and since $a = g$ for a freely falling body, the tension in the string

$$T = W - ma$$

becomes

$$T = mg - mg = 0$$

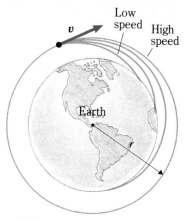

**FIGURE 7.16**

If an object is shot fast enough tangent to the earth, it will circle the earth. (Newton was probably the first to recognize this fact.)

The object appears weightless in a freely falling elevator! If we think about it a little, this is not strange at all. Since the elevator and everything in it are supposedly accelerating with the acceleration of free fall, by the very definition of free fall, there can be no force supporting the objects (elevator and everything in it) or in any way retarding their free fall. Hence all support forces on the elevator and everything in it must be zero. The tension in the cord supporting the object must be zero. All objects within the elevator appear to be weightless.

We see from these considerations that **in accelerating systems, the apparent weight of an object is not necessarily equal to its true weight.** In particular, if the system is falling freely,* all support forces must be zero and all objects appear to be weightless. This means that whenever a spaceship is falling freely in space, that is, when its rocket engines are not being operated, everything within this freely falling system appears weightless. It does not matter where the system is, or whether it is falling under the force of attraction of the earth, the sun, or some distant star; as long as it is falling freely, everything in it appears weightless.

A satellite circling the earth is simply an example of a freely falling object. At first this statement may surprise you, but it is easily seen to be correct. Consider the behavior of a projectile shot parallel to the horizontal surface of the earth in the absence of air friction. (At satellite altitudes, the air is so thin as to be almost negligible.) The situation is shown in Fig. 7.16. The various paths are for a projectile shot tangent to the earth at successively larger speeds. We see that, during the projectile's free fall to the earth, the curvature of the path decreases with increasing horizontal speed. If the projectile is shot fast enough tangent to the earth, the curvature of its path will match the curvature of the earth, as shown. In this case, the projectile (a satellite perhaps) will simply circle the earth. Since it circles the earth, the satellite is accelerating toward the earth's center. Its radial acceleration is simply $g$, the free-fall acceleration. In effect, the satellite is falling toward the center of the earth at all times, but the curvature of the earth prevents it from hitting. Since the satellite is in free fall, all objects within it appear weightless.

---

* Recall that a freely falling object is one that is subject to only one unbalanced external force: the pull of gravity.

---

## Minimum learning goals

When you finish this chapter, you should be able to

**1** Define (a) radian, (b) angular velocity, (c) angular acceleration, (d) tangential velocity, (e) tangential acceleration, (f) centripetal (radial) acceleration, (g) centripetal force, (h) apparent weight, (i) weightlessness.

**2** Convert an angle in degrees, radians, or revolutions to each of the other units.

**3** State the five angular-motion equations and use them to solve problems.

**4** Convert between tangential, angular, and linear quantities.

**5** Relate angular quantities to linear quantities involved in rolling wheels and a cord wound on a spool.

**6** Explain why a particle moving around a circle at constant speed is accelerating. Give the direction and magnitude of the acceleration.

**7** Compute the centripetal force needed to hold an object in a circular path.

**8** Compute the gravitational force one pointlike object exerts on another.

**9** Calculate the supporting force needed on an object of known mass if it is (a) moving with constant speed, (b) accelerating upward, (c) accelerating downward. Explain what is meant by apparent weight in such circumstances, and show why it differs from the object's weight.

**10** Explain why an object orbiting the earth (or in some similar situation) is said to be falling freely. Use your explanation to point out why objects appear weightless under certain circumstances.

## Questions and guesstimates

**1** As a wheel spins on its axle at a constant speed $\omega$, describe the following for a point $P$ at a radius $r$ from the center and state how each varies with $r$: (*a*) tangential velocity, (*b*) angular velocity, (*c*) angular acceleration, (*d*) tangential acceleration, (*e*) centripetal acceleration.

**2** After the standard tires on a car are replaced by tires with a 15 percent larger diameter, the car's speedometer no longer reads correctly. Explain how to obtain the correct reading from the actual reading.

**3** When mud flies off the tire of a moving bicycle, in what direction does it fly? Explain.

**4** Figure P7.1 shows a simplified version of a cyclone-type dust remover used for purifying industrial waste gases before they are vented to the atmosphere. The gas is whirled at high speed around a curved path, and the dust particles collect at the outer edge and are removed by a water spray or by some other means. Explain the principle behind this method.

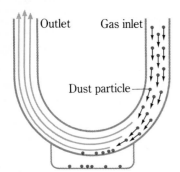

Outlet    Gas inlet

Dust particle

**FIGURE P7.1**

**5** Discuss the principle of the spin cycle in an automatic washing machine.

**6** An insect lands on a phonograph record that is on a turntable. Describe qualitatively the motion of the insect as the record begins to rotate. Assume that the insect is quite close to the axis and that there is some, but not much, friction between insect and record surface.

**7** On the moon, the acceleration due to gravity is 1.67 m/s². How would this make a person's life there different from what we are accustomed to on earth?

**8** To give a person a horizontal acceleration of $5g$, where $g = 9.8$ m/s², a force of "5 $g$'s" is required. What do we mean by this terminology? What do we mean when we say the pilot of an airplane experiences a force of several $g$'s as the plane pulls out of a steep dive? Why might the pilot "black out" if the pullout is too fast?

**9** From the fact that the moon circles the earth at a radius of $3.8 \times 10^8$ m, estimate the mass of the earth.

**10** Can we obtain the masses of the other planets of the Solar System if we know the radii of their orbits and the mass of the earth?

**11** About how fast can a car be going and still negotiate a turn from one street into a perpendicular street? Assume that each street is concrete with one lane in each direction.

**12** During the 1970 flight of Apollo 13 to the moon, serious trouble developed when the ship was about half-way there, and it returned to earth without executing its moon mission. After the trouble developed, however, the ship continued toward the moon, passed behind it, and only then returned to earth. Why didn't the astronauts simply turn around when the trouble developed?

**13** Suppose a huge mass, much larger than the mass of the Solar System or our galaxy, is created at this instant far away. The large gravitational force it would exert on the Solar System would cause us to accelerate toward this distant mass. After the first few seconds, what long-term effects would we notice on earth because of this acceleration? Assume that the earth's acceleraton due to this cause is of the order of 10 m/s².

## Problems

*Assume uniform acceleration unless stated otherwise.*

**1** Express each of the following angles in degrees, revolution, and radians: (*a*) 27°, (*b*) 3.2 rad, (*c*) 0.95 rev.

**2** Express each of the following angles in degrees, revolutions, and radians: (*a*) 0.48 rev, (*b*) 127°, (*c*) 0.45 rad.

**3** A roulette wheel of radius 120 cm has two numbers on its rim. The distance along the rim between the numbers is 4.0 cm. Find the angle subtended at the center of the wheel by the numbers. Give your answer in radians, degrees, and revolutions.

**4** A sphere of radius 20 cm has two dots on its surface. The distance along the surface between the dots is 3.0 cm. Find the angle subtended at the center of the sphere by the two dots. Express your answer in radians, degrees, and revolutions.

**5** Calculate the angular velocity of the second hand of a watch in radians per second and revolutions per minute.

**6** Calculate the angular velocity of the minute hand of a watch in degrees per second and radians per hour.

**7** A phonograph record rotates at 45 rev/min. (*a*) What is its angular speed in radians per second? (*b*) Through how many degrees does it rotate in 0.200 s?

**8** (*a*) What is the angular velocity of the hour hand of a clock in radians per second? (*b*) Through how many degrees does the hand turn in 5.0 s?

**9** A phonograph turntable accelerates from rest to an angular speed of 33 rev/min in 0.90 s. What is its average angular acceleration in revolutions per minute squared and radians per second squared?

**10** A phonograph turntable turning at 45 rev/min coasts to rest in 9.0 s. What is its average angular acceleration in revolutions per second squared and in radians per second squared?

■ **11** A merry-go-round takes 15 s to accelerate from rest to its operating speed of 3.0 rev/min. Find (*a*) its acceleration in revolutions per second squared and (*b*) the number of revolutions turned in this time.

■ **12** How large an angular acceleration (in radians per second squared) must be given to a wheel if it is to be accelerated from rest to a rotational speed of 460 rad/s after 5.8 rev?

■ **13** A certain roulette wheel coasts to rest in 12.0 s. If the wheel turns through 7.0 rev in that time, how fast was it originally turning?

■ **14** A wheel that is turning at 20 rev/min speeds up until its speed is 30 rev/min. The change takes 20 s. Find (*a*) its angular acceleration in radians per second squared and (*b*) the number of degrees through which it turned in this time.

**15** A ceiling fan is turning at 0.50 rev/s. The tip of its blade is 75 cm from the center. How fast, in centimeters per second, is the tip of the blade moving?

**16** A merry-go-round is rotating with a speed of 4.0 rev/min. How fast (in meters per second) is a child at a radius of 3.0 m moving?

**17** A bowling ball of diameter 22 cm rolls 15.0 m along the floor without slipping. Through how many revolutions does it roll?

**18** If its diameter is 60 cm, through how many revolutions must the wheel of a car turn as the car travels 300 m?

**19** What is the angular acceleration of a 50-cm-diameter wheel on a vehicle as the vehicle undergoes an acceleration of 0.50 m/s²?

**20** An object is being lifted by a cord wound on the rim of a wheel whose diameter is 30 cm. If the wheel is accelerating at 0.20 rad/s², what is the acceleration of the object in meters per second squared?

■ **21** The radius of the earth is $6.37 \times 10^6$ m. (*a*) How fast, in meters per second, is a tree at the equator moving because of the earth's rotation? (*b*) A polar bear at the North Pole?

■ **22** The earth orbits the sun in 365 days. What is the speed, in meters per second, of the earth in the orbit? The earth-sun distance is $1.50 \times 10^{11}$ m.

**23** A wheel of diameter 20.0 cm is turning at a rate of 0.40 rev/s and winds a string on its rim. How long a piece of string is wound in 30 s?

**24** A wheel turning at 1800 rev/min has a diameter of 6.0 cm. If a string is to be wound on the wheel, how much string is wound in 4.0 s?

**25** A vehicle is traveling along the road at 20 m/s. If the diameter of its wheels is 98 cm, how fast are the wheels rotating in revolutions per second, radians per second, and degrees per second?

■ **26** A 50-cm-diameter wheel comes loose from a car that is going 20 m/s, and the wheel rolls alongside the car. Find the angular speed of the wheel in revolutions per second, radians per second, and degrees per second.

■ **27** A bicycle with 60-cm-diameter wheels is coasting at 5.0 m/s. It decelerates uniformly and stops in 30 s. (*a*) How far does it go in this time? (*b*) Through how many revolutions does each wheel turn as the bicycle comes to a stop?

■ **28** A car with 80-cm-diameter wheels starts from rest and accelerates uniformly to 25 m/s in 30 s. Through how many revolutions does each wheel turn?

■ **29** A motor turning at 1400 rev/min coasts uniformly to rest in 15 s. (*a*) Find its angular deceleration and the number of revolutions it turns before stopping. (*b*) If the motor has a wheel of radius 5.0 cm attached to its shaft, what length of belt does the wheel wind in the 15 s?

■ **30** Two gear wheels that are meshed together have radii of 0.50 and 0.10 m. Through how many revolutions does the smaller wheel turn when the larger turns through 3 rev?

■ **31** A car accelerates uniformly from rest to 15 m/s in 20 s. Find the angular acceleration of one of its wheels and the number of revolutions turned by a wheel in the process. The radius of the car wheel is 0.33 m.

■ **32** A belt runs on a wheel of radius 30 cm. During the time the wheel takes to coast uniformly to rest from an initial speed of 2.0 rev/s, 25 m of belt length passes over the wheel. Find the deceleration of the wheel and the number of revolutions it turns while stopping.

**33** A 1200-kg car moving at 20 m/s is rounding a curve of radius 40 m. How large a horizontal force is needed to hold the car in this path?

**34** A 400-g mass at the end of a string is whirled in a horizontal circle of radius 80 cm. If its speed in the circle is 7.0 m/s, what must the tension in the string be? Neglect the force of gravity.

■ **35** A carton of eggs sits on the horizontal seat of a car as the car rounds a 20-m-radius bend at 15 m/s. What minimum coefficient of friction must exist between carton and seat if the eggs are not to slip?

■ **36** A 30-mg bug sits on the smooth edge of a 25-cm-radius phonograph record as the record is slowly brought up to its normal rotational speed of 33 rev/min. How large must the coefficient of friction between bug and record be if the bug is not to slip off? (It is a very compact bug, and so air friction can be ignored.)

■ **37** In a certain research device, a man is subjected to an acceleration of $5g$, that is, five times the acceleration due to gravity. This is done by rotating him in a horizontal circle at very high speed. The seat in which he is strapped is 9.0 m from the rotational axis. How fast is he rotating, in revolutions per second, if the centripetal force on him is five times his weight?

■ **38** An old trick is to swing a pail of water in a vertical circle. If the rotation rate is large enough, water will not fall out of the pail when the pail is upside down at the top of its path. What is the minimum speed your hand must have at the top of the circle if the trick is to succeed? Assume your arm is 0.60 m long.

■ **39** The designer of a roller coaster wishes the riders to experience weightlessness as they round the top of one hill. How fast must the car be going if the radius of curvature at the hilltop is 25 m?

■ **40** In an ultracentrifuge, a solution is rotated with an angular speed of 4000 rev/s at a radius of 10 cm. How large is the radial acceleration of each particle in the solution? Compare the centripetal force needed to hold a particle of mass $m$ in the circular path with the weight of the particle $mg$.

■ **41** The red blood cells and other particles suspended in blood are too light in weight to settle out easily when the blood is left standing. How fast (in revolutions per second) must a sample of blood be centrifuged at a radius of 8.0 cm if the centripetal force needed to hold one of the particles in a circular path is 10,000 times the weight of the particle $mg$? Why do the particles separate from the solution in a centrifuge?

■ **42** A certain car of mass $m$ has a maximum friction force of $0.7\,mg$ between it and the pavement as it rounds a curve

on a flat road ($\mu = 0.7$). How fast can the car be moving if it is to successfully negotiate a curve of radius 25 m?

**43** A neutron is an uncharged particle with a mass of $1.67 \times 10^{-27}$ kg and a radius of the order of $10^{-15}$ m. Find the gravitational attraction between two neutrons whose centers are $1.00 \times 10^{-12}$ m apart. Compare this with the weight of the neutron on earth.

**44** Find the force of gravity that the moon exerts on a 45-kg student sitting next to you on earth. The moon's mass is $7.3 \times 10^{22}$ kg, and its distance from the earth is $3.8 \times 10^5$ km. Compare this with the weight of the student on earth.

■ **45** Compare the gravitational pull on a spaceship at the surface of the earth with the gravitational pull when the ship is orbiting 1000 km above the surface of the earth. (The radius of the earth is 6370 km.)

■ **46** The planet Jupiter has a mass 314 times larger than that of the earth. Its radius is 11.3 times larger than the earth's radius. Find the acceleration due to gravity on Jupiter.

■ **47** A 50-kg astronaut is floating at rest in space 35 m from her stationary 150,000-kg spaceship. About how long will it take her to float to the ship under the action of the force of gravity?

■ **48** The acceleration due to gravity on the moon is only one-sixth that on earth. Assuming that the earth and the moon have the same average composition, what would you predict the moon's radius to be in terms of the earth's radius $R_e$? (In fact, the moon's radius is $0.27R_e$.)

■ **49** A satellite orbits the earth in about 80 min if its orbital radius is 6500 km. Use these data to find the mass of the earth.

■ **50** One satellite planet of Jupiter, called Callisto, circles Jupiter once each 16.8 days. Its orbital radius is $1.88 \times 10^9$ m. Use these data to find the mass of Jupiter.

■ **51** A 400-g ball at the end of a string is whirled in a horizontal circle of radius 80 cm. If its speed in the circle is 7.0 m/s, what must the tension in the string be? Do not neglect the weight of the ball; the string is not exactly horizontal.

■ **52** As shown in Fig. P7.2, a man on a rotating platform

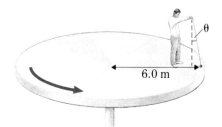

6.0 m

FIGURE P7.2

holds a pendulum in his hand. The pendulum ball is 6.0 m from the center of the platform. The rotational speed of the platform is 0.025 rev/s, and the pendulum hangs at an angle $\theta$ to the vertical. Find $\theta$.

■ **53** The bug in Fig. P7.3 has just lost its footing near the top of the bowling ball. It slides down the ball without appreciable friction. Show that it will leave the surface of the ball at the angle $\theta$ shown, where $\theta$ is given by $\cos \theta = 2/3$.

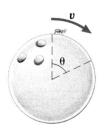

**FIGURE P7.3**

■ **54** Figure P7.4 shows a possible design for a space colony. It consists of a cylinder of diameter 7 km and length 30 km floating in space. Its interior is provided with an earthlike environment, and to simulate gravity, the cylinder spins on its axis. What should the rotation rate of the cylinder be, in revolutions per hour, so that a person standing on the landmass will press down on the ground with a force equal to his or her weight on earth? (For details, see G. K. O'Neill, *Physics Today,* September 1974, p. 32, and February 1977, p. 30.)

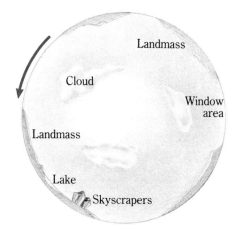

**FIGURE P7.4 (End view of cylinder)**

■ **55** A particle is to slide along the horizontal circular path on the inside of the funnel of Fig. P7.5. The surface of the funnel is frictionless. How fast must the particle be moving, in terms of $r$ and $\theta$, to execute this motion?

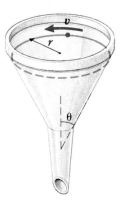

**FIGURE P7.5**

■ **56** A pendulum consisting of an 80-g ball at the end of a string 170 cm long is released at an angle of 50° to the vertical. Find the tension in the string when the angle is (*a*) 0° and (*b*) 30°.

■ **57** A hemispherical bowl of radius $b$ sits open side up on a table. A bead is shot horizontally tangent to the bowl's inner surface, and the bead follows a horizontal circular path of radius $R = 2b/3$. Assuming the motion to be frictionless, find the relation between $R$ and the bead's speed $v$ if the mass of the bead is $m$.

■ **58** The beads of masses $M$ and $m$ in Fig. P7.6 can slide freely on the wire circle. They are at rest in the positions shown when $M$ is released and collides elastically with $m$. What is the largest value possible for $m/M$ if bead $m$ is to then travel over the top of the circle without exerting a downward force on the wire there?

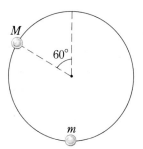

**FIGURE P7.6**

# 8 Rotational work, energy, and momentum

**N**EWTON'S second law relates the force acting on an object to the object's mass and linear acceleration: $F = ma$. When an object, such as a wheel, is rotating on an axis, torques can give it an angular acceleration. We see in this chapter that an analog to $F = ma$ exists for rotational motion. It relates the torque acting on an object to the product of the object's angular acceleration and a quantity that measures its rotational inertia. Moreover, we see that a rotating object has both kinetic energy and rotational momentum.

## 8.1 Rotational work and kinetic energy

It is easy to see that a rotating object has kinetic energy. For example, the rotating wheel in Fig. 8.1 is composed of many tiny bits of mass, each moving as the wheel turns. A typical bit of mass, such as the one labeled $m_1$, has a velocity $v_1$ and thus a kinetic energy $\frac{1}{2}m_1v_1^2$. Let us begin our study of the properties of rotating objects by seeing how a wheel might acquire kinetic energy.

The wheel in Fig. 8.2 is originally at rest, but it can rotate freely on an axle through its center. When a force $F$ pulls on the cord wound on the wheel's rim, the wheel begins to rotate. The work done by the force as it pulls the end of the string through a distance $s$ is

Work done by $F = Fs$

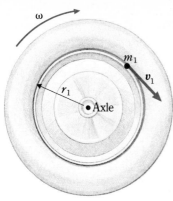

**FIGURE 8.1**

As the wheel rotates, each tiny bit of mass within it possesses kinetic energy, for $m_1$ it is $\frac{1}{2}m_1v_1^2$.

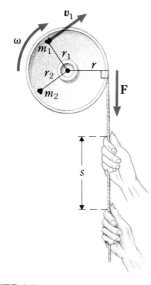

**FIGURE 8.2**

As the force **F** does work by pulling the string a distance $s$, the wheel gains an amount $Fs$ of kinetic energy.

As the length $s$ of string is unwound, the wheel turns through an angle $\theta$, which is related to $s$ through $s = r\theta$ (Eq. 7.1). Substitution of this value for $s$ yields the following expression for the work done:

Work done by $F = Fr\theta$

We can better understand this expression by noticing that $Fr$ is "force times lever arm" in Fig. 8.2, and this is simply the torque $\tau$ applied to the wheel.* Thus we find the relation between the work done on a wheel when it turns through an angle $\theta$ and the torque applied to it:

$$W = \tau\theta \tag{8.1}$$

It is interesting to notice that this is the result we would guess by analogy to linear motion. For linear motion, we had $W = F_x x$. In rotation, force is replaced by torque and linear distance by angular distance. Hence $F_x x$ becomes $\tau\theta$, as proved in Eq. 8.1.

According to the work-energy theorem, the work done on the wheel by the applied torque must appear as kinetic energy. We call the kinetic energy resident in a rotating object *kinetic energy of rotation* $KE_r$. You will recall that the kinetic energy of an object that is due to its linear motion is $\frac{1}{2}mv^2$, and we shall now refer to this as translational kinetic energy $KE_t$. Let us now compute the kinetic energy of a rotating object from a consideration of the kinetic energy of each of the little pieces of mass that compose the object.

Refer back to Fig. 8.1. As the wheel rotates, a typical tiny mass (such as $m_1$) that is part of the wheel possesses translational kinetic energy. For $m_1$, it is $\frac{1}{2}m_1v_1^2$. If there are $N$ tiny masses ($m_1, m_2, m_3, \ldots, m_N$) that compose the wheel, their combined kinetic energy is

$$KE \text{ of wheel} = \frac{1}{2}m_1v_1^2 + \frac{1}{2}m_2v_2^2 + \frac{1}{2}m_3v_3^2 + \cdots + \frac{1}{2}m_Nv_N^2$$

But $m_1$, for example, travels around a circle of radius $r_1$. Its tangential velocity on this circle is $v_1$. We know that the angular velocity of the wheel is related to this tangential velocity through $v_1 = \omega r_1$, and so we have

$$\frac{1}{2}m_1v_1^2 = \frac{1}{2}m_1\omega^2 r_1^2$$

Similar expressions exist for all the other tiny masses. We substitute these values in the kinetic energy equation and obtain

$$KE \text{ of wheel} = \frac{1}{2}m_1r_1^2\omega^2 + \frac{1}{2}m_2r_2^2\omega^2 + \cdots + \frac{1}{2}m_Nr_N^2\omega^2$$

Because $\omega$, the angular velocity of the wheel, can be factored out, we have

$$KE \text{ of wheel} = \frac{1}{2}\omega^2(m_1r_1^2 + m_2r_2^2 + \cdots + m_Nr_N^2)$$

The factor in parentheses is usually represented by the symbol $I$ and is called the **moment of inertia** of the rotating object:

$$I = \text{moment of inertia} = m_1r_1^2 + m_2r_2^2 + \cdots + m_Nr_N^2 \tag{8.2}$$

---

* You may wish to review the concept of torque, presented in Sec. 2.4.

We shall discuss the moment of inertia soon; then you will see why it is indeed a measure of the inertia of the wheel. Even now, however, you can see that it depends not only on the quantity of matter $m$ in the body, but also on how that matter is distributed.

Our expression for the kinetic energy of the rotating wheel can now be rewritten in terms of $I$:

$$KE_r = \text{rotational KE} = \tfrac{1}{2}I\omega^2 \tag{8.3}$$

This is the kinetic energy an object has because of its rotation. Notice that, once again, we could have guessed its general form. In analogy to $\tfrac{1}{2}mv^2$, we see that $v$ is replaced by $\omega$ and $I$ apparently is the rotational counterpart of the mass $m$.

As we mentioned earlier, the rotational energy is related to the work done on the wheel by the applied torque. To be specific, suppose the wheel is rotating with speed $\omega_o$ when a torque $\tau$ is suddenly applied to it. The torque acts while the wheel turns through an angle $\theta$ (so that the work the torque does is $\tau\theta$) and is then removed. At that time, the angular velocity of the wheel is $\omega_f$. The work-energy theorem describes this process by telling us that

Work done on wheel = change in wheel's KE

$$\tau\theta = \tfrac{1}{2}I\omega_f^2 - \tfrac{1}{2}I\omega_o^2$$
$$\tau\theta = \tfrac{1}{2}I(\omega_f^2 - \omega_o^2)$$

where we have used Eq. 8.3 for the wheel's rotational kinetic energy.

This relation between work and rotational kinetic energy can be simplified by using the angular motion equation (Eq. 7.5$d$),

$$\omega_f^2 - \omega_o^2 = 2\alpha\theta$$

After substituting this quantity, we can cancel $\theta$ from the equation and obtain

$$\tau = I\alpha \tag{8.4}$$

where $\alpha$ must be measured in radians per second squared. (Why?) We have thus arrived at a relation between the angular acceleration of the wheel and the torque that caused it. It is the rotational-motion analog to $F = ma$. As we expect, torque replaces $F$, angular acceleration replaces $a$, and the mysterious quantity $I$ replaces $m$. We investigate the meaning of $I$ in the next section.

## 8.2 Rotational inertia

We know that rotating objects have inertia. After you turn off an electric fan, the blade coasts for some time as the friction forces of the air slowly cause it to stop. If you try to decelerate it more quickly with your finger, its tendency to keep on rotating becomes obvious. The moment of inertia $I$ of the fan blade measures its rotational inertia, as we can understand in the following way.

In linear motion, the inertia of an object is represented by the object's mass. From $F = ma$, we have

$$m = \frac{F}{a}$$

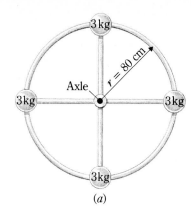

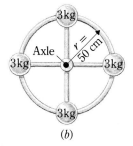

(a)

(b)

FIGURE 8.3

Which wheel is more difficult to set into rotation?

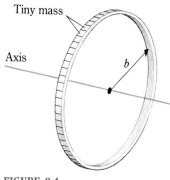

FIGURE 8.4

What is $I$ for the hoop about the axis shown?

Thus the mass tells us how large a force is required to produce a linear acceleration of $a = 1$ m/s². An object that has large inertia has large mass and requires a large force to give it an acceleration of 1 m/s².

The rotational analog of $F = ma$, which is $\tau = I\alpha$, gives similar information about $I$, the moment of inertia:

$$I = \frac{\tau}{\alpha}$$

which tells us how large a torque is needed to produce an angular acceleration of $\alpha = 1$ rad/s². Objects that have large values for $I$ require large torques to change their rotation rate. Clearly, $I$ measures the rotational inertia of any object.

Now let us examine the mathematical representation for moment of inertia. From Eq. 8.2,

$$I = m_1 r_1^2 + m_2 r_2^2 + \cdots + m_N r_N^2 = \sum_{i=1}^{N} m_i r_i^2$$

Let us apply this to the two wheels in Fig. 8.3. Each consists of four masses mounted on a frame of negligible mass. In part $a$, we have

$$I_a = m_1 r_1^2 + m_2 r_2^2 + m_3 r_3^2 + m_4 r_4^2$$
$$= (3 \text{ kg})(0.80 \text{ m})^2 + (3 \text{ kg})(0.80 \text{ m})^2 + (3 \text{ kg})(0.80 \text{ m})^2 + (3 \text{ kg})(0.80 \text{ m})^2$$
$$= 7.68 \text{ kg} \cdot \text{m}^2$$

For part $b$,

$$I_b = (3 \text{ kg})(0.50 \text{ m})^2 + (3 \text{ kg})(0.50 \text{ m})^2 + (3 \text{ kg})(0.50 \text{ m})^2 + (3 \text{ kg})(0.50 \text{ m})^2$$
$$= 3.00 \text{ kg} \cdot \text{m}^2$$

As we see, the wheel in $b$ has a much smaller moment of inertia than the one in $a$. Although the masses of the wheels are the same, their moments of inertia differ because the masses are farther from the axis of rotation in $a$ than in $b$. Because $I$ varies as $r^2$ (Eq. 8.2), the moment of inertia gets larger the farther the mass is from the axis. Thus, a greater torque is needed in $a$ than in $b$.

As a more practical example, let us compute the moment of inertia of the hoop of mass $M$ shown in Fig. 8.4. We assume it to rotate about an axis through its center and perpendicular to the plane of the hoop. In our mind's eye, we split the hoop into a large number of tiny masses as shown. Each mass is at a distance $b$ from the axis. The moment of inertia becomes

$$I_{\text{hoop}} = m_1 r_1^2 + m_2 r_2^2 + \cdots + m_N r_N^2$$
$$= m_1 b^2 + m_2 b^2 + \cdots + m_N b^2 = b^2(m_1 + m_2 + \cdots + m_N)$$

The sum of the tiny masses composing the hoop is simply its total mass $M$, however. Therefore

$$I_{\text{hoop}} = b^2 M$$

The moment of inertia for any object can be calculated in this same way. However, calculus is usually needed to carry out the summation. We list the values of $I$ for several simple objects in Table 8.1. In some cases, different axes of rotation are possible. For example, a cylinder might rotate about either of the two axes shown. Therefore one must state the axis being used when the moment of inertia is given.

■ **TABLE 8.1** **Moments of inertia**

| Object | Axis | $I$ | Radius of gyration $k$ |
|---|---|---|---|
| Hoop | | $mb^2$ | $b$ |
| Solid disk (radius $b$) | | $\frac{1}{2}mb^2$ | $b/\sqrt{2}$ |
| Solid sphere | | $\frac{2}{5}mb^2$ | $b\sqrt{\frac{2}{5}}$ |
| Solid cylinder (radius $b$) | | $\frac{1}{2}mb^2$ | $b/\sqrt{2}$ |
| Solid thin cylinder (length $L$) | | $\frac{1}{12}mL^2$ | $L/\sqrt{12}$ |

Another important feature concerning $I$ can be seen from Table 8.1. In all cases, $I$ is the product of the object's mass and the square of a length. For example, $I$ for a sphere is the mass of the sphere multiplied by $(\sqrt{\frac{2}{5}}\,b)^2$. Similarly, $I$ for a disk is $m(b/\sqrt{2})^2$. For a hoop, $I = mb^2$. In general, we can write

$$I = mk^2 \tag{8.5}$$

where $k$ is a length characteristic of the object. This length is called the object's **radius of gyration.** In a crude way, $k$ is the distance from the axis of an average bit of mass composing the object. For example, Table 8.1 shows that $k = b$ for a hoop, which is a reasonable value since each bit of mass composing the hoop is at a distance $b$ from the axis. For a sphere, however, $k = \sqrt{\frac{2}{5}}\,b = 0.63b$ because only the farthest points on the sphere are a distance $b$ from the axis. As another example, in Fig. 8.3a we have $k = 0.80$ m. For that object,

$$I = mk^2 = (12\ \text{kg})(0.80\ \text{m})^2 = 7.68\ \text{kg} \cdot \text{m}^2$$

as we found previously. Typical values for $k$ are given in Table 8.1.

Let us summarize what we have found.

**1** An object of mass $m$ possesses rotational inertia. We represent this quantity by $I$, the moment of inertia. In equation form, $I = mk^2$, where $k$ is an average distance from the axis of rotation to the various pieces of mass that compose the object.

**2** A rotating object has rotational kinetic energy, $KE_r = \frac{1}{2}I\omega^2$.

**3** A torque $\tau$ applied to an object that is free to rotate gives the object an angular acceleration: $\tau = I\alpha$.

**4** The work done by a torque $\tau$ when it acts through an angle $\theta$ is $\tau\theta$.

We now illustrate how we make practical use of these results.

### Example 8.1
Find the rotational kinetic energy of the earth due to its daily rotation on its axis. Assume it to be a uniform sphere, $m = 5.98 \times 10^{24}$ kg, $r = 6.37 \times 10^6$ m.

**Reasoning**   For a uniform sphere, Table 8.1 tells us that

$$I = \tfrac{2}{5}mr^2 = (\tfrac{2}{5})(5.98 \times 10^{24} \text{ kg})(6.37 \times 10^6 \text{ m})^2 = 9.71 \times 10^{37} \text{ kg} \cdot \text{m}^2$$

The angular velocity of the earth is

$$\omega = \left(1\,\frac{\text{rev}}{\text{day}}\right)\left(\frac{1}{86{,}400}\,\frac{\text{day}}{\text{s}}\right)\left(2\pi\,\frac{\text{rad}}{\text{rev}}\right) = 7.27 \times 10^{-5} \text{ rad/s}$$

Notice that we express $\omega$ in radians per second because all our fundamental equations are in radian measure. Then

$$KE_r = \tfrac{1}{2}I\omega^2 = \tfrac{1}{2}(9.71 \times 10^{37} \text{ kg} \cdot \text{m}^2)(7.27 \times 10^{-5} \text{ rad/s})^2 = 2.56 \times 10^{29} \text{ J} \quad \blacksquare$$

### Example 8.2
A certain wheel with a radius of 40 cm has a mass of 30 kg and a radius of gyration of 25 cm. A cord wound around its rim supplies a tangential force of 1.80 N to the wheel, which turns freely on an axle through its center. (See, for example, Fig. 8.2.) Find the angular acceleration of the wheel.

**Reasoning**   We use $\tau = I\alpha$ to find $\alpha$. In our case,

$$I = mk^2 = (30 \text{ kg})(0.25 \text{ m})^2 = 1.875 \text{ kg} \cdot \text{m}^2$$

where $k$ is the radius of gyration. Also

$$\tau = \text{force} \times \text{lever arm} = (1.80 \text{ N})(0.40 \text{ m}) = 0.720 \text{ N} \cdot \text{m}$$

Then we have

$$\alpha = \frac{\tau}{I} = \frac{0.720 \text{ N} \cdot \text{m}}{1.875 \text{ kg} \cdot \text{m}^2} = 0.384 \text{ rad/s}^2$$

We arrive at these units by noting that a newton is a kg $\cdot$ m/s$^2$ and by remembering that the "non-unit" radians may be inserted where necessary.   $\blacksquare$

### Example 8.3
The larger wheel in Fig. 8.5 has a mass of 80 kg and a radius $r$ of 25 cm. It is driven by a belt as shown. The tension in the upper part of the belt is 8.0 N, and that in the lower part is essentially zero. (a) How long does it take for the belt to accelerate the larger wheel from rest to a speed of 2.0 rev/s? (b) How far does the wheel turn in this time? (c) What is then the rotational kinetic energy of the wheel? Assume the wheel to be a uniform disk.

**Reasoning**   First let us change the final speed of the wheel to radians per second: 2.0 rev/s $= 4\pi$ rad/s. (a) We use $\tau = I\alpha$ to find the acceleration of the wheel. Because the applied force is 8.0 N with a lever arm of 0.25 m, the applied torque is

$$\tau = \text{force} \times \text{lever arm} = (8.0 \text{ N})(0.25 \text{ m}) = 2.0 \text{ N} \cdot \text{m}$$

The moment of inertia of a disk is (Table 8.1)

$$I_{\text{disk}} = \tfrac{1}{2}mb^2 = \tfrac{1}{2}(80 \text{ kg})(0.25 \text{ m})^2 = 2.50 \text{ kg} \cdot \text{m}^2$$

**FIGURE 8.5**

Angular acceleration is imparted to the large-radius wheel by the torque resulting from the tension $T$ in the upper part of the belt. Notice that the lower part of the belt is slack.

Then, from $\tau = I\alpha$,

$$\alpha = \frac{\tau}{I} = \frac{2.0 \text{ N} \cdot \text{m}}{2.50 \text{ kg} \cdot \text{m}^2} = 0.800 \text{ rad/s}^2$$

(You should be able to show that N/kg $\cdot$ m is 1/s².)
  Now we can use $\omega_f = \omega_o + \alpha t$ to find

$$t = \frac{\omega_f - \omega_o}{\alpha} = \frac{(4\pi - 0) \text{ rad/s}}{0.80 \text{ rad/s}^2} = 15.7 \text{ s}$$

(b) The angular distance turned in this time is given by $\theta = \overline{\omega}t$:

$$\theta = \tfrac{1}{2}(\omega_f + \omega_o)t = \tfrac{1}{2}(4\pi + 0 \text{ rad/s})(15.7 \text{ s}) = 98.6 \text{ rad}$$

(c) We know that $KE_r = \tfrac{1}{2}I\omega^2$, and so

$$KE_r = \tfrac{1}{2}(2.50 \text{ kg} \cdot \text{m}^2)(4\pi \text{ rad/s})^2 = 197 \text{ J}$$

(You should be able to show that kg $\cdot$ m² $\cdot$ rad/s² is joules.)  ■

---

*Exercise*  Find the rotational kinetic energy of the wheel using the work-energy theorem.  *Answer: 197 J*  ■

---

**Example 8.4**
The 3-kg object in Fig. 8.6a, hangs from a cord wound on a 40-kg wheel. The wheel has a radius of 0.75 m and a radius of gyration of 0.60 m. Find (a) the angular acceleration of the wheel and (b) the distance the object falls in the first 10 s after it is released. Notice that the 3-kg object weighs $3 \times 9.8 \text{ N} = 29.4 \text{ N}$.

*Reasoning*  Problems involving two bodies must be solved by isolating each in turn.
(a) The unbalanced force acting on the object hanging from the cord is $29.4 \text{ N} - T$, as shown in Fig. 8.6b. Using Newton's second law, $F = ma$, we have

$$29.4 \text{ N} - T = (3 \text{ kg})(a) \tag{a}$$

Isolating the wheel as in c, we have

$$\tau = I\alpha$$
$$T(0.75 \text{ m}) = (40 \text{ kg})(0.60 \text{ m})^2(\alpha)$$
$$T = (19.2 \text{ kg} \cdot \text{m})(\alpha) \tag{b}$$

The general idea is to solve Eqs. a and b simultaneously. To do this, we make use of the fact that $a = r\alpha$, which in this case gives $a = (0.75 \text{ m})(\alpha)$. Equation (a) then becomes

$$29.4 \text{ N} - T = (2.25 \text{ kg} \cdot \text{m})(\alpha)$$

Substitution from Eq. (b) gives

$$29.4 \text{ N} - (19.2 \text{ kg} \cdot \text{m})(\alpha) = (2.25 \text{ kg} \cdot \text{m})(\alpha)$$
$$29.4 \text{ N} = (21.45 \text{ kg} \cdot \text{m})(\alpha)$$
$$\alpha = 1.37 \text{ rad/s}^2$$

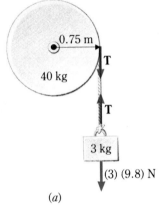

0.75 m

T

40 kg

T

3 kg

(3) (9.8) N

(a)

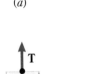

T

29.4 N

(b)

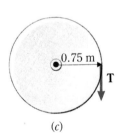

0.75 m

T

(c)

**FIGURE 8.6**
As the 3-kg object accelerates under the pull of gravity, the tension in the rope imparts an angular acceleration to the wheel.

You should verify the units of the answer.

(*b*) We make use of the usual linear-motion equations together with the fact that

$$a = r\alpha = 1.03 \text{ m/s}^2$$

Known are

$$a = 1.03 \text{ m/s}^2 \qquad v_o = 0 \qquad t = 10 \text{ s}$$

Using

$$y = v_o t + \tfrac{1}{2}at^2$$

gives

$$y = 51.5 \text{ m}$$

If the distance were known in a situation such as this, we could reverse the procedure and compute the moment of inertia of the wheel. This type of experiment is sometimes used to determine moments of inertia and radii of gyration. ∎

### Example 8.5

Find the speed of the object of Example 8.4 after it has dropped 80 cm.

*Reasoning*   To illustrate another approach, let us start anew using energy methods. As the object falls, it loses gravitational potential energy. At the same time, the wheel gains kinetic energy of rotation and the object gains translational kinetic energy. The law of conservation of energy tells us that

$$\text{Change in KE}_t + \text{change in KE}_r + \text{change in GPE} = 0$$

Therefore

$$(\tfrac{1}{2}mv^2 - 0) + (\tfrac{1}{2}I\omega^2 - 0) + (0 - mgh) = 0$$

But $\omega$ and $v$ are related through $v = \omega r$. Further, $m = 3$ kg, $h = 0.80$ m, and $I = Mk^2 = (40 \text{ kg})(0.60 \text{ m})^2 = 14.4 \text{ kg} \cdot \text{m}^2$. The energy equation thus becomes

$$\tfrac{1}{2}(3 \text{ kg})(v^2) + \tfrac{1}{2}(14.4 \text{ kg} \cdot \text{m}^2)\left(\frac{v}{0.75 \text{ m}}\right)^2 - (3 \text{ kg})(9.8 \text{ m/s}^2)(0.80 \text{ m}) = 0$$

Solving this equation for $v$, we find the object to be moving at 1.28 m/s. ∎

*Exercise*   How fast is the wheel turning in revolutions per second?   *Answer: 0.273 rev/s* ∎

## 8.3  Combined rotation and translation

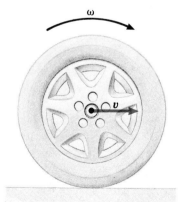

**FIGURE 8.7**

As the wheel rolls, it has both rotational and translational kinetic energy.

In Fig. 8.7 we see a wheel rolling without slipping. Each little piece of material within the wheel is undergoing two types of motion at the same time. The center of the wheel, which is the wheel's center of gravity, is moving horizontally with speed $v$. At the same time, however, the wheel is rotating about its center of gravity with angular speed $\omega$. Each piece of matter within the wheel experiences the combination of these two motions. Therefore, the wheel possesses both translational and rotational kinetic energy.

It is possible to show that the total kinetic energy of the wheel can be expressed quite simply if we restrict ourselves to rotation about one particular axis, that through the wheel's center of mass. (Except in rare cases,* the center of mass and the center of gravity are the same.) This is the usual axis for a rolling object. Then we can state that

The total kinetic energy of an object that is translating with speed $v$ as well as rotating with speed $\omega$ is $\frac{1}{2}mv^2 + \frac{1}{2}I\omega^2$.

Now let us exhibit the utility of this fact.

**FIGURE 8.8**

As the ball rolls to the bottom of the hill, its gravitational potential energy is changed to kinetic energy of rotation and translation.

### Example 8.6

A uniform sphere of radius $r$ and mass $m$ starts from rest at the top of an incline of height $h$ and rolls down (Fig. 8.8). How fast is the sphere moving when it reaches the bottom? (Assume that it rolls smoothly and that friction energy losses are negligible.)

**Reasoning** This problem is easily solved if we use the law of conservation of energy. In this instance, the sphere originally has gravitational potential energy, and this is changed to kinetic energy of translation and rotation. We have

$$\text{Change in KE}_t + \text{change in KE}_r + \text{change in GPE} = 0$$

which becomes, if we call the original and final heights of the ball $h_o$ and $h_f$, respectively,

$$\tfrac{1}{2}m(v_f^2 - v_o^2) + \tfrac{1}{2}I(\omega_f^2 - \omega_o^2) + mg(h_f - h_o) = 0$$

But $v_o$ and $\omega_o$ are zero. Further, $I = \frac{2}{5}mr^2$ for a uniform sphere, and so the above equation becomes

$$\tfrac{1}{2}mv_f^2 + \tfrac{1}{5}mr^2\omega_f^2 - mg(h_f - h_o) = 0$$

We wish to find $v_f$, and so we eliminate $\omega_f$ through the relation $v = r\omega$. Further, we are told that $h_f - h_o = h$, and so

$$\tfrac{1}{2}v_f^2 + \tfrac{1}{5}v_f^2 - gh = 0$$

$$v_f = \sqrt{\frac{10gh}{7}}$$

It is interesting to notice that the radius of the sphere cancels. Moreover, the translational kinetic energy term is $\frac{1}{2}v_f^2$ and the rotational kinetic energy term is $\frac{1}{5}v_f^2$. Therefore the translational kinetic energy of the sphere is 2.5 times larger than its rotational kinetic energy. If the rolling object had been a hoop, for which $I = mr^2$, the kinetic energy would have been equally apportioned between translational and rotational kinetic energy. ∎

* They are different if $g$ varies appreciably from point to point in the wheel, a rare occurrence indeed.

*Exercise* Suppose the sphere is hollow, a child's hollow rubber ball, for example. Is its speed the same, larger, or smaller than that calculated above *Answer: Smaller* ∎

## 8.4 Angular momentum

In view of the many analogies found thus far between linear and rotational phenomena, it should come as no surprise that linear momentum has a rotational counterpart. Rotational, or angular, momentum is associated with the fact that a rotating object persists in rotating. As you might expect from the fact that linear momentum is given by $mv$, the defining equation for **angular momentum** $L$ is

$$L = \text{angular momentum} = I\omega \tag{8.6}$$

We also give direction to angular momentum, and so it, like linear momentum, is a vector. Its direction is taken to be along the axis of rotation, as shown in Fig. 8.9.*

Angular momentum obeys a conservation law much like the one obeyed by linear momentum. The **law of conservation of angular momentum** can be stated as follows:

If no net torque acts on a body or system, its angular momentum remains constant in both magnitude and direction:

$$I\omega = \text{const} \qquad \text{if} \qquad \Sigma\tau = 0$$

Notice that not only is the magnitude of an object's angular momentum constant if no unbalanced torque acts on the object, but also the *direction* of the angular-momentum vector does not change. This is equivalent to saying that *the axis of rotation of a spinning object does not alter its orientation unless a torque acts on it to cause it to alter.* You can demonstrate this orientation effect using a simple gyroscope or a swiftly spinning wheel. For example, a large wheel set rotating about a north-south axis does not change its orientation readily unless very large forces are applied to it. When a torque is applied to a rotating system such as this, the resulting motion is interesting, since it appears to contradict what we expect to happen. Although the analysis of these effects is too complicated for us to pursue in this course, the effects are easily demonstrated, and your instructor may show you some.

*The axis of rotation of an object does not change its orientation unless an external torque causes it to do so.* This fact is of great importance for the earth as it circles the sun. No sizable torque is experienced by the earth since the major force on it, the pull of the sun, is a radial force. The earth's axis of rotation therefore remains fixed in direction with reference to the universe around us, that is, with respect to the distant stars. We can see this behavior in Fig. 8.10. Notice that the earth's path around the sun is nearly circular, but its rotation axis is not perpendicular to the plane defined by this orbit. Instead, the axis makes a fixed angle to the plane and, because of the conservation of angular momentum, maintains this orien-

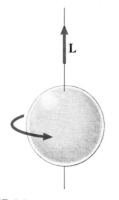

**FIGURE 8.9**
The rotating sphere has angular momentum.

---

* The following right-hand rule is used to determine the direction along the axis: grasp the axis in your right hand with your fingers circling it and pointing in the direction of rotation; your thumb then points along the axis in the direction of the angular momentum.

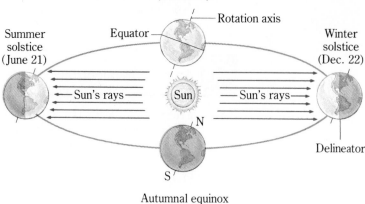

Vernal (spring) equinox
(March 21)

Rotation axis

Summer solstice (June 21)

Equator

Winter solstice (Dec. 22)

Sun's rays — Sun — Sun's rays

N

Delineator

S

Autumnal equinox
(Sept. 21)

**FIGURE 8.10**

As an example of the conservation of angular momentum, the earth's rotational axis retains its same orientation relative to the rest of the universe. The dates shown are approximate.

tation as the earth circles the sun. As you can see from the figure, the North Pole is in continuous daylight during summer and in continuous darkness in winter. The seasons experienced at lower latitudes are a less striking example of this same effect.

Many other examples of the conservation of angular momentum can be seen in the universe about us. This conservation law can be used to predict the rotation of planets in their orbits and the motions of the stars. It is influential in determining the behavior of atoms in molecules and of electrons in atoms. Its scope is unlimited. It applies to both the smallest and the largest objects in the universe.

*Example 8.7*

Consider a satellite circling the earth as shown in Fig. 8.11. Find the ratio of its speed at perihelion to that at aphelion.

*Reasoning* The satellite orbits the earth in an ellipse with the center of the earth at one focus. Since the earth's force on the satellite is radial, the angular momentum of the satellite about the earth's center must be conserved. When we denote perihelion and aphelion by subscripts $p$ and $a$, respectively, the conservation of angular momentum tells us that

$$I_p\omega_p = I_a\omega_a$$

The moment of inertia of a point mass $m$ (the satellite) at a distance $r$ from the rotational axis is simply $mr^2$, however. Therefore this relation becomes

$$mr_p^2\omega_p = mr_a^2\omega_a$$

This gives

$$\frac{\omega_p}{\omega_a} = \left(\frac{r_a}{r_p}\right)^2$$

We know that $v_T = \omega r$. Since the velocity is tangential at both perihelion and aphelion, we have

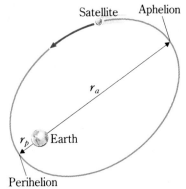

Satellite    Aphelion

$r_a$

$r_p$    Earth

Perihelion

**FIGURE 8.11**

Find the ratio of the satellite's speed at perihelion to that at aphelion.

$$\frac{v_p/r_p}{v_a/r_a} = \left(\frac{r_a}{r_p}\right)^2$$

which simplifies to the result requested,

$$\frac{v_p}{v_a} = \frac{r_a}{r_p}$$

Thus the satellite moves relatively more slowly at the greater distance.  ∎

---

### Example 8.8
Stars with mass greater than about 1.5 times that of our sun are unstable. Driven by gravitational forces, they sometimes collapse to form neutron stars. These are incredibly dense stars in which all atoms are collapsed; in effect, the atomic electrons and protons have combined to form neutrons. The final star has a radius of only about $10^{-5}$ that of the original. Our sun rotates on its axis about once each 25 days. How long would the sun take to complete 1 rev if it were to undergo such a collapse? Assume it to be a uniform sphere.

***Reasoning*** From the law of conservation of angular momentum,

$$(I\omega)_{\text{before}} = (I\omega)_{\text{after}}$$

For a sphere, $I = \frac{2}{5}Mr^2$, and so after substituting and canceling terms, we have

$$(r^2\omega)_{\text{before}} = (r^2\omega)_{\text{after}}$$

$$\omega_{\text{after}} = \omega_{\text{before}}\left(\frac{r_b}{r_a}\right)^2 = (\tfrac{1}{25}\text{ rev/day})(10^5)^2$$

$$= 4 \times 10^8 \text{ rev/day}$$

Because there are 86,400 s/day, the sun would complete 1 rev in $2 \times 10^{-4}$ s. Neutron stars rotate with periods of fractional seconds. Can you show that the surface speed of the collapsed sun is nearly the speed of light, $3 \times 10^8$ m/s?  ∎

---

***Exercise*** Find the ratio of the original rotational kinetic energy to the final rotational kinetic energy. *Answer:* $10^{-10}$  ∎

---

### Example 8.9
The beaker in Fig. 8.12 sits over the axis of a rotating turntable. The table is coasting on a frictionless bearing, and $I = 8.0 \times 10^{-4}$ kg · m² for the table-beaker combination. Water drips slowly into the beaker along the axis. If the beaker was rotating at 2.0 rev/s when empty, what is its rotational speed when it contains 300 g of water? The inner radius of the beaker is 3.5 cm.

***Reasoning*** Because the dripping water supplies neither an external torque nor angular momentum, the angular momentum of the system is conserved. Therefore,

Angular momentum before = angular momentum after

$$L_{\text{beaker+table}} \text{ at start} = L_{\text{beaker+table}} \text{ at end} + L_{\text{water}}$$

$$I_{b+t}\omega_o = I_{b+t}\omega_f + I_w\omega_f$$

**FIGURE 8.12**

Why does the coasting rotating system slow down as water drips into the beaker?

# How cats land feet first

Humans and other animals use the law of conservation of angular momentum in a seemingly instinctive way. Skilled athletes, however, make conscious use of this principle as they perform various maneuvers with their bodies. For example, the skater in the illustration controls her rotational speed by changing her moment of inertia. By bringing her arms and legs as close as possible to her rotation axis, she minimizes her moment of inertia and therefore maximizes her angular speed. Similarly, divers and acrobats change their body conformation as they fly through the air. A fast somersaulting motion is achieved by pulling the body into a tight, squat configuration, but when the body is straightened, the rotation rate decreases to compensate for the increased moment of inertia. The diver in the illustration can change his rotation rate by nearly a factor of four during the maneuver shown.

You have no doubt heard that a falling cat always lands feet first. Somehow the cat stops its tumbling motion in the air just as its feet are pointing toward the ground. Studies show that cats do this in the same way acrobats and divers control their rotation. By extending its paws as far as possible from its straightened body at just the proper time, the cat stops its tumbling and positions its body properly for landing. As you see, nature sometimes provides its creatures with the ability to use even its most abstruse laws.

(a) $I$ large, $\omega$ small

(b) $I$ small, $\omega$ large

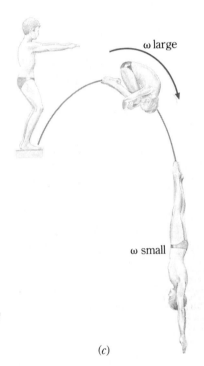

$\omega$ large

$\omega$ small

(c)

---

The water in the beaker takes the form of a disk with radius 3.5 cm. Because $I$ for a disk is $\frac{1}{2}mr^2$, we have

$$I_w = \tfrac{1}{2}(0.30 \text{ kg})(0.035 \text{ m})^2 = 1.84 \times 10^{-4} \text{ kg} \cdot \text{m}^2$$

Since we were told that $I_{b+t} = 8.0 \times 10^{-4}$ kg · m$^2$, the conservation equation becomes

$$(8.0 \times 10^{-4} \text{ kg} \cdot \text{m}^2)(\omega_o) = (8.0 \times 10^{-4} \text{ kg} \cdot \text{m}^2)(\omega_f)$$
$$+ (1.84 \times 10^{-4} \text{ kg} \cdot \text{m}^2)(\omega_f)$$

$$\omega_f = 0.813\omega_o = 0.813(2.0 \text{ rev/s}) = 1.63 \text{ rev/s} \quad ■$$

---

***Exercise*** Suppose the kinetic energy of the falling water can be neglected. Show that the final system has 18.7 percent less kinetic energy than the original system. What happened to this lost energy? ■

## Minimum learning goals

When you finish this chapter, you should be able to

**1** Define (*a*) moment of inertia, (*b*) radius of gyration, (*c*) rotational kinetic energy (*d*) angular momentum.

**2** Write the rotational analogs of $F = ma$, $KE_t = \frac{1}{2}mv^2$, $p = mv$, and $W = F_x x$.

**3** Compute the moment of inertia of simple objects such as a hoop or the object in Fig. 8.3.

**4** Use the relation $\tau = I\alpha$ in simple situations involving accelerated rotational motion.

**5** Relate the work done on an object by a torque to the object's change in rotational kinetic energy in simple situations.

**6** Find the total kinetic energy of an object that is rotating and translating simultaneously.

**7** Solve simple situations involving the conservation of energy for rolling objects.

**8** State the law of conservation of angular momentum and use it in simple problems.

## Questions and guesstimates

**1** Devise a demonstration that shows that a rotating wheel can do work because of its rotational kinetic energy.

**2** Two bicycle wheels are identical except that one has its rubber tire replaced by a solid ring of metal of the same size. They are mounted on identical stationary axles so that they can coast relatively freely. If they start with the same speed, which wheel will coast to rest first?

**3** Three wheels have the same mass and rim radius, but wheel *a* is a solid uniform disk, wheel *b* has a heavy rim with light spokes, and wheel *c* is a typical automobile wheel with tire. Compare their moments of inertia about their rotation axes.

**4** Estimate your moment of inertia when you are standing erect, using as axis (*a*) a vertical line that passes through the center of your body and (*b*) a horizontal line perpendicular to your stomach.

**5** One proposal for storing energy is by means of a massive, rapidly rotating flywheel. Discuss the pros and cons of this device as applied to (*a*) an automobile and (*b*) an electric power station.

**6** Refer to Fig. 8.6. Assume friction is negligible. (*a*) The tension in the connecting cord is less than *mg*. Why?

(*b*) What effect does changing the moment of inertia of the wheel have on the tension in the cord?

**7** A tiny bug is at rest on the rim of a turntable that rotates without friction. What happens to the turntable as the bug (*a*) runs radially in toward the center and (*b*) runs clockwise around the rim? Take into account when the bug starts, when it runs at constant speed, and when it stops.

**8** Which will roll down an incline faster, a hollow sphere or a solid sphere? Will the radius of the sphere affect its speed? Repeat for a hoop and a uniform solid disk.

**9** In order to keep a football or any other projectile from wobbling, the thrower causes it to spin about an axis in line with the direction of motion. Explain.

**10** A "do-it-yourselfer" builds a helicopter with a single propeller on a vertical axis. In the helicopter's maiden flight, the operator becomes sick because the whole helicopter tends to spin about a vertical axis. What is wrong? How is this difficulty overcome in more sophisticated machines?

**11** Suppose that the sun's attraction for the earth suddenly doubled. What effect would this have on the earth's rate of rotation and its orbit about the sun?

## Problems

**1** A force of 7.0 N is applied to a string wound on the rim of a 20-cm-diameter wheel. How much work is done by this force as it turns the wheel through 30°?

**2** The friction torque on a certain wheel-axle system is 0.050 N · m. How much work does this torque do on the system as the wheel turns through 3.5 rev?

**3** The 40-cm-long spokes of the object in Fig. P8.1 have negligible mass compared with the eight 3-kg masses. Find the moment of inertia of the object (*a*) about an axis through its center and perpendicular to the page and (*b*) about the line *AA'* as axis.

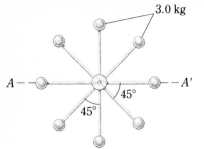

3.0 kg

45°

45°

A----                    ----A'

FIGURE P8.1

**4** Each of the four masses in Fig. P8.2 has a mass $m$. The connecting rods have negligible mass. Find the moment of inertia of the system about the axis ($a$) $AA'$ and ($b$) $BB'$. Consider the balls to be point masses.

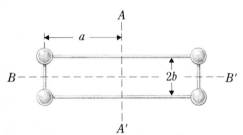

A

B----                    $2b$ ----B'

$a$

A'

FIGURE P8.2

**■ 5** The two hoops in Fig. P8.3 are held together by spokes of negligible mass. The mass of the inner hoop is

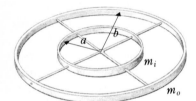

$a$    $b$

$m_i$

$m_o$

FIGURE P8.3

$m_i$, and its radius is $a$; for the outer hoop, the values are $m_o$ and $b$. Find the moment of inertia of the system about an axis through the center and perpendicular to the plane of the hoops.

**■ 6** The hoop of mass $M$ in Fig. P8.4 has a very light rod fastened through it with identical masses at its ends. Find the moment of inertia of the system about an axis through $C$ and perpendicular to the plane of the hoop.

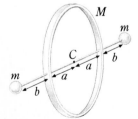

M

m

C

$a$  $b$

m

$a$

$b$

FIGURE P8.4

**■ 7** A wheel in the form of a uniform disk has a moment of inertia $I_d$. A rim of radius 30 cm and mass 1.50 kg is slipped onto the wheel. Find the moment of inertia of the combination.

**■ 8** Two wheels have moments of inertia $I_1$ and $I_2$. They are fastened together on the same axle so as to make a single wheel. Starting from the definition of $I$, prove that $I$ for the combined wheel is $I_1 + I_2$.

**9** How much work must be done on a wheel ($I = 0.25$ kg · m²) to accelerate it from rest to an angular speed of 2.0 rev/s?

**10** A wheel whose moment of inertia is 0.35 kg · m² is spinning at 0.80 rev/s when it begins to coast to rest. How much work do friction forces do in stopping it?

**11** How large a torque is required to give an angular acceleration of 3.0 rad/s² to a wheel that has a moment of inertia of 0.20 kg · m²?

**■ 12** A tangential force of 50 N applied to the rim of a 20-cm-radius wheel gives the wheel an acceleration of 0.150 rev/s². What is the moment of inertia of the wheel about its axis?

**■ 13** A wheel for which $I = 0.20$ kg · m² is spinning at 50 rev/min when the power source is shut off. It coasts uniformly to rest in 100 s. How large is the average torque that stops the wheel?

**■ 14** ($a$) A certain wheel has a mass of 50 kg and a radius of gyration of 25 cm. How large a torque is required to accelerate this wheel from rest to 0.40 rev/s in 20 s? ($b$) How far does the wheel turn in this time?

**■ 15** A force of 2.0 N acts tangential to the rim of a solid 50-kg disk that has a radius of 30 cm. ($a$) How long does it take to accelerate the disk (rotating about its usual axis) from rest to 4.0 rad/s? ($b$) Through how many revolutions does it rotate in this time?

**■ 16** A 6.0-cm-radius, 7.0-kg solid cylinder is mounted on an axle along its axis. A string wound around it supplies a tangential force of 3.5 N for 3.0 s. If the cylinder starts from rest, how fast is it rotating (in revolutions per second) at the end of the 3 s?

**■ 17** To determine the moment of inertia of a wheel, the experiment shown in Fig. 8.2 is performed. When the tension in the cord is 0.40 N, the wheel, starting from rest, rotates 3.0 rev in 8.0 s. The radius of the wheel is 7.0 cm. What is $I$ for the wheel?

**■ 18** When an apparatus such as that shown in Fig. 8.2 is used, how large a torque must the cord apply to the wheel in order to speed it up from rest to 0.70 rev/s in 20 s? The moment of inertia of the wheel is 1.60 kg · m².

**■ 19** A wheel mounted on an axle has a string wound on its rim. The moment of inertia of the wheel is 0.080 kg · m².

The wheel is accelerated from rest by a force of 30 N, which pulls the end of the string a distance of 75 cm. What is the wheel's final speed of rotation (in revolutions per second)?

■ **20** A wheel of radius 8.0 cm with $I = 0.070$ kg · m² is turning at 4.0 rev/s when a tangential friction force of 0.90 N is applied to the rim. Through how many revolutions does the wheel turn before stopping?

■ **21** A 6.0-cm-radius wheel is mounted on a horizontal axis. A string is wound on the rim, and a 50-g mass is suspended from the string. After the mass is released, the system accelerates in such a way that the mass drops 2.0 m in 4.0 s. What is the moment of inertia of the wheel? What is the tension in the cord as the mass is falling?

■ **22** A cylinder of radius 20 cm is mounted on a horizontal axis coincident with the cylinder's axis. A cord is wound on the cylinder, and a 90-g mass is hung from it. After being released, the mass drops 150 cm in 1.20 s. Find the moment of inertia of the cylinder and the tension in the cord while the mass is falling?

■ **23** A 90-g mass is suspended from the rim of a 60-cm-radius wheel by a string wound around the wheel. The wheel has $I = 0.070$ kg · m² and is mounted on a frictionless axle. It is accelerated from rest by letting the mass fall. How fast (in revolutions per second) is the wheel turning when the mass has fallen 120 cm?

■ **24** A wheel with $I = 800$ kg · cm² is spinning at 3.0 rev/s when a mechanism is engaged that causes a cord on the rim to lift a 5.0-kg mass as the wheel coasts to rest. How high can the wheel lift the mass? Neglect any change in total kinetic energy during the engagement. (See also Prob. 42.)

■ **25** The shaft of a 0.25-hp (output) motor rotates at 240 rev/min. (a) How much work can the motor do each second? (b) How large an output torque can the motor produce when it is operating at its rated speed?

■ **26** The output shaft of a gear system attached to a motor has a rated output power of 0.180 W. Mounted on the shaft is a wheel that has $I = 0.70$ kg · m². Estimate the minimum time it takes the motor to accelerate the wheel from rest to 0.50 rev/s.

■ **27** The system in Fig. P8.5 is released from rest. (a) How fast is the frictionless wheel ($I = 0.0050$ kg · m², $r = 7.0$ cm) turning after the 520-g mass has fallen 2.0 m? (b) How long does it take the mass to drop this far?

■ **28** The system in Fig. P8.6 is released from rest. There is no friction between the block and the table, and the pulley ($I = 0.0050$ kg · m², $r = 6.0$ cm) is frictionless. (a) How fast is the mass on the right moving after it has fallen 120 cm? (b) How long does it take the mass to fall this far?

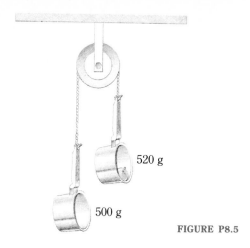

520 g

500 g

**FIGURE P8.5**

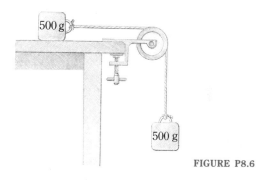

500 g

500 g

**FIGURE P8.6**

■ **29** A hoop of radius 5.0 cm starts from rest and rolls down a slope. (a) What is its linear speed when it reaches a point 40 cm lower than its starting point? (b) How fast is it rotating (in revolutions per second) at that time?

■ **30** Repeat Prob. 29 for a wheel that has a radius of 5.0 cm and a radius of gyration of 4.0 cm.

■ **31** A steel ball bearing of radius 0.70 cm is rolling along a table at 40 cm/s when it starts to roll up an incline. How high above the table level does it rise before stopping? Ignore friction losses.

■ **32** A uniform sphere and a uniform disk are rolled down an incline from the same point. Find the ratio of the disk's speed to that of the sphere at the bottom of the incline. Ignore friction losses.

■ **33** In Fig. P8.7, two identical small balls, each of mass 3.0 kg, are fastened to the ends of a light metal rod 1.00 m long. The rod is pivoted at its center and is rotating at 9.0 rev/s. An internal mechanism can move the balls in toward the pivot. (a) Find the moment of inertia of the original device. (b) If the balls are suddenly moved in until they are 25 cm from the pivot, what is the new speed of rotation?

0.50 m   0.50 m

FIGURE P8.7

$I_2$

$\omega_1$   $I_1$

FIGURE P8.8

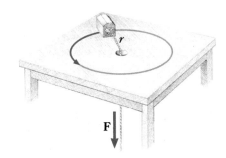

$r$

**F**

FIGURE P8.9

■ **34** It is surmised that our sun was formed in the gravitational collapse of a dust cloud that once filled the space now occupied by the Solar System and beyond. Assuming the original cloud to be a uniform sphere of radius $R_o$ with an average angular speed of $\omega_o$, how fast should the sun be rotating now? Ignore the small mass resident in the planets and assume the sun to be a uniform sphere of radius $R_s$.

■ **35** A woman stands over the center of a horizontal disk-like platform that is rotating freely at 1.5 rev/s about a vertical axis through its center and straight up through the woman. She holds two 5-kg masses in her hands close to her body. The combined moment of inertia of platform, woman, and masses is 1.2 kg · m². The woman now extends her arms so as to hold the masses far from her body. In so doing, she increases the moment of inertia of the system by 2.0 kg · m². (a) What is the final rotational speed of the platform? (b) Is the kinetic energy of the system changed during the process? Explain.

■ **36** An ice skater moving with speed $v_o$ past a post grabs the end of a rope tied to the post. The original length of the rope is $L_o$, but it shortens as the skater circles the post, thereby winding up the rope. Assuming that the skater coasts and does not try to stop, how fast will she be moving when the length of rope (the circle radius) is $L$? Assume that the post radius is much smaller than $L$.

■ **37** Figure P8.8 shows a disk and shaft (moment of inertia = $I_1$) coasting with angular speed $\omega_1$. A nonrotating disk (moment of inertia = $I_2$) is dropped onto the first disk and couples to it. (a) Find the angular speed of the disks after coupling. (b) Repeat if the dropped disk has an initial angular velocity $\omega_2$ in the same direction as $\omega_1$. (c) Repeat if $\omega_2 = \omega_1$ but is opposite in direction.

■ **38** The 20-g block in Fig. P8.9 revolves in a circle on a frictionless table. It is held by a cord that passes through a tiny hole at the center of the circle. The angular speed of the block is 3.00 rev/s, and $r = 80$ cm. (a) What is the magnitude of the force **F**? (b) If the cord is pulled down 15 cm, what is the new angular speed of the block? (c) How much work does the force **F** do to shorten the circle radius to 65 cm? Assume that the block is quite small relative to the radius of the circle.

■ **39** A children's merry-go-round consists of an essentially uniform, 150-kg solid disk rotating about a vertical axis. The radius of the disk is 7.0 m, and a 90-kg person is standing on it at its outer edge when it is coasting at 0.20 rev/s. How fast will the disk be rotating if the person walks 4.0 m toward the center?

■ **40** Suppose that the merry-go-round described in Prob. 39 has no one on it but is moving at 0.20 rev/s. If a 90-kg person quickly sits down on the edge, what is its new speed?

■ **41** A freely rotating turntable is turning with an angular speed $\omega_1$. Its moment of inertia is $I_1$. At the edge, at a radius $b$ from the center, sits a bug of mass $m$. The bug starts running with speed $v$ around the perimeter. What is the turntable's new angular speed if the bug runs (a) in the same direction the table is turning? (b) In the opposite direction?

■ **42** A wheel in the form of a uniform disk has a mass $M$ and is spinning with an angular speed $\omega_o$ while winding up a string on its rim. The slack string is attached to a mass $m$ sitting on the floor beneath the wheel. Eventually the slack is all taken up, and the mass is suddenly lifted from the floor. Show that the fraction of the total kinetic energy lost in the process of bringing the mass up to speed is $2m/(M + 2m)$. Neglect changes in gravitational potential energy.

# 9 Mechanical properties of matter

THE world around us is composed of three general types of materials: gases, liquids, and solids. In gases, the molecules are essentially independent of one another and fill the entire volume available to them. In liquids, the molecules are held to one another with forces strong enough to prevent them from escaping from one another, but they still can slip past one another fairly easily. In solids, very strong forces beween adjacent molecules hold them so tightly that the object appears rigid. In this chapter, we begin our study of the mechanical properties of these materials. We learn how they deform under external forces and how their properties can be described quantitatively.

## 9.1 Characterization of materials

To properly describe a material, we should describe the motion of each atom in it. Such a description would be worthless for most purposes, however, because it would be far too complicated and detailed for everyday use. The engineer who wishes to use a certain type of steel or the surgeon who needs a plastic filament for a suture neither wants nor requires an atomic description of the material. Only rarely are situations encountered in which more than the gross, overall properties of a material are needed. Usually, a nonatomic description of the material is sufficient. We therefore seek ways in which to characterize materials quantitatively in terms of their bulk properties.

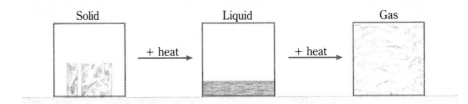

Solid      + heat →      Liquid      + heat →      Gas

**FIGURE 9.1**
Water can exist in three forms.

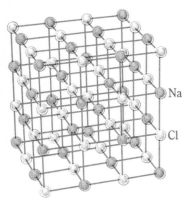

**FIGURE 9.2**
A schematic diagram of a very small portion of a salt (NaCl) crystal.

You are familiar with the classification method shown in Fig. 9.1. Perhaps the substance is water, originally in the form of solid ice. Heat melts the ice to form liquid water, and further heating changes the liquid to a gas, which fills the entire container. Sometimes the latter two forms are grouped together and called fluids. *Fluids are materials that flow readily under the action of applied forces.*

Solids are divided into two basic types. In **crystalline solids,** the atoms or molecules are arranged in a definite pattern throughout the solid. Table salt (sodium chloride) is an example of a crystalline solid. Its atoms are arranged in a cubic structure (or lattice), as shown schematically in Fig. 9.2. The exterior appearance of a crystalline material often indicates its internal regularity — as we see in Fig. 9.3 for a more complicated crystalline structure, that of ice.

In **amorphous solids,** the atoms are like grains of sand in a sandpile: they are not arranged in a repeating pattern. Typical amorphous solids are window glass and most transparent plastics.

These classifications are widely used, but they have definite limitations. For

**FIGURE 9.3**
The internal regularity of these snowflakes is manifest in their symmetric appearance. *(W. A. Bentley, NOAA)*

example, some amorphous solids flow appreciably over periods of many years. Are they fluids or solids? Scientists do not argue about the answer to such questions. Instead, they use quantitative methods to characterize materials in such borderline situations, because the material's numerical properties provide a less ambiguous description of it, as we see in the following sections.

## 9.2 Density

We frequently make use of a property called the *density* of a material, defined as

$$\text{Density} = \frac{\text{mass of substance}}{\text{volume of substance}}$$

Density is mass per unit volume and is represented by the Greek letter rho ($\rho$). Thus, if an object of volume $V$ has a mass $m$, its density is

$$\rho = \frac{m}{V} \tag{9.1}$$

The SI units of density are kilograms per cubic meter, but densities are sometimes given in grams per cubic centimeter. These units are related as follows:

$$1 \text{ g/cm}^3 = \left(\frac{1 \text{ g}}{\text{cm}^3}\right)\left(\frac{1 \text{ kg}}{1000 \text{ g}}\right)\left(\frac{10^6 \text{ cm}^3}{1 \text{ m}^3}\right) = 1000 \text{ kg/m}^3$$

Typical densities are given in Table 9.1. Because a material's volume changes with temperature, its density also depends on temperature. Unless indicated otherwise, the data in the table are for 20°C.

Because most materials expand as they are heated, densities usually decrease with increasing temperature. One notable exception is water in the range 0°C to 4.0°C. Water contracts with increasing temperature in this range only because ice molecules, and even molecules of liquid water at 0°C, exhibit a low-density ordered arrangement over short distances. The order breaks up with increasing temperature, and this allows the molecules to pack together more densely.

■ **TABLE 9.1**
**Densities**

| Material | Density $(kg/m^3)$* |
|---|---|
| Water (3.98°C) | 1,000 |
| Water | 998 |
| Air (0°C) | 1.29 |
| Air (20°C) | 1.20 |
| Aluminum | 2,700 |
| Bone | ~1,800 |
| Brass | 8,700 |
| Copper | 8,890 |
| Glass | 2,600 |
| Gold | 19,300 |
| Ice | 920 |
| Iron | 7,860 |
| Lead | 11,340 |
| Mercury (0°C) | 13,600 |
| Benzene | 879 |
| Gasoline | 680 |

* At 20°C unless specified otherwise.

### Example 9.1
A cube of uranium ($\rho_u = 18{,}680$ kg/m³) is 2 cm on each side. (*a*) Find its mass. (*b*) How large a cube of ice ($\rho_i = 920$ kg/cm³) has the same mass?

**Reasoning** (*a*) From the definition of density, $\rho = m/V$, we have that

$$m_u = \rho_u V_u = (18{,}680 \text{ kg/m}^3)(8.0 \times 10^{-6} \text{ m}^3) = 0.149 \text{ kg}$$

(*b*) Again from the definition,

$$V_i = \frac{m_i}{\rho_i} = \frac{0.149 \text{ kg}}{920 \text{ kg/m}^3} = 162 \times 10^{-6} \text{ m}^3$$

Taking the cube root of this gives the side of the ice cube to be 5.45 cm. ■

*Exercise* By what percentage does the volume change as ice melts to form liquid water ($\rho = 1000$ kg/m³)? Figure 9.4 shows the effect of the reverse process, water expanding as it freezes to ice. *Answer: — 8.0%* ∎

**FIGURE 9.4**

Because water expands as it freezes, the frozen beverage has forced off the bottle cap and broken the bottle.

## 9.3 Hooke's law; modulus

Objects such as a rubber band or a spring possess a property called *elasticity:* when stretched and allowed to retract, they return to their original length. Robert Hooke (1635 – 1703) investigated such behavior and formulated a rule, now known as **Hooke's law,** to describe his findings:

Hooke's law: When an object is stretched or otherwise distorted, the distortion is proportional to the distorting force that causes it.

For example, the spring of original length $L_o$ in Fig. 9.5 is stretched an amount $\Delta L$ by the applied force **F**. Hooke found that, provided the spring is not stretched too far, it stretches twice as far under twice the force. In general,

$$\Delta L \propto F$$

However, if the spring is stretched too much, beyond what is called its *elastic limit,* it deviates from this direct proportionality between $\Delta L$ and $F$. Furthermore, the spring does not retract to its original length when the force is removed.

When the spring in Fig. 9.5 is replaced by a solid rod, the rod also obeys Hooke's law. Although the rod's elongation is far smaller than that of the spring, the rod nonetheless stretches in conformity with Hooke's law at low elongations. The behavior observed in a typical experiment is shown in Fig. 9.6. Hooke's law is obeyed in the elastic region only. In the following discussion, we assume that the forces and elongations are small enough that the material is not distorted beyond its elastic limit.

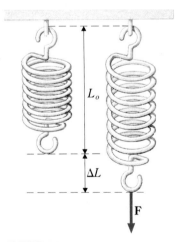

**FIGURE 9.5**

For a spring that obeys Hooke's law, the deformation $\Delta L$ varies in proportion to the applied force **F**.

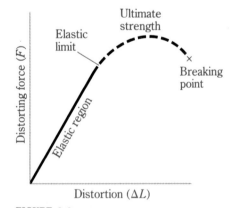

**FIGURE 9.6**

A typical stress-strain curve. Hooke's law applies in the linear (elastic) region. The maximum force that the rod can sustain is its ultimate strength. Often the material yields (softens) somewhat before breaking.

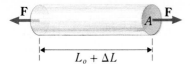

**FIGURE 9.7**

The stress is $F/A$, and the strain is $\Delta L/L_o$.

It is obvious that the elastic properties of rubber are very different from those of steel. To use Hooke's law in describing the elastic properties of solids, we use the terms *stress* and *strain*. Let us define these two quantities in terms of the stretching (or *tensile*) experiment of Fig. 9.7. A tensile (stretching) force **F** pulls on the end area $A$ of a rod of original length $L_o$ and causes the rod to stretch an amount $\Delta L$. We define the stress due to **F** as

$$\text{Stress} = \frac{\text{force}}{\text{area}} = \frac{F}{A} \tag{9.3}$$

The SI units of stress are newtons per square meter, the same as the units for pressure (Sec. 6.7). As we did for pressure, we call this combined unit the pascal (Pa): $1 \text{ Pa} = 1 \text{ N/m}^2$.

We define the strain in the rod of Fig. 9.7 to be

$$\text{Strain} = \frac{\text{elongation}}{\text{original length}} = \frac{\Delta L}{L_o} \tag{9.4}$$

Strain is defined as $\Delta L/L_o$, rather than just $\Delta L$, because a rod stretches in proportion to its original length. By dividing $\Delta L$ by $L_o$, we eliminate the effect of the length of the rod, an effect that tells us nothing about the material the rod is made of.

Because strain is the ratio of two lengths, it has no units. We see later in this section that there are many types of strain, depending on the geometry of the situation. In the present case, we are speaking about tensile strain. If the rod were being compressed lengthwise, the strain would still be the ratio of length change to the original length.

We are now in a position to restate Hooke's law. Stress is a measure of distorting force, and strain is a measure of distortion. Thus the law stating that a distorting force causes a proportional distortion, **Hooke's law,** becomes

$$\text{Stress} = (\text{constant})(\text{strain}) \tag{9.5}$$

In this form, the law may be applied to many situations other than the stretching of a rod. Hooke's experiments proved its applicability to the stretching, bending, and twisting of numerous springs and other objects. Of course, as was pointed out earlier, Hooke's law applies only in the elastic range of deformations.

The proportionality constant in Eq. 9.5 depends on the material and on the type of deformation involved. It is called the *elastic modulus* of the material. Thus, by definition,

$$\text{Modulus} = \frac{\text{stress}}{\text{strain}} \tag{9.6}$$

Because strain has no units, modulus has the units of stress. Notice that the modulus is large if a large stress produces only a small strain. Hence modulus is a measure of a material's rigidity. There are several different types of moduli, depending on the exact way in which the material is being stretched, bent, or otherwise distorted. We now discuss the most common of these moduli.

### Young's modulus

This modulus is used to describe situations such as the stretching of the rod in Fig. 9.7. Young's modulus would be of interest if you wished to compute how much a wire or rod would stretch under a tensile force. By definition,

$$\text{Young's modulus} = Y = \frac{F/A}{\Delta L/L_o} \tag{9.7}$$

Typical values of $Y$ are given in Table 9.2.

   In the elastic region, a bar compresses as much under a compressive force as it stretches under an equal stretching force. Therefore, Eq. 9.7 applies to compressions as well as elongations, provided the stress-strain ratio is always taken to be positive. However, compressed bars frequently buckle before the stress becomes very large.

### Shear modulus

Suppose we try to distort a cube of material in the manner shown in Fig. 9.8. A force **F** is applied parallel to the cube's top face, which has an area $A$. In this case, the stress is still $F/A$ and the strain is $\Delta L/L_o$, but observe how these symbols are defined in the figure. We have, for this situation in which the solid undergoes a deformation called *shear*,

$$\text{Shear modulus} = S = \frac{F/A}{\Delta L/L_o} \tag{9.8}$$

Although this equation is identical to Eq. 9.7 for Young's modulus, notice that the symbols are defined differently in the two cases. Typical values for $S$ are given in Table 9.2.

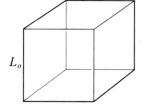

$L_o$

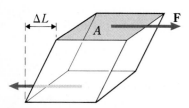

$\Delta L$    $A$    **F**

**FIGURE 9.8**
Here $\Delta L$ is exaggerated so that it can be seen. The shear modulus is given by $(F/A)/(\Delta L/L_o)$.

■ **TABLE 9.2    Approximate elastic properties**

| Material | Young's modulus (GPa) | Shear modulus (GPa) | Bulk modulus (GPa) | Elastic limit (GPa) | Tensile strength (GPa) |
|---|---|---|---|---|---|
| Aluminum | 70 | 23 | 70 | 0.13 | 0.14 |
| Brass | 90 | 36 | 60 | 0.35 | 0.45 |
| Copper | 110 | 42 | 140 | 0.16 | |
| Glass | 55 | 23 | 37 | | |
| Iron (wrought) | 90 | 70 | 100 | 0.17 | 0.32 |
| Lead (rolled) | 16 | 6 | 8 | | 0.02 |
| Polystyrene | 1.4 | 0.5 | 5 | | 0.05 |
| Rubber | 0.004 | 0.001 | 3 | | 0.03 |
| Steel | 200 | 80 | 160 | 0.24 | 0.48 |
| Tungsten | 350 | 120 | 20 | | 0.41 |
| Benzene | | | 1.0 | | |
| Mercury | | | 28 | | |
| Water | | | 2.2 | | |
| Air | | | $1 \times 10^{-4}$ | | |

1 GPa $= 10^9$ Pa and 1 Pa $= 1$ N/m². Hence $Y$ for aluminum is $70 \times 10^9$ N/m².

Because fluids flow when they are subjected to a shearing stress, we can state the following: *a fluid that is at rest, and remains so, is not subject to shear stress.* Therefore, in a fluid that remains at rest, shear forces of the type shown in Fig. 9.8 do not exist.

The reciprocals of the shear and Young's (tensile) moduli are frequently used. They measure how easily a material deforms rather than how difficult it is to deform it. They are given the descriptive names *shear compliance* and *tensile compliance.*

### Bulk modulus

Suppose a block of volume $V_o$ is subjected to a pressure increase $\Delta P$ on all sides (Fig. 9.9). The cube's volume change $\Delta V$ is a negative number since the volume shrinks. In this case, the strain is defined to be $-\Delta V/V_o$, and the stress, $F/A$, is the applied pressure increase $\Delta P$. As with the other types of moduli, bulk modulus $B$ is defined to be the ratio of stress to strain:*

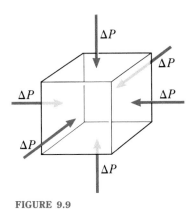

$$\text{Bulk modulus} = B = -\frac{\Delta P}{\Delta V/V_o} \qquad (9.9)$$

**FIGURE 9.9**
The cube, of original volume $V_o$, will contract by an amount $\Delta V$ under the action of the increased external pressure $\Delta P$.

### Bulk compressibility

The compressibility $k$ of a material is a measure of how easily the material is compressed. In other words, compressibility is just the reciprocal of bulk modulus. Usually, the equation by which compressibility is defined is written

$$-\frac{\Delta V}{V_o} = k\,\Delta P$$

It has the same units as reciprocal pressure. Liquids have a much higher compressibility than do solids. This reflects the fact that the molecules in a liquid are fairly widely separated. Compression simply pushes them closer together.

### Example 9.2

In a large lecture hall, a pendulum is to be made by suspending a 40-kg ball from the end of a steel wire 15 m long. (*a*) What cross-sectional area should the wire have if the applied stress in it is to be only 10 percent of its breaking stress? (*b*) How far will the ball stretch the wire?

**Reasoning** (*a*) From Table 9.2, the tensile strength of steel is $0.48 \times 10^9$ N/m$^2$, 10 percent of which is $0.48 \times 10^8$ N/m$^2$. This is to be the stress in the wire. From the definition of stress, $F/A$,

$$A = \frac{F}{\text{stress}} = \frac{mg}{\text{stress}} = \frac{(40 \text{ kg})(9.8 \text{ m/s}^2)}{0.48 \times 10^8 \text{ N/m}^2} = 8.17 \times 10^{-6} \text{ m}^2$$

which implies that the wire should have a radius of about 1.6 mm. (*b*) Young's modulus for steel is (Table 9.2) $200 \times 10^9$ Pa. From the definitions of strain and Young's modulus,

---

* The negative sign accounts for the fact that $\Delta V$ is negative when $\Delta P$ is positive.

$$\text{Strain} = \frac{\Delta L}{L_o} = \frac{\text{stress}}{Y} = \frac{0.48 \times 10^8 \text{ N/m}^2}{200 \times 10^9 \text{ N/m}^2} = 2.4 \times 10^{-4}$$

Using $L_o = 15.0$ m, we solve and find $\Delta L = 3.60$ mm. ∎

*Exercise* How large a stress is needed to produce a 0.020 percent elongation in an aluminum wire? *Answer: $1.40 \times 10^7$ Pa* ∎

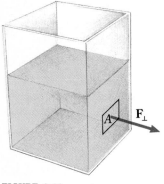

FIGURE 9.10

The average pressure on the area $A$ is $F_\perp / A$.

## 9.4 Pressure in fluids

Consider the fluid in the container of Fig. 9.10. It is at rest and exerts an outward force on the container wall. The outward force on the area $A$ is $\mathbf{F}_\perp$, where the subscript alerts you that the force is perpendicular to the wall. The average pressure on area $A$ is defined to be

$$\text{Average pressure} = \bar{P} = \frac{F_\perp}{A} \tag{9.10}$$

We often drop the subscript on $F_\perp$, and you are expected to know that the force is perpendicular to the area. As stated earlier, the SI unit for pressure is the pascal (Pa). Other frequently used units of pressure are listed on the inside cover of the text.

To measure the pressure within a fluid, the device shown in Fig. 9.11 could be used. The fluid exerts a force $\mathbf{F}$ on the piston, and the piston moves until the force exerted by the spring balances the force due to the fluid. When the device is suitably calibrated, the displacement of the piston can be used to measure $F$. If the area of the piston is $A$, then the pressure is simply $F/A$. By making the area of the piston very small, we can obtain the pressure very close to any point within the fluid. It is this quantity we refer to when we speak of the pressure at a point in the fluid. Even though force has direction, we define pressure to be a scalar quantity.

Let us now state and justify several important facts about pressure and forces within fluids.

**1** In a fluid at rest, the force exerted by the fluid on a surface placed in it is perpendicular to the surface.

Suppose this were not true. Then the fluid could exert a force tangential to the surface, but this is the type of force that occurs when shear is taking place in the fluid, as we saw in Fig. 9.8. Our fluid, assumed to be at rest, cannot give rise to shear forces, and so the force it exerts can only be perpendicular to the surface.

**2** The pressure on a surface is independent of the orientation of the surface.

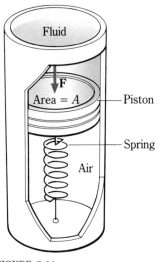

FIGURE 9.11

A simple device for measuring pressure.

To prove this, consider the triangular prism shown in cross section in Fig. 9.12. Suppose the prism is so tiny that it is little more than a point. Moreover, make it of a material that is identical in every way except rigidity to the fluid in which it is immersed. As a result, the prism remains at rest just as the fluid is at rest. The forces exerted by the fluid on the two upper sides of the prism are $F_1$ and $F_2$, perpendicular to the surfaces. Because the prism is at equilibrium, the horizontal forces that act on it must balance. Therefore $F_1 \sin \theta = F_2 \sin \theta$, and so $F_1 = F_2$. Since this means $P_1 A = P_2 A$, we conclude that $P_1 = P_2$. By placing the prism at arbitrary orienta-

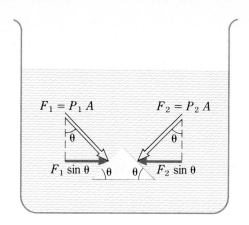

**FIGURE 9.12**

The triangular prism is much smaller than shown. Since it is in equilibrium, $F_1 \sin \theta = F_2 \sin \theta$.

tions and with arbitrary angle $\theta$, we conclude that, in the tiny pointlike region surrounding the prism, the pressure is the same no matter how the surfaces are oriented.

**3** In a fluid at rest, the pressure is constant along a horizontal plane.

We can prove this statement by referring to Fig. 9.13. Consider a small cubical volume element of the fluid similar to, but much smaller than, the volume $V_2$. Since the liquid in $V_2$ does not move, we know that the sum of the horizontal forces on it must be zero. The force toward the right on it is exerted by $V_1$ and is equal to the pressure in the gap at the left side of $V_2$ (call it $P_1$) multiplied by the side area of the cube (call it $A$). Thus

$$F \text{ to right on } V_2 = P_1 A$$

Similarly, letting $P_2$ be the pressure in the gap at the right of $V_2$, we have

$$F \text{ to left on } V_2 = P_2 A$$

Since the forces must balance, we see that $P_1 = P_2$.

Using the same reasoning, we can repeat this process over and over again to show that $P_1 = P_2 = P_3 = P_4$ and so on. This means that the pressure on a horizontal plane through $V_2$ must be the same everywhere. Notice that the pressure is independent of the shape of the container. Our only restriction is that the volume elements of the liquid be in equilibrium.

**4** A liquid at equilibrium in a series of open connected containers has its open surfaces at the same level (Fig. 9.14).

This follows directly from statement 3. The top of the liquid is at atmospheric pressure. To have the top higher at one point than at another would contradict statement 3. Why?

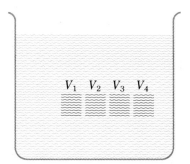

**FIGURE 9.13**

For a fluid at rest, the pressure is the same at all points on any given horizontal plane.

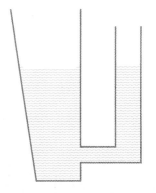

**FIGURE 9.14**

A liquid at equilibrium in open connected containers has its open surfaces at the same level.

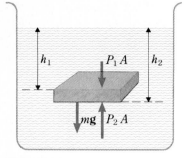

FIGURE 9.15

The force $P_2A$ balances the combined force $P_1A + mg$.

**5** The pressure difference between two levels in a fluid of density $\rho$ is

$$P_2 - P_1 = \rho g(h_2 - h_1) \tag{9.11a}$$

where $h_2$ and $h_1$ are the depths of the levels beneath the surface.

To prove this statement, refer to Fig. 9.15. The fluid in the container is at rest and remains so. Hence all the forces that act on the block of fluid shown in color must balance. Three forces act on it. If we call the top and bottom surface areas of the block $A$, these forces are (1) $P_1A$ down on the top, (2) $P_2A$ up on the bottom, and (3) $mg$ downward, the weight of the fluid in the block. The fluid is at equilibrium, and so $\Sigma F_y = 0$ for the block:

$$P_2A - P_1A - mg = 0$$

The volume of the block is $(h_2 - h_1)(A)$, and so its mass is (from $\rho = m/V$)

$$m = \rho V = \rho(h_2 - h_1)(A)$$

We substitute this in the pressure equation and find, after canceling $A$ from each term,

$$P_2 - P_1 = \rho g(h_2 - h_1) \tag{9.11a}$$

and so the proof is complete.

If we take the top of the block to be at the surface of the fluid, then $P_1$ is just the pressure of the atmosphere $P_a$. Further, $h_2 - h_1$ is then simply the depth $h$ of the lower surface of the block. Hence we find that the pressure at a depth $h$ in a fluid is

$$P = \rho g h + P_a \tag{9.11b}$$

where $\rho g h$ is the portion of the pressure that is due to the fluid and $P_a$ is the additional pressure due to the atmosphere above it (normally about $1 \times 10^5$ Pa).

In technical work, one often hears the terms **gauge pressure** and **absolute pressure**. The absolute pressure is the total pressure given by Eq. 9.11b. Many gauges, however, read only the difference between total pressure and $P_a$; that is, they read the quantity $\rho g h$ in Eq. 9.11b. This is what is called gauge pressure.

We next state what is called **Pascal's principle:**

**6** If an external force changes the equilibrium pressure of any point in a confined fluid, every point within the fluid experiences the same pressure change.

This fact was proved when we deduced Eq. 9.11b. Our derivation applied to any container shape and any external source of pressure, not just $P_a$. As Eq. 9.11b tells us, any change in the external pressure $P_a$ leads to a like change in $P$ everywhere in the fluid.

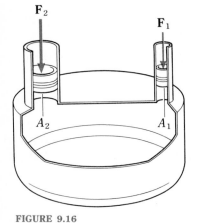

FIGURE 9.16

A small force on the small piston can balance a large force on the large piston.

---

*Example 9.3*

The apparatus shown in Fig. 9.16 is one version of a hydraulic press. When a force $F_1$ is applied to the piston of area $A_1$, how large a force on the other piston (of area $A_2$) is needed to balance it?

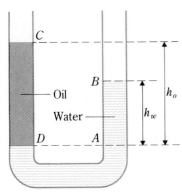

**FIGURE 9.17**

Because the water column $BA$ balances the oil column $CD$, the density of the oil can be determined.

**Reasoning** The force $F_1$ causes a pressure increase $\Delta P = F_1/A_1$. The same pressure increase occurs at piston 2, according to Pascal's principle. This furnishes an extra force

$$F_2 = A_2 \Delta P = F_1 \frac{A_2}{A_1}$$

on piston 2. This is the magnitude of the force we were asked to find. Notice that this device is a very effective force multiplier. Industrial presses often have $A_2/A_1$ ratios of thousands, and so tremendously large forces can be generated. ∎

---

**Example 9.4**

Water and oil are placed in the two arms of the glass U tube of Fig. 9.17. If they come to rest as shown, what is the density of the oil? The temperature is 20°C.

**Reasoning** Consider first the water below points $D$ and $A$. If the pressures at $D$ and $A$ were not equal, the water would flow. Since it does not, we conclude that the pressure at $D$ equals the pressure at $A$. These two pressures are, from Eq. 9.11b, $P_D = \rho_o g h_o + P_a$ and $P_A = \rho_w g h_w + P_a$. Equating, we find

$$\rho_o = \rho_w \frac{h_w}{h_o}$$

Because $\rho_w = 1000$ kg/m³, we can evaluate $\rho_o$ once $h_w$ and $h_o$ are measured.

As a point of interest, notice that the surfaces of the two liquids ($C$ and $B$) are not at the same level. Moreover, the pressure at a certain depth in one fluid is not equal to the pressure at an equal depth in the other fluid. The reasoning we used above for a single liquid breaks down in the case of two or more liquids. Can you point out where the reasoning fails in this latter case? ∎

---

**Exercise** If the oil has a density of 800 kg/m³, how long a column of water is needed to balance an oil column 30 cm long? *Answer: 24.0 cm* ∎

---

## 9.5 Archimedes' principle; buoyancy

The experiment shown in Fig. 9.18 is probably not new to you. It demonstrates the well-known fact that objects appear to weigh less when they are submerged in a fluid. If you have ever tried to support a person while you were swimming, you know full well that the supporting force required is far less than the weight of the person. Similarly, in Fig. 9.18, the supporting force $T$ is decreased when the object is placed in water. Apparently, the water exerts an upward force $F_B$ on the object. We call this the **buoyant force.**

The law of fluids that describes the buoyant force is called **Archimedes' principle.** To arrive at it, consider the object in Fig. 9.19. It is buoyed up by the fluid around it. Apparently, the net effect of the fluid forces acting on the object is to give it an upward force $F_B$. Basically $F_B$ is a result of the fact that pressure increases with depth, so that the upward force on the bottom of the object is larger than the downward force on its top.

To see how large the buoyant force is, notice what would happen if the object were made of the same material as the fluid. We would not be able to distinguish the object from the fluid, and so it would simply remain at rest with no outside support-

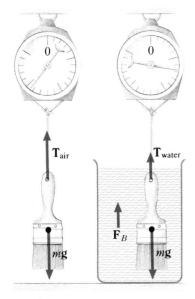

**FIGURE 9.18**

The fluid exerts an upward buoyant force $F_B$ on the object. The scale reads $T_{air}$ when the object is in air and $T_{water}$ when it is in water.

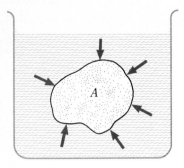

ing force required. This means that $F_B$ is just large enough to support the object in this case. That is, $F_B = mg$, where $mg$ is the weight of the fluid object.

Of course, the buoyant force due to the fluid cannot depend on what the object is made of. Hence $F_B$ is always the same, equal to the weight of the volume of fluid displaced by the object. We are therefore led to **Archimedes' principle:**

A body partially or wholly immersed in a fluid is buoyed up by a force equal to the weight of the fluid that it displaces.

You should go back through this reasoning to assure yourself that we made no use of the fact that the object in Fig. 9.19 is *wholly* immersed.

What does Archimedes' principle tell us about the buoyant force on the object?

**Example 9.5**

Suppose the object in Fig. 9.18 has a mass $M$ and a density $\rho$. Find its apparent weight (the scale reading $W_{app}$) when it is submerged in a fluid of density $\rho_f$.

**Reasoning** We know that the apparent weight is the object's actual weight decreased by the buoyant force $F_B$.

$$W_{app} = Mg - F_B$$

To find $F_B$, we first find the volume of the object. It is, from $\rho = m/V$,

$$V = \frac{M}{\rho}$$

This is also the volume of the displaced fluid. The mass of the displaced fluid is $\rho_f V$, and so

$$\text{Mass of displaced fluid} = M\frac{\rho_f}{\rho}$$

from which

$$F_B = \text{weight of displaced fluid} = Mg\frac{\rho_f}{\rho}$$

The apparent weight of the object is therefore

$$W_{app} = Mg - Mg\frac{\rho_f}{\rho} = W\left(1 - \frac{\rho_f}{\rho}\right) \qquad (9.12)$$

As a check, we notice that $W_{app} = 0$ when $\rho_f = \rho$; in other words, the object has an apparent weight of zero when its density is equal to the density of the fluid. ∎

**Exercise** Find the apparent weight of the object when only a fraction $k$ of its volume is submerged. *Answer:* $W(1 - k\rho_f/\rho)$ ∎

**Example 9.6**

A queen's gold crown has a mass of 1.30 kg. However, when it is weighed while it is completely immersed in water, its apparent mass is 1.14 kg. Is the crown solid gold?

**Reasoning** We easily compute $W$ for the crown:

$$W = mg = (1.30 \text{ kg})(9.8 \text{ m/s}^2)$$

Also, the apparent weight is

$$W_{\text{app}} = (1.14 \text{ kg})(9.8 \text{ m/s}^2)$$

Substituting in Eq. 9.12 with $\rho_f = 1000 \text{ kg/m}^3$ gives

$$\rho = 8120 \text{ kg/m}^3$$

From Table 9.1, the density of gold is 19,300 kg/m³. The crown's density is less than half this, and so it is either hollow or made of something that only looks like gold. ∎

## 9.6 Viscosity

Thick syrup and honey are examples of very viscous fluids. They flow quite slowly when poured. Water and alcohol, however, are much less viscous. We characterize the viscous nature of fluids by their *viscosity*. Very viscous, slow-flowing fluids have large viscosities. To give quantitative meaning to viscosity, we refer to the shearing experiment of Fig. 9.20. Two parallel plates, each of area $A$, are separated by a distance $L$, and the region between the plates is filled with a fluid whose viscosity we shall denote by $\eta$ (Greek eta). The force $F$ causes the top plate to move past the lower one with a speed $v$. We characterize the rapidity of the shearing motion by the *shear rate* of the two plates and the fluid beween them:

$$\text{Shear rate} = \frac{\text{speed of top plate past bottom plate}}{\text{distance between plates}} = \frac{v}{L}$$

Thus a shear stress $F/A$ applied to the upper plate causes a shear rate $v/L$ in the fluid.

We define the viscosity $\eta$ of the fluid as the shear stress required to produce unit shear rate:

$$\eta = \text{viscosity} = \frac{\text{shear stress}}{\text{shear rate}} \tag{9.13a}$$

As you see, a highly viscous fluid is one that requires a large shear stress to cause it to flow with a given shear rate.

In terms of the experiment of Fig. 9.20, we have stress $= F/A$ and rate $= v/L$. Using these measured quantities, we can compute the viscosity of the fluid:

$$\eta = \frac{\text{shear stress}}{\text{shear rate}} = \frac{F/A}{v/L} \tag{9.13b}$$

We see from its defining equation that the SI unit for viscosity is the pascal · second (Pa · s). This unit is given the special name *poiseuille* (Pl). Other common units for viscosity are the poise (P), where 1 P = 0.1 Pl, and the centipoise (cP). This

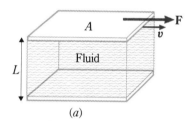

(a)

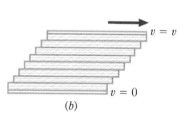

(b)

**FIGURE 9.20**

As the upper plate moves, layers of the fluid slide over one another. Viscous energy losses arise because of the friction forces that retard the motion of these layers.

**■ TABLE 9.3**
**Viscosities of liquids and gases at 30°C**

| Material | Viscosity (mPl)* |
|---|---|
| Air | 0.019 |
| Acetone | 0.295 |
| Methanol | 0.510 |
| Benzene | 0.564 |
| Water | 0.801 |
| Ethanol | 1.00 |
| Blood plasma | ~1.6 |
| SAE No. 10 oil | 200 |
| Glycerin | 629 |
| Glucose | $6.6 \times 10^{13}$ |

* 1 mPl = $10^{-3}$ Pa · s = 1 cP.

We can gain further insight into the meaning of viscosity by examining Fig. 9.20b. Notice that the fluid layers next to the two plates remain attached to the plates. We can think of the fluid beween the plates as consisting of many thin layers, many more than are shown. As the upper plate moves, these layers must slide over one another. In a high-viscosity fluid, the layers do not slide easily, and so a great amount of work must be done to shear the fluid. This friction-type work results in the production of thermal energy; it causes the fluid to become warmer. In industrial processes in which plastic objects are formed by huge shearing stresses, the plastic is often heated high above its initial temperature by this mechanism.

The flow of water and similar fluids through a pipe or tube is of particular interest, as we shall see. The volume $Q$ of fluid flowing through a pipe each second is called the *flow rate* in the pipe. For example, if 50 cm$^3$ of water flows out of the pipe in Fig. 9.21 each second, $Q = 50$ cm$^3$/s.

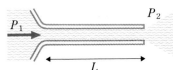

**FIGURE 9.21**
The flow rate through the tube is given by Poiseuille's law.

In Fig. 9.21, the pressures in the fluid at the two ends of the pipe are $P_1$ and $P_2$. We call $P_1 - P_2$ the *pressure differential,* and, as we might guess, the flow rate through the pipe is proportional to it for simple fluids. You would also expect the flow rate to be large for a pipe with a large radius $R$ and small length $L$. The equation for flow rate in a situation such as this was found by the French physician Jean Louis Marie Poiseuille (1799–1869). For flow rates that are not too large,

$$Q = \left( \frac{\pi R^4}{8\eta L} \right) (P_1 - P_2) \tag{9.14}$$

This is called **Poiseuille's law.** Notice that $Q$ increases as the fourth power of $R$, the pipe's radius.

---

### Example 9.7
Older people often develop blood-circulation difficulties because of deposits building up in their arteries. By what factor is the blood flow in an artery reduced if the artery's radius is cut in half? Assume the same pressure differential in the two cases.

*Reasoning* Poiseuille's law tells us that the volume of blood $Q$ flowing through the artery each second is related to $R$ by

$$Q \propto R^4$$

In the original artery, $Q_o = (\text{constant})(R_o^4)$, but in the constricted artery, $Q = (\text{constant})(R_o/2)^4$. Taking the ratio $Q/Q_o$, we find $Q/Q_o = 1/16$. The flow rate is reduced by a factor of 16. It is clear from this strong dependence of $Q$ on $R$ why blood-circulation difficulties result from arterial deposits. ■

---

*Exercise* Find the flow rate of water through a 20-cm-long capillary tube that has a diameter of 0.15 cm if the pressure differential across the tube is $4.0 \times 10^3$ Pa. Use 0.801 mPl for the viscosity of water. *Answer: 3.10 cm$^3$/s* ■

## 9.7 Bernoulli's equation

As we have seen, all liquids have a characteristic viscosity. If the viscosity is large, a great deal of work is needed to push the liquid through a pipe. Viscous friction forces between the layers of the liquid as they flow past each other cause the liquid to heat up. Many liquids, however, have such small viscosity that friction energy losses are negligible, at least for certain purposes. When this is the case, an important relationship called *Bernoulli's equation* can be found for the pressure in a moving fluid. It was first published by Daniel Bernoulli in 1738.

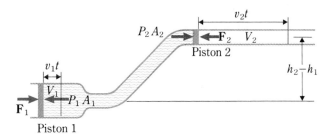

**FIGURE 9.22**

The work done by $F_1$ must equal the work done against $F_2$ plus the sum of the changes in the gravitational potential energy and kinetic energy of the fluid.

Consider the pipe system shown in Fig. 9.22. It is completely filled with an incompressible fluid between the two frictionless pistons. We shall say that piston 1 is being pushed to the right with constant speed $v_1$ and that piston 2 is moving to the right with speed $v_2$. The force $F_1$ on piston 1 is balanced by the force $P_1 A_1$ that results from the pressure of the fluid, where $A_1$ is the area of piston 1. (The forces on the piston must balance, or else it would be accelerating, and we have already specified that it is moving with constant speed.) Similarly, at piston 2, $F_2 = P_2 A_2$. In a time $t$, piston 1 moves a distance $v_1 t$, thereby displacing a volume of fluid $(v_1 t)(A_1)$. Since the fluid is incompressible, however, piston 2 must make way for an equal volume of fluid. Hence

$$(v_1 t)(A_1) = (v_2 t)(A_2)$$

Bernoulli asked what happens to the work done by piston 1. This work is just $F_1(v_1 t)$, and since $F_1 = P_1 A_1$, we have

Input work $= P_1 A_1 v_1 t$

Because piston 2 does an amount of work $F_2(v_2 t)$, some of the input work is used there.

In addition, the fluid pressed to the right by piston 1 is essentially transferred to the upper tube. As a consequence, that fluid (with mass $M$ and volume $A_1 v_1 t$) is given some gravitational potential energy. Moreover, since it is now traveling with a different speed $v_2$, its kinetic energy is changed. Of course, some energy is transformed to thermal energy by viscosity-related friction heating, but we are assuming this to be small. We therefore have the following equation that tells us what happened to the input energy:

Input work $=$ output work $+$ change in GPE $+$ change in KE

or, using the symbols of Fig. 9.22,

$$P_1 A_1 v_1 t = P_2 A_2 v_2 t + Mg(h_2 - h_1) + \tfrac{1}{2} Mv_2^2 - \tfrac{1}{2} Mv_1^2$$

where the volume of the fluid involved, $A_1 v_1 t$, has a mass $M$. From the definition of density,

$$M = \rho A_1 v_1 t = \rho A_2 v_2 t$$

Substituting this in the above equation gives, after rearrangement,

$$P_1 + \tfrac{1}{2}\rho v_1^2 + \rho g h_1 = P_2 + \tfrac{1}{2}\rho v_2^2 + \rho g h_2 \qquad (9.15)$$

This is **Bernoulli's equation.** Because the pistons need not really be present, points 1 and 2 can be any two points in the fluid. All that are needed are surfaces in the fluid; these surfaces can be imaginary, and the computation remains the same. Notice, however, that the equation is applicable only if friction forces can be neglected.

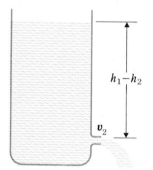

$h_1 - h_2$

$v_2$

**FIGURE 9.23**
Torricelli's theorem tells us how fast the liquid is moving as it flows out of the spigot.

### Example 9.8
A simple application of Bernoulli's equation is shown in Fig. 9.23. Suppose a large tank of water has a small spigot on it. Find the speed with which water flows from the spigot.

*Reasoning*   Since the spigot is so small, the efflux speed $v_2$ is much larger than the speed $v_1$ with which the top surface of the water drops. We can therefore approximate $v_1$ as zero. Bernoulli's equation can then be written

$$P_1 + \rho g h_1 = P_2 + \tfrac{1}{2}\rho v_2^2 + \rho g h_2$$

Because $P_1$ and $P_2$ are both nearly equal to atmospheric pressure, we can consider them equal. Solving for $v_2$ gives

$$v_2 = \sqrt{2g(h_1 - h_2)}$$

This is **Torricelli's theorem.** Notice that the efflux speed is the same that a ball falling through a height $h_1 - h_2$ would have. This points up the fact that when water flows from the spigot, it is as though the same amount of water had been taken from the top of the tank and dropped to the spigot level. The top level of the tank is a little lower, and the gravitational potential energy lost has gone to kinetic energy in the efflux water. If the spigot had been pointed upward, this kinetic energy would allow the ejected water to rise to the height of the top of the water in the tank before stopping. In practice, viscous energy losses would alter the result somewhat. ∎

*Exercise*   What is $v_2$ if the tank is closed at the top and the air pressure in it is $kP_t$, where $k$ is a constant?   *Answer:* $\sqrt{2g(h_1 - h_2) + 2(k-1)(P_t)/\rho}$ ∎

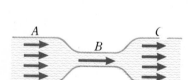

*A*   *B*   *C*

**FIGURE 9.24**
Since the fluid velocity is greatest at $B$, the pressure is lowest at that point.

### Example 9.9
Suppose water flows through the pipe system of Fig. 9.24. Because the same amount of water must flow past points $A$, $B$, and $C$ each second, the speed of the water in the narrow pipe at $B$ must be larger than the speed at $A$ and $C$. Assuming the flow speed at $A$ and $C$ to be 0.20 m/s and that at $B$ to be 2.0 m/s, compare the pressure at $B$ with that at $A$.

*Reasoning* Applying Bernoulli's equation and noting that the average gravitational potential energy is the same at both places, we have

$$P_A + \tfrac{1}{2}\rho v_A^2 = P_B + \tfrac{1}{2}\rho v_B^2$$

Substituting $v_A = 0.20$ m/s, $v_B = 2.0$ m/s, and $\rho = 1000$ kg/m³ gives $P_A - P_B = 1980$ Pa. Hence *the fluid pressure within the constriction is much less than that in the large pipes on either side.* This is probably opposite what you would guess at first. However, it is true and has wide application. Aspirators, for example, obtain a partial vacuum by forcing water through a constriction where the pressure is greatly reduced.

We can see qualitatively that the pressure at $A$ must be larger than that at $B$. Because each little volume of fluid is accelerated as it moves from $A$ to $B$, an unbalanced force to the right must act on it. To supply this unbalanced force, the pressure must decrease as one goes from $A$ to $B$. You should be able to reverse this line of reasoning to show that the pressure at $C$ is larger than that at $B$.

This result — that *pressure is low where speed is high* — provides an explanation of such diverse facts as the lift on an airplane wing and the curve ball pitched by a ballplayer. The flow around an airplane wing is shown in Fig. 9.25. Because the air has to travel farther over the top of the wing than under it, the air is moving faster on the upper side than on the lower. The pressure is therefore lower at the top, and the wing is forced upward. This effect is also used on race cars; winglike fins are used to produce a downward force, thereby increasing the normal force and hence the friction force between tires and track. This allows the car to travel around curves faster than would otherwise be possible. ■

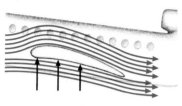

FIGURE 9.25

The airplane wing experiences a force that is directed from the region beneath the wing, where the air has low velocity and high pressure, to the high-velocity (low-pressure) region above the wing.

## 9.8 Laminar versus turbulent flow

Let us examine how a fluid flows through a pipe. Friction forces exerted on the fluid by the pipe walls tend to restrain the flow, as do the viscous forces within the fluid. As a result, the fluid close to the walls flows more slowly than that near the center of the pipe. We show this effect in Fig. 9.26$a$, where the lengths of the arrows indicate the magnitude of the velocity at various positions in the pipe. In Examples 9.8 and 9.9, $v$ was the average velocity on the pipe cross section.

Another feature of flow through a pipe is shown in Fig. 9.26$b$. Suppose a tiny speck of dust, like the one at point $A$, is flowing with the fluid. If the flow rate is low, the speck follows the line shown as it moves through the pipe. Other specks, and the fluid as well, follow similar smooth lines. We call these flow lines **streamlines,** and this is called **streamline,** or **laminar, flow.** In laminar flow, each element of the fluid follows a repeatable streamline.

If the speed of the fluid becomes high enough, the flow lines begin to behave erratically. At any instant, the flow lines may look like those in Fig. 9.26$c$. An instant later, the lines will take another form. This situation, in which the flow lines are contorted and vary widely with time, is called **turbulent flow.**

As you might guess, friction (or viscous) energy losses are nearly always larger in turbulent flow than in laminar flow. Turbulence causes rapid, chaotic motion that in turn increases distances moved and friction losses. Because of this, turbulence is to be avoided if friction losses are to be minimized. Automakers wish to minimize turbulent air flow around their cars, for example (Fig. 9.27). A means for predicting when turbulence occurs has obvious practical importance.

Although turbulence is very difficult to treat mathematically, there is a unifying

($a$) Fluid velocity

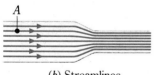

($b$) Streamlines

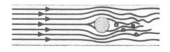

($c$) Turbulent flow

FIGURE 9.26

Examples of various features of flow in a pipe: ($a$) velocity profile; ($b$) laminar flow; ($c$) turbulent flow.

**FIGURE 9.27**
A wind tunnel test.

concept that simplifies the situation. Experiment shows that flow changes from laminar to turbulent when a critical value is reached for what is called the **Reynolds number** $N_R$, a dimensionless constant given by

$$N_R = \frac{\rho v D}{\eta} \qquad (9.16)$$

for a fluid with density $\rho$, viscosity $\eta$, and speed $v$ flowing through a pipe of diameter $D$. If $N_R$ exceeds about 2000, the flow usually becomes turbulent. This is not a precise rule because careful design can postpone the onset of turbulence. Reynolds numbers of 40,000 have been achieved for special laminar flow systems. (Equation 9.16 is also applicable to a sphere of diameter $D$ moving through a fluid. In that case, however, the critical value for $N_R$ is about 10.)

Despite its lack of precision, the critical value of 2000 is very useful, as is the Reynolds number itself. For example, two systems, one of which is a scale model of another, give rise to similar flow if $N_R$ is the same for both. Such systems are said to be *dynamically similar*. This concept forms the basis for small-scale wind tunnel tests of flow patterns around cars and planes. The flows are similar if $v$ is increased by the same factor by which $D$ is decreased because $N_R$ remains unchanged.

---

**Example 9.10**
How fast can a raindrop having a diameter of 3.0 mm fall before the flow of air around it becomes turbulent?

**Reasoning**  We use $N_R = 10 = \rho v_c D/\eta$, where $v_c$ is the critical speed we seek. Using $D = 3.0 \times 10^{-3}$ m, $\rho = 1.29$ kg/m³, and $\eta = 1.9 \times 10^{-5}$ Pl gives $v_c = 0.0491$ m/s. Notice how low this speed is. How far would a freely falling object have to fall before reaching this speed?  ■

---

**Exercise**  About what quantity of water can flow each second through a 2.0-cm-diameter pipe before turbulent flow sets in?  *Answer: 100 cm³*  ■

---

> ### 9.9 Viscous drag; Stokes' law

---

Because liquids and gases have nonzero viscosity, a force is required if an object is to be moved through them. Even the small viscosity of the air causes a large retarding force on an automobile traveling at high speed. If you stick your hand out the window of a fast-moving car, you easily recognize that considerable force must be exerted on your hand to move it through the air. You can probably think of several other examples of the following fact:

An object moving through a fluid experiences a retarding force called a *drag force*. The drag force increases with increasing speed of the object.

The exact value of the drag force is difficult to calculate, even in the simplest cases. Indeed, automobile companies devote a great amount of money to evaluating the drag forces on various car designs. Invariably, the force must be found experimentally by using models in a wind tunnel. However, the costly effort is fully justified by the increased efficiency of models that have the smallest drag forces.

The drag force a fluid exerts on a sphere moving through it was first computed by G. Stokes. Suppose a sphere of radius $r$ is being pulled with speed $v$ through a fluid of viscosity $\eta$. If the flow around the sphere is laminar, then the drag force $F_D$ on the sphere is given by **Stokes' law:**

$$F_D = 6\pi\eta rv \tag{9.17}$$

Other equations apply to objects of different shape. As long as the flow is laminar, however, the drag force is proportional to the speed of the object.

At speeds high enough to induce turbulence, the drag force is not simply proportional to speed. Instead, it becomes more complicated and varies as a power series in $v$. Usually, except in the simplest geometries, the drag force relation is found by experiment. Typical of these relations is that for an automobile traveling at moderate speeds. It is found that, for this case, the drag force $F_D$ is proportional to $v^2$:

$$F_D = \tfrac{1}{2}\rho C_d A_f v^2$$

where $\rho$ is the density of the air and $A_f$ is an equivalent frontal area for the car, about 2 to 3 m². The dimensionless constant $C_d$, called the *drag coefficient,* is about 0.35 for a very streamlined car and about 0.7 for a truck.*

We might also mention that, in the case of an automobile, rolling friction at the tires is important at low speeds. For speeds less than about 20 m/s, the retarding force due to this cause is not negligible. Experiment shows that this force, call it $F_r$, is given approximately by

$$F_r = \frac{M}{1000}(110 + 1.1v)$$

where $M$ is the mass of the car and all units are SI units. For a typical car, $F_r = F_D$ at a speed of about 15 m/s.

---

### Example 9.11

What horsepower is required to propel a car at 20 m/s on level pavement? Assume that $C_d = 0.45$, $A_f = 2.0$ m², and $M = 1000$ kg ($\rho_{air} = 1.29$ kg/m³).

*Reasoning*   The retarding force is

---

* You can determine the drag force for a car by means of simple experiments. See J. E. Farr, *The Physics Teacher,* **22:**320 (1983).

$$F = F_D + F_r = \tfrac{1}{2}\rho C_d A_f v^2 + \left(\frac{M}{1000}\right)(110 + 1.1v)$$

$$= 234\ \text{N} + 132\ \text{N} = 366\ \text{N}$$

The power expended against the force is

$$\text{Power} = Fv = (366\ \text{N})(20\ \text{m/s}) = 7320\ \text{W} = 9.8\ \text{hp}$$

Already at this speed, the term relating to air resistance is dominant. Because it increases as the square of $v$ and because $P = Fv$, the required horsepower varies approximately as $v^3$ at these speeds. Hence at twice this speed, the required power is about 80 hp. ■

## 9.10 Terminal velocity

There is a maximum speed a car can attain on a flat, straight roadway. The air friction forces that retard the car's motion increase with speed, as we saw in Sec. 9.9. When the car is traveling at a high enough speed, the friction force on it (the drag force) exactly balances the force driving it forward. Its acceleration is therefore zero, and it is traveling at the highest speed possible under these conditions. The car has reached its *terminal velocity.*

Terminal velocity is often used to describe the motion of falling objects (Fig. 9.28a). Gravity causes a downward force $mg$ on the object. An upward force (a drag force) is exerted by the friction effect of the air or other fluid in which the object is falling. Thus $F = ma$ becomes, for this situation,

$$mg - F_D = ma$$

$$a = g - \frac{F_D}{m}$$

Because the drag force increases with speed, the acceleration becomes zero when the speed reaches a high enough value. This maximum velocity of free fall is the object's terminal velocity. The approximate variation of speed with time for a falling object is shown in Fig. 9.28b. Let us now study a falling sphere when flow is not turbulent.

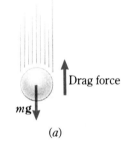

Drag force

$mg$

(a)

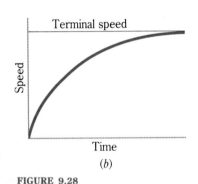

Terminal speed

Speed

Time

(b)

**FIGURE 9.28**
When the drag force equals $mg$, the terminal velocity has been reached.

### Example 9.12
Tiny particles in suspension in a liquid slowly fall through the liquid with a terminal velocity. This velocity is referred to as the *sedimentation rate.* Find the sedimentation rate for spherical particles of radius $b = 2.00 \times 10^{-3}$ cm as they fall through water at 20°C. The density of the material composing the particles is 1050 kg/m³, and the viscosity of water is 1.00 mPl.

**Reasoning** Three forces act on each particle: (1) its weight $m_p g$, (2) the buoyant force $F_B$, and (3) the drag force $F_D$. Application of $F = ma$ to a falling particle gives

$$\text{Weight} - \text{buoyant force} - \text{drag force} = m_p a$$

At the terminal velocity, $a = 0$. We then have

$$m_P g - F_B - 6\pi\eta b v_t = 0$$

The mass $m_P$ of the particle is given by $m = \rho V$ to be

$$m_P = \rho_P V_P = \rho_P(\tfrac{4}{3}\pi b^3)$$

The buoyant force is equal to the weight of the displaced liquid:

$$F_B = m_l g = (\rho_l V_P)(g) = \rho_l(\tfrac{4}{3}\pi b^3)(g)$$

Substitution of these values yields

$$\rho_P(\tfrac{4}{3}\pi b^3)(g) - \rho_l(\tfrac{4}{3}\pi b^3)(g) - 6\pi\eta b v_t = 0$$

$$v_t = \frac{2b^2 g}{9\eta}(\rho_P - \rho_l) \qquad\qquad (9.18)$$

Using the values indicated, we find $v_t = 4.36 \times 10^{-3}$ cm/s, a quite low velocity. Even so, a beaker containing this suspension would clarify in a few hours.

You can see from this example that sedimentation rates depend on the density difference between particle and liquid as well as on the size of the particle. Moreover, since the drag force depends upon the shape of the particle, shape also affects sedimentation rate. As a result of these dependencies, sedimentation rate measurements are often used in science, medicine, and industry to gain information concerning suspended particles. ∎

## 9.11 Blood-pressure measurement

One of the most important situations involving fluid flow is the flow of blood in our bodies. Because of the complex nature of the channels through which the blood flows, the detailed description of blood circulation is far from simple. Nevertheless, a very simple measurement of blood pressure is of great usefulness in medicine.

The pressure of the blood in your arteries and veins varies widely as a function of both time and position in your body. As the heart throbs, the blood pressure on the exit side alternately rises and falls. These pressure fluctuations persist throughout the arterial system. As the blood flows into smaller and smaller channels, however, friction effects and the elasticity of the channels tend to even out the flow pattern. By the time the blood flows into the veins for the return trip to the heart, the flow is almost uniform.

The wide pressure fluctuations in the arteries are important in several respects, although only two are mentioned here. (1) Extremely high maximum pressures can lead to rupture of the channel walls. Strokes are one evidence of such rupture. (2) The magnitudes of the pressure peaks and valleys during the heartbeat cycle provide information concerning constrictions in the channels and other factors that influence circulation.

Ordinary blood-pressure data give two numbers, the *systolic pressure* and the *diastolic pressure*. These are the pressure readings at the peak (systolic) and the low point (diastolic) of the blood flow cycle. To measure these values, doctors and nurses use a sphygmomanometer, as shown in Fig. 9.29.

An inflatable cuff is placed around the upper arm near heart level, and the pressure exerted by the inflated cuff is monitored with a mercury manometer. When

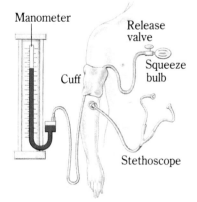

**FIGURE 9.29**
Apparatus used to measure blood pressure.

# The centrifuge

In scientific and technical work, it is frequently necessary to separate tiny particles from a solution even though the particles will not settle out by themselves. A centrifuge is used for this. The solution is whirled in a circle at high speed, and this causes the particles to separate from the solution. Let us see what scientific principles apply to this situation.

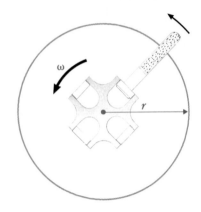

Consider a small particle of volume $V$ and density $\rho$ suspended in a fluid whose density is $\rho_o$. The mass of the particle is $m = \rho V$. Because its weight does not cause the particle to settle out, we are justified in neglecting gravity. Moreover, even when whirling in the centrifuge, the particle settles out very slowly. Hence we can approximate the situation by saying that the particle moves in a circle of radius $r$ with angular speed $\omega$. A centripetal force of $mv^2/r$, which is $m\omega^2 r$, is required to hold the particle in the circular path. Where does this force come from?

The particle is subject to a buoyant force $F_B$ due to the fluid in which it is immersed. In addition, the particle slowly settles out with a radial velocity $v$, the sedimentation velocity. It experiences a friction force due to this radial motion through the fluid, and we represent this force, as usual, by a constant times the sedimentation velocity, call it $kv$ in this case. Thus we see that the centripetal force $m\omega^2 r$ is supplied by the buoyant force and the friction force $kv$. We therefore can write

$$m\omega^2 r = F_B + kv$$

or

$$\rho V \omega^2 r = F_B + kv \qquad (a)$$

We can find $F_B$ easily by noticing the following fact. Suppose the particle is composed of exactly the same material that makes up the fluid. Since the fluid does not settle out of itself, $v$ is zero in that case. In addition, $m$ is simply $\rho_o V$, where $\rho_o$ is the density of the fluid. Therefore Eq. $(a)$ becomes, in this case,

$$\rho_o V \omega^2 r = F_B + 0$$
$$F_B = \rho_o V \omega^2 r$$

Since the buoyant force is supplied by the surrounding fluid, it is independent of the composition of the particle. Hence this value for $F_B$ applies to any particle of volume $V$. We can therefore write Eq. $(a)$ as

$$\rho V \omega^2 r = \rho_o V \omega^2 r + kv$$

which yields for the sedimentation velocity of the particle

$$v = \omega^2 \left( \frac{Vr}{k} \right) (\rho - \rho_o) \qquad (b)$$

Notice that the sedimentation velocity increases as the *square* of the angular velocity. Moreover, $v$ can be positive or negative, depending on whether the particle density is larger or smaller than the fluid density. Particles of very low density, such as air bubbles and the fat particles in whole milk, settle to the innermost end of the centrifuge tube. More dense particles, of course, settle to the outermost end of the tube.

air is pumped into the cuff, the pressure can be made to exceed the peak pressure in the artery. As a result, the flow of blood into the lower arm is cut off. A stethoscope (a listening device) placed on the artery below the cuff detects no sound because blood is not flowing through the artery.

The pressure in the cuff is slowly reduced by allowing the cuff to deflate. Suddenly, at the systolic pressure, the person taking the measurement begins to hear a pulse beat in the stethoscope. At that point, the pressure in the cuff is low enough for the blood to surge past it, through the artery, at the pressure peaks. This cuff pressure reading therefore gives the blood pressure at the peak of the heart-throb cycle. Actually, the sound heard is the result of turbulent flow past the constricted artery at the cuff.

As the cuff pressure is reduced further, the blood flow becomes less turbulent. The sound changes character, losing its sharpness, and eventually ceases. At that pressure, the diastolic pressure, the blood is able to flow past the cuff during all portions of the cycle. This reading therefore gives the lowest pressure during the pulse-beat cycle. The pressure gauge reads pressures in millimeters of mercury (mmHg), where 1 mmHg = 133 Pa. In a normal young person, the two pressures are about 120 and 80 mmHg (usually reported as 120/80). As a person ages, these pressures often change. Pressures above about 160 mmHg are abnormally high and usually require the attention of a physician.

## Minimum learning goals

When you finish this chapter, you should be able to

1 Define (*a*) fluid, (*b*) crystalline and amorphous solids, (*c*) density, (*d*) Hooke's law, (*e*) stress and strain, (*f*) modulus, (*g*) Young's modulus, (*h*) shear modulus, (*i*) bulk modulus, (*j*) pascal, (*k*) Pascal's principle, (*l*) buoyant force, (*m*) Archimedes' principle, (*n*) laminar and turbulent flow, (*o*) Bernoulli's equation, (*p*) drag force, (*q*) terminal velocity, (*r*) viscosity, (*s*) Reynolds number.

2 Use the definition of density in simple computations.

3 Use Hooke's law in its stress-strain form to compute the deformation of an elastic material in tension, shear, and bulk compression when the appropriate modulus is given.

4 Find force from pressure, and vice versa.

5 Given appropriate data, compute the absolute and gauge pressures at a given depth in a fluid.

6 Explain the theory of the hydraulic press.

7 Use Archimedes' principle to find the buoyant force on an object of known mass and density (or volume).

8 Identify each quantity in Poiseuille's equation and use it in simple calculations.

9 Use Bernoulli's equation to derive Torricelli's theorem and to show that pressure is least where flow rate is greatest.

10 Relate the drag force of an object to its terminal velocity in free fall.

## Questions and guesstimates

1 How could you determine the density of (*a*) a cubical block of metal, (*b*) a liquid, (*c*) an oddly shaped piece of rock?

2 How could you measure (*a*) the tensile modulus of the rubber in a rubber band? (*b*) The shear modulus of gelatin? (*c*) The bulk modulus of foam rubber?

3 Does the water pressure at the base of a dam depend on the size of the lake behind the dam?

4 A partly filled bottle of mercury has a screw cap. It is taken aloft on a spaceship. When the bottle is orbiting the earth in the ship, what is the pressure at a depth of 2.0 cm in the mercury? What is the pressure at this depth when the ship lands on the moon?

5 How can one determine the density of an irregular object that (*a*) sinks in water and (*b*) floats in water?

6 Estimate the average density of the human body. How

could you measure your density to within 1 percent using simple equipment at a swimming pool? Some people are able to float more easily than others. Explain what factors are involved.

**7** How is it possible for a ship made from steel to float? Won't steel always sink in water? How does a submarine move to various depths?

**8** A glass filled to the brim has an ice cube floating partly above the water. Does the water overflow as the ice cube melts?

**9** A glass filled to the brim with water sits on a scale. A block of wood is gently placed in the water so that it floats. Some water overflows, but it is wiped away. How does the final reading of the scale compare with the initial reading?

**10** Blood contains many tiny particles too small to be seen with a microscope. Sedimentation rates can be used to determine whether or not these particles have clumped into groups. Explain how this can be done and examine the assumptions you make.

**11** Using the facts that the density of air is 1.29 kg/m³ and atmospheric pressure at the ocean surface is $1 \times 10^5$ Pa, estimate the average height of the atmosphere above the earth. You will see in Chap. 10 how such an estimate differs from the true value.

**12** Discuss the physical meaning of Bernoulli's equation when the fluid is not moving.

**13** When you hold the end of a long strip of paper in your fingers, the strip will droop. If you blow horizontally over your fingers and above the strip, however, it will rise. Explain what lifts the paper against the force of gravity.

## Problems

**1** A solid sphere made of a certain material has a radius of 2.00 cm and a mass of 81.0 g. What is the density of the material?

**2** A solid cube made of a certain material is 1.500 cm on each edge. Its mass is 18.56 g. What is the density of the material?

**3** At 20°C, about what mass of air is there in a box-shaped room that is $8.0 \times 6.0 \times 2.8$ m³?

**4** The density within a neutron star is about $1 \times 10^{19}$ kg/m³. What would the diameter of the earth be if the earth were compressed to this density? $M_e = 5.98 \times 10^{24}$ kg.

**5** To determine the density of an unknown fluid, an empty 100-cm³ volumetric flask that has a mass of 58.71 g is filled to the mark with the fluid. The mass of the flask when full is 235.63 g. What is the density of the fluid?

▪ **6** To determine the approximate density of a 4.61-g stone, a student uses a 50-cm³ graduated cylinder that has a mass of 38.15 g. She places the stone in the cylinder and fills the cylinder with water to the 40-cm³ mark. The combined mass is then 81.26 g. What is the density of the stone?

**7** A wire 3.0 m long has a radius of 0.35 mm. When a 7.0-kg load is hung from it, the wire stretches 1.45 mm. What is Young's modulus for the material of the wire?

**8** A 20-kg load stretches a steel wire that is originally 150 cm long. The radius of the wire is 0.60 mm. How far does the wire stretch under this load?

**9** A cube of gelatin 3.5 cm on each edge is subjected to a shearing force of 0.40 N on its top surface. Because of this

force, the top surface displaces 2.5 mm. What is the shear modulus for gelatin?

**10** A rubber sheet 3.0 mm thick is cemented between two parallel metal plates. The rubber sheet is 8.0 cm × 8.0 cm and is the same size as the plates. A shear stress is applied to the rubber by pulling the two plates in opposite directions, each with a force of 40 N. How far does one plate move relative to the other if the shear modulus for rubber is 1.20 MPa?

**11** Atmospheric pressure is about $1.0 \times 10^5$ Pa. (*a*) By what fraction does the volume of a glass cube change as the air around it is removed in a vacuum chamber? (*b*) How large a pressure increase is needed to decrease a volume of mercury by 0.20 percent?

**12** A block of foam rubber shrinks by 10 percent as it is subjected to a pressure of 950 kPa. What is the bulk modulus for the rubber?

▪ **13** A 7.0-kg block is to be pulled along a horizontal surface by a horizontal steel wire whose cross section is 1.50 mm². If friction is ignored, what maximum acceleration can be given to the block? The tensile strength of steel is 0.50 GPa.

▪ **14** Repeat Prob. 13 if the block is being lifted straight up.

**15** Suppose there is a perfect vacuum inside a sealed coffee can. What force must the 6.0-cm-diameter cover support when it is exposed to the atmosphere? Use $P_a = 100$ kPa.

**16** How much force does the atmosphere exert on a person's back? Assume that $P_a = 100$ kPa and that the back has an area of 300 cm². Why doesn't this enormous force crush the person?

**17** What is the pressure due to the water at a depth of 10.4 m in a lake? Compare this with atmospheric pressure, about 100 kPa.

**18** What is the pressure due to the mercury at the base of a column of mercury 76 cm high? Compare this with atmospheric pressure, about 100 kPa.

**19** A car with all the windows closed runs off a bridge into a river. It comes to rest with the center of the driver's door 3.0 m below the surface. How large a force must the driver exert on the center of the door to push it open? The area of the door is 0.80 m$^2$.

**20** (*a*) What is the pressure due to the water 1609 m beneath the ocean's surface? Take the density of sea water to be 1025 kg/m$^3$. (*b*) If the bulk modulus is the same as for pure water, by what percent does the water density increase in going from the surface to this depth?

**21** A beaker has a 2.0-cm-thick layer of oil ($\rho_{oil} = 843$ kg/m$^3$) floating on 3.0 cm of water. What is the combined pressure due to the liquids at the bottom of the beaker?

**22** A 1200-kg automobile has its tires inflated to a gauge pressure of 180 kPa. How large an area of each tire is in contact with the pavement? Assume the wheels share the load equally.

**23** A glass tube is bent into a U shape, as in Fig. 9.17. Water is poured into the tube until it stands 10 cm high in each side. Benzene is added slowly to one side until the water on the other side rises 4 cm. What length is the benzene column?

**24** In Prob. 23, suppose that a 2.0-cm column of benzene is poured into one side. How far will the water in the other side rise?

**25** Hydraulic stamping machines exert tremendous forces on a sheet of metal to form it into the desired shape. Suppose the input force is 800 N on a piston that has a diameter of 1.40 cm. The output force is exerted on a piston that has a diameter of 35 cm. How large a force does the press exert on the sheet being formed?

**26** The plunger of a certain hypodermic needle has a cross-sectional area of 0.79 cm$^2$. How large a force must be applied to the plunger if liquid in the needle is to move into a vein where the pressure is 20 kPa above atmospheric pressure?

**27** Water is confined in a strong container by means of a piston with a 0.50-cm$^2$ cross-sectional area. How large a force on the piston is required to increase the density of the water by 0.0050 percent?

**28** A metal cube is 1.50 cm on each edge. What is the buoyant force on it when it is completely submerged in oil of density 850 kg/m$^3$?

**29** A 2.50-g object has an apparent mass of 1.63 g when it is completely submerged in water at 20°C. What is (*a*) the volume of the object and (*b*) its density?

**30** A 2.375-g object has an apparent mass of 1.942 g when it is completely submerged in water at 20°C. What is (*a*) the volume of the object and (*b*) its density?

**31** A 6.25-g object has an apparent mass of 5.41 g when completely submerged in oil of density 870 kg/m$^3$. Find the density of the object.

**32** A 4.873-g object has an apparent mass of 2.163 g when completely submerged in water and an apparent mass of 2.544 g when completely submerged in oil. What is the density of the oil?

**33** A woman weighs 475 N, and a downward force of 17 N must be applied to keep her completely submerged in water. What is the density of her body?

**34** A block of foam plastic has a volume of 24.0 cm$^3$ and a density of 820 kg/m$^3$. How large a force is required to hold it under water?

**35** The density of ice is 917 kg/m$^3$, and the approximate density of the seawater in which it floats is 1025 kg/m$^3$. What fraction of an iceberg is beneath the water surface?

**36** What must the minimum volume of a block of material ($\rho = 820$ kg/m$^3$) be if it is to hold a 60-kg man entirely above the surface of a lake when he stands on the block?

**37** When a beaker that is partly filled with water is placed on an accurate scale, the scale reads 20.00 g. If a piece of wood with a density of 920 kg/m$^3$ and a volume of 2.0 cm$^3$ is floated on the water, how much will the scale read?

**38** When a beaker that is partly filled with water is placed on an accurate scale, the scale reads 20.00 g. If a piece of metal with density 4000 kg/m$^3$ and volume 2.50 cm$^3$ is suspended by a thin string so that the metal is submerged in the water but does not rest on the bottom of the beaker, how much does the scale read?

**39** A 200-cm$^3$ block of foam plastic ($\rho = 620$ kg/m$^3$) is to be weighted with aluminum so that it just sinks in water. What mass of aluminum is required if the aluminum is to be hung from the block?

**40** A cube of metal ($\rho = 6250$ kg/m$^3$) has a cavity inside it. It weighs 2.5 times as much in air as it does when completely submerged in water. What fraction of the cube's volume is the cavity?

**41** By what factor does the quantity of fluid flowing through a capillary tube change when the length of the tube is increased to four times its original value and its diameter is increased to twice its original value? Assume the pressure difference across the tube remains unchanged.

**42** A hypodermic needle is replaced by another one that has half the length and one-third the diameter. By what factor must the pressure differential across the needle change if the flow rate is to remain unchanged?

■ **43** A hypodermic needle 4.0 cm long has an internal diameter of 0.25 mm. Its plunger has an area of 0.90 cm². When a force of 6.0 N is applied to the plunger, at what rate does water at 30°C flow through the needle?

■ **44** The blood pressure of a certain person is 120/80 mmHg. The average pressure is about 100 mmHg, which is $1.33 \times 10^4$ Pa. Assume that a 4.0-cm-long needle with internal radius 0.26 mm is inserted into the person's bloodstream where the pressure has this average value. At what rate does blood flow from the needle? Use $\eta_{blood} = 4$ mPl.

■ **45** A cubical block 2.0 cm on each edge sits on a flat plate. There is a 0.030-mm-thick oil film between block and plate ($\eta_{oil} = 0.40$ Pl). How large a force is required to pull the block across the plate at 0.25 cm/s? Assume the oil film remains intact.

**46** A pipe near the lower end of a large water-storage tank springs a small leak, and a stream of water shoots from it. The top of the water in the tank is 8 m above the leak. (*a*) With what speed does the water gush from the hole? (*b*) If the hole has an area of 0.060 cm², how much water flows out in 1 s?

■ **47** Water is flowing smoothly through a closed pipe system. At one point, the speed of the water is 2.5 m/s, and at a point 3.0 m higher than the first the speed is 4.0 m/s. (*a*) If the pressure is 80 kPa at the lower point, what is it at the upper point? (*b*) What would the pressure at the upper point be if the water were to stop flowing and the pressure at the lower point were 60 kPa?

■ **48** An airplane wing is designed so that the speed of the air below the wing is 400 m/s when the speed of the air across the top is 460 m/s. What is the pressure difference between the top and bottom of the wing?

**49** A scale model of a 1.80-m-high automobile is 9.0 cm high. If the model is tested in a wind tunnel, how fast should the air be moving to simulate the actual car's motion at 20 m/s?

**50** Show that the Reynolds number can be written as $N_R = 2Q\rho/\pi\eta r$ for flow through a cylindrical pipe of radius $r$.

**51** How fast can a 4.0-mm-diameter water drop fall through air before turbulent flow sets in? Use $N_R = 10$.

■ **52** If Stokes' law applies, what is the terminal speed of a 3.00-mm-diameter water drop falling through air? Does Stokes' law apply in this situation?

■ **53** A solid aluminum sphere of radius $b$ falls through water at 30°C. Find (*a*) the buoyant force on it and (*b*) its terminal speed. Assume laminar flow. (*c*) What value does $b$ have if $N_R$ for the sphere is to be 10?

■ **54** A piece of wood ($\rho = 840$ kg/m³) is fashioned into a sphere of radius 0.20 cm and released deep in a lake. What is its terminal velocity as it rises toward the surface? Assume laminar flow. Is this assumption justified?

■ **55** Show that the radius $b$ of a particle can be evaluated in a sedimentation experiment by use of the equation

$$b = \sqrt{\frac{9\eta v}{2g(\rho - \rho_f)}}$$

where $v$ is the sedimentation speed, $\rho$ and $\rho_f$ are the densities of particle and fluid, and $\eta$ is the viscosity of the fluid. Assume laminar flow for the spherical particle.

■ **56** An aluminum sphere of radius $b$ is being lifted with constant speed $v$ through a fluid of density $\rho_f$ and viscosity $\eta$ by means of a thin wire attached to it. Find the tension in the wire. Assume laminar flow.

■ **57** (*a*) Water is flowing upward in the pipe system of Fig. P9.1. At points 1 and 2, the pipe radii are $r_1$ and $r_2$, respectively. The water speed at point 1 is 40 cm/s. What is the pressure difference $P_2 - P_1$ between the two points? (*b*) Repeat if the flow direction is reversed.

■ **58** The vee-shaped hollow boat of length $L$ shown in cross section in Fig. P9.2 has sprung a leak (hole area = $A$) at a depth $D$ below the water line, and water is pouring into its empty hold. When the water has risen to the level shown, still below the hole, how fast is the boat sinking?

■ **59** A firefighter's water hose spews water from the nozzle at a rate of 0.0140 m³/s. When the nozzle is directed upward, the water shoots to a height of 30 m. Suppose the hose lies in a straight line along the ground when a firefighter tries to hold the nozzle end vertically. Describe the horizontal force that he or she must exert on the nozzle to hold it stationary.

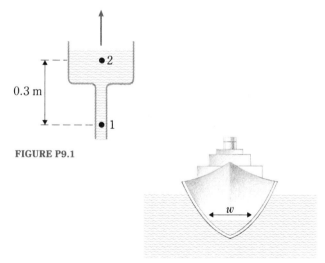

**FIGURE P9.1**

0.3 m

●2

●1

**FIGURE P9.2**

$w$

# 10 Gases and the kinetic theory

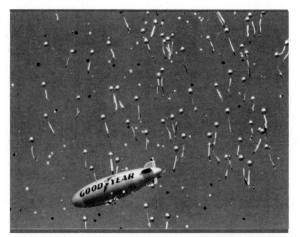

WE turn our attention now to the behavior of gases. As we saw in Chap. 6, the pressure of a gas on a surface is the result of the collisions of gas molecules with the surface. In a gas, the molecules have so much translational kinetic energy that the force of gravity and the attractions between molecules usually have negligible effects. Hence we can picture a gas as consisting of ball-like molecules shooting here and there; upon collision with each other or with the walls of a container, they rebound without sticking; their kinetic energy is so large that the amount they lose to gravitational potential energy as they rise is negligible. This model of a gas is basic to the *kinetic theory of gases*. In this chapter, we discuss the behavior of gases and how that behavior can be explained in terms of the kinetic theory, which assumes that gas molecules act like ball-like objects subject to the usual laws of mechanics.

## 10.1 Barometric pressure

The most important gas for humans is the atmosphere that envelopes the earth. We seldom stop to think of it, but we live near the bottom of a deep sea of air. This sea of air, the atmosphere, has no precise top; it simply thins out with increasing height above the earth. At 20 km above the earth, the density of the air is about 8 percent of its value at sea level. Because the atmosphere, like all other fluids, must obey the usual pressure-versus-depth relation, we can estimate the pressure at the surface of the earth due to the sea of air above us. Take the height of the atmosphere to be 20 km and its average density to be half its value at the surface of the earth, $\frac{1}{2} \times 1.29$ kg/m$^3$. Then the estimated pressure due to the atmosphere above us is

$$P_{\text{est}} = \rho g h \cong (0.65 \text{ kg/m}^3)(9.8 \text{ m/s}^2)(20 \times 10^3 \text{ m}) = 1.3 \times 10^5 \text{ Pa}$$

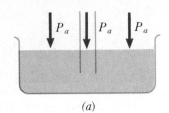

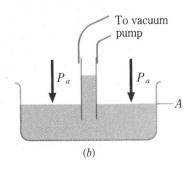

To vacuum pump

$P_a$  $P_a$

———A

(b)

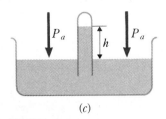

$P_a$  $h$  $P_a$

(c)

**FIGURE 10.1**

When the tube is evacuated, the mercury rises until $\rho gh = P_a$. Hence the device, a barometer, is capable of measuring pressure.

To measure atmospheric pressure, we use a *barometer*. There are many different devices used for this purpose, but the mercury barometer is one of the most fundamental. We can understand its operation by referring to Fig. 10.1. In part *a*, an open tube is partly immersed in a beaker of mercury. Because the air pressure inside the tube is equal to the air pressure outside, the mercury stands at the same level inside and outside.

Now suppose that a vacuum pump evacuates the air from the tube, as in Fig. 10.1*b*, and the tube is sealed off, as in *c*. Once all the air is removed, the pressure on the mercury surface inside the tube is zero. (Recall that the pressure exerted by a gas on a surface is the result of collisions of the gas molecules with the surface. If no molecules are present, we have a perfect vacuum and the pressure is zero.) Now the pressure at level *A* inside the tube is due only to the height *h* of mercury in the tube and is $\rho gh$, where $\rho$ is the density of mercury. Note that the pressure on the mercury surface at level *A* outside the tube is still atmospheric pressure $P_a$. Moreover, statement 3 in Sec. 9.4 tells us that the pressure inside the tube at level *A* is the same as it is outside the tube. Therefore

Pressure at *A* outside = pressure at *A* inside

$$P_a = \rho gh \tag{10.1}$$

Atmospheric pressure can support a column of mercury whose height is given by Eq. 10.1.

Because the density of mercury is known, atmospheric pressure is easily found from *h*. Pressure is often given in terms of the height of the mercury column. You have probably heard weather reports where the atmospheric pressure is stated to be about 760 mmHg or about 30 inHg. We can easily convert such a statement of pressure to SI units in the following way: an atmospheric pressure of 760 mmHg is

$$P_a = \rho gh = (13{,}600 \text{ kg/m}^3)(9.8 \text{ m/s}^2)(0.76 \text{ m}) = 1.01 \times 10^5 \text{ Pa}$$

where the density of mercury is 13,600 kg/m³. This value for $P_a$ is typical. *Standard atmospheric (barometric) pressure* is taken to be $1.01325 \times 10^5$ Pa. Note that our estimate for $P_a$ obtained earlier in this section has this order of magnitude. It is a very large pressure indeed, as evidenced by the experiment shown in Fig. 10.2.

**FIGURE 10.2**

As the pump removes the air inside the metal can, the can collapses under the unbalanced forces due to the atmospheric pressure outside it.

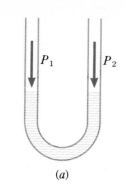

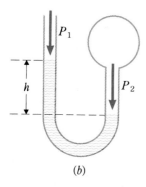

(a)

(b)

**FIGURE 10.3**

In a manometer, the pressure difference $P_2 - P_1$ is measured by the height difference $h$.

Commercial mercury barometers are more refined than the simple device shown in Fig. 10.1. They have an accurate scale beside the mercury column and special devices to adjust the level of the mercury in the beaker. There are other types of barometers based on different principles, but for accurate work the mercury barometer is preferred. However, it must be at least 76 cm long (why?), and so there is often good reason to replace it by a smaller, but less accurate, device.

Another device often used to measure gas pressures is a **manometer** (Fig. 10.3). Although it has many variations, a manometer is basically a U-shaped tube partly filled with a liquid, often mercury. If the mercury in a mercury manometer stands at equal levels in the two arms of the tube, as shown in $a$, we know that the two gas pressures $P_1$ and $P_2$ above the columns must be equal. However, if $P_2$ is larger than $P_1$, the columns will adjust as in $b$. The difference in heights $h$ of the two columns, measured in centimeters, gives the pressure difference $P_2 - P_1$ in centimeters of mercury. Usually one column is open to the atmosphere; let us say $P_1$ is atmospheric pressure. Then $P_2$ is found by adding the barometric pressure to $h$. For small pressure differences, it is often convenient to use a liquid that is less dense than mercury. If a liquid of density $\rho$ is used in place of mercury, the difference in levels is increased by a factor $13,600/\rho$, where $\rho$ is in kilograms per cubic meter. Thus standard atmospheric pressure is (76 cm) $\times$ (13,600/1000) = 1034 cm of water! Why isn't a water barometer used?

---

### Example 10.1

A simple test to determine the capabilities of a person's lungs is to have the person blow with full force into one end of a manometer, as shown in Fig. 10.4. Suppose that in a certain case a mercury manometer is used and the fluid level stands as shown. What is the pressure in the lungs? (Ordinarily, a mercury manometer would not be used because mercury vapor is dangerous on repeated exposure.)

*Reasoning* Let us call the air pressure in the lungs $P_L$. The pressure exerted on the left side of the manometer is then $P_L$. This pressure balances the pressure of the atmosphere plus the pressure due to the height $h = 6.0$ cmHg:

$$P_L = 6.0 \text{ cmHg} + P_a$$

**FIGURE 10.4**

By blowing into the manometer, the person is able to support a column of fluid 6.0 cm high. How large is $P_L$?

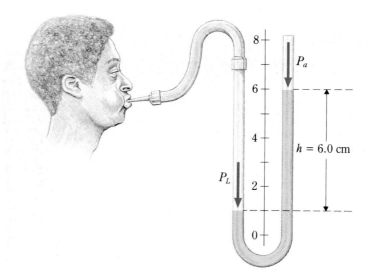

## Soft drinks and siphons

Those who have used a siphon know that you often start one by sucking on the siphon tube. They then come to the conclusion that the siphon operates on the same principle that applies to drinking a soft drink through a straw. The two situations are actually quite different from one another, however, as we shall see.

When you sip a fluid through a straw, the situation shown in (a) applies. By sucking on the straw, you reduce the air pressure inside to a value $P$ that is less than the atmospheric pressure $P_a$. Because of the imbalance between $P_a$ and $P$, the fluid rises in the straw until the fluid column, of height $h$, balances the pressure difference $P_a - P$. If you suck hard enough, atmospheric pressure pushes the fluid into your mouth.

In the siphon, we have the situation shown in (b). The fluid flows out the end at $B$ with speed $v_B$, and the pressure there is $P_a$, atmospheric pressure. At the surface of the reservoir at $A$, the pressure is $P_a$, and we can approximate the fluid velocity there as zero. Application of Bernoulli's equation at $A$ and $B$ gives

$$P_a + 0 + \rho g h_A = P_a + \tfrac{1}{2} v_B^2 + \rho g h_B$$
$$v_B = \sqrt{2g(h_B - h_A)}$$

which is Torricelli's theorem. Notice that $P_a$, atmospheric pressure, does not appear in the answer for $v_B$. Hence a siphon should operate in a vacuum (or on the moon) as well as in ordinary applications.

To gain further insight, notice that the fluid in the tube is not accelerating along its length. Therefore, if we ignore friction, the pressure in the tube is the same as it would be if the fluid were at rest. At $B$, the fluid pressure is $P_a$; at $C$, it is $P_a - \rho g(h_C - h_B)$; in the tube at $A$, it is $P_a - \rho g(h_A - h_B)$. Notice, however, that the pressure in the fluid outside the tube at $A$ is only $P_a$, larger than that inside the tube by an amount $\rho g(h_A - h_B)$. Because of this pressure differential at the mouth of the tube near $A$, the fluid is accelerated from the reservoir into the tube. It is this pressure differential that drives the fluid through the siphon. For a uniform-bore tube, this pressure differential arises because the fluid column on the right weighs more than the column on the left. In effect, the weight of the fluid on the right side lifts the fluid on the left. This is not as simple as it sounds, however, because the fluid flows even if the tube on the right is much narrower than the one on the left so that the fluid on the left weighs more. Why does the argument using weights break down in such a case?

(a) Sipping

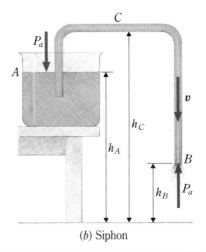

(b) Siphon

Normally, $P_a$ is about 76 cmHg, and so $P_L = 82$ cmHg. (Of course, in accurate work, $P_a$ would be read on a barometer.) To convert $P_L$ to SI units, we recall that the pressure due to a height $h$ of fluid is $\rho g h$; then

$$P_L = 82 \text{ cmHg} = (13{,}600 \text{ kg/m}^3)(9.8 \text{ m/s}^2)(0.82 \text{ m})$$
$$= 1.093 \times 10^5 \text{ Pa} \quad \blacksquare$$

**Exercise** A manometer that uses oil ($\rho = 840$ kg/m³) reads a pressure difference of 7.31 cm of oil. What is the pressure difference in SI units and in centimeters of mercury  *Answer: 0.602 Pa; 4.53 × 10⁻⁴ cmHg*  ■

## 10.2 Thermometers and temperature scales

The pressure of a gas depends on its temperature. This fact is well known to those who monitor the air pressure in their car tires. When the tires are hot, the pressure within the tires is higher than when the tires are cold. Of course, temperature is important in many other aspects of our lives as well. For this reason, we now discuss thermometers and the temperature scales associated with them.

The most common type of thermometer is shown in Fig. 10.5. A liquid — usually mercury or alcohol — is sealed in a glass capillary tube that has a bulb at one end. Because these liquids expand as the temperature increases, the liquid level in the capillary rises as the temperature increases. (The glass also expands, but much less than the liquid.) In olden times, the thermometer was marked into divisions in the following way.

**FIGURE 10.5**

The boiling and freezing points of water can be used to illustrate how the three usual temperature scales are related.

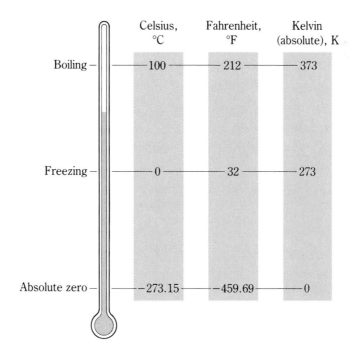

Two reference points are marked on the capillary. One is the position of the top of the liquid when the thermometer is at the temperature of ice and water at equilibrium under standard atmospheric pressure. This is the freezing level in Fig. 10.5. The second reference point is obtained when the thermometer is at the boiling point of water (under standard atmospheric pressure). This is the boiling level in the figure.

There are three temperature scales ordinarily encountered. The Celsius scale (formerly called centigrade) places the freezing point of water at 0°C (degrees Celsius) and the boiling point at 100°C. It is the temperature scale used by most of the world. The Fahrenheit scale assigns the values 32°F (degrees Fahrenheit) and 212°F to these two points. This scale is used mainly in the United States and is being consigned to oblivion as fast as the populace allows. The third scale, the Kelvin or absolute scale, is used by scientists everywhere. Its temperature unit is the *kelvin* (K). On it, the freezing and boiling points occur at 273.15 K and 373.15 K. Its advantages in science will become apparent as our studies progress.

These historical definitions of the temperature scales have been supplanted, as we shall see in Sec. 12.10. The new definitions, however, were chosen so as to preserve the scales essentially as they had been defined originally. As you can see from Fig. 10.5, there is a simple relationship between the Celsius temperature $t_C$ and the absolute (Kelvin) temperature $T$:

$$T = t_C + 273.15$$

Although we shall not use the Fahrenheit scale in this book, readings on it can be converted using the equations

$$t_C = (t_F - 32)(5/9)$$
$$T = 273.15 + (t_F - 32)(5/9)$$

We use thermometers routinely in our lives. Even so, there is a fundamental law of physics involving them that may have escaped your notice. When you place a thermometer in intimate contact with an object, the thermometer soon reaches a steady reading called the temperature of the object, and we say that the object and the thermometer are in *thermal equilibrium* with each other. If you now place this object in contact with an object that is at a higher temperature, the temperatures of the two objects will change and the objects will eventually reach thermal equilibrium at an intermediate temperature. We say that heat has flowed from the hotter to the colder object. These facts are well known. However, the following variation of this experiment is extremely important.

Suppose a thermometer reads the same temperature for two objects. What happens when these two objects are placed in intimate contact with each other? The answer is that nothing happens; the temperature of neither object changes. The two objects are in thermal equilibrium with each other. *Objects or systems that have the same temperature are in thermal equilibrium with each other.* This seemingly obvious statement is one form of the **zeroth law of thermodynamics,** which we state formally as follows:

Two bodies or systems that are in thermal equilibrium with a third body are in thermal equilibrium with each other.

As we see, temperatures are equal for objects that are in thermal equilibrium.

## 10.3 The mole and Avogadro's number

In the next section, we discuss how the pressure of a gas depends on its temperature and density. To facilitate that discussion, however, we need to use a few terms that are ordinarily learned in chemistry. Because you may be unfamiliar with these terms, let us now spend a short time discussing them.

The number of carbon atoms in 12 g of carbon* is called **Avogadro's number** $N_A$. Experiments show this number to be $6.0221 \times 10^{23}$ atoms per 12 g of carbon, and it is used to define a measure of the quantity of any substance, a quantity called the **mole**:

A mole of substance is that amount that contains $N_A$ particles.

For example, one mole of baseballs consists of $6.022 \times 10^{23}$ baseballs. Similarly, one mole of water consists of $N_A$ water molecules. As you see, the mole is a measure not of mass but of number of entities. To summarize,

Avogadro's number $= N_A = 6.0221 \times 10^{23}$ particles per mole

Because we shall most often use kilograms and kilomoles in discussions, we usually replace this value for $N_A$ by its equivalent:

$N_A = 6.0221 \times 10^{26}$ particles/kmol

Two related terms that we should be familiar with are **atomic mass** and **molecular mass** (popularly referred to as *atomic weight* and *molecular weight*). Both are represented by $M$.

The molecular (or atomic) mass $M$ of a substance is the mass in kilograms of one kilomole of the substance.

Because 12 kg of carbon contains $N_A$ atoms, one kilomole of carbon is 12 kg of it. Therefore $M$ for carbon is 12 kg/kmol. Similarly, 18 kg of water contains $N_A$ water molecules, and so $M$ for water ($H_2O$) is 18 kg/kmol. As other examples, $M$ for hydrogen (H) is 1 kg/kmol; for nitrogen gas ($N_2$), $M$ is 28 kg/kmol. The atomic masses of the elements are given in Appendixes 1 and 2.

---

### Example 10.2
The atomic mass of copper is 63.5 kg/kmol. Find the mass of a copper atom.

**Reasoning** Because $M = 63.5$ kg/kmol, 63.5 kg of copper contains $6.022 \times 10^{26}$ atoms. Therefore the mass of one atom is

$$\text{Mass per atom} = \frac{63.5 \text{ kg}}{6.022 \times 10^{26} \text{ atoms}} = 1.05 \times 10^{-25} \text{ kg/atom}$$

---

* Precisely, in 12 g of the isotope carbon 12.

This same method can be used to find the mass of any atom or molecule for which $M$ is known. Because $M$ kilograms contains $N_A$ entities, we have that

$$\text{Mass per entity} = \frac{M}{N_A} \quad \blacksquare$$

*Exercise* Find the mass of an oxygen molecule, $O_2$. For oxygen molecules, $M = 32.0$ kg/kmol. *Answer:* $5.31 \times 10^{-26}$ *kg* $\quad \blacksquare$

**Example 10.3**
Find the volume associated with a mercury atom in liquid mercury. For mercury, $\rho = 13{,}600$ kg/m$^3$ and $M = 201$ kg/kmol.

*Reasoning* The mass $m_o$ of a mercury atom is

$$m_o = \frac{201 \text{ kg/kmol}}{6.02 \times 10^{26} \text{ kmol}^{-1}} = 3.34 \times 10^{-25} \text{ kg}$$

Because there is 13,600 kg/m$^3$, we have

$$\text{Number of atoms per cubic meter} = \frac{13{,}600 \text{ kg/m}^3}{3.34 \times 10^{-25} \text{ kg/atom}}$$

$$= 4.07 \times 10^{28} \text{ atoms/m}^3$$

Taking the reciprocal of this, we find

$$\frac{\text{Volume}}{\text{Atom}} = 2.46 \times 10^{-29} \text{ m}^3/\text{atom}$$

If we consider the atom to be a sphere, this leads to a value of $1.8 \times 10^{-10}$ m for its radius. Thus we discover that even one of the largest atoms has a diameter of only about 0.4 nm. $\quad \blacksquare$

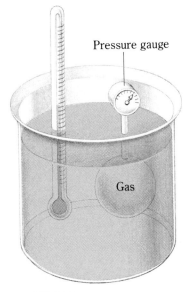

Pressure gauge

Gas

**FIGURE 10.6**
A simple device for measuring how temperature affects the pressure of a fixed volume of gas.

## 10.4 The ideal gas law

Some of the earliest investigators into the nature of temperature were concerned with how the pressure of a gas changes with temperature. Definitive experiments were carried out centuries ago, and today students still perform these basic experiments. A typical simple apparatus is shown in Fig. 10.6. The pressure of the gas is measured as a function of temperature while the volume of gas is held constant. When data from such an experiment are plotted, they lead to graphs such as the one shown in Fig. 10.7.

As you see from the graph, the data lead to a linear relation between absolute pressure (gauge pressure plus $P_a$) and temperature. Different straight lines are observed for different initial conditions within the container. In every case, however, *provided the gas is far from conditions under which it condenses, or liquefies,* a linear relation exists between temperature and pressure at constant volume.

Another informative experiment is to measure the volume of a gas as a function of temperature with the pressure held constant. Typical results are shown in Fig.

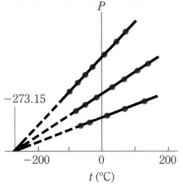

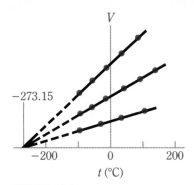

**FIGURE 10.7**
The pressure of a dilute gas at constant volume decreases as the temperature is lowered. The three curves are for the same gas, but with different amounts of gas in the volume.

**FIGURE 10.8**
The volume of a dilute gas varies linearly with temperature, provided $P$ is constant. The three curves are for the same gas, but at different pressures.

10.8. Here, too, a linear relation is found; the volume varies linearly with temperature. This is true for all gases if they are far removed from condensation conditions.

Figures 10.7 and 10.8 show another interesting feature: all the linear plots extrapolate to the same temperature intercept, $-273.15°$C. This unique temperature is called **absolute zero.** We cannot obtain data near this temperature for most gases because they condense at considerably higher temperatures. As condensation conditions are approached, the data deviate from a straight line. Even so, the fact that such a unique temperature exists indicates it may have fundamental importance. We shall discuss the meaning of absolute zero in more detail later.

All these data for noncondensing gases can be expressed in equation form. Experiment shows that the absolute pressure $P$ of a mass $m$ of gas confined to a volume $V$ at a Celsius temperature $t_C$ is given by

$$PV = \frac{m}{M} R(t_C + 273.15 \text{ C}°)$$

In this expression, $M$ is the molecular mass of the gas molecules and $R$ is a constant of nature called the *gas constant.* We shall give its value shortly. You can check this relation by noticing that, for constant $V$,

$$P = (\text{constant})(t_C + 273.15 \text{ C}°)$$

and for constant $P$,

$$V = (\text{constant})(t_C + 273.15 \text{ C}°)$$

When $t_C = -273.15°$C, both $P$ and $V$ go to zero. As we see, these relations for $P$ and $V$ check with the data of Figs. 10.7 and 10.8.

Apparently, nature reserves a special place for temperatures measured in the form $t_C + 273.15$ C$°$, that is, for a temperature scale with zero at absolute zero. We shall see further evidence for this preference as our studies progress. This scale is the temperature scale we have called the absolute or Kelvin temperature scale, and we represent temperatures on it by $T$. On it, the freezing point of water is 273.15 K, which we read as "273.15 kelvins." Notice that the word *degree* is *not* used.

In terms of temperatures on the Kelvin scale, the law we have found by experiment for gases becomes

$$PV = \frac{m}{M} RT \tag{10.2}$$

This is called the **ideal gas law,** and gases that obey it are called *ideal* gases. All gases that are far removed from conditions under which they condense are very nearly ideal. Precise measurements yield the following value for the gas constant:

$$R = 8314 \text{ J/kmol} \cdot \text{K} = 8.314 \text{ J/mol} \cdot \text{K}$$

## 10.5 Using the gas law

Now that we understand the meaning of the quantities in the gas law, we are in a position to apply it to various problems. The law is frequently stated as

$$PV = nRT \tag{10.2}$$

where $n = m/M$ is the number of kilomoles of gas in the volume $V$. In using this relation, it is very important that you pay attention to units. *Absolute* temperatures must be used for $T$. The pressure $P$ is in pascals (that is, $N/m^2$). Volumes are measured in cubic meters. To evaluate $n$, which is $m/M$, the mass $m$ of gas in the volume $V$ should be in kilograms. With these units, $R$ has the value $8314 \text{ J/kmol} \cdot \text{K}$.

---

**Example 10.4**
Standard atmospheric pressure and temperature are $1.01325 \times 10^5$ Pa and $0°C$. Find the volume that one kilomole of an ideal gas occupies at these values of $P$ and $T$.

*Reasoning*  For use in the gas law,

$$P = 1.013 \times 10^5 \text{ Pa}$$

$$T = 273.15 \text{ K} + 0 = 273.15 \text{ K}$$

The quantity of gas in the volume $V$ is given as 1 kmol, and so $m/M = n = 1$ kmol. Then $PV = nRT$ becomes

$$(1.013 \times 10^5 \text{ Pa})(V) = (1 \text{ kmol})(8314 \text{ J/kmol} \cdot \text{K})(273.15 \text{ K})$$

$$V = 22.4 \text{ m}^3/\text{kmol}$$

This is a convenient fact to remember:

One kilomole of any ideal gas occupies a volume of $22.4 \text{ m}^3$ under standard conditions.  ∎

---

**Example 10.5**
If 14 mg of nitrogen gas ($M = 28.0$ kg/kmol) is held at $27°C$ in a container that has a volume of $5000 \text{ cm}^3$, what is the gas pressure in the container?

*Reasoning* We simply substitute in the gas law, $PV = (m/M)RT$. In proper units, we have

$$P(5 \times 10^{-3} \text{ m}^3) = \left( \frac{14 \times 10^{-6} \text{ kg}}{28 \text{ kg/kmol}} \right) (8314 \text{ J/kmol} \cdot \text{K})(300 \text{ K})$$

$$P = 249 \text{ Pa}$$

Notice that 27°C must be changed to 300 K before use. ∎

*Exercise* What mass of nitrogen is present in the container under standard conditions? *Answer: 6.25 g* ∎

## Example 10.6

Use the gas law to determine the mass of air contained in a 50.0-cm³ flask at atmospheric pressure and 20°C.

*Reasoning* Because air is mostly nitrogen, we approximate the situation by setting $M = 28.0$ kg/kmol, the molecular mass of $N_2$. Also, $P = 1.013 \times 10^5$ Pa, $V = 50 \times 10^{-6}$ m³, and $T = 293$ K. Therefore, from $PV = (m/M)RT$,

$$m = \frac{PVM}{RT} = \frac{(1.013 \times 10^5 \text{ Pa})(50 \times 10^{-6} \text{ m}^3)(28 \text{ kg/kmol})}{(8314 \text{ J/kmol} \cdot \text{K})(293 \text{ K})}$$

$$= 5.82 \times 10^{-5} \text{ kg} \quad \blacksquare$$

*Exercise* Use the fact that 1 kmol occupies 22.4 m³ to find the mass of air in the flask at standard conditions. *Answer: 6.25 $\times$ 10⁻⁵ kg* ∎

## Example 10.7

An oil drum containing only air at 20°C is sealed off. It is then set out in the sun, where it heats up to 60°C. If the original pressure is 1.0 atm, what is the final pressure in the drum?

*Reasoning* Write the gas law twice, once for the original situation and once for the final situation:

$$P_1 V = nRT_1 \qquad P_2 V = nRT_2$$

where $V$ is the constant volume of the drum and $n$ is the number of kilomoles of gas in the drum. Dividing one equation by the other yields

$$\frac{P_1}{P_2} = \frac{T_1}{T_2}$$

where $T_1$ and $T_2$ are absolute temperatures. We have $P_1 = 1$ atm, $T_1 = 20 + 273 = 293$ K, and $T_2 = 333$ K. Therefore

$$P_2 = (1.0 \text{ atm}) \left( \frac{333 \text{ K}}{293 \text{ K}} \right) = 1.14 \text{ atm}$$

Notice that absolute temperatures must be used. The pressure units are not important as long as they are the same. ∎

**Example 10.8**

The gas in the piston of a diesel engine is at a temperaure of 27°C and a pressure of 74 cmHg when it is suddenly compressed. If the final pressure is 3700 cmHg and the final temperature is 547°C, what is the final volume of the gas in terms of the original volume?

**Reasoning**  Write the gas law twice,

$$P_1V_1 = nRT_1 \qquad P_2V_2 = nRT_2$$

Dividing yields

$$\frac{P_1V_1}{P_2V_2} = \frac{T_1}{T_2}$$

Substituting gives

$$\frac{V_1}{V_2}\frac{74\text{ cmHg}}{3700\text{ cmHg}} = \frac{273 + 27\ \text{K}}{273 + 547\ \text{K}}$$

from which $V_2 = 0.0547\,V_1$. Notice that the pressure units cancel, and so we do not need to convert to pascals.  ∎

---

**Example 10.9**

A car tire is filled to a pressure $P_1$ when the temperaure is 0°C. What is the pressure after the car has been driven at high speed and the gas in the tire reaches 35°C? Assume the volume of the tire does not change. Find $P_2$ if the tire gauge reads 190 kPa on a day when atmospheric pressure is 101 kPa at 0°C.

**Reasoning**  We write the gas law twice and obtain, as before,

$$\frac{P_1V_1}{P_2V_2} = \frac{T_1}{T_2}$$

In this case, $V_1 = V_2$, $T_1 = 273$ K, and $T_2 = 308$ K. Therefore

$$P_2 = \frac{P_1 T_2}{T_1} = P_1\left(\frac{308}{273}\right) = 1.13 P_1$$

Be very careful at this point. The gauge pressure is 190 kPa. However, $P$ must be the total pressure, defined as gauge pressure plus atmospheric pressure. Hence

$$P_2 = 1.13(190\text{ kPa} + 101\text{ kPa}) = 329\text{ kPa}$$

The final gauge reading, which is this value minus atmospheric pressure, is 228 kPa. Gauge pressures must always be changed to absolute pressures before they can be used in the gas law.  ∎

## 10.6 The molecular basis for the gas law

The gas law, $PV = nRT$, expresses the pressure of a gas in terms of the temperature. In Chap. 6, we used the kinetic theory to arrive at another expression for gas pressure. We computed the pressure resulting from molecular impacts with the container wall to be

$$P = \tfrac{2}{3} n_u (\tfrac{1}{2} m_o v^2)_{av} \tag{10.3}$$

where $m_o$ is the mass of a molecule, $n_u = N/V$ is the number of molecules per unit volume, and $(\tfrac{1}{2} m_o v^2)_{av}$ is the average translational kinetic energy of a gas molecule. If the theoretical approach we used in Chap. 6 is correct, it must yield the same pressure found by experiment, in other words, the pressure summarized by the gas law:

$$P = \frac{m}{VM} RT$$

where $m$ is the mass of gas (of molecular mass $M$) in the volume $V$. Notice that $m/V$ is the mass per unit volume. However, the number of molecules per unit volume multiplied by the mass of each, $n_u m_o$, is also the mass per unit volume. Hence we can replace $m/V$ by $n_u m_o$ to obtain

$$P = \frac{n_u m_o}{M} RT$$

We know, however, that $M$ is the mass per kilomole, and so $M/m_o$ is the number of molecules per kilomole — and this is simply Avogadro's number $N_A$. Therefore the gas law can be written

$$P = n_u \left( \frac{R}{N_A} \right) (T) = n_u k T \tag{10.4}$$

where $k$ is called **Boltzmann's constant:**

$$k = \frac{R}{N_A} = \frac{8314 \text{ J/kmol} \cdot \text{K}}{6.022 \times 10^{26} \text{ kmol}^{-1}} = 1.38 \times 10^{-23} \text{ J/K} \tag{10.5}$$

We now have two expressions for the pressure of an ideal gas. Equation 10.3 was arrived at using the kinetic theory of gases, while Eq. 10.4 came from the results of experiments as summarized in the gas law. Let us now equate these two expressions for $P$. The result is

$$n_u k T = \tfrac{2}{3} n_u (\tfrac{1}{2} m_o v^2)_{av}$$

After a little arithmetic, this yields

$$(\tfrac{1}{2} m_o v^2)_{av} = \tfrac{3}{2} k T \tag{10.6}$$

# Atoms: a historic struggle

The theory that all matter is composed of tiny pieces called atoms was postulated several times in the early history of physics. It was not until about 1800, however, that the concept of atoms took on a quantitative form. At that time, John Dalton showed that much of the chemistry then known could be explained if one assumed the existence of atoms that varied in character from element to element. He also inferred that the elements have masses that are integral multiples of the mass of the hydrogen atom. At the same time, Avogadro introduced the idea that the number that bears his name should have an important physical meaning. These concepts were used successfully and continuously by chemists.

In the mid-1800s, the kinetic theory of gases was developed by J. Joule (1848), R. Kronig (1856), R. Clausius (1857), J. C. Maxwell (1860), and especially L. E. Boltzmann (1872). The atomic theories proposed by these men fit the experimental data then available and gave a highly detailed picture of gas behavior. However, almost all experiments at that time were determinations of macroscopic properties of gases and solids. No direct evidence existed for the atoms and molecules that the theories assumed to exist. Because of this, many physicists prior to 1900 took the attitude that atoms did not exist. Or, if they did exist, they were too small to be observed and should therefore be of no concern to physicists.

Leaders in the opposition to the atomic approach were the noted scientist-philosopher of physics E. Mach and the highly regarded

physicist W. Ostwald. To sum up their reasoning in the words of Ostwald, one should try to free science "from hypothetical conceptions which lead to no immediately experimentally verifiable conclusions." He termed the atomic approach "those pernicious hypotheses" that place "hooks and points upon the atoms." Of course, chemists continued to use atomic concepts with great success.

In the years near 1900, Boltzmann struggled against these attacks on the kinetic theory he had so greatly advanced. It was discouraging work, and in 1898 he wrote, "I am conscious of being only an individual struggling weakly against the stream of time." At least partly as a result of this opposition to his work, he became severely depressed and, in 1906, committed suicide.

Shortly thereafter, the atomic theory was verified by direct experimental evidence. In 1908, J. B. Perrin showed that Brownian motion can be explained in terms of atomic concepts. A year later, R. A. Millikan confirmed the existence of the electron by showing that charge comes in unique packets. In the following years, a flood of experiments confirmed the validity of the atomic approach. By 1926, when Otto Stern first measured directly the distribution of atomic speeds in a gas and found perfect agreement with the predictions of atomic theory, the atomic approach had been so widely accepted as to make Stern's result an anticlimax.

We should not forget that the physics we know today has a very human past and involved the emotions as well as the cold scientific reasoning of those who have given us the presently accepted laws of physics.

This equation tells us that the average translational kinetic energy of a gas molecule is proportional to the absolute temperature of the gas. This is a very important equation because it tells us the physical meaning for absolute temperature as far as an ideal gas is concerned:

Absolute temperature is a measure of the average translational kinetic energy of a molecule of an ideal gas.

Notice that Eq. 10.6 applies to any ideal gas. It therefore tells us that

If two ideal gases have the same temperature, the average translational kinetic energy per molecule is the same in each.

Although these statements apply only to an ideal gas, we shall see in later chapters that absolute temperature is a measure of kinetic energy per molecule even in liquids and solids. However, it is not a simple measure.

Before leaving this section, we should point out that our results apply to real gases only at moderate and high temperatures. Near absolute zero, very strange things happen; some metals become resistanceless conductors of electricity, and certain fluids flow without friction (that is, their viscosity becomes zero). The behavior of molecules at low temperatures must be dealt with in terms of quantum mechanics, a topic we consider in the last few chapters of this book.

---

**Example 10.10**

On the average, how fast is a nitrogen molecule moving in air at $27\,°C$?

**Reasoning** We find $v$ for an average molecule from Eq. 10.6. For diatomic nitrogen gas, $M = 28.0$ kg/kmol, and so

$$m_o = \frac{M}{N_A} = \frac{28 \text{ kg/kmol}}{6.02 \times 10^{26} \text{ kmol}^{-1}} = 4.65 \times 10^{-26} \text{ kg}$$

Equation 10.6 can be solved for $v^2$ to give

$$v^2 = \frac{3kT}{m_o} = \frac{3(1.38 \times 10^{-23} \text{ J/K})(27 + 273 \text{ K})}{4.65 \times 10^{-26} \text{ kg}} = 2.67 \times 10^5 \text{ m}^2/\text{s}^2$$

from which the average velocity is 517 m/s. Notice how high a speed this is. In view of it, why does it take so long for the odor of a gas—perfume molecules, for example—to cross a room? ∎

---

## 10.7 Distribution of molecular speeds

In the previous sections, we ignored the fact that not all molecules in a gas are moving with the same speed. Let us now examine the speeds of the individual molecules. It is interesting to notice that this topic was treated theoretically long before vacuum techniques had been developed well enough to measure the speeds of the molecules in a gas. James Clerk Maxwell, whose name we will encounter again when we study electricity, was the theoretician who first successfully examined molecular speeds. In 1860, he derived an equation that describes the way in which

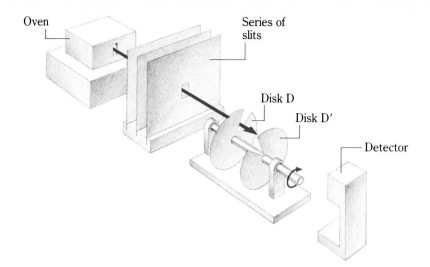

**FIGURE 10.9**

A schematic diagram of apparatus for measuring the speed of gas atoms. The entire apparatus is maintained in vacuum.

the speeds of gas molecules vary. This equation, together with the work of Boltzmann and other theoreticians, was the subject of a serious scientific controversy, as pointed out in the Special Note on page 216. Because experiments of the time could not observe molecules directly, theories involving these unproven entities were suspect. Not until 1926, when Otto Stern managed to devise and assemble the necessary equipment, was the prediction Maxwell had made 66 years earlier confirmed.

A diagram of an experiment to determine the speeds of gas atoms is shown in Fig. 10.9. The entire apparatus was confined in a vacuum chamber. Mercury was vaporized in the oven, and a beam of mercury gas atoms shot out of a hole in the oven. After passing through a series of slits, used to obtain a very thin beam of atoms, the atoms passed through a tiny slit in disk $D$. This disk was coupled to a disk $D'$, which was similar to $D$ but with the slit offset. Unless the disks were rotating, the beam could not pass through $D'$ and reach the detector.

If the disks were rotating, however, atoms that were traveling just fast enough to move from $D$ to $D'$ in the time taken for slit $D'$ to line up with the original position of slit $D$ would pass through $D'$ and reach the detector. In this way, the number of atoms having a given speed could be measured. The results obtained by Stern and others accurately confirmed the prediction Maxwell had made more than a half century earlier. These results are typified by the curves for mercury atoms shown in Fig. 10.10. Most of the mercury-gas atoms have speeds near 200 m/s at 460 K and near 400 m/s at 1700 K, although a wide range of speeds is found. As we expect, the speeds of gas atoms increase with increasing temperature. The nitrogen gas molecules in air have a similar speed distribution. However, the maxima in the curves for nitrogen at these two temperatures occur close to 500 m/s and 1000 m/s.

These experimental results, as well as many others, show that the average translational kinetic energy of a gas atom or molecule is $3kT/2$, the result predicted by the kinetic theory (Eq. 10.6). The concept of a gas consisting of individual particles darting here and there with an average energy $3kT/2$ was an important step in achieving an understanding of the thermal properties of all forms of matter, a topic we investigate in the next chapter.

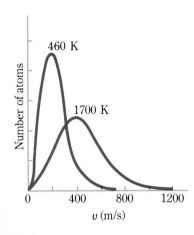

**FIGURE 10.10**

The distribution of molecular speeds in mercury gas at two temperatures.

## Minimum learning goals

When you finish this chapter, you should be able to

1 Define (*a*) barometer, (*b*) standard pressure, (*c*) standard temperature, (*d*) manometer, (*e*) thermal equilibrium, (*f*) absolute zero, (*g*) gas constant *R*, (*h*) ideal gas law, (*i*) mole, (*j*) kilomole, (*k*) Avogadro's number, (*l*) Boltzmann's constant.

2 Describe the construction of a mercury barometer and state how it is used to measure pressure.

3 Explain why the can in Fig. 10.2 collapses as the vacuum pump removes air from it.

4 Explain how a manometer measures pressure differences.

5 Sketch a diagram showing the three common temperature scales. On each, locate absolute zero and the freezing and boiling points of water.

6 Compute the mass of a molecule or atom if *M* for it is given.

7 Use the ideal gas law in simple computations.

8 State under what conditions a gas is likely to behave ideally.

9 Compute the average translational kinetic energy of the molecules of an ideal gas when the gas temperature is given.

10 Sketch a graph showing the distribution of speeds of gas molecules at two temperatures.

## Questions and guesstimates

1 Why don't people use water barometers? After all, mercury is poisonous and expensive.

2 We can imagine the molecules of an ideal gas as acting like tiny balls in continual motion. An ideal gas of colloidal-size particles can also exist. However, glass beads and pool balls do not act like ideal gases. At what size does the dividing line come, and to what is it due?

3 The composition of the atmosphere changes with altitude. As one goes farther above the earth, the percentage of hydrogen molecules in the air increases and the percentage of nitrogen molecules decreases. Why?

4 Compare the gravitational potential energy of a nitrogen molecule that is 1 m above the ground with its translational kinetic energy when the temperature is (*a*) $0°C$ and (*b*) $-270°C$.

5 Although the air is composed mostly of $N_2$ molecules, there is some $O_2$ present, of course. Do both kinds of molecules travel with the same average speed? What is the exact relation between the speeds?

6 To escape from the earth, an object must be shot out from it with a speed of at least 11,200 m/s. Explain why only a tiny amount of hydrogen exists in the atmosphere, even though billions of years ago there may have been more hydrogen than nitrogen in it.

7 **Boyle's law** for gases states that the volume of a gas varies inversely with pressure, provided the mass and temperature of the gas are maintained constant. Show that Boyle's law is a special case of the ideal gas law.

8 **Charles' law** for gases states that the volume of a gas increases in direct proportion to the absolute temperature, provided the pressure and mass of the gas are maintained constant. Show that this is a special case of the ideal gas law.

9 **Dalton's law of partial pressures** states that the total pressure of a mixture of gases is equal to the sum of the partial pressures of the gases in the mixture. Using the ideal gas law and kinetic theory, justify Dalton's law.

10 Hydrogen and oxygen gas are sealed off at atmospheric pressure in a strong glass jar containing two electrodes. A spark from the electrodes ignites the gases so that the reaction $2H_2 + O_2 \rightarrow 2H_2O$ results. Will the pressure in the tube have changed after the temperature has come back to its original value ($200°C$)? Explain. What if the original temperature is $200°C$ and the final temperature is $20°C$?

11 The earth's atmosphere contains about $10^{19}$ kg of gas. When Julius Caesar gasped, "Et tu, Brute!" as he lay mortally wounded, he exhaled a certain volume of nitrogen molecules. Estimate how many of these historic molecules you breathe in with each breath.

12 What would a mercury barometer read in a spaceship orbiting the earth if the air pressure in the spaceship is 75 cmHg?

# Problems

*Unless otherwise stated, assume that atmospheric pressure is 101 kPa.*

**1** Suppose a water barometer is used to measure atmospheric pressure. How tall is the water column if the pressure is 95 kPa?

**2** Denver, the "mile-high city," often has an atmospheric pressure of only 60 cmHg. How tall a column of oil (density = 850 kg/m³) will this pressure support?

**3** When atmospheric pressure is 100 kPa, with how large a force does the atmosphere press down on a 20 cm × 30 cm book resting on a table? If the book has a mass of 2.0 kg, what is the ratio of this force to the weight of the book?

**4** If the largest face of the can in Fig. 10.2 is 15 cm × 28 cm, how large a force does the atmosphere exert on that face? Assume atmospheric pressure to be 98 kPa. Compare this force with the weight of the can if the can's mass is 150 g.

**5** A mercury manometer is used to monitor the pressure in a reaction vessel. In the end that is open to the atmosphere, the mercury level is 2.64 cm higher than in the end leading to the vessel. The barometric reading is 74.83 cmHg. What is the pressure in the reaction vessel?

**6** A manometer that uses oil of density 878 kg/m³ is used to measure the pressure in an environmental testing chamber. In the end that is open to the atmosphere, the oil level is 12.5 cm higher than in the end leading to the chamber. The barometric reading is 73.54 cmHg. What is the pressure in the chamber?

**7** Convert to the other two temperature scales: (*a*) 68°F, (*b*) −37°C, (*c*) 870 K.

**8** Convert to the other two temperature scales: (*a*) 64°C, (*b*) −15°F, (*c*) 112 K.

▪ **9** At what temperature do the Celsius and Fahrenheit scales have the same numerical value?

▪ **10** At what temperature do the Fahrenheit and Kelvin scales have the same numerical value?

**11** What is the mass of a gold atom?

**12** What is the mass of a methane molecule, the chemical formula of which is $CH_4$?

**13** How many atoms are there in a 25-g cube of pure copper?

**14** How many molecules are there in 50 g of pure water, $H_2O$?

**15** The molecular mass of the nylon in a nylon thread is 10,000 kg/kmol. The density of nylon is 1100 kg/m³. (*a*) Find the mass of a nylon molecule. How many of these molecules exist in (*b*) 1 g of nylon and (*c*) 1 cm³ of nylon?

**16** Ethyl alcohol ($C_2H_5OH$) has a density of about 790 kg/m³. Find (*a*) the mass of an ethyl alcohol molecule and (*b*) the number of such molecules in 1 cm³.

**17** Consider a 50-kg woman to be one huge molecule. What is her molecular mass?

**18** Show that the number of ideal gas molecules in a container of volume $V$ and temperature $T$ is $N_A PV/RT$.

**19** A 500-cm³ tank contains oxygen ($O_2$) at 20°C and a gauge pressure of $2.50 \times 10^6$ Pa. What mass of oxygen is in the tank?

**20** A 800-cm³ tank contains monatomic argon gas (Ar) at 30°C and a gauge pressure of $1.32 \times 10^6$ Pa. What mass of argon is in the tank?

**21** What mass of air exists in a 5 × 6 × 8 m room at 20°C? Take the average molecular mass of air to be 28 kg/kmol.

▪ **22** The mass of a 250-cm³ flask is 287 mg more when it is filled with an unknown gas than when it is evacuated. When the flask is filled, the gas is at 20°C and $1.00 \times 10^5$ Pa. What is the molecular mass of the gas molecules.

**23** A test tube is filled with an ideal gas to a gauge pressure of 200 kPa at 20°C and sealed off. What is the gauge pressure in the tube when its temperature is 400°C?

**24** A sealed test tube contains nitrogen gas ($N_2$) at 20°C and a gauge pressure of 250 kPa. What is the gauge pressure of the gas when the temperature is −80°C?

**25** What volume of air at 100 kPa is needed to fill a tire of volume $V_o$ to a gauge pressure of 170 kPa? Assume the temperature of the air does not change.

**26** A 500-cm³ tank contains oxygen gas at a gauge pressure of 890 kPa. What volume would the oxygen occupy if it expanded until it reached atmospheric pressure, 100 kPa? Assume the gas temperature remains constant.

**27** A gas at 20°C and atmospheric pressure is compressed to a volume one-fifteenth as large as its original volume and an absolute pressure of 3000 kPa. What is the new temperature of the gas?

**28** In a diesel engine, the piston compresses air at 30°C from approximately atmospheric pressure to a pressure of about 5000 kPa and a volume about one-sixteenth the original volume. What is the temperature of the compressed air?

**29** One way to cool a gas is to let it expand. Typically, a gas at 20°C and 4000 kPa might be expanded to atmospheric pressure and a volume 33 times larger. Find the new temperature of the gas.

**30** A gas at 20°C and an absolute pressure of 800 kPa is suddenly expanded into a chamber having 15 times the volume. Its new temperature is −5°C. What is its new pressure?

■ **31** A vertical right circular cylinder of height 30.00 cm and base area 12.0 cm$^2$ is sitting open under standard temperature and pressure. A tight-fitting 5.0-kg piston is placed in the cylinder and allowed to fall to an equilibrium height. At equilibrium, what are the height of the piston and the pressure in the cylinder? Assume the final temperature to be 0°C.

■ **32** A fish at a depth of 8.0 m in a fresh-water lake exhales an air bubble whose volume is $V_o$. Find the volume of the air bubble just before it reaches the surface. Assume that the temperature of the bubble remains constant.

■ **33** A 20-cm-long cylindrical test tube is inverted and pushed vertically down into water. When the closed end is at the water surface, how high has the water risen inside the tube?

■ **34** A vertical capillary tube has a 5.0-cm column of mercury filling its lower part. The tube is sealed off (at atmospheric pressure) at a point 15 cm above the top of the mercury. If the tube is inverted, how long is the air column at its bottom?

**35** A piece of dry ice, $CO_2$, is placed in a test tube, which is then sealed off. If the mass of dry ice is 0.36 g and the sealed test tube has a volume of 20 cm$^3$, what is the final pressure of the $CO_2$ in the tube if all the $CO_2$ vaporizes and reaches thermal equilibrium with the surroundings at 27°C?

**36** When a 20-cm$^3$ test tube is sealed off at a very low temperature, a few drops of liquid nitrogen condense in the tube from the air (the boiling point of nitrogen is −210°C). What is the nitrogen pressure in the tube when the tube is warmed to 27°C if the mass of the drops is 0.075 g?

**37** The temperature of the sun's interior is estimated to be about $14 \times 10^6$ K. Protons ($m = 1.67 \times 10^{-27}$ kg) compose most of its mass. Compute the average speed of a proton by assuming that the protons act as particles in an ideal gas.

**38** At what temperature do the molecules of an ideal gas have an average speed that is ten times their average speed at 0°C?

**39** The escape velocity for a projectile on earth is about 11.2 km/s. (*a*) At what temperature do hydrogen molecules (H$_2$) have an average speed equal to this speed? (*b*) Repeat for nitrogen molecules (N$_2$).

■ **40** The temperature of outer space is about 3 K and consists mainly of a gas of single hydrogen atoms. On the average, there is about one such atom per cubic centimeter. (*a*) Find the pressure of this gas in outer space; express your answer in atmospheres. (*b*) Find the average kinetic energy of one of the hydrogen atoms. (*c*) What speed does an atom of this energy have?

■ **41** Find the density of water vapor (steam) at 1 atm and 100°C if it is considered an ideal gas. For comparison, the actual density is 0.598 kg/m$^3$. Try to justify any difference you may note.

■ **42** Show that the pressure of an ideal gas can be written $P = \rho v^2/3$.

■ **43** For many purposes, a pressure of $1 \times 10^{-8}$ mmHg is considered a reasonably good vacuum. At this pressure and room temperature, (*a*) find the number of kilomoles of nitrogen in 1 m$^3$ and (*b*) the number of nitrogen molecules in 1 cm$^3$.

■ **44** Use the graph in Fig. 10.10 to estimate the fraction of mercury atoms at 1700 K that have speeds in excess of 600 m/s.

■ **45** A beam of particles, each of mass $m_o$ and speed $v$, is directed along the $x$ axis. The beam strikes an area 1 mm square, with $1 \times 10^{15}$ particles striking per second. Find the pressure on the area due to the beam if the particles stick to the area when they hit. Evaluate for an electron beam in a television tube, where $m_o = 9.11 \times 10^{-31}$ kg and $v = 8 \times 10^7$ m/s.

■ **46** Assuming that the molecules of several ideal gases in the same container act independently, prove Dalton's law of partial pressures.

■ **47** Assume that the disks in Fig. 10.9 are 70 cm apart and that the angle between the two slits is 30°. When the disks rotate at 800 rev/s, what speed(s) may the gas molecules have if they are to reach the detector?

■ **48** In Fig. P10.1, a frictionless piston of area $A$ and mass $M$ separates two equal volumes $V_o$ of ideal gas in which the pressure is $P_o$. The cylindrical container is now set on end. Find the equilibrium upper volume in terms of $P_o$ and $V_o$.

Piston

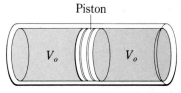

**FIGURE P10.1**

**49** A spherical balloon ($V = 4.20$ m³) is filled with helium gas ($M = 4.00$ kg/kmol) on a day on which atmospheric conditions have their standard value. (*a*) How many kilograms of helium are in the balloon if it floats in the air? Neglect the mass of the balloon. (*b*) What is the pressure of the helium?

**50** In Fig. P10.2, a uniform tube with an open stopcock is lowered into mercury so that 12 cm of the tube remains unfilled. After the stopcock is closed, the tube is lifted 8 cm. What is the height $y$ of the mercury in the tube? Assume standard atmospheric conditions.

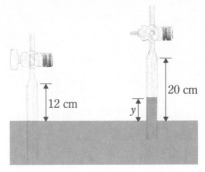

12 cm

20 cm

$y$

**FIGURE P10.2**

# 11 Thermal properties of matter

**T**HE discussion in the preceding chapter, which described the effect of heat on gases, considered gaseous atoms and molecules as acting like balls darting here and there. We ignored the fact that the atoms and molecules have internal structure and that they may possess forms of energy other than translational kinetic energy. Using these same simplifications, early investigators were able to achieve good agreement between theory and experiment for many gases. In liquids and solids, however, many other complications influence the thermal behavior of the atoms and molecules. The assumptions made in describing ideal gases no longer provide adequate descriptions of experimental results. Let us now see how we describe the thermal properties of these more complex systems.

## 11.1 The concept of heat

It has long been known that hot objects can be used to heat cooler objects. It has only been in the past century, however, that there has been any real understanding of the processes involved. Not surprisingly, our understanding of the nature of heat developed rapidly as the kinetic theory of gases evolved. As we saw in the preceding chapter, the kinetic theory leads at once to a physical meaning for temperature. The absolute temperature $T$ of a gas is proportional to the average translational kinetic energy of a molecule in the gas. We concluded that in a gas, the average translational

$$T_1 > T > T_2$$

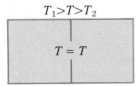

$$T = T$$

FIGURE 11.1

When two gases are brought into contact with each other, collisions between molecules originally at $T_1$ and molecules at $T_2$ cause the average molecular kinetic energy of both to change, thereby equalizing the temperatures.

kinetic energy of a molecule of mass $m_o$ can be found from the relation

$$(\tfrac{1}{2} m_o v^2)_{av} = \tfrac{3}{2} kT \tag{10.6}$$

where $k = 1.38 \times 10^{-23}$ J/K is Boltzmann's constant.

Suppose now that two containers of gas, originally at temperatures $T_1$ and $T_2$ ($T_1 > T_2$), are brought together and the gases allowed to mix, as in Fig. 11.1. We know from experiment that the final temperature $T$ of the gas is intermediate between $T_1$ and $T_2$. The interpretation of this on the basis of the kinetic theory is that the hot, high-energy molecules in container 1 collide with the cold, low-energy molecules in container 2; these collisions cause the molecules in container 1 to lose enough kinetic energy to those in container 2 to make the final average molecular translational kinetic energies in containers 1 and 2 the same, and thus the two gases are at the same temperature.

From this and many similar considerations, we conclude that, when two bodies at different temperatures are brought into contact with each other, energy is trans-ferred, or *flows,* from the hotter to the cooler body. *The energy that is transferred in a situation such as this is what we refer to as* **heat energy.**

Heat energy is the energy transferred from a warm body to a cooler one as a result of the temperature difference between the two.

Heat energy is therefore related to molecular energy. When an object is warmed, energy is added to it. When an object is cooled, it loses energy. The study of heat flow in matter is therefore the study of energy transfer.

## 11.2 Thermal energy

Let us now investigate what happens when heat energy flows into a substance. First let us consider a monatomic gas, such as helium. To a first approximation, each atom acts like a compact ball that darts here and there. Although it has some rotational kinetic energy because of its spinning motion on its axis, this energy ($\tfrac{1}{2} I \omega^2$) is negligibly small relative to the translational kinetic energy because the atom's moment of inertia is very, very small.

Diatomic molecules, such as $N_2$ and $O_2$, have much larger moments of inertia, however, and so they have comparable amounts of rotational and translational energy. We can picture such a molecule as shown in Fig. 11.2. Its two atoms are connected by a chemical bond that acts much like a spring. The atoms connected by this springlike bond vibrate relative to each other, and thus they have additional energy called vibrational energy. Unlike the situation in a monatomic gas, the added heat energy in a diatomic gas does not all appear as translational kinetic energy of the molecules.

As we go to gases composed of molecules that contain still more atoms, an even smaller fraction of the added heat energy becomes translational kinetic energy. Therefore, the more complex the molecules, the more heat energy must be added to provide a given increase in translational kinetic energy. Because only this latter energy controls the temperature of the gas, through Eq. 10.6,

$$KE_t = (\tfrac{1}{2} m_o v^2)_{av} = \tfrac{3}{2} kT$$

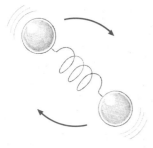

FIGURE 11.2

A diatomic gas molecule has both translational and rotational energy. It also has energy associated with the springlike bond between its atoms.

we see that, the more complex the gas molecule, the more heat energy must be added to increase its temperature a given amount.

In liquids and solids, the situation is very complex. Attractions and bonds between the often complex molecules give rise to additional possible destinations for added heat energy. The effect of the added energy is to cause the atoms and molecules of the substance to vibrate, rotate, and even translate in a multitude of motions. These motions are continually altered by random collisions of the moving atoms, and as a result have no set, repeatable direction; they are random and are called *thermal motions*. We term the energy resident in these random motions *thermal energy*.

Thermal energy is the energy associated with the random motion of atoms and molecules.

When heat energy flows into a substance, it most often becomes thermal energy; other possibilities are discussed later.

The thermal energy of a substance can be increased by mechanical means as well as by adding heat. For example, every Scout knows that you can start a fire by rubbing pieces of wood together. Friction work done on the wood sets molecules at the surface into violent random motion. These random motions constitute additional thermal energy. In general,

Friction-related losses of mechanical energy appear as thermal energy.

The law of conservation of energy assures us that the mechanical energy lost gives rise to an equal amount of thermal energy.

There is a technical difference between the terms *heat energy* and *thermal energy*. Heat energy is the energy that flows from one object to another because they are at different temperatures. Heat flows from a hot flame to a pan, for example. Thermal energy is the energy of random molecular motions within an object. The two are not the same, as the following example shows. When heat energy flows into a gas, it can appear as thermal energy (random-motion energy) of the gas molecules and thus cause a rise in temperature. However, the gas can expand and cause a piston to move so as to do work as the heat energy is added; at least part of the heat energy is then used to do work, and so not all of it is changed to thermal energy.

## 11.3 Heat units

Because heat and thermal energy are forms of energy, their basic unit in the SI is the joule. However, there are other units commonly used in heat measurements, units that were in widespread use long before it was known that heat is a form of energy. Because of their widespread use, we introduce them here.

A unit called the *calorie* (cal) was originally defined to be the heat required to raise the temperature of one gram of water one Celsius degree (1 C°). Today we define it in terms of the joule; one calorie is exactly 4.184 J. The "calorie" that nutritionists use in giving the number of calories in a portion of food is actually a kilocalorie. It is often distinguished from the calorie by calling it a "large calorie" or by writing it with a capital letter: Calorie (Cal). Another widely used unit for heat energy is the British thermal unit (Btu). It, too, is defined in relation to the joule; 1 Btu is exactly equal to 1054 J.

## James Prescott Joule (1818 – 1889)

An Englishman who inherited a prosperous brewery and pursued science as a hobby, James Prescott Joule was convinced that heat and energy are directly related. To prove his point, he carried out many experiments in which heat was generated from mechanical and electrical processes.

On his honeymoon, he and his bride measured the difference in water temperature between the top and bottom of a waterfall. He thus was able to calculate how much thermal energy was generated by the lost potential energy. Many more precise measurements led him to conclude the following: One unit of lost mechanical energy gives rise to a repeatable amount of thermal energy. The numerical comparison between the two is called the mechanical equivalent of heat. In today's units, the conversion is 1 cal = 4.184 J and is the definition of the calorie. In Joule's time, however, the calorie and the mechanical energy unit were considered unrelated.

The scientific community was slow to recognize Joule's achievements. When in 1847 he gave a lecture describing his discoveries, the scientific gathering ignored him, probably because he was an amateur. Later that year, he spoke before another group. Just when it seemed he would be ignored again, a young man (later to become Lord Kelvin) spoke up and pointed out the importance of Joule's results. Aided by Kelvin, Joule soon became a scientist of note. He is remembered not only for the work we have mentioned but also because he was an early advocate of the principle of conservation of energy.

## 11.4 Specific heat capacity

To increase the temperature of an object, we must increase the thermal energy of its molecules. We can do this by letting heat flow into the object from a hotter object. Similarly, if we wish to cool an object, we can allow heat energy to flow from it to a cooler object. To describe such processes as cooling and heating quantitatively, we must know how much energy is required to change the temperature of an object. *The quantity of heat that must flow into or out of a unit mass of a substance to change its temperature by one degree is called the* **specific heat capacity** *c of the substance.*

When a quantity $\Delta Q$ of heat flows into a mass $m$ of substance, the temperature of the mass will increase by an amount $\Delta T$. Then, by definition,*

$$\text{Specific heat capacity } c = \frac{\Delta Q}{m\Delta T}$$

After clearing fractions, we have

$$\Delta Q = cm \, \Delta T \tag{11.1}$$

---

* The notation $\Delta Q$ (read "delta Q") represents a change in $Q$. Similarly, $\Delta T$ represents a change in $T$. Increases are positive changes, and decreases are negative.

## ■ TABLE 11.1  Specific heat capacities

| Substance | $c(cal/g \cdot C°)$ | $c(J/kg \cdot C°)$ |
|---|---|---|
| Water | 1.000 | 4184 |
| Human body | 0.83 | 3470 |
| Ethanol | 0.55 | 2300 |
| Paraffin | 0.51 | 2100 |
| Ice (0°C) | 0.50 | 2100 |
| Steam (100°C)* | 0.46 | 1920 |
| Aluminum | 0.21 | 880 |
| Glass | 0.15 | 600 |
| Iron | 0.11 | 460 |
| Copper | 0.093 | 390 |
| Mercury | 0.033 | 140 |
| Lead | 0.031 | 130 |

* At constant volume.

From the definition, the units of specific heat capacity are J/kg · C°, although the unit cal/g · C° is in more common use. You should be able to show that

$$1 \text{ cal/g} \cdot C° = 4184 \text{ J/kg} \cdot C°$$

Typical values for $c$ are listed in Table 11.1. Notice that for water, $c = 1.000$ cal/g · C°. As we shall see, specific heat capacity varies somewhat with temperature. The values given in the table apply at temperatures near room temperature.

### Example 11.1
How much heat is required to change the temperature of (a) 400 g of water from 18 to 23°C? (b) 400 g of copper from 23 to 18°C?

*Reasoning*  We make use of the defining equation for $c$:

(a) $\Delta Q = cm \, \Delta T = (1.00 \text{ cal/g} \cdot C°)(400 \text{ g})(+5 \text{ C}°) = 2000 \text{ cal}$

or, in SI units,

$\Delta Q = (4184 \text{ J/kg} \cdot C°)(0.40 \text{ kg})(+5 \text{ C}°) = 8370 \text{ J}$

(b) In this case, $\Delta T = -5 \text{ C}°$, so that

$\Delta Q = (0.093 \text{ cal/g} \cdot C°)(400 \text{ g})(-5 \text{ C}°) = -186 \text{ cal} = -778 \text{ J}$

Here the substance loses heat as it cools, and so $\Delta Q$ is negative.  ■

*Exercise*  Determine the final temperature of 700 g of copper, originally at 16°C, to which 400 J of heat energy is added.  *Answer: 17.5° C*  ■

## 11.5 Vaporization and boiling

Let us now discuss what happens when a liquid evaporates. Molecules of a liquid are held rather tightly to each other by their mutual attractions. (These attractions are mostly electrical in nature.) For the most part, the molecules at the surface of the liquid are unable to escape into the region above the surface. As in a gas, however, thermal motion causes a few of the molecules to have extraordinarily high energy. A molecule with enough energy can escape from the surface and move from the liquid state to the gaseous state. In this way, the liquid *evaporates,* or *vaporizes.*

Because only the most energetic molecules can escape from the liquid, the average energy of the molecules left behind decreases as vaporization proceeds. Therefore, because temperature is a measure of energy, the temperature of an isolated liquid must decrease as vaporization proceeds. We thus arrive at an explanation for the well-known fact that evaporation can cool a liquid.

Obviously, energy must be furnished to the molecules of a liquid if they are to escape from the liquid surface. We define the **heat of vaporization** of a substance as follows:

The energy required to tear a unit mass of molecules loose from each other and thus change them from the liquid phase to the vapor phase is called the heat of vaporization $H_v$ of the liquid.

A like amount of energy is liberated as a unit mass of molecules condenses to form a liquid. Table 11.2 lists typical values of $H_v$. Because the molecules in a liquid possess more energy at high temperatures than at low temperatures, the additional energy required to tear them loose from each other varies with temperature; in other words, $H_v$ varies with temperature. It is 2470 kJ/kg for water at 10°C but only 2260 kJ/kg at 100°C. The values for $H_v$ given in Table 11.2 are those for the normal boiling temperature of the liquid.

A liquid boils when vapor bubbles form and grow within it. To see how this happens, we must first understand the term **vapor pressure.** Suppose you have a

■ **TABLE 11.2  Heats of vaporization and fusion**

| Substance | Melting point (°C) | Boiling point (°C) | $H_v$ kJ/kg | $H_v$ cal/g | $H_f$ kJ/kg | $H_f$ cal/g |
|---|---|---|---|---|---|---|
| Helium | −270 | −269 | 21 | 5.0 | 5.2 | 1.25 |
| Oxygen | −219 | −183 | 210 | 51 | 13.8 | 3.3 |
| Nitrogen | −210 | −196 | 200 | 48 | 25.5 | 6.1 |
| Ethanol | −114 | 78 | 854 | 204 | 105 | 25 |
| Mercury | −39 | 357 | 270 | 65 | 11.7 | 2.8 |
| Water | 0 | 100 | 2260 | 539 | 335 | 80 |
| Lead | 327 | 1750 | 858 | 205 | 23 | 5.9 |
| Aluminum | 660 | 2450 | 10500 | 2520 | 397 | 95 |
| Gold | 1063 | 2660 | 1580 | 377 | 64 | 15.4 |
| Copper | 1083 | 2595 | 4810 | 1150 | 205 | 49 |

**FIGURE 11.3**

When the vapor is saturated, the number of molecules evaporating from the liquid in a given time is exactly equal to the number condensing from the vapor.

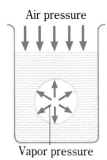

Air pressure

Vapor pressure

**FIGURE 11.4**

The boiling temperature is the temperature at which the vapor pressure in the bubble equals the external pressure on the liquid. (The size of the bubble is exaggerated.)

liquid and its vapor in a closed container, as in Fig. 11.3. At any given temperature, equilibrium occurs when the number of molecules evaporating from the liquid is balanced by the number returning from the vapor to the liquid. We call the pressure of the liquid's vapor under this equilibrium condition the vapor pressure of the liquid. It, of course, increases with increasing temperature. Why?

Because of the random motions of the molecules in a liquid, several molecules may pull away from each other and form an empty space, a hole, within the liquid. Molecules evaporate into the hole, and so the vapor pressure within it rises (Fig. 11.4). If time allows, the vapor pressure within the hole becomes equal to the vapor pressure of the liquid at this temperature. At low temperatures, the vapor pressure within the hole is small and the much larger atmospheric pressure applied to the liquid causes the hole to collapse. When the temperature of the liquid is high, however, the vapor pressure within the hole is large, larger perhaps than the pressure within the liquid due to the atmosphere. Then the excess pressure within the hole, now a vapor-filled bubble, causes the bubble to expand. The buoyancy of the bubble causes it, and many others like it, to rise to the surface of the liquid and explode, a phenomenon we recognize as boiling. Thus we see that a critical situation is reached when the temperature becomes high enough for the vapor pressure of the liquid to equal the pressure of the atmosphere above it. Then vapor-filled bubbles form and grow in the liquid, thereby producing boiling.

A liquid boils at a temperature where its vapor pressure just equals the external pressure on it.

Because its vapor pressure is 101 kPa at 100°C, water normally boils at 100°C. High in the mountains, however, where atmospheric pressure may be only 80 kPa, boiling occurs at about 94°C. In science and industry, a partial vacuum is often applied when liquids are distilled. The reduced external pressure allows the liquid to boil at a lower temperature and decreases the possibility of chemical reaction within it. Boiling points for typical liquids (at $P_a = 101$ kPa) are given in Table 11.2.

It is interesting to notice that vigorous heating of a boiling liquid does not increase its temperature. When heat is added to the boiling liquid, the energy is carried away by vaporization. More vigorous heating simply results in the formation of more vapor bubbles to carry away the energy. Thus the temperature remains at the boiling point until all the liquid has boiled away.

## 11.6 Heat of fusion and melting

Ice crystals melt at 0°C under standard pressure. Before melting occurs, the water molecules of the ice are ordered in a crystalline lattice. They are held in place by rather strong intermolecular forces. To melt the crystal, one must tear the molecules out of this tight arrangement and cause them to become disordered. This process requires energy, which is usually supplied by heat.

We therefore find that when a crystalline material is heated, it begins to melt at a certain temperature. As heat is slowly added to the crystal-liquid mixture, the temperature remains constant until all the crystals have melted. The substance has a definite melting temperature, and a definite amount of heat — called the **heat of fusion** — must be furnished to melt the crystals at this temperature.

The **heat of fusion** $H_f$ is the amount of heat energy required to melt a unit mass of a crystalline material. This same amount of heat is given off to the surroundings when a unit mass of the liquid crystallizes.

The heat of fusion for water is 335 J/kg (80 cal/g). Values for other materials are listed in Table 11.2. Notice that the hydrogen-bonded materials water and ethanol have much higher heats of fusion and vaporization than the others. Why?

The freezing point of a liquid can be altered somewhat by applying large pressures to the system. Materials that contract upon freezing have their melting points raised by increased pressure. Most materials behave in this way. A few materials, such as water, expand when they freeze. Increased pressure decreases the freezing point of such substances. The pressure of an ice skater's blade can cause the ice below it to melt. In this case, the skater is skating on ice lubricated with a thin film of water.

---

### Example 11.2

How much heat is released from 50 g of water as it (*a*) changes from liquid to the crystalline phase at 0°C and (*b*) changes from steam to liquid water at 100°C?

*Reasoning*   (*a*) When a mass *m* crystallizes, it liberates an energy $mH_f$. Thus

$$\Delta Q = mH_f = (50 \text{ g})(80 \text{ cal/g}) = 4000 \text{ cal} = 16,700 \text{ J}$$

(*b*) The heat liberated by a mass *m* of gas as it condenses is $mH_v$. Therefore

$$\Delta Q = mH_v = (50 \text{ g})(539 \text{ cal/g}) = 27,000 \text{ cal} = 113,000 \text{ J}$$

Notice that the steam-to-water transition liberates much more heat than does the water-to-ice transition.  ■

---

*Exercise*   How much heat is needed to melt 500 g of lead at 327°C?  *Answer: 4.29 × 10⁵ J*  ■

---

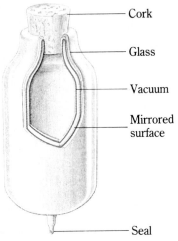

— Cork

— Glass

— Vacuum

— Mirrored surface

— Seal

**FIGURE 11.5**
A simple thermos flask.

## 11.7 Calorimetry

Many experiments involving heat are carried out in a container called a *calorimeter*. This is a device that thermally isolates materials so that heat cannot flow out of or into them from the surroundings or from the calorimeter. A perfect calorimeter completely prevents heat flow into or out of the system contained within it. The common vacuum thermos flask (Fig. 11.5) is a fairly good calorimeter. Heat is prevented from flowing through its double glass walls by the shiny metal coating on them and by the vacuum between them. We shall see in Secs. 11.9 to 11.11 why such a design is effective.

Suppose two or more materials at various temperatures are placed together inside a calorimeter. The materials will share thermal energies until they all reach the same temperature, that is, until thermal equilibrium is established. Because no energy flows into or out of the container, the law of conservation of energy leads us to a very important conclusion: if heat gains are taken as positive changes and heat losses are taken as negative changes, then

The sum of the heat changes within the calorimeter is zero.

In other words, the total energy of the isolated system inside the calorimeter is unchanged.

Before we apply this idea to various examples, let us review what types of heat changes we may encounter.

**1** When a mass $m$ undergoes a temperature change from $T_o$ to $T_f$, its heat change is

$$\Delta Q = mc(T_f - T_o)$$

where $c$ is its specific heat capacity.

**2** When a mass $m$ melts, its heat change is $\Delta Q_f = +mH_f$; when it crystallizes, its heat change is $\Delta Q_f = -mH_f$.

**3** When a mass $m$ vaporizes, its heat change is $\Delta Q_v = +mH_v$; when it condenses, its heat change is $\Delta Q_v = -mH_v$.

---

### Example 11.3

A cup contains 200 g of coffee at 98°C. What mass $M$ of ice at 0°C must be added to change the coffee temperature to 60°C? Neglect heat flow to the cup; that is, the cup is assumed to be a perfect calorimeter.

**Reasoning**  From the conservation of energy law,

Heat change of coffee + heat change of ice = 0

The temperature of the coffee (essentially water) changes from 98°C to 60°C. Therefore,

$$\text{Heat change of coffee} = mc(T_f - T_o) = (200 \text{ g})(1 \text{ cal/g} \cdot \text{C}°)(-38 \text{ C}°)$$
$$= -7600 \text{ cal}$$

The ice first melts and then warms from 0°C to 60°C. Therefore,

$$\text{Heat change of ice} = +MH_f + Mc(T_f - T_o)$$
$$= M(80 \text{ cal/g}) + M(1 \text{ cal/g} \cdot \text{C}°)(+60 \text{ C}°)$$
$$= M(140 \text{ cal/g})$$

Substitution of these values in the conservation of energy equation gives

$$-7600 \text{ cal} + M(140 \text{ cal/g}) = 0$$

which yields $M = 54.3$ g as the amount of ice required. Notice that we must be careful about signs in such a problem. If we are not careful, we will be confusing heat losses with heat gains, and this will invalidate our computation.  ∎

---

**Exercise**  Find the final temperature if only 40 g of ice is added.  *Answer: 68.3° C*

---

### Example 11.4

An 80-g piece of metal at 100°C is dropped into a calorimeter that contains 400 g of oil at 18.0°C. If the final temperature of the system is 23.1°C, what is the specific heat capacity of the metal? For the oil, $c = 0.65$ cal/g $\cdot$ C°. Call the specific heat capacity of the metal $c_m$, and neglect heat flow to the calorimeter.

**Reasoning**  From the conservation of energy law,

Heat change of metal + heat change of oil = 0

But

Heat change of oil $= mc(T_f - T_o) = (400 \text{ g})(0.65 \text{ cal/g} \cdot \text{C}°)(+5.1 \text{ C}°)$

$$= 1330 \text{ cal}$$

Heat change of metal $= mc(T_f - T_o) = (80 \text{ g})(c_m)(-76.9 \text{ C}°)$

$$= -6150(c_m) \text{ g} \cdot \text{C}°$$

After substituting in the energy equation, we have

$$-6150(c_m) \text{ g} \cdot \text{C}° + 1330 \text{ cal} = 0$$

$$c_m = 0.216 \text{ cal/g} \cdot \text{C}° \quad \blacksquare$$

### Example 11.5
A large glass container holds 500 g of mercury at 20°C. An electric heater immersed in the mercury delivers 70 W of power. How long does it take the heater to boil away 30 g of mercury? Ignore the mass of the heater, and assume that all the electric power is lost to heat.

*Reasoning*  The heater furnishes 70 J of heat to the mercury each second. So, in a time $t$,

Heat delivered to mercury in time $t = (70 \text{ J/s})(t)$

This heat changes the temperature of the 500 g of mercury from 20°C to its boiling point, 357°C. In addition, it vaporizes 30 g of the mercury. Hence, from $\Delta Q = mc(T_f - T_o)$ and $\Delta Q_v = mH_v$,

Heat change of Hg $= (500 \text{ g})(0.033 \text{ cal/g} \cdot \text{C}°)(357 - 20 \text{ C}°) + (30 \text{ g})(65 \text{ cal/g})$

$$= 5560 \text{ cal} + 1950 \text{ cal} = 7510 \text{ cal}$$

$$= 31,400 \text{ J}$$

Because all the heat delivered to the mercury must equal the heat change of the mercury, we equate these two expressions (notice that we changed them to like units, joules):

Heat delivered $=$ heat change
$$(70 \text{ J/s})(t) = 31,400 \text{ J}$$

from which the time $t$ taken is about 450 s.  $\blacksquare$

---

*Exercise*  How long would this same heater require to vaporize 50 g of water in a vat that is already at 100°C?  *Answer: 26.8 min*  $\blacksquare$

---

### Example 11.6
A 10.0-g lead bullet is traveling at 100 m/s when it strikes and embeds itself in a wooden block. By about how much does its temperature rise on impact? For this rough computation, assume that all the kinetic energy of the bullet changes to thermal energy in the bullet.

*Reasoning*   The thermal energy gained by the bullet equals the kinetic energy it loses:

$$\text{KE of bullet} = \tfrac{1}{2}mv^2$$
$$= (\tfrac{1}{2})(0.0100 \text{ kg})(100 \text{ m/s})^2 = 50 \text{ J} = 12.0 \text{ cal}$$

Notice the units conversion from joules to calories:

$$\text{Heat gained} = m_{Pb}c_{Pb}(T_f - T_o)$$
$$12 \text{ cal} = (10 \text{ g})(0.031 \text{ cal/g} \cdot \text{C}°)(\Delta T)$$
$$\Delta T = 38.7 \text{ C}°$$

Hence, if the original temperature of the bullet is 20°C, its final temperature is about 59°C. If the bullet were traveling at 600 m/s, we would find $\Delta T$ obtained in the above way to be 36 times as large, and the final temperature would be about 1430°C. Of course, the bullet would melt before this temperature was reached, and so the above computation would no longer be correct. How could the computation be carried out in this latter case?   ▪

---

### Example 11.7

When nutritionists state that 1 kg of bread has a food value of 2600 Cal, they mean that if the dried bread is burned in pure oxygen, it will give off 2600 kcal of heat energy. (Basically, the body generates heat from food in a somewhat similar chemical reaction.) Estimate how much heat energy a human body gives off each day.

*Reasoning*   Depending on the person, the nutritional calorie intake each day is 2000 to 3000 Cal. Since these are actually kilocalories, the body's metabolism generates on the order of $2 \times 10^6$ cal of heat. Because its temperature must remain nearly constant, the body must lose this energy as it is generated. The air we exhale and the evaporation of perspiration from the skin are well-known mechanisms for cooling the body, but others are important as well.   ▪

---

*Exercise*   If a 60-kg person retained all the heat energy from the 1800 Cal she consumed in a day, how much would her body temperature rise? Use $0.83 \text{ cal/g} \cdot \text{C}°$ for the specific heat of her body.   *Answer: 36.1 C°*   ▪

---

## 11.8 Thermal expansion

As we have seen, the temperature of a substance is a measure of the energy resident in its molecules. As the temperature of a liquid or solid is raised, the molecules, having greater energy, generally vibrate through larger distances. This increased amplitude of vibration of a given molecule forces its neighboring molecules to remain at a greater average distance from it. Hence, the solid or liquid expands. Although there are some notable exceptions to this rule over small temperature ranges (for example, water contracts in going from 0 to 4°C),* it is generally true

---

* In water, hydrogen bonding binds the molecules into groups of several molecules each in a definite structure even above the melting point of ice. As the temperature increases, these groups break up, causing a new, more compact arrangement of the molecules.

that substances expand with increasing temperature, provided a phase change does not occur.

Clearly, the thermal expansion of the metal in a building or bridge can be a matter of considerable practical importance. If provision were not made for thermal expansion, railway tracks and concrete highways would buckle under the action of the hot summer sun. You may know of situations where faulty design has caused difficulty in this respect. Therefore, it is necessary to know exactly how a material expands with temperature. To this end, a constant of linear thermal expansion $\alpha$ (alpha) and a constant of volume thermal expansion $\gamma$ (gamma) are defined and tabulated.

Suppose a bar of initial length $L$ undergoes a temperature change $\Delta T$ that causes its length to change by an amount $\Delta L$. The fractional change in the length of the bar is $\Delta L/L$. We define the **coefficient of linear thermal expansion** $\alpha$ for the material by

$$\alpha = \frac{\text{fractional change in length}}{\text{temperature change}} = \frac{\Delta L/L}{\Delta T}$$

which gives

$$\Delta L = \alpha L \, \Delta T \tag{11.2}$$

The units of $\alpha$ are reciprocal degrees, either $1/C^\circ$ or $1/K$. A few typical values for $\alpha$ are listed in Table 11.3.

As an example of the use of the linear expansion coefficient, suppose a brass rod 75 cm long is subjected to a temperature change of $+50$ C$^\circ$. Its length increase is (see Table 11.3 for $\alpha$)

$$\Delta L = \alpha L \, \Delta T = (19 \times 10^{-6}/C^\circ)(0.75 \text{ m})(50 \text{ C}^\circ) = 7.1 \times 10^{-4} \text{ m}$$

■ **TABLE 11.3** **Coefficients of thermal expansion per celsius degree at 20°C**

| *Substance* | $\alpha \times 10^6$ | $\gamma \times 10^6$ |
|---|---|---|
| Diamond | 1.2 | 3.5 |
| Glass (heat-resistant) | ~3 | ~9 |
| Glass (soft) | ~9 | ~27 |
| Iron and steel | 12 | 36 |
| Brick and concrete | ~10 | ~30 |
| Brass | 19 | 57 |
| Aluminum | 25 | 75 |
| Mercury | | 182 |
| Rubber | ~80 | ~240 |
| Glycerin | | 500 |
| Gasoline | | ~950 |
| Methanol | | 1200 |
| Benzene | | 1240 |
| Acetone | | 1490 |

Because this change in length is very small, the value of $L$ used to determine $\Delta L$ is not sufficiently temperature-dependent to cause worry about the temperature at which it is measured. Actually, $\alpha$ does vary somewhat with temperature, and for very precise work one should use the value appropriate to a given temperature range. In practice, however, this complication is seldom of any consequence.

Thermal expansion of volumes is also important. In a manner analogous to the way we defined the linear expansion coefficient, we define a **coefficient of volume thermal expansion** $\gamma$ as the relative change in volume per unit change in temperature:

$$\gamma = \frac{\Delta V/V}{\Delta T}$$

which yields

$$\Delta V = \gamma V \Delta T \tag{11.3}$$

The units of $\gamma$ are reciprocal degrees. As an example of its use, suppose that $100 \text{ cm}^3$ of benzene at $20°C$ is heated to $25°C$. According to Eq. 11.3, its volume will change by an amount (see Table 11.3 for $\gamma$)

$$\Delta V = (1.24 \times 10^{-3}/C°)(100 \text{ cm}^3)(5 \text{ C}°) = 0.62 \text{ cm}^3$$

This is a 0.6 percent change in volume, which is an appreciable change in $V$ for many purposes. It is therefore necessary to stipulate the temperature at which $V$ should be measured if the $\gamma$ coefficients in Table 11.3 are to apply. The values given there are for $T = 20°C$. Of course, for small temperature changes not too far removed from $20°C$, we can compute $\Delta V$ to fairly good precision using $V$ measured anywhere in this small temperature range.

Table 11.3 shows that the linear expansion coefficient for a solid is essentially one-third its volume expansion coefficient. This is a general rule for solids that expand to the same extent in all directions. Problem 45 asks you to show why this rule follows from the definitions of $\alpha$ and $\gamma$.

---

### Example 11.8
A slab of concrete in a highway is 20 m long. How much longer is it at $35°C$ than at $-15°C$?

*Reasoning*  We have

$$\Delta L = \alpha L \, \Delta T$$

From our tabulated values, $\alpha \approx 10 \times 10^{-6}/C°$, and so

$$\Delta L = (10^{-5}/C°)(20 \text{ m})(50 \text{ C}°) = 0.010 \text{ m}$$

An expansion crack of this magnitude would be required or else the concrete might buckle.  ∎

---

### Example 11.9
At $20°C$, the diameter of a circular hole in a brass sheet is 2.000 cm. How large is the diameter when the sheet is heated to $220°C$?

*Reasoning*   The sheet expands in the same way whether or not the hole is there. Thus the hole expands exactly as the circular piece of metal that originally filled the hole would expand. The hole's diameter obeys the relation

$$\frac{\Delta L}{L} = \alpha \, \Delta T$$

Because $L$ is given for 20°C, we can use the data of Table 11.3 directly:

$$\Delta L = (2.000 \text{ cm})(19 \times 10^{-6}/\text{C}°)(200 \text{ C}°) = 0.0076 \text{ cm}$$

The new diameter of the hole is 2.008 cm. Notice that, in this case, it is of no significance whether the value of $L$ at 20 or 220°C is used.   ∎

---

### Example 11.10

By how much does the volume of 100.0 cm³ of benzene (measured at 10°C) change as the benzene is heated to 30°C?

*Reasoning*   Because the constants in Table 11.3 are given for 20°C, we must first find the volume at that temperature, $V_{20}$:

$$\frac{\Delta V}{V_{20}} = \gamma \, \Delta T$$

Since $\Delta V = V_{20} - V_{10} = V_{20} - 100 \text{ cm}^3$, this becomes

$$V_{20} - 100 \text{ cm}^3 = V_{20}(1.24 \times 10^{-3}/\text{C}°)(10 \text{ C}°)$$
$$V_{20} = 101.26 \text{ cm}^3$$

Using this value for $V$ in $\Delta V/V = \gamma \, \Delta T$, we have

$$\Delta V = (101.3 \text{ cm}^3)(1.24 \times 10^{-3}/\text{C}°)(10 \text{ C}°) = 1.25 \text{ cm}^3$$

Therefore, the volume at 30°C is

$$V_{30} = 101.26 + 1.25 = 102.5 \text{ cm}^3$$

If we had not corrected $V$ to 20°C, we would have found $V_{30} = 102.4$ cm³. For most purposes, this difference in the two values is unimportant, and so it is usually not necessary to correct for the temperature at which $\alpha$ and $\gamma$ are measured, provided the temperature involved is not far removed from 20°C.   ∎

---

*Exercise*   How much benzene (at 30°C) must be removed from the flask to give a volume of 100.0 cm³ at 25°C?   *Answer: 1.88 cm³*   ∎

---

## 11.9  Transfer of heat: conduction

When you hold a metal spoon in hot water, heat flows along its handle to your hand. The interpretation of this is simple. Heat energy enters the spoon from the hot water and causes the atoms in that end of the spoon to acquire large thermal energy. They then vibrate with very large energy and, when they collide with their cooler

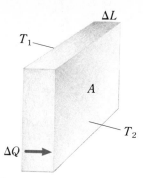

**FIGURE 11.6**

Because $T_1 > T_2$, heat flows through the slab in the direction shown.

neighbors, pass energy along to them. Thus thermal energy is passed along the spoon from the hot end to the cool end. Eventually the whole spoon becomes hot. This method of heat transfer is called **thermal conduction.**

In conduction, heat is transferred through a material by the collisions of adjacent atoms or molecules.

Conduction occurs at different rates in different materials. A wooden stick can burn at one end, and still the other end remains relatively cool, but a metal knife or spoon transmits heat rapidly from one end to the other. The ability of a material to conduct heat depends on its atomic structure. Metals have electrons within them that can move rather freely throughout the metal. As they move, they carry thermal energy from one part of the metal to another. Hence metals, because they contain many free electrons, are excellent heat conductors.

We use the experiment sketched in Fig. 11.6 to describe heat conduction quantitatively. A slab of material of thickness $\Delta L$ and face area $A$ has a temperature differential $T_1 - T_2 = \Delta T$ impressed across it. The rate of heat flow $\Delta Q/\Delta t$ through the slab depends on $\Delta T$, on $A$, and on $\Delta L$. It is directly proportional to $\Delta T$ and to $A$ (in other words, the rate of heat flow increases as $\Delta T$ and/or $A$ increases) and inversely proportional to $\Delta L$ (the rate decreases as $\Delta L$ increases). Experiment shows that

$$\frac{\Delta Q}{\Delta t} = k \frac{A \, \Delta T}{\Delta L} \tag{11.4}$$

The quantity $\Delta T/\Delta L$ is often called the *temperature gradient.* The constant $k$ depends on the material of which the slab is made and is called the *thermal conductivity* of the material. Typical values for $k$ are given in Table 11.4 for $\Delta Q/\Delta t$ in watts, $A$ in square meters, $\Delta L$ in meters, and $\Delta T$ in Celsius degrees or kelvins. Conversions to other units are given in the footnote. As you see, $k$ is large for good thermal conductors, such as metals, and small for thermal insulators.

The thermal conductivity of an object influences how hot the object feels to the touch. A hot metal burns your hand easily because heat can flow readily from it to your hand. A piece of wood at the same temperature does not burn your hand nearly as badly. Because of the much lower thermal conductivity of wood, only the thermal energy at the point of contact can easily reach your hand. In effect, your hand quickly cools the wood at the point of contact. Using similar reasoning, can you explain why a cold tile floor seems warmer to your bare feet if you stand on a carpet that lies on it?

■ **TABLE 11.4**
**Thermal conductivities***

| Material | $k$ $(W/K \cdot m)$† |
|---|---|
| Silver | 430 |
| Copper | 400 |
| Aluminum | 240 |
| Brass | 105 |
| Concrete | 0.8 |
| Glass | 0.8 |
| Brick | 0.6 |
| Asbestos paper | 0.2 |
| Rubber | 0.2 |
| Wood | 0.08 |
| Bone | 0.042 |
| Muscle | 0.042 |
| Glass wool (fiberglass) | 0.04 |
| Plastic foam | 0.03 |
| Fat | 0.021 |

\* These are approximate values; $k$ varies somewhat with temperature.

† 1 W/K · m = (1/418.4)(cal/s)/ C° · cm = 6.94 Btu · in/h · ft² · F°

### Example 11.11

The interior dimensions of a cubical soft drink cooler are $30 \times 30 \times 30$ cm. Each wall is 4.0 cm thick and is made of plastic for which $k = 0.032$ W/K · m. Ice inside the box maintains the interior temperature at 0°C. How much ice melts each hour when the outside temperature is 25°C?

**Reasoning** We approximate the situation by considering the heat flow through the combined area of six slabs, each of dimensions $30 \times 30 \times 4$ cm. The flow into the cooler is

$$\frac{\Delta Q}{\Delta t} = 6k \frac{A \, \Delta T}{\Delta L} = 6(0.032 \text{ W/K} \cdot \text{m}) \frac{(0.090 \text{ m}^2)(25 \text{ K})}{0.040 \text{ m}}$$

$$= 10.8 \text{ W} = 2.58 \text{ cal/s}$$

The heat flow in 1 h is 3600 times this value:

$$\Delta Q \text{ in } 1 \text{ h} = 9290 \text{ cal}$$

Because 80 cal melts 1 g of ice, the total ice melted in 1 h is

$$\text{Ice melted} = \frac{9280 \text{ cal}}{80 \text{ cal/g}} = 116 \text{ g} \quad \blacksquare$$

---

*Exercise* How much ice is required for walls 2.0 cm thick and an outside temperature of 18°C? *Answer: 167 g* ■

---

(a)

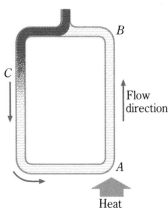

(b)

**FIGURE 11.7**

When heat is applied to the liquid in the tube, the dye shows that the liquid circulates counterclockwise. Heat is carried along by the liquid in a process known as convection.

## 11.10 Transfer of heat: convection

A simple experiment devised to illustrate convection is shown in Fig. 11.7. When the glass tube is filled with water, a little colored dye placed near the neck remains nearly motionless (part *a*). However, when the tube is heated at one corner, as in part *b*, the liquid begins to flow counterclockwise around the tube, carrying the dye with it.

The reason for this motion is quite simple. A heated liquid or gas expands, and so the water in the lower right corner of the tube at *A* expands when heated. It is now less dense than the rest of the liquid. The more dense column of liquid on the left can no longer be supported by the less dense column on the right. The left column falls, pushing the water along in the tube. Hence the liquid on the right flows upward. It cools as it moves along, and by the time it reaches point *C* it is cooler and denser than it was at point *A*. In summary, liquid heated at *A* rises to *B*. In so doing, it carries heat with it, and so heat is moved from *A* to *B* by the motion of the liquid. This means of heat transfer is called **convection.**

Fluid flow carries heat from place to place in convection.

Conduction does not involve the motion of the molecules over large distances. Heat is transferred from molecule to molecule by collision. In convection, however, the molecules of the transferring material move along with the heat. Only liquids and gases can transfer heat by convection because only in these materials can the molecules move over large distances.

Many homes are heated by air convection (Fig. 11.8). Even in heating systems without fans, the circulatory movement of the air is appreciable. For example, to a person standing over a hot-air register above an air furnace, the rush of hot air from the register is often quite noticeable. Proper design of such convection systems must allow the cool air to return to the furnace much as the cool liquid circulates back to point *A* in Fig. 11.7*b*. The purpose of the cold-air registers in such heating systems is to return the cool air to the furnace.

Weather phenomena are partly the result of convective air currents. Thermal currents that circulate air near the edges of mountain ranges are particularly interesting. Quite large effects are noticed at various fixed times of day as the cool air from the mountains flows down and causes the warm air on the nearby plains to rise. The Gulf Stream and the Japan current are other interesting examples of large-scale transfer of heat by convection.

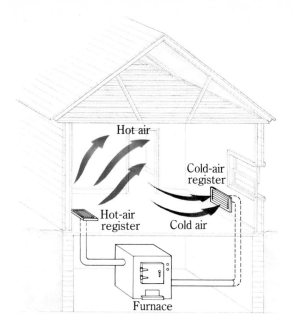

**FIGURE 11.8**
In convection, a moving fluid transports the heat energy.

**Transfer of heat: radiation**

We know that the sun warms the earth. It is, in fact, our major source of heat. We can easily see that the heat from the sun is not transferred to us by either conduction or convection. There are very few molecules in the vast reaches of space between us and the sun. Hence vibrational transfer by conduction and circulatory transfer by convection are impossible. We have here a case of heat transfer through a vacuum, that is, through empty space. This method of heat transfer is called **radiation.**

For many years, even to the first decade of this century, most scientists could not conceive of heat and light from the sun traveling through nothing. They therefore postulated that all space was filled by a "luminiferous ether." It was not until Einstein showed (in 1905) that the ether concept was neither useful nor verifiable that it was discarded. However, even before that time, the concept had run into formidable conceptual difficulties. We no longer consider the ether concept either necessary or convenient. As we shall see in our study of electromagnetic radiation, radiation can be understood without it.

Although an understanding of radiation requires knowledge that you will acquire in the study of electricity, we can even now obtain valuable information concerning it. All objects emit heat radiation, which is also referred to as infrared radiation. You have doubtless seen photographs taken by infrared radiation. The "snooperscope" used by the military allows one to see warm objects in the dark by means of the infrared radiation they emit. Hot objects emit heat radiation more intensely than colder objects. This fact is summarized in what is known as **Stefan's law:** the heat energy radiated per second by an object is

$$\frac{\Delta Q}{\Delta t} = e\sigma A T^4 \tag{11.5}$$

where $A$ is the object's surface area and $T$ is its absolute temperature. The con-

# Our fragile climate

The earth is a very special planet in that, unlike other planets known to us, its climate is able to sustain life. We sometimes fail to realize the tenuous grasp life has on earth. Life as we know it requires unique conditions. Plant life, without which we could not exist, requires an average temperature not far different from that of the earth. Moreover, if the earth's average temperature were to rise several degrees, most of the polar ice caps would melt and the seas would flood much of the land. Less obvious is the fact that ozone in the upper atmosphere shields us from cancer-causing ultraviolet radiation from the sun. We have only recently learned that the atmospheric ozone is being depleted by manufactured gases such as fluorocarbons. Once widely used as aerosol propellants and refrigerants, fluorocarbons are now being more wisely controlled. However, even the exhaust gases of jet aircraft place the atmospheric ozone in peril.

Carbon dioxide is one of the most critical gases in the earth's atmosphere. It acts as the earth's thermostat by means of what is called the greenhouse effect. As you know, the sun's rays keep the interior of a greenhouse warm even on a very cold day. Sunlight carries energy through the glass into the greenhouse; there it is absorbed as heat by the plants, soil, and so forth. But these objects, being at a much lower temperature than the sun, reradiate the energy at much longer wavelengths, wavelengths in the infrared. Since glass is not transparent to infrared, the energy received from the sun is trapped within the greenhouse; the greenhouse retains the heat energy and remains warm. For the earth as a whole, the carbon dioxide in the atmosphere plays the role of the greenhouse glass. It allows sunlight in but prevents heat radiation from escaping. We humans place additional carbon dioxide in the atmosphere every time we burn fossil fuels. As a result, the average temperature of the earth is slowly increasing. Although much research is being done on this potentially dangerous phenomenon, we do not yet know whether we are approaching the limits of disaster.

Although we often act as though our atmosphere is imperiled only by such obvious catastrophes as nuclear accidents and chemical releases, it is clear that we must protect against more subtle mechanisms as well. Life on earth is possible only because of a precarious balance in the gas layer that surrounds our planet.

---

stant $\sigma$ is a universal constant called the *Stefan-Boltzmann constant;* its value is $5.67 \times 10^{-8}$ W/m$^2 \cdot$ K$^4$. The constant $e$ depends on the nature of the radiating surface and ranges from zero to unity. It is called the *emissivity* of the surface. Dark, rough surfaces (such as a soot-blackened fireplace) have emissivities close to unity. (An object that has $e = 1.00$ is called a *blackbody.*) Shiny objects have lower $e$ values. For example, $e$ for polished copper is about 0.3. Human skin has $e$ values that range from 0.6 for light skin to about 0.8 for dark skin.

Notice how strongly the heat radiated depends on temperature; it varies as $T^4$. Although objects at all temperatures, even those near absolute zero, radiate heat, radiation increases rapidly with temperature. This is the basis for the use of thermograms (see Color Plate I) for determining temperature differences. As you look at the objects around you, each is radiating energy. Equilibrium is established when each object emits as much radiation as it absorbs from its surroundings.

**Example 11.12**

Compare the energy radiated per unit area of your body with that radiated by a chair with the same emissivity. Assume your body temperature to be 37°C and the chair temperature to be 15°C.

**Reasoning** We wish to compare the quantities

$$\left(\frac{1}{A}\frac{\Delta Q}{\Delta t}\right)_{\text{ch}} = e\sigma T_c^4 \quad \text{and} \quad \left(\frac{1}{A}\frac{\Delta Q}{\Delta t}\right)_{\text{b}} = e\sigma T_b^4$$

Therefore

$$\frac{\text{Body radiation}}{\text{Chair radiation}} = \frac{T_b^4}{T_c^4} = \left(\frac{310}{285}\right)^4 = 1.40$$

Even for such a small temperature difference, the radiation rates differ by 40 percent. ∎

---

## 11.12 Home insulation

Anyone who has to pay to heat or cool a home is interested in thermal insulation. It is obvious from a glance at Table 11.4 that metals are the worst insulators and plastic foam is among the best. Most modern buildings are heavily insulated by layers of foamed plastic or glass wool (fiberglass). These materials are very good insulators because they trap air, one of the very best insulators. Air by itself, however, undergoes convection, and its true potential is realized only when it is held in place by a porous material such as glass wool.

Most walls are layered structures. Suppose a wall consists of several layers whose conductivities and thicknesses are $k_1$ and $L_1$, $k_2$ and $L_2$, and so on. It turns out that the heat flow through a three-layer wall is

$$\frac{\Delta Q}{\Delta t} = \frac{A\,\Delta T}{(L_1/k_1) + (L_2/k_2) + (L_3/k_3)}$$

where $\Delta T$ is the temperature difference across the wall. Additional layers simply add more terms to the denominator. The quantities $L_1/k_1$ and so on measure each layer's resistance to heat flow and are referred to as the **R values** for each layer. In terms of them,

$$\frac{\Delta Q}{\Delta t} = \frac{A\,\Delta T}{R_1 + R_2 + \ldots + R_N} = \frac{A\,\Delta T}{R} \tag{11.6}$$

for $N$ layers. Typical $R$ values of interest are given in Table 11.5. We give $R$ in both SI and British units (ft² · F° · h/Btu), since the latter units are used frequently. The conversion factor is 1 ft² · F° · h/Btu = 0.176 m² · K/W.

To see the utility of $R$ values, consider a wall that consists of three layers: 2.00 cm of wood, 9 cm of glass wool, and 1.0 cm of gypsum board. Its total $R$ value is the sum of the individual values. From Table 11.5,

$$R = 0.185 + 1.95 + 0.060 = 2.20 \text{ m}^2 \cdot \text{K/W}$$

### ■ TABLE 11.5 *R* factors (approximate)

| Material | Thickness (cm) | R(m² · K/W) | R (ft² · F° · h/Btu) |
|---|---|---|---|
| Solid wood | 2.00 | 0.185 | 1.05 |
| Plywood | 1.30 | 0.111 | 0.63 |
| Fiberboard (insulating) | 1.90 | 0.370 | 2.1 |
| Gypsum board | 1.00 | 0.060 | 0.34 |
| Carpet plus pad | . . . . | 0.35 | 2.0 |
| Asphalt shingles | . . . . | 0.070 | 0.4 |
| Concrete (cast) | 20 | 0.11 | 0.64 |
| Concrete block: | | | |
| Normal | 20 | 0.20 | 1.1 |
| Lightweight | 20 | 0.35 | 2.0 |
| Glass wool (fiberglass) | 2.5 | 0.65 | 3.7 |
| | 9 | 1.95 | 11 |
| | 15 | 3.3 | 19 |
| Window (single-pane) | . . . . | 0.18 | 1 |
| Window (double-pane) | . . . . | 0.35 | 2 |

We can now use Eq. 11.6 with this total $R$ value to compute heat flow through the wall. Notice that most of the thermal resistance is due to the glass wool insulation. Even a layer of concrete blocks has much less effect than a 6-in-thick batt of glass wool.

Heat loss from buildings is vividly shown by thermograms such as the one shown in Color Plate I. An imaging device that is sensitive to heat radiation shows clearly the areas of a house that radiate extraordinary amounts of heat and thus provides a means for detecting the principal sites of heat leaks from a home or other building.

---

### Example 11.13
The wall for which we have just calculated an $R$ value of 2.20 m² · K/W has an area of 5 × 3 m². How much heat is lost through it each hour when the temperature is 20°C inside and −10°C outside?

*Reasoning*   We have that

$$\frac{\Delta Q}{\Delta t} = \frac{A \, \Delta T}{R_1 + R_2 + R_3} = \frac{(15 \text{ m}^2)(30 \text{ K})}{2.20 \text{ m}^2 \cdot \text{K/W}} = 204 \text{ W}$$

In 1 h, $\Delta t = 3600$ s, and so

$$\Delta Q = (204 \text{ J/s})(3600 \text{ s}) = 7.3 \times 10^5 \text{ J}$$

We have neglected the fact that there is usually a layer of stagnant air at the surface of the wall. This contributes an $R$ value of about 0.04 m² · K/W to each wall surface. However, a strong wind decreases this effect to nearly zero.   ■

## 11.13 Humidity

It is well known that on a day when the humidity is high, the air contains a great deal of water vapor. Humidity is a measure of the water content of the air. To be precise, **relative humidity** (RH) *is defined to be the ratio of the mass of water vapor per unit volume in the air to the mass per unit volume required to produce saturation at the same temperature.* As pointed out in Sec. 11.5, when a saturated vapor is in contact with a liquid, identical numbers of molecules leave and return to the surface of the liquid in a given time. Hence, at saturation, no net evaporation occurs. If the vapor is more than saturated, that is, **supersaturated,** drops condense from the vapor, and fog or rain is the result.

Because the pressure of an ideal gas is proportional to the number of molecules in it, the definition of relative humidity is often expressed in terms of pressures rather than masses. Water vapor is nearly an ideal gas, and so the two definitions are nearly identical. In equation form,

$$\text{Relative humidity} = \frac{m}{m_s} \approx \frac{P}{P_s} \tag{11.7}$$

where $m$ and $P$ are the mass per unit volume and pressure of the water vapor in the air and $m_s$ and $P_s$ are the respective values for saturated vapor. Some data for saturated water vapor at various temperatures are given in Table 11.6.

According to Table 11.6, saturated air at 20°C contains 17.1 g/m³ of water. Suppose that the air actually did contain 17.1 g/m³ of water vapor. If it were at any temperature above 20°C, it could contain still more vapor. However, if this amount of water vapor were present and the air were to cool to below 20°C (as it would, perhaps, after the sun went down), it would become supersaturated. Water drops would begin to fall out of the air in the form of fog, dew, or rain. *We term the temperature at which the air just becomes saturated the* **dew point.**

The dew point of air is a useful quantity. Suppose the temperature of the air on a certain day is 32°C. On that day, the meteorologist at the weather bureau cools some of the air down until fog or dew begins to settle out. Suppose she finds the dew point to be 16°C. She now knows from Table 11.6 that the air contains 13.50 g/m³ of water vapor, since this is the value for saturated vapor at 16°C. However, the actual air temperature is 32°C, and she knows that saturated air at this temperature holds 33.45 g/m³ of water. From this, she computes the relative humidity to be

$$\text{RH} = \frac{m}{m_s} = \frac{13.50}{33.45} \approx 0.40$$

We usually multiply this answer by 100 and say that the relative humidity is 40 percent.

The relative humidity can also be measured in other ways. One common method, the **wet-bulb–dry-bulb method,** makes use of the fact that liquids cool when they evaporate and that no evaporation occurs when the vapor is saturated. Hence, if the reading of a dry thermometer is compared with that of a thermometer that has a wet cloth wrapped around its bulb, the wet-bulb thermometer will usually read cooler than the dry-bulb thermometer. The difference in temperatures is a direct measure of relative humidity; the lower the relative humidity, the greater the difference. Tables have been compiled relating relative humidity to this temperature difference so that the relative humidity can be determined by reading the two thermometers.

■ **TABLE 11.6**
**Properties of saturated water vapor**

| Temperature (°C) | Water/volume (g/m³) |
|---|---|
| −8 | 2.74 |
| −4 | 3.66 |
| 0 | 4.84 |
| 4 | 6.33 |
| 8 | 8.21 |
| 12 | 10.57 |
| 16 | 13.50 |
| 20 | 17.12 |
| 24 | 21.54 |
| 28 | 26.93 |
| 32 | 33.45 |
| 36 | 41.82 |

High relative humidity in summer often causes us discomfort. This results from the fact that when we are hot, we perspire and the evaporation of the perspiration cools us. If the relative humidity is 100 percent, however, there can be no evaporation and therefore no cooling. That is why we do not feel the heat nearly so much in a hot, dry climate as in a hot, moist climate.

## Minimum learning goals

When you finish this chapter, you should be able to

1 Define (*a*) heat energy, (*b*) thermal energy, (*c*) calorie and Calorie, (*d*) specific heat capacity, (*e*) heat of vaporization, (*f*) vapor pressure, (*g*) heat of fusion, (*h*) calorimeter, (*i*) linear and volume thermal expansion coefficients, (*j*) heat conduction, (*k*) heat convection, (*l*) heat radiation, (*m*) Stefan's law, (*n*) *R* factor, (*o*) relative humidity, (*p*) dew point.

2 Use the equation $\Delta Q = cm\,\Delta T$ to solve simple problems involving heating and cooling.

3 Explain why the specific heat capacity of a substance depends upon its structure and complexity.

4 Explain why evaporation leads to cooling.

5 Explain why the boiling point of a liquid changes as the external pressure changes.

6 Describe qualitatively how the temperature of a crystalline substance changes as it is slowly heated, melted, heated further, and vaporized.

7 Solve simple problems in calorimetry. Explain why the law of conservation of energy is basic to the solution.

8 Use thermal expansion coefficients in simple situations.

9 Determine how much heat flows through a slab of material when the temperatures of the two faces of the slab are given.

10 Find the *R* factor for a wall consisting of several layers.

11 Calculate relative humidity when the dew point and air temperature are given.

## Questions and guesstimates

1 Which would you expect to have the larger specific heat capacity, 10 g of oxygen ($O_2$) gas or 10 g of argon (Ar) gas? Both contain about the same number of molecules. (Why?)

2 A student has a thermos jug containing an unknown substance at temperature $T_1$. After hot water at $T_2 > T_1$ is added, the temperature in the jug is still $T_1$. The student concludes that the material in the jug has an infinite specific heat capacity. Explain why the experiment implies that $c = \infty$. What is the probable explanation of these experimental results?

3 Can heat be added to something without its temperature changing? What if the "something" is a gas? A liquid? A solid?

4 A certain type of wax melts at 60°C. Describe an experiment by which you could determine its heat of fusion.

5 It is possible to make water boil furiously by cooling a flask of water that has been stoppered when the water was boiling at 100°C. Explain.

6 Why does cold metal feel cooler than wood at the same temperature?

7 When temperatures slightly below freezing are expected, farmers sometimes protect their fruits and vegetables by misting them with water. What is the principle behind this procedure?

8 Why is a burn caused by steam at 100°C apt to be much worse than one caused by hot water at 100°C?

9 Temperature fluctuations are much less pronounced on land close to large bodies of water than they are in the central regions of large land masses. Explain.

10 It is well known that a room filled with people becomes very warm unless it is properly ventilated. Assuming that a person gives off heat equivalent to the calories he or she burns throughout the day, estimate how much the temperature of your classroom would rise in 1 h if there were no heat loss out of the room.

11 About how much water would have to evaporate from the skin of an average-size person to cool his or her body

by 1 C°? How does this fit in with what you have heard about the effect of perspiration on the body? ($c_{body} \cong 0.83$ cal/g · C°)

**12** If ice is subjected to high pressure, its melting point is decreased to below 0°C. To a rough approximation, the melting temperature decreases by about 5 C° for each 6.0 × 10⁷ Pa of applied pressure. Estimate the melting temperature of ice beneath an ice skater's skate.

**13** Estimate the temperature of the sun's surface from the following facts: radiation reaching the earth from the sun is 1340 J/m² · s, the radius of the sun is 7 × 10⁸ m, and the sun-to-earth distance is 1.5 × 10¹¹ m.

## Problems

**1** How much heat (in calories and in joules) is needed to raise the temperature of 500 g of water from 12°C to 70°C?

**2** How much heat (in calories and in joules) must be removed from 150 g of water to cool it from 90°C to 55°C?

**3** How much heat (in calories and in joules) is removed in changing the temperature of 80 g of copper from 90°C to 20°C?

**4** How much heat (in calories and in joules) is required to heat 50 g of iron from 15°C to 85°C?

**5** How much heat is given off as 200 g of ethanol at 78°C is condensed from vapor to liquid and then cooled to 40°C?

**6** How much heat is required to heat 80 g of mercury from 20°C to 357°C and then vaporize it?

**7** How much heat must be removed from 200 g of water at 30°C to change it to ice at 0°C?

**8** How much heat must be added to 150 g of aluminum to change it from a solid at 20°C to a liquid at 660°C?

■ **9** An 18-g ice cube (at 0°C) is dropped into a glass containing 200 g of water at 25°C. If there is negligible heat exchange with the glass, what is the temperature after the ice melts?

■ **10** Molten lead at 327°C is poured into a hole in a block of ice at 0°C. How much ice is melted by 60 g of lead? Assume that thermal equilibrium with the block is achieved.

■ **11** Suppose that 40 g of solid mercury at its freezing point (−39°C) is dropped into a mixture of water and ice at 0°C. After equilibrium is achieved, the mercury–ice–water mixture is still at 0°C. How much additional ice is produced by the addition of the mercury?

■ **12** How much perspiration must evaporate from a 4.0-kg baby to reduce its temperature by 2 C°? The heat of vaporization for water at body temperature is about 580 cal/g.

■ **13** The average energy reaching us from the sun each second is 0.134 J/cm². Most of this energy is absorbed as it passes through the earth's atmosphere. Assume that 0.10 percent of it strikes the surface of a lake and evaporates the water. How much water evaporates from 1 m² in 1 h? Use $H_v = 590$ cal/g.

■ **14** When a volatile liquid, such as alcohol or ether, evaporates from your skin, noticeable cooling occurs. Suppose that 0.030 g of dichloroethane ($H_v \cong 85$ cal/g) evaporates from a 1-cm² area of your skin and cools a surface layer 0.035 cm thick. (*a*) By how much is the temperature of the skin lowered? Assume *c* for the skin to be 0.75 cal/g · C° and its density to be 0.95 g/cm³. Neglect the fact that the skin may crystallize. Also consider the value of *c* for dichloroethane to be negligibly small. (*b*) Why does the temperature of the skin not decrease as much as you calculate?

■ **15** If 200 g of lead at 100°C is dropped into 50 g of water at 20°C contained in a copper can of mass 50 g, what is the resulting temperature? Assume that the can maintains the same temperature as the water.

■ **16** To 100 g of water contained in a 50-g copper can at 35°C is added 20 g of ice at −10°C. What is the final temperature? Assume that the can maintains the same temperature as the water.

■ **17** How much steam at 100°C is needed to change 40 g of ice at −10°C to water at 20°C if the ice is in a 50-g copper can? Assume that the can maintains the same temperature as the ice and water.

■ **18** A 70-g can (*c* = 0.20 cal/g · C°) contains 400 g of water and 200 g of ice at equilibrium. To this is added a 300-g piece of hot metal (*c* = 0.10 cal/g · C°). The final temperature is 10°C. What was the original temperature of the metal?

■ **19** A certain calorimeter has a water equivalent of 4.9 g. That is, in heat exchanges, the calorimeter behaves like 4.9 g of water. It contains 40 g of oil at 50.0°C. When 100 g of lead at 30.0°C is added, the final temperature is 48.0°C. What is the specific heat capacity of the oil?

■ **20** Benzene boils at about 80°C. Benzene vapor at 80°C is bubbled into a calorimeter, the water equivalent of which is 20 g (see Prob. 19), containing 100 g of oil (*c* = 0.50 cal/g · C°) at 20°C. The final temperature after 7.0 g of benzene condenses is 30°C. What is the heat of vaporization of benzene? For benzene, $c = 0.40$ cal/g · C°.

■ **21** Assuming that the total heat of vaporization of water, 539 cal/g, can be used to supply the energy needed to separate 1 g of water molecules from one another, how much energy per molecule is needed for this purpose? Find the ratio of this energy to $kT$ at the boiling point.

■ **22** Water flows down a waterfall 20 m high in a continuous stream. If all of the water's lost gravitational potential energy appears as thermal energy in it, how much warmer is the water at the bottom than at the top?

■ **23** An electric heater supplies 1800 W of power in the form of heat to a tank of water. How long does it take to heat 200 kg of water in the tank from 20 to 70°C? Assume heat losses to the surroundings to be negligible.

■ **24** Cool water at 19.0°C enters a hot-water heater from which warm water at 80°C is being drawn at a rate of 300 g/min. How much electric power (in watts) does the heater consume in order to provide hot water at this rate? Assume that there is negligible heat loss to the surroundings.

■ **25** A 70-kg person consumes about 2500 Calories of food per day. If all this food energy were changed to heat and none of the heat escaped, what would be the temperature rise of the person's body? For the body, $c \cong 0.83$ cal/g · C°.

■ **26** Thirty people are sitting in a rectangular room that has dimensions 3.5 by 10 by 10 m. On the average, each person consumes 2500 Calories per day and loses one-twenty-fourth of this amount of heat each hour. By how much do the people raise the temperature of the air in the room in 1 h? For simplicity, assume the air to be $N_2$ at a density of 1.29 kg/m³. Neglect the volume occupied by the people. Assume the specific heat of air to be 0.20 cal/g · C°.

■ **27** In an extruder used to make synthetic fibers, a piston applies a pressure of 8.0 MPa to the molten plastic and forces it through a tiny nozzle at a rate of 0.00100 cm³/s. Assume that all the energy loss occurs as friction work in the plastic as it goes through the nozzle. (*a*) How much friction work is done each second? (*b*) By how much is the temperature of the plastic raised as it is forced through the nozzle? Assume $c_{plastic} = 0.20$ cal/g · C° and $\rho = 1000$ kg/m³. *Hint:* Think of the nozzle as a capillary tube of area $A$ and show that the work done in extruding a length $L$ is *PAL*.

■ **28** A 40-kg girl running at 5.0 m/s while playing basketball falls to the floor and skids along on her leg until she stops. How many calories of heat are generated between her leg and the floor? Assuming that all this heat energy is

confined to a volume of 2.0 cm³ of her flesh, what will be the temperature change of the flesh? Assume $c = 0.83$ cal/g · C° and $\rho = 950$ kg/m³ for flesh.

■ **29** A mass $m$ of lead rests at the bottom of a closed cardboard cylinder $L$ units long (Fig. P11.1). When the cylinder is quickly inverted by rotating it about its center, the lead falls through the length of the tube. Show that the rise in temperature of the lead after $n$ reversals is

$$\Delta T = \frac{ngL}{c_{Pb}}$$

What units must be used for $c_{Pb}$?

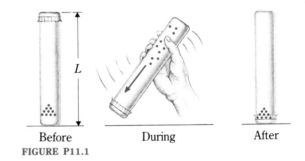

Before          During          After

**FIGURE P11.1**

■ **30** A certain 6-g bullet melts at 400°C and has a specific heat capacity of 0.20 cal/g · C° and a heat of fusion of 15 cal/g. (*a*) How much heat is needed to melt the bullet if it is originally at 0°C? (*b*) What is the slowest speed at which the bullet can travel if it is to just melt when it is stopped suddenly?

■ **31** A 2.2-g lead bullet is moving at 180 m/s when it strikes a bag of sand and is brought to rest. (*a*) Assuming that all the friction work is transferred to thermal energy in the bullet, what is the rise in the temperature of the bullet as it is brought to rest? (*b*) Repeat if the bullet lodges in a 50-g block of wood that is free to move.

■ **32** A hailstone made of ice at 0°C is falling at its terminal velocity. What fraction of it melts for each meter that it falls? Assume that 25 percent of the thermal energy generated goes into the hailstone.

**33** A meter stick made from aluminum is heated from 20°C to 35°C. What is its fractional change in length?

**34** A brass sphere has a radius of 2.0 cm at 20°C. What is its diameter at −80°C?

**35** A certain straight roadway is made of concrete slabs, each 25 m long. How large an expansion gap should be left between slabs at 20°C if they are to just touch at 45°C?

**36** The marks on an aluminum measuring tape are made when the tape is at 20°C. What is the percentage error due to contraction when the rule is used at −5°C?

**37** A salad-dressing jar has an aluminum screw-type lid. At 20°C, the lid fits so tightly that it does not screw off. By what factor will the lid's diameter expand if it is heated to 70°C? Why should you not heat the lid any longer than necessary?

**38** It is common machine-shop practice to "shrink-fit" cylindrical rods into holes in wheels, blocks, and plates. Suppose a 2.000-cm-diameter rod is to be fitted into a 1.985-cm-diameter hole in a brass block. How much must the block be heated for the cool rod to fit into the hole?

**39** A heat-resistant glass flask is calibrated to hold exactly 100 cm³ at 20°C. How much more will it hold at 30°C? *Hint:* The hollow flask expands as though it were a solid volume.

**40** A 250-cm³ volumetric flask is filled with benzene at 30°C. How much more benzene must be added if the flask is to be full after it is cooled to 20°C? (Ignore the expansion of the flask.)

**41** An iron beam 10 m long has its two ends embedded in substantial concrete pillars. If the structure is made at 5°C, what force does the beam exert on the pillars when the temperature is 30°C? The cross-sectional area of the beam is 60 cm². (Use $Y = 19 \times 10^{10}$ Pa.)

**42** A steel wire at 800°C is stretched tight between two fixed points. At what temperature does the wire break as it is cooled? Assume that the wire obeys Hooke's law. The strength of steel is $4.8 \times 10^8$ N/m².

**43** Show that the coefficient of volume thermal expansion for an ideal gas is $1/T$, provided the expansion is done at constant pressure.

**44** A uniform solid brass sphere of radius $b_o$ and mass $M$ is set spinning about a diameter with angular speed $\omega_o$. If its temperature is increased from 20 to 280°C without disturbing the sphere, what is its new (*a*) angular speed and (*b*) rotational kinetic energy?

**45** A cube of metal with original edge length $L_o$ has a volume $(L_o + \Delta L)^3$ after its temperature is increased by $\Delta T$. Use this fact to show that the volume thermal expansion coefficient for the material of the cube is, to first order, 3 times larger than its linear thermal expansion coefficient.

**46** It is desired to have a steel rod and an aluminum rod lengthen by the same amount for the same temperature change. What should be the ratio of the lengths of the two rods?

**47** How much heat flows through an 8.0-m² sheet of 1.25-cm-thick plywood ($k = 0.083$ W/K · m) in 1 h if the temperatures on its two faces are $-6$°C and 20°C?

**48** How much heat flows through a 25-cm-thick concrete wall (area = 35 m²) in 1 h if the temperature is 4.0°C on one side and 25°C on the other?

**49** An asbestos sheet 2.0 mm thick is used as a spacer between two brass plates, one at 80°C and the other at 20°C. How much heat flows through 60 cm² of area from one plate to the other in 1 h?

**50** Deep bore holes into the earth show that the temperature increases about 1 C° for each 30 m of depth. Assuming that the earth's crust has a thermal conductivity of about 1.5 W/K · m, how much heat flows out through the surface each second for each 1 m² of surface area?

**51** A brass pipe that is 10 cm in diameter and has a wall thickness of 0.25 cm carries steam at 100°C through a vat of circulating water at 20°C. How much heat is lost per meter of pipe in 1 s?

**52** A sheet of brass 0.50 cm thick has a 0.50-cm-thick rubber sheet sealed to one face. The other side of the brass sheet is connected to a bath maintained at 30°C. The other rubber surface is attached to a circulating bath at 80°C. Find the temperature at the rubber-brass junction.

**53** A cubical box has interior dimensions $40 \times 40 \times 40$ cm and a wall thickness of 3.0 cm. The box contains ice and water at 0°C when the outside temperature is 25°C. If 230 g of ice melt each hour, what is the thermal conductivity of the material the box is made of?

**54** A metal sphere with radius 2.0 cm is heated to 400°C and then suspended by a thin wire in a room at 20°C. At what rate is the sphere emitting heat energy if its emissivity is 0.65? Neglect any radiation the sphere receives from the room.

**55** A white-hot metal wire at 3000 K has a radius of 0.075 cm. Calculate the rate per unit length at which it emits radiation if its emissivity is 0.35. Ignore the radiation it receives from the surroundings.

**56** A solar collection panel for a hot water system receives solar radiation at a rate of 720 W/m². Its effective area is 4.0 m², and so it collects 2880 W. What flow of hot water at 70°C can it sustain if the original temperature of the water is 14°C?

**57** Two sheets of insulation have $R$ values of $R_1$ and $R_2$. Show that the $R$ value for the combination of one sheet on top of the other is $R_1 + R_2$.

**58** What is the $R$ value for a 3-cm-thick layer of (*a*) copper and (*b*) glass? Give your answer in both SI and British units.

**59** In Prob. 53, suppose that 90 g of ice melts in 1 h. What is the approximate $R$ value for the plastic foam from which the box is made? Give your answer in both SI and British units.

**60** What is the dew point on a day when the humidity is 75 percent and the temperature is 24°C?

**61** On a day when the air temperature is 20°C, condensation occurs on a soft drink glass when its temperature is lowered to 12°C. (*a*) What is the relative humidity? (*b*) What mass of water is present in a 1000-cm³ volume of air?

■ **62** A dehumidifier is used to change the relative humidity in a 7 × 4 × 3 m³ room from 95 percent to 60 percent on a day when the temperature is 20°C. How much water does the dehumidifier collect in the process? Assume that the room is sealed and that the temperature remains at about 20°C.

■ **63** A 70-kg aluminum disk that has a radius of 20 cm is spinning at 8.0 rev/s when a friction force applied to its rim causes it to coast to a stop. Assuming that two-thirds of the friction-generated heat appears in the disk, how much is the temperature of the disk increased?

■ **64** A cylindrical hot water tank (radius 18 cm, height 140 cm) is made of aluminum 4.0 mm thick. It is insulated by a 3.5-cm-thick sheath of glass wool ($k = 0.040$ W/K · m). Neglecting heat loss through the ends of the tank, how much average power must be furnished to maintain the water temperature at 65°C when the room temperature is 18°C?

■ **65** A vat contains 5.0 kg of molasses ($c = 4.0$ kJ/kg · K) that is being stirred by a paddle requiring a torque of 0.050 N · m to rotate it at 6.0 rev/s. Assuming no heat loss from the molasses, by how much does the temperature rise as a result of 8 min of stirring?

# 12 Thermodynamics

**L**ONG before the nature of atoms and molecules was known, a powerful way of discussing heat, work, and internal energy had been found. It involves the description of matter in terms of macroscopic properties* such as pressure, temperature, volume, and heat flow. This way of discussing the behavior of objects and substances is called **thermodynamics.** Today, even though we understand quite well how atoms and molecules behave, thermodynamics is still widely used in all branches of science. This chapter introduces this important and useful area of study.

## 12.1 State variables

In thermodynamics, we most frequently discuss the behavior of a definite group of molecules we call a **system.** A typical system may be the molecules in a gas-filled container or those in a solution, or even such a complex system as the molecules in a rubber band. For any very meaningful thermodynamic discussion, the system must be well specified. Only then can we give an unambiguous description of it.

To describe a system, we use quantities that apply either to the whole system or to some well-defined portion of it. Typical measurable quantities are pressure,

---

\* **Macroscopic** properties are those that involve the average effects of billions of molecules.

temperature, and volume. In thermodynamics, we also use such quantities as internal energy, heat, work, and a quantity we shall encounter later called entropy. As the condition of a system changes, these quantities may change. It is important that we know which quantities are suitable for representing the exact condition of the system. Let us now see what they are.

When a container of gas reaches equilibrium, the gas has a definite temperature, pressure, and volume. The ideal-gas law reflects this fact, since it tells us that $PV = nRT$. We see that when any two of these variables are given, the other can be calculated and is therefore known. This particular situation, where the gas (the system) has specified values of $P$, $V$, and $T$, is called a **state** of the system. Whenever the gas is returned to these same values of $P$, $V$, and $T$, its state will be the same. Even though each individual molecule within the system may not be doing exactly the same thing whenever the system is brought to this state, the system as a whole still appears the same when described in terms of macroscopic measurements.

*Certain features of a system are always the same when the system is in a given state. The variables that describe these features are called* **state variables.** For example, $P$, $V$, and $T$ are state variables for a system that consists of a gas. No matter how the system reaches a particular equilibrium state, that state is characterized by the same values for pressure, volume, and temperature. For example, a system consisting of the gas in a car tire can be specified by its $P$, $V$, and $T$ values. No matter what the history of the tire, a given state of the gas within the tire always has the same values for its state variables, $P$, $V$, and $T$.

Another important quantity used to characterize a system is its **internal energy:**

The internal energy $U$ of a system is the total energy of every type possessed by the system.

Internal energy can occur as thermal, chemical, nuclear, or electric energy; indeed, it encompasses all forms of energy. Because energy is conserved, you can keep a balance sheet for it. A system's internal energy is a well-defined quantity. If you know how much internal energy a system has today, you can (in principle, at least) tell how much internal energy it has at a later time. All you need to do is measure the energy changes that the system undergoes in the meantime. Then, like a good accountant, you can total the energy balance sheet and tell how much internal energy the system still possesses. *Internal energy is a state variable.*

Heat is *not* a state variable. There is no way we can evaluate how much heat a system contains. Even though we can compute how much heat must flow into a block of metal, for example, to cause a given temperature change, this same temperature change could have been caused by the action of friction without heat flow into the block. We cannot associate unambiguously a certain amount of heat with a given state of the system. The same change in states can be achieved by adding different quantities of heat. *Heat is not a state variable.*

## 12.2 The first law of thermodynamics

Early workers in thermodynamics developed the idea that energy is conserved. They became convinced that heat is a form of energy and therefore must be considered when accounting for energy gains and losses. Thus they were led to a fundamental relation between heat, work, and internal energy. Let us now see what this relation is.

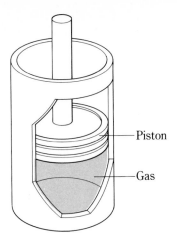

Piston

Gas

**FIGURE 12.1**

What does the first law state when the piston is immovable?

A system in a given state has a definite amount of internal energy. The system might be the gas molecules in the cylinder in Fig. 12.1. We now ask what happens to the system as an amount of heat $\Delta Q$ flows into it. This added energy can be utilized in only two ways: (1) it can lead to an increase in the internal energy of the system or (2) it can provide the system with the energy it needs to do an amount $\Delta W$ of work external to itself. For example, heat added to the gas in the cylinder of Fig. 12.1 can cause two changes: (1) it can raise the temperature of the gas and thus increase its internal energy and (2) it can cause the gas to expand, thereby lifting the piston and allowing the gas to do work on the piston, external work.

If you examine any system, you will find that adding heat to it results in a similar situation. We conclude that, for a system,

$$\left( \begin{array}{c} \text{Heat added} \\ \text{to system} \end{array} \right) = \left( \begin{array}{c} \text{increase in its} \\ \text{internal energy} \end{array} \right) + \left( \begin{array}{c} \text{external work} \\ \text{done by system} \end{array} \right)$$

This statement is called the **first law of thermodynamics.** In equation form,

$$\Delta Q = \Delta U + \Delta W \tag{12.1}$$

Notice that the first law is a result of the existence of the law of conservation of energy. We shall see that it is one of the two pedestals upon which thermodynamics is based.

In using the first law, we must be very careful about signs. The quantity $\Delta Q$ is always the heat that flows *into* the system. If heat flows out of the system, $\Delta Q$ is negative. The quantity $\Delta U$ is the *increase* in the internal energy of the system, and $\Delta W$ is the work done *by* the system. If the gas in Fig. 12.1 lifts the piston, then the gas does external work and $\Delta W$ is positive. If the piston is pushed down by an outside force, however, then $\Delta W$ is negative because the *gas* does negative work. To understand this latter statement, recall that work = force × displacement × cos $\theta$, where $\theta$ is the angle between the force vector and the displacement vector. In Fig. 12.1, $\mathbf{F}$ is the *upward* force the gas exerts on the piston. When the piston moves *downward* a distance $\Delta s$, the work done by the gas is

$$\Delta W = F \, \Delta s \cos 180° = - F \, \Delta s$$

Therefore, *when a gas is compressed, the work done* **by** *it is negative.*

The first law is applicable to all systems, no matter how complex. For example, consider your body as the system. (In order for the system to remain simple, you are not allowed to eat or excrete.) It loses internal energy as the day goes on. The food you have previously eaten is part of your internal energy. This and other forms of internal energy are used up as the day progresses. Much of the energy is lost as heat flows from your body to the surroundings. A smaller amount is lost as your body, the system, does work on the objects of the world about you. Solving the first law for $\Delta U$, we find that the following equation governs your body:

Change in internal energy $= \Delta Q -$ work done by body

Notice that in this case $\Delta Q$ is negative because heat flows from the system rather than into it. Hence $\Delta U$ for the system is negative; your body loses internal energy, and so you tire as the day goes along.

## 12.3 The work done by a system

To use the first law, we must calculate the work done by a system. A situation frequently encountered involves the work done during a volume change. Let us see how to compute it.

Consider a system consisting of a gas enclosed in a cylinder by a movable piston, as in Fig. 12.2. Suppose the gas just supports the weight of the piston, so that the pressure of the gas is maintained constant at a value given by

$$P = \frac{F}{A} = \frac{\text{piston weight}}{\text{piston area}}$$

When the gas is heated, it expands an amount $\Delta V$, as indicated in part b. During the expansion, the piston rises a distance $\Delta y$ and the work ($F \, \Delta y \cos \theta$) done on the piston by the gas is, because $\theta = 0°$ in this case,

$$\Delta W = F \, \Delta y = PA \, \Delta y$$

Since $A \, \Delta y$ is $\Delta V$, the increase in the volume of the gas, we find that

$$\Delta W = P \, \Delta V \tag{12.2}$$

In words,

The work a system does in an expansion $\Delta V$ against a constant pressure $P$ is $P \, \Delta V$.

If the system contracts, then $\Delta V$ is negative and the work done by the system is negative. In that case, the surroundings have done work on the system.

Often it is important to notice that the expansion work done by a system can be represented by an area on a graph called a $PV$ diagram. As a simple example of such a diagram, consider the thermal expansion of a solid metal object as it is heated. The pressure to which it is subjected is that of the atmosphere, $P_a$, which remains constant during the expansion. We can represent this expansion by the $PV$ graph in Fig. 12.3. The system starts at the point labeled "Cold." The object being heated has a volume $V_c$ at this point, and the pressure is $P_a$. During expansion, the system follows the graph line shown in color. Let us now find the work done by the system. From Eq. 12.2,

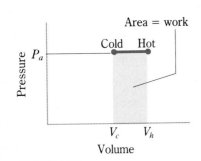

FIGURE 12.2

If the piston has an area $A$, then $\Delta V = A \, \Delta y$.

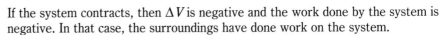

FIGURE 12.3

The work done by an object as it expands against the atmosphere equals the area under the $PV$ curve.

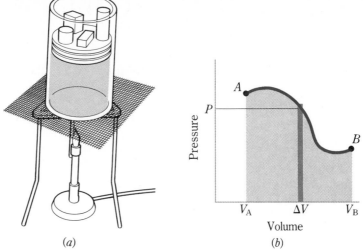

(a)                     (b)

**FIGURE 12.4**

The work done by a system in going from $A$ to $B$ is equal to the area under the curve.

$$\Delta W = P\,\Delta V = P_a(V_h - V_c)$$

Since $P_a$ is the height of the shaded rectangle in Fig. 12.3 and $V_h - V_c$ is its width, $P_a(V_h - V_c)$ is simply the shaded area. We therefore conclude that

The expansion work done by a system is equal to the area under its $PV$ curve.

Although we have derived this statement for an especially simple case, it is true in general. To see this, refer to Fig. 12.4. We see there a gas confined by a piston. The pressure on the piston can be changed by adding or removing weights. The pressure and volume can also be changed by heating or cooling the gas. Suppose the conditions are varied in such a way that the volume changes from $V_A$ to $V_B$, as represented by the colored path in part b.

Consider the most heavily shaded region, in which the pressure is approximately $P$ and the volume change is $\Delta V$. The work done in this small part of the total expansion is $P\,\Delta V$, which is equal to the area of the heavily shaded rectangle. The whole expansion from $A$ to $B$ can be thought of as a series of such tiny expansions. The work done during each is an area. The total work done is the sum of all these areas, and so it is the area under the curve from $A$ to $B$. As we see, *the expansion work is always equal to the area under the PV curve.*

---

### Example 12.1

Weights are added to the piston of Fig. 12.4a as its temperature changes so that the gas contracts in the way shown by the $PV$ diagram in Fig. 12.5. Find the work done by the gas in going from the situation represented by point $A$, through point $B$, to point $C$.

**Reasoning**   We must compute the area under the curve. The area under portion $AB$ is

$$(5 \times 10^5\ \text{Pa})[(800 - 500) \times 10^{-6}\ \text{m}^3)] = 150\ \text{J}$$

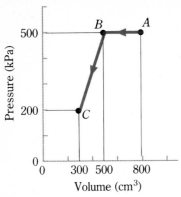

**FIGURE 12.5**
How much work is done by a gas as it moves from the conditions at point A to those at point C?

[Notice that 1 Pa = 1 N/m², and so (1 Pa)(1 m³) = 1 N · m = 1 J.] Similarly, the area under the curve from $B$ to $C$ is

$$(2 \times 10^5 \text{ Pa})(200 \times 10^{-6} \text{ m}^3) + \tfrac{1}{2}(3 \times 10^5 \text{ Pa})(200 \times 10^{-6} \text{ m}^3) = 70 \text{ J}$$

where we have made use of the fact that the area of a triangle is one-half the base times the height. Therefore

Area under $PV$ curve = 150 J + 70 J = 220 J

Because the process we are considering involves a decrease in volume, the work done by the gas is negative. We therefore conclude that the work done by the gas in going from $A$ to $C$ is − 220 J. ∎

---

***Exercise*** How much work would the system do if the $PV$ diagram were a straight line from $A$ to $C$? *Answer:* − 175 J ∎

---

## 12.4 Specific heats of ideal gases

You may have noticed in Chap. 11 that we did not discuss the specific heat capacities of gases. We postponed the topic until now so that you could appreciate more fully the following fact: the specific heat capacity of a gas maintained at constant volume ($c_V$) is smaller than the value for the same gas maintained at constant pressure ($c_P$). To understand this difference between $c_V$ and $c_P$, let us examine what the first law tells us about an ideal gas.

Consider first a mass $m$ of ideal gas maintained at a fixed volume $V$. When we add heat $\Delta Q$ to it, the first law tells us that

$$\Delta Q = \Delta U + \Delta W$$

Since the volume of the gas is constant, the gas can do no external work, and so $\Delta W = 0$. Therefore

$$\Delta Q_V = \Delta U \qquad \text{(volume constant)}$$

where the subscript $V$ reminds us that the volume of the gas is maintained constant. From the definition of specific heat capacity, we also have

$$\Delta Q_V = m c_V \Delta T$$

where $c_V$ is the specific heat at constant volume. Replacing $\Delta Q_V$ by $\Delta U$ and solving for $c_V$, we obtain

$$c_V = \frac{\Delta U}{m \, \Delta T}$$

Thus we see that

The quantity $c_V$ is the change in the internal energy of a gas per unit mass per degree temperature change.

Now let us look at the situation in which the gas is maintained at constant pressure because its volume is allowed to increase as it is heated. The first law becomes

$$\Delta Q_P = \Delta U + \Delta W = \Delta U + P\,\Delta V$$

The specific heat $c_P$ in this situation is for the gas at constant pressure:

$$\Delta Q_P = mc_P\,\Delta T$$

$$c_P = \frac{\Delta U + P\,\Delta V}{m\,\Delta T} = c_V + \frac{P\,\Delta V}{m\,\Delta T} \tag{12.3}$$

In other words, $c_P$ is larger than $c_V$ by an amount $P\,\Delta V/m\,\Delta T$.

It is nearly obvious why this difference occurs. At constant pressure, the added heat is expended in two ways: it increases the internal energy of the gas and, unlike the constant-volume case, it furnishes the energy for the gas to do external work. Therefore, for a given temperature change (that is, a given change in $\Delta U$), more heat must be supplied at constant pressure than at constant volume. Thus $c_P$ is always larger than $c_V$ for an ideal gas.

The ratio $c_P/c_V$ for a gas is usually represented by the symbol $\gamma$. Typical experimental data for $c_V$ and $\gamma$ are given in Table 12.1. As expected, $\gamma$ is always larger than unity.

Equation 12.3 relating $c_P$ and $c_V$ can be placed in a much more interesting form if we make use of the gas law. Let us write the gas law for a gas at constant pressure $P$ but at two temperatures, $T$ and $T + \Delta T$. The volumes at these temperatures are $V$ and $V + \Delta V$. For these two situations $PV = (m/M)\,RT$ becomes

■ TABLE 12.1   Specific heat capacities of gases

| Gas | $c_V(J/kg \cdot K)$ | $\dfrac{C_V}{R}$ | $\dfrac{C_P}{R}$ | $\dfrac{C_P - C_V}{R}$ | $\gamma = \dfrac{c_P}{c_V}$ |
|---|---|---|---|---|---|
| He | 3,130 | 1.50 | 2.49 | 0.99 | 1.66 |
| Ne | 620 | 1.50 | 2.46 | 0.96 | 1.64 |
| Ar | 310 | 1.50 | 2.50 | 1.00 | 1.67 |
| Kr | 150 | 1.50 | 2.52 | 1.02 | 1.68 |
| Xe | 95 | 1.50 | 2.49 | 0.99 | 1.66 |
| Hg (360°C) | 62 | 1.50 | 2.50 | 1.00 | 1.67 |
| $O_2$ | 650 | 2.48 | 3.48 | 1.00 | 1.40 |
| $N_2$ | 740 | 2.48 | 3.48 | 1.00 | 1.40 |
| $H_2$ | 10,000 | 2.40 | 3.39 | 0.99 | 1.41 |
| CO | 730 | 2.46 | 3.48 | 1.02 | 1.41 |
| HCl | 810 | 2.51 | 3.54 | 1.03 | 1.41 |
| $CO_2$ | 640 | 3.37 | 4.37 | 1.00 | 1.30 |
| $H_2O$ (200°C) | 1,500 | 3.23 | 4.23 | 1.00 | 1.31 |
| $CH_4$ | 1,690 | 3.24 | 4.24 | 1.00 | 1.31 |

All gases at 15°C unless stated otherwise.

$$PV = \frac{m}{M} RT$$

$$P(V + \Delta V) = \frac{m}{M} R(T + \Delta T)$$

Subtract the first equation from the second to obtain

$$P\Delta V = \frac{m}{M} R \Delta T$$

Rearrange to get

$$\frac{P \Delta V}{m \Delta T} = \frac{R}{M}$$

which is simply the last term of Eq. 12.3. Substituting this value in Eq. 12.3 leads to the following relation between $c_P$ and $c_V$:

$$c_P = c_V + \frac{R}{M} \tag{12.4}$$

This relation is often quoted in a somewhat different form. The quantity $cM$ is called the **molar specific heat** of the substance and is represented by $C$. Thus we have, in terms of molar specific units,

$$Mc_P = Mc_V + R$$

$$C_P - C_V = R \tag{12.5a}$$

The two molar specific heats differ by an amount equal to the gas constant.

This theoretical relation can be easily tested by comparing the measured values for $C_P - C_V$ with $R$. However, it is customary to eliminate the effects of units in this comparison by noting that both $R$ and the molar specific heats have the same units, joules/kilomole-kelvin. By dividing Eq. 12.5a by $R$, we obtain the following unitless equation:

$$\frac{C_P}{R} - \frac{C_V}{R} = 1 \tag{12.5b}$$

The pertinent experimental values are listed in Table 12.1. As theory predicts, the quantity $(C_P - C_V)/R$ is indeed very close to unity.

Another interesting feature of gases is shown in Table 12.1. Notice that the molar specific heats (divided by $R$) have unique values that depend on the structure of the molecule but not on the particular atom in the molecule. Monatomic gases, such as helium and neon, have $C_V/R = 1.50$. Diatomic gases, such as oxygen and nitrogen, have $C_V/R \cong 2.50$. More complex molecules have still larger values for $C_V/R$. Physicists are always intrigued by regularities such as these in fundamental quantities. Often such regularities lead to the discovery of fundamental laws. For example, the repetitive motions of the planets led Newton to discover the law of gravitation. Molar specific heat regularities provided one of the first clues that atomic processes require quantum mechanics for their description, a fact pointed out by Albert Einstein in 1906.

### Example 12.2

Estimate $c_V$ for nitric acid (NO) vapor. The molecular mass of NO is 30.0 kg/kmol.

**Reasoning** According to Table 12.1, this gas should have $C_V/R \cong 2.50$ because it is diatomic. We therefore suspect that $Mc_V/R \cong 2.50$ for nitric oxide gas, and so

$$c_V \cong 2.50 \frac{R}{M} = \frac{2.50(8314 \text{ J/kmol} \cdot \text{K})}{30 \text{ kg/kmol}} = 690 \text{ J/kg} \cdot \text{K}$$

The value found by experiment is 710 J/kg · K.  ■

---

**Exercise** For ethane gas ($C_2H_6$, $M = 30.0$ kg/kmol), $c_P = 1770$ J/kg · K. Find $c_V$. *Answer: 1490 J/kg · K*  ■

---

## 12.5 Typical processes in gases

Let us now return to the thermodynamic behavior of a gas as it is related to the $PV$ diagram. When we draw a $PV$ diagram, which is simply a graph showing how $P$ varies with $V$, we assume that the changes that are occurring in the system are slow enough for the pressure and temperature to be uniform throughout the whole system at any instant. Let us now examine several important ways in which a system composed of a gas can undergo change. *An* **isothermal process** *is one during which the temperature remains constant.* Because the temperature of an *ideal gas* is a measure of its internal energy, an isothermal process is a constant-internal-energy process. For an ideal gas, then, $\Delta U = 0$ during an isothermal process. The first law, $\Delta Q = \Delta U + \Delta W$, then becomes

$$\Delta Q = \Delta W \qquad \text{isothermal, ideal gas}$$

To examine the behavior of an ideal gas in an isothermal change, refer to Fig. 12.6a. We see there a container of gas in good thermal contact with a heat reservoir.

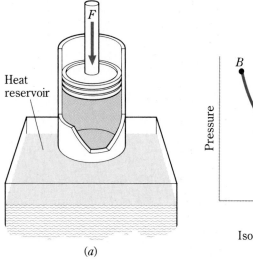

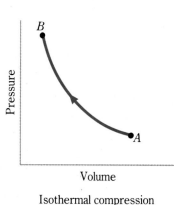

FIGURE 12.6

The $PV$ diagram for an isothermal compression.

Isothermal compression
(b)

(a)

The heat reservoir might be an oven, a cooling bath, or any other constant-temperature device. It maintains the container of gas at constant temperature provided the piston closing the container is not moved too rapidly.

When weights are slowly added to the piston, the pressure on the gas slowly increases and the volume decreases. The $PV$ diagram for this isothermal process is given in Fig. 12.6b. Its form is given by the ideal-gas law, $PV = nRT$, which becomes, when $T$ is constant,

$$P = \frac{\text{const}}{V}$$

Now suppose the gas has been compressed from point $A$ to point $B$ on the graph. If the force on the piston is reduced by slowly removing the weights, the above relation between $P$ and $V$ still applies. The system follows the same graph line as it moves from state $B$ back to state $A$. Such a process is said to be **reversible.** Notice that *in a* **reversible process,** *the state variables acquire the same values at all stages of the process regardless of the direction in which the process is being carried out.* Not all processes are reversible. For example, *a process that has appreciable friction losses cannot be reversible.* Why?

An **adiabatic process** *is one during which no heat is lost or gained by the system.* For example, if a system is well insulated from its surroundings, heat transfer is often negligible, and all processes taking place within the system are adiabatic. Or if a process is carried out extremely rapidly (such as a very sudden compression of a gas), no appreciable heat flows into or out of the system in that short time. Such a process is adiabatic.

For an adiabatic process, $\Delta Q = 0$ and the first law ($\Delta Q = \Delta U + \Delta W$) becomes

$$\Delta U = -\Delta W \qquad \text{adiabatic}$$

This relation, which is not restricted to an ideal gas, tells us that *if a system does adiabatic work, its internal energy must decrease.* The work is done at the expense of internal energy. If adiabatic work is done *on* the system, however, the internal energy increases. Examples 12.3 and 12.4 show two practical uses of these processes. First, however, we examine the adiabatic behavior of an ideal gas in more detail.

An adiabatic process for an ideal gas is not described in terms of $PV = nRT$ alone because all three state variables ($P$, $V$, and $T$) change during the process. To find how each changes, we need a second equation. We find this by noticing that the work done on the gas goes completely into increased *internal* energy. This increase in internal energy causes a temperature change in the system, but this same temperature change could have been carried out by adding heat to the system. Hence, a relation between heat, temperature change, and work can be found even for an adiabatic process. For an ideal gas, this line of thought leads to the following result: *if an ideal gas undergoes an adiabatic change from $P_1$, $V_1$, $T_1$ to $P_2$, $V_2$, $T_2$, then*

$$P_1 V_1^{\gamma} = P_2 V_2^{\gamma} \tag{12.6}$$

*where $\gamma = c_P/c_V$ for the gas.* Typical values for $c_P/c_V$ are listed in Table 12.1.

The $PV$ diagram graph line for an adiabatic change is shown by the solid line in Fig. 12.7. For comparison, the graph line for an isothermal change is shown by the dashed curve. As you might expect, the adiabatic graph line is steeper than the isothermal line at comparable values of $P$ and $V$.

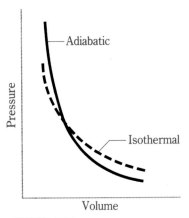

**FIGURE 12.7**

Comparison of the adiabatic and isothermal changes for an ideal gas.

## Example 12.3

In a cylinder of a diesel engine, the air is suddenly compressed by a piston and this causes a sudden rise in temperature. Fuel is then injected and ignites immediately because of the high temperature. Unlike a gasoline engine, a diesel engine does not require a spark plug to ignite the fuel. Suppose the air is compressed to one-fifteenth its original volume. Take its initial pressure to be $1 \times 10^5$ Pa, and its initial temperature to be 27°C. Find its final temperature after compression.

**Reasoning** This is essentially an adiabatic process, since the gas is compressed so rapidly. When we assume the gas to be $N_2$ and to act as an ideal gas, Table 12.1 gives $\gamma = 1.40$. We know from Eq. 12.6 that

$$P_1 V_1^\gamma = P_2 V_2^\gamma$$

$$\frac{P_1}{P_2} = \left(\frac{V_2}{V_1}\right)^\gamma$$

We are interested in the variation in temperature, however, not pressure. The ideal-gas law tells us that

$$P_1 V_1 = nRT_1 \qquad \text{and} \qquad P_2 V_2 = nRT_2$$

We divide one equation by the other to obtain

$$\frac{P_1}{P_2} = \left(\frac{T_1}{T_2}\right)\left(\frac{V_2}{V_1}\right)$$

Equating this value of $P_1/P_2$ to that found in the first equation gives

$$\left(\frac{V_2}{V_1}\right)^\gamma = \left(\frac{T_1}{T_2}\right)\left(\frac{V_2}{V_1}\right)$$

$$\frac{T_1}{T_2} = \left(\frac{V_2}{V_1}\right)^{\gamma-1}$$

In our case, $V_2/V_1 = \frac{1}{15}$, and so

$$T_2 = T_1(\tfrac{1}{15})^{-0.40}$$

We must use absolute temperatures for $T$, and so $T_1 = 300$ K. Taking logarithms of both sides of the equation gives*

$$\log T_2 = \log 300 - 0.40(\log 1 - \log 15)$$

$$= 2.477 - 0.40(0 - 1.176)$$

$$= 2.947$$

---

* You can use the $y^x$ key on your calculator to do this directly.

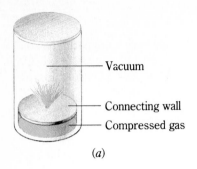

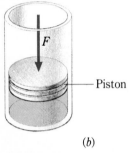

Vacuum

Connecting wall

Compressed gas

(a)

F

Piston

(b)

**FIGURE 12.8**

When a hole is made in the partition in (a), the gas expands into a vacuum. In (b), the partition has been replaced by a piston. The gas expands slowly as the piston moves upward. Under which condition will an ideal gas be cooled most?

Taking antilogs, we find $T_2 = 886$ K, which is $613\,^{\circ}$C. Notice how very hot the gas has become because of the adiabatic compression. ∎

---

*Exercise*   What is the pressure of the gas after compression?   *Answer: 4.43 MPa* ∎

---

### Example 12.4

As shown in Fig. 12.8a, a container is sectioned into two parts, with gas at high pressure in one part and vacuum in the other, much larger part. A small hole is opened in the connecting wall so that the gas expands into the vacuum chamber. Describe the temperature change of the gas, assuming the process to be adiabatic.

*Reasoning*   This type of process, in which a fluid expands through a small opening or porous disk, is called a *throttling process*. Since it is adiabatic, we can write, from the first law,

$$\Delta U = -\Delta W$$

where $\Delta W$ is the work done by the fluid in the process.

We are asked to deal with the special case of a gas. Suppose first that the gas is ideal. In Fig. 12.8a, the gas does no net work as it expands into the vacuum because the external pressure resisting the expansion is zero, and so $P\,\Delta V = 0$. Since the gas does no work, the fact that $\Delta U = -\Delta W$ tells us that its internal energy does not change. Because $T \sim U$ for an ideal gas, $T$ does not change. Therefore, since the expanding gas is ideal, the temperature remains unchanged.

However, the result is different when the fluid is not an ideal gas. Suppose the compressed material is a liquid, for example, butane, which vaporizes as it expands into the vacuum. Then energy must be furnished to the molecules in the liquid to separate them from one another; in other words, the heat of vaporization must be supplied.

Since the process is adiabatic, the required energy must come from the internal energy already resident in the fluid. As a consequence, the kinetic energy of the molecules decreases during the expansion. The temperature of the vapor is therefore lower than that of the original liquid. (In a sense, this is very much like cooling by evaporation.) You can see that many possibilities exist between the ideal gas, where no cooling occurs, and the volatile liquid, where a great deal of cooling occurs.

In certain cases, even an ideal gas cools upon adiabatic expansion. For example, suppose the partition is replaced by a movable piston, as in Fig. 12.8b. Then the gas does work as it expands against the piston. This work leads to a decrease in the internal energy of the gas, and the gas cools. ∎

---

### 12.6 | Nature and time's arrow

As we have seen, the first law of thermodynamics, like its cousin the law of conservation of energy, is a powerful tool for understanding physical processes. Any process that violates it is impossible. There are, however, many natural processes that are known to be impossible but do not violate the first law. We know that energy is conserved when a stone falls to the ground. As the stone strikes the ground, its kinetic energy is changed to thermal energy. However, a stone resting on the ground never changes the thermal energy in and near it to kinetic energy and goes shooting up into the air. The first law does not rule out such a possibility because the reverse process also conserves energy, but still the process does not occur.

There are many other processes in the universe that are not ruled out by the first law but still do not occur. For example, heat flows from hot to cold but not from cold to hot. Dry ice (solid carbon dioxide) vaporizes into the air, but carbon dioxide in the air does not by itself reform as a solid. A dead body decays and turns to dust, but the elements of the earth do not spontaneously form the body in the reverse process. Nature has a preferred direction for the course of spontaneous events. It is as though nature has decreed that time is not reversible. Time is like an arrow that points in only one direction, and all spontaneous natural processes must follow the path nature has chosen for them.

We shall see in what follows that a second law of thermodynamics governs the direction of time's arrow. This law tells us that order in the universe progresses relentlessly to disorder. In that sense, the second law of thermodynamics is sometimes stated in the following whimsical form: "Left to themselves, things go from good to bad to worse." Why this is true will become apparent as we examine the subjects of order and disorder.

## 12.7 Order versus disorder

As any gambler knows, the odds on an event happening are best if the event can occur in many different ways. To illustrate this fact, let us consider a game in which five identical coins are tossed onto a table after being well shaken. Only six possible events can result from such a toss. They are shown in Table 12.2.

At first guess, you might think that each event is equally likely to occur, but that is not correct. There is only one way in which event 1 or event 6 can occur. For event 1 to occur, *all* the coins must come up tails; for event 6, all must come up heads. However, there are five ways in which event 2 could occur. If we call the five coins A, B, C, D, and E, these ways are as listed in Table 12.3. Because there are 5 times as many ways for event 2 to occur, it is 5 times as likely to occur as event 1. Event 5 can also happen in five ways. As a result, events 2 and 5 are equally likely to occur. And both these events are 5 times as likely as events 1 and 6.

In the same way, we can show that events 3 and 4 are equally likely and each can occur in 10 ways. Therefore, events 3 and 4 are twice as likely to happen as events 2 and 5, and events 3 and 4 are 10 times as likely to happen as events 1 and 6. If you were a gambler, it is obvious which events you should lay your money on if no odds are given.

■ **TABLE 12.2**
**Six possible results (events) when five coins are tossed**

| Event | Heads | Tails |
|-------|-------|-------|
| 1 | 0 | 5 |
| 2 | 1 | 4 |
| 3 | 2 | 3 |
| 4 | 3 | 2 |
| 5 | 4 | 1 |
| 6 | 5 | 0 |

■ **TABLE 12.3**
**Possible ways for event 2 to occur**

| Way | A | B | C | D | E |
|-----|---|---|---|---|---|
| 1 | H | T | T | T | T |
| 2 | T | H | T | T | T |
| 3 | T | T | H | T | T |
| 4 | T | T | T | H | T |
| 5 | T | T | T | T | H |

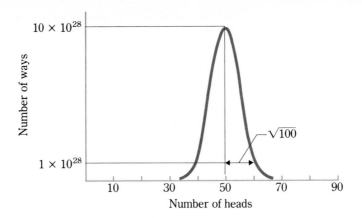

**FIGURE 12.9**

The number of ways in which the indicated number of heads come up when 100 coins are tossed. For less than 30 (or more than 70) heads, the number of ways is too small to be shown on this graph and can be approximated by zero.

We can extend this reasoning to a situation in which more coins are involved. Suppose 100 coins are tossed. As before, there is only one way in which all the coins can come up heads (or tails). The number of ways in which other combinations can occur becomes almost unbelievably large, however, as shown in Fig. 12.9. Notice that the number of ways in which 50 heads can come up is about $10 \times 10^{28}$. As you can see, the odds on all heads or all tails coming up are so small as to be negligible.

Indeed, the total number of ways for all combinations of heads and tails, that is, the sum of all the ways shown in the graph, is about $1 \times 10^{30}$. Therefore, the chance that all coins would come up heads is 1 in $10^{30}$. If you throw the coins once each 10 s for $10^{22}$ yr, your chance of all heads coming up once is about 10 percent. For all practical purposes, there is no chance at all of all heads or all tails occurring. As we see from Fig. 12.9, the only really likely occurrence is for nearly equal numbers of heads and tails to occur.

If we consider $10^{6}$ coins instead of $10^{2}$, the situation becomes even more striking. We can summarize all such results in a very simple way. Notice in Fig. 12.9 that the graph line decreases to about one-tenth of its maximum value at 40 and 60 heads. To give an estimate of the width of the peak, we could say that it extends from $50 - 10$ to $50 + 10$. In other words, if you throw 100 coins, the number of heads that should come up is $50 \pm 10$. The more general result is

If $N$ coins are tossed, the expected number of heads is about*

$$\tfrac{1}{2}N \pm \sqrt{N}$$

In the case of $10^{6}$ coins, we should expect $500,000 \pm 1000$ heads to come up. Notice how very precise this estimate is. It says that the expected number of heads lies between 501,000 and 499,000, a very narrow range indeed. As you can see, when the number of coins becomes very large, the percentage deviations from the average are very small.

This example with the coins is typical of our universe in general. When things are left to happen by themselves, they occur by chance. As a result, the probability laws applicable to tossed coins apply to these other situations as well. For example, suppose you have a box containing gas molecules, as in Fig. 12.10. In the air, there are about $3 \times 10^{19}$ molecules per cubic centimeter. Let us say the box has $10^{20}$ molecules in it. We now ask: What are the chances that the molecules will all bunch up in one half of the box?

---

* Precisely, if $N$ is large, the number of heads will lie in this range 96.3 percent of the time.

From our results with the coins, we can easily answer this question. To make the situations similar, call molecules on the left side of the box "heads" molecules and those on the right "tails" molecules. Our general result tells us that the number of heads will be about $\frac{1}{2}N \pm \sqrt{N}$, in this case about $(5{,}000{,}000{,}000 \pm 1) \times 10^{10}$.

Notice how small the expected deviation is: only $10^{10}/5 \times 10^{19}$, or about 1 part in $5 \times 10^9$. For all practical purposes, the number of molecules in one half of the box is the same as the number in the other half. And, of course, there is really no chance at all that all the molecules will spontaneously move to one side of the box.

These considerations have fundamental importance for all spontaneous processes. Reasoning from them, we can predict that thermal motion (as well as other random-type disturbances) causes systems to change from order to disorder. As a crude example, consider 100 coins again. Suppose we carefully arrange all with heads up. They then have a high degree of order. Now let us give them a type of motion similar to random thermal motion by shaking them. They quickly disorder and never return to their original state of order.

Similarly, we can give the gas molecules in Fig. 12.10 order by placing them all in one end of the box. If, however, we allow them to adjust with spontaneous thermal motion, they become disordered and fill the whole box. They will never spontaneously return to their original ordered state.

Basic to this discussion are the concepts of order and disorder. There is a simple method for comparing the disorder of two states. If a state can occur in only one way, it is highly ordered. In such a state, each molecule (or other particle) must be placed in a single exact way. In a disordered state, however, there are many possible ways for achieving the state. With these facts in mind, we can relate disorder and the probability (or number of ways) of achieving a state. That state having the highest disorder is the most probable state; it can occur in the largest number of ways. For example, the probability of $N$ coins all coming up heads is very small. This is a state of very low disorder. As we have seen, systems left to themselves move toward states of high disorder.

There are many examples that illustrate behavior of this type. We conclude from them that

If a system composed of many molecules is allowed to undergo spontaneous change, it changes in such a way that its disorder increases or, at best, does not decrease.

This law of nature, applicable to large numbers of molecules, is one form of the **second law of thermodynamics.** In the next section, we see another way in which the law can be stated.

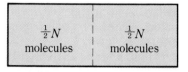

**FIGURE 12.10**

What is the likelihood that all the molecules will appear in one end of the box?

## 12.8 Entropy

The implications of order and disorder in a system can be approached in two quite different ways. Both approaches use a quantity called *entropy*. The concept of entropy was introduced in the mid-1800s by R. Clausius. Because the concept of atoms was still quite speculative at that time, Clausius followed conventional procedures and described the behavior of systems in terms of their macroscopic variables, namely, *P, V, T,* and *U.* To describe the consequences of the fact that heat flows preferentially from hot to cold, he found it convenient to define a quantity that he called entropy.

Suppose an amount of heat $\Delta Q$ is added in a reversible way to a system whose temperature is maintained at a value $T$. Then the attendant entropy change of the system, denoted by $\Delta S$, is defined to be

$$\Delta S = \frac{\Delta Q}{T} \tag{12.7}$$

As we see, the quantity $\Delta S$ is positive (the system experiences a gain in entropy) when heat flows into the system.

Clausius was able to show that entropy is a state variable. If two systems are macroscopically identical, they have the same entropy. In this respect, entropy is like $P$, $V$, $T$, and $U$. Because entropy is a state variable, it is an important, useful quantity in the description of thermodynamic processes.

As the understanding of atoms and molecules was advanced by such men as Boltzmann and Maxwell, it became apparent that entropy could be given a molecular meaning. About a quarter of a century after Clausius introduced the concept, Boltzmann was able to show that entropy could be given an equivalent definition expressed in terms of atoms and molecules. His definition related entropy to molecular disorder.

As we saw in the previous section, left to itself, thermal motion causes the disorder of a system to increase until maximum disorder is reached. Entropy behaves in a similar way, and so we expect disorder and entropy to be related. We can make this correspondence quantitative by measuring disorder in terms of the number of ways in which a system can exist in a given state. Each equilibrium state of a system can occur in a definite number of ways. Let us call that number $\Omega$ (Greek capital omega). A state for which $\Omega$ is very large is much more likely to occur than one for which $\Omega$ is small. Indeed, an equilibrium system composed of many molecules is always very close to the state that has the maximum value of $\Omega$.

Boltzmann proved that entropy and $\Omega$ are related as follows:

$$\text{Entropy} = S = k \ln \Omega \tag{12.8}$$

where $k$ is Boltzmann's constant and $\ln \Omega$ is the natural logarithm of $\Omega$. It is easy to see that entropy is a state variable, in other words, that it depends only on the equilibrium state of the system. The argument goes like this. Like $P$, $V$, $T$, and $U$, entropy describes an equilibrium state unambiguously. A system of many molecules reaches equilibrium when it has achieved the state with largest $\Omega$, that is, the most probable state, or the state of largest entropy. Hence, under a given set of constraints, a system at equilibrium has a well-defined value for $S$ as well as for $P$, $V$, $T$, and $U$. This fact is of great importance and gives entropy usefulness comparable to that of internal energy.

Notice what Eq. 12.8 tells us. If a state of the system can occur in only one way, then $\Omega = 1$. The logarithm of 1 is zero, however, and so the entropy of such a highly unlikely state is zero. However, if a state can occur in many ways, $\Omega$ is large. The entropy of a highly probable state is therefore large.

Recall that a highly ordered state can occur in only a few ways, whereas a disordered state can occur in many ways. Equation 12.8 then tells us that entropy is a measure of disorder. The more disordered the state of a system, the larger its entropy. As we see, *entropy is a state variable that measures disorder*. Because of this, we can give an alternative statement of the **second law**:

If an isolated system undergoes change, it changes in such a way that its entropy increases or at best remains constant.

The entropy of such a system increases until equilibrium is reached. Its entropy then remains constant. Example 12.5 shows how a change in entropy can be computed.

---

**Example 12.5**

By how much does the entropy of the system change as a 20-g cube of ice melts at 0°C? You should assume that heat can flow into the ice cube but that otherwise it is an isolated system.

**Reasoning** From Eq. 12.7 we have

$$\Delta S = \frac{\Delta Q}{T}$$

Notice that $T$ is the absolute temperature. In our particular case, $T = 273$ K. Because 80 cal of heat must be added for each gram of ice melted,

$$\Delta Q = (20 \text{ g})(80 \text{ cal/g})(4.184 \text{ J/cal}) \approx 6700 \text{ J}$$

Then

$$\Delta S = \frac{6700 \text{ J}}{273 \text{ K}} = 24.5 \text{ J/K}$$

Notice that the entropy of the system increases as the ice changes to water. The increase is a measure of the increase in disorder in the arrangement of the water molecules. The unit for entropy, joules per kelvin, is often referred to as the entropy unit, eu. Thus 1 eu = 1 J/K. ∎

---

**Exercise** What is the entropy change if the ice is originally at −10°C? *Answer: 24.7 J/K* ∎

---

## 12.9 Heat engines

The development of thermodynamics began in the late 1700s, the time of the Industrial Revolution. It was then that the invention of the steam engine led to a momentous change in our civilization. Because early steam engines were primitive devices that operated at low efficiency, scientists of the time were called upon to examine the physical laws governing these machines. It was this call that provided the major impetus for the early work in thermodynamics. The results of that work were far-reaching and today permeate both the physical and the biological sciences.

A steam engine is only one example of a heat engine:

A heat engine is any device that converts thermal energy to other useful forms of energy.

The steam engine obviously fits this description. So does a gasoline engine since it uses the thermal energy given off by the combustion of fuel. More exotic engines that use heat from the sun or from nuclear reactors are also heat engines. Let us now look at the laws all such engines obey.

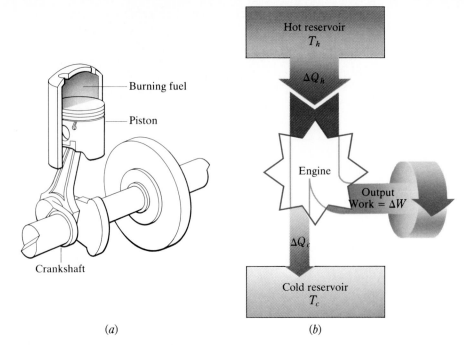

FIGURE 12.11

In a heat engine, the input energy $\Delta Q_h$ must equal the sum of the exhaust energy $\Delta Q_c$ and the output work $\Delta W$.

(a)                    (b)

A diagram of a simple engine is shown in Fig. 12.11a. The burning fuel creates high-pressure gas that causes the piston to move downward. This linear motion is changed to rotational motion by the crankshaft, and the engine goes through the same cycle of motion over and over. Of course, many details, such as the required valves and sparkplugs, are not shown. The essential feature, however, is that thermal energy is transformed to mechanical energy.

Figure 12.11b gives a more general representation of a heat engine. Heat $\Delta Q_h$ flows from a high-temperature (hot) reservoir to the engine. This is the input energy for the engine. Part of the input energy is used in doing work, and the remainder, $\Delta Q_c$ (the exhaust energy), flows to a low-temperature (cold) reservoir. The low-temperature reservoir is often simply the atmosphere, as when a car exhausts spent fuel through its tail pipe.

Because the engine must obey the law of conservation of energy, the first law of thermodynamics applied to it becomes, for one cycle of its motion,

$$\Delta Q = \Delta Q_h - \Delta Q_c = \Delta W + \Delta U$$

where $\Delta W$ is the output work of the engine per cycle. However, the engine goes through the same cycle over and over again with no net change in its internal energy. Hence, for a complete cycle, $\Delta U = 0$ and the above equation becomes

$$\Delta W = \Delta Q_h - \Delta Q_c$$

We shall use this relation to compute the efficiency of the engine. As with any machine, we define efficiency as the ratio of output work to input energy. In the present case,

$$\text{Efficiency} = \frac{\Delta W}{\Delta Q_h}$$

Substituting for $\Delta W$ the value given above, we have

$$\text{Efficiency} = \frac{\Delta Q_h - \Delta Q_c}{\Delta Q_h} = 1 - \frac{\Delta Q_c}{\Delta Q_h} \tag{12.9}$$

We see that the exhaust heat $\Delta Q_c$ is responsible for the inefficiency of a heat engine. If we could make $\Delta Q_c$ zero, the engine would be 100 percent efficient. However, we now use the concept of entropy to show that this is impossible. A very definite limit exists on the efficiency of any engine.

Let us compute the entropy change of the system of Fig. 12.11$b$ as heat flows. The engine is unchanged by the heat flow, and so its entropy does not change. However, the hot reservoir loses heat $\Delta Q_h$ while the cold reservoir gains heat $\Delta Q_c$. Therefore

$$\Delta S_h = \frac{-\Delta Q_h}{T_h} \qquad \text{and} \qquad \Delta S_c = \frac{\Delta Q_c}{T_c}$$

According to the second law, the total change in entropy must be greater than or equal to zero:

$$\Delta S_c + \Delta S_h \geqslant 0$$

$$\frac{\Delta Q_c}{T_c} - \frac{\Delta Q_h}{T_h} \geqslant 0$$

where $\geqslant$ is read as "is greater than or equal to." If we transpose the negative term, divide through by $\Delta Q_h$, and multiply through by $T_c$, we have

$$\frac{\Delta Q_c}{\Delta Q_h} \geqslant \frac{T_c}{T_h} \tag{12.10}$$

We can now substitute this value in Eq. 12.9 to obtain

$$\text{Efficiency} \leqslant 1 - \frac{T_c}{T_h} \tag{12.11}$$

According to Eq. 12.11, the maximum efficiency of any heat engine is

$$\text{Maximum efficiency} = 1 - \frac{T_c}{T_h} \tag{12.12}$$

Thus we find that efficiency is controlled by the ratio of the temperatures of the two reservoirs. The efficiency can be increased by using a very-high-temperature hot reservoir and a very-low-temperature cold exhaust. For example, if a steam engine has $T_h = 400$ K and an exhaust temperature $T_c = 300$ K, then its maximum efficiency is

$$\text{Maximum efficiency} = 1 - \frac{300}{400} = 0.25$$

Thus we arrive at the following startling result: even the best-designed heat engine has a very limited efficiency. Most engines, because of friction and extraneous heat loss, have efficiencies less than the maximum value. A heat engine can never have an efficiency greater than that given by Eq. 12.12. Gasoline — and, even

# Food, fat, and work

The human body is an engine that transforms food energy and oxygen to work and exhaust heat. Each day, you and I consume close to 1500 kcal (6.3 MJ) of energy in the form of food. That energy is used for various purposes, with the precise breakdown depending on our daily behavior. The body processes much of the food we eat to water, carbon dioxide, and energy. As it does so, waste heat is generated, and about 55 percent of the input food energy is released to the environment as heat. (See the diagram.) The remaining 45 percent is used to manufacture the materials that make up our bodies. In addition, an energy-rich compound called ATP (for adenosine triphosphate) is formed, and this material supplies energy for replenishing body cells and accomplishing the work associated with muscle contractions.

When the body is idling, so to speak, it expends only enough energy to maintain itself in an "as is" condition. The energy it uses each second to circulate blood and to carry out other required internal functions is called the basal metabolism rate. For mammals, including humans, this basic rate of energy consumption is about 4 MJ/day (960 kcal/day) for each square meter of surface area. For a 45-kg person, the average basal metabolism rate is about 6.7 MJ/day (1.6 Mcal/day). For a 90-kg person, this value is about 10 MJ/day. This energy expended by the body-engine as it idles appears mainly as heat exhausted to the surroundings. This is the waste heat indicated at the top left of the diagram below.

A relatively small percentage of the body's input energy is used in doing work that involves the use of our limb

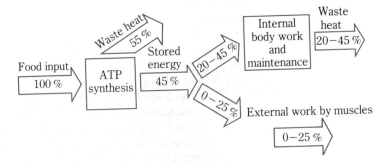

more, diesel — engines operate at temperatures higher than a steam engine does. A typical gasoline engine has an efficiency of about 30 percent, and the efficiency of a diesel engine may be as large as 40 percent.

One of the pioneers in thermodynamics, Sadi Carnot, was the first to show that a certain type of heat engine yields the maximum efficiency given by Eq. 12.12. It is an idealized engine that uses adiabatic and isothermal compressions and expansions. We refer to such an engine as a *Carnot engine*. For it, the inequalities of Eqs. 12.10 and 12.11 become equalities, so that

$$\frac{T_c}{T_h} = \frac{\Delta Q_c}{\Delta Q_h}$$

$$\text{Efficiency} = 1 - \frac{T_c}{T_h}$$

A family of four surrounded by the 2.5 tons of food they will probably consume in 1 year.

muscles. As indicated in the diagram, this amounts to less than about 25 percent of the food energy input. Typical energy-use rates for selected adult activities are 240 kcal/h for walking at a moderately brisk pace, 400 kcal/h for moderate exercise, and 600 kcal/h for strenuous exercise.

If the intake energy supplied each day by one's food exceeds the sum of the basal metabolism rate and the rate at which energy is used for additional work, the body stores the remaining energy as fat. Or, if food intake is too low, our body-engine consumes fat to meet its required supply of energy. As we see, three major factors control the production

or loss of fat: the basal metabolism rate, the muscle-related work we do, and food intake. Some people require more energy to keep their body-engine idling than do other people. To maintain their high basal metabolism rate, they must have a higher food intake. These are the people who, though relatively inactive, never seem to gain weight. Other people have a low basal metabolism rate and gain weight (fat) with a surprisingly low level of food input energy. For all of us, maintaining proper body weight requires balancing our input food energy with the energy required to do our daily work and to keep our body-engines idling.

## 12.10 The absolute temperature scale

The Carnot engine is used to define the absolute (Kelvin) temperature scale. We are now in a position to explain its definition. In the preceding section, we pointed out that the ratio of the heat flows at the two reservoirs ($\Delta Q_c/\Delta Q_h$) is related to the ratio of the temperatures of the reservoirs through

$$\frac{T_c}{T_h} = \frac{\Delta Q_c}{\Delta Q_h}$$

This relation is used to define absolute temperature.

Suppose you measure the heat flows for a Carnot engine operating between two reservoirs at temperatures $T_c$ and $T_h$. Suppose that the ratio of the two heats is 0.10. Then you know at once that

$$\frac{T_c}{T_h} = 0.10$$

Although this gives the ratio of the two temperatures, it does not tell us the value of either. If we knew one of these two temperatures, though, we could easily find the other.

We get around this difficulty by defining a value for one particular temperature. This temperature is the single unique temperature at which water, ice, and water vapor can coexist, a temperature called the **triple point** of water. It has been decided to take the triple point of water as having a defined temperature of 273.16 K. Water freezes at a temperature 0.01 C° lower than this, 273.15 K. Indeed, this definition was chosen so as to duplicate the familiar temperatures we indicated in Fig. 10.5. A reservoir maintained at the triple point of water thus has a temperature of 273.16 K. The temperature of any other reservoir can be obtained by operating a Carnot cycle between the two as described in the previous paragraph.

The absolute temperature scale as now defined facilitates measurements at very low and very high temperatures. Near absolute zero, where all gases condense and the liquids usually used in thermometers are solids, no ordinary thermometer is of value, but a type of Carnot engine can be used even at these temperatures. Thus we have a workable definition for temperature over the complete temperature range.

## 12.11 Refrigerators, air conditioners, and heat pumps

Ordinarily, heat flows from hot to cold. In refrigeration systems, however, we want the reverse to happen; heat is to be taken from the cold interior of the refrigerator and expelled into the warmer air exterior to it. Because this is not the direction in which heat would flow if left to itself, we must in some way force heat to flow in this "backward" direction. This requires the expenditure of energy by some outside source, as we shall now show.

A schematic diagram of a refrigeration system is shown in Fig. 12.12. Such a system makes use of a low-boiling-point fluid; Freon-12 is typical. This gas boils at $-30°C$ under atmospheric pressure. Let us begin with the compressor at the bottom of the unit. It compresses the Freon gas to a pressure high enough that it will liquefy when cooled somewhat. During this nearly adiabatic compression, the work done on the gas causes it to become quite hot. It is then cooled when it passes through the condenser cooling coils (usually at the rear of the refrigerator), over which a fan blows room air. When you are close to the refrigerator, you can often notice this warm air. As the highly compressed Freon cools, it condenses to a liquid in the condenser coils. The near-room-temperature liquid Freon under high pressure is then sent to the throttling valve.

The throttling valve (which is actually inside the cooling unit) has a tiny opening that allows the high-pressure Freon to spray into the low-pressure region within the evaporator tube. There two things happen to the Freon: it evaporates and it expands. Both these processes cool it, and so the evaporator tube becomes very cold. This coiled tubing is inside the refrigerator and is the refrigerator's cooling element. As the very cold Freon flows through the cooling coil, it cools the refrigerator by taking up heat. Eventually the (now warmer) Freon leaves the cooling unit

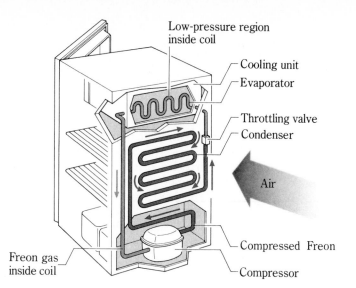

Low-pressure region
inside coil

Cooling unit

Evaporator

Throttling valve

Condenser

Air

Compressed Freon

Freon gas
inside coil

Compressor

**FIGURE 12.12**

A schematic diagram of a refrigerator.

and returns to the compressor. There it is compressed again, and both the heat it took up in the cooling unit and the heat generated during its compression are removed in the condenser. As you can see, the Freon takes up heat from the interior of the refrigerator, and this heat is ejected from the Freon through the action of the compressor and condenser.

Air-conditioning units operate in the same way. The cooling coil is inside the house, the condensing coil is outside, and heat is taken from the house and ejected to the outdoors. Let us now examine the operation of such a unit from a quantitative standpoint.

The refrigerator takes an amount of heat $\Delta Q_c$ from its interior, where the temperature is $T_c$, and expels an amount of heat $\Delta Q_h$ to the room, where the temperature is $T_h$. Notice that $\Delta Q_h$ is larger than $\Delta Q_c$ because the former includes the heat generated by the compressor. If the compressor does work $\Delta W$ in the process, then, from the fact that energy is conserved,

$$\Delta Q_h = \Delta Q_c + \Delta W$$

Thus we see that the input work needed to operate the refrigerator is

$$\Delta W = \Delta Q_h - \Delta Q_c$$

To be efficient, a refrigerator should remove a lot of heat $\Delta Q_c$ for a small amount of input work. We therefore define the **coefficient of performance** (COP) of the refrigerator:

$$\text{COP (refrigerator)} = \frac{\Delta Q_c}{\Delta W}$$

We can substitute the value of $\Delta W$ given above and obtain

$$\text{COP (refrigerator)} = \frac{\Delta Q_c}{\Delta Q_h - \Delta Q_c}$$

As we did for a heat engine, we can use entropy considerations to relate $\Delta Q_h$ and $\Delta Q_c$ to the temperatures of their respective heat reservoirs. Then we find that

$$\text{Maximum COP (refrigerator)} = \frac{T_c}{T_h - T_c} \qquad (12.13)$$

Notice that the performance is best when the temperature differential is small. This makes sense because we expect that less work needs to be done to force heat to flow to a reservoir that is at a slightly higher temperature than to one that is at a much higher temperature.

Heat pumps are much like refrigerators in that they pump heat from a cold region to a warm region. In fact, in their cooling mode, they act exactly like a refrigerator to cool the interior of a building. However, they can also use heat extracted from the colder air outside to heat the interior of a building. This is done by reversing the coils in Fig. 12.12 so that the cooling coil is outside the building and the condensing (heating) coil is inside. Notice that in this case the building is warmed by thermal energy taken from the often colder air outside the building. It may seem strange to you that heat energy from cold air is being used to heat warmer air, but that is what all refrigerators do. They take heat energy from the inside of the cold refrigerator and eject it to the warm room in which the refrigerator is sitting. Of course, heat does not by itself flow from cold to hot, but in these cases the electric energy used to run the compressor forces the heat to flow in this direction.

As the cost of energy increases, heat pumps are becoming more attractive for home heating because they make use of the waste thermal energy that surrounds us. To make use of that energy, however, heat must be forced to move from cold regions to warmer regions, and this requires external energy, the electric energy needed to run the compressor. Even so, much of this electric energy eventually appears as heat, and so it, too, contributes to the heating task. However, as you might guess, the electric energy required increases as the temperature differential between outside and inside increases. Because of this, heat pumps for home heating are currently restricted mostly to those areas where winters are not too severe. In the coldest climates, heat pumps can be used in conjunction with electric or other heating systems to provide suitable combined systems for heating the home.

---

**Example 12.6**

A dish of hot food is placed in a refrigerator maintained at 5°C. To cool to this temperature, the food must lose 210,000 J. How much electric energy is needed to operate the compressor if we assume that room temperaure is 20°C and that the system operates at its best possible performance level?

**Reasoning**  The maximum COP for the refrigerator is

$$\text{Maximum COP} = \frac{T_c}{T_h - T_c} = \frac{278 \text{ K}}{293 \text{ K} - 278 \text{ K}} = 18.5$$

From its definition,

$$\text{COP} = \frac{\Delta Q_c}{\Delta W}$$

from which

$$\Delta W = \frac{\Delta Q_c}{\text{COP}} = \frac{210{,}000 \text{ J}}{18.5} = 11{,}330 \text{ J}$$

Therefore the electric energy required would be about 11,000 J. If electric energy sells at perhaps 10 cents per kilowatthour (about $2.8 \times 10^{-6}$ cents per joule), the cost of cooling the food is about 0.03 cents.  ■

## 12.12 The heat death of the universe

Scientists believe that at the very beginning of time, the universe was highly compressed to a density unknown even in the densest stars. Although the extent of this compression is not known with certainty, it is likely that the portion of the universe we can see with our largest telescopes was compressed into a region with a diameter much smaller than that of the sun. At the time of that original compressed state, the beginning of time, the universe was an incredibly hot cauldron of energy. During the billions of years that have since elapsed, the universe has expanded adiabatically at speeds approaching the ultimate speed, the speed of light.

The laws of thermodynamics apply to this process, which has taken perhaps 15 billion years. During that time, heat energy has continuously flowed from hot to cold regions. As it has done so, the disorder and entropy of the universe have increased. The thermal energy — indeed, all energy — has become less useful on the average. The temperature of the original fireball has fallen continuously. There are still local hot spots (the sun and the stars), but most of the universe has long since cooled to temperatures far below those we experience on earth. The gas in the vast reaches of space has an average temperature of about 3 K.

We on earth are fortunate. Our nearby sun still floods us with energy, and we use this radiant energy to grow plants. These plants are then used as sources of energy for us and the other creatures on earth. Except for nuclear fuels, the sun is the ultimate source of the energy we use. Notice that the sun's usefulness to us is the result of its very high temperature. As it cools, it will radiate less energy. Over a few billion years, the earth will slowly lose its major energy source.

During all this time, the entropy of the universe will be increasing. The sun and other hot objects will be losing entropy at the rate of $\Delta Q/T_h$, where $\Delta Q$ is the heat lost in unit time by a hot object at temperature $T_h$. Although the cooler portions of the universe will receive this energy, they will be at a lower temperature $T_c$. As a result, they will gain an entropy $\Delta Q/T_c$ that is larger than $\Delta Q/T_h$. The entropy and disorder of the universe will continue to increase as hot objects cool and cold objects warm.

We can picture in our minds a time when everything in the universe has reached the same temperature. Then no heat flow can occur. The disorder of the universe (and its entropy) will have reached a maximum value. Even though all the original energy the universe once had will still be present, it will be useless. No plants will grow because there will be no hot object to light them. No engines will function because there will be no cooler place to which heat can be exhausted. No life will exist anywhere in the universe. The universe will have undergone what is known as its *heat death*.

Fortunately for us, this situation will not occur for billions of years. Indeed, we are not certain that it will ever happen. As you will learn in Chap. 28, there is a possibility that the universe will contract and once again become a fireball. That is another story, however, and we postpone discussion of it until the last chapter in this text.

## Minimum learning goals

When you finish this chapter, you should be able to

1 Define (*a*) state of a system, (*b*) state variable, (*c*) internal energy, (*d*) *PV* diagram, (*e*) molar specific heat, (*f*) isothermal process, (*g*) reversible process, (*h*) adiabatic process, (*i*) throttling process, (*j*) entropy, (*k*) Carnot engine.

2 State the first law of thermodynamics in equation form and explain the meaning of each term.

3 Use the relation $\Delta W = P\,\Delta V$ in simple situations. Compute the work done by a system during a given volume change if a graph of pressure versus volume is provided.

4 Explain why $c_P$ is larger than $c_V$ for a gas. Use Eqs. 12.4 and 12.5 in simple situations.

5 Apply the first law of thermodynamics to isothermal and adiabatic processes to describe the behavior of $\Delta Q$, $\Delta U$, and $\Delta W$.

6 Explain why a gas warms when it is compressed adiabatically.

7 Explain why a fluid is often cooled by a throttling process.

8 Give several examples of physical systems that, when left to themselves, become more disordered. Explain in each case why the reverse process is not observed.

9 Use the relation $\Delta S = \Delta Q/T$ to compute the entropy change of a simple system under an isothermal change.

10 State the second law of thermodynamics and explain how it leads to information not available from the first law.

11 Write the equation for the maximum efficiency of a heat engine and use it in simple situations.

12 Explain in a general way how a compressor-type refrigeration unit operates.

## Questions and guesstimates

1 An inventor claims that he has an engine that is started by a battery but then runs without an outside power source as it recharges the battery and does external work. What law of nature does this invention disprove? What does the first law tell you about perpetual motion machines?

2 The relation $\Delta U = \Delta Q - \Delta W$ is not always equivalent to the relation $\Delta U = \Delta Q - P\Delta V$. Give an example in which the latter relation does not apply even though the former does.

3 In each of the following processes, point out what is meant by each quantity in the equation $\Delta U = \Delta Q - \Delta W$: an ice cube slowly melts in water at $0°C$; ice heats from $-30°C$ to $-10°C$; steam in a closed boiler cools from $120°C$ to $110°C$; solid $CO_2$ (dry ice) sublimates (changes from a solid to a gas directly, without ever passing through the liquid phase) in dry air in a large container; a bottle of pop freezes and the bottle cracks.

4 Although $C_P - C_V = R$ for any ideal gas, $c_P - c_V$ varies from gas to gas. Why the difference?

5 How can one determine the molecular mass of an ideal gas by measuring $c_P$ and $c_V$ for it?

6 A container of ideal gas is to be compressed to half its original volume. Under which condition would the work done on the gas be larger: isothermal or adiabatic?

7 Two cylinders sit side by side and are enclosed by a movable piston. They are identical in all respects except that one contains oxygen ($O_2$) and the other contains helium (He). Both are compressed adiabatically to one-fifth their original volume. Which gas shows the larger temperature rise?

8 Suppose a box has such a good vacuum within it that it contains only five gas molecules. Sometimes all five will be in one half of the box. How can you reconcile this with the second law and our discussion of disorder?

9 Some people claim you can cool a watermelon by placing it in a wet blanket and letting it sit in a breeze even though the air temperature is very high. Doesn't this contradict the second law?

10 Estimate the rate at which a person's entropy changes while she or he is lounging around. The average person's metabolic rate under these circumstances is about 100 W (that is, the person uses stored energy at this rate).

11 Consider the simple heat engine in Fig. P12.1. The heated liquid on the right expands and is lifted by the cooler liquid on the left. As a result, the liquid circulates counterclockwise in the tube. As it does so, it rotates the paddle wheel, which is then coupled to external devices to do output work. Explain what factors affect the efficiency of this engine. What should be done to make it most efficient?

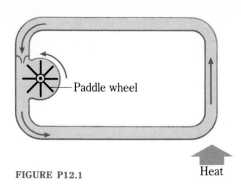

FIGURE P12.1

Heat

12 Each of a pair of dice has six sides, labeled 1 to 6. When a pair is tossed, what is the ratio of the chance that the up sides will total $x$ to the chance that they will total $y$ when (a) $x = 2$, $y = 3$ and (b) $x = 2$, $y = 4$?

13 A child who wishes to cool off the kitchen of a home opens the refrigerator door and leaves it open. Will this work? Answer the question from both short- and long-term considerations. Would the situation be any different if an old-fashioned ice box were used rather than a refrigerator?

14 At the present time, the sun is the major source of the energy we use here on earth. Trace this solar energy from its source through our uses of it and show that no conflict exists with the second law. Pay particular attention to the ordering process that occurs in photosynthesis.

## Problems

1 As a 20-g ice cube at $0°C$ melts to water at $0°C$, by how much does its internal energy change? Neglect its small volume change.

2 Find the change in internal energy of a 40-g block of metal ($c = 0.095$ cal/g · C°) as it is heated from $15°C$ to $400°C$. Neglect its small change in volume.

3 By how much must one lower the temperature of a 300-g block of metal ($c = 0.105$ cal/g · C°) to change its internal energy by 500 J? Neglect any volume change.

4 What is the internal energy change of 500 g of molten lead as it solidifies at its melting temperature? Neglect its small volume change.

5 Figure P12.2 shows the $PV$ diagram for a gas confined to a cylinder by a piston. How much work does the gas do as it expands from $A$ to $C$ along the curve?

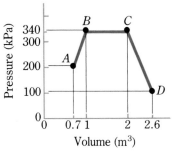

FIGURE P12.2

6 Figure P12.2 shows the $PV$ diagram for a gas confined to a cylinder by a piston. How much work does the gas do as it contracts along the curve from $D$ to $A$?

7 An ideal gas in a cylinder is slowly compressed to one-third its original volume. During this process, the temperature of the gas remains constant and the work done in compression is 75 J. (a) By how much does the internal energy of the gas change? (b) How much heat flows into the gas?

8 A certain quantity of helium gas in a closed container requires 25 J of heat to change its temperature from $20°C$ to $90°C$. What mass of helium is in the container? For helium, $c_V = 0.75$ cal/g · C° and $c_P = 1.25$ cal/g · C°.

■ 9 Suppose that 5 g of helium gas is heated from $-30°C$ to $120°C$. Find its change in internal energy and the work it does if the heating occurs (a) at constant volume and (b) at constant pressure. For helium, $c_V = 0.75$ cal/g · C° and $c_P = 1.25$ cal/g · C°.

■ 10 Helium at standard temperature and pressure fills a 500-cm³ closed container. How many joules of heat are required to raise its temperature to $70°C$? What is its pressure at $70°C$?

■ 11 Using the fact that air consists of approximately 78 percent $N_2$ and 22 percent $O_2$ by mass, calculate $c_V$ for air.

■ 12 Radon is a radioactive monatomic gas. Its atomic mass is 222 kg/kmol. Estimate its values for $c_V$ and $c_P$.

■ 13 How much heat must flow from an ideal monatomic gas if it is to be compressed from 800 cm³ to 300 cm³ under a constant pressure of 200 kPa? Its initial temperature is $127°C$.

■ 14 An ideal monatomic gas is maintained at a constant pressure of 300 kPa as it is slowly heated from an initial temperature of $27°C$. Its volume at $27°C$ is 250 cm³. How much heat must be added to raise its temperature to $300°C$?

■ 15 How much work is needed to compress an ideal gas

isothermally from 500 cm³ to 300 cm³ if, during the process, 25 cal of heat flows from the gas?

■ **16** How much work is required to adiabatically compress 25 g of helium in such a way that its temperature rises from 20°C to 500°C?

■ **17** An ideal gas in a cylinder is compressed adiabatically to one-third its original volume. During the process, 45 J of work is done on the gas by the compressing agent. (*a*) By how much does the internal energy of the gas change? (*b*) How much heat flows into the gas?

■ **18** A 10-g mass of ideal gas is compressed adiabatically in such a way that its temperature rises by 30 C° when 200 J of work is done on it by the compressive force. (*a*) How much does the internal energy of the gas change during the compression? (*b*) How much heat flows from the compressed gas as it cools to its original temperature? (*c*) What is $c_V$ for the gas?

■ **19** For air, $c_V = 0.177$ cal/g · C°. Suppose that air is confined to a cylinder by a movable piston under a constant pressure of 3.0 atm. How much heat must be added to the air if its temperature is to be changed from 27°C to 400°C? The mass of air in the cylinder is 20 g, and its original volume is 5860 cm³. *Hint:* Notice that $mc_V \Delta T$ is the internal energy one must add to the gas to change its temperature by $\Delta T$.

■ **20** A 16,000-cm³ cylinder closed at one end by a piston contains 20 g of air at 30°C. The piston is suddenly pushed in so as to change the gas volume to 1600 cm³. The compression is adiabatic, and the final temperature of the gas is 500°C. How much work is done? For air, $c_v = 0.177$ cal/g · C°.

■ **21** Suppose that 30 g of highly compressed air ($c_V = 0.177$ cal/g · C°) is confined to a cylinder by a piston. Its volume is 2400 cm³, its pressure is $10 \times 10^5$ Pa, and its temperature is 35°C. The air is expanded adiabatically until its volume is 24,000 cm³. During the process, 4100 J of work is done by the air. What is its final temperature?

■ **22** Helium gas at 20°C and a pressure of 2 atm is adiabatically compressed to one-fourth its original volume. What are (*a*) its final pressure and (*b*) its final temperature?

■ **23** Nitrogen gas is to be adiabatically compressed in such a way that its temperature rises from 20°C to 700°C. To what fraction of its original volume must the gas be compressed?

■ **24** Nitrogen gas is to be expanded adiabatically from an original pressure of $3.0 \times 10^6$ Pa and temperature of 27.0°C to such a volume that its temperature is −25°C. By what factor must its volume be expanded?

**25** Three pennies painted different colors are tossed at random. (*a*) In how many different heads-tails combinations can the pennies be arranged? (*b*) What is the chance

that all three will come up tails? (*c*) What is the chance for two heads and one tails?

**26** (*a*) When a pair of dice is thrown, in how many ways can the sum of the two up sides be 3? (*b*) In how many ways can the sum be 6? (*c*) What is the most likely sum for the up sides?

■ **27** When *N* labeled coins are tossed, $2^N$ combinations of heads and tails are possible. How many combinations are possible for (*a*) 3 coins, (*b*) 5 coins, (*c*) 50 coins? *Hint:* If you don't have a calculator or don't know how to use logarithms, try a method based on the idea that $a^{5q} = a^q a^q a^q a^q a^q$.

■ **28** Nine noninteracting ants have fallen into a box. (*a*) Using the explanation given in Prob. 27, determine the chance that all nine end up in the left half of the box. (*b*) What is the chance of eight ants ending up in the left half?

**29** What is the entropy change in 200 g of mercury as it changes from liquid to solid at its normal melting point, −39°C?

**30** When 500 g of lead is melted at its normal freezing point, 327°C, by how much does its entropy change?

**31** When watching television, a certain person has a metabolic rate of 108 W (that is, the person's body is emitting energy at this rate). At what rate is the person's entropy changing?

**32** During an isothermal expansion of an ideal gas at 40°C, the gas does 490 J of work. By how much does the entropy of the gas change?

**33** What is the approximate change in entropy of 140 g of water as it cools from 20°C to 18°C? (Assume *T* to be nearly constant.)

■ **34** Heat is slowly added to 2.0 m³ of nitrogen at 20°C and a constant pressure of 300 kPa. During the heating, the volume increases by 500 cm³. By about how much does the entropy of the gas change? The mass of the nitrogen is 6.9 kg (how can we calculate this?), and $c_P = 0.25$ cal/g · C°.

■ **35** Four coins can each come up heads or tails. (*a*) What is the entropy of the configuration in which all coins are heads? (*b*) For the configuration in which all but one are heads? (*c*) For two heads and two tails?

**36** A certain heat engine uses the interior of a hot furnace (900°C) as its hot reservoir and air at 80°C as its cold reservoir. What is its maximum efficiency?

**37** In modern high-pressure steam-turbine engines, the steam is heated to about 600°C and exhausted at close to 80°C. What is the highest efficiency of any engine operating between these two temperatures?

**38** Temperature differences between the surface water and bottom water in a large lake might be 5 C°. Assuming the surface water to be at 17°C, what is the highest efficiency a heat engine can have if it operates between these two temperatures?

**39** In a certain Carnot engine, the ratio of the exhaust heat to the input heat is 0.873 when the temperature of the hot reservoir is $-47°C$. What is the temperature of the cold reservoir?

■ **40** A scientist on an earthlike planet in a distant galaxy does not know that the earth exists, even though science on her planet is highly developed. She decides to set up a Carnot-engine-based temperature scale by using the condition that $T_h/T_c = \Delta Q_h/\Delta Q_c$, which is true throughout the entire universe. She wishes there to be 100 degrees between the boiling and freezing points of water. From measurements on a Carnot cycle at the boiling and freezing points of water, she finds $\Delta Q_c/\Delta Q_h = 0.732$. What temperatures must she assign to the boiling and freezing points?

■ **41** When gasoline is burned, it gives off 50,000 J/g, called its *heat of combustion*. A certain car uses 8.5 kg of gasoline per hour and has an efficiency of 25 percent. What horsepower does the car develop?

■ **42** A 1200-kg car is to accelerate from rest to 8.0 m/s in 7.0 s. (*a*) What is the minimum horsepower the engine must deliver if all friction losses are ignored? (*b*) By using this value and assuming that the car uses its fuel with an efficiency of 25 percent, determine how much gasoline the car burns in the 7.0 s. Gasoline has a heat of combustion of about 50,000 J/g; that is, it furnishes this much heat energy for each gram burned.

■ **43** A moderate-sized nuclear power plant might have an output of $8.0 \times 10^8$ W. (*a*) Assuming the plant to have an overall efficiency of 30 percent, how much energy is consumed by the plant each second? (*b*) How much heat does the plant exhaust to the cooling system each second? (*c*) Repeat parts *a* and *b* for a fossil-fuel plant with the same output and efficiency.

■ **44** A certain heat engine has an efficiency of 25 percent and an output of 300 hp. (*a*) How much input energy does it require per second? (*b*) How much heat does it give off each second?

■ **45** A certain refrigerator requires 0.75 hp for its operation. It is capable of removing 500 cal/s from the refrigerator. How much heat does it provide each second to the room in which it sits? What is its COP?

■ **46** An air conditioner that requires 0.75 hp for its operation exhausts 500 cal of heat to the outdoors each second.

How many calories does it take each second from the room it is cooling? What is its COP?

■ **47** Suppose a refrigerator has a COP of 6. (*a*) How much energy must be supplied to operate the refrigerator if it is to remove 2000 cal from its interior? (*b*) What horsepower rating must the refrigerator have to accomplish this in 60 s?

■ **48** Two Carnot engines are connected in series in such a way that the first operates between $T_1$ and $T_2$ and the second operates between $T_2$ and $T_3$, where $T_1 > T_2 > T_3$. If the second engine is run by the heat expelled from the first, show that the combined efficiency (output work/input) of the system is $1 - (T_3/T_1)$.

■ **49** An ideal gas is carried around the thermodynamic cycle shown in Fig. P12.3. How much work does the gas do in going around the complete cycle by way of (*a*) *ABCDA* and (*b*) *ADCBA*? (*c*) How much work is required to compress the gas from point *C* to point *A* along *CDA* of the cycle? (*d*) Along *CBA*?

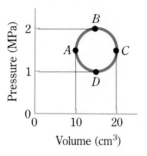

FIGURE P12.3

■ **50** For a certain diatomic ideal gas, $c_V = 920$ J/kg · K. Find the approximate values of (*a*) molecular mass $M$, (*b*) $C_P$, and (*c*) $\gamma$.

■ **51** An hourglass contains a mass $m$ of sand (of specific heat $c$) at a temperature $T_o$ near room temperature. It is inverted to count out one hour. During the hour, the center of gravity of the sand moves downward a distance $D$. Ignoring heat flow to the glass and surroundings, determine (*a*) by how much the sand warms up, (*b*) the change in its internal energy, and (*c*) the change in its entropy. (*d*) After a long time has passed, heat loss to the surroundings causes the temperature to return to $T_o$. How much more or less entropy does the sand now contain than it had when it originally had a temperature $T_o$?

# 13 Vibration and waves

$\mathbf{I}$N the preceding several chapters, we discussed mechanics and heat. Now we begin a new area of study, wave motion. Waves are generated by vibrating objects, such as a guitar string for sound waves and charges on an antenna for radio waves. In this chapter, we introduce waves and investigate the vibrating objects that act as their source. The following chapter is devoted to one special kind of wave, sound waves. Later we shall encounter other important types of waves, such as radio and light waves. Still later we shall learn that even a moving particle is in some ways like a wave. As you see, the topic of waves is wide-ranging and of great importance in the world around us.

## 13.1 Periodic motion

All vibrating systems undergo the same motion over and over again. For example, the pendulum in Fig. 13.1 vibrates (or oscillates) back and forth time after time after time. We say that such motion is periodic and define the **period** of the motion as follows:

The period of vibration $\tau$ (Greek tau) is the time taken to make one complete vibration.

In the case of the pendulum of Fig. 13.1, the period is the time it takes the pendulum to swing from $A$ to $C$ and back to $A$. Notice that the period is the *total* time the

**FIGURE 13.1**
A pendulum undergoes periodic motion. It executes one half cycle as it moves from the leftmost position to the rightmost position.

pendulum ball is away from $A$ during one complete vibration. We call the motion undergone in one period a *vibration cycle.*

Often we speak of the **frequency** of vibration, defined in the following way:

The frequency of vibration $f$ is the number of vibration cycles completed by the system in unit time.

People often express frequencies in cycles per second ($s^{-1}$). For example, a guitar string might undergo 330 vibration cycles in 1 second, and its frequency is thus $330\ s^{-1}$. The SI unit for frequency is the hertz (Hz), which is just another name for cycles per second: $1\ Hz = 1\ s^{-1}$.

There is an important relation between frequency $f$ and period $\tau$. Because frequency is the number of vibrations per unit time and because one vibration takes a time $\tau$,

$$f = \frac{\text{number of vibrations}}{\text{time taken}} = \frac{1\ \text{vibration}}{\text{period}}$$

Thus we have the general relation

$$f = \frac{1}{\tau} \tag{13.1}$$

This relation applies to all periodic motions. If the period of any motion is 0.020 s, say, then its frequency is 50 Hz.

We frequently speak of the **amplitude** of a periodic motion:

The amplitude is the maximum displacement from the position the object maintains when it is not vibrating.

For the pendulum in Fig. 13.1, the amplitude is the distance $AB$ or $BC$. Notice that the amplitude is only half of the full distance through which the system swings.

Another important feature of vibrating systems is the way they interchange energy between potential and kinetic. For example, when the pendulum ball in Fig. 13.1 is at $A$ or at $C$, it is at rest and thus has no kinetic energy. At these points, it possesses only gravitational potential energy. As it swings toward $B$, however, it loses potential energy and gains an equal amount of kinetic energy. Thus, as it swings back and forth, it retains a constant energy, but the energy keeps changing between kinetic and potential.

Another typical vibrating system is shown in Fig. 13.2. It consists of a mass at the end of a spring, and we assume that the mass can slide back and forth without friction. The spring-mass system is shown in its equilibrium position in part *a*. There is no horizontal force acting on the mass in this position. (The pull of gravity down is balanced by the push of the table up, and so the net vertical force on the mass is always zero.)

Suppose we compress the spring by moving the mass to the position shown in Fig. 13.2*b*. During this process, we do work on the spring, and so we store potential energy in it. The compressed spring exerts a force on the mass, a force that tends to push the mass back to the $x = 0$ position. If the mass is now released so that it can move freely under the force applied to it by the spring, the spring will accelerate it to the right until the position $x = 0$ is reached. The mass is now moving to the right quite swiftly, and the spring has lost all the potential energy stored in it when it was

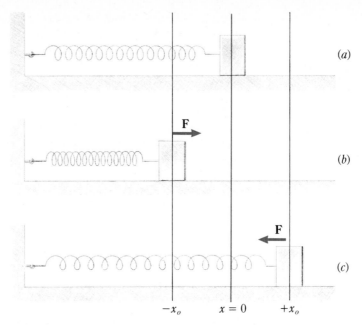

FIGURE 13.2

(*a*) The mass is at equilibrium at $x = 0$ before the system is set into motion. At that position, the spring exerts no force on the mass. (*b*) The compressed spring has potential energy stored in it and exerts a restoring force on the momentarily stationary mass. (*c*) The stretched spring has the same amount of potential energy stored in it as in (*b*). It exerts a restoring force on the momentarily stationary mass.

compressed. It is clear that the potential energy stored in the spring has been given to the mass and now appears as the kinetic energy of the moving mass.

The mass does not stop at $x = 0$, however, because it must lose its kinetic energy by doing work before it can come to rest. As the mass proceeds to the right of $x = 0$, it begins to stretch the spring and to store energy in it. By the time the mass reaches the position shown in Fig. 13.2*c*, it has lost all its kinetic energy doing work against the spring. The kinetic energy of the mass has been changed completely to potential energy in the stretched spring. Therefore, when the mass reaches $x = x_o$, its velocity becomes zero.

The spring, now stretched, accelerates the mass to the left. When the mass reaches $x = 0$, all the energy is once again in the form of kinetic energy. The mass again compresses the spring to the position $x = -x_o$, at which point all the kinetic energy is changed to potential energy stored in the compressed spring. Now the process repeats itself, and the mass will vibrate back and forth between $x = +x_o$ and $x = -x_o$ forever if there are no friction losses. Notice that as the mass oscillates, the energy oscillates back and forth between kinetic and potential but the total energy remains constant; energy is conserved.

The pendulum and spring-mass combination are only two examples of vibrating systems. All such systems involve a similar continuous interchange of energy. Because many of the systems of interest involve springs of one sort or another, let us take a moment to find the energy stored in a spring.

## 13.2 Energy in a Hooke's law spring

We saw in Chap. 9 that many elastic (springlike) systems obey Hooke's law, which states that a distorting force is proportional to the distortion it causes. For a spring being stretched by a force **F**, as in Fig. 13.3*a*, the distance $x$ the spring is stretched is related to $F$ through

$$F = kx \tag{13.2}$$

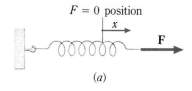

$F = 0$ position

$x$

**F**

(a)

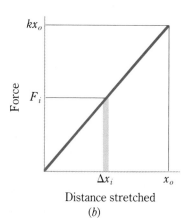

Force

$kx_o$

$F_i$

$\Delta x_i$      $x_o$

Distance stretched

(b)

**FIGURE 13.3**

The work done in stretching an elastic system is equal to the area under its graph of $F$ as a function of $x$.

where $k$ is called the **spring constant.** A stiff, hard-to-stretch spring has a large value for $k$. Because $k = F/x$, the spring constant has units of newtons per meter and is the stretching force per unit length of stretch. Figure 13.3$b$ shows how force varies with distortion for the Hooke's law spring of Fig. 13.3$a$. The graph is a straight line whose equation is given by Eq. 13.2. Let us now compute the energy stored in a stretched or compressed spring that obeys Hooke's law.

We can show that the work done in stretching the spring from $x = 0$ to $x = x_o$ is the area under the graph line in Fig. 13.3$b$. To do that, we examine the shaded rectangle shown. Its area is $F_i \Delta x_i$, where $F_i$ is the stretching force that prevails during the small increase in distortion $\Delta x_i$. Because $W = F_s \Delta s$, this area is also the work done by the stretching force during this small increase in displacement. Imagine the region under the line from $x = 0$ to $x = x_o$ to be filled with such rectangles. The sum of their areas gives the total work done in stretching the spring from x $= 0$ to $x = x_o$. Hence

The work done in stretching or compressing an elastic element is equal to the area under its $F$-versus-$x$ graph line.

You should be able to extend this discussion to confirm the compression portion of this statement.

Since the area of a triangle is one-half its base times its height, we see from Fig. 13.3 that the area under the graph line is $(\frac{1}{2}x_o)(kx_o)$. However, this equals the work done in stretching the spring, and so it is equal to the potential energy stored in the spring. We therefore conclude that the potential energy stored in a spring with constant $k$ that has been stretched or compressed a distance $x$ is

Spring $PE = \frac{1}{2}kx^2$        (13.3)

Now that we know how much elastic energy is stored in a spring (or any other system that obeys Hooke's law), we can use the law of energy conservation to learn much about the vibration of the system shown in Fig. 13.2. Because we are assuming friction losses to be negligible, the sum of the potential energy stored in the spring and the kinetic energy of the mass must remain constant. To place this in equation form, consider once again the system in Fig. 13.2. The spring is stretched to $x = x_o$ and released. Because its total energy is $\frac{1}{2}kx_o^2$, its energy at any time is

$PE + KE = \frac{1}{2}kx_o^2$

Substituting gives

$\frac{1}{2}kx^2 + \frac{1}{2}mv^2 = \frac{1}{2}kx_o^2$       (13.4)

where $m$ and $v$ pertain to the mass at the end of the spring, which itself has negligible mass.

Equation 13.4, despite its simplicity, is a very powerful tool. Using it, we can describe the velocity of the mass at any point in the motion:

$$v = \pm \sqrt{\frac{k}{m}(x_o^2 - x^2)}$$

Do not memorize this equation because it is simply Eq. 13.4 rearranged. Notice that $v = 0$ at $x = x_o$, when the mass is at the end of its swing; when $x = 0$, the velocity has its largest value, $x_o \sqrt{k/m}$. We already know these facts from our qualitative

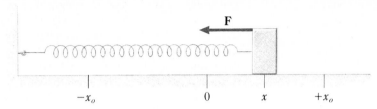

discussion of energy interchange. Now, however, we are able to find the velocity of the vibrating mass at any position $x$.

It remains yet to find the acceleration of the vibrating mass. When the system is vibrating freely, the situation is as shown in Fig. 13.4. The only unbalanced force acting on the mass is **F**, the pull of the spring on it. We call this a restoring force because it is always in such a direction as to pull or push the system back to its equilibrium position. The magnitude of **F** is $kx$, the force required to stretch the spring an amount $x$. However, its direction is *opposite* the direction of stretch, and so its value is $-kx$; the negative sign signifies that it is a restoring force, that is, one directed opposite the displacement $x$. Since **F** is the unbalanced force on the mass, the acceleration of the mass is, from $F = ma$,

$$a = -\frac{k}{m}x \qquad (13.5)$$

Notice that the magnitude of the acceleration is maximum when $x = \pm x_o$ because there the restoring force is maximum; when $x = 0$, the restoring force is zero and so is the acceleration. As you see, we can use Eqs. 13.4 and 13.5 to find the velocity and acceleration of the mass at any displacement $x$.

---

### Example 13.1

A particular spring stretches 20 cm when a 500-g mass is hung from it. Suppose a 2.0-kg mass is attached to the end of this spring and the system is then vibrated horizontally as in Fig. 13.4 by displacing the mass 40 cm from its equilibrium position and releasing it. Find (*a*) the maximum velocity of the mass, (*b*) its maximum acceleration, and (*c*) its velocity and acceleration when $x = 10$ cm.

*Reasoning* Let us first find $k$, the spring constant. Because the stretching force $F = mg = 4.90$ N, we have $k = F/x = (4.90 \text{ N})/(0.20 \text{ m}) = 24.5$ N/m. The law of energy conservation tells us that

$$\tfrac{1}{2}kx^2 + \tfrac{1}{2}mv^2 = \tfrac{1}{2}kx_o^2$$

from which

$$v = \pm\sqrt{\frac{k}{m}(x_o^2 - x^2)}$$

where $x_o$ is the maximum amount of stretch.
(*a*) The maximum velocity occurs when the mass goes through the equilibrium position, that is, at $x = 0$. Therefore,

$$v_{\text{max}} = x_o\sqrt{\frac{k}{m}} = 0.40 \text{ m}\sqrt{\frac{24.5 \text{ N/m}}{2.0 \text{ kg}}} = 1.40 \text{ m/s}$$

(*b*) Because $a = F/m$ and $F = -kx$, the magnitude of the acceleration is maximum when $x = \pm x_o$:

$$a_{\text{max}} = \frac{kx_o}{m} = \frac{(24.5 \text{ N/m})(0.40 \text{ m})}{2.0 \text{ kg}} = 4.90 \text{ m/s}^2$$

(c)   When $x = 0.10$ m,

$$v = \pm \sqrt{\frac{k}{m}(x_o^2 - x^2)} = \pm \sqrt{\frac{24.5 \text{ N/m}}{2.0 \text{ kg}}(0.16 \text{ m}^2 - 0.01 \text{ m}^2)} = \pm 1.36 \text{ m/s}$$

What is the meaning of the $\pm$ sign?

$$a = -\frac{kx}{m} = -\frac{(24.5 \text{ N/m})(0.10 \text{ m})}{2.0 \text{ kg}} = -1.22 \text{ m/s}^2 \quad \blacksquare$$

---

**Exercise**   Find $v$ and $a$ when $x = -5$ cm.   *Answer: $\pm 1.39$ m/s; 0.613 m/s²*   ■

---

## 13.3 Simple harmonic motion

Many vibrating systems appear to vibrate in the same way a mass at the end of a spring vibrates. There is a simple test we can apply to a vibrating mass to see whether it duplicates this type of vibration. The essential feature of such a vibration is that a restoring force that obeys Hooke's law governs the system's motion. In the case of the spring-mass system, the mass is subject to a restoring force of the form $F = -kx$:

Restoring force $= -$(spring constant)(distortion)

If the restoring force that acts on a mass has this form (that is, if it varies linearly with distortion), then the acceleration of the vibrating mass is, from $F = ma$,

$$a = \frac{F}{m} = \frac{\text{restoring force}}{m}$$

or

$$a = -\frac{(\text{spring constant})(\text{distortion})}{m} = -\frac{k}{m}x \qquad (13.6)$$

We call motion that is governed by Eq. 13.6 **simple harmonic motion** (SHM). A mass vibrates at the end of a spring with simple harmonic motion.

To test whether or not an object vibrates with simple harmonic motion, we note whether the object's acceleration obeys Eq. 13.6. If it does, the system is vibrating back and forth with the motion exemplified by the spring-mass system, simple harmonic motion. Because all objects subject to a restoring force that obeys Hooke's law have an acceleration given by Eq. 13.6, we can state at once that

Objects subject only to a restoring force that obeys Hooke's law undergo simple harmonic motion.

Thus we know at once that a system will exhibit simple harmonic motion if either of the following is true: it obeys Hooke's law or its acceleration is given by Eq. 13.6.

## 13.4 Frequency of vibration in simple harmonic motion

It is easy to find the frequency of vibration for a simple harmonic oscillator if calculus is used. We shall not use that method here, however, because many of you are not yet comfortable with that branch of mathematics. Instead, we shall use a simple method based upon what is called a **reference circle.**

The reference circle shown in Fig. 13.5 is a circular path of radius $x_o$ along which a *reference particle* $Q$ moves with constant angular speed $\omega$. We now show that the *x*-coordinate position of this particle, point $P$, undergoes simple harmonic motion. To do this, we show that the acceleration of point $P$ obeys Eq. 13.6. First, notice that point $P$ appears qualitatively to undergo simple harmonic motion. As $Q$ travels around the circle at constant speed, point $P$ moves back and forth from $+x_o$ to $-x_o$ to $+x_o$ over and over again. Let us now prove that the motion of $P$ is simple harmonic.

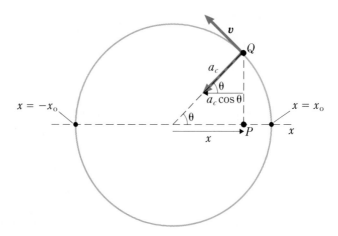

**FIGURE 13.5**

As $Q$ moves around the circle of radius $x_o$ with constant speed $v$, point $P$ undergoes simple harmonic motion from $-x_o \leq x \leq x_o$. Because the radius of the circle is $x_o$, $x = x_o \cos \theta$.

Because $Q$ travels around the circle with constant angular speed $\omega$ (or tangential speed $v = \omega r = \omega x_o$), it has a centripetal acceleration (from Eq. 7.9)

$$a_c = \frac{v^2}{x_o} = \frac{\omega^2 x_o^2}{x_o} = \omega^2 x_o$$

The radial direction of $a_c$ is shown in Fig. 13.5. Point $P$ is also accelerating, but its acceleration is along the $x$ axis. To find $a_P$, we notice that it is simply the $x$ component of the acceleration of $a$ since $P$ and $Q$ move in unison. As we see in Fig. 13.5, the $x$ component of $a_c$ is $a_c \cos \theta$, and so

$$a_P = - a_c \cos \theta$$

The negative sign shows the direction of the acceleration.

We can put this in more meaningful form by using the facts that $\cos \theta = x/x_o$ and $a_c = \omega^2 x_o$:

$$a_P = - \omega^2 x_o \frac{x}{x_o} = - \omega^2 x \tag{13.7}$$

where $\omega$ is constant. This is an important result because it shows that the acceleration of point $P$ is of the same form as that in Eq. 13.6. We know, however, that this acceleration is characteristic of simple harmonic motion. Therefore we have proved that point $P$ undergoes simple harmonic motion.

Now that we know this, we can find the frequency of the motion. To do so, we notice that both Eqs. 13.6 and 13.7 apply to a mass undergoing simple harmonic motion. Comparing the two equations, we see that $\omega^2$ must be the same as $k/m$. Hence

$$\omega = \sqrt{\frac{k}{m}} \tag{13.8}$$

This is the angular speed with which particle $Q$ in Fig. 13.5 must move if point $P$ is to represent the simple harmonic motion of a mass $m$ under the action of a restoring force whose force constant is $k$.

We are really more interested in the motion of the mass itself, represented by point $P$ in Fig. 13.5. The frequency of its back-and-forth motion, $f$, is related to the angular frequency $\omega$ through $\omega = 2\pi f$ because $f$ is equal to the angular frequency expressed in revolutions per second. Equation 13.8 therefore becomes

$$f = \frac{1}{2\pi} \sqrt{\frac{k}{m}} \tag{13.9}$$

and we conclude that *a mass m subject to a Hooke's law restoring force with spring constant k undergoes simple harmonic motion with a frequency given by Eq. 13.9.*

---

**Example 13.2**

Find the frequency of vibration for the system of Example 13.1.

**Reasoning**  In that example, the spring constant was 24.5 N/m and the mass at the end of the spring was 2.0 kg. From Eq. 13.9,

$$f = \frac{1}{2\pi} \sqrt{\frac{24.5 \text{ N/m}}{2.0 \text{ kg}}} = 0.56 \text{ s}^{-1} = 0.56 \text{ Hz} \quad \blacksquare$$

---

## 13.5 Sinusoidal motion

We shall now see that a simple mathematical equation can be written for an object vibrating with simple harmonic motion. Point $P$ in Fig. 13.5 has an $x$-coordinate position given by

$$x = x_o \cos \theta$$

However, the angle $\theta$ is continuously changing as $Q$ moves around the reference circle. Since our motion equations state that (Eq. 7.2)

$$\theta = \omega t$$

we have at once that

$$x = x_o \cos (\omega t)$$

where $\omega$ is the angular speed, in radians per second, of $Q$ on the reference circle. We know, however, that $\omega = 2\pi f$, where $f$ is the angular speed in revolutions per second (Secs. 7.1 and 7.2). We also know that as $Q$ makes one complete revolution around the circle, point $P$ vibrates through one cycle. Hence $f$ is also the vibration frequency of the harmonic oscillator represented by $P$. Therefore the equation that describes the motion of the harmonic oscillator $P$ is

$$x = x_o \cos (2\pi ft) \tag{13.10}$$

where $f$ is the frequency of oscillation.

Equation 13.10 states that an object vibrating with simple harmonic motion has a displacement that varies as $\cos (2\pi ft)$. To see exactly what this type of vibration entails, consider the experiment illustrated in Fig. 13.6. An object with an attached pen is suspended from a spring. When the object is raised to $y_o$ and released, it undergoes simple harmonic motion with amplitude $y_o$. Behind the vibrating object is a sheet of paper moving to the left at constant speed. The pen marks on this paper the position of the mass as it vibrates up and down.

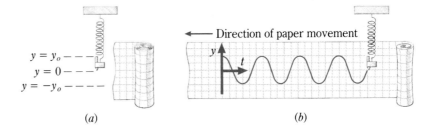

**FIGURE 13.6**

The vibrating mass traces a cosine curve as a function of time.

(a)    (b)

Let us count time from the instant the object is released. This is the left end of the trace in *b*. We take this point as $t = 0$. The position of the object at the instant shown in the figure occurs at some later time. Hence, this can be considered a plot of the displacement $y$ of the object as a function of time. According to Eq. 13.10, the equation of this trace is

$$y = y_o \cos (2\pi ft) = y_o \cos (\omega t) = y_o \cos \left(\frac{2\pi t}{\tau}\right) \tag{13.11}$$

This general type of trace, the graph of $y$ versus time, is called a **sinusoidal curve,** and the motion that causes it is called **sinusoidal motion.** Clearly, *simple harmonic motion and sinusoidal motion are the same.*

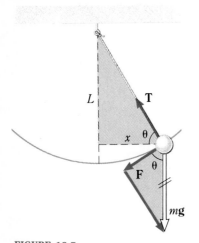

**FIGURE 13.7**

The two $\theta$s are equal. Notice also that if the swing is not too large, the component of **F** along $x$ (which is the $x$-direction restoring force) is nearly equal to **F**.

## 13.6 The simple pendulum

We know that a simple pendulum like that shown in Fig. 13.7 vibrates with what appears at first glance to be simple harmonic motion. If the restoring force on a body is proportional to the displacement, Hooke's law applies and the vibration is simple harmonic motion, as we saw previously. Let us now examine the restoring force for the pendulum and see whether it obeys Hooke's law.

The pendulum ball moves along the arc of the circle shown. Since the tension $T$ in the pendulum cord always pulls perpendicularly to this arc, it neither speeds nor slows the motion. It is the weight $mg$ of the ball that causes the motion. However, only $F$, the component of the weight tangential to the circular arc, is effective in producing the motion. From Fig. 13.7,

$$F = -mg \cos \theta$$

where the two angles labeled $\theta$ in the figure are equal. (Why?) The minus sign shows that $F$ is a restoring force. We can also see from the figure that $\cos \theta = x/L$, where $x$ is the horizontal displacement of the ball and $L$ is the length of the pendulum. Therefore, the equation for $F$ becomes

$$F = -\frac{mg}{L} x \qquad (13.12)$$

This is similar in form to Hooke's law, $F = -kx$, except in one detail. In Hooke's law, the displacement $x$ should be in the direction opposite the force. We see from Fig. 13.7 that this is not true in this case. Although $F$ and $x$ are not exactly in line, they are very nearly so if the angle of swing is very small. Therefore, *the pendulum does approximate Hooke's law if it is not swinging too widely. To that approximation, the pendulum undergoes simple harmonic motion.* Comparing Eq. 13.12 with Hooke's law, $F = -kx$, we see that the spring constant $k$ for the pendulum is just

$$k = \frac{mg}{L}$$

When this value for $k$ is substituted in Eq. 13.9, which is the formula for the frequency of a body in simple harmonic motion, we have

$$f = \frac{1}{2\pi} \sqrt{\frac{g}{L}} \qquad (13.13)$$

Notice that *the frequency of a simple pendulum does not depend on the mass of the ball.* It depends only on the length of the pendulum and on the acceleration due to gravity $g$. This offers a precise means for measuring $g$ in a simple experiment. If a pendulum of known length is timed so that its frequency of vibration is known, $g$ can be computed at once from Eq. 13.13. With proper precautions, this method for determining $g$ is extremely accurate.

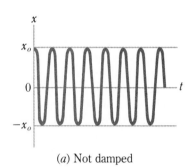

(a) Not damped

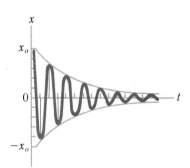

(b) Slightly damped

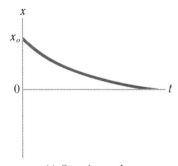

(c) Overdamped

**FIGURE 13.8**

The way a system vibrates depends on the extent of the energy losses within it.

## 13.7 Forced vibrations

In any vibrating system, there is always some loss of energy to friction forces. As a result, a vibrating pendulum or mass at the end of a spring vibrates with constantly decreasing amplitude as time goes by. This fact is illustrated in Fig. 13.8. Part *a* shows the vibration in the ideal case with no friction. It is the situation we discussed in preceding sections. A more realistic situation is shown in *b*, where the vibration is fairly strongly influenced by friction forces. We say that such a system is *damped* and that, in this case, the vibration *damps down* fairly quickly.

When the friction forces are very large, the system does not vibrate at all; instead, it simply returns slowly to its equilibrium position, as shown in Fig. 13.8c. Such a system is said to be *overdamped*. This situation exists, for example, when the mass at the end of a spring is immersed in a very viscous liquid. The mass does not move beyond the equilibrium position in such a case. When the friction forces are just large

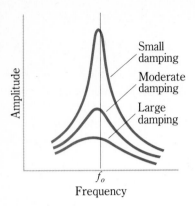

A graph showing the amplitude of the driven vibration as a function of equal driving forces at various frequencies $f$. The resonance frequency of the undamped vibration is $f_o$. The three curves are for the same oscillator at different extents of damping.

enough that the system returns to the equilibrium position without overshooting it, we say that the system is *critically damped*.

If any system is to vibrate for an extended time, energy must be added continually to replace the energy lost doing work against friction forces. For example, to keep a child swinging at constant amplitude on a swing, you must push the swing from time to time to add energy to the system. This is typical of all vibrating systems. An outside agent must feed energy into the system if the vibration is not to damp down.

Everyone knows that there is a right and a wrong way to push a swing if it is to swing high. You must push with the motion of the swing and not against it. Only in that way can energy be added effectively. In fact, if you push against the motion, the vibration can be brought to a stop, since the vibrating object must then do work on the pushing agent. These simple facts have importance in all forced, or *driven*, vibrating systems.

In a driven system, such as a grandfather clock, the vibration is usually sustained by a repetitive force acting on the system. This force has a frequency $f$, which may or may not be the same as the natural frequency of vibration of the system $f_o$. When $f = f_o$, the driving agent is most effective in adding energy to the system. At all other frequencies, the driving force is not quite in step with the motion of the system, and so its action is less effective in adding energy. How the amplitude of a vibrating system varies with the frequency of the applied force is shown in Fig. 13.9. Notice, as we said above, that the driving force is most effective when its frequency $f$ equals the natural frequency $f_o$ of the system. In that case, we say that the force is in **resonance** with the system. More is said about $f_o$, the **resonance frequency** of the system, in Sec. 13.10, where we discuss the resonance vibration of strings.

## 13.8 Wave terminology

Many vibrating objects act as sources for waves. For example, sound waves can originate from a vibrating tuning fork or from a vibrating guitar string. We begin our study of waves by referring to one easily visualized type: a wave on a string. A disturbance can be sent down a string as shown in Fig. 13.10. The disturbance, or

A pulse carries energy down the string. What is the speed of the pulse?

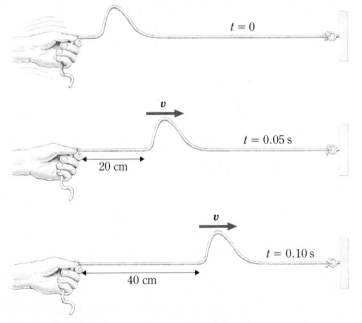

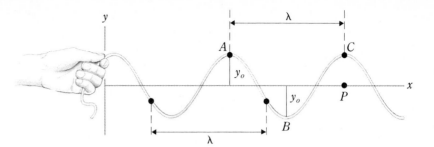

**FIGURE 13.11**

A source that vibrates with simple harmonic motion sends a sinusoidal wave down the string.

pulse, is initiated by a sudden up-and-down motion of the hand holding the string and travels with speed $v$ along the string. Note two very impotant features of such a pulse. First, it carries energy down the string. When the pulse strikes a given point on the string, it causes that portion of the string to momentarily acquire both kinetic and potential energy. This energy was given to the pulse by the source that initiated it. The energy moves with speed $v$ down the string along with the pulse.

Second, the pulse on the string is a record of the history of what the source has done. We can see in Fig. 13.10 that the hand moved to initiate the pulse at a definite time in the past. Indeed, what the source was doing at any past time $t$ is shown by the string at a distance $x = vt$ from the source. In other words, the string at a distance $x$ from the source is performing the same motion that the source initiated at a time $t = x/v$ previously.

Let us now see what happens when the source vibrates with simple harmonic motion. The situation is shown in Fig. 13.11. As we expect, the string shows the past history of the way its end was vibrated. The up-and-down motion of the string's end is transmitted down the string with speed $v$. As a result, the string has a sinusoidal form at any given instant, and this sinusoidal pattern on the string travels to the right with speed $v$. As it moves, the pattern carries energy down the string, energy furnished by the source of vibration.

There are certain words we use to describe such a wave. The points $A$ and $C$, the tops of the wave, are called **wave crests.** Points such as $B$ are called **wave troughs.** We call the maximum displacement of the string from its equilibrium position the **amplitude** of the wave. The amplitude of the wave in Fig. 13.11 is $y_o$. Notice that it is only half the total vertical displacement of the string.

The distance between two adjacent crests on the wave — between $A$ and $C$, for example — is equal to a distance called the **wavelength** of the wave, indicated by $\lambda$ (Greek lower-case lambda) in Fig. 13.11. The wavelength of a wave is the distance between any two equivalent adjacent points along the wave. One wavelength of a wave is sent out by the wave source as it executes one complete vibration.

At a point such as $P$, the string repeatedly moves up and down as the wave moves to the right. During the time it takes the source to send out one wavelength, one wavelength must pass through $P$ to make way for a new wavelength. As a result, $P$ undergoes one complete cycle of motion in the same time the source takes to undergo one complete vibration. We see from this that the period of the vibrating source is the same as the period of vibration of a point in the path of the wave. This time taken for a complete vibration of a point in the path of the wave is called the **period** of the wave and is represented by $\tau$. As for an oscillator, the frequency of the wave is related to its period through $f = 1/\tau$. Furthermore, the frequency is equal to the number of wave crests passing through point $P$ each second.

A very important relation exists between the wavelength of the wave and the frequency. Referring again to Fig. 13.11, it is apparent that a length $\lambda$ of the wave is sent out during the time $\tau$ that the wave source takes to undergo one complete vibration. Therefore, the wave moves a distance $\lambda$ in a time $\tau$, and so we find, from $v = x/t$, that $v = \lambda/\tau$, where $v$ is the speed of the wave. Thus

$$\lambda = v\tau \qquad \text{and} \qquad \lambda = \frac{v}{f} \tag{13.14}$$

This relation is true for all waves, not just for waves on a string.

The speed of a wave on a string is given by a particularly simple relation that we state without derivation. If the tension in the string is $T$ and if the mass of a length $L$ is $m$, then the speed of the wave along the string is

$$v = \sqrt{\frac{T}{m/L}} \tag{13.15}$$

We can easily justify that speed should depend on tension and mass per unit length. The tension in the string is responsible for the force that accelerates a piece of the string as the pulse passes through the region. The greater the tension, the greater the acceleration, and so the motion of the pulse is swift if the tension is high. On the other hand, the more massive the string, the more inertia it has. The mass per unit length therefore affects the speed with which the pulse moves. A massive string has large inertia, and the speed of a pulse on it will be low.

---

**Example 13.3**
A certain guitar string has a mass of 2.0 g and a length of 60 cm. What must the tension in the string be if the speed of a wave on it is to be 300 m/s?

**Reasoning** From Eq. 13.15, $T = (m/L)(v^2)$, and since $v = 300$ m/s, $m = 0.0020$ kg, and $L = 0.60$ m, $T = 300$ N. Notice that this tension is really quite large: it is equivalent to the weight of a mass of about 30 kg pulling on the string. ∎

---

**13.9** **Reflection of a wave**

The string in Fig. 13.11 must be held taut by a support at its right-hand end. When the wave traveling to the right along the string strikes the support, the energy carried by the wave is stopped. We now wish to discuss what happens to this energy. There are two extreme possibilities: all the energy could be absorbed in the support as thermal energy or all the energy could be reflected back to the left along the string. Between these two extremes, of course, is the possibility that part of the energy is absorbed and the rest is reflected. For the purpose of this discussion, we shall assume that all the energy is reflected, an assumption that is approximately correct in many cases.

To study this effect, let us consider only a single wave crest propagating along the string, as illustrated in Fig. 13.12a. When this crest reaches the wall, it cannot pull the wall upward in the same way it pulls the string upward. Because the wall does not move, it exerts a downward pull on the string. This pull accelerates the string downward to such an extent that its momentum carries it below the zero line. The result is that the pulse is turned upside down as it hits the wall, and the reflected wave appears as shown in Fig. 13.12b. If the string had been completely free to move up and down at that end, the wave would not have been turned over, although it would still have been reflected (because the energy in the wave could not just disappear at the end of the string). In summary, *a pulse is inverted by reflection at a fixed end. It is reflected, but not inverted, by a free end.*

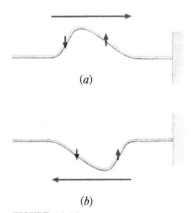

(a)

(b)

FIGURE 13.12

A pulse on a string is inverted when it is reflected from a fixed end.

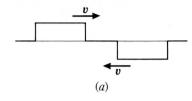

(a)

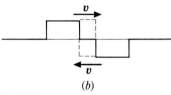

(b)

**FIGURE 13.13**

According to the superposition principle, two waves on the same string add as shown when they meet.

Next, let us consider what happens when a reflected pulse traveling to the left along the string meets a pulse moving to the right. Suppose two rectangular pulses are going in opposite directions on the same string, as in Fig. 13.13*a*. After they meet and begin to overlap, the situation is that shown in *b*. The positions of the original pulses in the region of overlap are shown in color. It is found from experience that the string displaces as shown by the black line: its displacement is the vector sum of the individual wave displacements. This is an example of the **principle of superposition:**

A point subjected to two or more waves simultaneously is displaced an amount equal to the vector sum of the individual disturbances.

All waves we deal with in this text conform to this principle.

We are now ready to see what happens when a sinusoidal wave traveling down a string is reflected by a rigid support at its end. The situation is shown in Fig. 13.14. When the incident wave in *a* reaches the support, it is inverted and reflected. Part *b* shows the hypothetical incident and reflected waves. We term them "hypothetical" because the string itself obeys neither wave. Instead, it sums the two waves and

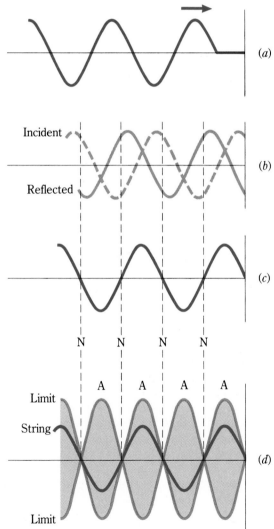

**FIGURE 13.14**

The combined incident and reflected waves give rise to nodes and antinodes on the string. (The amplitudes of the waves are much exaggerated.)

takes the form shown in *c* at the instant at which the incident and reflected waves are positioned as in *b*. Notice that the displacement of the string at the support is zero, as it must always be. Moreover, the displacement at this instant is zero at several other points as well.

Now comes the really interesting part. Suppose we redraw Fig. 13.14*b* for any other instant. Even though the incident and reflected waves are at different positions in the new sketch, their sum is still zero at the same points as in *c*. In other words, the string never moves at all at the positions labeled N. If you were to watch the string, it would appear blurred as it oscillates back and forth between the limits shown in *d*. The points N along the string, which never move, are called **nodes.** The points A, which are midway between the nodes and move most, are called **antinodes.** This type of vibration, in which the string vibrates back and forth within a well-defined envelope (or limiting curve), is called a **standing wave.** We shall have more to say about standing waves in a moment.

If you look at the instantaneous position of the string in *d*, you can see that the nodes are half a wavelength apart. Similarly, the distance between adjacent antinodes is $\frac{1}{2}\lambda$. This is a convenient fact to remember:

The distance between adjacent nodes or adjacent antinodes is $\frac{1}{2}\lambda$.

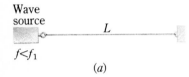

Wave source

$L$

$f < f_1$

(*a*)

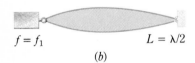

$f = f_1$  $L = \lambda/2$

(*b*)

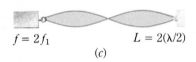

$f = 2f_1$  $L = 2(\lambda/2)$

(*c*)

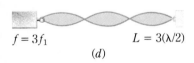

$f = 3f_1$  $L = 3(\lambda/2)$

(*d*)

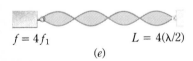

$f = 4f_1$  $L = 4(\lambda/2)$

(*e*)

**FIGURE 13.15**

Resonances of a string.

## 13.10 Wave resonance

When a pendulum, a child on a swing, or a mass at the end of a spring is set into motion by a periodic force, the system moves with largest amplitude when the frequency of the force equals the natural vibration frequency of the free system. In Sec. 13.7, we used the example of pushing a child on a swing to show that a system undergoes resonance, that is, it vibrates most strongly when the frequency of the driving force is equal to the frequency of the system's free vibration. A similar situation exists for vibration of a string, as shown in Fig. 13.15. If you vibrate it with too low a frequency, the string vibrates so little that it appears motionless, as in Fig. 13.15*a*. If you slowly increase the frequency of vibration, the string begins to vibrate widely at a certain frequency, as in Fig. 13.15*b*. At this fundamental resonance frequency, $f_1$, the string vibrates widely and appears as a blur between the limits shown. This is an impressive example of the standing-wave phenomenon mentioned earlier. Experiment shows that the string also resonates to other higher frequencies, as shown in *c*–*e* of the figure. In summary, the string resonates to the fundamental frequency $f_1$ and to higher frequencies $2f_1$, $3f_1$, $4f_1$, and so on.

There is an easy way to specify the conditions under which resonance occurs. Looking at Fig. 13.15, we see that the string always resonates in whole segments, where a *segment* is the distance between adjacent nodes or antinodes. The fixed ends are always at nodes. Therefore the string will resonate only if it is one segment long, two segments long, and so on. Since the length of a segment is $\frac{1}{2}\lambda$, however, the string can resonate only if it is $\lambda/2$ long, or $2(\lambda/2)$ long, or $3(\lambda/2)$ long, and so on. Indeed, we can state in general that a string fastened firmly at its two ends resonates only if it is a whole number of half wavelengths long. For example, in Fig. 13.15*b*–*e*, the length of the string is equal to $\lambda/2$, $2(\lambda/2)$, $3(\lambda/2)$, and $4(\lambda/2)$. In general, then, for resonance of a string fastened at both ends,

$$L = n\frac{\lambda_n}{2} \quad \text{where } n = 1, 2, 3, \ldots \tag{13.16}$$

and where $\lambda_n$ is the wavelength when the string resonates in $n$ segments. Since wavelength is related to frequency by Eq. 13.14, we see at once that a string of fixed length resonates to only certain very special frequencies. We say that the resonant frequencies $f_n$ of the string are *quantized*, meaning that they are separated by frequency gaps; in other words, quantum jumps in frequency exist between the resonance frequencies:

$$f_n = \frac{v}{\lambda_n} = \frac{v}{2L/n} = n\frac{v}{2L} = nf_1$$

The resonance frequencies are integer multiples of the fundamental resonance frequency $f_1$.

---

### Example 13.4

The speed of a wave on a particular string is 24 m/s. If the string is 6.0 m long, to what driving frequencies will it resonate? Draw a picture of the string for the first three resonance frequencies.

*Reasoning*  The possible resonance wavelengths are given by Eq. 13.16. For $L = 6.0$ m,

$$\lambda_n = \frac{2L}{n} \qquad \lambda_1 = 12 \text{ m} \qquad \lambda_2 = 6 \text{ m} \qquad \lambda_3 = 4 \text{ m} \qquad \lambda_n = \frac{12}{n} \text{ m}$$

Now we can use Eq. 13.14 to find the frequency; from $f = v/\lambda$,

$$f_1 = \tfrac{24}{12} = 2 \text{ Hz} \qquad f_2 = \tfrac{24}{6} = 4 \text{ Hz} \qquad f_3 = \tfrac{24}{4} = 6 \text{ Hz} \qquad f_n = \frac{24}{12/n} = 2n \text{ Hz}$$

The modes of vibration, that is, the various standing waves, of the string are as shown in Fig. 13.15$b$–$d$ when the frequency is $f_1$, $f_2$, and $f_3$.  ∎

---

*Exercise*  If the string resonates in three segments to $f = 11$ Hz, what is the speed of the waves?  *Answer: 44.0 m/s*  ∎

---

### 13.11  Transverse waves

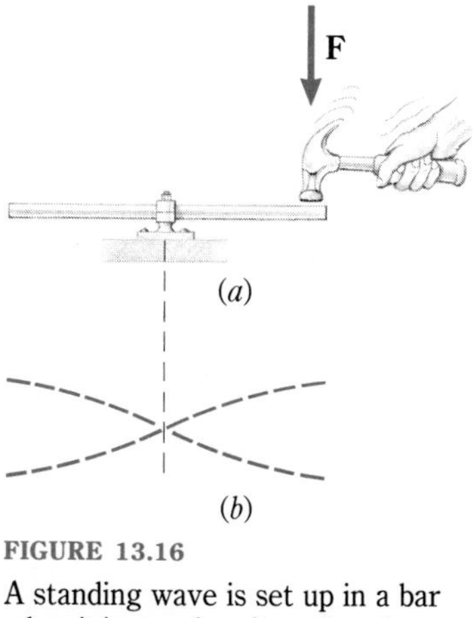

We have spent a great deal of space discussing waves on a string because the principles that apply to them apply to many other vibrating systems.

For example, if a metal bar clamped at its center is struck at its end, as shown in Fig. 13.16, it vibrates. The mode of vibration of the bar is indicated in Fig. 13.16$b$. The center of the bar must be a node because it is tightly clamped in place and therefore cannot move. Since the ends of the bar are not held rigidly, we expect antinodes near them. If we assume the ends to be antinodes, the length $L$ of the bar is one-half wavelength. This follows from the fact that the distance between two successive antinodes is $\lambda/2$. Since we know that $\lambda = 2L$, we can measure the frequency of vibration of the bar and use $f = v/\lambda$ to compute the speed of a wave traveling in it.

The waves we have been considering so far are called **transverse** waves. They

**FIGURE 13.16**

A standing wave is set up in a bar when it is struck on its end as shown.

are given this name because the particles of the string or bar move across (or perpendicular to) the direction in which the wave is propagating. For example, the wave on a string propagates from left to right down the string while the string itself moves up and down. Because "transverse" means "crosswise" or "across," we term waves such as this transverse waves.

We shall encounter other transverse waves as our studies progress. Water waves are approximately transverse, as are the waves on a drumhead and cymbal. Perhaps the most important of all transverse waves is the electromagnetic wave. As we shall see, this type of wave encompasses not only light and radio waves but heat radiation and x-rays as well. All of these transverse waves behave much like waves on a string, and so what we have learned here is applicable to these other wave types as well.

## 13.12 Longitudinal waves

An interesting experiment can be done with a very long spring placed on a smooth tabletop and tied at one end. One form of this experiment (not the best from a practical standpoint) is illustrated in Fig. 13.17. In part *a*, the spring is shown at equilibrium on the tabletop. If it is suddenly compressed as in part *b*, the loops near the end at which the compression force is applied are compressed before the rest of the spring experiences the disturbance. The compressed loops then exert a force on the loops to the right of them, and the compression travels down the spring. At the fixed end, the compressional energy is reflected; thus the compression is reversed and ends up traveling to the left, as shown in *d*.

This type of wave is not a transverse wave since the particles of the spring vibrate back and forth in the direction in which the wave is propagated, along the spring. A compressional wave such as this, where the motion of the particles is along the direction of wave propagation, is called a **longitudinal wave.**

**FIGURE 13.17**
A longitudinal pulse travels down the spring and is reflected at the fixed end.

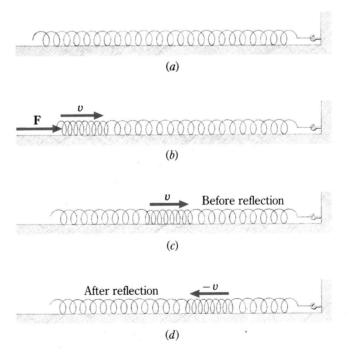

## 13.13 Standing compressional waves on a spring

It is clear that a longitudinal wave on a spring has many features in common with a transverse wave on a string. If a compressional wave is sent down a spring, the wave and its energy are usually reflected at the end of the spring. This reflected wave can interfere with the later waves being sent down the spring from the source. If the proper relation is maintained between the frequency of the source and the various parameters of the spring, resonance occurs. It is this feature of the spring system that we now investigate.

As with resonance on a string, the driving source in a spring system is usually close to a node (a point of zero motion) since at resonance the spring moves much more than the source. Also, if the other end of the spring is held motionless, it too must be at a node. We represent the resonance motion of a spring by the diagrams in Fig. 13.18. Although the displacement of the spring is longitudinal, we still use the same type of diagram used for transverse vibrations to show the nodes and anti-nodes. Now, however, we cannot interpret the diagrams as showing the actual shape of the vibrating system. Despite that fact, analysis of the resonance forms in Fig. 13.18 proceeds much as for resonance in a string.

We notice in Fig. 13.18 that the distance between adjacent nodes is $\lambda/2$ and that the resonating spring is an integral number of segments long. Therefore, if the spring has a length $L$, it resonates to waves that have the following wavelengths:

$$n\frac{\lambda_n}{2} = L \qquad \text{where } n = 1, 2, \ldots$$

This relation, when combined with the one for wavelength and frequency, $\lambda = v/f$, tells us at once that the spring resonance frequencies are

$$f_n = n\frac{v}{2L} \qquad \text{where } n = 1, 2, \ldots$$

### Example 13.5

A spring 300 cm long resonates in three segments (nodes at both ends) when the driving frequency is 20 Hz. What is the speed of the wave in the spring?

*Reasoning* The spring is vibrating as in Fig. 13.18d. In this case, the spring resonates in three segments, so $n = 3$. Then, because the length of a segment is $\lambda/2$,

$$3\frac{\lambda_3}{2} = L$$

$$\lambda_3 = 2.00 \text{ m}$$

Since we know that $\lambda = v/f$ and $f_3 = 20$ Hz, we have

$$v = f\lambda = (20 \text{ s}^{-1})(2.0 \text{ m}) = 40.0 \text{ m/s}$$

We could, of course, have obtained this result by simply substituting in the relation

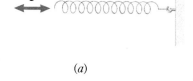

(a)

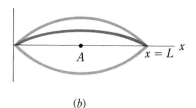

(b)

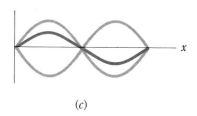

(c)

(d)

**FIGURE 13.18**
Standing waves result from longitudinal vibration of a spring.

$$f_n = n \frac{v}{2L}$$

using $n = 3$. However, most physicists prefer not to memorize a different relation for each case. They ordinarily use the number of half wavelengths on the total spring to find $\lambda$ and then the relation $f = v/\lambda$ to find the unknown. As a matter of fact, almost all the resonance situations we encounter can be described by the use of this relation and an examination of the resonant system. It is not necessary to memorize an equation for each case. ■

---

*Exercise*  What is the wave speed when the wave vibrates at the same frequency in five segments, rather than three?  *Answer: 24.0 m/s*  ■

---

## 13.14 Compressional waves on a bar

We have already discussed the transverse vibrations of a bar. It is possible to set up longitudinal vibrations in such a bar as well. There are many possible ways of doing this. One simple method is to strike the end of the bar, as shown in Fig. 13.19a.

Because the bar acts like a very stiff spring, the blow on its end sends a compressional wave down the bar. This wave is quite complex in form and is actually a large group of waves of various frequencies. The bar resonates only to certain frequencies, and hence it selects the proper frequency to which it will resonate.

In this particular case, the wave must have a node at the center of the bar and antinodes at the ends. The lowest resonant-frequency motion is shown in Fig. 13.19b. Remember that the motion of the bar is longitudinal even though the graph of the motion looks like that for the similar case for transverse vibration. Clearly in this case the bar is one-half wavelength long. Since the length $L$ of the bar is known, the wavelength is also known. If the frequency of vibration is known as well, the velocity of compressional waves in the material can be computed by using $\lambda = v/f$. (A simple way to measure the frequency of vibration is to compare the sound given off by the bar with the sound given off by tuning forks of known frequency. More accurate methods are discussed later.) Some typical values for the speed of compressional waves in various materials are given in Table 13.1.

Other resonance vibrations of the bar in Fig. 13.19 are also possible. Any vibration that has a node at the center of the bar and antinodes at the ends is allowed. You should examine this case to see what the next highest resonant frequency is. It will have three nodes along the bar. (Why is the resonance with two nodes along the bar not allowed in this case?) Since friction losses between molecular planes within the vibrating bar are more serious at higher frequencies, these resonance motions usually die out rapidly — that is, their energy is lost to heat — and the bar ceases to vibrate.

The speed of compressional waves in various materials can be related to the density and elasticity of the materials. The speed in a rod is

$$v = \sqrt{\frac{Y}{\rho}} \tag{13.17}$$

where $\rho$ is the density and $Y$ is Young's modulus for the material of the rod. Notice that consistent units must always be used in this relation. In the case of extended solids and liquids, $Y$ must be replaced by the bulk modulus $B$, defined in Eq. 9.4.

It seems reasonable that speed increases with increasing modulus because a pulse

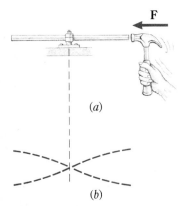

**FIGURE 13.19**
When the bar is struck as shown, a longitudinal standing wave is set up in it.

---

■ **TABLE 13.1**
**Speed of compressional waves**

| Material | Speed (m/s) |
|---|---|
| Air (0°C) | 331 |
| Water (15°C) | 1447 |
| Copper | 3500 |
| Glass | 4000–5500 |
| Wrought iron | 4900–5100 |
| Steel | 5000 |

should be transmitted most swiftly down the stiffest spring. A stiff spring has very little give, and so an impulse is felt almost at once through the entire length. The density factor in the expression for $v$ represents the inertia of the material. As we would expect, large inertia (high density) gives rise to slowly moving pulses.

---

### Example 13.6

If the thin metal bar of Fig. 13.19 is 0.925 m long and resonates as shown with a frequency of 2700 Hz, what is Young's modulus of the metal? The density of the metal is 7.86 g/cm³.

*Reasoning* We first need to determine $v$ so that we can use Eq. 13.17. From Fig. 13.19, we see that 0.925 m is equivalent to $\lambda/2$. Hence $\lambda = 1.85$ m. Using $v = \lambda f$ gives

$$v = (1.85 \text{ m})(2700 \text{ s}^{-1}) = 5000 \text{ m/s}$$

Then, from Eq. 13.19,

$$(5.00)(10^3 \text{ m/s}) = \sqrt{\frac{Y}{7.86 \times 10^3 \text{ kg/m}^3}}$$

$$Y \cong 2.0 \times 10^{11} \text{ N/m}^2$$

Examination of Table 9.2 indicates that this bar might well be made of steel. ∎

---

*Exercise* Determine the speed of compressional waves in a large block of this metal. Its bulk modulus is $1.50 \times 10^{11}$ N/m². *Answer: 4370 m/s* ∎

---

## Minimum learning goals

When you finish this chapter, you should be able to

1 Define or explain (*a*) amplitude, period, and frequency of vibration, (*b*) hertz, (*c*) spring constant, (*d*) simple harmonic motion, (*e*) sinusoidal motion, (*f*) damping, (*g*) resonance, (*h*) sinusoidal wave, (*i*) wavelength, (*j*) wave crest and wave trough, (*k*) amplitude, period, and frequency of a wave, (*l*) node and antinode, (*m*) standing wave, (*n*) wave resonance, (*o*) relationship between segment length and $\lambda$, (*p*) transverse wave, (*q*) longitudinal wave.

2 Use energy considerations to find the speed of a simple harmonic motion oscillator at any position in its path. State where the speed is greatest and where it is least.

3 Use $F = ma$ to find the acceleration of a simple harmonic motion oscillator at any point in its path. State where the acceleration is largest and where it is smallest.

4 Explain how one can ascertain whether or not a motion is simple harmonic and how the test method is related to Hooke's law.

5 Explain how motion in a reference circle is related to simple harmonic motion.

6 Given sufficient data, find the natural frequency of vibration of (*a*) a spring-mass system and (*b*) a pendulum.

7 Explain why simple harmonic motion is called sinusoidal motion. Give the equation for a sinusoidal curve and explain the quantities in it.

8 Point out what causes the restoring force in the case of a simple pendulum and explain why the motion is only approximately simple harmonic. Give the equation for the period of the motion.

9 Sketch several standing-wave forms for a string fixed at both ends. Given the length of the string and either $f$ or $v$, use the standing-wave pattern to compute $v$ or $f$.

10 Draw the compressional standing-wave resonance forms for (*a*) a spring fixed at both ends and (*b*) a rod held rigidly at one point, with that point being at an end, or $0.5L$ from one end, or $0.25L$ from one end. Compute the resonance frequency for each situation.

## Questions and guesstimates

**1** On the same horizontal axis, sketch graph lines that relate the horizontal position of a pendulum bob to (*a*) its kinetic energy, (*b*) its potential energy, and (*c*) its total energy.

**2** On the same horizontal axis, sketch graph lines that relate the position of the mass in a spring-mass system to (*a*) its velocity and (*b*) its acceleration.

**3** Two equal-weight masses hang together at the end of the same spring, and the system is set vibrating. What happens to the amplitude, frequency, and maximum speed of the end of the spring if one of the masses falls off when (*a*) the spring is at its largest extension and (*b*) the mass is passing through the equilibrium position?

**4** A precocious student says that she can predict the frequency of a spring-mass system even though she knows neither the spring constant nor the mass. All she needs to know is how far the spring stretches when the mass is hung from it. Should you bet money that she cannot do this?

**5** How does the period of a pendulum change if the pendulum is in an accelerating elevator? Consider both upward and downward acceleration.

**6** How could you compute the up-and-down resonance frequency of a car by using data on how the car lowers as the load on it is increased? Estimate this frequency for an automobile. When might it be important?

**7** Sometimes an automatic washer vibrates strongly during the spin-dry cycle. Why? Is an unbalanced load the whole story? What should a designer do to minimize this problem?

**8** The value of *g* on the moon is one-sixth that on earth. How would the frequency of vibration of each of the following change as each is moved from earth to the moon: (*a*) a horizontal spring-mass system; (*b*) a vertical spring-mass system; (*c*) a simple pendulum? How would each behave in a spaceship orbiting the earth?

**9** The two idealized pulses in Fig. P13.1 are moving down the string at 20 m/s. Sketch how the string will look 0.40 s later. Repeat for 0.20 s later.

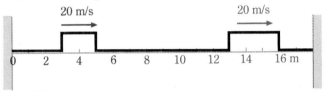

**FIGURE P13.1**

**10** Is it possible for two identical waves traveling in the same direction down a string to give rise to a standing wave?

**11** If you watch people trying to carry a large pan full of water, you will see that some are quite successful at it, but for others, who are equally careful, the water sloshes badly. What makes the difference?

## Problems

**1** A mass hanging at the end of a spring is 20 cm above the floor in its equilibrium position. When the mass is pulled down 8.0 cm and released, it reaches its lowest point 23 more times in the first 127 s after release. What are the (*a*) frequency, (*b*) period, and (*c*) amplitude of the motion?

**2** A long pendulum held initially at the highest point of its arc swings between two points 6.70 cm apart. It takes 385 s for the pendulum to reach the starting point for the 100th time after its release. What are the (*a*) period, (*b*) frequency, and (*c*) amplitude of the motion?

**3** A certain spring that obeys Hooke's law stretches 30 cm when a load of 0.35 N is added to it. How much energy is stored in the spring when it is compressed 5.0 cm?

**4** The spring on a popgun obeys Hooke's law and requires a force of 220 N to compress it 12.0 cm to its cocked position. How much energy is stored in the cocked spring?

■ **5** A 400-g mass is attached to the end of a spring ($k = 80$ N/m). The spring is then stretched 3.5 cm from its equilibrium position and released. Find (*a*) the speed of the mass as it passes through the equilibrium position and (*b*) its acceleration just as it is released.

■ **6** A 200-g mass hangs at the end of a spring ($k = 6.0$ N/m). The mass is pulled down 4.0 cm from its equilibrium position and released. What are (*a*) its initial acceleration and (*b*) its speed as it passes through the equilibrium position?

**7** A spring-mass system slides on a horizontal, frictionless surface. The mass is 60 g. A horizontal force of 0.80 N applied to it stretches the spring 4.0 cm. (*a*) What is the acceleration of the mass when the system is released? (*b*) What is its speed as it passes through its equilibrium position?

**8** The force constant for a spring in a popgun is 1200 N/m, and the spring is compressed 7.0 cm when the gun is cocked. If the pellet shot by the gun has a mass of 15.0 g, what is the maximum speed with which the pellet can leave the gun?

**9** A 2.0-kg mass oscillates with simple harmonic motion at the end of a spring. The amplitude of motion is 30 cm, and the spring has a constant of 500 N/m. Find the speed and acceleration of the mass when its displacement is (*a*) 30 cm, (*b*) 0 cm, (*c*) 15 cm.

**10** The 300-g mass in a spring-mass system has a maximum speed of 20 cm/s as it vibrates with an amplitude of 8.0 cm. Find (*a*) the spring constant, (*b*) the maximum acceleration of the mass, and (*c*) the speed and acceleration of the mass when it is 3.0 cm from its equilibrium position.

**11** A pendulum 80 cm long is pulled aside to an angle of 15° and released. Its ball has a mass of 300 g. Find the speed and acceleration of the ball when the pendulum angle is (*a*) 0°, (*b*) 15°, (*c*) 8°. *Hint:* 15° is not a very small angle.

**12** Show that the speed and acceleration of a pendulum ball released at an angle $\theta_o$ to the vertical are

$$v = \sqrt{2gL(\cos\theta - \cos\theta_o)} \quad \text{and} \quad a = g\sin\theta$$

**13** A circle of radius 20 m is marked out in the center of a football field, and a girl runs around it at 4.0 m/s. Meanwhile a boy runs up and down the sideline so as to keep pace with her motion in that direction. Find (*a*) the frequency of his motion, (*b*) his acceleration at the end of his path and (*c*) his maximum speed.

**14** A satellite circles the earth at 3100 m/s in an orbit that passes over both poles and has a radius of $4.2 \times 10^7$ m. Consider a spot on the earth's axis that keeps pace with the satellite's component of motion along the axis. Find (*a*) the frequency of the spot's motion, (*b*) the acceleration of the spot at the end of its path and (*c*) the maximum speed of the spot.

**15** A 40-g mass hangs at the end of a spring. When 20 g is added, the spring stretches an additional 5.0 cm. What is the frequency with which the combined 60 g vibrates?

**16** When a 100-g mass hangs at its end, a certain spring vibrates in such a way that it completes 25 cycles in 83 s. What is the spring constant?

**17** Two children notice that they can make a car vibrate up and down by periodically pushing downward on it. The car vibrates through nine cycles in 13.0 s. (*a*) Assuming that the car has a mass of 1200 kg, find the spring constant for its suspension system. (*b*) If this value is correct, about how much should the car lower as a 70-kg person gets in and sits in the front seat?

**18** A mass $m$ is hung at the end of a length $L$ of wire that has a cross-sectional area $A$ and a Young's modulus $Y$. Show that the mass will vibrate up and down with a frequency $f = (1/2\pi)\sqrt{AY/Lm}$.

**19** A mass vibrates back and forth obeying the equation $x = 20\sin(3.0t)$ cm. Find its (*a*) amplitude, (*b*) frequency, and (*c*) period of motion. (*d*) If the mass is 400 g, what is the spring constant?

**20** As it vibrates up and down at the end of a spring, a 300-g mass obeys the relation $y = 8.0\sin(7.0t)$ cm. Find (*a*) the spring constant and the (*b*) amplitude, (*c*) frequency, and (*d*) period of the motion.

**21** How long must a pendulum be if its period is to be 1 s (*a*) on earth and (*b*) on the moon? Objects weigh one-sixth as much on the moon as on earth.

**22** Two pendulums are to have frequencies such that $f_1 = 2f_2$. What should be the ratio $L_1/L_2$ of their lengths?

**23** A pendulum is drawn aside to a certain angle and released. When the ball passes the low point of the arc, the tension in the string is twice the weight of the ball. Show that the original displacement angle is 60°.

**24** A 6.0-g bullet moving horizontally at 90 m/s strikes a 200-g block at rest and lodges in the block (Fig. P13.2). The block ends up vibrating with an amplitude of 8.0 cm. What is the spring constant?

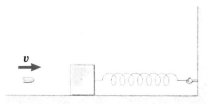

FIGURE P13.2

**25** A piston undergoes vertical simple harmonic motion with amplitude 18.0 cm and frequency *f*. A washer sits freely on top of the piston. At low piston frequencies, the washer moves up and down with the piston. However, at very high frequencies, the washer momentarily floats above the piston as the piston starts its downward motion. (*a*) What is the maximum acceleration of the piston when the washer begins to separate from it? (*b*) What is the lowest frequency at which separation occurs?

**26** Show that the following statements are true for a spring that obeys Hooke's law in the region of concern. (*a*) A force applied to the spring causes it to stretch (or compress) the same amount whether or not it is subject to a prior additional stretching force. (*b*) The frequency of vibration of a mass at the end of the spring is the same for the spring hanging vertically as it is for the spring lying horizontally with the mass supported on a frictionless surface.

**27** A spring that has a constant of 20 N/m is mounted vertically in a tube, and a 80-g mass is set on top of the spring and allowed to come to rest. We refer to this position as the zero position. The mass is now pushed 8.0 cm farther down the tube. How high above the zero position does the mass fly when the spring is released?

**28** A compressed spring with a mass attached to its end is immersed in a container of water at 20.000°C. The container, spring, mass, and water together are equivalent to 80 g of water. After the spring is released, it vibrates back and forth with decreasing amplitude as a result of friction (viscous) forces imposed by the liquid. When the system stops vibrating, the temperature is 20.100°C. (*a*) How much energy was stored in the spring? (*b*) If the spring was compressed 7.0 cm, what is its constant?

**29** Refer to Fig. P13.1. Sketch the situation 1.8 s later.

**30** In Fig. P13.1, how long does it take for the pulses to return to the same positions?

**31** The wave in Fig. P13.3 is moving to the right at 20 cm/s. Find its (*a*) wavelength, (*b*) amplitude, (*c*) frequency, and (*d*) period.

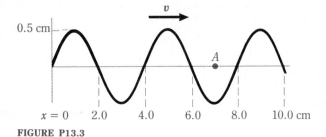

FIGURE P13.3

**32** When the wave on a string passes through *A* in Fig. P13.3, the string vibrates as $y = y_o \sin(2\pi ft)$. What are $y_o$ and $f$ if the speed of the wave is 30 cm/s?

**33** All radio waves travel through air with a speed of $3.0 \times 10^8$ m/s. What is the wavelength of a typical wave from a station that emits waves of frequency 1200 kHz?

**34** Light waves travel through air with a speed of $3.0 \times 10^8$ m/s. Blue light has a wavelength of about 450 nm. What is the frequency of these waves?

**35** How large a mass must be hung on the end of a thread 200 cm long if the speed of transverse waves in the thread is to be 4000 cm/s and if 300 cm of the thread has a mass of 0.50 g?

**36** An electric wire that has a mass of 5.0 kg for a 60-m length is strung between two poles 20 m apart. What must the tension in the wire be if it is to take 0.80 s for a pulse to travel along the wire from one pole to the other?

**37** Refer to Fig. P13.1. What does the disturbance on the string look like 0.40 s after the instant shown?

**38** Refer to Fig. P13.1. What does the disturbance look like when the leading edge of the (reflected) first pulse is at the position of the rear edge of the second (unreflected) pulse?

**39** A standing wave is set up on a string by an oscillator that has a frequency of 120 Hz. The distance between nodes in the pattern is 37 cm. Find (*a*) the wavelength of the waves and (*b*) their speed.

**40** A string 280 cm long is set into vibration in such a way that a standing wave consisting of three segments exists on it. A variable-frequency strobe light reveals that the string is vibrating at 82 Hz. Find (*a*) the wavelength of the waves and (*b*) their speed.

**41** A wire 130 cm long vibrates with four nodes, two of them at the two ends. Find (*a*) the wavelength and (*b*) the speed of the wave in the wire if the wire is being vibrated at 500 Hz.

**42** A string 160 cm long vibrates as a standing wave with four segments. The speed of the waves on the string is 18.0 m/s. Find (*a*) the wavelength and (*b*) the frequency of the waves.

**43** A certain string held at its two ends resonates to several frequencies, the lowest of which is 200 Hz. What are the next three higher frequencies to which it resonates?

**44** A certain string resonates in three segments to a frequency of 80 Hz. Give four other frequencies to which it will resonate.

**45** A resonating string has one of its resonances at 320 Hz and the next one at 400 hz. What is its lowest resonance frequency?

**46** A wire held firmly at its two ends vibrates in four segments to a frequency $f_1$ and in five segments to a frequency $f_2$. What is the ratio $f_1/f_2$?

**47** A violinist changes the tone from a string by moving a finger along the string. If the free string's lowest vibration frequency is 264 Hz, what are its three lowest vibration frequencies when the violinist's finger is placed one-fourth of the way from the upper end?

**48** A string 160 cm long has two adjacent resonances at 85 and 102 Hz. (*a*) What is its lowest resonance frequency? (*b*) What is the length of a segment at the 85-Hz resonance? (*c*) What is the speed of the waves on the string?

■ **49** An iron bar clamped at its center is set into longitudinal vibration when a person pulls outward on its end. (*a*) If the bar is 200 cm long, find the lowest frequency at which it resonates. (*b*) Repeat if it is clamped at one end. ($v = 5 \times 10^3$ m/s)

■ **50** An iron bar 100 cm long is clamped 25 cm from one end. Give the lowest resonance frequency for longitudinal waves in it. ($v = 5 \times 10^3$ m/s)

■ **51** A spring is stretched to a length of 4.00 m and set into longitudinal vibration by an oscillator at one end. When the driving frequency is 3.0 Hz, the spring vibrates with five antinodes along its length. What is the speed of compressional waves on the spring? Assume that the two ends are nodes.

■ **52** A steel bar 75 cm long is clamped at one end. What are the first three resonance frequencies for compressional waves? ($v = 5000$ m/s)

■ **53** A brass rod 45 cm long is dropped one end first onto a hard floor, but it is caught before it topples over. According to a bystander who claims the gift of perfect pitch, the rod emits a tone of frequency 2800 Hz. If he is right, what is the speed of sound in brass?

■ **54** A stiff wire forms a circular loop of diameter *D*. It is clamped by knife edges at two points opposite each other. A transverse wave is sent around the loop by a small oscillator that acts close to one clamp. Show that the resonance frequencies of the loop are $nv/\pi D$, where *n* is any integer and *v* is the speed of the waves.

■ **55** A cylinder of length *L* and density $\rho$ floats in water, as shown in Fig. P13.4. When it is pushed downward a small distance and released, it oscillates up and down. Prove that the motion is simple harmonic and that its frequency is $(1/2\pi)\sqrt{g\rho_w/\rho L}$.

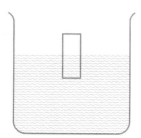

**FIGURE P13.4**

■ **56** Find the frequency with which the pendulum in Fig. P13.5 oscillates for small oscillations.

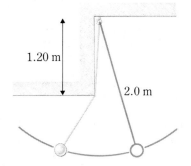

1.20 m

2.0 m

**FIGURE P13.5**

■ **57** A nonviscous liquid is placed in an open-ended U tube of uniform bore, as shown in Fig. P13.6. The total distance from *A* to *B* through the tube is *L*. The liquid is set in oscillation by a person blowing momentarily on the end above *A*. Show that the motion is simple harmonic and has a frequency $(1/\pi)\sqrt{g/2L}$.

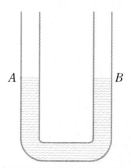

*A*                    *B*

**FIGURE P13.6**

■ **58** A thin horizontal wire of length $L_o$ and density $\rho$ is subjected to a stretching force large enough to stretch it a distance $\Delta L$. Show that the speed of transverse waves on the wire is $\sqrt{Y\Delta L/\rho L_o}$, where *Y* is Young's modulus for the material of the wire.

# 14 Sound

THE concepts involving wave motion discussed in the preceding chapter are now applied to one particular form of wave motion, sound. Not only is a study of sound important in its own right, but also it affords us a valuable means of consolidating our knowledge of wave motion in general. We shall find that many of the ideas discussed in connection with sound will also be important in our study of light and other types of wave motion.

## 14.1 The origin of sound

Sound is usually defined as any compressional disturbance traveling through a material in such a way that it is capable of setting the human eardrum into motion, thereby giving rise to the sensation of hearing. Notice that sound waves, because they are compressional waves in a material, require a substance in which to travel. They cannot travel through a vacuum because there is nothing there to transmit the wave compressions. A common demonstration is to show that the ringing of a bell cannot be heard if the bell is in a vacuum chamber. The bell is vibrating, but there is no surrounding material that can transmit the vibration.

A source of sound waves must be capable of sending out a compressional wave into an adjacent sound-carrying substance. Whether it is a vibrating string on the sound box of a violin or an exploding firecracker, the sound source generates vibrations in the air or other sound-carrying material. Although sound waves, being compressional, can be transmitted through almost any substance, solid or fluid, we are usually concerned with sound waves in air.

## 14.2 Sound waves in air

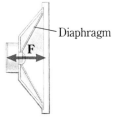

**FIGURE 14.1**

The flexible loudspeaker diaphragm vibrates back and forth to send compressions out into the air.

Let us now consider the action of a loudspeaker generating simple sounds. A simple loudspeaker consists of a cone-shaped sheet of flexible material, called a diaphragm, that can be oscillated back and forth by means of an applied force **F**, as in Fig. 14.1. (We shall see how the force is obtained when we study magnetic forces in Chap. 18.)

When the diaphragm in Fig. 14.1 moves to the right, it compresses the air in front of it and a **compression** travels out through the air. An instant later, the diaphragm is moving to the left, leaving a region of decreased air pressure in its wake, called a **rarefaction.** This disturbance also travels out from the loudspeaker. Hence a series of pressure disturbances, the compressions and rarefactions, travel out from the loudspeaker. This situation is sketched in Fig. 14.2, where the compressions are shown as $A$, $B$, and $C$ and the rarefactions as $P$, $Q$, and $R$. A plot of the air pressure along this sound wave at a given instant is also given in Fig. 14.2. The pressure at the horizontal line in this graph is average atmospheric pressure. The compressions and rarefactions cause only slight variations in air pressure. Even for very loud sounds, the pressure variations are only about 0.01 percent of atmospheric pressure.

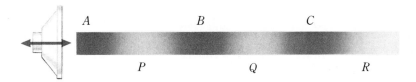

**FIGURE 14.2**

The sound wave sent out by the loudspeaker consists of alternate high- and low-pressure regions in the air. In practice, $P$ changes by only about 0.01 percent or less.

The sound waves sent out by a loudspeaker or any other sound source are not usually confined to a straight-line path. Instead, they spread out in all directions from the source. To better understand this feature of wave motion, refer to Fig. 14.3$a$, which shows a water wave traveling out from a source. A diagrammatic way of representing this situation is shown in Fig. 14.3$b$. As we see, the wave crests (called **wavefronts** in this context) take the form of larger and larger circles as they move away from the source. At great distances from the source, the circles are so large that they have very little curvature. Such a wave crest from a very distant source appears as a nearly straight line as it sweeps past a point on the water. Waves far from their source possess wavefronts with negligible curvature and are called **plane waves,** a terminology carried over from three-dimensional waves, as we shall soon see.

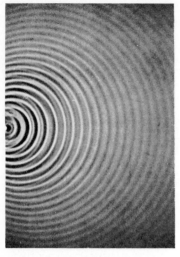

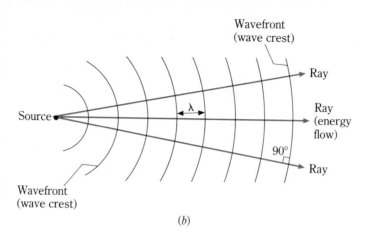

Wavefront
(wave crest)

Ray

Source

$\lambda$

Ray
(energy
flow)

90°

Ray

Wavefront
(wave crest)

(b)

(a)

**FIGURE 14.3**

(a) A wave source sends circular waves across the surface of the water. (b) A diagram used to represent a situation such as (a). (*Education Development Center*)

The waves in Fig. 14.3 carry energy away from the source. Because energy moves in the direction of wave propagation, the energy carried by the wave travels out along radial lines, such as those labeled **ray** in the figure. Notice that the *rays are of necessity perpendicular to the wavefronts.* Because wavefronts far from a source form nearly straight lines and because rays are perpendicular to wavefronts, *rays are parallel to each other when they are far from a wave source, that is, in a plane wave.*

An analogous situation exists for sound waves in air. Because this is a three-dimensional situation, however, the wavefronts are spherical surfaces centered on the sound source. These spherical waves have decreasing curvature as they move farther away from the source and appear essentially flat at a distant point; the waves are then essentially flat planes, and they, too, are called plane waves. As before, the rays are perpendicular to the wavefronts, and so the rays are parallel to each other in plane waves.

We should notice one other feature of the waves in Fig. 14.3*a*: their amplitude decreases with increasing distance from the source. This simply reflects the fact that the energy carried by the wave is spread out over an increasingly larger wavefront as the wave proceeds. Hence a unit length of a wavefront contains less energy as the front moves away from the source. This feature did not concern us with a wave on a string, spring, or rod because all the energy propagated along the same line, in only one dimension. With water waves, which propagate in two dimensions, and three-dimensional sound waves, however, the energy is spread ever more thinly as the surface area of the wavefront increases. Only in plane waves can this decrease in amplitude be ignored. In that case, the rays are parallel to each other, and so the energy is carried forward along a single direction; it does not spread over a larger wavefront as the wave moves along.

## 14.3 The speed of sound

Sound waves are compressional waves in matter. We know from Eq. 13.17 that the speed of such waves in a fluid is

$$v = \sqrt{\frac{B}{\rho}}$$

(13.17)

## ■ TABLE 14.1
### Speed of sound*

| Material | $v$ (m/s) |
|---|---|
| Air† | 331.45 |
| Oxygen | 316 |
| Helium | 965 |
| Hydrogen | 1284 |
| Water | 1402 |
| Water (20°C) | 1482 |
| Water (50°C) | 1543 |

* Values are at 0°C unless noted otherwise.

† For air near room temperature, $v = 331.45 + 0.61t$ m/s, where $t$ is the Celsius temperature.

where $\rho$ is the density of the fluid and $B$ is its bulk modulus. It turns out that the bulk modulus for an ideal gas* is simply $\gamma P$, where $\gamma = c_P/c_V$ and $P$ is the pressure of the gas. Thus the speed of sound (compressional waves) in an ideal gas is

$$v = \sqrt{\frac{\gamma P}{\rho}} \qquad \text{(ideal gas)}$$

From the ideal gas law, however — $PV = (m/M)RT$ — the pressure of an ideal gas is

$$P = \frac{m}{V}\frac{RT}{M} = \rho\frac{RT}{M}$$

because $m/V$ is simply the gas density. Substitution of this in the velocity equation yields

$$v = \sqrt{\frac{\gamma RT}{M}} \qquad (14.1)$$

This is a very interesting result because it tells us that the speed of sound in an ideal gas is independent of pressure or density. Typical values for the speed of sound at 0°C are given in Table 14.1. The footnote to the table states how the velocity in air changes with temperature.

### Example 14.1
Find the speed of sound in neon gas at 0°C.

*Reasoning*    We can use Eq. 14.1 with $M = 20.18$ kg/kmol. For a monatomic gas, $\gamma = 1.66$ (Table 12.1). Then

$$v = \sqrt{\frac{\gamma RT}{M}} = \sqrt{\frac{(1.66)(8314 \text{ J/kmol} \cdot \text{K})(273 \text{ K})}{20.18 \text{ kg/kmol}}} = 432 \text{ m/s} \quad ■$$

*Exercise*    Two ideal gases have the same molecular mass $M$, but gas $A$ is monatomic and gas $B$ is diatomic. Find the ratio $v_A/v_B$.    *Answer: 1.09*    ■

## 14.4 Intensity and intensity level

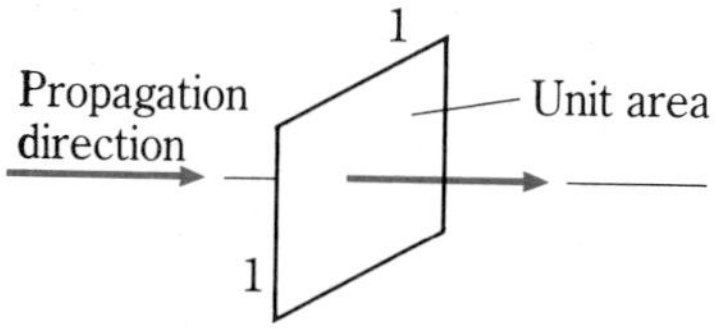

We saw in Chap. 13 that the source that sends a wave down a string also sends energy with the wave. Indeed, all waves carry energy along with them. Sound waves are no exception. For example, the loudspeaker in Fig. 14.2 sends out sound-wave energy, which flows in the direction of wave propagation.

Suppose a sound wave is traveling in the propagation direction shown in Fig. 14.4. We define the intensity of the wave in terms of the energy it carries. To be precise,

FIGURE 14.4
Sound intensity is measured as the amount of energy flowing through a unit area per second. The area must be perpendicular to the direction of propagation, as shown.

---

* This is true only under adiabatic conditions. For isothermal conditions, $B = P$.

**■ TABLE 14.2  Approximate sound intensities and intensity levels**

| Type of sound | Intensity (W/m²) | Intensity level (dB) |
|---|---|---|
| Pain-producing | 1 | 120 |
| Jackhammer or riveter* | $10^{-2}$ | 100 |
| Busy street traffic* | $10^{-5}$ | 70 |
| Ordinary conversation* | $10^{-6}$ | 60 |
| Average whisper* | $10^{-10}$ | 20 |
| Rustle of leaves* | $10^{-11}$ | 10 |
| Barely audible sound | $10^{-12}$ | 0 |

\* For a person near the sound source.

we erect a unit area perpendicular to the direction of propagation, as shown. We then define the intensity $I$ of the wave to be the energy the wave carries per second through this unit area. Since power is energy per second,

Sound intensity $I$ is the power passing through a unit area perpendicular to the direction of wave propagation.

$$I = \frac{\text{power}}{\text{area}}$$

**■ TABLE 14.3**
**The decibel* scale**

| Intensity (W/m²) | Intensity level (dB) |
|---|---|
| $10^{-12}$ | 0 |
| $10^{-11}$ | 10 |
| $10^{-10}$ | 20 |
| $10^{-9}$ | 30 |
| · | · |
| · | · |
| · | · |
| $10^{-1}$ | 110 |
| 1 | 120 |
| 10 | 130 |

\* 1 B (bel) = 10 dB and is named after Alexander Graham Bell, the inventor of the telephone.

The units of sound intensity are typically watts per square meter. Representative sound intensities are listed in Table 14.2. Notice what a wide range of intensities can be heard by the ear, a truly remarkable measuring device.

Notice in Table 14.2 that the sound intensities we usually hear span a range of at least $10^{12}$. It is convenient in many situations to use a different scale to express the loudness of sounds, a scale that uses logarithms to compress the range of intensities that extends from $10^0$ to $10^{12}$ W/m² into a scale that ranges from 0 to 120. We call it the **decibel scale,** and readings on it are called *intensity levels* or *sound levels* (Table 14.3). The relation between the two scales is

$$\text{Intensity level in decibels (dB)} = 10 \log \frac{I}{I_o} \tag{14.2}$$

where $I$ is the intensity (in watts per square meter) of the sound under consideration and $I_o$ is often, but not always, taken as the intensity of the least audible sound, $10^{-12}$ W/m². Notice that the intensity level of the least audible sound is

$$10 \log \frac{I}{I_o} = 10 \log \frac{10^{-12}}{10^{-12}} = 10 \log 1 = 0 \text{ dB}$$

while a pain-producing sound whose intensity is 1 W/m² has an intensity level of

$$10 \log \frac{I}{I_o} = 10 \log \frac{1}{10^{-12}} = 10 \log 10^{12} = 120 \text{ dB}$$

***Example 14.2***

Find the sound level in decibels of a sound wave that has an intensity of $10^{-5}$ W/m².

*Reasoning*   From Eq. 14.2

$$\text{Sound level (in dB)} = 10 \log \frac{I}{I_o} = 10 \log \frac{10^{-5}}{10^{-12}} = 10 \log 10^7 = (10)(7) = 70 \text{ dB} \quad \blacksquare$$

*Exercise*   Find the sound level equivalent to an intensity of $4 \times 10^{-8}$ W/m². *Answer: 46 dB* ∎

***Example 14.3***

Find the intensity of a sound that has an intensity level of 35 dB.

*Reasoning*   From Eq. 14.2,

$$35 \text{ dB} = 10 \log \frac{I}{10^{-12} \text{ W/m}^2}$$

If we divide by 10 and recall that

$$\log \frac{a}{b} = \log a - \log b$$

we have

$$3.5 \text{ dB} = \log I - \log 10^{-12} = \log I - (-12)$$

$$\log I = -8.5$$

The antilog of $-8.5$ is $3.16 \times 10^{-9}$, and so the intensity of the sound is $3.16 \times 10^{-9}$ W/m². ∎

*Exercise*   Find the intensity of a 75-dB sound. *Answer: 3.16 × 10⁻⁵ W/m²* ∎

## 14.5 The frequency response of the ear

People vary in their ability to hear sounds. We all know persons whose hearing has been in some way impaired. The sensitivity of their ears has decreased to considerably below that of a person with normal hearing. However, most people agree fairly well on the intensity of a sound that is just audible and also on how loud a sound must be before it causes pain. We can therefore set average limits of audibility for the human ear. The lower limit is the intensity of a just-audible sound, and the upper limit is a sound so intense that it hurts the ear.

The response of the ear to sound depends on the frequency of the sound as well as its intensity. The ear is more sensitive to some frequencies than to others. Most people cannot hear sound waves that have a frequency higher than about 20,000 Hz. Waves of higher frequency are called **ultrasonic waves** — meaning "beyond" or "above" sound in the sense of frequency. Similarly, most people are not able to hear sounds below a frequency of about 20 Hz. The ear is most sensitive near 3000 Hz. At

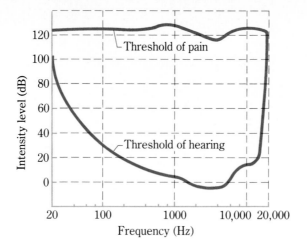

**FIGURE 14.5**

The normal ear can hear sounds that have intensities above the lower curve.

frequencies other than this, the sound must be made more intense before it is audible. This variation of ear sensitivity with frequency is shown in Fig. 14.5. The lower curve shows the minimum audible intensity level as a function of frequency. For example, a sound wave with a frequency of 1000 Hz can be heard when it has an intensity level of about 5 dB. At a frequency of 100 Hz, however, the sound level must be about 30 dB if the sound is to be audible. Of course, near the frequency limits of audible sound (20 and 20,000 Hz), the intensity must be very large for a sound to be heard.

The upper curve in Fig. 14.5 shows how intense a sound must be to produce pain. It is not very frequency-dependent and shows that an intensity level of about 120 dB is painful. Such high sound levels can cause permanent damage to the ear. Indeed, long-term exposure to sounds in the range down to about 90 dB can cause hearing impairment. Of course, loss of hearing can be caused by other factors as well.

There are some people who have an unusual hearing impairment, often unknown to themselves: they are unable to hear sound frequencies above perhaps 6000 Hz. Since most of the sounds we hear consist, partly at least, of frequencies below this, these people are still able to hear sounds that are audible to other people. However, the *quality* of the sounds they hear is quite unlike that of sounds heard by a person with normal hearing. Quality and pitch of sound are complex, subjective properties of sounds discussed in the next section.

## 14.6 Sound pitch and quality

**Pitch** is our qualitative conception of whether a musical sound (a tone) is high, like that of a soprano, or low, like that of a bass singer. To investigate pitch and its relation to other features of sound, it is convenient to use a simple experiment. If a high-quality loudspeaker is driven by an electrical system that generates a sinusoidal force, the sound wave given off is an almost pure sine wave that has the frequency being generated by the electrical system. Anyone who can carry a tune can compare the pitch of this sound with that of another sound. If the frequency of the driving force is increased so as to increase the frequency of the sound given off by the loudspeaker, the listener will notice at once that the new sound has a higher pitch than the first sound. In such cases, the terms *pitch* and *frequency* are nearly synonymous.

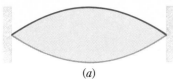

*(a)*

Fundamental, 1st harmonic: $f_1$

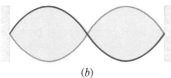

*(b)*

2nd harmonic, 1st overtone: $2f_1$

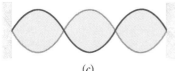

*(c)*

3rd harmonic, 2nd overtone: $3f_1$

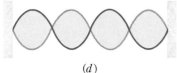

*(d)*

4th harmonic, 3rd overtone: $4f_1$

**FIGURE 14.6**

The four simplest vibration patterns for standing waves on a string.

**FIGURE 14.7**

Each musical instrument has its own characteristic sound. The quality of the sound is influenced by the number and strengths of the harmonics that compose it. The heights of the vertical bars show the relative strength (amplitude) of each harmonic wave. Are the instruments sounding the same note?

Single-frequency sound waves are not very common, however. When a violin string is bowed, for example, the sound wave given off is not a pure sine wave. This is readily apparent to anyone who has ever compared the tone from a violin being played by an expert with that from one being played by a beginner. In the former case, the tone is full and melodious, whereas the beginner may obtain rasping sounds on the same string. We say that the **quality** of the tone is different in the two cases.

As we saw in Sec. 13.10, a string may resonate in more than one way. Typical simple vibration patterns are shown and named in Fig. 14.6. Since the wavelengths in the cases shown are in the ratio $1 : \frac{1}{2} : \frac{1}{3} : \frac{1}{4}$ and since $f = v/\lambda$, the vibration frequencies are in the ratio $1 : 2 : 3 : 4$.

It is very difficult, however, to cause a string to vibrate exactly as shown in any single pattern of Fig. 14.6. Instead, if the string is bowed near one end, as usually occurs, it vibrates in several ways at once, causing several harmonics to occur at the same time. To find the resulting vibration, it is necessary to add the waves for the various harmonics excited. Of course, because each harmonic is excited to a different amplitude, the proper amplitude must be used for each harmonic when they are added.

A typical example for a vibrating violin string is shown in Fig. 14.7. The amplitudes of vibration of the various harmonics are indicated by the lengths of the vertical bars. In this case, all but the first two harmonics are relatively weak. Even so, the tone heard by the ear is different from what it would be if only the first or the second harmonic alone were present.

Figure 14.7 also shows similar diagrams for the sounds of various other instruments. The piano string shows many more harmonics than the violin string. This is probably the result of the way in which the strings are set into vibration. The violinist pulls a bow slowly across the string, whereas the piano string is excited by a hammer blow.

*The quality of a sound depends on the number and type of harmonics occurring in it.* If all sounds were pure sine waves, much of the variety of sound would be lost. The tone of all human voices would be the same, and a voice could be recognized only by a characteristic frequency or inflection. Much of the beauty of music would be lost if the qualities of all sounds were the same.

In a complex sound, such as that of a piano or clarinet, pitch is not always easily defined. It can no longer be taken as identical to frequency since the sound contains several waves that are nearly equal in amplitude but vary in frequency. In such cases, it is difficult to match tones accurately. For this reason, it is quite common for an inexperienced singer to sing a tone that is twice the frequency of the fundamental tone of a violin and be unaware that the tones are not the same. Moreover, many listeners would be unable to recognize that the singer was not sounding the desired tone.

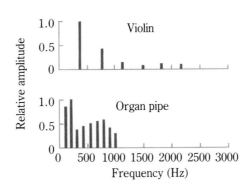

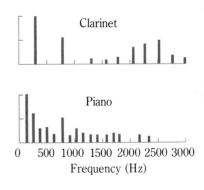

## 14.7 Interference of sound waves

Suppose we have a pipe system such as that shown in Fig. 14.8. A pure sine wave is sent into the pipe at the left by a loudspeaker. The sound splits, with half the sound intensity going up through the upper pipe and the remaining half through the lower pipe. Each pipe carries half the sound, and this sound is a wave motion in the air, that is, a series of compressions and rarefactions.

Eventually the two waves are reunited at the outlet on the right ($D$), where a sound detector, such as an ear or microphone, is placed. The sound emitted at $D$ can be made loud or very faint, depending on the position of the sliding upper pipe $EAF$. Moreover, as this pipe is slowly pulled upward, the sound intensity at $D$ becomes alternately large and small. We now investigate the reasons for this phenomenon, called interference.

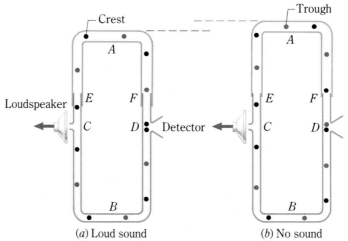

(a) Loud sound           (b) No sound

**FIGURE 14.8**

A sound wave from the loudspeaker is split into two parts. When they are reunited at the detector ($D$), either a loud or a weak sound results, depending on the path lengths traveled by the two parts.

When the air is compressed by a rightward movement of the loudspeaker diaphragm, a region of high pressure (a compression) starts into the pipe at $C$. This compression causes compressions to move in both pipes, toward $A$ and toward $B$. In other words, the original compression at $C$ splits into two equal parts; one part goes up toward $A$, and the other goes down toward $B$. Since the compressions, represented by the black dots, propagate through the pipes with the speed of sound, they reach $D$ at the same time, provided that $L_A$, the pipe length from $C$ to $D$ through $A$, is the same as $L_B$, the pipe length from $C$ to $D$ through $B$. They are reunited at $D$, giving the original compression, and this is what exits at $D$. This situation is shown in Fig. 14.8$a$. The colored dots represent the rarefactions.

Of course, the loudspeaker is sending out a pure sound, a sinusoidal wave, consisting of alternate compressions and rarefactions. If $L_A = L_B$, the two portions of the original compression always meet at $D$. The same is true for the two halves of the rarefactions. If $L_A = L_B$, compressions and rarefactions identical to the originals exist at $D$, and the sound is loud. Examine Fig. 14.8$a$ to convince yourself that,

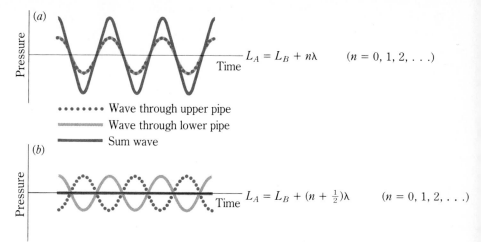

$L_A = L_B + n\lambda \qquad (n = 0, 1, 2, \ldots)$

........ Wave through upper pipe
———— Wave through lower pipe
———— Sum wave

$L_A = L_B + (n + \tfrac{1}{2})\lambda \qquad (n = 0, 1, 2, \ldots)$

**FIGURE 14.9**

Waves $A$ and $B$ can reinforce or cancel depending on their positions relative to each other. The waves in (a) are in phase, while in (b) waves $A$ and $B$ are 180° (or one-half wavelength) out of phase.

as time progresses, the wave through the upper pipe and that through the lower pipe reinforce each other when they meet at the detector.

We can represent the situation at the detector in Fig. 14.8a by the graph in Fig. 14.9a. Plotted there are the wave through the upper pipe, the wave through the lower pipe, and the sum of the two waves at the detector. Notice that, as we have already concluded, crest falls on crest and trough falls on trough so as to cause a large-amplitude wave (loud sound) at the detector.

Let us now look at Fig. 14.8b, where the path through $A$ has been lengthened by sliding the upper pipe away from the source and detector. The path through $A$ is now one-half wavelength longer than the path through $B$. Consequently, when the $A$ and $B$ waves rejoin at the receiver, a crest of the $B$ wave joins a trough of the $A$ wave. Furthermore, as time goes on, crests always meet troughs when the two waves meet. Because a crest is canceled by a trough, we conclude that the two waves in Fig. 14.8b rejoin so as to cancel each other at the detector and no sound is heard. This cancellation is also shown in Fig. 14.9b. We are therefore led to the following conclusion: *if two identical waves are displaced one-half wavelength relative to each other, they cancel each other.*

We should mention another terminology used to describe situations such as this. It is based on the fact that a sinusoidal wave is related to a sinusoidal vibration and that the vibration varies as $\sin \omega t$ or as $\sin \theta$. During one complete cycle of vibration, $\theta$ varies from zero to 360° (or zero to $2\pi$). We thus associate one wavelength with 360°, half a wavelength with 180° ($\pi$), and one-fourth wavelength with 90° ($\pi/2$). In view of this, we often say that a path length that is one wavelength long is equivalent to a **phase** of 360° or of $2\pi$. Similarly, a path length of one-half wavelength is equivalent to a phase of 180° or of $\pi$.

We extend this terminology to path length differences. For example, when one wave is half a wavelength behind another wave, we say the waves are a half wavelength, or 180°, out of phase with each other. Hence a retardation of $\tfrac{1}{2}\lambda$ is equivalent to a phase difference of 180°. Similarly, 90° is equivalent to $\tfrac{1}{4}\lambda$, and so on. Referring to Figs. 14.8 and 14.9, we see that the waves are in phase in *a* but 180° out of phase in *b*.

Returning now to the apparatus of Fig. 14.8b, we see that if $L_A$ is increased still further by lifting the upper pipe an additional amount, the wave through that section is retarded still more. When $L_A$ is one whole wavelength longer than $L_B$, a crest of the wave through the upper pipe again reaches $D$ at the same time as a crest of the wave through the lower pipe. Although these crests did not start together at $C$ (the upper crest occurred one compression prior to the lower crest), this is of no concern

# The acoustics of large rooms

Have you ever noticed that certain restaurants and meeting rooms are particularly noisy even though the people in them do not seem to be especially boisterous? Often the cause of this annoying clamor is the lack of sound-absorbing materials on the inner surfaces of the rooms. We do not often stop to think that the sound level in a classroom would be a deafening roar if it were not for sound absorption. Even with only one person speaking, the sound energy within the room would keep increasing until the walls would be shaking. Since this does not happen, sound energy must be absorbed by the surfaces within the room. Obviously the surfaces of a room are very important in determining the room's acoustic properties.

Large concert halls and lecture rooms are partly characterized by their reverberation time, the time it takes a sound to die out. To make the reverberation time of a room small, we can open its windows, put velvet curtains on its walls, use acoustic tile for its ceiling, cover its floor with plush carpet, or fill it with people wearing soft clothing. All of these measures either absorb sound energy or let it escape from the room. In a room that lacks these features, sound reflects efficiently from surface to surface and permeates the room for a long time before it is completely dissipated.

The most desirable reverberation time for a room depends on the purpose for which the room is to be used. Reverberation is undesirable in a lecture room because it obscures spoken syllables and words; here a reverberation time of even 1 s is usually considered too long. In a concert hall, however, we want a rather long reverberation time, perhaps double that for a lecture hall, because then the listener is effectively immersed in sound reflected from all sides.

Reverberation time is only one of many important features of a lecture or concert hall. The architect is faced with a delicate design situation. Poor designs abound and are often readily discerned as such by both performer and audience, but the factors that contribute to poor design are usually very difficult to pinpoint. Obviously, major reflections that cause conflicting, delayed echoes are to be avoided. Therefore a concert hall should have features that break up the sound; columns, wall niches and protruding design elements, hanging panels, and even huge chandeliers are of value in dispersing sound, thereby reducing distant echoes and bathing the listener in an ocean of closely related sounds.

With the advent of electronic amplification of sound, another complication has been added. A listener at the rear of a large lecture hall should, in an ideal situation, hear the words simultaneously from the various loudspeakers and from the mouth of the person speaking. Otherwise, speech is garbled. Although this ideal is impossible to realize fully, good systems phase the sounds from the various loudspeakers so as to minimize the problem.

As you see, the acoustic features of a modern hall are complex. Without proper design, the hall will be inferior for many of its intended uses.

when they reach $D$. The wave once again appears as in Fig. 14.9$a$, and so sound of the original intensity is produced at $D$.

Similarly, if $L_A$ is increased until it is $1\frac{1}{2}$ wavelengths longer than $L_B$, the situation in Fig. 14.9$b$ arises once again. No sound is heard at $D$. Moreover, it is clear that no sound is heard at $D$ whenever $L_A = L_B + (n + \frac{1}{2})\lambda$, where $n$ can be any integer, including zero. When two waves exactly cancel each other in this way, we say that there is complete *destructive interference*. Moreover, reinforcement, which we call *constructive interference*, of the waves occurs whenever $L_A = L_B + n\lambda$, where $n$ is any integer, including zero. Of course, if $L_B$ is greater than $L_A$, these interferences also occur.

It is not necessary to have a pipe system such as this to obtain interference. We need only obtain two waves that are exactly the same in frequency and shape. If these waves are combined after traveling different distances, they will interfere with each other. The following example explains another situation involving interference. We shall see in Chap. 24 that the interference of light waves is of great importance.

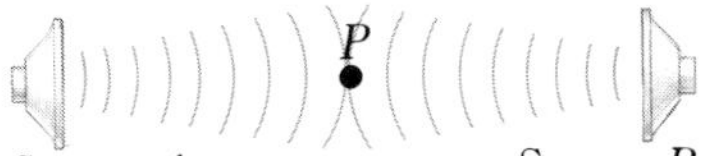

---

## Example 14.4

When the two identical sound sources in Fig. 14.10 vibrate in phase to send identical waves ($\lambda = 70$ cm) toward each other, a loud sound is heard at $P$, midway between them. Now source $B$ is slowly moved to the right. How far must it be moved before the sound at $P$ becomes very weak?

*Reasoning*   Originally the sound at $P$ is loud because the waves reach there in phase and give rise to constructive interference. (The path lengths from the two sources are the same.) When source $B$ is moved to the right, the waves are no longer in phase at $P$. If the distance from $P$ to $B$ ($\overline{PB}$) is increased until it is $\frac{1}{2}\lambda$ longer than the distance from $A$ to $P$ ($\overline{AP}$), then a crest from $A$ will meet a trough from $B$ and the result is destructive interference (and very weak* sound) at $P$. Hence to obtain minimum sound at $P$, source $B$ must be moved to the right a distance $\frac{1}{2}\lambda$, which is $\frac{1}{2} \times 70$ cm $= 35$ cm. Can you show that minimum sound is also obtained for a distance $\lambda + \frac{1}{2}\lambda$?  ∎

---

*Exercise*   What must the relation between the lengths $\overline{PB}$ and $\overline{AP}$ be if loud sound is to occur at $P$?  *Answer:* $\overline{PB} - \overline{AP} = \pm\, n\lambda$, where $n = 0, 1, 2, \ldots$  ∎

---

Source $A$          Source $B$

**FIGURE 14.10**
When $\overline{AP} = \overline{PB}$, the waves reinforce at $P$.

---

## 14.8 Beats

When people tune a string on a piano, they do not merely listen to see whether the tone of the string is the same as that of the standard tuning fork used for comparison. Instead, they use a much more precise way of judging the accuracy to which the string is adjusted. They listen for *beats* between the sounds of the two vibrating objects. This is a very sensitive method for obtaining agreement of frequency and is widely used for that purpose.

---

* The sound intensity is usually not exactly zero because the amplitudes of the waves decrease somewhat with distance. Why?

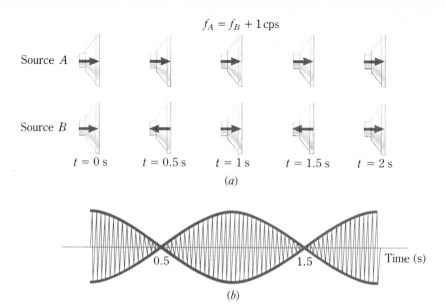

$$f_A = f_B + 1\,\text{cps}$$

Source $A$

Source $B$

$t = 0$ s    $t = 0.5$ s    $t = 1$ s    $t = 1.5$ s    $t = 2$ s

(a)

0.5              1.5        Time (s)

(b)

**FIGURE 14.11**

Beats result when two similar sources vibrate with slightly different frequencies.

Suppose that two vibrating bodies, $A$ and $B$, vibrate with slightly different frequencies. For example, they might be the two loudspeakers in Fig. 14.11$a$. Consider what happens when source $A$ is vibrating with frequency of 1000 Hz and source $B$ with $f = 999$ Hz. At time $t = 0$, the loudspeakers are in phase, that is, both are sending out a compression, as indicated by the rightward-pointing arrow. If a person's ear is equidistant from the speakers, the compressions arrive at the ear together, and a large compression results; a loud sound is heard.

As time goes on, $B$, vibrating at a slightly lower frequency than $A$, begins to fall behind. After $\frac{1}{2}$ s, $A$ has vibrated 500.00 times and is just sending out a compression, as shown for $t = \frac{1}{2}$ s in Fig. 14.11. Loudspeaker $B$, however, has vibrated only 499.50 times and is exactly one-half cycle behind $A$. It is sending out a rarefaction (the leftward-pointing arrow). The compression from $A$ now reaches the ear at the same time as the rarefaction from $B$, and they exactly cancel each other. Hence, at this instant, no sound is heard.

As time continues, loudspeaker $B$ falls still farther behind $A$. After 1 s, $B$ has vibrated exactly 999 times while $A$ has vibrated exactly 1000 times. Source $B$ has now fallen exactly one cycle behind $A$. Hence once again they are both sending out compressions together, and a loud sound is heard.

This process continues, as shown in the succeeding portions of Fig. 14.11$a$. At times of 0, 1, 2, 3, . . . s, the sources are in phase and the sound heard is loudest. At times of $\frac{1}{2}$, $1\frac{1}{2}$, $2\frac{1}{2}$, . . . s, nothing is heard because the sources are 180° out of phase. Hence the ear hears a series of sound pulses, or beats, one beat each second.

The beating sound wave is shown in Fig. 14.11$b$, where the combined wave from the two sources and its envelope are shown as a function of time. The combined wave has maximum intensity at $t = 0, 1, 2$ s, as we just discussed. As this combined wave strikes the ear, a beat is heard each time a maximum in the sound envelope reaches the ear.

Notice that the frequencies differ by exactly 1 Hz in this case, and exactly one beat is heard each second. You should carry through the above analysis for the case in which the sources differ by, say, 3 Hz. In this case, there will be 3 beats per second. In fact, *the number of beats per second equals the difference in frequencies*

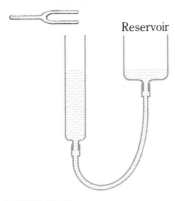

FIGURE 14.12

When the water in the tube is just the right height, resonance will occur.

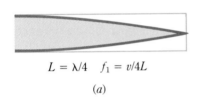

$L = \lambda/4 \quad f_1 = v/4L$

(a)

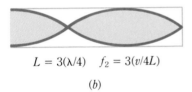

$L = 3(\lambda/4) \quad f_2 = 3(v/4L)$

(b)

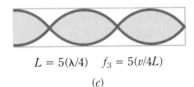

$L = 5(\lambda/4) \quad f_3 = 5(v/4L)$

(c)

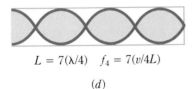

$L = 7(\lambda/4) \quad f_4 = 7(v/4L)$

(d)

FIGURE 14.13

Simple modes of vibration for a resonating pipe closed at one end.

*of the two sources.* The piano tuner tries to adjust the tension in a piano string until the beats between the sound from the string and the sound from the tuning fork become extremely far apart in time.

Sometimes the beat frequency (that is, the number of beats per second) between two sound waves gives rise to a third distinct sound. For example, suppose the frequencies of two sounds are 1000 Hz and 1200 Hz. Their beat frequency is 200 Hz, and the listener will hear a sound of this frequency in addition to the two original sounds.

This phenomenon of beats occurs in any type of vibration that is a combination of identical vibrations from two sources. It may therefore be used to compare frequencies of vibrations other than sound. As in the case discussed here, this is a sensitive means of comparing frequencies. We shall see another example of this in Sec. 14.10 when we discuss how sound and radar waves are used to measure speed. In general, two waves, identical except for frequency, give rise to beats when they are combined, if their frequencies differ. The beat frequency equals the difference between the frequencies of the two waves.

## 14.9 Resonance of air columns

If you hold a vibrating tuning fork over the open end of a glass tube that is partly filled with water, the sound of the tuning fork can be greatly amplified under certain conditions. While the fork is held as shown in Fig. 14.12, the reservoir is lowered so that the water level in the tube falls. When the tube water level is at a certain height, the air column above the water in the tube resonates loudly to the sound being sent into it by the tuning fork. In fact, there are usually several heights at which the air column in the tube resonates.

The situation here is much like the case of a vibrating string. In place of the string excited by an oscillator at one end, we have an air column with a sound source at its open end. The oscillator sends a wave down the string, and the wave is reflected at the other end. Similarly, the sound source sends a sound wave down the air column, and the wave is reflected by the surface of the water. As we learned in Chap. 13, a string resonates when the wavelength of the wave is properly related to the length of the string. In particular, nodes must occur at the two ends of the string, and so the string resonates only if its length is $n(\frac{1}{2}\lambda)$, where $n$ is an integer and $\frac{1}{2}\lambda$ is the distance between nodes.

Resonance of the air column (or air-filled tube) in Fig. 14.12 differs in a major respect from that of a string. The water surface in the air-filled tube acts as a fixed, closed end for the tube. Longitudinal motion of the air in the tube is prevented at this closed end by the water. Hence the closed end of the air column (the water surface) must be the location of a node in the sound-wave resonance pattern. At the open end of the tube, however, the air is free to move out into the open space above, and so at that point there is maximum motion, that is, an antinode.* Therefore, if the air column in Fig. 14.12 is to resonate, it must have a node at its closed end and an antinode at its open end. The air column will resonate only if the sound wavelength is such that the wave fits into the column with a node at one end and an antinode at the other. Several of these resonating waves are shown schematically in Fig. 14.13.

---

* The antinode is not precisely at the end of the tube. However, for tubes with radii much smaller than $\lambda$, this complication can usually be ignored.

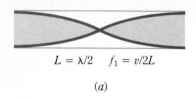

$L = \lambda/2 \quad f_1 = v/2L$

(a)

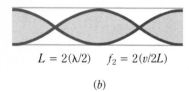

$L = 2(\lambda/2) \quad f_2 = 2(v/2L)$

(b)

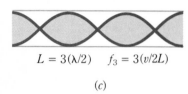

$L = 3(\lambda/2) \quad f_3 = 3(v/2L)$

(c)

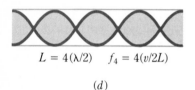

$L = 4(\lambda/2) \quad f_4 = 4(v/2L)$

(d)

**FIGURE 14.14**

Simple modes of vibration for a resonating pipe open at both ends.

Remember that the distance between two nodes or two antinodes is $\lambda/2$. Hence, the distance from a node to an antinode is $\lambda/4$. If we call the length of the air-filled tube $L$, the length in Fig. 14.13$a$ is from a node to an antinode, or $L = \lambda/4$. In $b$, the tube is three node-to-antinode lengths long; hence $L = 3(\lambda/4)$, and so on.

The resonance frequencies shown in Fig. 14.13 can be found from $f = v/\lambda$. These frequencies are easily computed by using the values found for $\lambda$ in terms of $L$. Notice that the first overtone $f_2$ is just $3f_1$. The second overtone $f_3$ is $5f_1$, and so on. The customary terminology is to call the frequency $2f_1$ the second harmonic, $3f_1$ the third harmonic, and so forth. In this case, the tube resonates only to the odd harmonic frequencies. Since the frequency of a tuning fork is usually known, resonances in a tube such as the one in Fig. 14.12 can be used to measure the velocity of sound.

A pipe or tube need not be closed at one end to resonate. For example, you can use a piece of glass tube as a whistle by blowing across either of its ends. The simplest possible resonances of a tube that is open at both ends are shown in Fig. 14.14. In each case, the ends of the tube are antinodes because the air is free to move there. The resonance frequencies are computed in the usual way by using the fact that $f = v/\lambda$, with $\lambda$ being determined as indicated in the figure.

When you blow across the end of a tube, this very complex process sends waves of many frequencies down the tube. Of this multitude of frequencies, the tube resonates to only one or two. For this reason, the resonating tube usually gives off a loud sound of a single frequency. However, if you try hard enough, you can often cause the tube to resonate to two frequencies at the same time and thereby give off two tones simultaneously.

Many musical instruments make use of resonating air columns. The flute and piccolo are basically tubes whose length can be changed by means of holes along the tube. A clarinet is similar, but in that case the sound waves are generated by the vibration of a reed in the mouthpiece. More complex tube resonance systems are seen in the trumpet, trombone, and tuba. In these instruments, the player elicits various resonance tones by changing the length of the resonating pipe. In addition, the sound waves in these instruments are generated by the vibration of the player's lips in the mouthpiece.

## 14.10 The Doppler effect

We now turn to an entirely different aspect of waves of all kinds and of sound waves in particular, a phenomenon referred to as the **Doppler effect.** You have certainly experienced this effect, although you may have been unaware of it. Whenever a fast-moving car approaches you, the sound of its horn (or even of the car itself) behaves in a peculiar way. The pitch of the sound is higher when the car is approaching you than when it is moving away from you. This same effect can be observed when an airplane flies by. The sound changes as the plane approaches and then recedes. A similar effect occurs with light and electromagnetic waves. Indeed, a highway speed trap uses radar in this way to measure speeds.

The basic idea behind the Doppler effect can be seen by referring to Fig. 14.15. Shown there is a water wave source that is moving to the right through the water. Although the source sends out circular waves, the centers of successive circles move to the right with the source. This causes the wave crests to be closer together at the right than at the left. The moving source, in effect, causes the wavelength of the waves to be different in various directions.

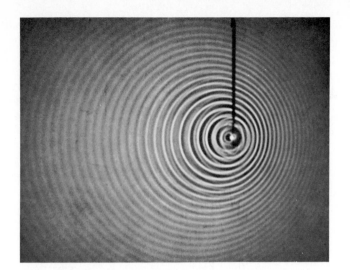

FIGURE 14.15
The vertical rod vibrates in such a way that its lower end acts as a source of water waves. Because the source is moving to the right, it causes the wavelengths in that direction to be shortened.
*(Education Development Center)*

A similar phenomenon occurs with sound waves, as we can see from Fig. 14.16. The wave source in Fig. 14.16*a* is stationary. If a listener at *P* is also stationary, the frequency of the disturbance striking the listener's ear is identical to the source frequency *f*.

Figure 14.16*b* shows what happens when the source is moving and the listener is stationary. The moving source causes the wavelengths of the waves it sends out to differ in different directions. A listener in front of the source hears a frequency higher than *f*, but a listener behind the source hears a lower frequency. In general,

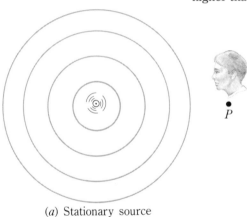

(*a*) Stationary source

FIGURE 14.16
The frequency of sound that a person hears depends on the velocity of both the source and the listener. In (*b*), the source is moving to the right. When it is at *A*, it sends out the crest labeled *A*; when at *B*, it sends out crest *B*; and so on.

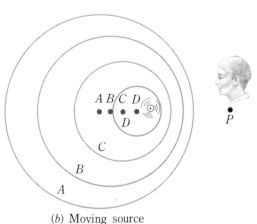

(*b*) Moving source

14.10 The Doppler effect    **317**

The observed frequency of a sound is raised when the distance between observer and source is decreasing and lowered when the separation distance is increasing.

A different situation exists when the source is stationary and the listener is moving. Suppose in Fig. 14.16a that the listener is moving toward the source. Because of this motion, the listener runs into some of the crests sooner than he would have if he had remained stationary at $P$. As a result, a listener approaching the source hears a frequency larger than $f$. By similar reasoning, a listener moving away from the source hears a frequency less than $f$. Indeed, if the listener is moving away with the speed of sound, only the crest accompanying him acts on his ear. In effect, the frequency heard is reduced to zero.

Let us now examine this phenomenon quantitatively. Taking velocities to the right in Fig. 14.16 as positive, we define $v_w$ = velocity of wave, $v_L$ = velocity of listener, and $v_s$ = velocity of source, all relative to the earth. Further, let $\tau$ be the period and $f$ the frequency of the wave emitted by the stationary source and $\tau'$ and $f'$ the period and frequency of the vibration detected by the listener.

The wavelength of the wave emitted by the stationary source is given by our usual equation:

$$\lambda = v_w \tau$$

As we see in Fig. 14.16, however, the wavelength changes when the source is moving. Along the line of forward motion, the crests are closer together by an amount $v_s \tau$; this is the distance the source moves during the time $\tau$ between the emission of successive crests. As a result, the actual wavelength (distance between crests) is

$$\lambda - v_s \tau$$

or, after replacing $\lambda$ by $v_w \tau$,

Distance between crests $= (v_w - v_s)(\tau)$

To find $\tau'$, the period detected by the listener, we must find the time observed between the arrival of successive crests. If the listener were stationary, this time would be, from $x = vt$,

$$\tau' = \frac{\text{distance between crests}}{v_w} \qquad \text{(stationary observer)}$$

If the listener is also moving to the right with a velocity $v_L$, the speed of the wave relative to the listener is $v_w - v_L$, and so $\tau'$ is

$$\tau' = \frac{\text{distance between crests}}{v_w - v_L}$$

The distance between crests is $(v_w - v_s)(\tau)$, and so

$$\tau' = \frac{(v_w - v_s)(\tau)}{v_w - v_L}$$

Usually this is expressed in terms of the frequencies $f = 1/\tau$ and $f' = 1/\tau'$. We then find for the ratio of the frequency heard $f'$ to the frequency of the emitter $f$:

$$\frac{f'}{f} = \frac{v_w - v_L}{v_w - v_s} = \frac{1 - (v_L/v_w)}{1 - (v_s/v_w)} \tag{14.3}$$

In using Eq. 14.3, we must be careful about signs. All the velocities are assumed to be along the $x$ axis with the source to the left of the listener, and $v_w$ is always positive. **The velocities must be given appropriate plus or minus signs depending on whether they are in the positive or negative direction.** Some people prefer to write Eq. 14.3 in the following way because it helps to keep track of signs: for sound waves,

$$\frac{f'}{f} = \frac{\text{speed of wave relative to listener}}{\text{speed of wave relative to source}} \tag{14.4}$$

This form shows clearly that $f'$ is increased if the listener is moving against the waves and if the source is moving in the direction of the waves.

This phenomenon is called the *Doppler effect,* after Christian Johann Doppler, who showed in 1842 that it should be observed for sound and light waves. Einstein showed, however, that Eq. 14.3 is not correct for light waves when the source or the observer is moving with a speed close to the speed of light. The difficulty arises because of the relativity principle, which states that the speed of light in vacuum is independent of the motion of the observer or the light source.

An interesting situation arises when the speed of the sound source approaches or equals the speed of sound. Then, from Eq. 14.3, we see that the sound frequency $f'$ approaches infinity. This simply means that a nearly infinite number of wave crests reach the listener in a very short time. We can easily understand this by referring once again to Fig. 14.16*b*.

Suppose the moving source has a speed equal to the speed of sound. Then all the wave crests in front of the source lie upon one another. They, together with the source itself, all pass a given point at the same time. All the energy of the sound waves is compressed into a very small region in front of the source. Consequently, this very concentrated region of sound energy, a *shock wave,* causes an extremely loud sound as it passes the point. Basically this is the origin of the sonic boom that accompanies the flight of supersonic aircraft.

When an airplane moves through the air with a speed near that of sound, the noise and air disturbances originating from the plane are built up into a shock wave, which, as we have just seen, is simply a region of very dense sound energy. The exact shape of the shock wave depends upon the speed of the airplane. In general, the shock zone covers the surface of a cone, as illustrated in Fig. 14.17. As we shall see in Prob. 46, the angle $\theta$ of the cone depends upon the ratio of the speed of the plane $v_p$ to the speed of sound waves $v_w$ in the following way (provided $v_w < v_p$):

$$\sin \theta = \frac{v_w}{v_p}$$

As $v_p$ becomes larger with respect to $v_w$, the angle of the cone decreases.

The ratio $v_p/v_w$ is often called the *Mach number.* In this terminology, a plane traveling at Mach 2 is moving at twice the speed of sound.

Since the shock wave is a region of very concentrated sound energy, it can cause severe damage when it strikes something. The familiar sonic boom is the result of the conical shock-wave surface passing over the earth. For example, in Fig. 14.17,

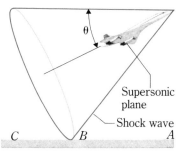

**FIGURE 14.17**

The sonic boom has already hit point $C$ and is moving through point $B$ toward point $A$.

## The speed of blood and baseballs

The advent of modern electronics and microcomputers has made the Doppler effect a very practical tool. Today we can measure the speed of inaccessible objects by bouncing sound or radar waves off them. When a sports announcer states the speed with which a pitcher has just thrown a ball across the plate, the measurement was probably made by means of an ultrasonic wave and the Doppler effect. Let us see how this is done.

A sound source in the stands behind the catcher sends a sound beam of frequency $f_o$ toward the pitcher. The beam is reflected by the pitched ball and returns to the source with a frequency that has been shifted by the Doppler effect. Using the difference in frequency between the sound source and the reflected wave, a microcomputer reads out the speed of the ball. The formula for the effect is found as follows.

The original sound beam has a frequency $f_o$ and a speed $v_w$. The frequency with which it strikes the ball is given by Eq. 14.3 or 14.4:

$$\frac{f'}{f_o} = \left( \frac{\text{wave speed relative to ball}}{\text{wave speed relative to source}} \right)$$

$$= \left( \frac{v_w + v_b}{v_w} \right)$$

where $v_b$ is the speed of the ball. This is the frequency with which the reflected wave leaves the ball. Since the ball acts like a moving sound source, the frequency of the wave returning to the receiver in the stands is

$$\frac{f}{f'} = \left( \frac{\text{wave speed relative to receiver}}{\text{wave speed relative to source}} \right)$$

$$= \left( \frac{v_w}{v_w - v_b} \right)$$

After substituting for $f'$, we have

$$f = f_o \left( \frac{v_w + v_b}{v_w} \right) \left( \frac{v_w}{v_w - v_b} \right)$$

$$= f_o \left( \frac{v_w + v_b}{v_w - v_b} \right)$$

This can be solved to find the speed of the ball:

$$v_b = v_w \left( \frac{f - f_o}{f + f_o} \right)$$

A microphone receives the reflected wave; then, by the use of appropriate circuitry, the beat frequency between $f$ and $f_o$ is measured, thereby determining $f$. Using this value for $f$, a microcomputer calculates $v_b$ and displays it on a readout device.

This same general method is used in medicine to measure the speed of blood moving through a vein or artery. The particles carried along by the blood plasma play the role of the baseball. Similarly, a police officer measures the speed of a car using radar waves. As you see, the Doppler effect, which until about 1950 was of only limited practical importance, is now in widespread use.

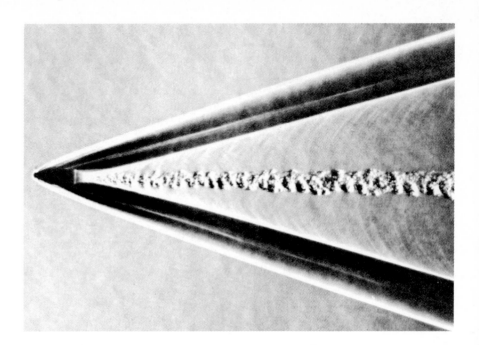

Shock waves generated by a projectile shooting through the air. *(Peter B. Wegener)*

the sonic boom will soon strike point *A*; it is currently striking point *B* and has already hit point *C*. Depending upon the intensity of the wave, its effects will be more or less damaging. Figure 14.18 shows the shock wave near a high-speed projectile.

---

**Example 14.5**

A car is moving at 20 m/s along a straight road with its 500-Hz horn sounding. You are standing at the side of the road. What frequency do you hear as the car is (*a*) approaching you and (*b*) receding from you?

***Reasoning*** We know that the speed of sound $v_w$ is about 340 m/s, $v_c = 20$ m/s, and $f = 500$ Hz. Using Eq. 14.4, we have

(*a*) Car approaching:

$$f' = f\left(\frac{\text{wave speed relative to listener}}{\text{wave speed relative to source}}\right) = f\left(\frac{v_w}{v_w - 20 \text{ m/s}}\right)$$

$$= (500 \text{ Hz})\left(\frac{340 \text{ m/s}}{320 \text{ m/s}}\right) = 531 \text{ Hz}$$

(*b*) Car receding:

$$f' = f\left(\frac{\text{wave speed relative to listener}}{\text{wave speed relative to source}}\right) = f\left(\frac{v_w}{v_w + 20 \text{ m/s}}\right)$$

$$= (500 \text{ Hz})\left(\frac{340 \text{ m/s}}{360 \text{ m/s}}\right) = 472 \text{ Hz} \quad \blacksquare$$

## Example 14.6

A stationary police officer is using radar to monitor car speeds. The radar sends out a wave ($v_w = 3 \times 10^8$ m/s and $f_o = 1 \times 10^{10}$ Hz) that strikes an approaching automobile and is reflected back to the police officer. What is the difference between the frequency of the wave reaching the officer and $1 \times 10^{10}$ Hz if the car's speed is 25 m/s?

*Reasoning*  From the special note "The Speed of Blood and Baseballs," the frequency of the reflected wave reaching the police radar is*

$$f = f_o \left( \frac{v_w + v_c}{v_w - v_c} \right) = (1 \times 10^{10} \text{ Hz}) \left( \frac{3 \times 10^8 + 25 \text{ m/s}}{3 \times 10^8 - 25 \text{ m/s}} \right)$$

To solve this with minimum arithmetic, divide the numerator and denominator by $3 \times 10^8$ m/s. Then we have

$$f = (1 \times 10^{10} \text{ Hz}) \left( \frac{1 + 8.3 \times 10^{-8}}{1 - 8.3 \times 10^{-8}} \right)$$

$$\cong (1 \times 10^{10} \text{ Hz})(1 + 8.3 \times 10^{-8})(1 + 8.3 \times 10^8)$$

where we have used the expansion formula that $1/(1 - x) \cong 1 + x$ if $x$ is small. Simplification gives

$$f = 1 \times 10^{10} + 1670 + 0.000064 \cong 1 \times 10^{10} + 1670 \text{ Hz}$$

Although the change in frequency (1670 Hz) is very small compared with $f_o$, the beat frequency between $f_o$ and $f$ is also 1670 Hz, an easily measured quantity. The radar device uses a microcomputer to compute the car's speed from the beat frequency. ∎

------

* If the car had been going at nearly the speed of light, an impossible situation, relativistic effects would alter this computation somewhat.

## Minimum learning goals

When you finish this chapter, you should be able to

1  Define (*a*) sound wave, (*b*) wavefront, (*c*) plane wave, (*d*) sound intensity, (*e*) sound level, (*f*) decibel, (*g*) quality of sound, (*h*) beats, (*i*) overtones versus harmonics, (*j*) Doppler effect, (*k*) shock wave.

2  Explain what is meant by a sound wave and why sound cannot travel through vacuum.

3  Recall the approximate velocity of sound in air.

4  Convert from sound intensity in watts per square meter to sound level (intensity level) in decibels.

5  Sketch the approximate response curve for the normal ear as a function of frequency. Give the approximate decibel levels for very weak and very loud sounds. Identify the ultrasonic region.

6  Explain the concept of sound quality and point out why it is different from frequency.

7  Combine two waves of the same frequency and amplitude but of different phase so as to obtain destructive and/or constructive interference.

8  Use beats to find the difference in frequency between two sound sources.

9  Find the resonance frequencies of sound in given pipes.

10  Explain what is meant by the Doppler effect and compute the frequency shift noticed for an approaching or receding sound source.

11  Explain how a shock wave originates and why it gives rise to a sonic boom.

## Questions and guesstimates

1 Explain why a bell ringing inside a vacuum chamber cannot be heard on the outside.

2 Would you expect a sound heard under water to have the same frequency as a sound heard in air if the sources vibrate identically? Explain.

3 When a deep-voiced man inhales helium and then speaks, his voice sounds high-pitched. Why?

4 Suppose a few pipes of a pipe organ were mounted close to a hot heater. Would this affect the performance of the organ? Explain.

5 A siren can be made by drilling equally spaced holes on a circle concentric to the axis of a solid metal disk. When the disk is rotated while a jet of air is blowing against it near the holes, a sirenlike tone is given off. Explain how this gives the sensation of sound to the ear, and state what factors influence the pitch and quality of the tone.

6 A singer claims to be able to shatter a wine glass by singing a particular note. Could this be true? Explain.

7 There is on the market a device that uses intense ultrasonic waves in water to wash dirt loose from cloth and other objects. Explain how this works, and list its advantages and disadvantages.

8 Suppose that on some distant planet there exist humanoids whose hearing mechanism is designed as follows. From the outside, their heads look like our own. However, a 1-cm-diameter, hard-surfaced cylindrical hole passes through the head from ear to ear. At the midpoint of the channel, a thin, circular membrane acts like a drumhead, separating the two halves of the channel. These beings experience the sensation of sound when this drumhead vibrates. What can you infer about their hearing abilities and about the way they communicate orally with one another?

9 In an adult human, the ear canal, which is the hollow tube leading from the outer ear to the eardrum, has a length of about 2.5 cm. How does the resonance frequency of such a tube correspond with the sensitivity curve of the ear?

10 Estimate the frequencies at which a test tube 15 cm long resonates when you blow across its lip.

11 A sound source emits a 1000-Hz sound in all directions while a strong wind blows to the east past the source. How do the frequency, speed, and wavelength of the observed sound depend on the position of the observer?

12 Two speakers are to be connected to a stereo. The directions say, "Set the speakers side by side and connect the two red wires from the amplifier to the two terminals at the back of one speaker. Connect the two gray wires from the amplifier to the two terminals at the back of the other speaker. Listen to the sound. Reverse the red wires at the speaker so that the wire that was in the left speaker terminal is now in the right terminal and vice versa. Listen to the sound again. The final connections should be in the way that gives maximum sound." Explain the physical reasons for these instructions.

## Problems

*Unless otherwise stated, use 340 m/s for the speed of sound in air.*

1 The thunder from a lightning flash is heard 8.0 s after the flash is seen. How far away is the lightning? Assume the speed of light to be nearly infinite (it is $3 \times 10^8$ m/s).

2 The sound from a carpenter's hammer is heard by someone far away considerably after the hammer impact. What is the delay time if the hammer is 800 m from the observer? Assume the speed of light ($3 \times 10^8$ m/s) to be nearly infinite.

3 A loudspeaker vibrates with a frequency of 400 Hz. What is the wavelength of the sound waves it emits? Repeat for a frequency of 15,000 Hz.

4 In order to locate objects in the dark, a bat screams at an ultrasonic frequency near 60,000 Hz. To what wavelength is this equivalent? What frequency is required if the sound wavelength is 1.0 mm?

5 Use Eq. 14.1 to compute the speed of sound in nitrogen gas at 0°C.

6 Compute the speed of sound at 0°C in a hypothetical ideal diatomic gas that has a molecular mass of 70 kg/kmol.

7 From the fact that the speed of sound in mercury is 1451 m/s, find the bulk modulus of mercury.

8 The velocity of sound in bone at 1 MHz is 3450 m/s, and the density of bone is about 1.82 g/cm³. Compute the bulk modulus of bone at this frequency.

9 By about what percentage does the speed of sound in air change as the air temperature changes from 15°C to 21°C?

**10** A depth finder on a fishing boat sends sound waves downward and detects the reflected wave. The instrument shows a school of fish at a depth of 8.5 m. How long after a sound pulse is sent out is it received at the boat?

**11** A certain stereo set consumes power at a rate of 30 W. It has two speakers, each of which has a 40-cm² area from which the sound comes. If 0.030 W of sound power comes from each speaker, what is the sound intensity at the speaker? With what efficiency does the set convert electric energy to sound energy?

**12** A certain loudspeaker has a circular opening with an area of 30 cm². Assume that the sound it emits is uniform and outward through this entire opening. If the sound intensity at the opening is $2 \times 10^{-4}$ W/m², how much power is being radiated as sound?

■ **13** A beam of sound has an intensity of $3 \times 10^{-6}$ W/m². What is the intensity level in decibels?

■ **14** What is the intensity level in decibels for a sound that has an intensity of 0.70 W/m²?

■ **15** (a) What is the intensity of a 40-dB sound? (b) If the sound level is 40 dB close to a speaker that has an area of 120 cm², how much sound energy comes from the speaker each second?

■ **16** Eight typists in a room give rise to an average sound intensity level of 60 dB. What is the intensity level in the room when an additional three typists, each generating the same amount of noise, begin to type?

■ **17** The sound level in a room in which 40 people are conversing is 70 dB. How many people must leave to lower the sound level to 60 dB? Assume that the people speak identically.

■ **18** A tiny sound source sends sound equally in all directions. If the intensity is $5 \times 10^{-3}$ W/m² 4.0 m from the source, (a) how much sound energy does the source emit each second? (b) What is the intensity 3.0 m from the source? (c) Prove that the intensities at radii $r_1$ and $r_2$ from the source are related through $I_1/I_2 = r_2^2/r_1^2$.

■ **19** What is the sound intensity at a location where the sound level is 23 dB?

■ **20** Find the sound intensity in a room where the intensity level is 78 dB.

■ **21** A sound beam has a cross-sectional area of 2.0 cm² and a sound level of 95 dB. The beam strikes a sheet of sound absorber that absorbs essentially all of it. How many calories are generated in the absorber in 1 min?

■ **22** When the decibel level of a certain sound is increased by a factor of 5, its intensity also increases fivefold. What is the original intensity of the sound?

■ **23** The tiny loudspeaker in Fig. P14.1 is in an air-filled tube that is bent into a circle. If the circumference of the

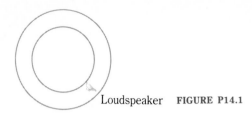

Loudspeaker    FIGURE P14.1

circle is 7.0 m, what are the three lowest vibration frequencies of the loudspeaker for which intense sound exists in the tube? (The drawing is not to scale.)

■ **24** The loudspeaker in Fig. P14.1 sends sound through the air-filled interior of a hollow tube that is bent into a circle. The tube resonates to loudspeaker frequencies of 87, 174, 261, and 348 Hz, as well as to higher frequencies. What is the circumference of the circle? Assume the loudspeaker to be much tinier than shown.

■ **25** The two identical loudspeakers in Fig. P14.2 vibrate in phase with a frequency of 4000 Hz. At what values of $x$ are (a) loud sound heard at $P$? (b) Weak sound heard at $P$?

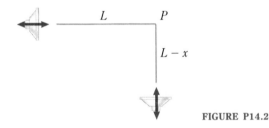

FIGURE P14.2

■ **26** In Fig. P14.2, the two identical loudspeakers vibrate in phase with the same frequency. The length $L$ is 20 cm longer than the length $L - x$. The frequency of the waves sent out from the speakers is slowly increased from 30 Hz. At what frequency does an observer at $P$ hear (a) maximum loudness and (b) minimum loudness?

■ **27** Two tiny sound sources face each other and send out identical, in-phase waves. One source is at $x = 0$, and the other is at $x = 3.0$ m. If the wavelength of the waves sent out by the sources is 60 cm, at what points along the line from $x = 0$ to $x = 3.0$ m does a detector register minimum sound?

■ **28** Assume the situation described in Prob. 27 but with variable-frequency sources. The source frequency starts low and then is increased slowly. At what frequencies is minimum sound heard at the point $x = 2.0$ m?

**29** Two violins are slightly out of tune. When they are bowed so as to send out the supposedly same note, a beat frequency of 0.90 Hz is noticed. If the frequency emitted by one violin is $f_o$, what are the possible frequencies emitted by the other?

**30** Two pianos sound the same note, but the vibration from one is 221.60 Hz and that of the other is 221.40 Hz. What is the beat frequency between the two tones?

**31** A pipe open at one end and closed at the other is 82 cm long. What are the three lowest frequencies to which it will resonate? Draw the wave within the tube for each frequency. Repeat for a pipe open at both ends.

**32** What are the three lowest frequencies to which a pipe open at both ends will resonate? The pipe is 65 cm long. Draw the wave within the tube for each frequency. Repeat for a pipe closed at one end.

**33** In an experiment like that shown in Fig. 14.12, resonance occurs when the water is 33.00 cm high in the tube and again when it is 41.00 cm high. If no resonance occurs at any height between these two values, find the frequency of the tuning fork.

**34** A man wishes to find out how far down the water level is in the iron pipe leading into an old well. Being blessed with perfect pitch, he merely hums musical sounds at the mouth of the pipe and notices that the lowest-frequency resonance is 94 Hz. About how far from the top of the pipe is the water level?

**35** The Lincoln Tunnel under the Hudson River in New York City is about 2600 m long. To what sound frequencies does it resonate? What practical importance, if any, do you think this has?

■ **36** A certain organ pipe resonates to a frequency of 600 Hz when its temperature is 20°C. When this pipe at 20°C and an identical pipe that is at 25°C because it is near a heater are sounded in unison, what is the beat frequency between their two sounds?

■ **37** A cylindrical tube resonates at the following consecutive frequencies: 375 Hz, 525 Hz, and 675 Hz. (*a*) What is its fundamental resonance frequency? (*b*) Is it open at just one end or at both ends?

■ **38** The speed of sound in hydrogen is about 1270 m/s. If a pipe that resonates in its fundamental frequency to 700 Hz in air is filled with hydrogen, what will its fundamental resonance frequency be?

**39** How fast must a car be coming toward you for its horn to appear to be 5 percent higher in frequency than when the car is standing still?

**40** A sound source vibrates at 200 Hz and is receding from an observer at 18 m/s. If the speed of sound is 331.0 m/s, what frequency does the observer hear?

■ **41** A plane moving at 500 km/h flies in a straight line that passes directly over your head. Assuming that the frequency of sound you hear as the plane approaches from far away is $f_a$, find the frequency of sound $f_r$ that you hear as it recedes into the distance.

■ **42** At the coordinate origin, a sound source of frequency $f$ sends waves out along the positive $x$ axis. There is a 20-m/s wind blowing in the positive $x$ direction. (*a*) Find the frequency and wavelength of the sound heard by a stationary observer on the positive $x$ axis. Represent the speed of sound in still air by $v$. (*b*) Repeat for the wind blowing in the negative $x$ direction.

■ **43** A source of sound waves whose frequency is $f$ is approaching a wall with speed $v_s$. The sound reflects back to an observer traveling with the source. Find the frequency of sound heard by the observer. Represent the speed of sound by $v_w$.

■ **44** A narrow tube open at one end and closed at the other is 40 cm long. It resonates under the action of a 220-Hz sound source provided the source is receding from it. How fast must the source be moving?

■ **45** Two identical sound sources vibrate at 150 Hz. Suppose that the two sources and a detector are placed along a straight line, with the detector at $x = 0$, one source at $x = 20$ m, and the other source at $x = 30$ m. If the 30-m source begins moving toward larger $x$ values at 7.0 m/s, what is the beat frequency heard by the detector?

■ **46** Show that the apex angle of a shock wave cone is given by $\sin \theta = v_w/v_p$, where $v_w$ is the wave speed and $v_p$ is the plane speed. *Hint:* Draw the spherical waves generated by the plane at successive times and consider their envelope.

■ **47** A car blowing a 2000-Hz horn is moving at 100 km/h along the negative $x$ axis toward the coordinate origin. What frequency does a listener on the $y$ axis at $y = 30$ m hear when the car is at $x = -40$ m?

# 15 Electric forces and fields

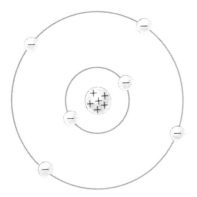

I T is difficult for us to visualize the world of a century ago, when the use of electricity was just beginning. A few people had electric lights, but the electrical appliances and machines that we are so accustomed to today were totally missing. Primitive electric motors and batteries were curiosities just beginning to show practical importance. What a contrast to today, when nearly everything we do makes use of electricity in some way. Because of its widespread use, electricity is a tool that all educated people must understand. We shall spend the next several chapters learning about the ways in which electricity plays an influential role in the world about us.*

**FIGURE 15.1**

A schematic representation of a carbon atom. The negative charges on its six electrons are balanced by the positive nuclear charge. (The nucleus and electrons are much smaller than shown.)

## 15.1 Atoms as the source of charge

An atom is composed of a tiny positively charged nucleus around which are negatively charged particles called **electrons.** This is illustrated for an atom of carbon in Fig. 15.1. You may recall from courses in chemistry that all atoms are electrically neutral. That is, the quantity of positive charge on the nucleus exactly equals the quantity of total negative charge on the electrons about the nucleus. In the case of the carbon atom, if $-e$ is the charge on each electron, the charge on the nucleus is exactly $+6e$. We shall postpone a detailed discussion of the atom to a later chapter and merely make use of its electrical constitution here.

It appears that the universe as a whole is nearly, if not completely, neutral from an electrical standpoint. The earth has very little, if any, excess of either positive or

---

* The material in this chapter and the next is basic to the entire study of electricity and the many facets of nature that involve electrical effects. It is therefore imperative that you understand this material thoroughly before passing on to the following chapters.

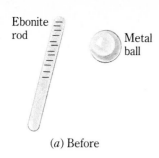

Ebonite rod        Metal ball

(*a*) Before

(*b*) During

(*c*) After

FIGURE 15.2

When the negatively charged ebonite rod touches the uncharged metal ball, electrons are conducted off the rod onto the ball.

negative charge. For nearly all practical purposes, it can be considered to have zero net charge. The vast majority of the charges on and in the earth reside in atoms. When free negative or positive charges are found, they are usually assumed to have come from the tearing apart of an atom.

It is not at all difficult to remove an electron from an atom — under certain circumstances. For example, if a rod of ebonite (hard rubber) is rubbed with animal fur, some of the electrons from the atoms in the fur are rubbed off onto the ebonite rod. (The reason for this charge transfer is not simple to explain. It is covered in courses dealing with solid-state physics.) Hence, the rod acquires a net excess of electrons. When it is touched to a metal object, some of the excess electrons are transferred to the metal, as illustrated in Fig. 15.2.

Similarly, if a glass rod is rubbed with silk, some of the electrons leave the atoms in the rod and give rise to an excess of positive charge on it. If the positively charged rod is touched to a neutral metal ball, electrons leave some of the atoms of the metal and replace those electrons lost by the atoms in the glass. As a result, the metal ball acquires a net positive charge. Many other materials give rise to a separation of charge when rubbed together. The ones we have described were used to define positive and negative charge before the existence of the electron was even known.

## 15.2 Forces between charges

Now that we know how to obtain charged bodies, it is possible to examine the forces between charges. One of the simplest ways of doing this is to make use of very light metal-coated balls. Such balls can be charged by touching them with a glass or ebonite rod. If the balls are suspended by light threads, four interesting experiments can be performed. These are illustrated in Fig. 15.3, and from them we can conclude the following:

**1** Like charges repel one another; that is, two positive charges repel each other, as do two negative charges.

**2** Unlike charges attract one another; that is, positive charges attract negative charges, and vice versa.

**3** The magnitude of the electric force between two charged objects often exceeds the gravitational attraction between them. (For example, the gravitational force between the two balls is far too small to affect the way they hang.)

FIGURE 15.3

In (*a*) the balls are uncharged. The charged balls in (*b*), (*c*), and (*d*) show that like charges repel one another while unlike charges attract one another.

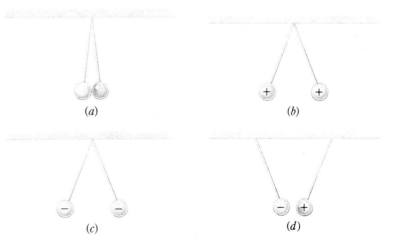

(*a*)    (*b*)

(*c*)    (*d*)

## 15.3 Insulators and conductors

Although all materials are made of atoms and all atoms are made of electrons and nuclei, we are well aware that the electrical properties of substances vary widely. There are two basic groups into which substances can be divided according to their electrical properties: *conductors* and *nonconductors* (or *insulators*). Materials intermediate between these two groups are called *semiconductors*.

In insulators, the electrons of any given atom are bound tightly to that atom and cannot wander through the material. Hence, even if a charged rod is brought close to an insulator, the electrons and nuclei in the atoms of the insulator cannot move under the attraction or repulsion of the rod's charge. Charges cannot move freely through insulators.

Electrical conductors behave quite differently. These substances contain charges that are free to move throughout the material. Metals are familiar conductors; though each atom of the metal is normally neutral (that is, uncharged), the electrons farthest from the nucleus are easily freed from the atom. They then move through the metal, carrying their negative charge from place to place in the process. Therefore, when a negatively charged rod is brought close to a piece of metal, the rod repels some of the free electrons in the metal to the most distant regions of the metal. Similarly, a positively charged rod attracts free electrons to the portion of the metal nearest the rod.

Metals are not the only electrical conductors. Many substances — ionic solutions, for example — contain ions (charged atoms) that can move relatively freely through the substance. All electrical conductors contain charges that can move over long distances when repelled or attracted by nearby charged objects.

## 15.4 The electroscope

The *electroscope* (Fig. 15.4) is a simple device used for detecting and measuring charges of small magnitude. A metal rod from which are suspended two very thin leaves of gold foil is held inside a metal case by an insulator that keeps the rod from touching the case. The two faces of the case are covered with glass, so that the positions of the leaves can be seen.

Suppose some negative charge is placed on the electroscope by touching a charged piece of ebonite to its metal sphere. The charge is confined entirely to the sphere, rod, and leaves since they are insulated. Because like charges repel one another, the negative charges distribute themselves more or less uniformly over the sphere, rod, and leaves. As a result, the leaves, being free to bend and being repelled by the charges on one another, take the position shown in Fig. 15.5a.

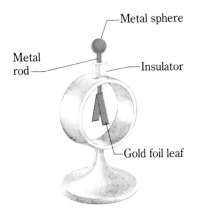

— Metal sphere

Metal rod —

— Insulator

— Gold foil leaf

**FIGURE 15.4**

One type of gold-leaf electroscope. The portion consisting of the metal sphere, rod, and gold foil is insulated from the case.

**FIGURE 15.5**

The uncharged electroscope is used to determine the sign and approximate magnitude of the charge on an object.

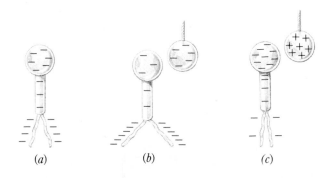

(a)        (b)        (c)

(a) Before

(b) During

(c) After

**FIGURE 15.6**

A metal ball can be charged by conduction. Note that the rod gives some of its charge to the ball, and therefore the ball and the rod have like charges.

**FIGURE 15.7**

Charging a metal sphere by induction. Note that the rod and the ball end up with unlike charges.

If a negatively charged ball is now brought close to the metal sphere of the electroscope, as shown in Fig. 15.5b, many of the negative charges in the metal sphere are repelled down the rod, causing the leaves to separate farther. An opposite effect is observed if a positively charged ball is brought close to the electroscope (Fig. 15.5c). In addition, a ball with no net charge does not disturb the electroscope to any great extent. With this apparatus, the sign of a charge and its approximate magnitude can be determined. You should convince yourself that a similar procedure can be followed if the electroscope is charged positively.

## 15.5 Charging by conduction and by induction

There are two general ways of placing charge on a metal object using a second object that is already charged. As an example, consider the ways in which you might use a negatively charged ebonite rod to charge a metal ball. One way is to touch the ball to the rod. On contact, some of the excess negative charge on the rod moves onto the ball. This process, shown in Fig. 15.6, is called **charging by conduction.**

The same rod can also be used in a different way to charge the ball. This is shown in Fig. 15.7. In this process, **charging by induction,** the rod is not touched to the ball at all. When the rod is brought close to the left side of the ball, some of the electrons in the metal are repelled to the right side of the ball, leaving a positive charge on the left side. Since no charge has been added to or subtracted from the ball, it is still neutral, of course. Now, suppose you touch the ball with an object other than the charged ebonite rod, such as your finger. Because your body is a conductor (though not a good one), charge moves from the ball through your body to the earth. Thus the nearby negatively charged rod induces negative charge to leave the ball and move to the ground. (We say that the ball is **grounded,** and the symbol —|ı· is used to show this. Usually, to ground an object, we attach it by a wire to a water pipe or some other object that goes into the earth.) Once negative charge has moved from the ball to the ground, the ball is no longer neutral. After the conducting path to ground is removed, the negative rod can be taken away and the ball will have a positive charge. (Why must the grounding object be removed *before* the rod is?)

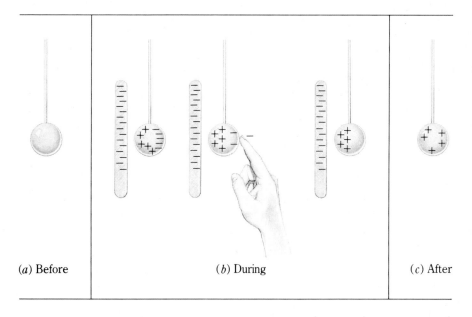

(a) Before       (b) During       (c) After

If you compare Figs. 15.6 and 15.7, you can see that an ebonite rod can charge a metal object *negatively* by *conduction* but it charges the same object *positively* by *induction*. You may find it interesting to work out similar diagrams using a positively charged glass rod. The charges are reversed in that case.

## 15.6 Faraday's ice-pail experiment

In 1843, Michael Faraday carried out a simple but highly instructive experiment. He attached a metal ice pail to an electroscope, as shown in Fig. 15.8a. When a positively charged metal ball suspended from a thread was lowered into the pail (without touching it), as in (b), the leaves of the electroscope spread apart (diverged). Moreover, when the ball was moved from place to place within the pail, the leaves remained in the same divergent position. Only when the ball was removed did the leaves return to their original collapsed position, as in (a).

Faraday further noticed that if the charged metal ball was touched to the inside of the pail as in (c), the electroscope leaves remained in the divergent position. Now, however, when the ball was removed from the pail, the leaves remained divergent, as in (d). When the ball was brought close to a second electroscope, it was found to be no longer charged. Apparently, when the ball touched the inside of the pail, the excess charge on the ball was completely neutralized. Since the leaves of the electroscope attached to the outside of the pail did not move when the ball touched the pail, Faraday concluded that the inner surface of the pail had just enough charge on it to neutralize the ball.

FIGURE 15.8
Faraday's ice-pail experiment.

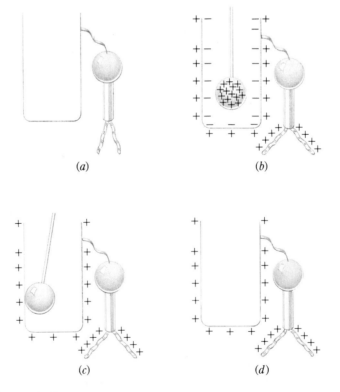

(a)          (b)

(c)          (d)

From these experiments, we can draw the following conclusions:

**1** A charged metal object suspended inside a neutral metal container induces an equal and opposite charge on the inside of the container.

**2** When the charged metal object is touched to the inside of the container, the induced charge exactly neutralizes the excess charge on the object.

**3** When a charged object is placed within a metal container, an equal charge of the same sign is forced to the outer surface of the container.

**4** All the charge on any metal object will reside on its outer surface if a conducting path is provided so that the charge can move there.

These are important facts concerning electric charges on metal. We shall interpret them more fully when we examine Coulomb's law and the concept of electric fields in Secs. 15.8 to 15.12.

## 15.7 Conservation of charge

In mechanics, we learned that nature conserves certain quantities. Among these are energy, linear momentum, and angular momentum. Each of these quantities obeys a conservation law, and, as we have seen, that fact is of great importance in our universe.

There are also conservation laws that apply to electrical quantities. One of these is the law of conservation of electric charge. By this we mean that the algebraic sum of all the charges in the universe remains constant. This fact has become clear to us only in this century. When physicists became able to create new particles by bombarding one high-energy particle with another in huge accelerators, they found that charges are always created (or annihilated) in pairs. Any reaction that creates an electron (charge $-e$) also creates a particle whose charge is $+e$. Similarly, when a particle with charge $+e$, such as a positron (a positive electron), combines with a particle of charge $-e$, both charges disappear; their algebraic sum at the outset was zero, and it remains zero after the reaction is completed. In every experiment, the algebraic sum of the charges before the reaction is the same as that afterwards. There appears to be no way in which net charge can be created or destroyed. We conclude that

The net charge of the universe does not change.

This is called the **law of conservation of charge.**

Notice that the law does not state that the number of electrons or protons in the universe is constant. We know of many reactions in which pairs of oppositely charged particles are created or destroyed, as we shall see in Chap. 28. Even though we do not yet know the net charge of the universe, or of our galaxy, or even of the earth and its atmosphere (it is possible that these net charges are close to zero), our knowledge of the conservation of charge is still of great use to us. We shall use this concept when we discuss electric circuits. Moreover, in these days, when particle physicists are trying to understand the particles that may be created in high-energy reactions, the conservation law guides them in deciding which reactions are possible.

# Hunting quarks

Discoveries in physics often result from the interplay of the experiments and theories of several persons. A still continuing example of this fact is the current search for particles with charge smaller than the charge carried by the electron and proton.

Freed from their war-related duties after the completion of World War II, physicists began a search for a new particle predicted in 1935 by the Japanese theorist Hideki Yukawa. These physicists were surprised to learn that collisions produced in high-energy accelerators yielded a myriad of previously unknown particles. The electron, proton, and neutron were found to be only a few of the many particles nature can be teased to reveal.

It was suspected that there must be some relation between these various particles, and so simplifications of current theories were sought. One theory, proposed in 1963 by Murray Gell-Mann and George Zweig, postulated the existence of still other particles, which they called quarks. These particles, they reasoned, must have only one-third the charge of the proton and electron. Their theory "explained" many of the newly discovered particles, as well as the neutron and proton, by assuming them to be combinations of quarks. Predictions could be made from their theory, and so it could be tested.

In the following years, experiments confirmed Gell-Mann and Zweig's theory in broad outline. In order to obtain perfect agreement with experimental results, however, the theory has been modified in its details. Still, the basic idea that quarks are the building blocks for the proton, neutron, and many other particles seems to be correct. Even as this is written, however, 24 years after they were proposed, individual quarks still elude us. Despite what now appear to be "false alarms," no particle with charge smaller than that of the electron and proton has yet been isolated, but the search continues. Moreover, theorists are struggling to understand why quarks, which are almost certain to exist, are thus far impossible to isolate.

## 15.8 Coulomb's law

The mathematical law that describes how like charges repel and unlike charges attract each other was discovered in 1785 by Charles Augustin de Coulomb (1736–1806) and is called **Coulomb's law.** By means of a very sensitive balance, similar to that used by Cavendish in his study of gravitation, Coulomb was able to measure accurately the force between two small charged balls (Fig. 15.9). Two small balls,

**FIGURE 15.9**
The two unequal, unlike charges attract each other with equal force.

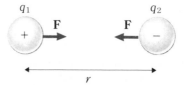

much smaller than those shown with a distance $r$ between centers, carry charges $+q_1$ and $-q_2$. After a number of experiments, Coulomb concluded that the force on ball 1 varied in proportion to the product of the two charges and inversely as the square of the distance between centers:

$$F \propto \frac{q_1 q_2}{r^2}$$

$$F = \text{(constant)} \left( \frac{q_1 q_2}{r^2} \right) \tag{15.1}$$

and that the force was in the direction shown in Fig. 15.9. According to Newton's law of action and reaction, the force on ball 2 must be identical in magnitude but oppositely directed. Coulomb's law applies only to point charges and to charges whose dimensions are small relative to $r$. Otherwise, the value to be used for $r$ in Eq. 15.1 is open to question.

To make the above statement of Coulomb's law into an equation, we must first decide on a unit in which to measure charge. The SI unit of charge is defined in terms of electric current, as we shall see in Sec. 21.8. For now, we simply state that the SI charge unit is the *coulomb* (C). When we use this unit for $q_1$ and $q_2$, Coulomb's law can be written

$$F = k \frac{q_1 q_2}{r^2} \tag{15.2}$$

where $F$ is in newtons and $r$ is in meters. The constant of proportionality $k$, determined by experiment, is $8.9874 \times 10^9$ N · m²/C² when the experiment is performed in vacuum (or air, to good approximation). We shall usually take $k$ to be $9.0 \times 10^9$ N · m²/C².*

The coulomb is a very large quantity of charge relative to the charge on an electron. If we call the electron's charge $-e$, the experimentally determined value is

$$e = 1.60218 \times 10^{-19} \text{ C}$$

The proton has a charge $+e$. In fact, all the charged particles thus far discovered have a charge of $\pm e$ or integral multiples of this value. We therefore conclude that *there appears to be a basic unit of charge, the* **quantum of charge,** *of magnitude e.* However, we are still in doubt on this point. As explained in the accompanying note, recent theories on fundamental particles hold that many particles, such as the proton and neutron (but not the electron), consist of combinations of even more fundamental particles with charge less than $e$.

Experiment shows another interesting feature of the Coulomb force. When several charged particles exert forces on one another, the forces are additive. For example, suppose two charges are close to a third charge. Each of the two exerts a Coulomb's law force on the third, and the total force on the third is simply the vector sum of the two separate forces. We call this fact the *superposition principle* for Coulomb's law forces. Its use will become evident in some of the examples that follow.

---

* The constant $k$ is often written as $1/4\pi\varepsilon_0$, where $\varepsilon_0$, the *permittivity of free space,* has an experimentally determined value of $8.85 \times 10^{-12}$ C²/N · m².

## Example 15.1

A copper penny has a mass of about 3 g and contains about $3 \times 10^{22}$ copper atoms. Suppose two pennies are 2.0 m apart and carry equal charges $q$. (*a*) How large must $q$ be if the force of repulsion on one penny due to the other is equal to the weight of a penny? (*b*) How many electrons must be removed from a penny to give it this charge? (*c*) What fraction of the atoms have lost electrons in such a case?

**FIGURE 15.10**

As shown in Example 15.1, only a tiny fraction of the electrons need to be removed from a penny to give rise to large electric forces.

$$F \quad +q \qquad\qquad +q$$

2 m

**Reasoning** (*a*) The situation is shown in Fig. 15.10. Each penny has a mass of $3 \times 10^{-3}$ kg. The weight of a penny is $mg = 0.0294$ N. Because the diameter of a penny is much smaller than 2.0 m, we can use Coulomb's law, $F = kq_1q_2/r^2$, to give the repulsion force:

$$F = (9 \times 10^9 \ \text{N} \cdot \text{m}^2/\text{C}^2) \frac{q^2}{(2.0 \ \text{m})^2}$$

Solving for $q$ when $F = 0.0294$ N gives $q = 3.61 \times 10^{-6}$ C.

(*b*) Since each electron removed from a penny leaves an unbalanced charge of $1.60 \times 10^{-19}$ C behind, the number of electrons removed must be

$$\text{Number of electrons} = \frac{\text{charge}}{\text{charge/electron}} = \frac{3.61 \times 10^{-6} \ \text{C}}{1.60 \times 10^{-19} \ \text{C}} = 2.26 \times 10^{13}$$

(*c*) There are about $3 \times 10^{22}$ atoms in the penny. So the fraction that have lost a single electron is

$$\text{Fraction} = \frac{2.3 \times 10^{13}}{3 \times 10^{22}} = 7.7 \times 10^{-10}$$

an exceedingly small fraction of the total number of electrons. Note that charges as small as $10^{-6}$ C (a microcoulomb, $\mu$C) give rise to easily measurable forces between ordinary-sized objects. ∎

## Example 15.2

Find the force on the center charge in Fig. 15.11.

**FIGURE 15.11**

The central charge is attracted to $q_1$ by force $\mathbf{F}_1$ and to $q_3$ by force $\mathbf{F}_3$.

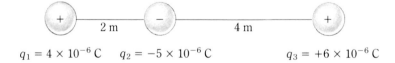

$$+ \qquad\qquad - \qquad\qquad\qquad\qquad +$$
$$\text{2 m} \qquad\qquad\qquad \text{4 m}$$

$q_1 = 4 \times 10^{-6}$ C     $q_2 = -5 \times 10^{-6}$ C     $q_3 = +6 \times 10^{-6}$ C

$$\mathbf{F}_1 \qquad\qquad \mathbf{F}_3$$

***Reasoning*** Charge $q_2$ is attracted to $q_1$ by force $F_1$ and to charge $q_3$ by $F_3$. Approximating the charges as point charges, we use Coulomb's law, $F = kq_1q_2/r^2$, to give

$$F_1 = (9 \times 10^9 \text{ N} \cdot \text{m}^2/\text{C}^2) \frac{(4 \times 10^{-6} \text{ C})(5 \times 10^{-6} \text{ C})}{(2 \text{ m})^2} = 0.0450 \text{ N}$$

$$F_3 = (9 \times 10^9 \text{ N} \cdot \text{m}^2/\text{C}^2) \frac{(5 \times 10^{-6} \text{ C})(6 \times 10^{-6} \text{ C})}{(4 \text{ m})^2} = 0.0169 \text{ N}$$

where the directions of the forces are shown in the figure. Using the superposition principle, we find the net force on $q_3$ by taking the vector sum of $\mathbf{F}_1$ and $\mathbf{F}_2$. The result is

$$F = F_1 - F_2 = 0.0281 \text{ N}$$

directed to the left. Because we were computing only the magnitudes of $\mathbf{F}_1$ and $\mathbf{F}_2$ in the above equations, we did not carry the signs of the charges in our computation. ∎

***Exercise*** Find the force on the 4-$\mu$C charge. *Answer: $3.90 \times 10^{-2}$ N* ∎

## Example 15.3
Find the resultant force on the $+10$-$\mu$C charge of Fig. 15.12.

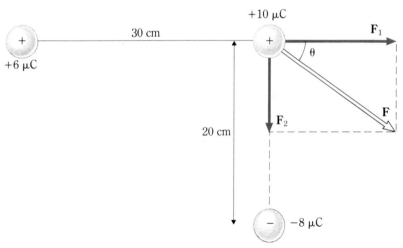

**FIGURE 15.12**
To find the resultant force on the $+10$-$\mu$C charge, we must add the forces exerted on it by the other two charges.

***Reasoning*** The $+6$-$\mu$C and $-8$-$\mu$C charges exert forces $\mathbf{F}_1$ and $\mathbf{F}_2$, respectively, on the 10-$\mu$C charge, as shown. Use Coulomb's law to show that

$$F_1 = 6.00 \text{ N} \quad \text{and} \quad F_2 = 18.0 \text{ N}$$

Because these two forces are at right angles, the resultant force $F$ is

$$F = \sqrt{F_1^2 + F_2^2} = 19.0 \text{ N} \qquad \text{and} \qquad \tan \theta = \frac{F_2}{F_1} = 3.00$$

from which

$$\theta = 71.5° \quad \blacksquare$$

### Example 15.4
Find the force on the $+20$-$\mu$C charge in Fig. 15.13.

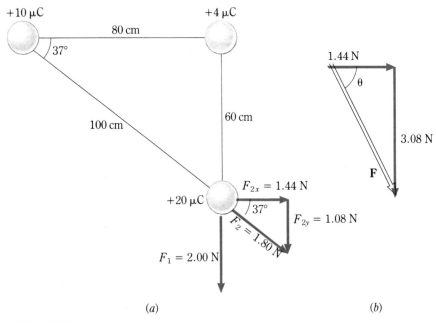

(a)                                                                  (b)

**FIGURE 15.13**
The vector forces acting on the 20-$\mu$C charge produce
the resultant force in (b).

*Reasoning*   From Coulomb's law,

$$F_1 = 2.00 \text{ N} \qquad \text{and} \qquad F_2 = 1.80 \text{ N}$$

Resolving $\mathbf{F}_2$ into components, we have $F_{2x} = F_2 \cos 37° = 1.44$ N and $F_{2y} = 1.08$ N. The total $y$ force on the charge is $F_y = 2.00 + 1.08 = 3.08$ N downward, and the total $x$ force is 1.44 N to the right (Fig. 15.13$b$). The resultant force is

$$F = \sqrt{(1.44)^2 + (3.08)^2} = 3.40 \text{ N} \qquad \text{and} \qquad \tan \theta = \frac{3.08}{1.44} = 2.14$$

from which $\theta \approx 65°$ in the direction shown.  $\blacksquare$

*Exercise*   Find the magnitude of the force on the 10-$\mu$C charge.  *Answer:*
2.27 N  $\blacksquare$

We find it convenient to discuss electric forces in terms of a concept called the *electric field.* It serves much the same purpose in electricity that the concept of the gravitational field serves in mechanics. Before we discuss this new concept in detail, let us review the more familiar situation of the gravitational field.

We are familiar with the fact that the gravitational force of the earth pulls downward on objects on and above its surface. Similarly, objects on and near the moon are pulled downward to its surface by the gravitational force of the moon. We say that a gravitational field exists in these regions. A gravitational field is said to exist in a region if an object experiences a gravitational force there. Where the gravitational force is strong, the field is strong.

It is convenient to sketch gravitational fields. That of the earth is shown in Fig. 15.14. We interpret this sketch in the following way. If an object is placed at point *A*,

**FIGURE 15.14**
The gravitational field of the earth is directed radially inward and becomes stronger as one approaches the earth.

it will experience a force in the direction of the arrowhead, toward the earth's center. The lines, called *lines of force* (or, sometimes, *field lines*), show the direction of the earth's gravitational pull; this is taken to be the direction of the gravitational field. (To be truly representational, of course, Fig. 15.14 should be drawn in three dimensions, with lines of force directed from all sides toward earth's center.)

Lines of force not only represent the direction of the force but also indicate its relative magnitude. You can see this in Fig. 15.14 by noticing that the lines are closer together near the earth, where the force is strong, than they are farther away from the earth, where the force is weaker. We shall return to this feature of field lines after we discuss the electric field.

The electric field represents the electric force a stationary positive charge experiences. Consider how you might go about determining the electric field in a region. You could simply place a charged object (call it a test charge) in the region and determine the force on it due to all other charges. However, your test charge exerts forces on all other charges in the vicinity and, if these charges are in metals, could cause them to move. To eliminate this difficulty, we imagine that the test charge has a very special property: *the* **test charge** *is a fictitious charge that exerts no forces on nearby charges.* We represent it by $q_t$. In practice, we can approximate the test charge by using a very small charge that disturbs other charges in the vicinity by only a negligible amount.

We take the direction of the electric field at a point to be the same as the direction of the force on a *positive* test charge placed at that point. For example, suppose a positive test charge is placed at point $A$ in Fig. 15.15$a$. It is attracted radially inward, as shown by the arrow at $A$. Indeed, the force on the positive test charge is directed radially inward no matter where in the neighborhood of the central negative charge it is placed. We therefore surmise that the electric field is directed as shown by the arrows: *the electric field near a negative charge is directed radially into the charge.*

We can obtain the field near a positive charge in the same way, as shown in Fig. 15.15$b$. The positive test charge is repelled radially outward by the central positive charge. Hence, *the electric field near a positive charge is directed radially away from the charge.*

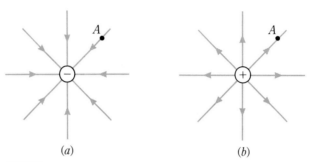

(a)           (b)

**FIGURE 15.15**
The electric field is directed radially inward toward a small negative charge and radially outward from a small positive charge.

The directed lines we have drawn in Fig. 15.15 to show the direction of the electric field are called **electric field lines** or **electric lines of force.** As we have seen, electric field lines originate on and are directed away from positive charges, and end on and are directed toward negative charges.

To make the electric field concept quantitative, we define a quantity called the **electric field strength E.** The direction of **E**, a vector quantity, is taken to be the same as that of the electric field. The magnitude of **E** is that of the force experienced by a unit positive test charge. Thus, if the electric field strength is measured using a test charge of magnitude $q_t$,

$$\mathbf{E} = \frac{\mathbf{F}}{q_t} \tag{15.3}$$

where **F** is the force of electrical origin experienced by the stationary test charge $q_t$. The units of **E** are clearly newtons per coulomb. Because **E** is a force per unit charge, we frequently state that it is the force per unit positive test charge. However, we should realize that, in measuring the strength of an electric field, we would use a charge much smaller than 1 C so as not to disturb the other charges present.

The relative strength of the electric field can be estimated by examining the field-line diagram. For example, the field lines in Fig. 15.15 are closest together near the charges. The force on a unit positive test charge (the electric field strength) is also largest close to the charges. Electric field strength is largest where the field lines are closest together. We often estimate the field strength in a region by noticing the closeness of the field lines in that region in a sketch of the electric field.

## 15.10 The electric field of a point charge

We are often interested in the electric field generated by an ion or some other charged atomic-size particle. For most purposes, we can consider these to be point charges. Even a charged sphere acts like a point charge under certain circumstances, as we point out shortly. It is therefore important for us to know the electric field due to a point charge.

Suppose we wish to find the electric field strength at point $P$ in Fig. 15.16, which is a distance $r$ away from a positive point charge $q$. We know that the electric field due to $q$ is radially outward, as we saw in Fig. 15.15$b$. Hence $\mathbf{E}$ at point $P$ is in the direction shown. If we imagine a test charge $q_t$ placed at $P$, the force on it is given by Coulomb's law:

$$F = k\frac{qq_t}{r^2}$$

Dividing through by $q_t$ to obtain $F/q_t$, the electric field strength, we have

$$\frac{F}{q_t} = k\frac{q}{r^2}$$

from which

$$E = k\frac{q}{r^2} \qquad \text{for a point charge} \tag{15.4}$$

If $q$ is positive, the electric field is directed radially outward; if $q$ is negative, the field is directed radially inward.

We can extend this relation to another important situation, the field of a uniformly charged sphere. From a large distance away, a charged sphere (let it be positive) appears as a point charge, and so the field lines it generates extend radially from it into space. Because the charge on the sphere is uniform, the lines are uniformly spaced around the sphere. And, as we approach closer to the sphere, the lines must remain uniformly spaced. Thus even close to the sphere, they remain radial and similar to those of a point charge. Hence, for a uniformly charged sphere, the field looks like that in Fig. 15.15 for a point charge. We therefore conclude that

Outside a uniformly charged sphere, the field is that of an equal point charge placed at its center.

Thus Eq. 15.4 applies to a uniformly charged sphere as well as to a point charge. Notice, however, that it applies only to the region outside the sphere.

---

### Example 15.5

Find the electric field strength 50 cm from a positive point charge of $1 \times 10^{-4}$ C.

**Reasoning**  We wish to find $\mathbf{E}$ at point $P$ in Fig. 15.16 with $r = 0.50$ m and $q = 1 \times 10^{-4}$ C. Because $q$ is positive, the test charge we mentally place at $P$ is repelled outward by $q$. Hence the direction of $\mathbf{E}$ is as shown. To find the magnitude of $\mathbf{E}$, we use Eq. 15.4:

**FIGURE 15.16**

To find the electric field $\mathbf{E}$ at point $P$, we must compute the force a unit positive test charge would experience if it were placed at that point.

$$E = k\frac{q}{r^2} = (9 \times 10^9 \text{ N} \cdot \text{m}^2/\text{C}^2)\frac{(1 \times 10^{-4} \text{ C})}{(0.50 \text{ m})^2} = 3.60 \times 10^6 \text{ N/C} \quad \blacksquare$$

*Exercise* What would the field strength at $P$ be if the point charge is a uniformly charged sphere of radius 3.0 cm? *Answer: $3.60 \times 10^6$ N/C* ∎

### Example 15.6
Find the magnitude of **E** at point $B$ in Fig. 15.17 due to the two point charges.

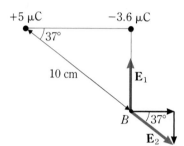

*Reasoning* A test charge at $B$ is attracted upward by the $-3.6$-$\mu$C charge. Call this portion of the field $\mathbf{E}_1$. The test charge is repelled by the $+5$-$\mu$C charge, and so the portion of the field at $B$ due to it is $\mathbf{E}_2$. Using Eq. 15.4, we find that

$$E_1 = (9 \times 10^9 \text{ N} \cdot \text{m}^2/\text{C}^2)\frac{(3.6 \times 10^{-6} \text{ C})}{(0.060 \text{ m})^2} = 9.0 \times 10^6 \text{ N/C}$$

$$E_2 = 4.5 \times 10^6 \text{ N/C}$$

These are vector quantities, forces per unit charge, and must be added as such. We see that

$$E_{1x} = 0 \quad \text{and} \quad E_{1y} = +9.0 \times 10^6 \text{ N/C}$$
$$E_{2x} = E_2 \cos 37° = 3.6 \times 10^6 \text{ N/C}$$
$$E_{2y} = -E_2 \sin 37° = -2.7 \times 10^6 \text{ N/C}$$

Therefore the resultant field at $B$ has components

$$E_x = 3.6 \times 10^6 \text{ N/C}$$
$$E_y = 9.0 \times 10^6 - 2.7 \times 10^6 = 6.3 \times 10^6 \text{ N/C}$$

Then

$$E = \sqrt{E_x^2 + E_y^2} = 7.26 \times 10^6 \text{ N/C}$$

How would you find the direction of **E**? ∎

*Exercise* Find the field components at $B$ if both charges are positive. *Answer: $3.6 \times 10^6$ N/C, $-12.0 \times 10^6$ N/C* ∎

### Example 15.7

Sparking occurs through air when the electric field strength exceeds about $3 \times 10^6$ N/C. (We call this the *electric strength* of air.) About how much charge can a 10.0-cm-diameter metal sphere hold before sparking occurs?

***Reasoning*** We know that Eq. 15.4 applies to the sphere and that $q$ is the charge on the sphere. At the sphere's surface, $E$ cannot exceed $3 \times 10^6$ N/C, and $r = 0.050$ m in this case. Because the field strength outside a uniform sphere is $kq/r^2$,

$$E = 3 \times 10^6 \text{ N/C} = (9 \times 10^9 \text{ N} \cdot \text{m}^2/\text{C}^2) \frac{q}{(0.050 \text{ m})^2}$$

from which $q = 8.3 \times 10^{-7}$ C. As we see, a sphere this size can hold a charge of about 1 $\mu$C in air. ∎

***Exercise*** What is the field strength 75 cm from the center of the sphere if its charge is 0.50 $\mu$C? *Answer: 8000 N/C* ∎

## 15.11 The electric field in various systems

We can obtain a great deal of insight into a problem by examining the pertinent electric-field drawing. To illustrate this point, let us examine several electric fields and the charge distributions that give rise to them. In doing so, we recall that we defined the electric field strength as the force per unit positive test charge. Lines of force must always be drawn out of positive charges and into negative charges.

Consider the electric field about two equal charges, one positive, the other negative. The electric field is similar about each except that it is directed into the negative charge and out of the positive charge. This field is shown in Fig. 15.18. You

**FIGURE 15.18**

The electric field lines originate on the positive charge and end on the negative one.

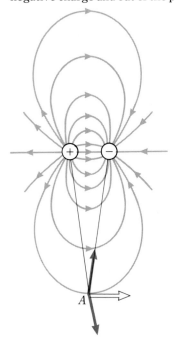

should examine several points in the figure to convince yourself that a positive test charge placed there would experience a force in the direction indicated by the force lines. To see how this is done, consider point $A$. A positive test charge at $A$ is repelled by the positive charge and attracted by the negative charge. The attractive force equals the repulsive force because the test charge is as close to the positive charge as it is to the negative charge. The resultant of these two forces is tangent to the line of force at $A$.

The field in the neighborhood of two like charges is shown in Fig. 15.19. Try to justify the way the lines are drawn. Where is the field strongest? Why is it weak midway between the charges?

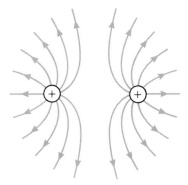

Suppose a positive, uniformly charged ball is held above a large metal plate, as in Fig. 15.20. If the plate is attached to the ground, charge may run onto it from the earth or off of it to the earth. We see at once that the electrons in the metal plate are attracted by the positive charge. Although they cannot leave the plate, they congregate at its surface just below the positive charge. Negative charge flows onto the plate from the ground in order to replace the electrons induced to take up the positions closest to the charged ball.

A positive test charge used to determine the direction of the field between ball and plate experiences forces as shown by the lines of force in Fig. 15.20. You should verify that these lines represent the force on a positive test charge. Similarly, the density of the lines indicates the field strength. As one would expect, the field is strongest close to the positive charge.

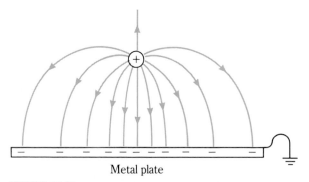

Metal plate

**FIGURE 15.20**

The positive charge attracts negative charges to the top of the metal plate. Why are the field lines perpendicular to the plate at its surface?

Two subtle but important features of Fig. 15.20 should be pointed out. First, the lines of force do not penetrate into the metal plate; all stop at the surface. This indicates that the electric field strength is zero within the metal; that is, there is no force on a charge in the metal. This is readily understood from the following reasoning. We are considering only electrostatic conditions, that is, no continuing flow of charge. If there were a field within the metal, it would exert a force on the metal's free electrons. The electrons would then move, and charge would flow; but no charges flow *under electrostatic conditions,* and hence *the electric field strength must be zero inside a metal.*

The second point has to do with the fact that the lines of force are perpendicular to the plate surface. First, we point out the lines represent the force on a positive charge. A negative charge would experience a force in the opposite direction. Thus, in Fig. 15.20, the negatively charged electrons at the plate surface experience a force trying to pull them upward. This is the result of the attraction by the positive charge suspended above the plate. However, the electrons are held within the body of the plate and cannot come loose under this force.

Suppose the force were not exactly perpendicular to the metal plate. Then there would be a component of the force along the surface. Since the electrons of the plate can move in this direction, they would flow under the action of this force, and a current would be present. However, since this is an electrostatic situation, there is no charge flow. Hence, there can be no component of the field parallel to the metal surface. We conclude that the field is directed perpendicular to the surface.

We shall use these facts later. They are quite general in any electrostatic problem. For this reason, we state them once again.

Under electrostatic conditions:

**1** The electric field strength within a metal is zero.

**2** The electric field just outside the surface of a metal is perpendicular to the surface.

Since these facts were proved by reference to the movement of free electrons within the metal, the proofs do not hold true for insulators, which contain no free electrons. Hence the rules apply only to metals and not to insulators.

---

### Example 15.8

Show by means of a diagram using lines of force that a charge suspended within a cavity in a hollow metal object induces an equal and opposite charge on the surface of the cavity.

*Reasoning*   Let us suppose the object to be a hollow metal sphere, as shown in cross section in Fig. 15.21. Lines of force come out of the positive charge $q$ suspended in the cavity. Since the field strength within the metal must be zero in electrostatic situations such as this, the lines from $q$ must stop when they hit the cavity surface. However, this means that the cavity surface must possess a negative charge, since lines of force go into and terminate only on negative charges. Moreover, since the number of lines originating on $q$ is equal to the number terminating on the negative charge, we infer that the charge on the cavity surface is $-q$.

If the metal object was neutral before $q$ was placed in the cavity and if no charge was allowed to run onto or off the object, it must still have no net charge. Therefore, a charge $+q$ must exist on the outer portions. Can you prove, using lines of force, that the charge must be on the outer surface? Can you show that the results obtained here are independent of the shape of the object and cavity? How does the proof proceed if $q$ is negative?   ∎

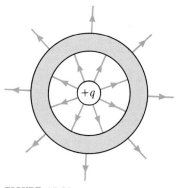

**FIGURE 15.21**

A charge $-q$ is induced on the inner surface of the hollow metal sphere. Is this true even in the absence of the high degree of symmetry assumed here?

## 15.12 Parallel metal plates

The electric field between two oppositely charged metal plates is of particular importance in electricity, as we shall see as our studies progress. We show a typical situation in Fig. 15.22a. The charges on the plates come from a battery (discussed in the next chapter). The battery gives one plate a positive charge and the other a negative charge. Because the charges attract one another, they reside for the most part on the inner surfaces of the plates. (Notice the symbol ⊣⊢ , commonly used to signify a battery.)

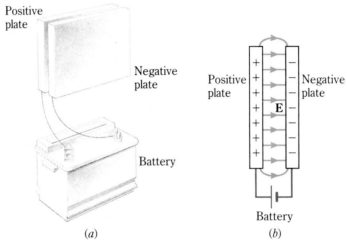

FIGURE 15.22

The battery places equal charges of opposite sign on the two metal plates.

We are interested in the electric field between the plates. A positive test charge placed in this region experiences a force from left to right because of the repulsion of the positive plate and the attraction of the negative plate. The electric field lines are therefore directed as indicated. Although some "fringing" of the lines occurs near the edges, the field lines are mostly directed straight across the gap between the plates. Moreover, as we see in the next chapter, the field is of constant strength throughout the region between the plates (indicated by the equal spacing of the lines).

Suppose a particle with positive charge $q$ and mass $m$ is placed between the plates. It experiences a force to the right because of $\mathbf{E}$. Since $\mathbf{E}$ is defined to be the force per unit positive test charge, our particle experiences a force $q$ times larger, or $q\mathbf{E}$, toward the right-hand plate. We can use this same reasoning to find the force on a charge $q$ in any electric field. Thus we arrive at the following result:

A charge $q$ in an electric field $\mathbf{E}$ experiences a force

$$\mathbf{F} = q\mathbf{E} \tag{15.3a}$$

Notice that $\mathbf{F}$ is in the same direction as $\mathbf{E}$ if $q$ is positive. Why is the direction of $\mathbf{F}$ opposite that of $\mathbf{E}$ if $q$ is negative?

## Example 15.9

The gap between the plates in Fig. 15.22 is 0.20 cm wide, and the constant field strength in the gap is 6000 N/C. A proton ($q = e$ and $m = 1.67 \times 10^{-27}$ kg) is released at the positive plate. What is its speed just before it strikes the negative plate? Assume vacuum between the plates.

**Reasoning**  The force on the proton is

$$F = qE = (1.60 \times 10^{-19} \text{ C})(6000 \text{ N/C}) = 9.60 \times 10^{-16} \text{ N}$$

It therefore experiences an acceleration

$$a = \frac{F}{m} = \frac{9.60 \times 10^{-16} \text{ N}}{1.67 \times 10^{-27} \text{ kg}} = 5.75 \times 10^{11} \text{ m/s}^2$$

To find its speed after it falls through the gap, a distance of $2.0 \times 10^{-3}$ m, we use

$$v^2 = v_o^2 + 2ax = 0 + 2(5.75 \times 10^{11} \text{ m/s}^2)(2.0 \times 10^{-3} \text{ m})$$

$$v = 48,000 \text{ m/s} \quad \blacksquare$$

**Exercise**  Suppose the proton is shot to the left from the negative plate with an initial speed of 60,000 m/s. How fast is it going just before it reaches the positive plate? *Answer: 36,000 m/s* ■

---

## Minimum learning goals

When you finish this chapter, you should be able to

1  Define (*a*) conductor, (*b*) insulator, (*c*) free electron, (*d*) electrical ground, (*e*) induced charge, (*f*) Coulomb's law, (*g*) electric field lines, (*h*) electric field strength **E**.

2  Give the magnitude and sign of the charge on the electron and proton.

3  Describe qualitatively how charges on a metal object redistribute when a charged object is brought nearby. Explain how an object can be charged by conduction and by induction.

4  State the conclusions that can be drawn from the Faraday ice-pail experiment.

5  Use Coulomb's law to find the force on a charge due to nearby point charges.

6  Find the electric field strength at a point due to several specified point charges.

7  Sketch the electric field lines in the vicinity of simple charged objects.

8  Specify the following under electrostatic conditions: (*a*) field in a metal, (*b*) origin of field lines, (*c*) termination points of field lines, (*d*) angle at which field lines strike metal surfaces.

9  Use the relation **F** = *q***E** in simple situations.

10  Find the field strength outside a uniform spherical charge.

---

## Questions and guesstimates

1  A tiny charged ball hangs from a thread. How can you tell whether the charge on the ball is positive or negative?

2  You can place a static charge on nearly any dry piece of plastic by rubbing it with a piece of fabric, fur, or plastic wrap. How can you determine the sign of the charge placed on the plastic?

3  Static electricity produces sparks that can cause some volatile gases to explode. This used to be a real danger in hospital operating rooms because the anesthetic that was then used, ether, is combustible. What measures can be taken to minimize this danger?

4  The electric strength of air is about $3 \times 10^6$ N/C.

That is, a spark will jump through the air if the electric field strength exceeds this value. Why do sparks jump preferentially from sharp metal points and edges? When your body becomes highly charged as you walk across a deep-pile carpet in dry weather, why does a spark jump from your fingernail to a metal object, such as a stove or doorknob?

**5** Clothes often cling together when they are removed from the dryer. Why? What is often done to eliminate this effect?

**6** Never try to wipe the dust off a phonograph record with an ordinary cotton or wool cloth. Why?

**7** In dry climates, one frequently sees (or hears) sparks jump when hair is combed or when clothes are removed in darkness. Why?

**8** Two equal-magnitude positive point charges are a distance $D$ apart. Where can you place a third charge so that

the resultant force on it is zero? Is it in stable equilibrium there?

**9** A positive point charge and a much larger negative point charge are a distance $D$ apart. Is there any place a third point charge can be placed where the resultant force on it is zero?

**10** A tiny ball with charge $q$ is suspended between two very large parallel metal plates that are grounded. Sketch the electric field between the plates. What can you infer about the induced charges on the plates?

**11** Properly drawn electric field lines never cross each other. Why?

**12** Sensitive apparatus is frequently shielded from unwanted electric fields by placing it inside a metal can or a fine-mesh wire box that is grounded. Explain why the field of a charge placed outside such a shield does not affect the interior region.

## Problems

**1** Two point charges, $q_1 = +5.0\,\mu C$ and $q_2 = -3.0\,\mu C$, are 80 cm apart. Find the magnitude and direction of the force on both.

**2** Two protons are brought to a distance of $3 \times 10^{-14}$ m from each other. (*a*) Find the magnitude and direction of the force on both. (*b*) What is the ratio of the magnitude of this force to the weight of a proton on earth? The proton can be considered a point charge. Its mass is $1.67 \times 10^{-27}$ kg.

**3** Two point charges are placed on the $x$ axis: $+5.0\,\mu C$ at $x = 0$ m and $+7.0\,\mu C$ at $x = +25$ cm. Find the magnitude and direction of the force on the 7.0-$\mu C$ charge.

**4** A point charge of $+1.60\,\mu C$ is placed on the $x$ axis at $x = 20$ cm, and an unknown charge $q$ is placed at $x = 50$ cm on the axis. The force on the 1.60-$\mu C$ charge is 0.80 N in the positive $x$ direction. What are the magnitude and sign of $q$?

**5** Two point charges $q_1$ and $q_2$ are 50 cm apart, and they repel each other with a force of 0.30 N. The algebraic sum of the two charges is $+6.2\,\mu C$. Find $q_1$ and $q_2$.

**6** Repeat Prob. 5 for the two charges attracting rather than repelling.

**7** Two identical 200-g balls of diameter 3.0 cm are 5.0 m apart. Each carries a charge of $+6.0\,\mu C$. One of the balls is released. Find its acceleration. Neglect gravitation.

**8** Two identical pointlike balls, each with a mass of 50 g, are 200 cm apart. They carry equal charges $q$. How large is $q$ if the electrostatic repulsion between the balls equals their gravitational attraction?

**9** A medium-size person contains about $3 \times 10^{28}$ protons and a like number of electrons. Suppose two people are 50 m apart and 0.10 percent of the electrons on one person are transferred to the other. What force is required to hold the two apart at this distance?

**10** In the Bohr model of the hydrogen atom, an electron orbits a stationary proton at a radius of $0.53 \times 10^{-10}$ m. (*a*) How large a force does the proton exert on the orbiting electron? This force holds the electron in orbit. (*b*) How fast is the electron moving? ($m_e = 9.1 \times 10^{-31}$ kg)

**11** The following three point charges are placed on the $x$ axis: $+4\,\mu C$ at $x = 0$, $+3\,\mu C$ at $x = 30$ cm, and $+5\,\mu C$ at $x = 60$ cm. Find the force on (*a*) the 5-$\mu C$ charge and (*b*) the 3-$\mu C$ charge.

**12** Three point charges with values $-3$, $+4$, and $-6\,\mu C$ are placed at $x = 0$, $x = 40$, and $x = 120$ cm, respectively. Find the force on (*a*) the $-3$-$\mu C$ charge and (*b*) the $+4$-$\mu C$ charge.

**13** Two point charges are placed on the $x$ axis: a $-4$-$\mu C$ charge at $x = 0$ and a $+3$-$\mu C$ charge at $x = 80$ cm. At what point(s) in the vicinity of these two charges can a $+5$-$\mu C$ charge be placed so that it experiences no resultant force?

**14** Two point charges are placed on the $x$ axis: a $+6$-$\mu C$ charge at $x = 0$ and a $+5$-$\mu C$ charge at $x = 200$ cm. At what point(s) in the vicinity of these two charges is the resultant force on a $+8$-$\mu C$ charge zero?

**15** Two point charges are placed on the $x$ axis: $+5.0$-$\mu C$ charge at $x = 0$ and an $+8.0$-$\mu C$ charge at $x = 90$ cm.

Where on the $x$ axis can a third charge be placed so that the net force on all three charges is zero? Evaluate the third charge.

■ 16 A 4.0-$\mu$C point charge is placed at the coordinate origin. Two other point charges are placed on the $x$ axis: $q_1$ at $x = 30$ cm and $q_2$ at $x = 50$ cm. Find the magnitude and sign of $q_1$ and $q_2$ if the net force on all three charges is zero.

■ 17 The four point charges in Fig. P15.1 are each $+3.0$ $\mu$C. Find the magnitude and direction of the force on $q_2$ due to the other charges. Use $a = 30$ cm and $b = 60$ cm.

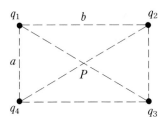

FIGURE P15.1

■ 18 The three point charges in Fig. P15.2 are each $7.0\,\mu$C. Find the magnitude and direction of the force on $q_3$ due to the other charges. Use $a = 50$ cm.

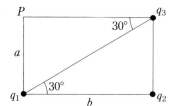

FIGURE P15.2

■ 19 In Fig. P15.2, $q_1 = q_2 = +6.0\,\mu$C and $q_3 = -8.0\,\mu$C. Find the magnitude and direction of the force on $q_1$. Use $a = 2.0$ m.

■ 20 In Fig. P15.1, $q_1 = q_2 = +6.0\ \mu$C and $q_3 = q_4 = -8.0\,\mu$C. Find the magnitude and direction of the force on $q_1$. Use $a = 30$ cm and $b = 60$ cm.

■ 21 Two balls hang from a single support, as shown in Fig. P15.3. Each has a mass of 0.50 g and carries a charge $q$. The length of each string is 35 cm, and the balls come to equilbrium with $\theta = 40°$. Find the charge on each ball.

■ 22 Repeat Prob. 21 if the balls carry unequal charges, with the ball on the left having one-third the charge of that on the right.

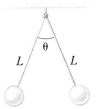

FIGURE P15.3

■ 23 Two point charges $q_1$ and $q_2$ 90 cm apart exert a force of 0.050 N on each other. It is known that $q_1 + q_2 = 4.5$ $\mu$C. Find the magnitudes of $q_1$ and $q_2$. Is the force attractive or repulsive?

■ 24 Radium nuclei are radioactive and emit alpha particles ($m = 4 \times 1.66 \times 10^{-27}$ kg, $q = +2e$). The nucleus left behind has a charge of $+86e$ and a very large mass. Find (a) the force exerted on the alpha particle by the nucleus when they are $5 \times 10^{-14}$ m apart and (b) the acceleration of the alpha particle at that instant.

25 A tiny ball carrying a charge of $-3.0 \times 10^{-12}$ C experiences an eastward force of $8.0 \times 10^{-7}$ N due to its charge when it is suspended at a certain point in space. What are the magnitude and direction of **E** at that point?

26 The electric field in a certain region is directed eastward and has a strength of 3000 N/C. Find the magnitude and direction of the force experienced by a $-8.0 \times 10^{-5}$ C charge placed in this region.

■ 27 An electron is released in a region where the electric field is in the positive $x$ direction and has a strength of 4000 N/C. Find the magnitude and direction of the electron's acceleration. ($m_e = 9.11 \times 10^{-31}$ kg)

28 A tiny drop has mass $m$ and charge $+q$. It is to be supported against gravity by an electric field. Find the magnitude and direction of **E** in terms of $q$ and $m$.

■ 29 A 0.200-g sphere is suspended by a thread in an electric field of 5000 N/C that is directed straight upward. The tension in the thread is $2.16 \times 10^{-3}$ N. Find the charge on the sphere.

■ 30 In Fig. P15.4, the electric field causes the ball of mass $m$ and charge $q$ to hang at the angle shown. Find **E** in terms of $m$ and $q$.

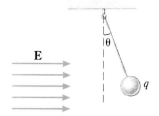

FIGURE P15.4

31 Find the electric field strength due to a point charge $q = -5.0\ \mu$C at a distance of 80 cm from the charge. Is the field directed radially outward or inward?

32 Two charges are placed on the $x$ axis: a $+6$-$\mu$C charge at $x = 80$ cm and a $-3$-$\mu$C charge at $x = 0$. Find **E** at (a) $x = 30$ cm and (b) $x = 100$ cm.

■ **33** Find **E** at the center of the rectangle in Fig. P15.1 if (*a*) $q_1 = q_2 = q_3 = q_4$ and (*b*) $q_1 = q_2 = 2\ \mu C$ and $q_3 = q_4 = -4\ \mu C$. Take $a = 30$ cm and $b = 60$ cm.

■ **34** If $q_1 = q_3 = -6.4\ \mu C$ and $q_2 = 2\ \mu C$ in Fig. P15.2, find the electric field strength at *P*. Take $a = 30$ cm.

■ **35** Two charges are placed on the *x* axis: a $+20$-$\mu C$ charge at $x = 50$ cm and a $-30$-$\mu C$ charge at $x = 80$ cm. Where, if anywhere, is $E = 0$ in this region?

■ **36** An isolated hollow metal sphere of radius 20 cm carries a charge of $-5.0\ \mu C$. What is the magnitude of **E** (*a*) in the empty region inside the sphere and (*b*) 30 cm from the center of the sphere?

■ **37** A proton traveling along the *x* axis is slowed by a uniform electric field **E**. At $x = 20$ cm, the proton has a speed of $3.5 \times 10^6$ m/s, and at $x = 80$ cm, its speed is zero. Find the magnitude and direction of **E**. ($m_p = 1.67 \times 10^{-27}$ kg)

■ **38** At a certain instant, an electron is traveling out from the origin along the *x* axis with a speed of $7.0 \times 10^6$ m/s. An electric field **E** along the *x* axis causes the electron to slow, stop, reverse its direction, and return to the same spot in 35.0 $\mu s$. What are the magnitude and direction of **E**? ($m_e = 9.11 \times 10^{-31}$ kg)

■ **39** An electron is shot from the coordinate origin out along the positive *x* axis with a speed $v_{xo}$. There is a *y*-directed electric field **E** in this region. (*a*) Show that the *y* coordinate of the electron a time *t* later is $y = -eEt^2/2m_e$. (*b*) Show that the electron's *x* and *y* coordinates are related through $y = (eE/2m_e v_{xo}^2)(x^2)$.

# 16 Electrical potential

**I**N our study of mechanics, we found the concepts of work and potential energy to be of great utility. Even though many situations were too involved to be solved in detail by using forces, the energy approach often allowed us to obtain useful results quickly. We see in this chapter that the concept of electrical potential energy is extremely useful in applications of electricity. It is indispensable for an understanding of such widely diverse topics as electric circuits and nuclear accelerators.

## 16.1 Electrical potential energy

When we discussed the movement of an object from one place to another within a gravitational field, we used the concept of gravitational potential energy. To lift an object of mass $m$, we must apply an upward force $mg$ to it in order to balance the downward pull of gravity. The work done in lifting the object through a distance $h$ is simply force times distance, which is $mgh$. Because of this lifting work done on the object, a like amount of gravitational potential energy is stored in it. When the object is allowed to fall freely through the height $h$, it acquires kinetic energy, and from the law of energy conservation, we can write

GPE at height $h$ = KE gained in falling through $h$

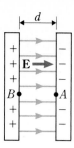

FIGURE 16.1

The electric field between the two oppositely charged parallel plates is uniform.

FIGURE 16.2

A force $\mathbf{F} = -\mathbf{E}q$ is required if the object with charge $q$ is to be moved from $A$ to $B$ in Fig. 16.1.

We made extensive use of gravitational potential energy and its interconversion with kinetic energy in our study of mechanics.

A similar situation exists in electricity. Charged objects often possess electrical potential energy that can be transformed to kinetic energy. To show this, consider a charged object between two charged parallel plates. (We ignore the gravitational force in this discussion because it is negligibly small compared with the electric forces we are concerned with.) The electric field in the central region between the plates is shown in Fig. 16.1; it has a constant value $\mathbf{E}$ and is directed as shown. Figure 16.2 shows the forces acting on a charged object between the plates. Because of the electric field, the object with charge $q$ experiences a force $\mathbf{E}q$ directed to the right. If we are to hold the charged object in place, we must exert a force $\mathbf{F} = -\mathbf{E}q$ on it.

Suppose the charged object (which is much tinier than shown) is originally at point $A$ in Fig. 16.1. If we are to move it to point $B$, we must pull it the entire way with the force $\mathbf{F}$. Hence we do work on the object as we pull it from $A$ to $B$. As a result, we store energy in it, electrical potential energy. After we get the object to point $B$, we can release it and recover this potential energy in the form of kinetic energy. The charged object at $B$ will be pulled toward $A$ by the (now unbalanced) force $\mathbf{E}q$ that acts on it. Therefore, when it is released at $B$, the object accelerates toward $A$.

We see, therefore, that it is possible to store electrical potential energy in a charged object. Just as we store gravitational potential energy in an object by lifting it against the pull of gravity, we store electrical potential energy in a charged object by pulling it against the electric-field force that acts on it. Remember that only differences in gravitational potential energy are important, and so we defined the gravitational potential energy of an object at one position relative to another. The gravitational potential energy of an object at point $B$ relative to point $A$ is the work one must do against gravity to move the object from $A$ to $B$. Similarly, in electricity we define the electrical potential energy of a charged object at a point $B$ relative to another point $A$:

The electrical potential energy of a charge at point $B$ relative to point $A$ is equal to the work against electric forces required to move the charge from $A$ to $B$.

As the charge is allowed to move from $B$ to $A$, the electrical potential energy stored in it is transformed to kinetic energy.

Our discussion thus far has been in terms of positively charged objects. Consider what would happen to a negatively charged object placed between the plates of Fig. 16.1. It would be attracted by the positive plate and repelled by the negative one. Hence a negative charge would move from $A$ to $B$ rather than from $B$ to $A$. Therefore a negatively charged object has more electrical potential energy at point $A$ than at point $B$; this is just the opposite of the situation for positively charged objects.

In electricity, we go one step further than we did in mechanics by making extensive use of the potential energy per unit charge. As you will see, this quantity is closely related to what is colloquially referred to as *voltage*.

Suppose we pull a positive charge $q$ from point $A$ to point $B$ in Fig. 16.1. This requires a certain amount of work on our part. We define the **potential difference** (or voltage) between $A$ and $B$ as that work divided by the charge $q$. In other words, the potential difference between two points is the work one must do to carry a $+1$-C test charge from one point to the other. In effect, it is the potential energy difference between the two points for a unit positive test charge. To summarize:

The potential difference from point $A$ to point $B$ is the work required to carry a unit positive test charge from $A$ to $B$; it is work per unit positive charge.

The unit for potential difference (or voltage) is work divided by charge, which is joules per coulomb; we call this unit the **volt** (V) and represent the potential difference from $A$ to $B$ as $V_{AB}$. Often, when there is no ambiguity as to the two points involved, we drop the subscripts and represent the potential difference by $V$.

Let us now compute the potential difference between points $A$ and $B$ in Fig. 16.1. To pull a charge $q$ through the distance $d$ from $A$ to $B$, we have to pull with a constant force of magnitude $Eq$, as we saw in Fig. 16.2. Thus the work needed to move a charge $q$ from $A$ to $B$ is

Work to move $q$ = force $\times$ distance $= (Eq)(d)$

The potential difference from $A$ to $B$ is work per unit charge, the work we have computed divided by $q$. Therefore

$$V_{AB} = Ed \qquad \text{(constant field)} \tag{16.1}$$

Note that this relation applies only to the special case shown in Fig. 16.1, a situation in which the electric field is constant. In more complicated cases in which $\mathbf{E}$ is not constant, we must return to the definition of potential difference:

$V_{AB}$ is the work one must do to carry a unit positive test charge from $A$ to $B$.

We can use this definition to obtain a simple relation between work and potential difference. Because $V_{AB}$ is the work per unit charge, the work for a charge $q$ is $q V_{AB}$. To carry a charge $q$ from $A$ to $B$ requires

$$\text{Work} = q V_{AB} \tag{16.2}$$

As discussed earlier, this work is stored as electrical potential energy.

---

### Example 16.1
Suppose the potential difference between the two plates in Fig. 16.1 is 12.0 V. What is the magnitude of the electric field between the two plates if their separation is 0.50 cm?

***Reasoning*** We know that $V_{AB} = 12.0$ V and $d = 0.0050$ m. Using Eq. 16.1 for the potential difference between parallel plates, we have

$$V_{AB} = Ed$$

$$12.0 \text{ V} = E(5.0 \times 10^{-3} \text{ m})$$

from which $E = 2400$ V/m. We have previously expressed fields in units of newtons per coulomb. You should be able to show that volts per meter are equivalent to newtons per coulomb. ∎

### Example 16.2

If a proton ($q = e$, $m = 1.67 \times 10^{-27}$ kg) is released from point $B$ in Fig. 16.1, what is its speed just before it strikes the plate at $A$? Take $V_{AB} = 45$ V.

***Reasoning*** When a proton, which is a positive charge, is released from $B$, it accelerates toward $A$. The electrical potential energy it loses is changed to kinetic energy. The change in potential energy is equal to the work done in moving the charge from $A$ to $B$ in the first place, namely $qV_{AB}$. Therefore

$$\text{Change in KE} = \text{change in PE}$$

$$\tfrac{1}{2}mv_A^2 = qV_{AB}$$

$$\tfrac{1}{2}(1.67 \times 10^{-27} \text{ kg}) (v_A^2) = (1.60 \times 10^{-19} \text{ C})(45 \text{ V})$$

which gives $v_A = 9.29 \times 10^4$ m/s. ∎

### Example 16.3

Find the speed of an electron ($q = -e$, $m = 9.11 \times 10^{-31}$ kg) just as it reaches $B$ in Fig. 16.1 after being released from $A$. Assume $V_{AB} = 45$ V.

***Reasoning*** Notice that, unlike the positive proton, which moves from $B$ to $A$, the negative electron moves from $A$ to $B$ because it is attracted by the positive plate and repelled by the negative plate. We have

$$\text{Change in KE} = \text{change in PE}$$

$$\tfrac{1}{2}mv_B^2 = eV_{AB}$$

$$\tfrac{1}{2}(9.11 \times 10^{-31} \text{ kg})(v_B^2) = (1.60 \times 10^{-19} \text{ C})(45 \text{ V})$$

which gives $v_B = 3.98 \times 10^6$ m/s. Notice that the electron moves much faster than the proton of Example 16.2. Why? ∎

***Exercise*** With what speed must an electron be shot from $B$ toward $A$ if it is to just reach $A$ before stopping? *Answer: $3.98 \times 10^6$ m/s* ∎

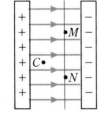

**FIGURE 16.3**
Points $M$ and $N$ are on an equipotential line.

## 16.3 Equipotentials

Let us now look at points other than $A$ and $B$ in the region between two charged plates. For example, we might ask for the potential difference between points $M$ and $N$ in Fig. 16.3. Because potential difference is simply work per unit charge, we must find the work required to move a unit positive test charge from $M$ to $N$. Note that to hold the test charge in place, we must exert a force to the left on it. This force is needed to counterbalance the effect of the electric field on the test charge. If we move the charge from $M$ to $N$, our balancing force does no work since the direction of motion is perpendicular to the force. Indeed, we see that no work is ever needed to move the test charge in a direction perpendicular to the electric field. Therefore, there is no potential difference between points $M$ and $N$ in Fig. 16.3. In fact, it should be clear that all points on the line passing through $M$ and $N$ are at the same potential; there is no potential difference between them. We call this line of constant potential an **equipotential line.** Moreover, the plane that lies through this line and is parallel to the plates is a constant-potential plane, which we call an equipotential plane. No work is done in moving a charge along an equipotential line or equipotential plane since such motion is always perpendicular to the lines of force, that is, the electric field. Conversely, *lines of force are always perpendicular to equipotential lines.*

There are an infinite number of equipotential lines and planes between the plates in Fig. 16.3. Can you show that no total work is done on a charge in moving it along a path that begins and ends on the same equipotential line or plane? The same reasoning can be used to prove that the work done in carrying a charge from $M$ to a point such as $C$ in Fig. 16.3 is independent of the path taken. We therefore conclude that *the static electric field is a conservative field. The electrical potential difference between two points is a constant independent of the path used for its computation.*

Before leaving our discussion of equipotentials, we should point out that *metal objects are equipotential volumes under electrostatic conditions.* For example, suppose you are concerned with a solid piece of metal of any shape. We know that under electrostatic conditions, the electric field everywhere within it is zero. As a result, no force exists on a test charge within the metal, and so no work is required to move the test charge from one place to another. Hence there can be no potential difference between any two points within the metal; the metal constitutes an equipotential volume.

---

### Example 16.4

Sketch the equipotentials and electric field lines near a charged solid metal object.

*Reasoning* Consider the charged metal object shown in cross section in Fig. 16.4. The object is an equipotential volume, and so its surface is an equipotential surface. Because lines of force must be perpendicular to equipotential lines and

FIGURE 16.4

The equipotentials are perpendicular to the field lines.

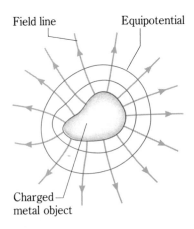

Field line    Equipotential

Charged metal object

surfaces, the electric field lines must be perpendicular to the object's surface. In addition, the equipotentials must be perpendicular to the field lines. ∎

*Exercise* Suppose the object in Fig. 16.4 is viewed from a great distance, so that it appears pointlike. Draw the equipotentials and field lines as they are now observed. *Answer: The field lines are radial, and the equipotentials are circles.* ∎

## 16.4 Batteries as sources of energy

One of the easiest ways to supply a potential difference between two points is by using a battery. There are many kinds of batteries. Although most are essentially chemical devices, other types are now becoming common. The lead-cell battery in an automobile uses a chemical reaction to supply energy. This is likewise true of the "dry cell," which is not dry inside despite its name. Perhaps you have heard of solar cells, which are used to supply energy to solar-powered watches and hand calculators as well as for more exotic purposes. Solar batteries, which operate on quite different principles from chemical batteries, transform light directly into electric energy. Other types of nonchemical batteries are currently being developed. Despite this diversity, the purpose of any battery is to supply electric energy.

Every simple battery has two terminals (metal posts) that provide a means for connecting wires to the battery. The quantity we commonly call the voltage of a battery is the potential difference between its two terminals, typically 1.5 V for a dry cell and 12 V for an automobile battery. When the battery terminals are connected by wires to two metal plates, as in Fig. 16.5, one plate acquires a positive

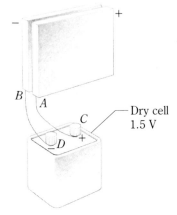

FIGURE 16.5

The potential difference from *B* to *A* is 1.5 V, the emf of the battery. Terminal *C* is positive and 1.5 V higher in potential than *D*.

charge and the other acquires an equal-magnitude negative charge. The terminal that supplies the positive charge is said to be positive, and the other is said to be negative. As noted in Sec. 15.12, the symbol used for a battery is ──+│┠── , where the longer vertical line marked "+" is the positive terminal and the shorter vertical line marked "−" is the negative terminal. Usually the plus and minus signs are left off the symbol, and you are expected to know that the longer line represents the positive terminal. Often the positive terminal of a battery is stamped with a plus sign or painted red.

The potential difference between the terminals of a battery depends somewhat on whether or not charge is flowing from the battery. Its potential difference when no charge is flowing is called the **electromotive force (emf)** of the battery. For many purposes, the emf of a battery and the potential difference between its terminals, even when charge is flowing from it, can be considered to be the same. We denote emf by the symbol $\mathcal{E}$. Do not confuse it with the symbol $E$ used for electric field strength.

Let us examine the situation in Fig. 16.5 in more detail. When the originally uncharged metal plates are attached to the battery by metal wires, charge flows for a tiny instant as the battery establishes the charges on the plates. Thereafter, no charge flows and the situation is electrostatic. You will recall that metals are equipotential volumes under electrostatic conditions. Hence the wire from terminal $C$ to plate $A$ and the plate are at the same potential. Similarly, terminal $D$, which is at a potential 1.5 V lower than $C$, is at the same potential as plate $B$. Therefore, the potential difference between plates $A$ and $B$ is 1.5 V, with plate $A$ being at the higher potential because it is positive. We thus conclude that, under electrostatic conditions, *the potential difference between a metal object connected to one battery terminal and another metal object connected to the other terminal is equal to the terminal potential difference of the battery.*

We saw in Sec. 16.2 that the charges on charged plates have electrical potential energy. Because the plates in Fig. 16.5 acquire their charge from the battery, the battery is the source of the energy that the charges on the plates possess. This is but one of many ways in which a battery acts as an energy source. When a flashlight battery lights a bulb, the heat and light energy that the bulb gives off is furnished by the battery. When a battery causes a motor to run, the mechanical energy output of the motor is furnished by the battery. As our study of electricity progresses, we shall learn of still other sources of electric energy.

### Example 16.5

(*a*) How much work is done by an external force in carrying a proton from the negative terminal of a 9.0-V battery to the positive terminal? (*b*) Repeat for an electron.

*Reasoning*   The potential difference from the negative terminal, point $A$, to the positive terminal, point $B$, is $V_{AB} = +9.0$ V. The potential difference from $A$ to $B$ is positive because the positive terminal ($B$) is at the higher potential; a positive test charge is repelled by $B$ and attracted by $A$.

(*a*) For a proton, $q = 1.60 \times 10^{-19}$ C, and so

$$\text{Work}_{AB} = qV_{AB} = (1.60 \times 10^{-19} \text{ C})(9.0 \text{ V}) = 1.44 \times 10^{-18} \text{ J}$$

(*b*) For an electron, $q = -1.60 \times 10^{-19}$ C, and so

$$\text{Work}_{AB} = qV_{AB} = (-1.60 \times 10^{-19} \text{ C})(9.0 \text{ V}) = -1.44 \times 10^{-18} \text{ J}$$

This work is negative because the electron, a negatively charged particle, is attracted to point $B$. We had a similar situation in mechanics when we lowered an object toward the earth, in which case the force supporting the object did negative work. ∎

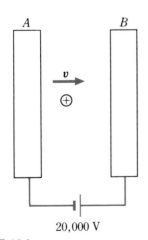

*Exercise*   How much work is required to move a proton from the positive to the negative terminal? *Answer: $-1.44 \times 10^{-18}$ J* ∎

FIGURE 16.6

Will the proton speed up or slow down as it moves toward plate $B$?

### Example 16.6

The proton ($q = 1.60 \times 10^{-19}$ C, $m = 1.67 \times 10^{-27}$ kg) in Fig. 16.6 is shot from plate $A$ toward plate $B$. A 20,000-V battery is attached to the plates. If the proton's initial speed at $A$ is $8 \times 10^6$ m/s, what is its speed just before it strikes $B$?

*Reasoning*   Plate *B* is positive (why?) and thus repels the proton. Therefore, if it is to move from *A* to *B*, the proton must do work equal to $qV_{AB}$. During the process, it loses an equal amount of kinetic energy. We can therefore write, from energy conservation,

Change in KE + change in PE = 0

$$(\tfrac{1}{2}mv_B^2 - \tfrac{1}{2}mv_A^2) + qV_{AB} = 0$$

$$v_B^2 = v_A^2 - \frac{2qV_{AB}}{m}$$

which yields $v_B = 7.76 \times 10^6$ m/s. Note that $v_B$ is less than $v_A$, as it must be.   ∎

*Exercise*   What must the potential difference between the plates be if the proton is to stop just before it reaches *B*?   *Answer: 3.34 × 10⁵ V*   ∎

---

## 16.5  The electronvolt energy unit

The SI unit for energy is the joule. In atomic and nuclear physics, however, there is another unit for energy that is so widely used that we must be familiar with it. This unit is defined in terms of the energy a charge of magnitude *e* gains as it moves through a potential difference of one volt:

One electronvolt (eV) is the energy acquired by a charge of magnitude *e* as it moves through a potential difference of one volt.

To see how the electronvolt is related to the joule, we recall that the energy a charge *q* acquires as it moves through a potential difference $V_{AB}$ is

Energy in joules = $qV_{AB}$

This energy is 1 eV when $q = e = 1.602 \times 10^{-19}$ C and $V_{AB} = 1$ V. Therefore

$1 \text{ eV} = (1.602 \times 10^{-19} \text{ C})(1 \text{ V}) = 1.602 \times 10^{-19} \text{ J}$

You can easily see the convenience of this unit if we consider a charge *q* that moves through a potential difference $V_{AB}$. Its energy is

Energy in joules = $qV_{AB}$

or, using the conversion factor $1.602 \times 10^{-19}$ J/eV,

$$\text{Energy in eV} = \left( \frac{q}{1.602 \times 10^{-19} \text{ J/eV}} \right)(V_{AB}) \tag{16.3}$$

In atomic and nuclear physics, the particles carry charges that are integral multiples of $1.602 \times 10^{-19}$ C, and so the ratio in parentheses is unity or some other small integer.

When a proton moves through a potential difference of 1000 V, say, its energy is, from Eq. 16.3,

$$\text{Energy} = \left(\frac{1.602 \times 10^{-19}}{1.602 \times 10^{-19}}\right)(1000)\ \text{eV}$$

$$= 1000\ \text{eV}$$

Similarly, if a particle that has a charge $3e$ moves through 1000 V, the energy it acquires is $3 \times 1000 = 3000$ eV. Even though the electronvolt cannot be used in our SI-based equations, its convenience in other respects has established it firmly in science.

---

***Example 16.7***
To tear the single electron loose from a hydrogen atom requires an energy of 13.6 eV. Suppose we wish to knock an electron loose by bombarding hydrogen atoms with protons that have been accelerated through a potential difference $V_{AB}$. What is the minimum value for $V_{AB}$ needed?

***Reasoning*** Each proton must have an energy of at least 13.6 eV. Since each has a charge of $1.602 \times 10^{-19}$ C, its energy in electronvolts is numerically equal to the potential difference through which it moves. Hence the required potential difference is 13.6 V. ∎

---

***Exercise*** Repeat if the bombarding particles are ions that have a charge $3e$.
*Answer: 4.53 V* ∎

---

## 16.6 Absolute potentials

So far, we have been concerned only with differences in potential because, as in gravitational potential, the choice of a position for zero potential energy is merely a matter of convenience. Gravitational potential energy can be measured with respect to any point we choose: a tabletop, the ground, the top of a building, or wherever. Similarly, in electrical potential energy problems, the zero potential energy location is a matter of choice. In electric circuit theory, a particular wire in the circuit may be attached to the ground (perhaps connected to a water pipe). This point is usually taken to have zero potential energy. However, a different zero for electrical potential is frequently taken, as we shall now see.

When dealing with atoms and molecules, we frequently specify the zero of electrical potential energy in a different way. To illustrate, let us refer to Fig. 16.7, in

**FIGURE 16.7**

We define the absolute potential at $B$ to be the work done in carrying a unit positive test charge from infinity up to $B$.

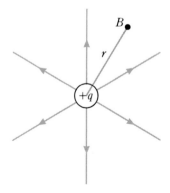

which we see a sphere carrying charge $+q$. This sphere might represent a proton, for example. It is customary in such cases to take the zero of electrical potential energy to be an infinite distance away from the charge. Let us now see the consequences of such a choice.

Consider the point $B$ in Fig. 16.7, which is at a radius $r$ from the center of the charge.

The **absolute potential** at point $B$ is the work done in carrying a unit positive test charge from infinity up to $B$.

In effect, what we are doing is as follows. So far, we have discussed situations in terms of potential differences $V_{AB}$. Now, however, we specify that point $A$ is to be taken at infinity. Further, we specify that the potential at infinity is to be taken as zero, so that the potential at point $B$ becomes what we refer to as the absolute potential at $B$. Note carefully that *when we speak of the absolute potential at a point, we are really speaking about the potential difference between that point and infinity.*

A particularly simple expression can be found for the absolute potential outside a single isolated charge, such as the one shown in Fig. 16.7. We need to compute the work done in carrying our unit test charge from infinity up to point $B$, for example. This computation is not as simple as finding the potential difference between two parallel plates. In the parallel-plate case, $E$ is constant and so the force on the test charge is constant. Hence the potential difference is simply $Ed$.

In Fig. 16.7, however, $E$ is zero at infinity and gets stronger and stronger as we come closer to $B$. Since the value of $E$ changes, so does the force on our test charge as we bring it in from infinity. In spite of that fact, we can easily find the work done in bringing the test charge from infinity to point $B$ by using calculus. We find that the absolute potential ($V$) at point $B$ in Fig. 16.7 due to the charge $+q$ (in other words, the work needed to carry a unit positive test charge from infinity up to $B$) is

$$V = k\frac{q}{r} \cong (9 \times 10^9 \text{ N} \cdot \text{m}^2/\text{C}^2)\left(\frac{q}{r}\right) \tag{16.4}$$

where $k$ is the same constant that appears in Coulomb's law. This equation, like Coulomb's law, *applies only to point charges and uniformly charged spheres.* In other cases, the distance $r$ is not usually easily determined.

---

*Example 16.8*
Suppose in Fig. 16.7 that $r = 50$ cm and $q = 5 \times 10^{-6}$ C. If a proton is released at point $B$, how fast will it be moving when it gets far away?

*Reasoning*  The proton starts at point $B$, where the absolute potential is

$$V = k\frac{q}{r} = (9 \times 10^9 \text{ N} \cdot \text{m}^2/\text{C}^2)\left(\frac{5 \times 10^{-6} \text{ C}}{0.50 \text{ m}}\right) = 90{,}000 \text{ V}$$

It is then repelled to infinity, where its absolute potential is zero. Hence it moves through a potential difference of 90,000 V. It loses $(90{,}000)(1.6 \times 10^{-19})$ J of potential energy and gains a like amount of kinetic energy. Therefore

$$\tfrac{1}{2}mv^2 = (90{,}000 \text{ V})(1.60 \times 10^{-19} \text{ C})$$

from which $v = 4.15 \times 10^6$ m/s.  ∎

## Example 16.9

Compute the absolute potential at point $B$ in the vicinity of the three point charges in Fig. 16.8.

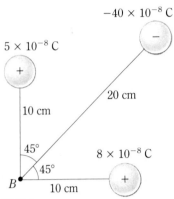

**FIGURE 16.8**

Find the absolute potential at point $B$.

*Reasoning* When a unit positive test charge is carried from infinity to $B$, work is done against the repulsions of the two positive charges, and the negative charge tends to pull the positive test charge in toward $B$. Hence, the work done per unit charge consists of two positive amounts of work (those done against the repulsions of the two positive charges) and a negative amount (the work done *on* the test charge by the attraction of the negative charge):

Due to $+5 \times 10^{-8}$:  $V_1 = (9 \times 10^9) \left( \dfrac{5 \times 10^{-8}}{0.10} \right) = 4500$ V

Due to $+8 \times 10^{-8}$:  $V_2 = (9 \times 10^9) \left( \dfrac{8 \times 10^{-8}}{0.10} \right) = 7200$ V

Due to $-40 \times 10^{-8}$:  $V_3 = -(9 \times 10^9) \left( \dfrac{40 \times 10^{-8}}{0.20} \right) = -18{,}000$ V

Note that $V_3$ is negative; absolute potentials due to negative charges are always negative. Unlike forces and fields, work and potential are scalars, not vectors. Hence, they are numbers without direction, merely plus or minus, and may be added directly. The result for the absolute potential at $B$ is $V_B = V_1 + V_2 + V_3 = -6300$ V. Since the result is negative, we conclude that the unit positive charge actually has less potential energy at $B$ than it did at infinity. This must mean that the effect of the attraction of the negative charge for it is stronger than the repulsion of the two positive charges. ■

*Exercise* Sketch the electric field lines and equipotentials at very large distances from $B$ in Fig. 16.8. Estimate the absolute potential for $r \gg 20$ cm. *Answer:* $V = -k \, (27 \times 10^{-8} \text{ C})/r$ ■

# Millikan's oil-drop experiment

The magnitude of the quantum of charge, the electronic charge $e$, was first measured accurately by R. A. Millikan and his coworkers (1909 to 1913). Millikan's experiment has a simplicity and directness that established without doubt that charge is quantized.

His apparatus consisted of two parallel metal plates (separation $\approx$ 1 mm and radius $\approx$ 6 cm). A variable-voltage source placed charges on the plates, giving rise to a known uniform field $E$ between them. By varying the voltage in this way, Millikan was able to change the value of $E$.

An atomizer was used to generate tiny oil drops, which fell through a hole in the upper plate. Friction effects in the atomizer often led to charges on the drops. Consider a drop of mass $m$, radius $a$, and charge $q$ between the plates. The drop could be observed through a microscope as it sparkled in an intense light used to illuminate it. Millikan adjusted the electric field between the plates until the charged drop remained motionless. He then knew that the gravitational force $mg$ on the drop was balanced by the electric force $Eq$ on its charge:

$$mg = Eq$$

from which

$$q = \frac{mg}{E}$$

Therefore the charge on the drop could be computed from a knowledge of $E$, $m$, and $g$.

The electric field strength $E$ can be found easily from the voltage and the distance between the plates by use of Eq. 16.1, $V = Ed$. To find the mass of the drop, Millikan measured $v$, the terminal velocity of fall for the drop when $E = 0$. Then, from Stokes' law,

$$mg = 6\pi\eta av$$

where $a$ is the radius of the drop and $\eta$ is the viscosity of the air. Because $m = \rho(4\pi a^3/3)$, where $\rho$ is the density of the oil, Millikan could find the radius, and thus the mass, of the oil drop. He then could substitute in the equation for the charge to find the charge on the drop.

Millikan carried out thousands of measurements of this general type (slightly modified for higher accuracy). His result for $q$ was always an integral multiple of $1.60 \times 10^{-19}$ C. He therefore concluded that charge is quantized, the magnitude of the charge quantum being $1.60 \times 10^{-19}$ C, which we designate by $e$.

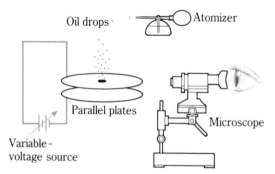

Oil drops · Atomizer

Parallel plates

Variable - voltage source

Microscope

## Example 16.10

In the Bohr model of the hydrogen atom, sketched in Fig. 16.9, the pointlike electron ($q = -e$) moves in a circular orbit ($r = 0.053$ nm) with the positive nucleus ($q = +e$) at the center. (a) Find the absolute potential due to the nucleus at the electron's position. (b) How much energy is needed to pull the electron loose from the atom?

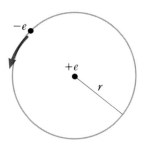

**FIGURE 16.9**
The Bohr model of the hydrogen atom. The electron moves in a circular orbit of 0.053 nm radius about the center of the atom.

**Reasoning** (a) The absolute potential due to the nucleus at the position of the electron is $V = kq/r$, with $r = 5.3 \times 10^{-11}$ m and $q = 1.60 \times 10^{-19}$ C. Using these values, we find that

$$V = 27.2 \text{ V}$$

(b) The potential due to the nucleus at a point very far away from the nucleus is zero because $r \to \infty$ in $kq/r$. Therefore, if the atom's electron is to be removed from the orbit and carried to infinity (that is, the atom is to be ionized), it must be carried through a potential difference

$$\Delta V = V_\infty - V = 0 - 27.2 \text{ V} = -27.2 \text{ V}$$

The negative sign tells us it is a potential drop from the orbit to infinity. However, because the electron is negatively charged, it is attracted by the nucleus, and so work must be done on it to pull it to infinity. The work done is

$$\text{Work} = q(\Delta V) = (1.60 \times 10^{-19} \text{ C})(27.2 \text{ V}) = 4.35 \times 10^{-18} \text{ J}$$

In atomic physics, this work would often be expressed as 27.2 eV. In practice, only half this amount of energy must be furnished from outside the atom to ionize it. The electron's kinetic energy in orbit furnishes the other half of the required energy, as we shall see in Chap. 26. ∎

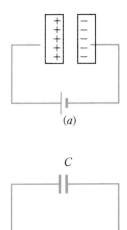

**FIGURE 16.10**
Equal and opposite charges reside on the inner faces of the capacitor plates. Notice the symbol used for a capacitor in *b*.

## 16.7 Capacitors

We have made frequent reference to two oppositely charged metal plates. This is one form of a device that is of considerable practical importance for storage of electric charge and energy, as we shall see in later chapters. It is called a **capacitor** (or *condenser*). Such a device connected to a battery is shown in Fig. 16.10. The positive terminal of the battery places a positive charge on one plate, and the negative terminal places an equal-magnitude negative charge on the other plate.

These charges attract one another, and so they reside on the inner surfaces of the plates, as shown. Such a device is capable of storing charge. As shown in part b, the symbol used for a capacitor is —||— or sometimes —|(—.

Let us represent the charge on the positive plate in Fig. 16.10 by $+q$. The charge on the negative plate is then $-q$. Because the electric field $E$ between the plates is proportional to $q$ and because the potential difference $V$ between the plates is proportional to $E$ (remember that $V = Ed$) and because things proportional to the same thing are proportional to each other, we find that $q$ is proportional to $V$. Therefore

$$q = (\text{constant})(V) \qquad \text{or} \qquad \frac{q}{V} = \text{constant}$$

This constant is simply the charge that the capacitor holds for each volt of potential difference between the plates. It is called the **capacitance** $C$ of the capacitor:

$$\text{Capacitance} = C = \frac{q}{V} \tag{16.5}$$

The unit of capacitance is coulombs per volt, called the **farad** (F). The capacitance $C$ depends on the geometry of the capacitor. It is large if the plates have a large surface area and are very close together. Can you explain why this is true?

There is a particularly simple expression for the capacitance of a parallel-plate capacitor. If each plate has a surface area $A$ and the plates are separated by only a tiny distance $d$, then the capacitance is

$$C = \frac{\epsilon_0 A}{d} \qquad \text{parallel plates} \tag{16.6}$$

where $\epsilon_0$ (read "epsilon sub zero") is the permittivity of free space ($8.85 \times 10^{-12}$ F/m), mentioned in Sec. 15.8 in connection with Coulomb's law. You are asked in Prob. 16.40 to show that the units of $\epsilon_0$ given there ($C^2/N \cdot m^2$) are the same as those given here (F/m).

Capacitances are usually of the order of microfarads or smaller, as we can see from a simple computation. Suppose two parallel plates each have a face area of 10 cm $\times$ 10 cm $= 100$ cm$^2$ and are 0.10 mm apart. Their capacitance is given by Eq. 16.6:

$$C = \frac{(8.85 \times 10^{-12} \text{ F/m})(100 \times 10^{-4} \text{ m}^2)}{1 \times 10^{-4} \text{ m}} = 8.85 \times 10^{-10} \text{ F}$$

which is $8.85 \times 10^{-4} \mu\text{F}$.

In practice, most parallel-plate capacitors contain a sheet of nonconducting material between the plates. This sheet allows the plates to be placed very close together with no fear that they will touch and permit the charges to join together. Many commercial capacitors are formed by taking two thin sheets of metal foil and laying one on top of the other with a thin plastic film between them to keep them from touching. The layered sheets are then rolled up into a tight cylinder and packaged for convenience. The device is essentially a parallel-plate capacitor, but it looks very different from the sketch in Fig. 16.10. Capacitors with a capacitance of 0.1 $\mu$F, a common size, occupy a volume of about 1 cm$^3$ when made this way. Figure 16.11 shows two common capacitors.

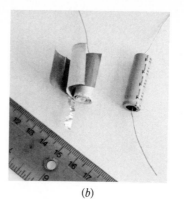

(a)                (b)

**FIGURE 16.11**

Two sheets of metal foil separated by an insulator act as the plates of a commercial capacitor. If the sheets are rolled or folded into a compact package, a parallel-plate capacitor can be reduced to a convenient size. We show two types of capacitor in both original and partly disassembled form. (*a*) A 100-pF capacitor that uses a thin plastic sheet as insulator. (*b*) A 470-$\mu$F electrolytic capacitor that uses a thin oxide coating on the metal foil as insulator. A paper spacer impregnated with moist electrolyte separates the metal sheets. Although they provide large capacitance, electrolytic capacitors usually cannot withstand high voltages.

## 16.8 Dielectrics

Despite the fact that nonconductors contain no free charges, they have a marked effect on electric fields in which they are placed. These materials, called **dielectrics** in this context, tend to cancel the electric fields established by charged objects. We now see how they do this.

We can divide dielectrics into two groups, those that contain molecular dipoles and those that do not. A **dipole** consists of two equal-magnitude charges of opposite sign separated by a small distance, as shown in Fig. 16.12*a*. Many molecules,

$+q$      $-q$

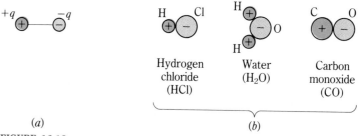

Hydrogen chloride (HCl)      Water ($H_2O$)      Carbon monoxide (CO)

(*a*)                          (*b*)

**FIGURE 16.12**

The dipolar molecules in (*b*) act like the dipole in (*a*).

although electrically neutral (that is, uncharged), are in effect tiny dipoles. Examples are shown in Fig. 16.12b. Molecules such as these are called *dipolar molecules*. When a dipolar molecule is placed in an electric field, as in Fig. 16.13, its oppositely charged ends experience equal, oppositely directed forces ($\mathbf{E}q$ and $-\mathbf{E}q$). The

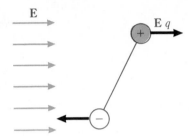

**FIGURE 16.13**

An electric field causes a dipole to experience a torque that tends to align it in the field.

resultant torque on the molecule tends to align it in the electric field. As a result, dipolar molecules between charged plates tend to align as shown in Fig. 16.14. In practice, thermal motion prevents them from becoming fully aligned except in extremely strong fields.

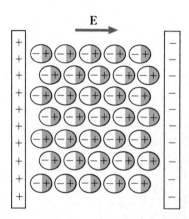

**FIGURE 16.14**

Dipoles align along the field lines.

Atoms and many molecules do not possess dipoles. Although they have negatively charged electrons and positively charged nuclei, the effective centers of the two types of charge coincide, as shown at the top of Fig. 16.15. As a result, these atoms and molecules behave as though the negative and positive charges were not separated, and so they possess no permanent dipole. However, when such an atom or molecule is placed in an electric field, as shown in the lower part of Fig. 16.15, the negatively charged electron cloud is attracted slightly to the left and the positively charged nucleus is repelled slightly to the right. This slight shift in the charges causes the atom (or molecule) to become a dipole; we say that it has become *polarized* and now possesses an *induced dipole*.

We see then that all materials, when placed in an electric field, possess dipoles aligned with the field, as in Fig. 16.14. Notice how the positive plate induces the negative ends of the dipoles to come close to it and the negative plate attracts the positive ends. Notice further in Fig. 16.14 that the dipole alignment causes a layer of positive charges (the positive ends of the dipoles) to exist near the plate on the right.

**(a)**

**(b)**

**Thermograms.** Because the heat radiated from a surface varies as $T^4$, infrared radiation is a sensitive measure of surface temperature. Infrared-detecting cameras were used to obtain these two thermograms. Electronic processing of the image obtained results in a display of temperatures in terms of colors. In these photographs, the brighter colors represent warmer areas. To interpret the photographs, we use the fact that the temperature difference between the top and bottom of the scale in *b* is 10 C°. *(a)* Why should the eaves of these houses (blue and purple) be cooler than the siding (green and yellow) and the windows (white and red)? *(Daedulus)* *(b)* The dark blue (cooler) side of the jacket is made from fibers that have one-tenth the diameter of those used in its bright green (warmer) side. The thinner fibers trap air much better within the fabric than do the heavier fibers and thereby insulate the body more effectively. *(3M)*

# VISIBLE SPECTRUM

# EMISSION (BRIGHT LINE) SPECTRA

**Refraction, interference, and reflection.** White light can be separated into its colors in a variety of ways. *(a)* A prism makes use of refraction and the fact that the refractive index of the material the prism is made of changes with wavelength. *(Bausch & Lomb)* *(b)* The colors seen in the soap bubbles are due to interference between the two beams of light reflected from the two faces of the water film that forms the bubbles. *(AIP/T. Young)* *(c)* A rainbow owes its colors to refraction and to the total internal reflection of light beams from the sun as they are scattered by water droplets in fog or rain clouds. This rainbow was photographed over Isaac Newton's birthplace at Woolfthorpe Manor in England. *(AIP/R. Bishop)*

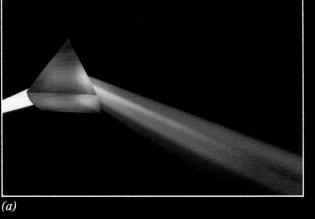

*(a)*

*(b)*

*(c)*

**Lasers.** The technological uses of lasers in today's world cover a broad range of applications, from the powerful to the elegantly delicate. *(a)* The two large cylindrical blocks of metal are being welded by a powerful laser beam. *(Photo Researchers/D. Luria)* *(b)* This helium-neon laser beam has a power of only a fraction of a milliwatt. It is being used to diagnose an eye disease by using the Doppler effect to monitor blood flow in the retina. In applications such as this, the laser beam is carried to the site of its use by an optical fiber conduit. *(Science Source/A. Tsiaras)*

*(a)*

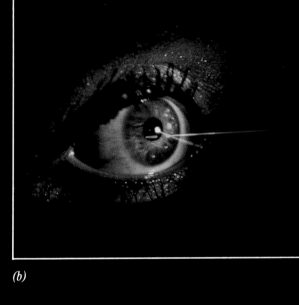

*(b)*

**Fiber optics.** A handful of optical fibers. These light "pipes," which work on the principle of total internal reflectance, have had a profound effect on fields ranging from medicine to communications. One of the medical applications is described above. In the field of telecommunications, kilometer after kilometer of traditional copper cable is being replaced by optical fibers as telephone companies update their systems to handle the "information revolution." *(W. Sproul)*

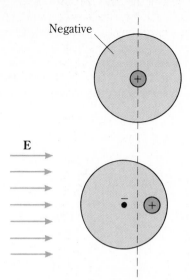

**FIGURE 16.15**

An impressed electric field causes
a nonpolar atom or molecule to
become a dipole, an induced dipole.

Similarly, there is a layer of negative charges near the plate on the left. When a slab
of dielectric material is placed between the plates, as in Fig. 16.16, the dipole
alignment causes charges to appear on the two faces of the slab. These charges are
simply the charged ends of the dipoles at the surfaces of the dielectric. We refer to
this type of charge as *induced polarization charge* or *bound charge*. The latter
name reflects the fact that this charge is bound to atoms and molecules within the
dielectric; it is not free to move from its parent atom or molecule.

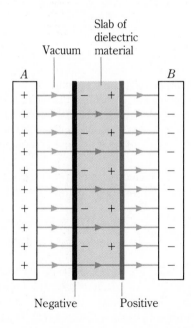

**FIGURE 16.16**

The electric field induces bound
charges on the surface of the
dielectric. They cause the field to
be less inside the dielectric than
outside.

The induced charge on the dielectric influences the electric field within it. We can
see why this is so by considering a somewhat different situation, one in which a

metal slab replaces the dielectric slab. Then the situation becomes that shown in Fig. 16.17. Because the induced charge on the left side of the metal slab equals in magnitude the charge on plate $A$, all the field lines that originate on plate $A$ must end on the induced charges on the left side of the slab. A like number of lines originate on the induced charge on the right side of the slab, and these lines end on plate $B$. Most important, the field within the metal is zero. The induced charges on the slab cancel the field within the region occupied by the metal.

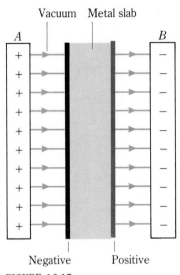

**FIGURE 16.17**
When the dielectric slab of Fig. 16.16 is replaced by a metal plate, the field within the plate is zero.

■ **TABLE 16.1**
**Dielectric constants (20°C)**

| Material | $K$ |
| --- | --- |
| Vacuum | 1.00000 |
| Air | 1.006 |
| Paraffin | 2.1 |
| Petroleum oil | 2.2 |
| Benzene | 2.29 |
| Polystyrene | 2.6 |
| Ice (−5°C) | 2.9 |
| Mica | 6 |
| Acetone | 27 |
| Methyl alcohol | 38 |
| Water | 81 |
| Metal | ∞ |

For the dielectric slab in Fig. 16.16, the induced charges on its surfaces are smaller than those on the metal slab of Fig. 16.17. Therefore, fewer of the field lines emanating from plate $A$ can terminate on the face of the slab. As a result, many of the field lines continue on through the dielectric. However, because not all the lines go through the dielectric, it is clear that the field inside the slab is less than that in the vacuum region outside it. Moreover, the field is decreased most by a dielectric that is easily polarized, that is, one in which polarization charge is most easily induced.

The ability of a dielectric to decrease the electric field strength is characterized by its **dielectric constant $K$**, defined by reference to Fig. 16.16:

$$\text{Dielectric constant } K = \frac{\text{electric field in vacuum}}{\text{electric field in dielectric}}$$

The electric field is only $1/K$ as large inside the dielectric as outside it. Typical dielectric constants are given in Table 16.1. Notice that vacuum does not alter the field at all, and so its dielectric constant is unity. Because air has so few molecules per unit volume, its constant differs only slightly from that of vacuum. For most solids, $K$ is in the range from 2 to 10. Although we do not consider metals to be dielectrics, you should be able to show that the dielectric constant for a metal is infinite.

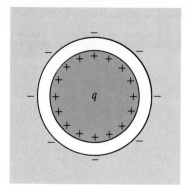

**FIGURE 16.18**

A charged sphere in an infinite dielectric. Why is the electric field decreased by the dielectric?

**The effects of dielectrics**

Coulomb's law is altered when charges are immersed in a dielectric. To see why this occurs, refer to Fig. 16.18. We see there a sphere with charge $q$ immersed in a dielectric that extends a great distance in all directions — in other words, an essentially infinite dielectric. Notice how the sphere induces a charge on the dielectric surface next to it. This induced charge in effect cancels some of the charge on the sphere. Thus the electric field in the dielectric is reduced from the value $E = kq/r^2$ that applies in a vacuum. The dielectric reduces the field by a factor of $1/K$, so that the field in the dielectric is

$$E = k \frac{q}{K r^2} \qquad \text{(point charge)} \tag{16.7}$$

This is the electric field of a point charge immersed in a dielectric.

Suppose two point charges $q_1$ and $q_2$ that are a distance $r$ apart are immersed in an infinite dielectric. The field due to $q_1$ at the position of $q_2$ is given by Eq. 16.7 after $q$ has been replaced by $q_1$. This field causes a force $Eq_2$ on $q_2$, and so we find that the force on $q_2$ due to $q_1$ is

$$F = k \frac{q_1 q_2}{K r^2} \qquad \text{(Coulomb's law)} \tag{16.8}$$

This is Coulomb's law for point charges in an infinite dielectric.

Because a dielectric greatly influences the forces between charges, chemical and biological reactions are markedly solvent-dependent. For example, two ions in solution exert forces given by Eq. 16.8 on each other. Water has $K = 81$, and so the force between two ions is much less in water than it is in a liquid such as benzene, which has $K = 2.3$. As a result, the $Na^+$ and $Cl^-$ ions of sodium chloride can escape from each other in water, whereas they cannot do so in benzene. Hence water dissolves NaCl and benzene does not. There are many other similar situations in chemical and biological systems where the dielectric constant of the solvent is the controlling factor in chemical reactions.

There is a simple way to use a capacitor to measure the dielectric constant of a material. Consider the parallel-plate capacitor shown in Fig. 16.16. It has a charge $q$ on its positive plate, and the distance between plates is $d$. Let us say that the electric field between the plates is $E_{vac}$ when vacuum exists there, that is, when the dielectric is absent. Then the voltage difference between the plates is

$$V_{vac} = E_{vac} d$$

the usual parallel-plate relation. When a dielectric fills the entire region between the plates, however, the field between them is reduced by a factor $1/K$, to become $E_{vac}/K$. Then the potential difference between the plates is

$$V_{di} = Ed = \frac{E_{vac}}{K} d$$

We can find the dielectric constant by taking the ratio of these two voltages:

$$\frac{V_{vac}}{V_{di}} = \frac{E_{vac} d}{(E_{vac}/K)(d)} = K$$

This result is often stated in terms of capacitances. Because $C = q/V$, we have

$$C_{vac} = \frac{q}{V_{vac}} \quad \text{and} \quad C_{di} = \frac{q}{V_{di}}$$

where $C_{vac}$ and $C_{di}$ are the capacitances with vacuum and dielectric between the plates. Division of one equation by the other gives

$$\frac{C_{di}}{C_{vac}} = \frac{V_{vac}}{V_{di}} = K \tag{16.9}$$

Thus we see that the capacitance of a capacitor is increased by a factor $K$ when a dielectric is placed between its plates. We can therefore determine the dielectric constant of a material by measuring the capacitance of an empty capacitor and comparing it with the capacitance when the material fills the capacitor.

## 16.10 The energy stored in a capacitor

A charged capacitor has electrical potential energy stored in it. We know this to be true because one of its charges, when released from one plate, gains kinetic energy as it moves to the other plate. We now compute how much energy is stored in a charged capacitor.

Consider a hypothetical experiment in which we charge an initially uncharged parallel-plate capacitor in the following way. We take tiny bits of charge $\Delta q$ from one plate, carry them to the other plate, and deposit them there. Initially, there is no charge on the plates and the potential difference between them is zero. Later, however, after we have deposited considerable charge on one plate, the potential difference is sizable, and we need to expend a great deal of work to carry an additional charge $\Delta q$ across. Finally, after a total charge $q$ has been deposited and the potential difference is $V$, the work required to transport $\Delta q$ is $V \Delta q$. The total work done is easily found by using calculus. However, it appears reasonable that the total work done should be equivalent to the work done in carrying the whole charge $q$ across the average potential difference during the charging process, $\frac{1}{2} V$. This turns out to be a correct assumption. Therefore, the energy stored in a capacitor with charge $q$ and potential difference $V$ is

$$\text{Energy} = \tfrac{1}{2} q V \tag{16.10}$$

By use of the defining equation for capacitance, $q = CV$, Eq. 16.10 can be written in the alternative forms

$$\text{Energy} = \tfrac{1}{2} q V = \tfrac{1}{2} C V^2 = \frac{q^2}{2C}$$

## 16.11 The energy stored in an electric field

In the preceding section, we saw that the energy stored in a charged capacitor is $\frac{1}{2} C V^2$, where $V$ is the voltage difference across a capacitor having a capacitance $C$. Although it is not necessary to specify exactly how and where this energy is stored, it is sometimes convenient to think of it as being stored in the electric field between the capacitor plates. With this in mind, it would be well to express the equation for the stored energy in terms of the electric field $E$ between the plates. We can do this by recalling that, for a parallel-plate capacitor, $V = Ed$, where $d$ is the separation of the plates.

We therefore have for the energy stored in a parallel-plate capacitor

$$\text{Energy} = \tfrac{1}{2}CV^2 = \tfrac{1}{2}CE^2d^2$$

From Eq. 16.6, however, the capacitance of a parallel-plate capacitor with plate area $A$ is

$$C = \frac{\epsilon_0 A}{d}$$

provided the capacitor has vacuum between its plates. If it is filled with a dielectric with a constant $K$, the equation becomes

$$C = \frac{K\epsilon_0 A}{d}$$

Substituting this value for $C$ in the energy equation yields

$$\text{Energy} = (\tfrac{1}{2}\epsilon_0 KE^2)(Ad)$$

The term $Ad$ is the volume of the space between the capacitor plates — in other words, the volume in which the constant electric field $E$ exists. Dividing both sides of the equation by the volume gives us an expression for the energy per unit volume, that is, the energy we picture as being stored in a unit volume of the region of space where the electric field is $E$:

$$\text{Energy density} = \text{energy per unit volume} = \tfrac{1}{2}\epsilon_0 KE^2 \tag{16.11}$$

Notice that the energy stored in a unit volume of space is proportional to the square of the electric field strength. It is often convenient to use Eq. 16.11 for assigning energy to an electric field. Although this expression was derived for a very special case, it is shown in more advanced texts that it has general validity.

## Minimum learning goals

When you finish this chapter, you should be able to

1 Define (a) potential difference, (b) volt, (c) equipotential lines, surfaces, and volumes, (d) emf, (e) electronvolt, (f) absolute potential, (g) capacitor, (h) capacitance, (i) farad, (j) dielectric, (k) dipole, (l) dielectric constant.

2 Find the potential difference between two points when the work required to carry a charge $q$ from one point to the other is given (or vice versa).

3 Find the potential difference between any two points in a region in which a known uniform electric field exists.

4 Sketch the equipotentials and field lines in simple situations.

5 Use the relation $W = qV_{AB}$ in simple specified situations.

6 Find the energy change in electronvolts of a particle of known charge due to its movement through a given potential difference. Convert energies between electronvolts and joules.

7 Find the absolute potential at a point due to several specified point charges near the point.

**8** Find the change in kinetic energy of a charged particle due to its motion through a given potential difference. If either the initial or final speed is known, find the other speed.

**9** Draw a diagram of a parallel-plate capacitor and give the relation between $q$, $V$, and $C$.

**10** Explain the difference between a liquid or solid that has a large dielectric constant and one that has a small dielectric constant.

**11** Compute the energy stored in a given capacitor charged to a known potential difference.

## Questions and guesstimates

**1** Two points $A$ and $B$ are at the same potential. Does this necessarily mean that no work is done in carrying a positive test charge from one point to the other? Does it mean that no force has to be exerted to carry the test charge from one point to the other? Explain.

**2** Can two equipotential surfaces intersect? Explain.

**3** The absolute potential midway between two equal but oppositely charged point charges is zero. Can you find an obvious path along which no work would be done in carrying a positive test charge from infinity up to this point? Explain.

**4** Starting from the fact that a piece of metal is an equipotential body under electrostatic conditions, prove that the electric field inside a hollow piece of metal is zero.

**5** If the absolute potential is zero at a point, must the electric field be zero there as well?

**6** What can be said about the electric field in a region in which the absolute potential is constant?

**7** Prove that all points of a metal object are at the same potential under electrostatic conditions. Does this also apply in a hole inside the object? Does it matter if there is a charge suspended in the hole?

**8** A parallel-plate capacitor has a fixed charge $q$ on its plates. The plates are now pulled farther apart. The puller must do work. Why? Does the potential difference change during the process? What happens to the work done by the puller?

**9** A hollow uniformly charged spherical metal ball has a charge $+q$. Where is the charge located? Is the absolute potential zero inside the sphere? Is it constant? What is it? Repeat for a charge $-q$.

**10** Electrostatic methods are frequently used in industry to spray-paint metal objects. The sprayer is attached to one terminal of a high-voltage source, and the metal object to be painted is attached to the other. Explain the principle of operation for this method. Why does it generate less air pollution and use less paint than conventional methods?

**11** Two identical metal spheres carry charges $+q$ and $-2q$. They are touched together and again separated.

What are their final charges? If the two spheres have different radii, which has the larger final charge?

**12** The electric strength of air is about 30,000 V/cm. By this we mean that when the electric field intensity exceeds this value, a spark will jump through the air. We say that "electric breakdown" has occurred. Using this value, estimate the potential difference between two objects where a spark jumps. A typical situation might be the spark that jumps between your body and a metal door handle after you have walked on a deep carpet or slid across a plastic car seat in very dry weather.

**13** Refer to the data given in the previous question. About how much charge could you place on a metal sphere that has a diameter of 50 cm?

**14** A simple electrostatic precipitator for removing smoke from air can be constructed as shown in Fig. P16.1. A very thin wire is placed along the axis of a much larger metal tube, and a high voltage is applied to these two elements, with the wire being made the negative terminal. If the wire is very thin and the voltage high, the electric field near the wire will be very high. Why? Tiny sparks (called corona) are formed near the wire due to electric breakdown (see Question 12), and electrons shoot away from the wire. Why? They charge the smoke particles negatively. How? These particles then move to the tube and precipitate there. Why? As a result, the smoke is removed from the air.

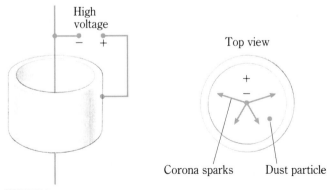

**FIGURE P16.1**

## Problems

**1** How much work is required to carry a $+5.0\text{-}\mu\text{C}$ charge from the negative to the positive terminal of a 12.0-V battery? From the positive to the negative terminal?

**2** How much work is required to carry an electron from the positive to the negative terminal of a 9.0-V battery? Repeat for a proton.

**3** Two parallel metal plates 0.50 mm apart are connected to the terminals of a 1.55-V battery. (*a*) What is the electric field strength between the plates? (*b*) How large a force would an electron experience if it were between them?

**4** Two parallel metal plates are 0.20 mm apart, and the electric field strength between them is 4000 V/m. (*a*) What is the potential difference between the plates? (*b*) How large a force would a proton experience if it were between them?

■ **5** Points $A$ and $B$ are on the $x$ axis. Point $A$ is at $x = 50$ cm, and $B$ is at $x = 80$ cm. The potential difference from $B$ to $A$ is 70 V, with $A$ being at the higher potential. (*a*) Find $E_x$, the constant electric field in the $x$ direction, in this region. (*b*) Repeat if $B$ is at the higher potential.

■ **6** In a certain region of space, the electric field is directed in the negative $x$ direction and has a magnitude of 5000 V/m. What is the potential difference between the origin of coordinates and the point (*a*) $x = 0, y = 0, z = 10$ cm; (*b*) $x = 20$ cm, $y = 0, z = 0$; (*c*) $x = -30$ cm, $y = 0$, $z = 0$; (*d*) $x = -30$ cm, $y = 10$ cm, $z = 0$?

■ **7** An electron is released at the coordinate origin in a region in which there is an electric field of 3000 V/m in the positive $y$ direction. (*a*) Find the time the electron takes to reach a speed of $8.0 \times 10^6$ m/s. (*b*) Where is it at that time?

■ **8** A proton is traveling in the positive $x$ direction with a speed of $7.0 \times 10^5$ m/s. An electric field is now switched on with $E_x = -400$ V/m, $E_y = E_z = 0$. What is the speed of the proton after it has traveled 2.0 m?

■ **9** A tiny ball carrying a charge of $+25$ nC is held by a thread between two horizontal parallel plates that are 3.0 cm apart. (*a*) When the potential difference between the plates is 6000 V, the tension in the thread is zero. What is the mass of the ball? (*b*) What is the tension in the thread when the polarity of the plates is reversed?

■ **10** Two vertical parallel plates are 4.0 cm apart, and the potential difference between them is 8000 V. A tiny ball ($m = 2.5 \times 10^{-4}$ g) is suspended as a pendulum between them. The very thin thread holding the ball reaches equilibrium at an angle to the vertical of 17.0°. What is the charge on the ball?

**11** A proton is released from rest and accelerates through a 45-V potential difference. What is its final speed?

**12** Through how large a potential difference must an electron move if it is to be accelerated from rest to $7.0 \times 10^5$ m/s?

■ **13** An electron is shot from one large metal plate toward a parallel plate. If the electron's initial speed is $5 \times 10^6$ m/s and its speed just before it hits the second plate is $3 \times 10^6$ m/s, what is the potential difference between the plates? Is the second plate at a higher or lower potential than the first?

■ **14** A proton is shot with speed $v_o$ from one metal plate toward a second plate parallel to the first. If the potential difference between the plates is $V$, find the speed of the proton just before it strikes the second plate. Two answers are possible; give both.

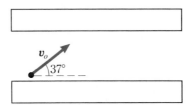

**FIGURE P16.2**

■ **15** A proton is shot from the lower plate in Fig. P16.2 with a velocity $v_o = 3 \times 10^4$ m/s at the angle shown. What must the potential difference between the plates be if the proton is to just miss the upper plate?

■ **16** An electron is shot from the lower plate in Fig. P16.2 at the angle shown. The potential difference between the plates is 4000 V. What must the initial speed of the electron be if it is to just miss the upper plate? Should the top plate be positive or negative?

**17** An alpha particle ($m = 4 \times 1.66 \times 10^{-27}$ kg, $q = 2e$) has a kinetic energy of 6.2 MeV. (*a*) What is its energy in joules? (*b*) What is its speed? (*c*) Through how large a potential difference must it move to attain this energy?

**18** A triply ionized lithium atom ($m = 6.94 \times 1.66 \times 10^{-27}$ kg, $q = 3e$) is accelerated through a potential difference of 8000 V. What is its kinetic energy in electronvolts? What is its speed?

■ **19** Two parallel plates have a potential difference of 60 V. (*a*) A proton is shot from the negative plate toward the positive plate with an initial energy of 80 eV. What is its kinetic energy just before it strikes the positive plate? (*b*) Repeat if it is shot from the positive plate toward the negative plate.

■ **20** A proton is shot with a kinetic energy of 4000 eV from a negative plate toward a positive plate. The potential difference between the plates is 1500 V. (*a*) How much kinetic energy (in electronvolts) does the proton lose as it shoots to the positive plate? (*b*) What is its kinetic energy (in electronvolts) just before it hits the plate? (*c*) Repeat for an alpha particle with the same initial kinetic energy. (The charge on an alpha particle, a helium nucleus, is $+2e$.)

■ **21** An electron moving with a speed of $3.0 \times 10^6$ m/s is accelerated through a potential difference of 20 V. What is its new speed?

■ **22** A proton moving with a speed of $5.0 \times 10^7$ m/s is to be slowed to $1.2 \times 10^7$ m/s. How large a potential difference must it move through in order to slow this much?

**23** What is the absolute potential $3 \times 10^{-14}$ m from the center of an atomic nucleus if the nuclear charge is $80e$? Ignore the electrons of the atom. If a proton is released at this radius, what will its kinetic energy be (in millions of electronvolts) when it is far from the nucleus?

**24** A metal sphere of radius 20 cm carries a uniform charge of $3.0 \times 10^{-8}$ C. Assume that it is far from all other objects. What is the absolute potential at its surface due to its charge?

**25** Two point charges are placed on the $x$ axis: a $+5.0\text{-}\mu$C charge at $x = 3.0$ cm and a $-6.0\text{-}\mu$C charge at $x = 18.0$ cm. Find the absolute potential at (*a*) $x = 10$ cm and (*b*) $x = -5$ cm due to the two charges.

**26** Four identical $-4.0\text{-}\mu$C point charges are placed at the four corners of a square that is 30 cm on each side. What is the absolute potential at the center of the square due to the four charges?

**27** Repeat Prob. 26 if three of the charges are positive and the fourth is negative.

■ **28** A metal sphere of radius 4.0 cm hangs from a thin thread in the center of a very large room. It carries a charge of $-5 \times 10^{-9}$ C. What is the approximate potential difference between the sphere and the walls of the room?

**29** What is the charge on a 30-nF capacitor subjected to a potential difference of 800 V? How much energy is stored in this capacitor?

**30** What potential difference must be applied to a $6.0\text{-}\mu$F capacitor if the charge on it is to be 7.0 mC? How much energy is then stored in the capacitor?

**31** A parallel-plate capacitor has a plate spacing of 0.025 mm. What must the area of each plate be if the capacitance is to be $0.50\ \mu$F and the material between the plates is (*a*) vacuum and (*b*) plastic with $K = 4.0$?

**32** Two identical metal plates are placed parallel to each other. The gap between them is 0.040 mm, and the area of each is 300 cm². (*a*) Find the capacitance of the plates if vacuum exists between them. (*b*) How much charge exists on the capacitor when it is connected to a 12.0-V battery? (*c*) Repeat parts *a* and *b* if the space between the plates is filled with a material of dielectric constant 3.8.

■ **33** The gap in a certain parallel-plate capacitor can be changed without otherwise disturbing the electrical system. In position $A$ the capacitance is $4.0 \times 10^{-9}$ F, and in position $B$ it is $3.7 \times 10^{-9}$ F. The capacitor is charged by a 12-V battery when in position $A$. The battery is then removed, and the capacitor is changed to position $B$ without changing the charge on it. (*a*) How much charge is on the capacitor in position $A$? (*b*) What is the voltage across it in position $B$? (*c*) By how much does its stored energy change as it goes from $A$ to $B$? (*d*) What is the minimum amount of work that the person holding the plates must have done to change the capacitor from $A$ to $B$?

■ **34** Repeat the previous problem if the battery is left attached to the plates as the plate separation is changed.

■ **35** Sparking will occur through air if the electric field exceeds about $3.0 \times 10^6$ V/m. How large a charge can be placed on a $20 \times 10^{-12}$ F parallel-plate capacitor having air between the plates before sparking takes place? Assume the area of each plate is 20 cm².

■ **36** A parallel-plate air capacitor holds a charge of 30 nC when subjected to a potential difference $V_o$. When a fluid is placed between the plates, the charge on the capacitor increases to 87 nC if the potential difference is maintained at $V_o$. What is the dielectric constant of the fluid?

■ **37** A parallel-plate air capacitor is charged by placing a 90-V battery across it. The battery is then removed, and an insulating liquid is poured between the plates to fill the air gap. The voltage across the capacitor is now 28 V. What is the dielectric constant of the liquid?

■ **38** Two point charges $q_1$ and $q_2$ are on the $x$ axis at $x_1$ and $x_2$, respectively. Where, if anywhere, on the $x$ axis is the absolute potential zero if (*a*) both $q_1$ and $q_2$ are positive and (*b*) $q_1 = -3q_2$?

■ **39** A pendulum of length $L$ hangs from the ceiling of a room in which a downward electric field $\mathbf{E}$ exists. The pendulum ball has a mass $m$ and a positive charge $q$. Find the frequency of the pendulum for small-angle oscillation.

■ **40** Show that C²/N · m² and F/m, the two units used for $\epsilon_0$, are equivalent.

# 17 Direct-current circuits

MOST practical applications of electricity involve charges that are in motion. Light bulbs, for instance, are lighted by charges flowing through them. Charge flowing through the coils of a motor causes its shaft to rotate. When we turn on a radio or television, we allow charge to flow through the device so that it can operate. Although common household devices are designed to operate with alternating current (ac), we shall begin our study of them by discussing direct-current (dc) circuits, circuits in which the flow of charge is in one direction only.

## 17.1 Electric current

We begin our discussion of charges in motion by defining a quantity called electric current. Suppose we have a device, called a *charge gun*, that can shoot out a stream of charged particles, such as ions or electrons. (A television set uses a gun to shoot an electron beam at the screen.) For our discussion, consider a gun that shoots a beam of positive particles through a hole in a plate, as in Fig. 17.1.

The beam passing through the hole constitutes a flow of charge, and we now wish to characterize the magnitude of this flow. We do so by defining a quantity called **electric current,** which we designate by the symbol $I$:

**FIGURE 17.1**
The beam of moving charges passes through the hole in the plate. If a charge $\Delta q$ passes through the hole in time $\Delta t$, the current is $\Delta q/\Delta t$.

If in a time $\Delta t$ a beam carries a charge $\Delta q$ past a given point (the plate in this case), then the current carried by the beam is

$$I = \frac{\Delta q}{\Delta t} \qquad (17.1)$$

The unit of current, the coulomb per second, is called the **ampere** (A).

If the charges in the beam are positive, then both $\Delta q$ and $I$ are positive. If the beam is composed of negative charges, however, then both $\Delta q$ and $I$ are negative. For this reason, a flow of negative charge in one direction is equivalent to a positive current in the opposite direction. You might object that currents in metals are carried by electrons, which are negative charges, and so current should be defined in terms of negative-charge flow. Historically, however, the sign of the charge carriers was not known. In fact, it is very difficult to determine experimentally which type of charge is moving. Hence there was no great compulsion to change terminology once the nature of the charge carriers in metals was learned.

To see what our definition means for currents in wires, refer to Fig. 17.2, which shows positive charge carriers. If a charge $\Delta q$ flows through the cross section at $A$ in a time $\Delta t$, then the current in the wire is defined by Eq. 17.1 to be

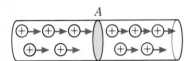

$$I = \frac{\Delta q}{\Delta t}$$

**FIGURE 17.2**
The current in amperes in the wire is defined to be the quantity of positive charge in coulombs flowing through a cross section such as $A$ in 1 s.

just as in Fig. 17.1. The current is again in the direction of charge flow because the charge carriers are positive. In metals, however, the current carriers are electrons (negative), and so the current and charge motion have opposite directions. As you will see, in practical circuit electricity, we speak only in terms of the current and seldom examine whether the charge carriers are positive or negative.

---

***Example 17.1***
The current through a flashlight bulb is 0.150 A. How many electrons flow through the bulb each second?

***Reasoning*** Since current is the charge per second that flows past a point, we know that 0.150 C of charge flows through the bulb each second. Each electron carries a charge of magnitude $1.60 \times 10^{-19}$ C. The number of electrons needed to make up a charge of 0.150 C is

$$\text{Number of electrons} = \frac{0.150\ \text{C}}{1.60 \times 10^{-19}\ \text{C/electron}} = 9.36 \times 10^{17}\ \text{electrons}$$

As we shall soon see, this tremendously large number is what causes electric currents in wires to be analogous to water flow in pipes. ∎

---

## 17.2 A simple electric circuit

Before we discuss the way an electric circuit behaves, let us look at a more easily visualized situation, the flow of water molecules through a pipe. Figure 17.3 shows a pipe system completely filled with water. A pump provides energy to the water

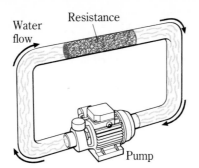

FIGURE 17.3

The water pump furnishes energy to the water and causes it to flow through the tightly packed glass wool in the resistance section.

molecules and causes them to flow through the pipes. Because water fills the entire pipe system and is incompressible, all the pipes carry the same water current. The pipes are large enough so that little viscous loss occurs, but the pipe section labeled "resistance" is packed with glass wool so that the water has great difficulty flowing through it. Obviously, the resistance section is the major obstacle to flow; nearly all the energy furnished to the water by the pump appears as viscous energy losses (that is, heat) in the resistance section. In effect, the water simply carries energy from the pump to the resistance section, where the energy is liberated as heat.

An analogous electrical system is shown in Fig. 17.4a. A battery is connected to metal wires to form what is called an *electric circuit*. The colored wire is much

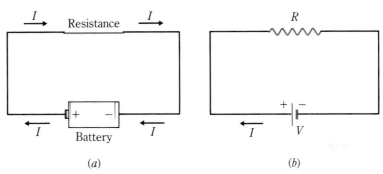

FIGURE 17.4

The battery causes charges to flow in the circuit. Energy given to the charges by the battery is released as heat in the resistance. Part (*b*) is a schematic diagram for the circuit in (*a*).

thinner than the black wires, and so it offers a very large resistance to charge flow through it.* These metal wires contain a multitude of free electrons, which we can liken to water molecules in the pipes of Fig. 17.3. Just as the pump furnishes energy to the water molecules, the battery furnishes energy to the free charges in the metal and causes them to flow.

Most of the energy furnished by the battery is lost as heat when the charges flow through the high-resistance wire. Thus the flowing charges simply carry energy from the battery to the resistance; there the energy is liberated as heat. In fact, if

---

* Alternatively, the colored wire could be made of a metal that offers much more resistance to charge flow than does the metal used to construct the remainder of the circuit. An example of such a system is one using iron for the colored wire and copper for the black ones.

the amount of heat liberated is large enough, the wire will become white-hot. An example of this is shown in Fig. 17.5, where a battery causes charge to flow through a flashlight bulb. The filament of the bulb, a hair-thin wire, glows white-hot as it releases the energy furnished by the battery.

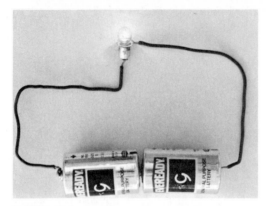

**FIGURE 17.5**

The two 1.55-V batteries shown here are connected together in such a way as to act as a single 3.10-V battery. Where does the heat and light energy emitted by the bulb come from?

Figure 17.4*b* shows the diagram used to represent the circuit in *a*. Notice the symbol ——∿∿—— used for the resistance wire. We call it a **resistor.** All other wires are assumed to have negligible resistance to flow, and so no appreciable heat is generated in them. The energy furnished by the battery of voltage *V* is delivered to the resistor *R* and is there released as heat.

Before leaving this section, we should point out another similarity between water flow in a pipe and charge flow in an electric circuit. In the water circuit, it is obvious that when a certain amount of water enters one end of the pump, an equal amount flows out the other end. Because the pipes are filled, water cannot flow in one section unless it flows in all sections. Like the water molecules, the free charges in the metal of the wires in the electric circuit fill their confining "pipes," the wires. When any amount of charge flows into one end of the battery, a like amount must flow out the other end. Thus the current (the flow of charge per second) is the same everywhere in the circuit of Fig. 17.4. In the next section, we find the relationship between the current, the battery voltage, and the resisting effect of the resistor.

## 17.3 Ohm's law

Let us examine the circuit shown in Fig. 17.6. Because we assume that only negligible energy losses occur in the wire from *p* to *a*, the energy of the charges does not change as they move through this section of the wire. Hence wire *pa* is an equipotential, and so point *a* is at the same electrical potential as point *p*. Similarly, point *b* is at the same potential as point *n*. Hence we arrive at the fact that the potential difference across the resistor is the same as the potential difference across the battery, namely *V*.

Because end *a* of the resistor is connected to the positive terminal of the battery, point *a* is at a higher potential than point *b*. Any positive charge free to move through

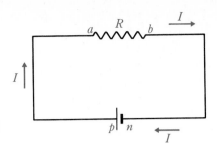

FIGURE 17.6

The direction of the current is always from high to low potential through a resistance.

the resistor moves from $a$ to $b$—in other words, from high to low potential. Hence the direction of the current through the resistor is from $a$ to $b$. In fact,

The direction of the current through a resistor is always from its high-potential end to its low-potential end.

We characterize a resistor by its resistance $R$. If a potential difference $V$ across the resistor causes a current $I$ through it, then the resistance $R$ is defined to be

$$R = \frac{V}{I} \quad \text{or} \quad V = IR \quad\quad\quad (17.2)$$

The unit for resistance is the volt per ampere, which is called the ohm ($\Omega$). The relation expressed in Eq. 17.2 was first proposed by Georg Simon Ohm (1787–1854), whose experiments showed that $I$ is proportional to $V$. Consequently, Eq. 17.2 is often called *Ohm's law*. However, Ohm also believed that the resistance of a wire is the same under all conditions. This is not always true; for example, $R$ varies with temperature, as we shall soon learn. Even so, we still refer to Eq. 17.2 as Ohm's law, but we recognize that $R$, as defined by it, can vary.

---

*Example 17.2*

A certain flashlight bulb draws a current of 0.080 A when the potential difference across it is 1.55 V. What is the resistance of the bulb?

*Reasoning*  This situation is shown in Fig. 17.5. We are told that $V = 1.55$ V across the resistor (the filament of the bulb) and that $I$ through it is 0.080 A. Using Ohm's law, $V = IR$, we have

$$R = \frac{V}{I} = \frac{1.55 \text{ V}}{0.080 \text{ A}} = 19.4 \ \Omega$$

We shall see in the next section that the bulb's resistance is much lower if its filament is not white-hot.  ∎

---

## 17.4 Resistivity and its temperature dependence

Wires that are identical in size but made from different metals have different resistances. For example, a copper wire has less resistance than an iron one of the same size. We therefore need a way to characterize the resistance properties of a material. To do this, let us consider the wire of length $L$ and cross-sectional area $A$ shown in Fig. 17.7. As you might guess, the resistance of the wire increases as $L$ is made larger and decreases as $A$ is made larger. Indeed, experiment shows that

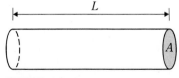

FIGURE 17.7

The resistance of a uniform wire varies directly with $L$ and inversely with $A$.

$$R \sim \frac{L}{A}$$

We can change this proportionality to an equation by introducing a constant of proportionality $\rho$ (Greek rho):

$$R = \rho \frac{L}{A} \quad \text{or} \quad \rho = R \frac{A}{L} \tag{17.3}$$

where $\rho$ has the units ohm-meters and depends on the material from which the wire is made. We call $\rho$ the **resistivity** of the material. It is numerically equal to the resistance between opposite faces of a cube (1 m $\times$ 1 m $\times$ 1 m) made from the material. For very good electrical conductors, such as copper, $\rho$ is small. Typical values for resistivity are given in Table 17.1. Notice that values are given for insulators as well as metals. Electrically insulating materials, such as wood and glass, contain a few ions (usually impurities) that give rise to charge motion when a voltage is impressed across them. Therefore, the resistivity of these materials is not infinite.

Resistivity changes with temperature. For example, the resistance of a metal filament in an incandescent light bulb increases more than tenfold as the filament changes from room temperature to white-hot. Although the resistance of ordinary metals increases with temperature, the opposite occurs for graphite and most semiconductors. We shall see the reason for this latter behavior in Chap. 26.

The variation of resistance (and resistivity) with temperature can be represented over a limited range as follows:

$$\Delta R = R_o \alpha \Delta T \quad \text{and} \quad \Delta \rho = \rho_o \alpha \Delta T \tag{17.4}$$

In this expression, $\Delta R$ (or $\Delta \rho$) is the change in resistance (or resistivity) undergone by a resistor of resistance $R_o$ (or resistivity $\rho_o$) when its temperature is changed by an amount $\Delta T$. The constant $\alpha$, which depends upon the material, is called the *temperature coefficient of resistivity*. The typical values given in Table 17.2 are

■ TABLE 17.1
Resistivity at 20°C

| Material | $\rho \, (\Omega \cdot m)$ |
|---|---|
| Silver | $1.6 \times 10^{-8}$ |
| Copper | $1.7 \times 10^{-8}$ |
| Aluminum | $2.8 \times 10^{-8}$ |
| Tungsten | $5.6 \times 10^{-8}$ |
| Iron | $10 \times 10^{-8}$ |
| Graphite | $3.5 \times 10^{-5}$ |
| Blood | 1.5 |
| Fat | 25 |
| Wood | $10^8 - 10^{12}$ |
| Glass | $10^{12}$ |
| Polystyrene | $10^{15} - 10^{19}$ |

■ TABLE 17.2
Temperature coefficients of resistivity at 20°C

| Material | $\alpha \, (per \, C°)$ |
|---|---|
| Silver | 0.0038 |
| Copper | 0.0039 |
| Aluminum | 0.0040 |
| Tungsten | 0.0045 |
| Iron | 0.0050 |
| Graphite | $-0.0005$ |
| Germanium | $-0.05$ |
| Silicon | $-0.07$ |

correct only for small temperature changes near the temperature at which $R_o$ and $\rho_o$ are measured.

Because resistance varies with temperature, it can be used to measure temperature. The small electronic probes now widely used as fever thermometers make use of this fact. However, these devices use semiconductor resistors, materials that have exceptionally large temperature coefficients of resistivity.

---

**Example 17.3**

Number 12 copper wire has a cross-sectional area of 0.0331 cm². What is the resistance of a 40-m length?

**Reasoning**  We make use of $R = \rho\,(L/A)$ with $L = 40$ m, $A = 0.0331 \times 10^{-4}$ m², and $\rho = 1.7 \times 10^{-8}\ \Omega \cdot$ m:

$$R = \rho \frac{L}{A} = \frac{(1.7 \times 10^{-8}\ \Omega \cdot \text{m})(40\ \text{m})}{0.0331 \times 10^{-4}\ \text{m}^2} = 0.205\ \Omega$$

This wire has a diameter typical of electrical connecting wires, and so you see why we can usually ignore the resistance of such wires.  ■

---

**Example 17.4**

A light bulb whose filament is made of tungsten has a resistance of 240 Ω when white-hot (about 1800°C). Find the approximate resistance of the bulb at room temperature (20°C).

**Reasoning**  The temperature range here is far too large for Eq. 17.4 to be precise. However, as an approximation,

$$\Delta R = R_o \alpha \Delta T$$

becomes

$$\Delta R = R_o\,(0.0045/\text{C}°)(1780°\text{C})$$

where $R_o$ is the resistance at 20°C. Since $\Delta R = 240\ \Omega - R_o$, we have

$$240\ \Omega - R_o = 8.0 R_o$$

from which we find $R_o = 26.7\ \Omega$.  ■

---

**Exercise**  At what temperature is the light bulb's resistance 40.0 Ω?  *Answer: 131° C*  ■

---

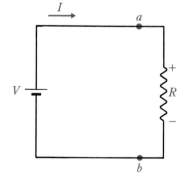

**FIGURE 17.8**

The energy lost by the battery appears as heat in the resistor.

## 17.5 Power and electrical heating

When a battery sends a current through a resistor, as in Fig. 17.8, the battery is furnishing energy to the resistor. In effect, the battery lifts charge from the low potential at $b$ to the higher potential at $a$. The potential difference between these two points is simply the voltage of the battery. The work done by the battery in the process of lifting a charge $\Delta q$ from $b$ to $a$ is, from Eq. 16.2,

# Super-conductors

The resistance of metals decreases as their temperature is lowered. This was known as early as 1835, from the measurements of Heinrich Lentz. There was some speculation in later years that perhaps the resistance of a metal is zero at absolute zero ($-273\,°C = 0$ K). However, no one was able to measure resistance at very low temperatures until much later than 1835 because there was no way to cool objects to such low temperatures.

The attainment of very low temperatures received considerable impetus in 1883, when Wroblewski and Olzewski succeeded in liquefying air. With liquid air as a cooling agent, it was possible to carry out experiments at the boiling point of liquid oxygen ($-183\,°C$) and liquid nitrogen ($-196\,°C$). In 1898, James Dewar succeeded in liquefying hydrogen, which has a boiling point of $-263\,°C$, or 10 K. There remained only one other gas to liquefy, helium. This was finally done in 1908 by Heike Kamerlingh Onnes, and it was found that liquid helium boiled at 4.2 K.

Using liquid helium to achieve lower temperatures than had ever been possible before, Kamerlingh Onnes set out to measure the resistance of metals at very low temperatures. The measuring technique was quite simple in principle but quite complex in practice because of the difficulty of maintaining the temperature constant. Since resistance is defined by Ohm's law, Kamerlingh Onnes needed only to measure the voltage drop across a cylinder of metal with a known current through it. Then $R = V/I$. For his measurements, Kamerlingh Onnes used pure mercury as his metal. (It solidifies at $-39\,°C$ and so was in its solid form.) When he carried out the requisite measurements in 1911, he obtained the astonishing data shown in the accompanying graph.

Although the metal was steadily approaching a very low resistance as the temperature was lowered, the resistance suddenly decreased to zero at 4.2 K. The mercury had become what we now refer to as a *superconductor*. Subsequent

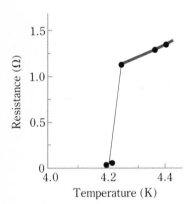

---

$$\text{Work} = V\Delta q$$

Thus the charge $\Delta q$ is given an energy $V\Delta q$ as it is raised by the battery from the low-potential end of the resistor to the high-potential end. The charge loses this energy as it generates heat while flowing through the resistor.

Let us next compute the power output of the battery as it drives a current through the circuit. We know from Eq. 5.2 that

$$\text{Power} = \frac{\text{work done}}{\text{time taken}}$$

If the battery transports a charge $\Delta q$ through a potential difference $V$ in a time $\Delta t$, we know that the work done in the time $\Delta t$ is $V \Delta q$, and so the equation for power becomes

$$\text{Power} = \frac{V\Delta q}{\Delta t}$$

experiments indicated that charge flows essentially forever in a ring made of a superconductor, even though no source of emf is maintained. As far as experiment has shown, the resistance of a superconductor is essentially zero.

Kamerlingh Onnes later succeeded in showing that lead, tin, and indium also become superconductors at 7.2, 3.7, and 3.4 K, respectively. Since that time, many other superconductors have been found. Until 1987, however, no material was known to be superconducting above about 20 K, a temperature too low to be reached easily. But in 1987, certain metal compounds were discovered that become superconducting at temperatures above the boiling point of nitrogen, 77 K. Because liquid nitrogen is cheap and readily available, these new materials are easily cooled to their superconducting state. As a result, it appears that widespread use of superconductors will become a reality in the near future.

Since no heat is generated (and no energy is lost) as charge flows through a superconductor, important applications have already been found for superconductivity. For example, the large accelerators ("atom smashers") used to create new fundamental particles require large magnetic fields; the electromagnets used for this purpose use coils of superconducting wire. Superconducting magnets are also used to provide the huge magnetic fields needed to counterbalance the force of gravity in the magnetically levitated trains now being developed. In medicine, superconducting magnets and detectors are necessary components of the apparatus used in making magnetic field scans of the body. Similar apparatus is used in such widely diverse fields as geology and computer technology. The discovery of new materials that become superconducting at easily accessible temperatures is almost certain to spark a major revolution in science and technology.

---

But $\Delta q/\Delta t$ is simply the current $I$ in the circuit. Hence we find that the power delivered by a voltage source $V$ as it furnishes a current $I$ is

$$\text{Power} = VI \tag{17.5}$$

As the charges pass through the resistor, they *fall* through a potential difference $V$. Consequently, Eq. 17.5 also gives the electric power *lost* in the resistor. Thus we have the following relation for the electric power loss for a current $I$ through a resistor $R$:

$$\text{Power loss in resistor} = VI = I^2R = \frac{V^2}{R} \tag{17.6}$$

where the latter forms are obtained by use of Ohm's law, $V = IR$.

We learned in Sec. 5.2 that the unit of power is the joule per second, which is given the name watt (W). We are all familiar with the use of this unit in electricity because we see it on light bulbs and electrical appliances. For example, if you examine a 60-W bulb, you will see stamped on it "60 W, 120 V." This means that the bulb consumes 60 W of power when 120 V is impressed across it. Another example is a 1500-W electric space heater designed for use on 120 V. Because power is work per unit time, the space heater furnishes 1500 J of heat each second when operated on 120 V.

Electrical energy to operate appliances is furnished by the power company, which charges us for this energy at a rate of perhaps 10 cents per kilowatthour (kWh). The energy unit the company uses, the kilowatthour, is arrived at by using the defining equation for power, which may be rewritten as

Work = power × time taken

If we measure power in kilowatts and time in hours, then the work (or energy) furnished is in kilowatthours. For example, a 1500-W heater operated for 6 h uses the following amount of electrical energy:

Energy = power × time = (1.50 kW)(6 h) = 9.0 kWh

At a cost of $0.10 per kilowatthour, it would cost $0.90 to run such a heater for 6 h.

This non-SI energy unit is related to the joule through

$$1 \text{ kWh} = 3.60 \times 10^6 \text{ J}$$

You are asked to obtain this conversion in Prob. 53 at the end of this chapter.

---

**Example 17.5**

How much heat does a 40-W light bulb generate in 20 min?

*Reasoning*   Because power is work per unit time, a 40-W bulb generates 40 J of heat per second. Therefore, in (20)(60) s it develops

Heat = (20)(60)(40) J = 48,000 J   ■

---

*Exercise*   How many calories of heat is this?   *Answer: 11,500 cal*   ■

---

**Example 17.6**

If electrical energy costs 10 cents per kilowatthour, how much does it cost to operate a 700-W coffeemaker for 30 min?

*Reasoning*   Because work (or energy) is power multiplied by time,

Energy = power × time = (0.700 kW)(0.50 h) = 0.350 kWh

At $0.10 per kilowatthour, the cost is $0.0350.   ■

---

*Exercise*   How much would it cost to heat 200 g of coffee from 20 to 90°C using this coffeemaker? Neglect heat losses.   *Answer: 0.163 cent*   ■

## 17.6 Kirchhoff's point rule

Until now, we have been discussing current in a single wire, in which, of course, each moving charge has to follow the same path. We call the path along which the charge moves an *electric circuit*. Many electric circuits are more complicated than a single wire, however, because the moving charges can follow any one of several paths between two points. Analysis of these more involved circuits requires the use of two basic rules, called Kirchhoff's rules. They are very easy to understand and remember because they are almost obvious.

To see what the first rule is, refer to Fig. 17.9. We see there a situation in which several wires meet at a junction point. Consider point $a$. The current into this point

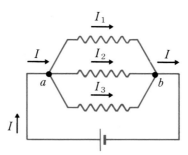

**FIGURE 17.9**

Kirchhoff's point rule tells us that $I = I_1 + I_2 + I_3$.

is $I$, and the currents out of it are $I_1$, $I_2$, and $I_3$. What can we say about these currents? We can easily answer this question if we think of the water-pipe analogy, in which the quantity $I$ represents a flow of so many cubic centimeters of water into point $a$ each second. If the pipes do not leak and because water cannot disappear, this exact same amount of water must flow out of point $a$ each second. In other words, since point $a$ cannot store water, as much must flow out of the point as flows in. We can therefore state that the water current into point $a$ must equal the current out of point $a$.

A similar situation applies to the flow of charge. Because the amount of charge is conserved, the current out of point $a$ must equal the current that enters point $a$. Therefore, we can state that $I = I_1 + I_2 + I_3$. Moreover, this same rule must apply to any point in the circuit. At $b$, for example,

Current into $b$ = current out of $b$

becomes

$$I_1 + I_2 + I_3 = I$$

This is identical to the equation found for point $a$. We can summarize this result in what is called **Kirchhoff's point rule:**

The sum of all the currents into a point must equal the sum of all the currents leaving the point.

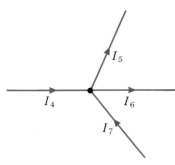

**FIGURE 17.10**

According to the point rule, $I_4 + I_7 = I_5 + I_6$.

We shall find that this simple rule is of very great importance. As another example of its use, notice that when the rule is applied to Fig. 17.10, it gives $I_4 + I_7 = I_5 + I_6$.

## 17.7 Kirchhoff's loop rule

To understand the second of Kirchhoff's rules, consider a circuit in which the current is either steady or changing very slowly. In these cases, the electric field near the circuit is essentially electrostatic. Under such conditions, the electric field is a conservative field. By this we mean that the work done in carrying a positive test charge from one point to another in the field is independent of the path followed. Let us see what this tells us about the electric circuit of Fig. 17.11.

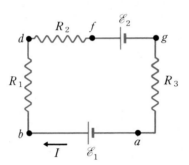

FIGURE 17.11
What does Kirchhoff's loop rule tell us about this circuit?

Suppose we start at point $a$ in the circuit and carry a positive test charge along it through points $b$, $d$, $f$, $g$, and back to $a$. How much work have we done in carrying the test charge *around the circuit and back to the starting point*? Since the charge has been moved around a closed path in a conservative field, the answer is zero.

The situation here is much like a similar situation involving the gravitational field, another conservative field. Suppose you rise in the morning from your bed and return to it at night. The net amount of work you have done for the whole day on your body against the gravitational field is zero. Since the starting and ending points are the same, the gravitational potential energy of your body is unchanged. The total work done against gravity is zero.

In the electric circuit of Fig. 17.11, we see that the following is true. If we carry a positive test charge around the circuit and back to the starting point, zero net work is done. This must mean that the charge was carried through an equal amount of voltage rises and voltage drops. The net effect of all these voltage changes (with rises taken as positive and drops as negative) is that their algebraic sum is zero. This fact is summarized in **Kirchhoff's loop rule:**

The algebraic sum of the voltage changes around a closed circuit must equal zero.

As we see, the loop rule is intimately connected with potential rises and drops. For that reason, let us review what happens to the potential as we move across a resistor, a battery, and a capacitor.

Suppose we move from $a$ to $b$ through the resistor in Fig. 17.12. We know that the current direction is always from high to low potential through a resistor. Hence we know that the change from $a$ to $b$ is a drop in potential; therefore its sign is negative. Ohm's law tells us that its magnitude is $IR$. The change in potential in going from $a$ to $b$ is $-IR$.

The battery symbol tells us that the left side of the battery in Fig. 17.12 is positive. (Remember, the long side of the symbol is positive.) Therefore point $a$ is $\mathcal{E}$ volts higher than point $b$. Going from $a$ to $b$, the potential change is $-\mathcal{E}$.

Resistor: $a$ to $b \rightarrow -IR$

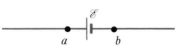

Battery: $a$ to $b \rightarrow -\mathcal{E}$

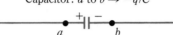

Capacitor: $a$ to $b \rightarrow -q/C$

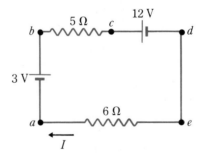

**FIGURE 17.12**

In each case, going from $a$ to $b$ is a voltage drop, that is, a negative voltage change. Going from $b$ to $a$, the voltage change would be positive.

For the capacitor, we must be told which plate is positively charged. According to the diagram, plate $a$ is positive. It therefore is at the higher potential. Since the potential difference across a capacitor is given by Eq. 16.5, $q = CV$, to be $q/C$, the potential change in going from $a$ to $b$ is $-q/C$.

The potential change in each of these three cases is negative when we go from $a$ to $b$. If we were going from $b$ to $a$, the change would be positive. Let us now use the loop rule in a few simple circuits before moving on to its more serious applications.

---

**Example 17.7**

Find the current in the circuit of Fig. 17.13.

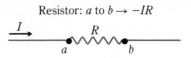

**FIGURE 17.13**

When we solve this circuit, how will our answers tell us that we have chosen $I$ in the wrong direction?

**Reasoning**  Let us guess that the current is in the direction shown. (You might protest that this is wrong since the 12-V battery will certainly have more effect than the 3-V battery; but one of the nice things about Kirchhoff's rules is that even a poor guesser can use them, as we shall see.) We pick a point such as $a$ as a starting point and move around the circuit. The voltage changes are

$$a \rightarrow b \quad +3 \text{ V}$$
$$b \rightarrow c \quad -(5 \ \Omega)(I)$$
$$c \rightarrow d \quad -12 \text{ V}$$
$$d \rightarrow e \quad \ \ 0 \text{ V}$$
$$e \rightarrow a \quad -(6 \ \Omega)(I)$$

Check these values so that you are sure about the signs we have used. The sum of these voltage changes must be zero. Therefore, in volts,

$$3 - (5 \ \Omega)(I) - 12 - (6 \ \Omega)(I) = 0$$

Solving for $I$, we find $I = -\frac{9}{11}$ A. The negative sign tells us we guessed wrongly the direction of the current. No harm is done. The current is $\frac{9}{11}$ A in the opposite direction.

Suppose we had circled the circuit in the reverse direction. Then our equation would have been

$$+(6 \ \Omega)(I) + 12 + (5 \ \Omega)(I) - 3 = 0$$

from which $I = -\frac{9}{11}$ A, as before.

In solving this circuit, be sure you understand our choice of signs for the voltage changes. Also note that the current is the same at all points in the circuit. Why? ∎

---

*Exercise*   Find $I$ if the 3-V battery is reversed.   *Answer: $-1.36$ A* ∎

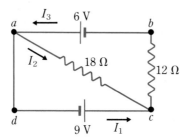

**FIGURE 17.14**

Determine the value of the currents in all three wires.

---

*Example 17.8*

Find the currents in all the wires of the circuit shown in Fig. 17.14.

*Reasoning*   We assign currents to all the wires and give each a symbol and direction. Once again, we waste little time trying to guess proper direction because our answer will indicate direction.

Starting at $a$, let us follow the loop $acda$ and write the loop rule. We have, in volts (be sure you understand the signs used),

$$-(18 \ \Omega)(I_2) - 9 + 0 = 0$$
$$I_2 = -0.50 \ \text{A}$$

The current $I_2$ is therefore in the direction opposite that shown.

Now let us move around the loop $abcda$. We have

$$-6 + (12 \ \Omega)(I_3) - 9 = 0$$
$$I_3 = 1.25 \ \text{A}$$

We could also write a loop equation for loop $abca$, but no new voltage changes would appear in it. Therefore, this equation would contain no new information and would be redundant. Instead, we write the point rule for point $c$:

$$I_1 + I_2 = I_3$$

Substituting, we have, in amperes,

$$I_1 - 0.50 = 1.25$$

Notice that we carry along the signs we found for $I_2$ and $I_3$. Solving, we find that $I_1 = 1.75$ A. ∎

---

**Exercise**  Find $I_2$ and $I_3$ if the 9-V battery is reversed.  *Answer: 0.500 A, − 0.250 A* ∎

---

## 17.8 Resistors in series and in parallel

There are two configurations of resistor circuits that can be simplified very easily. If we recognize them, we can often greatly decrease the amount of work needed to solve a problem. In the configuration shown in Fig. 17.15a, the resistors between points *a* and *b* are said to be in **series.** *To move from point a to point b, there is only one possible path. That path goes through all the resistors. Each resistor carries the same current.*

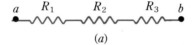

(a)

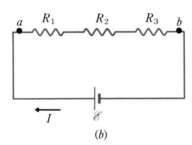

(b)

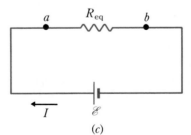

**FIGURE 17.15**

The three resistors are in series. They are equivalent to $R_{eq} = R_1 + R_2 + R_3$.

(c)

If we connect this combination across a battery, as in part *b*, we might guess it would act like a single resistor equal to the sum $R_1 + R_2 + R_3$. In what follows, we prove this guess to be correct. What we wish to do is find an equivalent resistor $R_{eq}$

that draws the same current as the three-resistor combination. To do this, we make use of the loop rule for the circuits of parts $b$ and $c$.

Going clockwise around the circuit in part $b$, we have the following loop equation:

$$+\mathcal{E} - IR_1 - IR_2 - IR_3 = 0$$

This can be rewritten as

$$\frac{\mathcal{E}}{I} = R_1 + R_2 + R_3$$

Writing a similar loop equation for part $c$ gives

$$+\mathcal{E} - IR_{eq} = 0$$

$$\frac{\mathcal{E}}{I} = R_{eq}$$

From these two equations, we see that

$$R_{eq} = R_1 + R_2 + R_3 \qquad \text{series} \qquad (17.7)$$

In general,

Several resistors in series are equivalent to a single resistor equal to their sum.

Another common configuration is shown in Fig. 17.16$a$. These resistors are said to be in **parallel.** *In a parallel configuration, one end of each resistor is connected to a point a and the other end of each resistor is connected to a point b.* When a

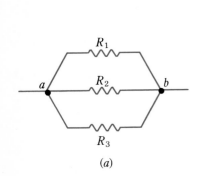
(a)

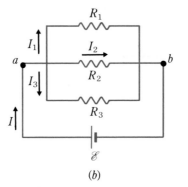
(b)

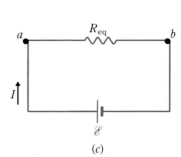
(c)

**FIGURE 17.16**

The three resistors are in parallel. Their equivalent is given by $1/R_{eq} = 1/R_1 + 1/R_2 + 1/R_3$.

potential difference is placed across the combination, as in Fig. 17.16$b$, each resistor has this same potential difference across it. We wish to find the equivalent resistor $R_{eq}$ that can be used to replace the combination and still draw the same current.

Notice in Fig. 17.16$b$ that the potential difference across each resistor is $\mathcal{E}$. Therefore, Ohm's law tells us that

$$I_1 = \frac{\mathcal{E}}{R_1} \qquad I_2 = \frac{\mathcal{E}}{R_2} \qquad I_3 = \frac{\mathcal{E}}{R_3}$$

But the point rule tells us that

$$I = I_1 + I_2 + I_3$$

and so

$$I = \frac{\mathcal{E}}{R_1} + \frac{\mathcal{E}}{R_2} + \frac{\mathcal{E}}{R_3}$$

If we divide both sides of this equation by $\mathcal{E}$, it becomes

$$\frac{I}{\mathcal{E}} = \frac{1}{R_1} + \frac{1}{R_2} + \frac{1}{R_3}$$

Now let us look at the circuit in part $c$. The loop rule tells us that

$$+\mathcal{E} - IR_{eq} = 0$$

from which

$$\frac{I}{\mathcal{E}} = \frac{1}{R_{eq}}$$

We can now equate these two expressions for $I/\mathcal{E}$ to give

$$\frac{1}{R_{eq}} = \frac{1}{R_1} + \frac{1}{R_2} + \frac{1}{R_3} \qquad \text{parallel} \tag{17.8}$$

In general,

For several resistors in parallel, the reciprocal of their equivalent resistance is equal to the sum of the reciprocals of the resistances.

---

**FIGURE 17.17**

The parallel resistors between $b$ and $c$ are equivalent to 2 Ω, as shown in (*b*). The two series resistors in (*b*) can be combined, as in (*c*).

### Example 17.9

Find the current $I$ through the battery in Fig. 17.17$a$.

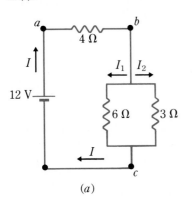

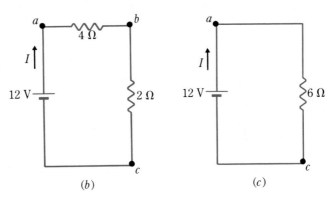

(*a*)  (*b*)  (*c*)

**Reasoning** We could solve this using Kirchhoff's rules. However, it is usually simpler to combine obvious series and parallel resistors before writing the loop equations. The resistances may be reduced as shown in parts *b* and *c*. Let us first combine the two parallel resistors between points *b* and *c*. We have

$$\frac{1}{R_{bc}} = \frac{1}{6} + \frac{1}{3} = \frac{1}{6} + \frac{2}{6} = \frac{3}{6}$$

$$R_{bc} = 2 \ \Omega$$

An equivalent circuit is now drawn in Fig. 17.17*b*, with the parallel combination replaced by its equivalent resistance. We see that the 4- and 2-$\Omega$ resistors are connected in series between points *a* and *c*. Their equivalent is

$$R_{ac} = 4 \ \Omega + 2 \ \Omega = 6 \ \Omega$$

A new equivalent circuit, shown in part *c*, is now drawn. This is a situation to which Ohm's law can be applied. The voltage difference across the 6-$\Omega$ resistor is 12 V. Hence we have

$$I = \frac{V}{R} = \frac{12 \ \text{V}}{6 \ \Omega} = 2 \ \text{A} \quad \blacksquare$$

**Example 17.10**
Find the current through the battery in Fig. 17.18*a*.

**FIGURE 17.18**

The complex circuit of (*a*) can be reduced to the simple equivalent circuit in (*d*).

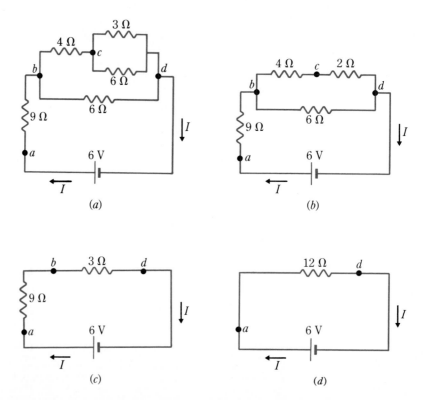

*Reasoning* Here we have two parallel resistors again: 3 Ω and 6 Ω. They can be replaced by a 2-Ω resistor, as in part *b*. The 2- and 4-Ω resistors are in series and so are equivalent to a 6-Ω resistor. This equivalent 6-Ω resistor is in parallel with the actual 6-Ω resistor. By using the reciprocal formula, we know we can replace the two 6-Ω resistors in parallel by a 3-Ω resistor. This is drawn in part *c*. Finally, the circuit can be reduced to the form shown in part *d*. Using Ohm's law for it, we have

$$I = \frac{V}{R} = \frac{6\text{ V}}{12\text{ Ω}} = 0.500\text{ A} \quad \blacksquare$$

We can treat series and parallel combinations of capacitors in much the same way we treat resistors. We show the two configurations in Fig. 17.19*a* and *b*. The results for the equivalent capacitors are shown in each case. These results are obtained in problems at the end of the chapter. Notice that the relations are just the opposite of those we found for resistors. *Parallel capacitors add directly, while the reciprocals of series capacitances add.*

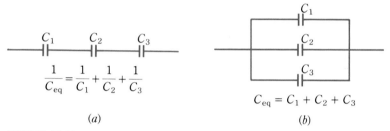

$$\frac{1}{C_{eq}} = \frac{1}{C_1} + \frac{1}{C_2} + \frac{1}{C_3}$$

(*a*)

$$C_{eq} = C_1 + C_2 + C_3$$

(*b*)

**FIGURE 17.19**

Reduction of series and parallel capacitances: (*a*) series; (*b*) parallel.

## 17.9 Solving circuit problems

We now have at our disposal the tools needed to solve most direct-current (dc) circuit problems. Before we use these tools in several examples, let us state a few facts we should remember. Although each problem has its own peculiar features, the following general approach is most often useful.

**1** Draw the circuit.

**2** Assign a current (both symbol and direction) to each important wire. Be careful to use only one current designation for a given wire even though it may contain several elements. If one wire branches off from another, the currents in the two branches must usually be considered different from each other.

**3** Reduce series and parallel resistance systems whenever possible and convenient.

**4** Write the loop equations for the simplified circuit. Remember that an equation is redundant unless it contains a new voltage change.

**5** Write the point equations for the junction points, remembering that any equation that does not use a new current is redundant.

**6** Solve these equations for the unknowns.

## Example 17.11

Find the currents in the wires of the circuit shown in Fig. 17.20.

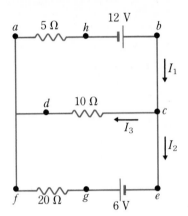

**FIGURE 17.20**

A circuit that is easily solved using Kirchhoff's rules.

*Reasoning* Steps 1 and 2 of our procedure are already completed in the diagram. There are no simple resistance systems. (The resistors in *ab* and *ef* are *not* in simple parallel with the center resistor. Notice that the right-hand ends are connected to batteries, not to a common point.)

Let us now start at point *a* and write the loop equation for loop *abcda*:

$$-(5\ \Omega)(I_1) + 12\ \text{V} - (10\ \Omega)(I_3) = 0 \tag{1}$$

Similarly, for loop *dcefd*,

$$+(10\ \Omega)(I_3) + 6\ \text{V} - (20\ \Omega)(I_2) = 0 \tag{2}$$

We have three unknowns, $I_1$, $I_2$, and $I_3$, but only two equations. The loop equation *abcefa* includes no new voltage changes, and so we ignore it.

To obtain a third equation, we write the point equation for point *c*:

$$I_1 = I_3 + I_2 \tag{3}$$

Substitution of this value for $I_1$ in Eq. 1 gives

$$-(5\ \Omega)(I_3) - (5\ \Omega)(I_2) + 12\ \text{V} - (10\ \Omega)(I_3) = 0$$
$$-(5\ \Omega)(I_2) - (15\ \Omega)(I_3) + 12\ \text{V} = 0 \tag{4}$$

We now can solve Eqs. 2 and 4 simultaneously. Solving Eq. 2 for $I_3$, we find

$$I_3 = 2I_2 - 0.60\ \text{A} \tag{5}$$

Substituting this in Eq. 4 gives

$$-(5\ \Omega)(I_2) - (30\ \Omega)(I_2) + 9\ \text{V} + 12\ \text{V} = 0$$
$$I_2 = 0.600\ \text{A}$$

Using this value in Eq. 5 gives

$I_3 = 0.600$ A

Now, by use of Eq. 3, we find

$I_1 = 1.20$ A  ∎

---

**Example 17.12**
Find $I_1$, $I_2$, and $I_3$ for the circuit in Fig. 17.21.

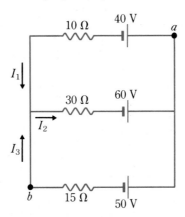

**FIGURE 17.21**
Find the three unknown currents.

*Reasoning*  Choosing the currents in the directions shown and noticing that $I_1 + I_3 = I_2$, we can write the appropriate circuit or loop equations. Taking the loop through the 40- and 60-V batteries, starting at $a$ and going counterclockwise, we have, in volts,

$$-40 - (10\ \Omega)(I_1) - (30\ \Omega)(I_2) + 60 = 0$$
$$I_1 + 3I_2 = 2\ \text{A}$$

Next, starting from $a$ and going clockwise through the 40- and 50-V batteries gives

$$-40 - (10\ \Omega)(I_1) + (15\ \Omega)(I_2 - I_1) + 50 = 0$$
$$2.5I_1 - 1.5I_2 = 1\ \text{A}$$

After multiplying the second equation by 2 and adding it to the first, we have

$6I_1 = 4$ A

$I_1 = \frac{2}{3}$ A

Using this in the first equation gives

$3I_2 = \frac{4}{3}$ A

$I_2 = \frac{4}{9}$ A

And since $I_3 = I_2 - I_1$, we have

$I_3 = -\frac{2}{9}$ A

Clearly, the direction of $I_3$ is opposite what is shown in the diagram.  ∎

*Exercise* Find $V_{ab}$ in Fig. 17.21. *Answer:* $-46.7$ V ∎

## Example 17.13

Find the values of $\mathcal{E}$, $R$, and $I$ in Fig. 17.22. The ammeter reads 0.50 A, and the voltmeter reads 16 V. The polarity of the 8-$\Omega$ resistor is as shown, and the current in the center branch is in the direction indicated. Assume the meters to be ideal; that is, the ammeter resistance is zero and the voltmeter resistance is infinite. This being true, the meters can be ignored.

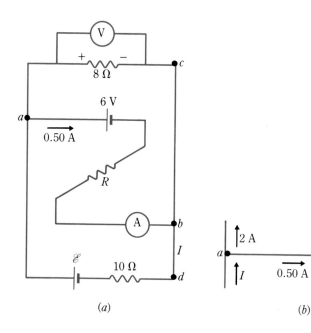

**FIGURE 17.22**

The ammeter and voltmeter readings are known. We wish to find $I$, $X$, and $\mathcal{E}$.

*(a)*          *(b)*

*Reasoning*  Since the voltmeter reads 16 V across the 8-$\Omega$ resistor, Ohm's law tells us that the current in the wire on the top is 2 A. From the polarity of the 8-$\Omega$ resistor, we know that the current is traveling to the right through the wire.

The currents at point *a* must be as shown in Fig. 17.22*b*. From the point rule,

$$I = 2.00 + 0.50 = 2.50 \text{ A}$$

Write the circuit equation for loop *abca*. In doing so, recall that an ideal ammeter has negligible resistance, and so the voltage change across it is zero. In volts,

$$-6 - 0.50R + 16 = 0$$
$$R = 20.0 \ \Omega$$

The circuit equation for loop *acbda* gives

$$-16 - 25 + \mathcal{E} = 0$$
$$\mathcal{E} = 41.0 \text{ V} \quad ∎$$

## Example 17.14

For the circuit in Fig. 17.23, find $I_1$, $I_2$, $I_3$, and the charge on the capacitor.

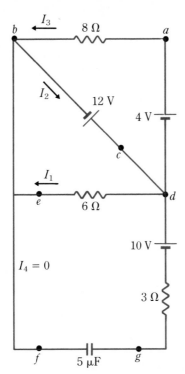

**FIGURE 17.23**

When the capacitor is fully charged, the current through the bottom wire is zero, and that portion of the circuit can be neglected.

**Reasoning**  Notice first that the current in wire *fg* is zero once the capacitor is charged. Hence, $I_4 = 0$, and we can ignore this wire when solving the rest of the circuit. At point *d*,

$$I_2 = I_1 + I_3$$

Writing the loop equation for *dcbed* gives, in volts,

$$-12 + (6\ \Omega)(I_1) = 0$$
$$I_1 = 2.00\ \text{A}$$

Using the loop *dabcd* gives

$$+4 - (8\ \Omega)(I_3) + 12 = 0$$
$$I_3 = 2.00\ \text{A}$$
$$I_2 = I_1 + I_3 = 4.00\ \text{A}$$

If we knew the potential difference between points *f* and *g*, we could use the relation $q = CV_{fg}$ to find *q*. We can easily find $V_{fg}$ by starting at point *g* and moving around the circuit *gdefg*, adding the voltage rises and falls. Doing this gives

$$0 + 10 - (6\ \Omega)(I_1) + V_{fg} = 0$$
$$V_{fg} = -10 + 12 = +2.00\ \text{V}$$

17.9 Solving circuit problems     **395**

The positive sign shows that we have gone uphill in potential in going from $f$ to $g$ and so the $g$ side of the capacitor is positive. We now have, for the charge on the capacitor,

$$q = CV_{fg} = (5 \times 10^{-6} \text{ F})(2 \text{ V}) = 1.00 \times 10^{-5} \text{ C} \quad \blacksquare$$

## 17.10 Ammeters and voltmeters

In Example 17.13, we saw a typical situation in which an ammeter and a voltmeter are used in a circuit. Although we learn how these meters are constructed in Chap. 18, let us not postpone a discussion of how they are used, since you will be using them in the laboratory.

An ammeter is used to measure the current through a wire. We connect it directly in line with the wire, as shown in Fig. 17.24a. Notice that the current to be measured passes through the meter. If the meter had much resistance, it would disturb the circuit. Therefore, an ideal ammeter has zero resistance. The ammeters you will be using in the laboratory will usually have a resistance of only a fraction of an ohm.

Voltmeters are used to measure potential difference. To measure the potential difference $V = IR$ across the resistor in Fig. 17.24b, the voltmeter terminals must be connected to the two ends of the resistor as shown. Ideally, we want the voltmeter to leave the circuit undisturbed. This is possible only if the voltmeter resistance is very large. An ideal voltmeter has infinite resistance and therefore provides no path for the current.

Students who confuse the ammeter and the voltmeter in situations such as those shown in Fig. 17.24 face serious danger to life and happiness because of the severe displeasure of their laboratory teacher. An ideal voltmeter has an infinite resistance. As such, no current passes through it when its terminals are connected to two points that differ appreciably in potential. An ideal ammeter has zero resistance, however. If its terminals were inadvertently connected to two points of different potential, the current through the ammeter would be given by

$$I = \frac{V}{R} = \frac{\text{something}}{\text{zero}} \rightarrow \infty$$

This student error is accompanied by smoke issuing from the meter case, irreparable harm to the meter, and an antagonistic attitude on the part of the instructor. So beware.

## 17.11 House circuits

We are all familiar with the ordinary electric circuits that extend throughout our houses. The power company runs at least two wires to each house to provide a potential difference of about 120 V.* These lead-in wires usually have a large

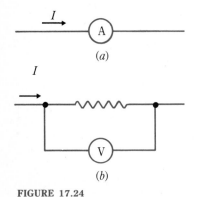

$I$

(a)

$I$

(b)

**FIGURE 17.24**

Why should an ammeter have little resistance and a voltmeter have nearly infinite resistance?

---

* The potential difference supplied by the power company is constantly reversing in a sinusoidal way. We discuss this type of voltage in detail in Chap. 20. For the purposes of the present discussion, the alternating voltage has the same effect as a direct-current voltage.

diameter so that they can carry considerable current without heating up. (The larger the cross-sectional area of the wire, the less its resistance. Since heat generated is proportional to $I^2R$, the low resistance ensures low heat dissipation.)

In most newer houses, the wires inside the house are capable of carrying about 20 A without undue heating. However, to protect against too large a current, a fuse or a circuit breaker is placed in series with the wire. Its purpose is to disconnect the wire from the voltage source if greater than the allowed current is drawn from it. This procedure automatically disconnects any wire that is accidentally called on to carry more than the safe current.

A typical house circuit consists of two parallel wires strung through the house from the 120-V source provided by the lead-in wires (Fig. 17.25). One terminal of each light bulb, appliance, and so on, is connected to the high-potential wire, and the

**FIGURE 17.25**

When a switch S is closed, current passes through the device that the switch controls. Will the lamp light when its cord is plugged in?

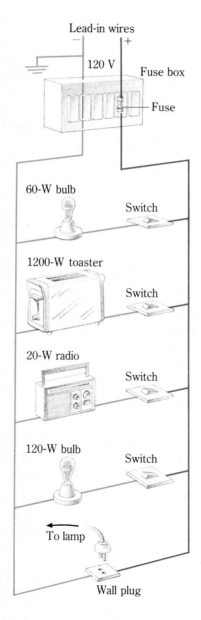

other terminal is connected to the low-potential wire. When the switch to that appliance is closed, charge can flow through the device. The low-potential wire is usually grounded.

Many 120-V appliances have a third prong on the plug. This furnishes a connection between a ground wire and the metal frame of the appliance. If, by accident, the high-voltage wire touches the metal frame of the appliance, a direct connection to ground is made. The effect is the same as connecting the high- and low-voltage wires directly. There is a large current through the high-voltage wire to ground, and the fuse in the high-voltage wire will blow. If the ground wire is absent, such a malfunction will leave the whole appliance "floating" at high potential. Anyone touching the metal frame will then suffer a shock.

Let us compute how much current is drawn by the 60-W bulb of Fig. 17.25 when it is turned on. Since power $= VI$ and since $P = 60$ W and $V = 120$ V in this case, the current through the bulb is $I = 0.500$ A. Similarly, when turned on, the toaster draws 10.0 A, the radio draws 0.167 A, and the 120-W bulb draws 1.00 A. If they are all turned on at once, a total of 11.667 A passes through the fuse. Usually a house circuit is fused for no less than 15 A, and so there is no danger in this case.

A house with many electrical appliances requires more than one circuit. Most houses have several separate circuits, each with its own fuse, similar to the one shown in Fig. 17.25.

It is interesting to compute the resistance of a light bulb. When the bulb is cool, its resistance is not very large. However, when it is connected across the rated voltage, usually 120 V, its resistance element becomes white-hot. As discussed previously, its resistance increases considerably when the bulb heats up. When it is hot, it is operating at the wattage stamped on it. Suppose we have a 60-W, 120-V bulb. We know that

$$P = VI = \frac{V^2}{R}$$

$$60 \text{ W} = \frac{(120 \text{ V})^2}{R}$$

$$R = 240 \ \Omega$$

We saw in Example 17.4 that the resistance of this bulb at room temperature is about 30 $\Omega$.

---

## 17.12 Electrical safety

Since we use electrical appliances daily, we should understand the elements of electrical safety. There are two ways in which electricity can kill a person. It can cause the muscles of the heart and lungs (or other vital organs) to malfunction, or it can cause fatal burns.

Even a small electric current can seriously disrupt cell functions in that portion of the body through which it passes. When the electric current is 0.001 A or higher, a person can feel the sensation of shock. At currents 10 times larger, 0.01 A, a person is unable to release an electric wire held in the hand because the current causes the hand muscles to contract violently. Currents larger than 0.02 A through the torso paralyze the respiratory muscles and stop breathing. Unless artificial respiration is started at once, the victim will suffocate. Of course, the victim must be freed from the voltage source before he or she can be touched safely; otherwise the rescuer, too, will be in great danger. A current of about 0.1 A passing through the region of the heart will shock the heart muscles into rapid, erratic contractions (ventricular fibrillation) so that the heart can no longer function. Finally, currents of 1 A and higher through body tissue cause serious burns.

To prevent injury, *the important quantity to control is current. Voltage is important only because it can cause charge to flow.* Even though your body can be charged to a potential thousands of volts higher than the metal of an automobile when you slide across the car seat, you feel only a harmless shock as you touch the door handle. Your body cannot hold much charge on itself, and so the current through your hand to the door handle is short-lived and the effect on your body cells is negligible.

In some circumstances, a 120-V house circuit is almost sure to cause death. One of the two wires of the circuit is usually attached to the ground, and so it is always at the same potential as the water pipes in the house. Suppose you are soaking in a bathtub; your body is effectively connected to the ground through the water and piping. If your hand accidentally touches the high-potential wire of the house circuit (by touching an exposed wire on a radio or heater, for example), charge will flow through your body to the ground. Because of the large, efficient contact your body makes with the ground, the resistance of the body circuit is low. Consequently the current through your body is so large that you are in danger of being electrocuted.

Similar situations exist elsewhere. For example, if you accidentally touch an exposed wire while you are standing on the ground with wet feet, you are in far greater danger than if you are on a dry, insulating surface. The electric circuit through your body to the ground has a much higher resistance if your feet are dry. Similarly, if you sustain an electric shock by touching a bare wire or a faulty appliance, the shock is greater if your other hand is touching a faucet or is in water.

As you can see from these examples, the danger from electric shock can be eliminated by avoiding a current path through the body. When the voltage is greater than about 50 V, avoid touching any exposed metal portion of the circuit. If a high-voltage wire must be touched, for example, in a power-line accident when help is not immediately available, use a dry stick or some other substantial piece of insulating material to move it. When in doubt about safety, avoid all contact or close approach to metal or to the wet earth. *Above all, do not let your body become the connecting link between two points that are at widely different potentials.*

## 17.13 The emf and terminal potential of a battery

Probably everyone has noticed at one time or another that the lights on an automobile dim when the motor is started. The electric starter used on an automobile draws considerable current from the battery. In so doing, it lowers the potential between the battery terminals, and the car lights dim. We now investigate this nonconstancy of the terminal potential difference of a battery.

As pointed out in Chap. 16, the emf of a battery is generated by the chemical action in the battery. When no current is being drawn from the battery, the difference in potential between the terminals is equal to the emf. However, a battery is a very complex chemical device, and it behaves like an emf in series with a resistor. An equivalent circuit for a battery is shown in Fig. 17.26.

Notice that when no current is being drawn from the battery, there is no potential drop across the internal battery resistance $r$. Hence the potential difference between the terminals is equal to the emf. However, if the battery is connected across

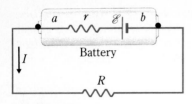

**FIGURE 17.27**
The terminal voltage of the
battery is $\mathcal{E} - Ir$.

a resistor, as in Fig. 17.27, the current is $I$ and the terminal potential difference is
$\mathcal{E} - Ir$.

For a good 12-V battery, the internal resistance is only of the order of 0.01 Ω. If
this battery is connected across a 3-Ω resistor, we have

$$I = \frac{12 \text{ V}}{3\,\Omega + 0.01}\,\Omega \approx 4 \text{ A}$$

The terminal potential is the voltage difference between points $a$ and $b$:

Terminal potential $= 12 \text{ V} - (4 \text{ A})(0.01 \text{ Ω}) = 11.96 \text{ V}$

In this case, the terminal potential is nearly equal to the emf.

As a battery becomes older, however, its internal resistance increases. If the
resistance of the battery in Fig. 17.27 is 1.0 Ω, the current is

$$I = \frac{12 \text{ V}}{4\,\Omega} = 3.0 \text{ A}$$

and the terminal potential is

Terminal potential $= 12 \text{ V} - 3.0 \text{ V} = 9.0 \text{ V}$

It should be clear that when a starter on a car draws 100 A from the battery, the
terminal potential of even a new battery will decrease noticeably.

---

**Example 17.15**
What is the terminal potential of each battery in Fig. 17.28?

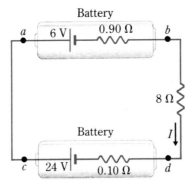

**FIGURE 17.28**
The 24-V battery is charging the
6-V battery. We find that the
terminal potential difference of a
discharging battery is less than its
emf, while the reverse is true for a
battery that is being charged.

**Reasoning** The two batteries are opposing each other, and so the effective
driving emf in the circuit is 18 V. The current is in the direction shown and is

$$I = \frac{18 \text{ V}}{9\,\Omega} = 2 \text{ A}$$

The potential difference from $d$ to $c$ is

$$V = -0.2 \text{ V} + 24 \text{ V} = 23.8 \text{ V}$$

Hence the terminal potential of the 24-V battery is less than its emf. For the terminal potential of the other battery from $b$ to $a$, we have

$$V = +1.8 \text{ V} + 6 \text{ V} = 7.8 \text{ V}$$

Notice that the terminal potential of this battery is larger than the emf. This is always the case when a battery is being charged, as the 6-V battery is in this example. ∎

## Example 17.16

Electrical contacts placed at various positions on a person's head show that potential differences exist and that they fluctuate in characteristic ways. These potential differences can be recorded by an *electroencephalograph*. Typically these differences are of the order of $5 \times 10^{-4}$ V and fluctuate in times of the order of 0.10 s. Suppose the head's resistance between two of these contacts is 10,000 Ω. How large a current can the voltage-recording device draw from the voltage being measured if the voltage read is to be in error by less than 1 percent? How large a resistance must the recording device have?

*Reasoning* The portion of the head between the two electrodes acts as a battery with emf $= V_o$ and internal resistance $= 10,000$ Ω. When a current $I$ is being drawn, the internal resistance causes a voltage error $(10,000 \text{ Ω})(I)$. We wish

$$\frac{(10,000 \text{ Ω})(I)}{V_o} < 0.01$$

Therefore the recorder cannot draw a current in excess of $0.01 V_o/10,000$, or $10^{-6} V_o$ amperes. Since the current through the voltage recorder is caused by $V_o$, the voltage source being measured, we can use Ohm's law to find the resistance $R_o$ of the recorder. Thus $V = IR$ becomes

$$V_o = (10^{-6} V_o)(R_o)$$

from which the recorder resistance must be at least

$$R_o = 10^6 \text{ Ω} \quad \blacksquare$$

## Minimum learning goals

When you finish this chapter, you should be able to

1 Define (*a*) dc circuit, (*b*) current, (*c*) ampere, (*d*) Ohm's law, (*e*) resistance, (*f*) resistivity, (*g*) ohm, (*h*) temperature coefficient of resistivity, (*i*) watt, (*j*) electric power, (*k*) kilowatthour, (*l*) Kirchhoff's rules, (*m*) series and parallel circuits, (*n*) equivalent resistance, (*o*) terminal potential difference and emf.

2 Use the relation $I = \Delta q/\Delta t$ in simple situations.

3 Interpret a simple circuit diagram. State the potential difference between various points of the circuit.

4 State which end of a resistor is at the higher potential when the direction of current through the resistor is given.

5 Use Ohm's law in simple situations.

6 Compute the resistance of a given piece of wire if the resistivity of the wire material is known.

7 Find the resistance of a wire at a given temperature when its resistance and temperature coefficient at some reference temperature are known.

8 Use the power equation $P = VI$ to find the power loss or gain in a resistor, battery, and capacitor under dc conditions.

9 Apply Kirchhoff's point rule.

10 Write the loop equation for a series circuit that contains batteries and resistors.

**11** Reduce a given set of series and parallel resistors to a single equivalent resistor.

**12** Use Kirchhoff's rules to solve dc circuits that contain batteries, resistances, and capacitances.

**13** Sketch a typical house circuit and point out the various elements in it. Compute the current drawn by various portions of a house circuit when the appliances it is running are given.

**14** Analyze a given electrical situation from the viewpoint of safety.

**15** Explain why the terminal potential of a battery is not always equal to the emf. Find the terminal potential if $\mathcal{E}$, $I$, and $r$ are known.

## Questions and guesstimates

**1** Sometimes a student insists that current is used up in a resistor. Arguing from the water analogy, how would you convince such a student that current is not lost in a resistor?

**2** How do we know which end of a battery is at the higher potential, that is, positive, in a schematic diagram of a circuit? How do we know which end of a resistor is at the higher potential?

**3** Fluorescent light bulbs are usually more efficient light emitters than incandescent bulbs. That is, for the same input energy, the fluorescent bulb gives off more light than the incandescent bulb. Carefully touch a fluorescent bulb and an incandescent one after each has been lit for a few minutes. Explain why the incandescent bulb is a less efficient light emitter.

**4** In Fig. P17.1$a$, the pump lifts water to the upper reservoir at a rate such that the water level remains constant. The water slowly trickles out of the narrow tube into the lower reservoir. Point out the similarities between this water circuit and the electric circuit shown in part $b$.

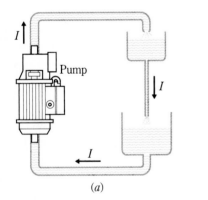 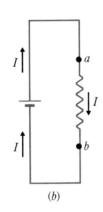

(a)                    (b)

**FIGURE P17.1**

**5** A resistor is connected from point $a$ to point $b$. How does one tell whether there is a potential drop or a potential rise from $a$ to $b$? Repeat for a battery and for a capacitor.

**6** Explain the following statement: For series resistors, the equivalent resistance is always larger than the largest resistance in the series; for parallel resistors, the equivalent resistance is always smaller than the smallest resistance in the combination.

**7** Using an ohmmeter (basically a battery in series with a very sensitive ammeter), measure your resistance from one hand to the other. A current of about 0.02 A through the midsection of the body is sufficient to paralyze the breathing mechanism. About how large a voltage difference between your hands is needed to electrocute you?

**8** If you grasp the two wires leading from the two plates of a charged capacitor, you may feel a shock. The effect is much greater for a 2-$\mu$F capacitor than for a 0.02-$\mu$F capacitor, even though both are charged to the same potential difference. Why?

**9** Birds perch on high-tension wires all the time. Why are they not electrocuted, even when they perch on a part of the wire where the insulation has worn off?

**10** If there is a current of only a small fraction of an ampere into one of a person's hands and out the other, the person will probably be electrocuted. If the current path is into one hand and out the elbow on the same arm, the person can survive even if the current is large enough to burn the flesh. Explain.

**11** Parents frequently worry about children playing near electrical outlets. Discuss the various factors that determine how badly shocked a child could be. What would happen if a child were to cut a lamp cord in two with a pair of noninsulated, wire-cutting pliers when the cord is plugged in? Is the child in any danger?

**12** Explain why touching an exposed circuit wire when you are in a damp basement is much more dangerous than touching the same wire when you are on the second floor.

**13** It is extremely dangerous to use a plug-in radio near a bathtub when you are taking a bath. Why? Does the same reasoning apply to a battery-operated radio?

# Problems

*Ignore internal resistances in batteries unless they are given.*

**1** There is current of 0.40 A through a light bulb. (*a*) How much charge is transported through the bulb in 3 h? (*b*) How many electrons flow through the bulb in this time?

**2** The beam current in a certain television tube is 35 $\mu$A. How many electrons strike the screen in 1 min?

**3** A battery charger sends a current of 4.0 A through a battery for 6.0 h. How much charge is furnished to the battery in this time?

**4** A certain car battery maintains a current of 2.0 A for 7.0 h. How much charge flows from the battery during this time?

**5** When a 1.55-V flashlight battery is connected across a bulb, the current is 25 mA. What is the bulb's resistance?

**6** Suppose the resistance through your body from one hand to the other is 30,000 $\Omega$. How large is the current through your body as your hands grasp the terminals of a 12-V battery?

**7** Find the resistance at 20°C of a 20-m length of aluminum wire that has a diameter of 2.0 mm.

**8** Find the resistance at 20°C of a 50-cm length of silver wire that has a radius of 0.050 mm.

**9** The resistance of a spool of insulated copper wire is measured by noticing that a 12.0-V battery causes a current of 0.60 A in the entire length of wire. The diameter of the metal part of the wire is 0.100 cm. What length of wire is on the spool?

**10** Number 10 copper wire has a diameter of 2.59 mm. The approximate safe current for this size wire is 30 A. (At higher currents, the wire becomes too hot.) (*a*) Find the resistance of a 25-m length of this wire at 20°C. (*b*) How large a potential difference exists between the ends of a wire this long when it carries 30 A?

**11** A coil of tungsten wire that has a resistance of 25.0 $\Omega$ at 20.0°C is to be used to measure temperature. How much does its resistance change for a 5-C° temperature change near 20°C?

**12** The resistance of a coil of wire is 165.3 $\Omega$ at 20°C and 173.6 $\Omega$ at 45°C. What is the temperature coefficient of resistance for the material from which the wire is made?

■ **13** What is the percentage change in the resistance of a copper wire as its temperature changes from 14 to 33°C?

■ **14** At what temperature is the resistivity of iron the same as that of aluminum at 20°C?

■ **15** A 30-$\Omega$ resistor that is temperature-independent is to be made by using a graphite resistor in series with an iron resistor. What should the resistance of each be at 20°C?

■ **16** A graphite resistor is placed in series with a 7.0-$\Omega$ (at 20°C) iron resistor. How large should the resistance of the graphite resistor be in order for the combined resistance to be temperature-independent?

**17** A light bulb is marked 60 W/120 V. (*a*) What current does it draw? (*b*) What is its resistance when operating on 120 V?

**18** A 40-W fluorescent lamp is designed to operate on 120 V. In operation, (*a*) how much current does it draw and (*b*) what is its resistance?

■ **19** (*a*) How much energy (mostly heat) does a lighted 100-W bulb emit in 3 min? (*b*) How many kilowatthours does it consume in this time?

■ **20** How long will it take a 500-W submersible heater to heat 200 g of water from 18 to 80°C? Assume no heat loss to the surroundings.

**21** Find the equivalent resistance of the following resistances (*a*) in series and (*b*) in parallel: 5 $\Omega$, 7 $\Omega$, 9 $\Omega$, 3 $\Omega$.

**22** Find the equivalent resistance of the following resistances (*a*) in series and (*b*) in parallel: 14 $\Omega$, 11 $\Omega$, 20 $\Omega$, 18 $\Omega$.

■ **23** A resistor system that consists of 3-, 4-, and 12-$\Omega$ resistors in parallel is placed in series with a 4.5-$\Omega$ resistor. Find the equivalent resistance of the combination.

■ **24** In Fig. P17.2, each resistor is 7.0 $\Omega$ and $\mathcal{E} = 1.50$ V. Find (*a*) the equivalent resistance of the resistor combination and (*b*) the current drawn from the battery.

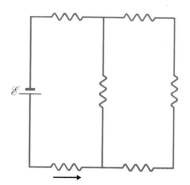

**FIGURE P17.2**

■ **25** In Fig. P17.2, all the resistors are 4.0 $\Omega$ except for the two at the top. Each of these is 6.0 $\Omega$. Find (*a*) the equivalent resistance of the combination and (*b*) the current drawn from the battery if $\varepsilon = 15.0$ V.

■ **26** Each resistor in Fig. P17.2 is 8.0 Ω, and $\varepsilon = 45$ V. Find the current (*a*) from the battery and (*b*) through the resistor at the far right.

■ **27** Find the equivalent resistance seen by the battery in Fig. P17.3 (*a*) when the switch S is open and (*b*) when the switch is closed. (*c*) What is the current in the 4-Ω resistor when the switch is open?

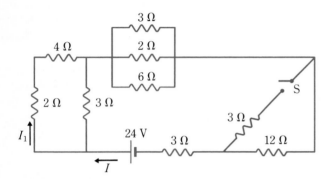

**FIGURE P17.3**

■ **28** How many resistance values can you obtain using the following three resistors: 4, 5, and 6 Ω? What are they?

■ **29** (*a*) Find the equivalent resistance for the circuit in Fig. P17.4 if each resistor is 3 Ω. The circuit is shown in its entirety. (*b*) Find $I$. (*c*) What is $I_2$?

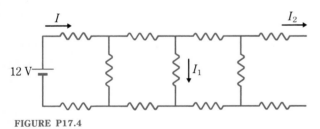

**FIGURE P17.4**

■ **30** Prove the relation for parallel capacitors given in Fig. 17.19.

■ **31** Prove the relation for series capacitors given in Fig. 17.19.

■ **32** How many capacitance values can you obtain using the following three capacitors: 6, 8, and 12 μF? What are they?

■ **33** Find the equivalent capacitance of the system shown in Fig. P17.5 when switch S is open.

■ **34** Find the equivalent capacitance of the system shown in Fig. P17.5 when the switch S is closed.

■ **35** Find $I_1$, $I_2$, and $I_3$ in Fig. P17.6.

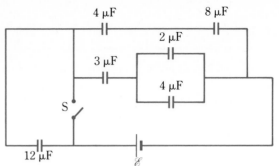

**FIGURE P17.5**

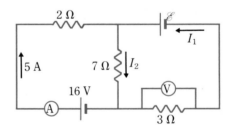

**FIGURE P17.6**

■ **36** Suppose the polarity of the 12-V battery in Fig. P17.6 is reversed. What are the new values of $I_1$, $I_2$, and $I_3$?

■ **37** In Fig. P17.7, the ammeter reads 5 A. Find $I_1$, $I_2$, $\varepsilon$, and the voltmeter reading.

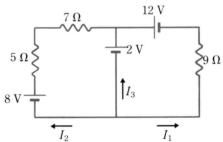

**FIGURE P17.7**

■ **38** In Fig. P17.7, suppose that the ammeter reads 5 A but the direction of the current is opposite that shown. Find $\varepsilon$ and the voltmeter reading.

■ **39** In Fig. P17.8, find $I$, $I_1$, and the charge on the 7-μF capacitor. *Hint:* Note that the current in many of the wires is zero.

■ **40** In Fig. P17.8, suppose that the 7-μF capacitor is replaced by a wire of zero resistance. Find $I$, $I_1$, and the charge on the 2-μF capacitor. See the hint in Prob. 39.

■ **41** The voltmeter in Fig. P17.9 reads 4.0 V, and the ammeter reads 2.0 A with the current direction as indicated. Find (*a*) $R$ and (*b*) $\varepsilon$. (In writing the circuit equation, notice that the voltage drop across $R$ is 4 V.)

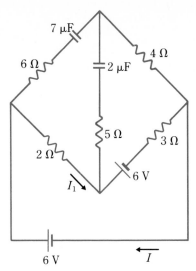

FIGURE P17.8

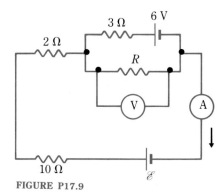

FIGURE P17.9

**42** In Fig. P17.9, how large must $\mathcal{E}$ be if the current through the 6-V battery is to be zero and $R$ is 12 $\Omega$?

**43** In Fig. P17.9, what would (a) the ammeter and (b) the voltmeter read if $\mathcal{E}$ were 30 V and R were 7.0 $\Omega$?

**44** In Fig. P17.8, the 6-V battery at the bottom is replaced by an unknown emf $\mathcal{E}$ with the same polarity. The charge on the 7-$\mu$ F capacitor is then 56 $\mu$C. Find $\mathcal{E}$, $I$, and $I_1$.

**45** A particular 120-V circuit has operating on it a 1000-W toaster, a 90-W lamp, and a 500-W soldering iron. The fuse in the circuit blows when a 60-W bulb is turned on. What is the rating of the fuse?

**46** A person plans to operate a 1200-W dryer, a washer that requires 460 W, four 60-W bulbs, and a 40-W radio from the same 120-V line. For what minimum current must this line be fused?

**47** Two capacitors, one 3 $\mu$F and the other 4 $\mu$F, are individually charged to a potential difference of 12 V by connecting them, one at a time, across a battery. After they are removed from the battery, they are connected, with the positive plate of one connected to the positive plate of the other and the negative plate of one connected to the negative plate of the other. Find (a) the potential across each and (b) the resultant charge on each. *Hint:* After they are connected, the potential drop is the same across the two capacitors.

**48** Repeat Prob. 47 for the capacitors connected with the positive plate of one connected to the negative plate of the other.

**49** Each resistor in Fig. P17.10 is 4 $\Omega$. Find the equivalent resistance between $a$ and $b$. *Hint:* From symmetry, many of the wires contain identical currents.

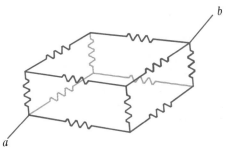

FIGURE P17.10

**50** When a current of 3.5 A is drawn from a certain battery, its terminal voltage drops from its zero-current value of 1.57 V to 1.26 V. Find the internal resistance of the battery.

**51** What is the maximum current that can be drawn from a battery that has an emf of 1.57 V and an internal resistance of 0.20 $\Omega$?

**52** The terminal potential of a battery is 11.40 V when a 20.0-$\Omega$ resistor is connected across its terminals and 10.16 V when a 4.0-$\Omega$ resistor is connected. Find the emf and internal resistance of the battery.

**53** Prove that 1 kWh = $3.60 \times 10^6$ J.

# 18 Magnetism

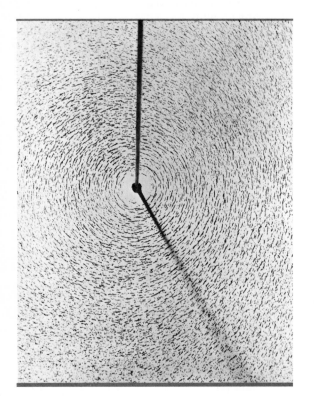

AS children in grade school, we performed simple experiments that involved magnetism. We learned that a bar magnet has two poles, a north pole and a south pole. Further, we found that unlike poles attract each other and like poles repel each other. By sprinkling iron filings on a glass plate placed above a magnet, we discovered that the filings formed a picture of the magnetic field that surrounds the magnet. Most of these facts were known thousands of years ago. Not until 1819, however, did scientists learn that magnetism is closely related to electric currents and fields. Even today, scientists are still making discoveries concerning magnetism and the materials from which magnets are made. As we shall see in the following chapters, magnets and their effects are only a small facet of magnetism.

## 18.1 Magnetic field plotting

Much of the terminology of magnetism was developed centuries ago by those who first investigated the behavior of magnets. The first magnets were simply pieces of iron-bearing rock called lodestone. We find it convenient to describe the terminology in reference to bar magnets and compass needles. If a bar magnet is suspended by a delicate fiber, as in Fig. 18.1, one particular end always points approximately northward on the earth. This end is called the north pole of the magnet; the other

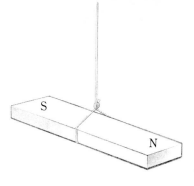

FIGURE 18.1
The north pole of a magnet is defined to be the pole that points northward on the earth when the magnet is freely suspended.

end is the south pole. A device such as this is nothing more than a simple compass. Note that the arrowhead end of a compass needle is its north pole.

Further studies with magnets show that the north poles of two magnets repel each other and that the south pole of one magnet is always attracted by the north pole of another magnet. If one tries to separate the north pole of a bar magnet from the south pole by breaking the magnet in two, the effort proves unsuccessful. The broken magnet becomes two new bar magnets, each with a north and a south pole. These are familiar qualitative features of magnets.

Strange things happen in the vicinity of magnets. Iron filings and nails experience an attraction to the magnet and, if allowed to do so, dart to the magnet. Compass needles deflect when a magnet is brought near. We shall learn of other phenomena that occur in the vicinity of magnets. All these phenomena are explained in terms of what we call the *magnetic field* of the magnet.

We define the direction of the magnetic field at a point as the direction a compass needle points when placed at the point. For example, suppose we wish to sketch the magnetic field near the bar magnet shown in Fig. 18.2. We can do so by placing many tiny compass needles in the vicinity and observing their orientation. Because the arrowhead end of a compass needle is a north pole, it is repelled by the north pole of

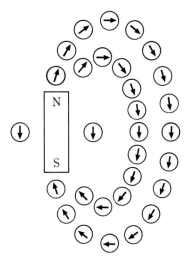

FIGURE 18.2
The direction of the magnetic field in the vicinity of a magnet can be found by using a compass needle.

the magnet; hence a compass needle near the north pole of the magnet points away from the magnet. Similarly, a needle near the south pole points toward the magnet because unlike poles attract. To sketch a magnetic field, we draw a series of lines about the magnet in such a way that the arrows on the lines show the direction in which a compass needle would point. These lines, called *magnetic field lines,* are shown for three types of magnet in Fig. 18.3. Like the compass needles that define them,

Magnetic field lines point out from the north pole of a magnet and enter the south pole.

Sketches such as those in Fig. 18.3 show not only the direction of the field but its strength as well. As with electric fields, the magnetic field lines are closest together where the field is strongest.

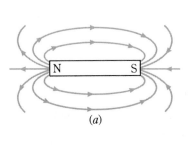
(a)

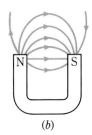
(b)

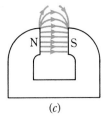
(c)

**FIGURE 18.3**
The magnetic field points away from the north pole of a magnet and to the south pole.

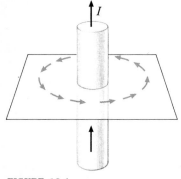

**FIGURE 18.4**
The magnetic field forms circles concentric to the current-carrying wire.

## 18.2 The magnetic field of an electric current

Magnets are not the only source of magnetic fields. In 1820, Hans Christian Oersted discovered that an electric current in a wire causes a nearby compass needle to deflect. This indicates *that a current in a wire is capable of generating a magnetic field.* We now know from many other types of experiments that this is indeed the case. Further, the magnetic field of a magnet is also the result of the motion of charges, as we shall show later.

Oersted investigated the nature of the magnetic field about a long, straight wire carrying current in the direction indicated in Fig. 18.4. When a compass is placed near the wire, the needle lies with its length tangent to a circle concentric with the wire, and the inference is that a magnetic field exists in a circular form about the wire. As is to be expected, the strength of the field is greatest close to the wire. A three-dimensional representation of the magnetic field is shown in Fig. 18.5. (In this and later diagrams, the symbol · indicates an arrow coming toward the reader and the ✕ represents an arrow going away from the reader. The symbols are meant to suggest the point and tail of the arrow that shows the direction of the magnetic field.)

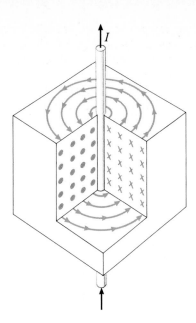

FIGURE 18.5

The magnetic field circles about the long, straight wire. Its magnitude decreases inversely with the distance from the wire.

There is a simple **right-hand rule** for remembering the direction of the magnetic field about a wire. If you grasp the wire in your right hand with your thumb pointing in the direction of the current, your fingers will circle the wire in the direction of the field (Fig. 18.6).

FIGURE 18.6

When you grasp a current-carrying wire in your right hand with the thumb pointing in the direction of the current, your fingers circle the wire in the same direction as the magnetic field.

## 18.3 The force on a current in a magnetic field

Thus far we have discussed only the qualitative features of the magnetic field. We now seek some means of precisely measuring it. The measurement method we shall explain makes use of another important feature of magnetic fields: they cause forces on electric currents.

A wire carrying a current through a region in which a magnetic field exists experiences a force due to the field.

As an example of this phenomenom, consider the situation shown in Fig. 18.7a. A magnet causes the horizontal magnetic field shown by the colored lines. In this field is a wire carrying a current directed vertically upward. Experiment shows that the current-carrying wire experiences a force resulting from the magnetic field. The force **F** on the wire is not in the direction of the magnetic field but is directed into the page, as shown. If the direction of the current is reversed, the direction of the force on the wire also reverses; it is then directed out of the page. We can see this more clearly if we redraw the situation in two dimensions, as in Fig. 18.7b. Notice there that the line of the wire and a magnetic field line intersecting it determine a plane, the plane of the page. The force the wire experiences is always perpendicular to this plane; in this case, the force is into the page.

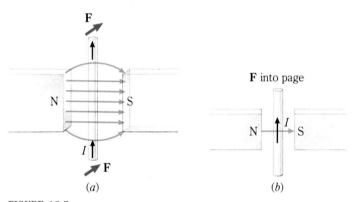

**FIGURE 18.7**
The magnetic field (colored lines) furnished by the poles of the bar magnets causes the current-carrying wire to experience a force.

The magnitude of the force caused by the magnetic field depends on several factors. As you would perhaps guess, it is proportional to both the current in the wire and the length of the wire in the magnetic field. Moreover, the magnitude (or strength) of the magnetic field also affects the magnitude of **F**. For a very weak field, **F** is very small.

We characterize the strength of the magnetic field by the magnetic field vector **B**. It is defined in the following way. Suppose that, as in Fig. 18.7, a length $L$ of wire carries a current $I$ through a magnetic field; further, suppose that the field lines are perpendicular to the wire. The wire experiences a force **F** into the page because of the magnetic field. We define the magnetic field vector **B** to be in the direction of the field lines and to have a magnitude given by

$$B = \frac{F}{IL} \qquad \textbf{(B perpendicular to wire)}$$

This is the force per unit length of wire per ampere of current through the wire. The SI unit for **B** is the newton per ampere-meter, which is called the tesla (T). This unit is sometimes called the weber per square meter (Wb/m²) for reasons we shall explain in the next chapter. A non-SI unit used for **B** is the gauss (G), where $1 \text{ G} = 10^{-4}$ T. For comparison purposes, the magnitude of **B** in the earth's magnetic field is about $5 \times 10^{-5}$ T, while $B$ near the end of a strong bar magnet is of the order of 0.1 T. The magnetic field vector **B** is commonly referred to by any of the following names: *magnetic field strength, magnetic induction,* and *magnetic flux density.*

In Fig. 18.7, we have assumed that the field lines (and therefore **B**) are perpendicular to the current direction (that is, to the wire). Let us now see what happens when the two are not perpendicular. Suppose the field lines and the wire are parallel to each other rather than perpendicular, as in Fig. 18.8a. Then the wire experiences no force. *A current parallel to the field lines experiences no force due to a magnetic field.* Clearly, the relative orientation of the field lines and the current direction is very influential.

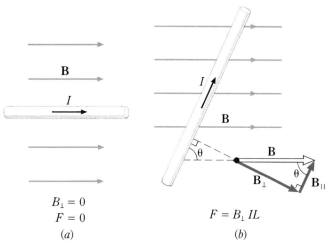

**FIGURE 18.8**
When a current-carrying wire is immersed in an impressed magnetic field, the force on it is proportional to the component of **B** that is perpendicular to the wire.

There is a simple way to describe the effect of the wire's orientation in the field. Picture **B** as having two components, one perpendicular to the wire and the other parallel to it (Fig. 18.8b). The parallel component $B_\parallel$ exerts no force on the wire. The perpendicular component $B_\perp$ exerts the following force on the wire, as we can see by referring to the definition of the magnitude of **B**:

$$F = B_\perp IL \tag{18.1}$$

We therefore have an expression for the force on a current-carrying wire immersed in a magnetic field.

---

**Example 18.1**
In Fig. 18.8b, suppose that $B = 2.0$ G, $\theta = 53°$, and $I = 20$ A. Find the force on a 30-cm length of the wire.

**Reasoning**   We know that $B_\perp = B \sin \theta = B(0.799)$. Converting $B$ to SI units, we have $B = 2.0$ G $= 2.0 \times 10^{-4}$ T. Then

$$F = B_\perp IL = (2.0 \times 10^{-4} \text{ T})(0.799)(20 \text{ A})(0.30 \text{ m}) = 9.58 \times 10^{-4} \text{ N} \quad \blacksquare$$

---

**Exercise**   Find $F$ if the wire is perpendicular to the field lines.   *Answer: $12.0 \times 10^{-4}$ N* ∎

## 18.4 An extension of the right-hand rule

In the preceding section, it was pointed out that the direction of the force experienced by a wire carrying a current in a magnetic field is perpendicular to the plane defined by the wire and the field. We now consider a simple intuitive extension of the right-hand rule (Sec. 18.2) that helps us state the direction of the force experienced by the wire. It is purely an intuitive aid for remembering the direction of the force. No real physical significance should be attached to the rule; it is simply a memory device.

The rule is shown in Fig. 18.9. *Point the fingers of your right hand along the lines of the magnetic field and your thumb in the direction of the current. The force on the wire is in the direction in which your palm would push.*

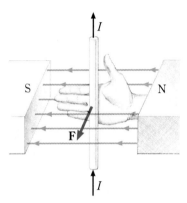

**FIGURE 18.9**

The right-hand rule: the fingers point in the direction of **B**, the thumb points in the general direction of the current, and the palm pushes in the direction of **F**.

Let there be no confusion on this point. The line of the magnetic field vector **B** and the line of the wire together define a plane (the plane of the page in Figs. 18.7 and 18.8*b*). The force on the wire is always perpendicular to this plane. Once you know this, a pure guess allows you a 50 percent chance of obtaining the proper direction for the force. It must be either into or out of the plane. To find which is the proper alternative, use the rule illustrated in Fig. 18.9. The direction of the force in Fig. 18.9 is toward you, out of the page. Using the same rule, you see that the direction of the force in Figs. 18.7 and 18.8*b* is into the page.

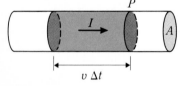

**FIGURE 18.10**

All the charge in the length $v \, \Delta t$ will pass through the area at $P$ in a time $\Delta t$.

## 18.5 Forces on moving charges

A current is the result of the motion of charged particles, and a wire carrying a current experiences a force in a magnetic field. Is it really necessary that these particles be in a wire in order for them to experience a force from a magnetic field? It would appear that if the force were exerted directly on the charge carriers, even a beam of charge from a charge gun would experience a force due to a magnetic field. Indeed, this turns out to be the case. Let us now find the force on individual moving charges by investigating further the force on a current-carrying wire.

To find the force caused by a magnetic field on a single charge carrier moving in a wire, we must divide the total force on a length $L$ of wire by the number of charge carriers in this length. If the wire has a cross-sectional area $A$, as in Fig. 18.10, the volume of a length $L$ is $AL$. If there are $n_u$ charge carriers per unit volume, the number of charge carriers in length $L$ is $n_u AL$. Therefore

$$\text{Force per charge} = \frac{\text{force on wire}}{\text{number of carriers}} = \frac{B_\perp IL}{n_u AL} = \frac{B_\perp I}{n_u A}$$

We still need to express the current in terms of the individual charges that cause it. A charge carrier moves a certain distance in the direction of the current in a time $\Delta t$. If the average speed of the carrier is $v$, then the distance moved in time $\Delta t$ is $v\Delta t$. Hence, in a time $\Delta t$, all the charge carriers in a length $v\Delta t$ to the left of point $P$ in Fig. 18.10 move through the cross section at $P$. Because the volume of this length of wire is $Av\Delta t$ and because there are $n_u$ charge carriers per unit volume, the number of charge carriers that pass $P$ in time $\Delta t$ is $n_u Av\Delta t$. Each carries a charge $q$, and so

$$I = \frac{\text{charge past } P \text{ in time } \Delta t}{\Delta t} = \frac{qn_u Av \, \Delta t}{\Delta t} = qn_u Av$$

We can now use this value for $I$ in the expression for the force per unit charge:

$$F = \frac{B_\perp I}{n_u A} = qvB_\perp$$

We therefore conclude the following:

A charge $q$ moving with speed $v$ perpendicular to a magnetic field of magnitude $B_\perp$ experiences a force of magnitude

$$F = qvB_\perp \tag{18.2}$$

As with the force on a wire, the direction of the force on the moving charge is given by the right-hand rule. Moreover, the component of $\mathbf{B}$ parallel to the particle velocity $\mathbf{v}$ causes no force on the particle.

As an example of such a situation, refer to Fig. 18.11. We see there a charge $q$ moving with velocity $\mathbf{v}$ through a magnetic field $\mathbf{B}$ directed out of the page. The intersecting vectors $\mathbf{B}$ and $\mathbf{v}$ define a plane (horizontal), and the force $\mathbf{F}$ on $q$ is perpendicular to this plane. Using the right-hand rule (Fig. 18.9 with $I$ replaced by $v$), we find that $\mathbf{F}$ has the direction shown in Fig. 18.11. Notice that Eq. 18.2 tells us that the direction of $\mathbf{F}$ reverses when the particle is charged negatively. Hence if the particle in Fig. 18.11 had been negative, $\mathbf{F}$ would be directed upward instead of downward.

Notice that the force is perpendicular to the velocity vector of the charge and therefore has no component in the direction of motion. As a result, it does no work on the charge.

**FIGURE 18.11**

Use the right-hand rule to find the direction of $\mathbf{F}$.

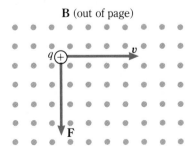

A steady magnetic field does no work on a charge moving through it.

Despite this fact, the force deflects the particle. However, it does not change its kinetic energy, since no work is done.

## 18.6 Particle motion in a magnetic field

Suppose a positively charged particle is moving with velocity **v** in a magnetic field **B**, as shown in Fig. 18.12. The field is directed into the page, and so **v** and **B** are

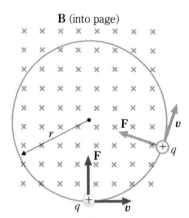

**FIGURE 18.12**

The charged particle follows a circular path in the uniform magnetic field.

mutually perpendicular. We know that the particle, which we shall say has charge $q$, experiences a force whose magnitude is given by Eq. 18.2:

$$F = qvB$$

The right-hand rule tells you that the force on the charge is in the direction shown in Fig. 18.12. As pointed out before, the force is perpendicular to **v**, and so $v$ does not change.

   The mechanical analogy of this situation is well known to you. It occurs when a ball at the end of a string is swung in a circle and when the gravitational force of the sun on the earth causes the earth to orbit the sun. In each case, a force perpendicular to **v** causes the object to follow a circular path. In the present case, the force $qvB$ due to the magnetic field furnishes the centripetal force needed for circular motion. Equating $qvB$ to the centripetal force $mv^2/r$ gives

$$qvB = \frac{mv^2}{r}$$

$$r = \frac{mv}{qB} \qquad (18.3)$$

A particle moving perpendicular to a uniform magnetic field follows a circular path with this radius. For a practical example of this type of motion, see Fig. 18.13.

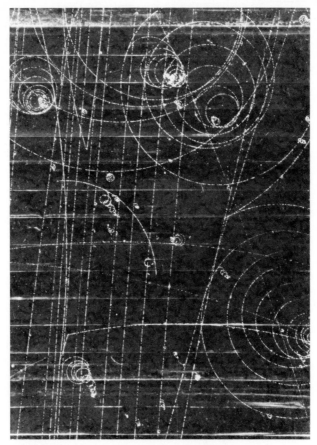

**FIGURE 18.13**
Particle paths in a magnetic field as exhibited in a bubble chamber. Which way are the spiraling particles moving? If the magnetic field is out of the page, which particles are positive? *(Courtesy of Lawrence Livermore Laboratory, University of California, Berkeley)*

---

***Example 18.2***
If the charge in Fig. 18.12 is a proton ($m = 1.67 \times 10^{-27}$ kg, $q = +e$) moving at a speed of $10^6$ m/s and if the radius of the circle is 4.0 cm, what is the magnitude of the magnetic field?

***Reasoning*** The force on the charge is $qvB$, and this force must furnish the centripetal force $mv^2/r$. Equating magnitudes,

$$\frac{mv^2}{r} = qvB$$

$$B = \frac{mv}{rq}$$

Using the given values, we find $B = 0.260$ T.  ∎

*Exercise* Repeat if the particle is an electron. How would Fig. 18.12 be different? *Answer: B = 1.42 × 10⁻⁴ T; the direction of motion would be clockwise.*  ∎

### Example 18.3

Figure 18.14 is a sketch of a **velocity selector.** A charged particle is shot into a region in which uniform electric and magnetic fields exist, with **E** and **B** perpendicular to each other. (Notice that a vertical electric field is generated by the parallel plates.) Show that the particle does not deflect from its straight-line path if its speed is equal to $E/B$.

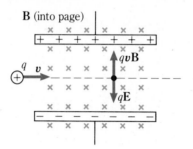

**FIGURE 18.14**

The velocity selector passes undeflected those particles for which the electric force $qE$ equals the magnetic force $qvB$.

*Reasoning* The magnetic field into the page exerts a force $qvB$ *upward* on the particle. (Check this by using the right-hand rule.) The electric field is directed downward and thus exerts a force $qE$ *downward* on the particle. If the particle is not to deflect, these two forces must be equal. Hence

$$Eq = qvB$$

$$v = \frac{E}{B}$$

A particle with this speed will pass through the region of the crossed fields and emerge undeflected. Notice that the charge cancels. Can you show that a negative particle would behave in the same way? Clearly, this device can be used to select particles of speed $E/B$ from a beam of heterogeneous particles that has a distribution of velocities.  ∎

## 18.7 The Hall effect

There are very few electrical phenomena that show clearly the sign of the charge carriers. Most experiments can be explained equally well by positive charges flowing in one direction or by negative charges flowing in the reverse direction. The experiment we describe in this section is one of the few that distinguish between positive and negative charge carriers.

Consider the circuit in Fig. 18.15a. A battery is connected to the ends of a thin, uniform conducting ribbon made perhaps from a metal. The two symmetric points $m$ and $n$ are at the same potential, and so there is no voltage difference between them. When a magnetic field is impressed perpendicular to the ribbon, however, as in b, points $m$ and $n$ differ in potential. That is, a potential difference appears between $m$ and $n$ when a magnetic field is present. Let us see how it arises.

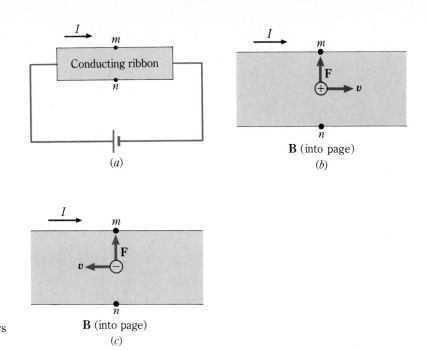

(a)

(b)

(c)

**FIGURE 18.15**

Can you show that the voltage reverses sign if the charge carriers are negative instead of positive?

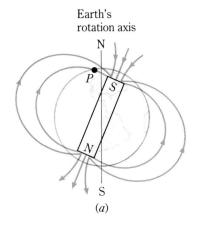

(a)

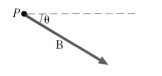

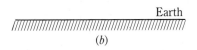

Earth

(b)

**FIGURE 18.16**

(a) The earth's magnetic field. (b) The dip angle $\theta$ is the angle between **B** and the horizontal.

Suppose that the charges flowing through the ribbon are positive. One such charge is shown in Fig. 18.15b. The right-hand rule tells us that the charge is forced upward, toward $m$. Hence point $m$ becomes positively charged and a potential difference appears between $m$ and $n$. We repeat, point $m$ is positive when the charge carriers are positive.

Suppose, alternatively, that the current consists of negative charges moving to the left, as in Fig. 18.15c. Using the right-hand rule, we find that these negative charges experience an upward force, toward $m$. Thus in this case point $m$ becomes *negatively* charged.

Here, then, we have a clear-cut way of determining the sign of the charge carriers in a material. This effect was discovered by Edwin Hall in 1879, and it is called the **Hall effect.** Using it, present-day scientists are able to tell the sign of the charge carriers in newly developed electronic materials for use in solid-state electronics. The Hall effect also forms the basis for a commercially produced device for measuring magnetic fields.

## 18.8 The earth's magnetic field

Experiments such as the one outlined in the preceding section are complicated by the fact that the earth acts like a very large magnet (Fig. 18.16a). Only the approximate location of the poles within the earth can be given, but they are indicated roughly on the figure. Notice, in particular, that the geographic poles, defined by the axis of rotation of the earth, do not coincide with the magnetic poles.

How do we locate the position of the magnetic pole on the north end of the earth? We use a compass, the north pole of which is, by definition, the end of the needle that points north. Therefore, *the pole at the north end of the earth must actually be the south pole of a magnet* since it attracts the north pole of the compass needle. This is, of course, purely the result of the fact that we choose to call the end of a magnet that is attracted to the geographic north a north pole. Although this definition of north and south poles causes some confusion, it should cause no difficulty for the careful student.

As can be seen in Fig. 18.16a, the earth's magnetic field is parallel to the surface of the earth only near the equator. Near the poles, the field is almost perpendicular to the earth. The quantities needed to specify the magnetic field of the earth at a given place are the direction and magnitude of the field. Usually the values given are the magnitude of **B** and the angle of dip ($\theta$ in Fig. 18.16b).

**18.9** ## The magnetic field of a current

We pointed out in Sec. 18.2 that electric currents generate magnetic fields. Let us now see what these fields look like for various wire configurations of importance. The relations we present can be derived from theory, and the results are confirmed by experiment.

### Long, straight wire

The magnetic field generated by a current in a long, straight wire circles the wire, as we saw in Fig. 18.6. The direction of **B** is given by the right-hand rule. Its magnitude as a function of the radial distance $r$ from the axis of the wire is

$$B = \frac{\mu_o I}{2\pi r} \tag{18.4}$$

where $I$ is the current in the wire. By definition, the constant $\mu_o$ is exactly $4\pi \times 10^{-7}$ T · m/A and is called the *permeability of free space*. We shall learn more about $\mu_o$ in Sec. 21.7.

### Circular loop of wire

Suppose a circular loop of wire carries a current $I$, as shown in Fig. 18.17a. Apply

**FIGURE 18.17**

Two views of the magnetic field about a current-carrying loop. (*a*) Perspective view. (*b*) A cross section as seen from directly above in (*a*).

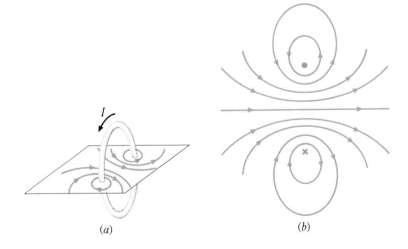

(a)                    (b)

the right-hand rule to the wire by grasping it with your right hand with your thumb pointing in the direction of the current. Notice that your fingers circle through the loop in the same sense as do the magnetic field lines in Fig. 18.17a. We show the field lines in more detail in b. If the radius of the loop is a, then the magnitude of the field at the loop's center is

$$B = \frac{\mu_o I}{2a} \tag{18.5}$$

A coil consisting of $N$ loops tightly packed together produces a field $N$ times larger at its center.

### Solenoid

The coils in Fig. 18.18 are called solenoids. The one in $a$ is more loosely wound than most. However, you can see from it, using the right-hand rule, how the wire loops generate the magnetic field. Part $b$ shows in cross section a portion of a much more closely wound solenoid. As indicated, the magnetic field within it is nearly uniform. Inside a long, hollow solenoid that carries a current $I$ and has $n$ loops of wire per meter length, the magnetic field magnitude is

$$B = \mu_o n I \tag{18.6}$$

This relation applies throughout the interior of the hollow solenoid, except near its ends. A solenoid is often used to produce a field that is approximately uniform. Solenoids wound on iron cores are used as electromagnets in doorbells and many other devices.

**FIGURE 18.18**

(*a*) A perspective view of a loosely wound solenoid. (*b*) Cross-sectional view of a solenoid that has many loops of wire. Far from the ends, the fields are essentially uniform inside the solenoids.

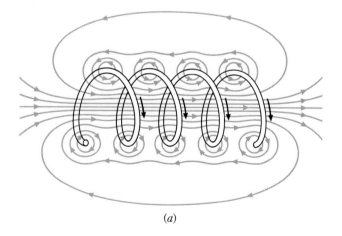

(*a*)

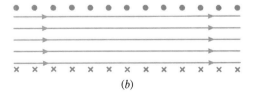

(*b*)

# Geomagnetism and plate tectonics

Geologists studying the magnetic properties of rocks have discovered an amazing fact: the earth's magnetic field has reversed direction many times through the ages. This fact is most clearly shown by magnetic rocks in the ocean floor, where rocks of different ages have opposite polarity. It is believed that 171 field reversals have occurred in the past $76 \times 10^6$ years, with a span of about 400,000 yr between reversals. Presumably the earth's field is the result of vast currents in the molten iron of the earth's interior. Why these currents should reverse direction is still unknown. We can only speculate on the turmoil that accompanies such a reversal.

These magnetic field reversals provide striking evidence for a theory (called plate tectonics) first presented in 1912 by Alfred Wegener: the earth's continents float from place to place on the surface of the earth's sphere. Rejected by nearly everyone until the 1960s, Wegener's theory received new life from a magnetic history imprinted on the ocean floor. We now know that the earth's surface is created continuously at crack lines (called ridges) that thread through the ocean centers. Molten rock (lava) from the earth's interior rises through these cracks and forms new ocean floor. The new floor pushes aside the older material and causes the distant continents to slowly separate. To make way for this new surface, colliding continents in some cases cause mountain ranges (such as the Himalayas) to rise; in other cases, old ocean floor is forced to move beneath the continents, where it descends to become a part of the earth's molten interior.

At the mid-Atlantic ridge, new surface is being generated at such a rate that Europe and North America are separating at a rate of about 3 cm/yr. The upswelling lava that forms the new ocean floor at the midocean ridges solidifies to form rock, which often contains small magnetic regions. As the lava solidifies, the magnetic material of these regions is magnetized along the direction of the earth's magnetic field. Geologists have found that the magnetization direction in the sea-floor rocks varies with distance away from the mid-ocean ridges. Thus we have clear evidence for both the magnetic history of the earth and the continuous growth of the ocean floor. Today we believe that nearly 200 million years ago all the continents were joined in one continuous mass, called Pangaea. Driven by upswell from the earth's molten interior, Pangaea broke up, and the earth's continents drifted slowly to the positions they now occupy.

---

**Example 18.4**
Find $B$ at a radial distance of 5 cm from the center of a wire that carries 30 A.

*Reasoning*  The field of a long, straight wire is, from Eq. 18.4,

$$B = \frac{\mu_o I}{2\pi r} = \frac{(4\pi \times 10^{-7} \text{ T} \cdot \text{m/A})(30 \text{ A})}{2\pi(0.050 \text{ m})} = 1.20 \times 10^{-4} \text{ T} = 1.20 \text{ G}$$

Notice how very small the field is.  ■

*Exercise*   A solenoid 40 cm long has 800 loops and carries a current of 5.0 A. Find (*a*) the number of loops per meter, *n*, and (*b*) the magnitude of the magnetic field in the center of the solenoid.   *Answer: 2000; 12.6 mT*   ■

## 18.10 The definition of the ampere unit

An interesting situation of fundamental importance is shown in Fig. 18.19. We see there two parallel wires a distance *b* apart carrying currents $I_1$ and $I_2$ in the same direction. Wire 1 generates a magnetic field

$$B_1 = \frac{\mu_o I_1}{2\pi r}$$

and this field is imposed on wire 2. Thus $I_2$ experiences a force due to $B_1$. You should be able to use the right-hand rule to show that this force is such as to push wire 2 toward wire 1.

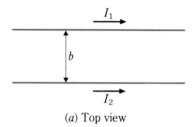

(*a*) Top view

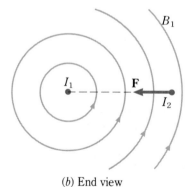

**FIGURE 18.19**
Two parallel currents attract each other. What would happen if they were antiparallel?

(*b*) End view

To find the force caused by wire 1 on a unit length of wire 2, we use $F = B_\perp IL$. Because $B_1$ is perpendicular to wire 2 and because $r = b$ in this case, we have

$$\frac{F}{L} = B_1 I_2 = \frac{\mu_o I_1 I_2}{2\pi b}$$

as the force $I_1$ exerts on unit length of $I_2$. The action-reaction law tells us that an equal-magnitude force is exerted on $I_1$ by $I_2$.

Suppose the same current exists in both wires, so that $I_1 = I_2 = I$. Then the above equation can be solved for $I$ to give

$$I = \sqrt{\frac{2\pi b F}{\mu_o L}}$$

We can measure $F$, $b$, and $L$ in SI units that we have already defined. As we learned in Sec. 18.9, $\mu_o$ is defined to be exactly $4\pi \times 10^{-7}$ T · m/A in the SI. Therefore $I$ in amperes is measurable in terms of basic quantities. Thus this basic experiment furnishes us with a way of defining the ampere.

## 18.11 Magnetic materials

We learned in grade school that magnets attract iron but not most other materials. It turns out that only a few *ferromagnetic* materials (iron, nickel, cobalt, gadolinium, dysprosium, and their alloys) are greatly influenced by a steady magnetic field.

Some atoms act like tiny bar magnets. To see why, refer to the often-used atom model shown in Fig. 18.20, which pictures the electrons as orbiting the nucleus.

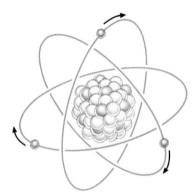

**FIGURE 18.20**

The orbiting electrons, acting like current loops, generate magnetic fields.

Since an electron in its orbit constitutes a circular current loop, each orbiting electron in Fig. 18.20 generates a magnetic field similar to that of the loop in Fig. 18.21*a*. Notice that this field is very similar to that of the short bar magnet shown in *b*. We thus conclude that each orbiting electron in an atom acts like a small bar magnet.

**FIGURE 18.21**

Notice how the current loop acts like a short bar magnet.

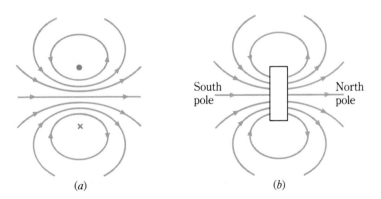

(a)    (b)

There is also another phenomenon that causes atoms to act like magnets. Particles such as the electron and proton act as though they are spinning on an axis through their centers; we say that such particles have *spin*. Any spinning charge acts like a current loop and generates a magnetic field. As a result, these particles act like tiny bar magnets.

In many atoms, the magnetic effects of the electrons cancel. In other atoms, the cancellation is nearly complete. Only in the transition-element atoms is cancellation small enough for the atom to act like a fairly powerful bar magnet. These atoms, the ferromagnetic atoms mentioned earlier, act like tiny compass needles, each having a north and a south pole. Such atomic magnets are referred to as *magnetic dipoles*.

We know that if we place a group of tiny magnets close to one another, insofar as possible they arrange themselves in such a way that each south pole is close to a north pole. This is the result of the attraction of unlike poles and the repulsion of like poles. The lowest potential energy of the system is reached when the magnets are arranged somewhat as shown in Fig. 18.22a. Notice that magnets arranged in this way are equivalent to a single large magnet.

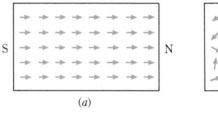
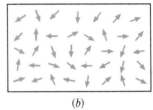

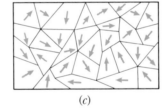

**FIGURE 18.22**

(*a*) A magnetized piece of iron; (*b*) the disordered, unmagnetized iron shown schematically; (*c*) a more realistic sketch of the domains.

However, if the magnets are strongly agitated (perhaps by someone shaking the board on which they rest), they will break loose from the alignment shown in *a*, as shown in Fig. 18.22*b*. Notice now that the individual magnets no longer align so as to produce a strong bar magnet.

An analogous situation exists with atoms in a solid. Thermal vibrations agitate the system and prevent the atoms from ordering themselves as in Fig. 18.22*a*. Only certain atomic magnets — iron and the other ferromagnetic materials — can preserve the pattern shown in *a* at ordinary temperatures. Even these atoms, when heated enough, acquire enough thermal energy to break loose and disorient as in part *b*. The temperature at which this happens is quite definite for each atomic species and is called the **Curie temperature.** There are, in addition to the magnetic forces between the ferromagnetic atoms, other forces that are much more complex. These forces can be understood only in terms of quantum mechanics, and so we are unable to discuss them further here. They play a major role in the aligning of atomic magnets.

Most materials have their atomic magnets, if there are any, randomly oriented, as in Fig. 18.22*b*. The ferromagnetic materials, however, consist of little regions in which the atoms are all aligned as in Fig. 18.22*a*. Each of these oriented regions is called a **domain.** In an ordinary piece of iron, each domain may contain as many as $10^{16}$ atoms and consist of a region a small fraction of a millimeter in linear dimension. However, the domains in an unmagnetized piece of iron are randomly oriented, and so the situation is much as in Fig. 18.22*b*, where now the arrows represent domains instead of atoms. A more realistic picture of the domains is given in part *c*. For a bar of iron to be magnetized, the domains within it must be lined up. This can be done in the following way.

Suppose you start with the unmagnetized bar of iron shown in Fig. 18.23*a*. A solenoid with a current in it, such as that shown in *b*, possesses a weak magnetic field within itself. If now you place an iron core in the solenoid, the magnetic field of the solenoid will exert forces on the domains. Those domains that are oriented along the field grow, and those oriented in other directions decrease in size. In effect, the domains orient under the action of the field, as shown in *c*. The iron is now a bar magnet, with north and south poles. In what is referred to as soft iron, the domains are easily oriented, but in what is called hard iron the field must be made quite strong or the domains must be agitated by heat or mechanical means to allow them to turn into the direction of the field. (The designations *hard* and *soft* refer only to the magnetic properties, not to the physical hardness.) It is possible, however, to align the domains nearly perfectly and form a large bar magnet.

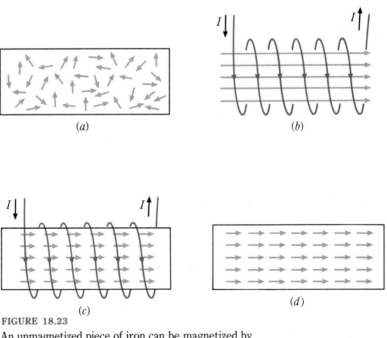

**FIGURE 18.23**
An unmagnetized piece of iron can be magnetized by using the field of a solenoid to line up the domains.

Once the domains have been aligned, the magnetic field consists of two parts: the original small field of the solenoid plus the field produced by the bar magnet, which is usually hundreds of times larger than the field of the solenoid. *The combination of a solenoid and a piece of very soft iron is called an* **electromagnet.**

If the current in the solenoid is turned off, the domains in a bar of soft iron return nearly to their original random state. Thermal motion causes them to disarrange. This is a desirable situation in an electromagnet because it makes it possible to turn it on or off at will. A piece of hard iron used in the solenoid, on the other hand, would retain most of its alignment when it was removed from the solenoid and would be a permanent bar magnet, as shown in Fig. 18.23$d$.

Let us summarize what we have found about the magnetic properties of materials. Most materials, when placed in a magnetic field, change the field scarcely at all. A very few substances, however, chiefly iron and its alloys, increase the magnitude of the magnetic field in which they are placed; often the field is strengthened by factors of hundreds. As we shall see, the ability of iron to greatly augment a magnetic field is of prime importance in many applications of magnetism.

## 18.12 The torque on a current loop

Many practical devices, including motors and many meters, make use of the torque that a current loop experiences when it is placed in a magnetic field. To see how such a torque originates, refer to Fig. 18.24$a$, where we show a current-carrying coil in a magnetic field. The coil is mounted on an axle and can rotate. Using the right-hand rule, we find the forces on the various sides to be those illustrated. Notice that only the two forces $\mathbf{F}_h$ cause a torque about the axis of rotation. Even these two forces cause no torque when the plane of the coil is perpendicular to the field of the magnet. Maximum torque occurs when the magnetic field lines skim past the surface of the coil, that is, when they lie in the plane of the coil, because then the lever arm for $\mathbf{F}_h$ is maximum.

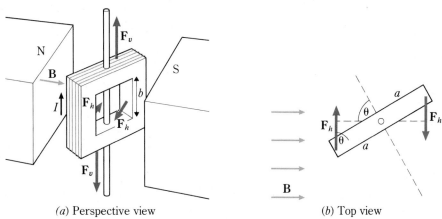

(a) Perspective view　　　　　　　　　　(b) Top view

**FIGURE 18.24**
The magnetic field causes the current-carrying coil to experience a torque.

To obtain a quantitative expression for the torque on the coil, we note that each of the two forces $\mathbf{F}_h$ gives a torque

$$(F_h)(\text{lever arm})$$

From Fig. 18.24$b$ we see that the lever arm is $a \sin \theta$. Therefore the torque on the coil is

$$\text{Torque} = 2F_h a \sin \theta$$

where $\theta$ is the angle between **B** and the perpendicular to the surface area of the coil. But $F_h$ is simply the force on the vertical side of the coil. If the vertical side has a length $b$ and if the current is $I$, each vertical wire contributes a force $BIb$ to $F_h$. There are $N$ loops on the coil, however, and so $F_h = NBIb$ and the torque becomes

$$\text{Torque} = (2ab)(NI)(B \sin \theta)$$

Notice that $2ab$ is simply the area of the coil. We can therefore write

$$\text{Torque} = (\text{area})(NI)(B \sin \theta) \qquad (18.7)$$

Although we have derived Eq. 18.7 for a very specially shaped coil, it turns out that it is true for all flat coils. Because $NI$ is the current around the coil, the important features of the coil (aside from its orientation) are its area and the current in it. In view of this, it is customary to define a quantity called the **magnetic moment** of a current loop:

$$\mu = \text{magnetic moment} = (\text{area})(I)$$

There is a definite advantage to thinking of a current loop as a bar magnet characterized by its magnetic moment, as we now shall see.

It has been pointed out (Fig. 18.21) that a current loop has a magnetic field similar to that of a bar magnet. This is shown once again in Fig. 18.25. Notice that the coil acts like a short, fat bar magnet with north and south poles. This is also true for the single current loop in Fig. 18.25c. Moreover, when placed in a magnetic field, the coil and loop experience a torque in the same direction as the torque on a bar

**FIGURE 18.25**

The coil in (a) acts like the bar magnet in (b). Notice how the direction of the magnetic moment $\mu$ is assigned in (c).

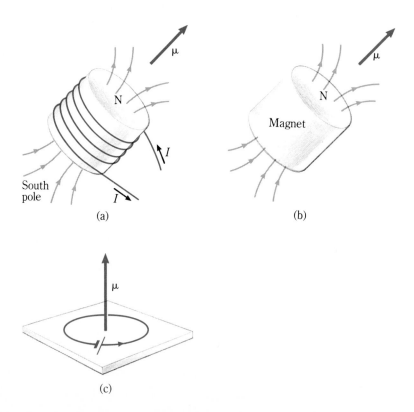

(a)

(b)

(c)

magnet. For example, if **B** is directed from left to right in Fig. 18.25, all three devices experience a clockwise torque.

We can obtain maximum usefulness from the concept of magnetic moment if we assign it direction. The direction assigned to $\mu$ is shown in Fig. 18.25$b$ and $c$. Notice that it is directed along the axis of the magnet, loop, or coil in such a way that it follows the central field line through the coil. As a result, the magnetic-moment vector $\mu$ points out of the north pole of the equivalent magnet. This has the following important consequence:

A current loop placed in a magnetic field rotates so as to align its magnetic-moment vector with the magnetic-field vector. The torque on it is

$$\text{Torque} = \mu B \sin \theta$$

where $\theta$ is the angle between $\mu$ and **B**.

You can appreciate why this is true by recalling that a compass needle is simply a bar magnet and that the field direction is defined to be the direction along which the needle aligns. We shall find it convenient from time to time to think of a current loop as a magnet with magnetic moment $\mu$.

---

### Example 18.5

In 1913 Niels Bohr postulated the following reasonably successful picture for the hydrogen atom. At its center is a proton of charge $+1.6 \times 10^{-19}$ C and mass $1.67 \times 10^{-27}$ kg. The diameter of the proton is of the order of $10^{-15}$ m. Circling the proton at a radius of $0.53 \times 10^{-10}$ m is an electron of charge $-1.6 \times 10^{-19}$ C and mass $\frac{1}{1840}$ that of the proton. Bohr concluded that the electron circled the nucleus $6.6 \times 10^{15}$ times each second. This motion of the charged electron around a circle is equivalent to a current in a loop of wire. Find the magnetic moment of the hydrogen atom resulting from the orbital motion of its electron.

*Reasoning* By definition, the magnetic moment $\mu$ is $IA$, where $A$ is the area of the loop:

$$A = \pi r^2 = (\pi)(0.53 \times 10^{-10})^2 \text{ m}^2 = 0.88 \times 10^{-20} \text{ m}^2$$

To find $I$, we note that the current in a loop is equal to the charge passing a given point in a second. For the present case, the electron circles the atom $6.6 \times 10^{15}$ times each second, carrying a charge of $1.6 \times 10^{-19}$ C past a given point that many times each second. Hence

$$I = (1.6 \times 10^{-19} \text{ C})(6.6 \times 10^{15} \text{ s}^{-1}) = 1.05 \times 10^{-3} \text{ A}$$

We therefore find that the atom acts as a small magnet with magnetic moment

$$\mu = IA = (1.05 \times 10^{-3} \text{ A})(0.88 \times 10^{-20} \text{ m}^2)$$
$$= 9.3 \times 10^{-24} \text{ A} \cdot \text{m}^2$$

When techniques for measuring this magnetic moment became available, the predicted value was confirmed. In addition, the electron *itself* was found to act as a small magnet, and this was pictured as being the result of a spinning motion of the charged electron about an axis through its own center, as we mentioned previously. ∎

As we just saw, a current-carrying coil in a magnetic field experiences a torque. Because the torque is proportional to the current in the coil, this effect can be used to measure current. In fact, most common electric meters are based on the movement of a coil in a magnetic field.

To see how this effect is utilized, let us refer to Fig. 18.26, where we see a current-carrying coil between the poles of a magnet. When the current is in the direction shown, the coil acts as a magnet with its north pole on the back side. (Check this and see.) We indicate this fact by the magnetic-moment vector $\mu$.

Because the magnetic-moment vector tries to align with the field, the coil rotates so as to point this vector toward the south pole of the permanent magnet. However, a spring attached to the coil provides a torque to stop this rotation. As a result, the coil rotates an amount proportional to the current in it. Therefore, the rotation of the coil, indicated by the pointer attached to it, can be used as a measure of the current in the coil.

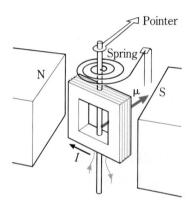

**FIGURE 18.26**

The current in the galvanometer coil causes it to rotate in the magnetic field provided by the permanent magnet.

We call the device sketched in Fig. 18.26 a *meter movement*. In practice, the coil often has an iron core to intensify the field and the torque. Many very sensitive ammeters, called *galvanometers,* are simply a movement such as this placed in an appropriate case. For that reason, it is common to use the terms *meter movement* and *galvanometer* interchangeably.

The sensitivity of a meter movement, that is, how large a deflection results from a given amount of current, depends on several factors. Of course, the stiffness of the restoring spring is of primary importance. The spring must be responsive enough to measure reasonably small currents, but on the other hand, it must not be too delicate if the instrument is to be rugged and portable. The sensitivity also depends on the number of turns of wire on the coil. If the number of turns is doubled, the torque is doubled as well.

A very sensitive ammeter gives full-scale deflection for a current of only a fraction of a microampere. Such a highly sensitive meter must have a large number of turns of wire on its coil, and so its resistance might easily be 100 $\Omega$. Even so, a voltage of $10^{-4}$ V across its terminals would cause a current of $10^{-6}$ A. Hence, it could be used as a very sensitive voltmeter as well as an ammeter. The table-model galvanometers used widely for student work give a full-scale deflection for a current of about 1 mA ($10^{-3}$ A). They have a resistance of about 20 $\Omega$.

The core of an ammeter is the meter movement. If it is to be very sensitive, the ammeter may be a meter movement without any modification (in which case it is most often called a galvanometer). However, most meter movements give full-scale deflection for a few milliamperes or less, and so they cannot be used directly to measure larger currents. Suppose, as an example, that we wish to make an ammeter that deflects full scale for a current of 2 A. The movement to be used has a resistance of 20 Ω and deflects full scale for a current of 3 mA. (This is a 60-mV movement; that is, $60 \times 10^{-3}$ V across its terminals causes a current of 3 mA and a full-scale deflection.)

In order to make the desired ammeter from this movement, we must find some way of allowing 2 A through the meter while there is only 0.003 A through the movement. This can be done with the meter design shown in Fig. 18.27. The meter movement is the circular device with resistance $R_m$.

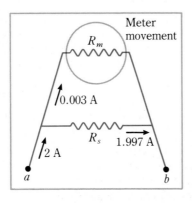

FIGURE 18.27

Only a small portion of the current goes through the movement of an ammeter. Most of it goes through the shunt resistor $R_s$.

When there is a current of 2 A into this meter through terminal $a$, we want only 0.003 A through the movement, which has a resistance $R_m$. This means that we must place a small resistance in parallel with the movement so that 1.997 A goes through it. We call this small resistor a **shunt** resistance, and it is indicated as $R_s$ in Fig. 18.27. To find how large $R_s$ should be, we proceed as follows.

The movement deflects full scale when there is a potential of $60 \times 10^{-3}$ V across its terminals from $a$ to $b$, but under these same conditions, the shunt resistor must carry a current of 1.997 A. Using Ohm's law, we find

$$V = IR$$
$$60 \times 10^{-3} = 1.997 \, R_s$$
$$R_s \approx 0.030 \ \Omega$$

Notice that the shunt resistor is extremely small. Often it is just a piece of copper wire. Moreover, the resistance of the ammeter is less than 0.030 Ω since $R_s$ is in parallel with $R_m$. This is as it should be since we do not wish the ammeter to disturb the circuit when it is connected in series. (What would happen if this ammeter were accidentally connected across a potential difference of 1 V?)

Many ammeters have more than one range because they contain several alternative shunts. The sensitivity of the meter is determined by which shunt is connected across the meter terminals.

## 18.15 Voltmeters

We can construct a voltmeter from the same movement we used to make an ammeter. The movement deflects full scale for a current of 0.003 A and has a resistance of 20 Ω. Hence, the potential difference across its terminals at full-scale deflection is (0.003)(20), or 60 mV. Let us construct a 90-V voltmeter from this movement.

An appropriate circuit for this purpose is shown in Fig. 18.28. When the potential across the terminals is 90 V, we want a current of 0.003 A through the meter. To find the resistance $R_x$ that must be placed in series with the movement, we apply Ohm's law to the circuit of Fig. 18.28:

$$V = IR$$

$$90 = (0.003)(20 + R_x)$$

$$R_x = 29{,}980 \ \Omega$$

Clearly, this meter possesses a very high resistance. Since we use a voltmeter by connecting it across a potential difference, we require it to have a high resistance so that it does not disturb the circuit. The best voltmeters have a much higher resistance than the one we designed. Our meter is said to have a resistance of 29,980/90, or 333, Ω/V. This does not mean that the resistance varies depending on the scale reading. It is merely the ratio of the resistance of the meter to its maximum scale reading. A similar meter with a more sensitive movement would have a higher resistance. (Why?)

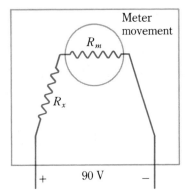

**FIGURE 18.28**

To make a voltmeter from a sensitive movement, we place a large resistor in series with the movement.

## Minimum learning goals

When you finish this chapter, you should be able to

**1** Define (*a*) right-hand rule for a magnetic field, (*b*) right-hand rule for a force, (*c*) magnetic field strength, induction, and flux density, (*d*) tesla, weber/m², and gauss, (*e*) velocity selector, (*f*) solenoid, (*g*) Hall effect, (*h*) ferromagnetic material, (*i*) domain, (*j*) electromagnet, (*k*) magnetic moment, (*l*) meter movement, (*m*) shunt resistor.

**2** Sketch the magnetic field in the vicinity of (*a*) various magnets, (*b*) a straight current-carrying wire, (*c*) a loop of current-carrying wire, (*d*) a solenoid.

**3** Use a compass to determine the direction of the magnetic field lines in a given region.

**4** Find the magnitude and direction of the force on a given straight-wire current in a specified magnetic field.

**5** Use $F = B_\perp IL$ to find one of the quantities when the others are given.

**6** Use $F = qvB_\perp$ to find one of the quantities when the others are given. Describe quantitatively the path followed by a particle of known charge and speed moving perpendicular to a given magnetic field.

**7** Compute the magnetic field for a long, straight wire carrying a current and for a solenoid when sufficient data are given.

**8** Given a list of common materials, select those that greatly alter the magnetic field into which they are placed.

**9** Describe in terms of domains what happens when a bar of ferromagnetic material is magnetized or demagnetized.

**10** Explain why the Hall effect allows us to determine the sign of charge carriers.

**11** State which way a current-carrying coil will turn when placed in a given position in a magnetic field. Compute the torque on the coil when sufficient data are given.

**12** Point out where the effective north and south poles are for a current-carrying loop. Explain what is meant by the magnetic-moment vector for a current loop.

**13** Explain the major features of a meter movement. Tell how it is used to make an ammeter or a voltmeter.

**14** Compute the shunt resistance needed to make a given movement into an ammeter of stated range.

**15** Compute the series resistance needed to make a given movement into a voltmeter of stated range.

## Questions and guesstimates

**1** The north pole of a bar magnet is brought close to an unmagnetized iron nail. What does the field of the magnet do to the nail? Why is the nail attracted by the magnet?

**2** Two concentric circular loops lie on a table. The larger loop carries a current of 10 A counterclockwise, and the smaller carries a current of 5 A clockwise. Describe the forces on each loop.

**3** As shown in Fig. P18.1, two high-voltage leads cause a beam of charged particles to shoot to the right through a partially evacuated tube. Their path is shown by a fluorescent screen placed along the length of the tube. When a magnet is brought close, the beam deflects. How could you determine the sign of the charge on the particles?

**4** Describe the motion of an electron shot into a long solenoid at a small angle to the solenoid axis.

**5** It is sometimes said that the earth's North Pole is a south pole, and vice versa. What does this mean?

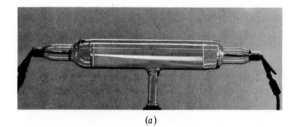

(*a*)

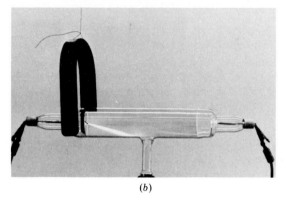

(*b*)

**FIGURE P18.1**

**6** When a beam of electrons is shot into a certain region of space, the electrons travel through the region in a straight line. Can we conclude there is no electric field in the region? No magnetic field?

**7** In a certain experiment, a beam of electrons shot out along the positive *x* axis deflects toward positive *y* values in the *xy* plane. If this deflection is the result of a magnetic field, in what direction is the field? Repeat for an electric field.

**8** A beam of charged particles is deflected as it passes through a certain region of space. By taking measurements on the beam, how could you determine which caused the deflection, a magnetic or an electric field?

**9** A proton is shot from the coordinate origin out along the positive *x* axis. There is a uniform magnetic field in the positive *y* direction. (*a*) Describe the motion of the proton, paying particular attention to the quadrants in which it travels. (*b*) Repeat for an electron. (*c*) Repeat if the proton velocity is such that $v_x = v_y$ and $v_z = 0$. (*d*) Repeat if $v_x = v_y = v_z \neq 0$.

**10** We know that in a television tube electrons are shot from one end of the tube to the other, where they strike the fluorescent screen. Suppose your little brother insists that his general science teacher says that protons are

used, not electrons. How could you prove to him that he is wrong without dismantling the set?

**11** Suppose you are given a material that is a poor conductor but conducts enough to obtain a measurable current through it. How could you decide whether the current is made up of positive or negative charges, or both? Give as many ways as you can.

**12** It is proposed to furnish the propulsion force to a spaceship in the following way. Electricity is furnished by a nuclear reactor or some other means. Large currents are sent through copper bars in the ship, and the forces exerted on these bars by the earth's magnetic field propel the ship. What objections do you see to such a plan?

**13** Cosmic rays (charged particles coming to the earth from outer space) are unable to reach the surface of the earth unless they have very high energy. One reason is that they have to penetrate the earth's atmosphere. However, for particles coming toward the equator along a radius of the earth, magnetic effects are also important. Explain why, being careful to point out why particles can reach the poles of the earth without encountering this difficulty.

**14** Give an order-of-magnitude estimate of the displacement of the electron beam on a television screen as a result of the earth's magnetic field.

## Problems

**1** A power line carries a current of 30 A straight east in a region where the magnetic field is parallel to the earth's surface and directed straight north, with $B = 8.5 \times 10^{-4}$ T. Find the magnitude and direction of the force due to the field on a 5.0-m length of the power line.

**2** A wire carries a current of 25 A straight up from the surface of the earth. The earth's field there is horizontal and directed straight north, and for it $B = 8.0 \times 10^{-4}$ T. Find the magnitude and direction of the force due to the field on a 30-cm length of the wire.

**3** A circular loop of wire, $r = 8.0$ cm, lies on a table and carries a current of 2.0 A. Impressed on the loop is a vertical magnetic field of magnitude 0.065 T and directed downward. (*a*) Find the total force on the loop because of the field. (*b*) Find the approximate force on a 3.0-mm length of the loop.

**4** A certain horizontal east-west wire has a mass of 0.175 g per meter of length and carries a current *I*. Impressed on the wire is a northward-directed horizontal magnetic field of magnitude 0.40 T. Find *I* if the magnetic force is to support the weight of the wire.

■ **5** A rectangular wire loop lies on a table and carries a current *I*. Impressed on it is a magnetic field that makes an angle $\theta$ with the vertical. Prove that the net force on the loop due to the field is zero.

■ **6** Repeat Prob. 5 if the loop is circular.

■ **7** A wire in the *xy* plane makes an angle of 60° with the positive *x* axis and carries a 5.0-A current toward positive *xy* values. Impressed on it is a 0.035-T magnetic field. Find the magnitude and direction of the force on a 90-cm length of the wire if the field is directed along (*a*) the positive *x* axis, (*b*) the positive *y* axis, (*c*) the positive *z* axis.

■ **8** A wire in the *xy* plane makes an angle of 20° with the positive *y* axis and carries a current of 8.0 A toward negative *xy* values. Impressed on it is a 25-mT field. Find the magnitude and direction of the force on a 120-cm length of the wire if the field is directed along (*a*) the positive *y* axis, (*b*) the positive *x* axis, (*c*) the positive *z* axis.

■ **9** A proton is shot out along the positive *x* axis with a speed of $6.0 \times 10^4$ m/s. Find the direction and magnitude of the force exerted on it by a 0.035-T magnetic field if the field is in (a) the negative y direction, (b) the positive z direction, (c) the negative x direction.

**10** An electron is shot out along the positive y axis with a speed of $7.0 \times 10^5$ m/s. Find the direction and magnitude of the force exerted on it by a 6.0-mT magnetic field if the field is in (a) the negative y direction, (b) the positive x direction, (c) the positive z direction.

■ **11** An electron is moving in the xy plane with a velocity of $3.0 \times 10^5$ m/s directed at an angle of 20° above the positive x axis. Impressed on the electron is a 7.5-mT field. Find the magnitude and direction of the force on the electron if the direction of the field is along (a) the positive x axis, (b) the negative y axis, (c) the positive z axis.

■ **12** A proton is moving in the xy plane with a velocity of $5.0 \times 10^4$ m/s directed at an angle of 70° above the $+x$ axis. Impressed on it is a 0.0035-T field. Find the magnitude and direction of the force on the proton if the direction of the field is along (a) the positive y axis, (b) the negative x axis, (c) the positive z axis.

**13** A proton with speed $3.5 \times 10^5$ m/s is traveling perpendicular to a 25-mT field. Describe quantitatively the path it follows.

**14** An electron with speed $7.0 \times 10^6$ m/s is traveling perpendicular to a 2.5-mT field. Describe quantitatively the path it follows.

■ **15** A proton falls from rest through a potential difference of $2.5 \times 10^5$ V. It then enters a 0.50-T magnetic field perpendicular to the field lines. How large is the radius of the circle in which the proton travels?

■ **16** A proton is accelerated through an unknown potential difference, enters a $3.00 \times 10^{-2}$ T magnetic field perpendicular to the field lines, and describes a circle with a radius of 30.0 cm. What is the energy of the proton in electron-volts?

■ **17** When a fast-moving particle such as an electron shoots through superheated liquid hydrogen, bubbles form along the path of the particle. Figure P18.2 shows the

**FIGURE P18.2**

paths of several particles in such a "bubble chamber." The paths are curved because of a magnetic field perpendicular to and into the page. Assume this field to have a magnitude of 3.0 mT. Is the particle that is leaving point a and moving to the right positive or negative? If you assume it to be an electron, approximately what is its speed? (The tracks are actual size. Assume them to be in the plane of the page.)

■ **18** The particle that starts at point b in Fig. P18.2 slows as it moves through the liquid hydrogen. As a result, it spirals inward. Assume the same data as in Prob. 17 and assume the particle to be an electron. Find its speed at point c.

■ **19** An electron is shot from the origin of coordinates with a velocity of $2.0 \times 10^7$ m/s at an angle of 30° with the positive x axis. There is in this region a 4.0-mT magnetic field in the positive x direction. Describe the motion of the electron quantitatively. *Hint:* Separate the particle velocity into components parallel and perpendicular to the x axis.

■ **20** A particle of charge q is traveling with a velocity **v** perpendicular to a uniform magnetic field **B**. Relate the radius of its path to the kinetic energy of the particle.

■ **21** A particular beam of electrons travels in a straight line through crossed magnetic and electric fields in a velocity selector. The value of B is 0.050 T, and the plates are separated by 5 cm and have a voltage of 100 V across them. Find (a) the speed of the electrons and (b) the radius of the circle in which the electrons travel when the plate voltage is zero.

■ **22** When moving perpendicular to a 40-mT field, a proton beam follows a circle of radius 120 cm. How large an electric field perpendicular to both **v** and **B** will cause the proton to move in a straight line?

■ **23** Two parallel straight wires are 20 cm apart, and each carries a current of 15 A. Find the magnetic field at a point in the plane of the wires that is 4.0 cm from one wire and 16.0 cm from the other if the currents are (a) in the same direction and (b) in opposite directions.

**24** A long, straight wire carries 5.0 A out along the positive x axis, and a second wire carries 7.0 A out along the positive y axis. Find their combined field at the point $x = 4$ cm, $y = 6$ cm.

**25** A long solenoid has 5000 loops of wire on its 60-cm length. The solenoid diameter is 2.0 cm. Find B inside the solenoid when the current is 300 mA.

**26** How large is B inside a long solenoid that has 2000 turns on its 50-cm length if it carries a current of 3 A?

**27** A solenoid 62.8 cm long has 200 loops on it. How large must the current be if B is to be 0.50 T inside?

**28** A straight wire carrying a current of 10 A lies along the axis of a long solenoid in which $B = 3 \times 10^{-4}$ T. (*a*) How large is the force on a 1.0-cm length of the wire? (*b*) What is the value of $B$ inside the solenoid at 0.50 cm from its axis?

**29** A long, straight wire carries a current of 20 A long the axis of a solenoid in which the magnetic field (in the absence of the wire) is 5.0 G. Find the magnitude of the total magnetic field 0.40 cm from the axis of the solenoid. The solenoid radius is 0.50 cm.

**30** A certain solenoid has 70 loops per centimeter length and carries a current of 8.0 A. A proton is shot from a point on the solenoid axis with a velocity of $3 \times 10^5$ m/s at an angle of 15° to the axis. Describe quantitatively the path it follows. *Hint:* Separate the initial velocity into components parallel and perpendicular to the axis.

**31** A flat coil of wire with 30 loops lies on a table. Its area is 150 cm², and it carries a current of 25 A. Find the torque on it due to a 70-mT field if the field lines are (*a*) parallel to the tabletop, (*b*) perpendicular to the tabletop, (*c*) at an angle of 20° to the horizontal. (*d*) What is the magnetic moment of the coil?

**32** A flat coil of wire with 50 loops lies flat against the north wall of a room. The coil's area is 200 cm², and it carries a current of 30 A. Find the torque on the coil due to a 90-mT field if the field lines are directed (*a*) eastward, (*b*) northward, (*c*) vertically, (*d*) in the plane of the wall at an angle of 75° to the vertical. (*e*) What is the magnetic moment of the coil?

**33** A square coil of wire 12 cm on an edge has 50 loops and a mass of 80 g. It lies flat on a table and is subject to a horizontal field of 0.043 T parallel to one edge. How large must the current in the coil be if one edge of the coil is to lift from the table?

**34** A circular coil of wire of radius 8.0 cm has 30 loops and a mass of 40 g. It lies flat on a table and is subject to a field of 70 mT that makes an angle of 60° to the vertical. How large must the current in the coil be if one part of the coil is to lift from the table?

**35** A certain meter movement has a resistance of 40 Ω and deflects full scale for a voltage of 200 mV across its terminals. How can it be made into a 3-A ammeter?

**36** If a meter movement deflects full scale to a current of 0.010 A and has a resistance of 80 Ω, how can it be made into a 4-A ammeter?

**37** How can the meter movement described in Prob. 36 be made into a 30-V voltmeter?

**38** How can the movement of Prob. 35 be made into a 20-V voltmeter?

**39** How can the movement of Prob. 36 be made into an ammeter with two ranges, 10 and 1.0 A?

**40** How can the movement of Prob. 36 be made into a voltmeter with two ranges, 12 and 120 V?

**41** The uniform magnetic field within a long solenoid has a magnitude $B$ and is directed along the solenoid's axis. What is the largest speed an electron shot radially from the axis can have if it is to escape hitting the inner surface of the solenoid? The inner radius of the solenoid is $R$.

**42** A hollow, straight plastic pipe has charge uniformly distributed over its cylindrical face. The charge per unit length is $Q$, and the pipe is very long. If it is rotating on its axis with a frequency $f$, what is the magnetic field inside the pipe due to the charge on its surface?

# 19 Electromagnetic induction

$\mathbf{T}$HE industrial revolution that transformed the world more than a century ago was based on three major scientific advances: the invention of the steam engine and the further development of heat engines through the use of thermodynamics, the discovery that forces to turn motors can be produced by the interaction of electric currents with magnetic fields, and the discovery that currents can be produced by changing magnetic fields. We have already discussed the first two advances. In this chapter, we investigate the third.

## 19.1 Induced emf

The discovery that electric currents generate magnetic fields was made in 1820. As so often happens in science, this newly found facet of nature led to intense investigations of related phenomena. One avenue of experimentation was followed by those who attempted to answer the question "If currents produce magnetic fields, is it not possible that magnetic fields produce currents?" The answer to this question was shown to be affirmative about ten years later by Michael Faraday (1791–1867) in England and independently by Joseph Henry (1797–1878) in the United States.* Let us now discuss an experiment that shows this effect vividly.

---

* Henry's work, carried out in relative obscurity in Albany, New York, was published only in the United States, and few people knew of it. As a result, his experiments had very little influence on scientific progress at that time.

The experiment makes use of the simple equipment shown in Fig. 19.1*a*. We see there two simple series circuits. One consists of a battery and switch connected in series by a long wire coiled around an iron rod. We call this coil the **primary coil** because it is attached to the battery. A second, independent wire is also coiled around the rod. This coil is in series with a galvanometer, but has no battery in its circuit. It is called the **secondary coil.**

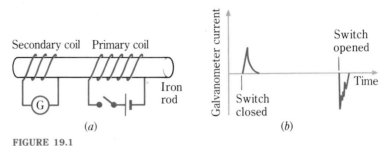

FIGURE 19.1

An induced current exists in the secondary coil only when the current in the primary coil is changing. The current pulses are actually much narrower than in (*b*).

Since there is no battery in the secondary-coil circuit, one might guess that the current through it is always zero. But a startling fact emerges if the switch in the primary circuit is suddenly either closed or opened. At that exact instant, the galvanometer suddenly deflects and then returns to zero. In other words, a current exists in the secondary-coil circuit for an instant. It is as though the secondary circuit possessed a battery (a source of emf) only for the short time it takes the switch to open or to close. We say that an **induced emf** exists in the secondary coil during that instant.

Figure 19.1*b* shows another feature of this induced current and emf: the short-lived current is in one direction when the switch is pushed closed, but the emf causes a current pulse in the opposite direction when the switch is pulled open. This tells us that the direction of the induced emf depends on whether the current in the primary coil is increasing or decreasing.

A second, somewhat similar experiment is shown in Fig. 19.2. It involves a bar magnet and a coil in series with a galvanometer. When the magnet is stationary beside the coil, as in *a* and *c*, there is no current in the coil. However, when the magnet is moving relative to the coil, a current exists in the coil, as indicated in *b*, *d*, and *e*. As we see, an induced emf exists in the coil only when the magnet and coil are in relative motion. No induced emf exists when conditions are not changing.

There are two ways to analyze this effect. One uses the fact that a charge moving through a magnetic field experiences a force. Although Fig. 19.2 shows the magnet in motion, exactly the same effects occur if the magnet is held fixed and the coil is moved.* Consider what happens when the coil moves toward the magnet. The free charges in the wire, since they are moving in the magnetic field of the magnet,

---

* This is an example of the fact that motion is a relative quantity. When relative motion occurs between two objects, the effect of one on the other is a function only of the relative motion. It makes no difference which object is considered to be at rest. More is said about this in Chap. 25 when we discuss the theory of relativity.

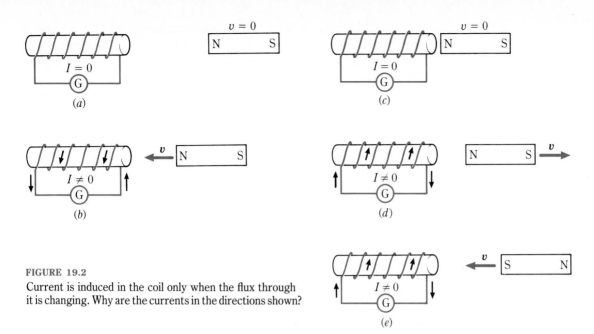

FIGURE 19.2

Current is induced in the coil only when the flux through it is changing. Why are the currents in the directions shown?

experience a force $qvB_\perp$, as discussed in Chap. 18. They flow under the action of this force and thereby give rise to the induced current.

This approach shows us at once how induced emf is related to phenomena we have already studied, and we shall use this way of looking at the situation from time to time. In most practical situations, however, another approach is more useful. It involves the concept of magnetic flux, as we shall see in the next sections.

## 19.2 Magnetic flux

Faraday explained the induced emf in a coil in terms of a quantity called magnetic flux. To do so, he made a rule about how to draw magnetic field lines. If the magnetic field in a certain region of space has a magnitude $B$, we agree to draw a number of field lines equal to $B$ through a unit area erected perpendicular to the field lines. Thus, in Fig. 19.3, the field has a magnitude $B = 16$ T because 16 field lines pass through the 1-m² area erected perpendicular to the field. In general, if an area $A$ is erected perpendicular to the field lines, the number of lines that pass through it is $BA$.

**FIGURE 19.3**

We agree to draw a number of field lines equal to $B$ through a unit area erected perpendicular to the field lines.

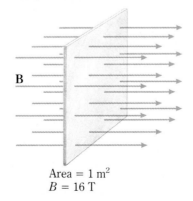

Area = 1 m²
$B = 16$ T

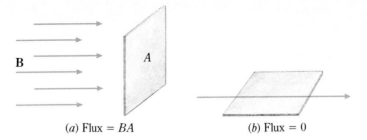

FIGURE 19.4

The flux through an area depends
on the relative orientation of the
area and the field lines.

(a) Flux = BA          (b) Flux = 0

We define the **magnetic flux** through an area to be the number of field lines that pass through the area. For example, the flux through the area shown in Fig. 19.3 is 16. Let us now look at Fig. 19.4. The area $A$ in part $a$ is erected perpendicular to the field lines, and so $BA$ lines pass through it; the flux has a numerical value $BA$ in this case. The situation in $b$ is quite different, however. We see there that no lines pass through the area, and so the flux through it is zero.

We can approach these situations most easily by considering the components of **B**, as shown in Fig. 19.5. The component of **B** parallel to the surface gives rise to

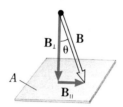

FIGURE 19.5

The flux through an area is $B_{\perp}A$.

field lines that simply skim the area and thus contribute zero flux. Only the component perpendicular to the area gives rise to field lines through the area. Hence the flux through an area $A$ is

$$\text{Magnetic flux} = \phi = B_{\perp}A \qquad (19.1)$$

where $\phi$ is the Greek letter phi. In the SI, $B$ is in tesla and $A$ is in square meters, and so the unit of flux is tesla-meters². This unit is often called the *weber* (Wb):

$$1 \text{ Wb} = 1 \text{ T} \cdot \text{m}^2$$

From this we see that the unit for $B$, the tesla, is simply

$$1 \text{ T} = 1 \text{ Wb/m}^2$$

a unit for $B$ that we mentioned earlier.

---

### Example 19.1

In a certain room, the magnetic field has a magnitude of $4.0 \times 10^{-5}$ T and is directed at an angle of 70° below the horizontal. Find the flux through a 400 cm × 80 cm tabletop in the room.

***Reasoning*** We can use Fig. 19.5 to represent this, with $\theta = 20°$ and $A = (4.0 \text{ m})(0.80 \text{ m}) = 3.2 \text{ m}^2$. Then, since $B_{\perp} = B \cos \theta$, we have

$$\phi = B_\perp A = (4.0 \times 10^{-5} \text{ T})(\cos 20°)(3.2 \text{ m}^2) = 1.20 \times 10^{-4} \text{ T} \cdot \text{m}^2 \quad \blacksquare$$

*Exercise* How much flux goes through the 15-m² north wall of the room if the field has no component in the east-west direction? *Answer: 2.05 × 10⁻⁴ T · m²* ∎

## 19.3 Faraday's law

After performing many experiments similar to those in Figs. 19.1 and 19.2, Faraday concluded that an induced emf exists in a coil only if the flux through the coil is changing. As another example, look at the experiment shown in Fig. 19.6. When the switch is closed, the current in the primary coil generates the magnetic field shown. Since the field lines follow the iron rod, considerable flux passes through the secondary coil. If the switch is pulled open, this flux decreases to zero because the current that generates it stops. During the time in which this change in flux is occurring, an induced emf exists in the secondary coil, but no induced emf exists when the flux is not changing.

**FIGURE 19.6**

What happens in the secondary coil when the current in the primary coil is steady? What happens when the switch is pulled open? When it is pushed closed?

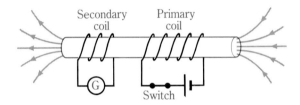

The experiment shown in Fig. 19.7 also confirms Faraday's conclusion. Notice that the flux through the coil increases as the magnet is brought closer to it. When the magnet is stationary, no flux change is occurring and so there is no induced emf

**FIGURE 19.7**

When the magnet is moved from the position in (*a*) to that in (*b*), the current in the coil is in the direction shown in (*c*). Why?

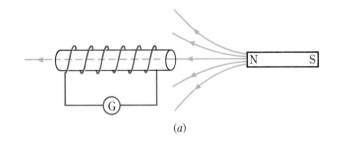

(*a*)

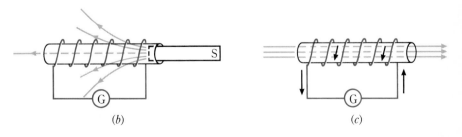

(*b*)          (*c*)

in the coil. However, while the magnet is being moved from the position in *a* to that in *b*, the flux through the coil is changing and so there is then an emf induced in the coil.

Let us now give a statement of Faraday's quantitative results. Suppose that a coil with *N* loops has the flux through it changed from $\phi_1$ to $\phi_2$ in a time $\Delta t$. Then the change in flux through the coil is simply $\Delta\phi = \phi_2 - \phi_1$. Faraday found that the emf induced in the coil while the flux is changing is

$$\text{emf} = N\frac{\Delta\phi}{\Delta t} \qquad (19.2)$$

This is called **Faraday's law.**

The direction of the induced emf is often important and can be determined from experiments such as those shown in Figs. 19.2 and 19.7. As the magnet in Fig. 19.7 is moved from the position in *a* to that in *b*, the induced current shown in *c* exists. The colored lines in *c* show the magnetic field produced in the coil by this induced current. Notice that the flux produced in the coil by the coil's own induced current is directed to the *right*. However, as the magnet was moved from the position in *a* to that in *b*, it furnished flux directed to the *left* through the coil. In other words, the flux due to the induced current tends to cancel the change in flux that caused the induced current. If you examine Fig. 19.2, you will notice similar behavior. The induced current is in such a direction as to try to cancel the change in flux through the coil. This rule is embodied in **Lenz's law:**

A change in flux through a loop of wire induces an emf in the loop. The direction of the current produced by the induced emf is such that the flux generated by the current tends to counterbalance the original change in flux through the loop.*

Again, if the flux through a coil is to the *left* and *increasing,* the induced emf will try to put flux through the coil to the *right*. If the flux is to the *left* and *decreasing,* the induced emf will try to put flux through the coil to the *left*. The induced emf always tries to counterbalance the change in flux.

---

### Example 19.2

A solenoid with 100 loops has a cross-sectional area of 4.0 cm². It is suddenly transferred from a region where there is no magnetic field to one where the field is 0.50 T directed down its length. If the transfer requires 0.020 s, how large is the average emf induced in the solenoid?

***Reasoning*** From our defining equation for flux, Eq. 19.1, $\phi = B_\perp A$. Then

$$\phi_1 = 0 \qquad \text{and} \qquad \phi_2 = (0.50 \text{ T})(4.0 \times 10^{-4} \text{ m}^2) = 2.0 \times 10^{-4} \text{ Wb}$$

We therefore have that

$$\Delta\phi = \phi_2 - \phi_1 = 2.0 \times 10^{-4} \text{ Wb}$$

---

* This is a consequence of the conservation of energy. If the flux generated by the induced emf were in such a direction as to augment the change in flux, the induced emf could continue to induce a larger and larger emf, without end.

Faraday's law yields

$$\text{emf} = N\frac{\Delta\phi}{\Delta t} = 100\,\frac{2.0 \times 10^{-4}\ \text{Wb}}{2.0 \times 10^{-2}\ \text{s}} = 1.0\ \text{V}$$

You should be able to show that one weber is one volt-second, and so the unit of the answer is volts. ∎

## 19.4 Mutual induction

Faraday's law for the induced emf in a coil applies to any method for changing the flux through the coil. Suppose we have two coils placed side by side, as in Fig. 19.8. When the switch is open, both coils have zero flux through them. When the switch is suddenly closed, the primary coil will act as an electromagnet and will generate flux

**FIGURE 19.8**

Why does the current flow from $a$ to $b$ through the resistor at the instant the switch is opened?

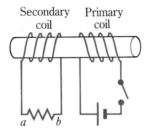

Secondary coil    Primary coil

in the region near it. Some of the flux from the primary coil will go through the secondary coil. Hence, the flux through the secondary will change when the switch is suddenly closed. According to Faraday's law, an induced emf will be generated in the secondary for an instant as the current in the primary rises from zero to its final value. You should be able to show that the direction of the induced current through the resistor in Fig. 19.8 will be from $b$ to $a$ just as the switch is closed. The current will be in the opposite direction just as the switch is opened.

The magnitude of the induced emf generated in the secondary will depend on many geometric factors. Among these are the number of turns of wire on each coil, how close together the coils are to each other, and their cross-sectional area. (Why?) In addition, since the flux through the secondary will be proportional to the current in the primary, the induced emf in the secondary will be proportional to the rate of change of current in the primary, $\Delta I_p/\Delta t$. We therefore write the following equation for the induced emf in the secondary:

$$\text{emf}_{\text{sec}} = M\frac{\Delta I_p}{\Delta t} \tag{19.3}$$

where the proportionality constant $M$ contains the effects of the geometry of the two coils. It is called the **mutual inductance** of the two coils. If the emf is in volts, $I$ in amperes, and $t$ in seconds, the unit $M$ is defined to be the **henry** (H), or V · s/A.

---

***Example 19.3***
Two coils of wire wound on an iron core have a mutual inductance of 0.50 H. How large an average emf is generated in the secondary by the primary as the current in the primary is increased from 2.0 to 3.0 A in 0.010 s?

# Michael Faraday (1791–1867)

Faraday, the son of an English blacksmith, received only a rudimentary education before finding work as a bookbinder's errand boy in 1804. Soon he became the bookbinder's apprentice, and this gave him the opportunity to read the books passing through his hands. Growing fascinated by science, he carried out such simple experiments as his meager salary allowed and attended public scientific lectures, the most inspiring of which was the series of lectures given by the great chemist Sir Humphry Davy. In 1813, Faraday took the bold step of writing to Davy to request employment in his laboratory. Davy was impressed by the notes Faraday enclosed, which he had taken at the lecture series, and offered Faraday a job as a servant in his laboratory. From these humble beginnings, tutored by Davy, he advanced swiftly as a result of his obvious experimental skill and insight. In 1825, he was made director of the laboratory of The Royal Institution. His whole life was devoted to experimental investigations, particularly in electricity. He is recognized as one of the greatest experimental scientists of the nineteenth century.

*Reasoning* Let us first point out that the flux when two coils are wound on a core of iron is much larger than the flux when they are wound on a nonmagnetic core. Hence, the value of $M$ given here is much larger than the value one would ordinarily have for coils with a noniron core.

Making use of Eq. 19.3, we have

$$\text{emf} = M\frac{\Delta I}{\Delta t} = (0.50 \text{ H})\left(\frac{1.0 \text{ A}}{0.010 \text{ s}}\right) = 50 \text{ V}$$

Notice that this emf exists in the secondary for only an instant. As soon as the current in the primary becomes steady, the flux is no longer changing and there is no induced emf. ■

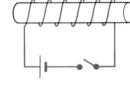

**FIGURE 19.9**

When the switch is first closed, the coil will induce an emf in itself. Will it aid or oppose the battery?

## 19.5 Self-inductance

Faraday's law tells us that any change in flux through a coil induces an emf in the coil. This means that when a current through a coil changes, the coil induces an emf in itself. Suppose the current in the coil shown in Fig. 19.9 changes from zero to some finite value when the switch is first closed. A flux is generated by the current and is directed to the left through the coil. By Faraday's law, an emf is induced in the coil, and it tries to produce flux to the right through the coil. Hence, the induced emf must be opposed to the emf of the battery. However, if the switch is suddenly opened, the induced emf will aid, rather than oppose, the battery. (You should be able to show this.)

Here, too, the geometry of the coil as well as the core material determines how large the induced emf will be. If $\Delta I / \Delta t$ is the rate of change of current through the coil, we can write for the average induced emf

$$\text{emf} = L \frac{\Delta I}{\Delta t} \tag{19.4}$$

The constant of proportionality $L$ is called the **self-inductance** of the coil. It has the same units as mutual inductance: henrys.

Obviously, if the coil is wound on an iron core, the flux through it is much greater than if no magnetic material is present. Hence, if a large self-inductance is desired, the inductor should be wound on an iron core. We return to the behavior of mutual and self-inductance in later sections. They are of particular importance in alternating-current (ac) circuits, where the current, and thus the flux, are changing continually.

---

### Example 19.4

A certain solenoid has cross-sectional area $A$, length $D$, and $n$ loops per unit length. What is its self-inductance?

**Reasoning**  We know from Eq. 18.6 that the field in the solenoid is given by $\mu_o n I$, where $I$ is the current in the solenoid. Since the field is directed straight down the solenoid, the flux through each loop is

$$\phi = B_\perp A = \mu_o n I A$$

Let us consider what happens as the current in the solenoid changes from $I$ to $I + \Delta I$. The new flux will be an amount $\Delta \phi$ larger than before, and so

$$\phi + \Delta \phi = \mu_o n A (I + \Delta I)$$

Subtracting the previous equation from this gives

$$\Delta \phi = \mu_o n A \, \Delta I$$

According to Faraday's law, the induced emf in the solenoid is

$$\text{emf} = N \frac{\Delta \phi}{\Delta t} = (nD) \mu_o n A \frac{\Delta I}{\Delta t}$$

where $N$, the number of loops on the solenoid coil, has been replaced by its value, $nD$. We compare this with the defining equation for self-inductance,

$$\text{emf} = L \frac{\Delta I}{\Delta t}$$

From the comparison, we see that the self-inductance is

$$L = \mu_o n^2 DA$$

This is the inductance we set out to compute. If the solenoid had an iron core, the field within it, and therefore its inductance, would be hundreds of times larger than the value we have computed. ∎

---

*Exercise* Evaluate $L$ for an 80-cm-long air-filled coil that has 500 loops and a 1.20-cm diameter. *Answer: $4.44 \times 10^{-5}$ H* ∎

---

## 19.6 Inductance-resistance circuits

Some very interesting and useful properties of a self-inductance coil are explored in detail in Chap. 20. At present, we are concerned with only one facet of the inductor's behavior — its ability to store energy.

Consider the circuit shown in Fig. 19.10. It consists of an inductance coil (represented by the symbol ∿∿), a resistor, a battery, and a switch. If the inductor were

FIGURE 19.10

Why doesn't the current rise to a value $\mathscr{E}/R$ immediately after the switch is closed?

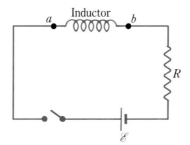

not in the circuit, the current would rise quickly just as the switch was closed. The final current would be given by Ohm's law to be $\mathscr{E}/R$. However, as the current rises, flux is generated in the inductance coil. This induces an emf in the coil in such a direction as to oppose the increasing current. In other words, the inductor acts as a battery whose polarity is the reverse of that of the actual battery in the circuit.

The upshot of this is that the coil keeps the current from rising too rapidly. Despite this fact, the current continues to rise until it finally achieves its normal Ohm's law value, $\mathscr{E}/R$. This behavior is shown in Fig. 19.11. As we see there, the current rises to 0.63 of its maximum value in a time $L/R$. This time $L/R$ is called the **time constant** of the circuit.* As you would expect, the current rise time is longest for a large inductance (because the larger the inductance, the larger the opposition to flow) and a small resistance (because the smaller the resistance, the higher the final current).

It is important to notice that the inductor acts as a transitory battery of reverse polarity in the circuit. As the current rises, it is, in effect, charging this emf source in the inductor. Let us now compute how much work is done against the opposing emf of the coil.

The emf induced in the coil is, from Eq. 19.4, $L(\Delta I/\Delta t)$. Therefore, when there is a current through the coil, the charges flow through a potential difference

---

\* The factor 0.63 is $1 - (1/e) = 1 - (1/2.718)$.

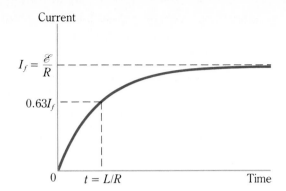

FIGURE 19.11
After the switch in Fig. 19.10 is
closed, the current rises as shown.

Current

$I_f = \dfrac{\mathscr{E}}{R}$

$0.63I_f$

$0 \qquad t = L/R \qquad\qquad\qquad$ Time

$L(\Delta I/\Delta t)$. The work the current does as it carries a charge $\Delta q$ through the inductor and through the potential difference $L(\Delta I/\Delta t)$ is

$$\Delta W = (\Delta q)(V) = (\Delta q)\left( L\,\frac{\Delta I}{\Delta t}\right)$$

This can be simplified if we notice that $\Delta q/\Delta t$ is simply $I$. Then,

$$\Delta W = LI\,\Delta I$$

To summarize, this is the work done on the coil as the current with present value $I$ is increased by an amount $\Delta I$.

We wish to sum all these small quantities of work as the current in the circuit changes from zero to its final maximum value $I_f$. This sum is evaluated in one of the exercises at the end of this chapter. The result for the work expended on the coil while the current in it is changed from $I = 0$ to $I = I_f$ is found to be

$$W = \tfrac{1}{2}LI_f^2$$

This supplied work can be thought of as energy stored in the coil. Vivid evidence for this stored energy is shown when the switch in a circuit such as that of Fig. 19.10 is pulled open. If the inductance is large, a large spark jumps across the gap of the switch. Moreover, a very high voltage is induced in the coil as it tries unsuccessfully to oppose the loss of flux through itself. We therefore have found that

An inductor $L$ through which there is a current $I$ has stored in it an energy $\tfrac{1}{2}LI^2$.

## 19.7 The energy in a magnetic field

You will recall that we computed the energy stored in an electric field by considering the energy stored in a capacitor (Sec. 16.11). Let us now find the energy stored in a magnetic field by considering the energy stored in an inductor. We assume the inductor to be a long solenoid. As was shown in Chap. 18, the magnetic field is confined mainly to the core of the solenoid and has a uniform value, $B = \mu_o nI$.

We computed the inductance of a solenoid in Example 19.4:

$$L = \mu_o n^2 DA$$

where $D$ is the length of the solenoid and $A$ is its cross-sectional area. Notice, however, that $DA$ is simply the volume of the core region of the solenoid.

The energy stored inside the solenoid is

$$\text{Energy} = \tfrac{1}{2}LI^2 = \tfrac{1}{2}\mu_o n^2 I^2 DA$$

from which the energy per unit volume is

$$\frac{\text{Energy}}{\text{Volume}} = \frac{\tfrac{1}{2}\mu_o n^2 I^2 DA}{DA} = \tfrac{1}{2}\mu_o n^2 I^2$$

The field in the solenoid is $B = \mu_o nI$, however, from which $I = B/(\mu_o n)$. Substituting this value in the above equation gives

$$\frac{\text{Energy}}{\text{Volume}} = \tfrac{1}{2}\mu_o n^2 \frac{B^2}{\mu_o^2 n^2}$$

$$\text{Energy per unit volume} = \frac{B^2}{2\mu_o} \tag{19.5}$$

for the energy density in a magnetic field of magnitude $B$. This is to be compared with the value $\tfrac{1}{2}\epsilon_o E^2$ we found for the energy density in an electric field in vacuum. Of course, Eq. 19.5 applies only to magnetic fields in vacuum and to nonmagnetic materials. Although this result was obtained for a solenoid, it applies to all other situations as well. The concept of energy stored in a magnetic field will be important when we study the way in which light and other electromagnetic waves carry energy.

---

**Example 19.5**

A certain inductor has an inductance of 0.50 H and a resistance of 2.0 $\Omega$. It is placed in series with a switch, a 12.0-V battery, and a 4.0-$\Omega$ resistor. Find the time constant of the circuit and the final energy stored in the inductor.

*Reasoning*  The time constant is

$$\frac{L}{R} = \frac{0.50 \text{ H}}{(2.0 + 4.0)\ \Omega} = 0.083 \text{ s}$$

Although the rise of the current is delayed somewhat in this circuit, the delay is not large.

The final current in the circuit is $I_f = V/R = (12 \text{ V})/(6\ \Omega) = 2.0$ A. Therefore, the energy stored in the inductor is

$$\text{Energy} = \tfrac{1}{2}LI^2 = \tfrac{1}{2}(0.50 \text{ H})(2 \text{ A})^2 = 1.0 \text{ J} \quad \blacksquare$$

---

### 19.8 Motional emf

An induced emf can occur in many ways. Until now, we have been dealing mainly with flux changes through stationary coils and the attendant induced emf. Sometimes, however, the induced emf is the result of the motion of a wire through the magnetic field. In such cases, it is often more convenient to use an approach that is not based directly on the concept of flux change through a loop.

Let us begin our discussion by referring to the simple experiment shown in Fig. 19.12. A rod of approximate length $d$ rolls with velocity $\mathbf{v}$ along parallel wires that form a U-shaped loop from $m$ through $r$ and $s$ back to $n$. Notice that the rod and wires form a loop *(pqrsp)* to the left of the rod. As the rod moves to the right, the area of this loop increases.

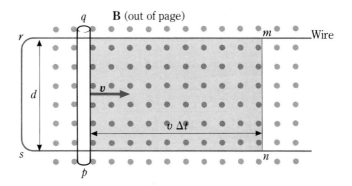

FIGURE 19.12

The induced emf can be treated as the result of the change of flux through the loop.

Suppose a magnetic field $\mathbf{B}$ directed out of the page exists in this region. As the rod moves along, the flux through the area of the loop increases because the area increases. Hence an emf is induced in the loop. To find this emf, we note that in time $\Delta t$ the rod rolls a distance $v \Delta t$. As a result, the loop area increases by an amount $\Delta A = d(v \Delta t)$, the shaded area in the figure. The accompanying flux change is

$$\Delta \phi = B_\perp \Delta A = B_\perp dv \Delta t$$

Then, according to Faraday's law, the magnitude of the induced emf in the loop is

$$\text{Induced emf} = \frac{\Delta \phi}{\Delta t} = B_\perp vd$$

You should convince yourself that this induced emf is directed clockwise around the loop.

There is another way in which we can approach this situation. Consider a positive charge $q$ in the moving rod. This charge, because of its motion with velocity $\mathbf{v}$ through the field $\mathbf{B}$, experiences a force $qvB_\perp$. In this case, the total field is perpendicular to the velocity of the charge and so $B = B_\perp$. We conclude that

$$F = \text{force on } q = qvB_\perp$$

Using the right-hand rule, you can see that the force on $q$ is directed from $q$ to $p$ along the rod.

From the definition of electric field as force per unit charge, however, we conclude that the charges moving with the rod experience an electric field directed from $q$ to $p$ along the rod. Thus*

$$E = \frac{F}{q} = vB$$

---

* Strictly speaking, this value for $E$ applies only in a reference frame moving with the charge.

If we recall that the potential difference between two points equals the work done in carrying a unit test charge from one point to the other, then we know that the potential difference from $p$ to $q$ due to the electric field $E$ is

$$V = Ed = B_\perp vd$$

Notice that this is exactly equal to the induced emf in the loop that we found using Faraday's law. Moreover, the electric field induced by the charge motion causes a clockwise current in the loop, the same direction we found from Faraday's law.

We can summarize these latter results in the following way:

A wire (or rod) of length $d$ moving with velocity $\mathbf{v}$ perpendicular to both a field $\mathbf{B}$ and its own length has an induced emf along its length of

$$\text{Induced emf} = B_\perp vd \qquad (19.6)$$

This is called a *motional emf.*

In the more general case in which $\mathbf{B}$, $\mathbf{v}$, and the wire are not mutually perpendicular, the components of $\mathbf{B}$ and $\mathbf{v}$ that are perpendicular to each other and to the wire must be used.

This statement is often paraphrased in terms of lines of flux. When the rod in Fig. 19.12 moves so as to change the flux through the loop by an amount $\Delta\phi$, the rod cuts through $\Delta\phi$ flux lines. But the induced emf in the rod is simply $\Delta\phi/\Delta t$, the rate at which the rod is cutting the lines of flux. We can therefore state:

A moving wire has within itself an induced emf equal to the rate at which the wire is cutting magnetic flux lines.

As we shall see, the concept of a motional emf is convenient in certain situations.

---

### Example 19.6

A rod of length 5.0 m is held horizontal with its axis in an east-west direction. It is allowed to fall straight down. What is the emf induced in it when its speed is 3.0 m/s if the earth's magnetic field is 0.60 G at an angle of 53° below the horizontal?

*Reasoning* We use Eq. 19.6. Because $B_\perp = B\cos 53°$ and because $\mathbf{v}$ is already perpendicular to the rod, we have

$$\text{emf} = B_\perp vd = (B\cos 53°)vd$$

$$= (0.60 \times 10^{-4}\text{ T} \times 0.60)(3.0\text{ m/s})(5.0\text{ m}) = 5.4 \times 10^{-4}\text{ V}$$

You should be able to show that the same result is obtained by finding the rate at which the rod is cutting flux lines. ∎

---

### 19.9 AC generators

A generator is a device that converts mechanical energy to electrical energy. It does this by changing the flux through a coil, thereby inducing an emf between the two terminals of the coil. In theory, the flux could be changed either by moving a magnet

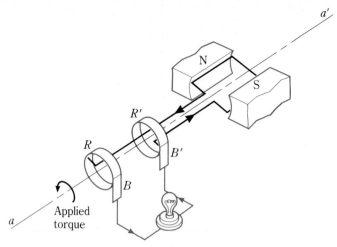

**FIGURE 19.13**

An alternating emf is produced between terminals $B$ and $B'$ as the loop rotates in an external magnetic field.

with respect to the coil or by moving the coil with respect to a magnet. The latter procedure is more easily realized in practice and is the one ordinarily used.

A schematic representation of a simple generator is shown in Fig. 19.13. An external energy source causes the loop of wire to rotate in the magnetic field of the magnet. (In practice, the loop is replaced by a coil wound on an iron core so as to intensify the effects we discuss.) As the loop rotates, the flux through it changes continually. This changing flux induces an emf in the loop, and the emf causes a current through the loop in the direction indicated. The current can be used to do useful work, perhaps to light a bulb as shown.

The purpose of a generator is to produce electrical energy that can be used to do work. Let us now determine from where this energy comes.

In a good generator, the external energy source that rotates the coil does minimal work against friction. Work it must do, however, because the generator produces current that can do work. To see how this interchange of input work with output work occurs, notice what happens to the current-carrying wire of the coil. Since the wire passes through a magnetic field, the current experiences a force due to the magnetic field. This force turns out to be in a direction that opposes the rotation of the coil; the larger the current, the larger the opposing force. Thus we see that the external energy source must do work to rotate the coil, and the more current drawn from the generator to do useful work, the more work the external energy source must do to rotate the coil. In this way, the energy source that drives the generator furnishes the energy that the generator current uses to do useful work. In practice, the external source might be a diesel motor or a waterfall. Let us now look at the operation of the generator in detail so that we can ascertain the form of the emf it produces.

In your mind's eye, replace the single loop of wire in Fig. 19.13 by a coil containing $N$ loops. The coil rotates on axis $aa'$ in a uniform magnetic field. Notice that one end of the coil is attached to ring $R$ and the other end to ring $R'$. These rings, called *slip rings,* are fastened rigidly to the coil and rotate as a unit with it. Contact between the rotating rings and the stationary outside terminals is made by means of the brushes $B$ and $B'$ that slide along the rings. In a very simple motor, the brushes might be short ribbons of spring steel.

To see how the induced emf between the terminals of the coil is generated, let us

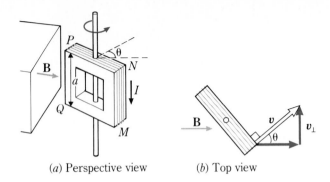

FIGURE 19.14
An induced emf is produced in the
rotating coil.

(a) Perspective view　　(b) Top view

refer to Fig. 19.14$a$. The coil is assumed to be rotating in the direction indicated. As you see, it is moving from a position in which the field lines are perpendicular to its plane to a position in which the lines just skim past it. In other words, the flux to the right through the coil is decreasing. Because of this change in flux, an emf is induced in the coil.

Using Lenz's law, we see that the induced emf in the coil because of this changing flux is directed from $M$ to $Q$ in the lower wire. Only then is the induced current in the direction indicated, so as to replenish the flux through the coil and counteract the change.

Notice, however, what happens once the coil has turned through 180° from the position shown. Everything in Fig. 19.14$a$ is unchanged except that points $M$ and $N$ are interchanged with points $Q$ and $P$. As a result, the induced emf is now directed from $Q$ to $M$, the reverse of what it was before. It is clear that the induced current in the coil keeps reversing as the coil continues to turn.

To analyze this situation quantitatively, we could find $\Delta\phi/\Delta t$ for the coil and use Faraday's law to compute the induced emf. When calculus is used, this approach is very convenient. However, to avoid the use of calculus, we analyze the system in terms of motional emfs.

You will recall that when a wire cuts through the flux lines, an emf equal to $Bvd$ is induced in it, where all three quantities must be mutually perpendicular. In Fig. 19.14$a$, only sides $MN$ and $PQ$ of the loop cut through the flux lines. So only these two sides have an induced emf.

We shall compute the induced emf in side $MN$, which is moving with velocity $\mathbf{v}$ through the field, by first computing $v_\perp$, the magnitude of the component of $\mathbf{v}$ that is perpendicular to $\mathbf{B}$. From Fig. 19.14$b$, we see that $v_\perp = v\sin\theta$. Because the wire $MN$, the field $\mathbf{B}$, and $v_\perp$ are all mutually perpendicular, the induced emf in $MN$ is

$$\text{emf}_{MN} = B(v\sin\theta)(a)$$

Using the right-hand rule for the deflection of moving positive charges, you should satisfy yourself that the direction of the induced current is from $N$ to $M$ in side $MN$ and from $Q$ to $P$ in side $PQ$. Therefore, an identical induced emf in $PQ$ will augment the induced emf in $MN$. Hence we find that

$$\text{Induced emf in loop} = 2Bva\sin\theta$$

We can put this in a more convenient form if we note that $v$ is the tangential speed of point $M$ as it describes a circle about the axis of rotation. Calling the radius of this circle $r$ (where $r = \frac{1}{2}MQ$), we have

$$v = \omega r = 2\pi f r$$

where $f$ is the frequency of rotation of the loop. Moreover, $\theta$ is simply the rotation angle of the loop, and it increases continuously by the relation

$$\theta = \omega t = 2\pi f t$$

When these substitutions are made, the induced emf becomes

$$\text{emf} = 2\pi f B(2ra) \sin 2\pi f t$$

but $2ra$ is simply the area $A$ of the loop, and so we have as our final result

$$\text{emf} = 2\pi f A B \sin 2\pi f t \qquad (19.7)$$

If instead of a single loop we had a coil consisting of $N$ loops, the emf would be $N$ times as large.

As we see, *the induced emf in a rotating coil varies sinusoidally with time.* A graph of this behavior is given in Fig. 19.15. The induced emf (or voltage) has its

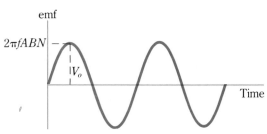

FIGURE 19.15

An alternating emf is induced in a coil rotating in a uniform magnetic field.

maximum value when the sine is unity. Therefore the maximum emf is $2\pi f A B N$. It is reasonable that the induced voltage should be large for large $f$ (flux is changing fastest), for large $A$ and $B$ (the flux is large), and for a large number of loops on the coil.

Equation 19.7 is frequently written in the alternative form

$$V = V_o \sin 2\pi f t$$

where $V$ is the voltage at any instant and $V_o$ is the maximum voltage. Clearly, the voltage in the rotating coil varies sinusoidally and reverses its direction twice during each rotation cycle.

From the above considerations, it becomes evident that a coil of wire rotating in a magnetic field has an alternating emf generated between its terminals. If such a generator is used as the power source in the simple circuit shown in Fig. 19.16, the current through the resistor will reverse its direction $2f$ times per second. (Notice that the symbol for an alternating-voltage generator is ⊙.)

The ac generators used by power companies are usually more complex than the one discussed here, but their basic operation is the same. Mechanical energy to rotate the coil is usually furnished by steam turbines or by water power. Let us

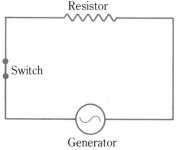

FIGURE 19.16

A simple ac circuit.

briefly consider the conversion of energy in a system such as that shown in Fig. 19.16.

If the circuit is open so that there is no current in the generator coil, very little force need be exerted to rotate the coil. As soon as current is drawn from the generator (the coil), however, the magnetic field exerts a force on the current-carrying wires of the generator, and these forces are in such a direction as to stop the coil from rotating. Hence, the mechanical energy fed into the generator is dependent on the current drawn from the generator—more current requires more mechanical energy.

At an instant when the voltage of the generator is $V$, the power being delivered to the resistor in Fig. 19.16 is $VI$. Clearly, if $I$ is very small, the power consumed by the resistor is small and the mechanical energy needed to operate the generator is small. We therefore see that *the energy needed to operate the generator depends directly on the energy being drawn from it. The mechanical energy is transformed to electrical energy by means of the interactions between magnetic field and charge motion within the coil of the generator.*

## 19.10 | Motors

An electric motor is a device that converts electrical energy to mechanical energy. A schematic diagram of a simple motor is shown in Fig. 19.17. A source of emf (a battery in this case) sends current through the loop of wire. The magnetic field causes the current-carrying loop to experience a torque that rotates the loop on its

FIGURE 19.17

A simple dc motor. With the slip ring as shown, which way should the motor rotate?

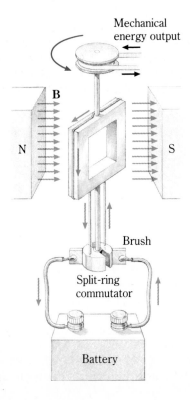

axis. (Use the right-hand rule to convince yourself that the loop rotates in the direction indicated.) Thus the energy furnished to the loop by the battery causes the loop to turn and thereby causes it to do external work by means of a pulley on its axle. The more work the motor performs, the harder it is to turn and so the more energy the battery must furnish to it.

In your mind's eye, replace the single loop in Fig. 19.17 by a coil wound on an iron core so as to make the motor more realistic. As we discussed previously, a current-carrying coil wound on an iron core acts as an electromagnet, a bar magnet. Use the right-hand rule to convince yourself that the front side of the coil (as shown in the figure) is its north pole and the rear side is its south pole. Because of the nearby permanent-magnet poles, forces on the poles of the coil cause it to turn in the direction indicated. However, when the plane of the coil is perpendicular to the page, its south pole is as close as possible to the permanent north pole; the coil will then stop rotating if nothing further is done.

To keep the coil rotating, we reverse the direction of the current in it, thereby reversing its north and south poles. This reversal is accomplished by means of the split-ring commutator. Electrical contact is made with the separate halves of the ring through stationary brushes that slide on the ring as it and the coil turn. (The brushes are typically slippery conducting blocks of graphite and are pushed against the ring by springs.) Notice that, as the coil turns, current enters it first through one half of the ring and then through the other half. In this way, the direction of the current through the coil is reversed at just the proper instant to keep the coil turning.

There are many kinds of electric motors. Many use electromagnets instead of permanent ones. Most have more than one coil so as to produce a more constant torque. Some run on both ac and dc voltage, whereas others run on only one or the other. In all motors, however, the source of emf furnishes energy to the coil by means of a current. It is this energy that the motor uses to do work.

Before leaving the subject of motors, we should point out that a motor is much like a generator running in reverse. The rotating coil of the motor acts like the coil of the generator and has an emf generated in it. This emf is in such a direction as to oppose the emf running the motor. For this reason, it is called a *back* or *counter emf*. Since the resistance of a motor is usually quite small, the chief limitation on the current through it is the back emf. When a motor is overloaded, it slows down. This in turn decreases the back emf (why?) and causes the motor to draw more current. This increased current through the overloaded motor may, on occasion, become large enough to burn it out. To protect against this, many motors have a thermal switch that inactivates them when they become too hot.

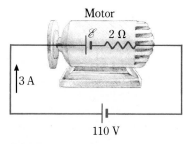

Motor

3 A

110 V

**FIGURE 19.18**
A motor acts as though it were a resistance in series with a counter emf.

### Example 19.7
A particular permanent-magnet motor has a resistance of 2.0 Ω. It draws 3.0 A when operating normally on a 110-V line. How large is the back emf it develops?

***Reasoning*** The motor can be thought of as a battery in series with a resistance. Since the battery is to represent the back emf, it must oppose the operating power source. The situation is shown in Fig. 19.18. Writing the loop equation, we find

$$+110\text{ V} - \mathcal{E} - (3\text{ A})(2\text{ Ω}) = 0$$

$$\mathcal{E} = 104\text{ V}$$

Notice how very large the back emf is. ∎

## *Tape recorders*

The tape recorder is one of the many practical devices that make use of magnetization and induced emf. At the heart of a tape deck are the recording and playback heads, illustrated schematically in the figure. The sound to be recorded is transformed to an electric current by a microphone and amplifier. This current is then sent through the coil of an electromagnet, the recording head. The tape is pulled across a gap in the magnet and, since the tape is coated with tiny magnetic particles, it provides a pathway across the gap for the field lines. These field lines magnetize the tape in proportion to the current in the coil and thus in proportion to the sound signal being recorded.

As the magnetized tape moves across the gap in the playback head, it causes a varying flux through the playback coil. Hence an emf is generated in this coil. By use of appropriate circuitry, this induced emf yields a signal that is an amplified duplicate of the original current in the microphone. When this current is fed to a loudspeaker, it reproduces the sound originally used to magnetize the tape. (A video recorder, although somewhat more complex, employs these same principles.)

To achieve faithful sound reproduction, the tape must contain very small magnetic particles that respond readily to the magnetizing field. Moreover, once it is magnetized, the tape must preserve the magnetic record well. Of course, exposure to strong stray magnetic fields will disrupt the information on the tape and must be avoided.

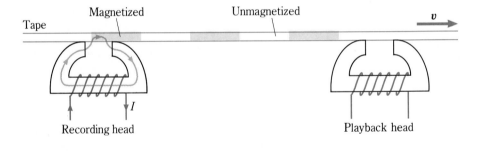

## 19.11 Transformers

One of the most important applications of electromagnetic induction takes place in the transformer, a device that changes (transforms) one ac voltage to another ac voltage. For example, in the typical television set, a transformer changes the 120-V ac input voltage to the about 15,000 V needed to operate the picture tube. As another example, the common doorbell requires a voltage of about 9 V, and a transformer is used to obtain this voltage from the usual house-line voltage of about 120 V. Transformers cannot be used on dc voltages because a continually changing flux is basic to their operation.

A diagram of a typical transformer is shown in Fig. 19.19. The transformer consists of an iron core onto which are wound two coils, the primary (with $N_p$ loops)

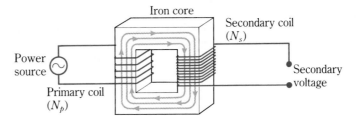

FIGURE 19.19
An iron core step-up transformer.

and the secondary (with $N_s$ loops). The primary coil is connected to the ac power source, and the alternating current in this coil sets up a changing magnetic field in the iron core. Because flux lines tend to follow iron, the lines circle through the secondary coil as indicated. Therefore the flux $\phi$ through the primary and secondary coils is the same.

The changing flux through the secondary coil gives rise to an induced emf in it, given by

$$\text{Secondary emf} = -N_s \frac{\Delta\phi}{\Delta t}$$

In most transformers, the resistance of the coils is negligible, and so the current-limiting factor in the primary is the back emf the primary coil induces in itself. In other words, the induced emf in the primary is equal to the voltage of the power source. We can therefore write that

$$\text{Primary emf} = -N_p \frac{\Delta\phi}{\Delta t}$$

where $\phi$ is the same flux that flows through the secondary coil.

Taking the ratio of these two induced emfs, we find

$$\frac{\text{Secondary emf}}{\text{Primary emf}} = \frac{N_s}{N_p} \tag{19.8}$$

This is the **transformer equation,** and it tells us how the secondary emf is related to the primary emf. The two are in the same ratio as the number of loops on the coils. A transformer that raises the input emf ($N_s > N_p$) is called a *step-up transformer* and one that lowers it ($N_p > N_s$) is called a *step-down transformer*. Notice carefully that transformers make use of ac, not dc, voltages.

If the secondary circuit is not closed, the current in it must be zero. Hence, there is no power loss in the secondary coil when it is not in use. Moreover, we show in the next chapter that there is also no power loss in an inductor that has no resistance. This fact makes it possible for the power company to keep transformers running throughout a city even when no one is using the electricity they are providing. The transformers themselves consume very little energy.

However, if current is drawn from the secondary — to run an electric heater, for example — energy is consumed by the heater. This energy must be fed into the primary of the transformer so that it can be delivered to the secondary. Under these conditions, the loss in power at the secondary causes the primary to act as though it had resistance.

One of the most important uses of transformers has to do with power transmission. Many power companies provide power to cities that are perhaps 100 km from the generators. This proves to be quite a problem. Suppose that each person in a city of 100,000 people is using 150 W of power. This is the equivalent of one or two lighted light bulbs for each person. The power consumed is (150)(100,000) W, and at a voltage of 120 V (the usual house voltage) we have

$$P_{\text{total}} = VI$$

$$(150 \text{ W})(100{,}000) = (120 \text{ V})(I)$$

$$I = 125{,}000 \text{ A}$$

Since an ordinary house wire can safely carry only about 20 A without overheating, the power company would need the equivalent of about 6500 of these wires to carry power to the city at this level of current. Although this is not impossible, the cost of the copper alone would be tremendous. The power companies get around this difficulty quite nicely by noticing that the important quantity in determining power is $VI$ and not $I$ alone. They therefore transmit power over long distances at very high voltages instead of very high currents. In the above example, if $V$ is 100,000 V, we now have

$$(150 \text{ W})(100{,}000) = (100{,}000 \text{ V})(I)$$

$$I = 150 \text{ A}$$

As you see, the required current is much less in this case. Therefore, power companies use high-voltage-difference lines (usually referred to as high-tension lines) to transmit power over long distances.

Of course, they would not dare to have such high voltages wired directly to a house because the danger from electrocution and fire would be tremendous. Instead, they use step-down transformers to convert these high voltages to about 120 V.

Many houses also have 240-V lines because large appliances (air conditioners, dryers, stoves) are usually run on 240 V rather than 120 V. This is for essentially the same reason that the power companies use high voltages. You should be able to explain why these large power-consuming devices are more profitably run on 240 V than on 120 V.

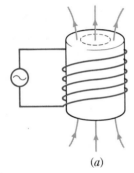

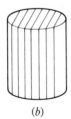

*(a)*

*(b)*

**FIGURE 19.20**

In (*a*) the changing flux induces eddy currents to flow in the metal core. To prevent these currents, the core can be laminated, as in (*b*).

## 19.12 Eddy currents

In our discussion of transformers, we neglected one very important feature of their construction. If the core of the transformer were solid metal, it would become extremely hot and considerable energy loss would occur. Let us now see why this would happen and what can be done to prevent it.

Consider the coil of wire wound on a metal core in Fig. 19.20*a*. When an ac source is used to drive current through the coil, an oscillating flux is set up through the metal core. Because the flux through the dotted path shown in *a* keeps changing, an emf is induced around this path. This induced emf, similar to other emfs induced throughout the core, causes circular currents within the core. We therefore see that

There are induced currents called **eddy currents** in metal objects subjected to a changing magnetic field.

Although eddy currents are sometimes advantageous (as in the induction heating of metals), they are more often undesirable. In particular, they waste energy by causing unwanted heating in transformers, motors, and generators. To eliminate eddy currents, the metal cores of these devices are *laminated,* that is, slit into thin slices that are insulated from each other, as shown in Fig. 19.20*b*. Because of the insulating barriers, charge can no longer flow around paths such as the dotted circle in *a*. As a result, unwanted heating of the metal is greatly reduced.

## Minimum learning goals

When you finish this chapter, you should be able to

1  Define (*a*) induced emf, (*b*) Faraday's law, (*c*) Lenz's law, (*d*) mutual and self-inductance, (*e*) time constant $= L/R$, (*f*) motional emf, (*g*) ac voltage, (*h*) back (counter) emf, (*i*) transformer, (*j*) eddy current, (*k*) lamination.

2  When presented with a simple experiment that involves a change in flux through a coil, explain qualitatively how the induced emf in the coil behaves.

3  Make quantitative use of Faraday's law in simple situations.

4  Explain how the induced emf arises in a mutual inductance and describe what geometric variables influence the observed mutual inductance. Repeat for a self-inductance.

5  Sketch the graph of current versus time for a circuit

consisting of an inductor, a resistor, and a battery just after the circuit is closed. Locate the time constant on the graph.

6  Explain qualitatively why a wire that is cutting through magnetic field lines has an emf generated between its ends. Compute this emf in the case of a wire moving perpendicular to the field lines.

7  Sketch a simple ac generator. Explain how it gives rise to an ac voltage. Sketch a graph of voltage versus time.

8  Explain why the back emf of a motor depends on the speed of the motor.

9  Explain how a transformer changes voltage.

10  Show how eddy currents arise and explain how lamination minimizes them.

## Questions and guesstimates

1  Two circular current loops lie on a table. Loop 1 has a battery and switch in it, and loop 2 is just a closed wire loop. Describe what happens in loop 2 when the switch in loop 1 is suddenly closed and suddenly opened (*a*) when the loops overlap and (*b*) when they do not overlap. Sketch a graph of current versus time in each case.

2  A long, straight wire carries a current along the top of a table. A rectangular loop of wire lies on the table. If the current in the straight wire is suddenly shut off, what is the direction of the current induced in the loop? Draw a diagram for several positions of the loop relative to the wire, showing in each case the direction of the induced current in the loop.

3  A copper ring lies on a table. There is a hole through the table at the center of the ring. If a bar magnet is held vertically by its south pole high above the table and is then released so that it drops through the hole, describe the induced emf in the ring and the forces that act on the magnet.

4  What happens in the secondary coil in Fig. P19.1 when the switch in the primary-coil circuit is (*a*) pushed closed and (*b*) pulled open? Repeat for Fig. 19.1.

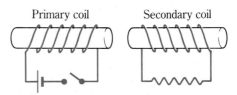

FIGURE P19.1

5  Suppose you are given two identical flat coils. How should the coils be placed so that their mutual inductance is (*a*) largest and (*b*) smallest? When the coils are connected in series by a flexible wire, how should they be positioned to make the self-inductance (*c*) largest and (*d*) smallest?

6  A small coil is placed inside a long solenoid. How does the mutual inductance of the two change with the orientation of the coil?

**7** A very long copper pipe is oriented vertically. Describe the motion of a bar magnet that is dropped lengthwise down the pipe. Why does the magnet reach a terminal speed?

**8** Discuss the possibility of using induced emfs in earth satellites to power the various pieces of electronic equipment. Satellites travel with very high speed through the earth's magnetic field.

**9** A closed wire loop experiences a large stopping force as it falls into a magnetic field. Justify this assertion by reference to Fig. P19.2. Does the same effect occur when a solid piece of metal supported by a string swings in a magnetic field? This general effect is referred to as *magnetic damping* of motion.

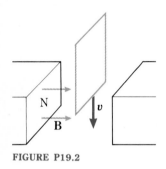

FIGURE P19.2

**10** In Fig. P19.3, the metal ring sits on the end of a solenoid and is held in place there. An alternating current (produced by an alternating emf) is sent through the solenoid. The ring becomes hot. Why? A metal plate also becomes hot when it is held above the solenoid. Explain how eddy currents are induced in this plate and cause it to heat up.

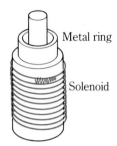

Metal ring

Solenoid

FIGURE P19.3

**11** The metal ring in Fig. P19.3 is made of copper, and the solenoid has an iron core to increase its field. When the current is turned on in the solenoid, the ring flies upward. Explain. Be particularly careful about directions.

**12** Motors do work on external objects. Explain clearly how energy is transferred from the electric current to the rotating portion of the motor.

**13** Explain how electric generators transform mechanical work into electric energy.

## Problems

**1** A flat rectangular piece of cardboard with an area of 200 cm² is placed in a magnetic field of 30 mT. The perpendicular to the surface of the cardboard makes an angle $\theta$ to the field lines. Find the magnetic flux through the area if $\theta$ is (a) 0°, (b) 90°, (c) 25°.

**2** A flat circular loop that has an area of 150 cm² is placed in a 40-mT magnetic field. The perpendicular to the area of the loop makes an angle $\theta$ to the field lines. Find the magnetic flux through the loop if $\theta$ is (a) 90°, (b) 0°, (c) 70°.

■ **3** A hollow solenoid that has 500 loops on its 80-cm length carries a current of 3.0 A. Suspended within the central region of the solenoid is a loop of wire that has a cross-sectional area $A$. How much flux goes through the loop if the angle between the loop axis and the solenoid axis is (a) 0°, (b) 90°, (c) 50°?

■ **4** In Fig. P19.4, the cross-sectional area of the solenoid is $A_s$ and that of coil 1 is $A_c$. A current through the solenoid causes a magnetic field of 25 mT inside it and essentially

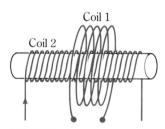

Coil 1

Coil 2

FIGURE P19.4

zero field outside. There is no current in coil 1. Find the flux through each loop of (a) the solenoid and (b) coil 1.

**5** A 20-loop coil lies on a tabletop in a region in which the magnetic field is vertically upward and has a value of 0.50 mT. The field is reduced to zero in 0.50 s. If the radius of each loop is 7.0 cm, what is the average induced emf in the coil (a) while the field is changing and (b) before the field begins to change? (c) As viewed from above, is the induced emf in the coil clockwise or counterclockwise?

**6** A 16-loop square coil has sides 3.0 cm long. It rests flat against the north pole of a large electromagnet. The current in the electromagnet is slowly increased, so that the magnetic field increases from zero to 0.40 T in 3.5 s. (*a*) Find the average induced emf in the coil while the current is being changed. (*b*) If you look toward the north pole, is the induced emf in the coil clockwise or counterclockwise?

**7** At a certain place, the earth's magnetic field is $7.0 \times 10^{-5}$ T and directed at an angle of 65° below the horizontal. (*a*) A man places his 2.0-cm-diameter wedding band flat on a table and slides it across the table with a speed of 50 cm/s. What is the average induced emf in the ring? (*b*) Repeat if the man rolls the band in a straight line across the table at this same speed.

**8** As a laboratory experiment, a student attempts to measure the earth's magnetic field by connecting the two ends of a horizontal loop (area = 0.70 m²) to the terminals of a sensitive voltmeter. She then moves the loop parallel to the earth at 5.0 m/s. (*a*) If the vertical component of the earth's field there is $4.5 \times 10^{-5}$ T, what does the voltmeter read? (*b*) Suppose instead that she flips the coil completely over in 1.50 s. What is the average voltage induced in the loop?

■ **9** The flexible wire loop in Fig. P19.5 has an area of 80 cm² and is in a magnetic field of 0.70 mT. By grasping the loop at *m* and *n* and pulling radially outward, a student collapses it to a straight line in 0.15 s. Find the average induced emf in the loop as it is collapsed.

**FIGURE P19.5**

■ **10** When the wire loop in Fig. P19.5 is suddenly rotated through 90° about an axis through *m* and *n* in 0.20 s, an average emf of 55 $\mu$V is induced in it. The area of the loop is 30 cm². (*a*) What is the magnitude of the magnetic field? (*b*) What is the average emf if the rotation angle is 180°?

■ **11** The wire loop in Fig. P19.6 is being pulled out of the magnetic field at a constant rate. If the field is uniform (1.50 mT) in the region shown and zero elsewhere, (*a*) what is the induced emf in the loop and (*b*) what is its direction?

■ **12** Magnetic transducers are often used to monitor small vibrations. For example, the end of a vibrating bar is attached to a coil that vibrates in and out of a uniform mag-

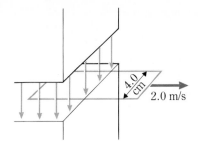

**FIGURE P19.6**

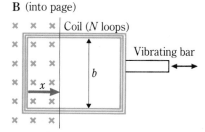

**FIGURE P19.7**

netic field **B**, as shown in Fig. P19.7. Show that the speed of the end of the bar is related to the emf induced in the coil by emf = *NBbv*.

■ **13** The average field produced within a solenoid is 0.85 T. A secondary coil of 100 turns is wound on the solenoid. If the cross-sectional area of the solenoid is 0.50 cm², how large an average emf is induced in the secondary coil if the field in the solenoid is reduced to zero in 0.010 s?

■ **14** A bar magnet can be moved into or out of a 200-turn coil into which it fits snugly. An average emf of 0.30 V is induced in the coil when the magnet is suddenly brought up and inserted in 0.30 s. If the cross-sectional area of the magnet is 2.0 cm², find the value of *B* in it.

■ **15** In Fig. P19.4, the radius of coil 1 is *b* and the radius of the solenoid is *a*. The solenoid is actually much longer than shown. (*a*) If the magnetic field in the solenoid is changing at a rate of 0.030 T/s, find the induced emf in coil 1. This coil has *N* loops on it, and the solenoid has *n* loops per meter of length. (*b*) If a bar of iron that intensifies the field by a factor of 300 is placed inside the solenoid so as to fill it, what is the emf induced in coil 1?

■ **16** The current in an air-core solenoid is increasing at a rate of 2.0 A/s. There are $10^6$ turns of wire on the solenoid for each meter of its length, and its cross-sectional area is 1.80 cm². A secondary coil of $10^4$ turns is wound over the solenoid. How large an emf is induced in the secondary?

■ **17** Two coils are wound tightly on the same iron core. The cross-sectional area of both is about 4.0 cm². When there is a current of 2.0 A in the primary, $B = 0.30$ T. There are 100 turns on the secondary. (*a*) How large an emf is induced in the secondary if the current in the primary drops uniformly to zero in 0.050 s? (*b*) What is the mutual inductance of the coils?

■ **18** A long, iron-core solenoid with 2000 turns has a cross-sectional area of 4.0 cm². When there is a current of 3.0 A in it, $B = 0.50$ T. (*a*) How large an average emf is induced in the solenoid if the current is reduced to zero in 0.10 s? (*b*) How large is the self-inductance?

■ **19** A 20-mH self-inductance carries a current of 2.50 A. (*a*) How much energy is stored in the inductance? (*b*) If this energy is stored in 2.0 cm³ of air, what is the average value of $B$ in the air?

■ **20** The coil of a certain solenoid is wound uniformly on a wooden rod in such a way that the interior of the solenoid has a volume of 20 cm³. When there is a current of 0.30 A in the solenoid, the field within it is 0.060 T. (*a*) Find the energy per unit volume of the magnetic field. (*b*) How much energy is stored in the solenoid? (*c*) What is the self-inductance of the solenoid?

■ **21** We saw in Sec. 19.6 that a change $\Delta I$ in the current through an inductor increases the energy stored in the inductor by $\Delta W = LI\Delta I$. Sketch a graph of $I$ versus $I$. (It is a straight line at an angle of 45°.) Notice that $I\Delta I$ is the area of a vertical rectangle of height $I$ and width $\Delta I$. Use this fact to show that an inductor carrying a current $I_f$ has an energy $\frac{1}{2}LI_f^2$ stored in it.

■ **22** An inductor $L$ is in series with a resistor $R$ and a battery. The steady current in the circuit is $I_o$. Without interrupting the current, the battery is replaced by a wire. The current in the remaining circuit decreases to zero in time $T$. (*a*) How much energy was originally stored in the inductor? (*b*) How much heat is dissipated in the resistor during the time $T$? (*c*) What average current $I_a$ through the resistor in time $T$ will produce this amount of heat? (*d*) How many time constants long must $T$ be if $I_a = \frac{1}{2}I_o$?

■ **23** A metal airplane flies parallel to the ground in a westerly direction at 300 m/s. (*a*) If the downward vertical component of the earth's magnetic field is 0.80 G, what is the potential difference between the tips of the wings, which are 25 m apart? (*b*) Which wing tip is positive, the north or the south? (*c*) Can this voltage be measured. (*d*) If so, how?

**24** An engineer decides to light a train station by utilizing the emf induced in the axles of the trains running on the tracks. (*a*) If the vertical component of the earth's field is $0.80 \times 10^{-4}$ T and the tracks are 1.5 m apart, how large an emf is produced between the tracks by a train traveling 35 m/s? (*b*) Could this voltage be utilized on the moving train? (*c*) Could it be utilized in the train station at the distant end of the tracks? Explain your answers to *b* and *c*.

■ **25** The metal rod in Fig. P19.8 slides down the incline while a vertical magnetic field **B** is present. (*a*) Find the induced emf in the rod when its speed is $v$. (*b*) If the resistance of the loop is $R$, what is the current in the loop? (*c*) Is the current clockwise or counterclockwise? (*d*) How large is the force acting up the incline on the rod because of the current in the magnetic field? (*e*) Does this force tend to speed up or slow down the rod?

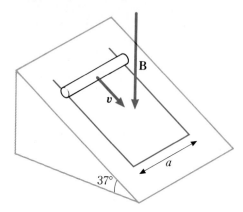

**FIGURE P19.8**

■ **26** Suppose that the speed with which the loop in Fig. P19.6 is being pulled is $v$. The width of the loop is $d$, and the magnetic field has a constant value $B$ between the poles and zero elsewhere. (*a*) Find the induced emf in the loop. (*b*) If the resistance of the loop is $R$, what is the current in it at the instant shown? (*c*) With how large a force must the loop be pulled to keep its speed constant?

■ **27** For the situation shown in Fig. P19.8 and described in Prob. 25, find the terminal speed of the bar as it slides without friction down the incline. The mass of the bar is $m$, and the resistance of the loop is nearly zero except for the resistance of the bar, which is $R$.

■ **28** The square loop of resistance $R$ in Fig. P19.2 has an edge length $a$ and a mass $m$. Assuming the magnitude of the magnetic field to be $B$ between the poles and zero elsewhere, find the terminal speed of the loop as it enters the region between the poles. Assume that the loop is in about the position shown when its terminal speed is reached and that it is narrower than drawn.

**29** A coil consisting of 200 loops, each with a 4.0-cm² area, rotates with a frequency of 120 rev/s in a 0.035-T field. Write the induced voltage in the coil in the form $V = V_o \sin \omega t$.

**30** If a single coil rotating in a magnetic field gives a voltage $V = 20 \cos(800t)$ V, what is the frequency of rotation of the coil?

**31** A 140-loop coil that has an area of 80 cm² rotates at a frequency of 30 rotations/s in a magnetic field. If the maximum induced emf in the coil is 4.0 V, what is the value of $B$?

**32** If a 200-loop coil in a generator has an area of 500 cm² and rotates in a field with $B = 30$ mT, how fast must the coil be rotating in order to generate a maximum voltage of 1.50 V?

**33** The coil of a motor has a resistance of 4.0 Ω. When the motor is turning at rated speed, it draws a current of 3.0 A from 120 V. (*a*) How large is the counter emf of the motor? (*b*) How much current would it draw if the coil were stopped from rotating?

**34** Very large motors take nearly a minute to get up to speed after they are turned on. One such motor has a resistance of 0.75 Ω and normally draws 8.0 A on 120 V. (*a*) What resistance (called the starting resistance) must be placed in series with the motor if it is not to draw more than 20 A when it is first turned on? (This resistance is later removed, of course.) (*b*) What is the back emf of this motor when it is operating at normal speed?

**35** A certain transformer in a radio changes the 120-V line voltage to 8.0 V. (*a*) What is the turns ratio $N_p/N_s$ for this transformer? (*b*) By mistake, the transformer is connected into the circuit backward. About what output voltage does it deliver before everything burns out?

**36** A transformer in a neon sign is designed to change 120 V ac to 18,000 V ac. (*a*) What is the turns ratio $N_p/N_s$ for this transformer? (*b*) If the transformer is connected backward (120 V to the secondary), what voltage appears across the primary?

**37** The resistance of no. 10 copper wire is about $5.2 \times 10^{-3}$ Ω/m. It can carry a current of only about 30 A without overheating. A power company wants to use wires of this size to deliver 30 MW of power to a city 40 km from a generating station. What fraction of the power sent from the station is lost along the transmission lines if the transmission voltage is (*a*) 200 V and (*b*) 100,000 V? Assume that the 30-A restriction is not exceeded.

# 20 Alternating currents

IN the past few chapters, we were concerned mainly with direct currents, that is, those in which the charges flow continuously in one direction. We saw in Chap. 19, however, that a voltage source of alternating polarity is obtained by rotating a coil in a magnetic field. An alternating-voltage source such as this gives rise to alternating currents, and these, too, are of great importance. We see in this chapter how such currents behave when they are sent through resistances, capacitances, and inductances.

## 20.1 Charging and discharging a capacitor

Let us begin our study of circuits in which the current varies with time by examining the simple circuit in Fig. 20.1$a$. Suppose that the switch is open initially and that no charge exists on the capacitor, whose value we represent by $C$. We wish to know what happens when the switch is suddenly closed.

When the switch is closed, the battery tries to send current clockwise around the circuit. Since there is initially no charge on the capacitor, the current $i$ is limited only by the resistor $R$. Therefore, just after the switch is closed (at time $t = 0$), the current is $i_o = V_o/R$, as shown in $b$. As time goes on, however, the capacitor becomes charged and the current decreases. The current must drop to zero when the capacitor's charge becomes $q = CV_o$.

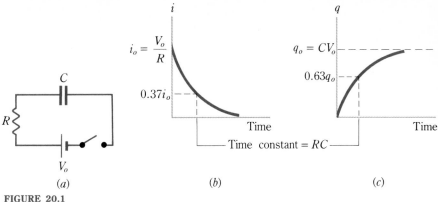

**FIGURE 20.1**

The time constant $RC$ is a convenient measure of the time a capacitor takes to charge or discharge.

The exact way the current behaves in this circuit is shown in Fig. 20.1b. The curve followed by the current is called an **exponential decay curve.*** Analysis of the circuit in part *a* shows that the current drops to a value of $0.3679i_o$ in a time equal to the product $RC$. (It is an interesting problem in unit manipulation to show that ohm-farads are equivalent to seconds.) *We call the product $RC$ the* **time constant** *of this circuit. It is the time in seconds required for the current to decrease to about 0.37 times its original value.*

As long as there is a current in the circuit, the capacitor charges. When the current finally stops, the capacitor's charge is $q_o = CV_o$. The charge $q$ as a function of time is shown in Fig. 20.1c. Notice that here the time constant measures the time in seconds it takes the capacitor to become about two-thirds charged. As we see, *the time constant is a rough measure of the time needed to charge a capacitor.*

If a resistor $R$ is connected directly across a charged capacitor $C$, the capacitor will discharge through the resistor. If we assume the initial potential difference between the capacitor terminals to be $V_o$, the current from the capacitor as it discharges will vary as in Fig. 20.1b. The capacitor discharge current behaves in the same way as the charging current. It turns out that the capacitor is about two-thirds discharged in one time constant. Here, too, the time constant $RC$ is a rough measure of the time required for the process.

---

### Example 20.1

In most television sets, a capacitor is charged to a potential difference of about 20,000 V. As a safety measure, a resistor called a bleeder is connected across the capacitor's terminals so that the capacitor will discharge after the set has been turned off. Suppose a bleeder resistor is $10^6\ \Omega$ and $C = 10\ \mu F$. About how long must you wait after turning off the set before it is safe to touch the capacitor?

***Reasoning*** The time constant for this $RC$ circuit is $RC$. Therefore,

$$\text{Time constant} = (10^6\ \Omega)(10^{-5}\ \text{F}) = 10\ \text{s}$$

---

* The equation for the curve of Fig. 20.1b is $i = i_o e^{-t/(RC)}$, where $e$ is the base for natural logarithms, $e \cong 2.718$, and $e^{-1} = 1/e \cong 0.3679$. A function of the form $e^x$ or $e^{-x}$ is called an exponential function.

As a guess, we might say that it would be safe to touch the capacitor after 10 time constants have passed. However, we can be more precise if we know the following fact about the exponential decay curve of Fig. 20.1$b$. The value plotted on such a curve decreases by a factor of 0.37 during each time constant. As a result, in the figure we see that $i = 0.37 i_o$ at $t = RC$. At $t = 2RC$, $i = (0.37)(0.37) i_o$. At $t = 3RC$, $i = (0.37)^3 i_o$, and so on. After 10 time constants, $i = i_o (0.37)^{10}$, which is $4.5 \times 10^{-5} i_o$. At the time $t = 10RC$, the current and charge have been reduced to $4.5 \times 10^{-5}$ times their original values. As we see, the current is extremely small after 10 time constants have passed; the charge and potential are reduced by a factor of $4.5 \times 10^{-5}$, and so V is then 0.90 V, a perfectly safe value. ∎

## 20.2 AC quantities; rms values

Electric power companies furnish what is known as ac (alternating-current) voltages. They generate these with rotating-coil generators, and the voltage $v$ thus provided is similar to the ac voltage shown in Fig. 19.15. You will recall that it is a sinusoidal voltage and is given by

$$v = v_o \sin 2\pi ft$$

where $f$ is the rotation frequency of the generator coil (60 Hz in the United States). This type of voltage, when impressed across a resistor, gives rise to a current such as the one shown in Fig. 20.2, a sinusoidal current. Its equation is

$$i = i_o \sin 2\pi ft$$

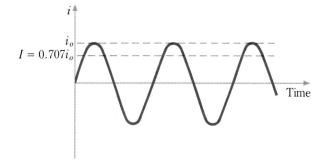

FIGURE 20.2
For an alternating current, the effective, or rms, current is $i_o/\sqrt{2} \cong 0.707 i_o$.

As you see, we have changed notation from that used in previous chapters. In the remainder of this text, we shall use small letters, $v$ and $i$, for voltages and currents that vary with time. We shall soon see that capital letters, $V$ and $I$, are reserved for other quantities when discussing alternating voltages and currents.

It is interesting to notice that, over one cycle, an ac voltage or current has an average value of zero. As you can see from Fig. 20.2, a sine function (as well as a cosine function) is negative as much as it is positive. Hence its average value is zero. Therefore, for an ac voltage and current,

$$v_{av} = i_{av} = 0$$

Because of this, alternating currents cannot be used to charge batteries or for similar applications. If the battery is being charged when the current is positive, it will undergo an equal discharge when the current is negative.

There was considerable controversy in the late 1800s concerning which was more practical, ac or dc electricity. Both can be used for lighting and to run motors. In the end, ac triumphed because it could be used to run transformers, which are very important in modern technology.

The electric power furnished to our homes is often used to operate an electric stove or an incandescent lamp. Uses such as these involve the heat generated by the current in a resistor. Since the electric power provided in such situations is given by $i^2R$, it does not matter whether the current is negative or positive because $i^2$ is always positive. Alternating current is as useful in these applications as direct current is. Because $i_{av}$ and $v_{av}$ are zero for ac conditions, however, we need a special way to describe currents and voltages in ac circuits.

Consider an alternating current $i = i_o \sin 2\pi ft$ that is delivering power to a resistor $R$. The power given to the resistor at any instant is

$$\text{Power} = i^2R = Ri_o^2 \sin^2 2\pi ft$$

In most applications, we are interested only in the average power:

$$\text{Average power} = Ri_o^2 (\sin^2 2\pi ft)_{av}$$

It turns out* that the average value of $\sin^2 \theta$ is 0.50. Therefore

$$\text{Average power} = R \left( \frac{i_o}{\sqrt{2}} \right)^2$$

Thus it appears that the important value for an alternating current is $i_o/\sqrt{2}$, which is $0.707 i_o$. We call this the *root-mean-square* (or *effective*) current and represent it by the symbol $I$. Similarly, we define the root-mean-square (rms) voltage as $v_o/\sqrt{2}$. In summary:

The rms current $I$ and rms voltage $V$ are given by

$$I = \frac{i_o}{\sqrt{2}} \qquad \text{and} \qquad V = \frac{v_o}{\sqrt{2}}$$

Notice that the value of $I$ is shown in Fig. 20.2.

Most ac voltmeters and ammeters read rms voltage or current. From time to time, you may see a meter calibrated to read the peak voltage $v_o$ or peak current $i_o$. Because most dc meters read average values, they will not deflect if they are connected to ac systems. From the way in which the rms current is defined, the power loss in a resistor is merely $I^2R$, where $I$ is the rms value. Why, in a dc system, are the rms current, average current, and instantaneous current all equal?

---

* Recall that $\sin^2 \theta + \cos^2 \theta = 1$. Because the graphs of $\sin \theta$ and $\cos \theta$ have the same shape, $(\sin^2 \theta)_{av} = (\cos^2 \theta)_{av}$. Therefore $(\sin^2 \theta)_{av} + (\cos^2 \theta)_{av} = 1$ becomes $2 (\sin^2 \theta)_{av} = 1$, from which $(\sin^2 \theta)_{av} = 0.50$.

## 20.3 Resistance circuits

We introduce the subject of ac circuits by considering in turn three different circuit elements connected in series with an ac voltage source. First let us consider the simple resistance circuit shown in Fig. 20.3a. At any particular instant, Ohm's law applied to the resistor tells us that $v = iR$. Because $v = v_o \sin 2\pi ft$, we have

$$i = \frac{v}{R} = \frac{v_o}{R} \sin 2\pi ft$$

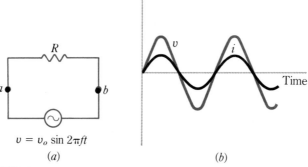

$v = v_o \sin 2\pi ft$

*(a)*            *(b)*

**FIGURE 20.3**
The current in a resistor is in phase with the voltage across its terminals.

Hence both the voltage and the current vary with time in the same way because they are both sine functions. We plot them in Fig. 20.3b. Notice that $i$ and $v$ vary in unison. We say that the current and voltage in this circuit are *in phase*.

As outlined in the previous section, the power loss in the resistor is $I^2R$. In this particular case, where only a resistance is present, $I = V/R$, and so the power loss could also be written $IV$, where $I$ and $V$ are the rms meter readings. We see in the next sections that there is no average power loss in pure capacitors or pure inductors. *All power losses in simple ac circuits occur in resistors.*

## 20.4 Capacitance circuits

Let us now consider the capacitance circuit in Fig. 20.4a. We know that the potential from $a$ to $b$ is equal to the voltage of the ac source, $v_o \sin 2\pi ft$. However, we recall that the potential across a capacitor is $q/C$. If we equate these two expressions for the same voltage, we have

$$\frac{q}{C} = v_o \sin 2\pi ft$$

Because $C$ is a constant, the charge on the capacitor oscillates in value in the same way as the source voltage. When the voltage source makes point $a$ positive and point $b$ negative, plate $a$ of the capacitor is positively charged. When the source voltage is

**FIGURE 20.4**

The voltage across a capacitor reaches its maximum $\frac{1}{4}$ cycle later than the current. The charge on the capacitor is in phase with the voltage across the capacitor.

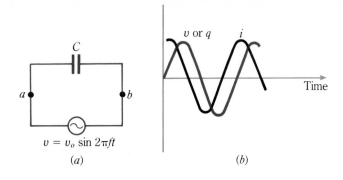

$v = v_o \sin 2\pi ft$

(a)

(b)

reversed, point $a$ and plate $a$ of the capacitor are negative. Moreover, the capacitor charge reaches its maximum at the same time the voltage does. Hence the graph of voltage versus time looks the same as the graph of charge versus time, as shown in Fig. 20.4$b$.

To see how the current in the circuit varies with time, we recall that current is the rate of change of charge. Hence, the rate of change of charge on the capacitor, $\Delta q/\Delta t$, is the current in the circuit. But $\Delta q/\Delta t$ is merely the slope* of the curve obtained when $q$ is plotted against $t$. This curve is plotted in Fig. 20.4$b$, and so we need only plot its slope to have a graph of the current in the circuit. This is shown as the black curve in Fig. 20.4$b$.

Notice that now the current is not in step with the voltage. The current has its maximum at $t = 0$, but the voltage does not reach its maximum until $\frac{1}{4}$ cycle later. We say that the current leads the voltage by $90°$ (or that the current and the voltage are $90°$ out of phase) in such a circuit. We shall now see what this implies for the power dissipated in the circuit.

From Fig. 20.4$b$, we see that the voltage across the capacitor follows a sine curve while the current obeys a cosine curve. We have

$$v = v_o \sin 2\pi ft \qquad \text{and} \qquad i = i_o \cos 2\pi ft$$

Now the instantaneous power furnished to the capacitor is given by the usual relation:

$$\text{Power} = vi = v_o i_o \sin 2\pi ft \cos 2\pi ft$$

This can be put in a much nicer form if you recall that

$$\sin 2\theta = 2 \sin \theta \cos \theta$$

Hence we find that

$$\text{Instantaneous power} = \tfrac{1}{2} v_o i_o \sin 4\pi ft$$

---

* The **slope** of a curve is exactly what the word implies. It is the rate at which the curve is rising. If the curve is flat and horizontal, it is not rising at all, and so its slope is zero. If the curve is rising rapidly, its slope is large. If the curve is decreasing, that is, going down, it is, in effect, rising at a negative rate. Its slope is therefore negative.

This means that the instantaneous power furnished to the capacitor varies sinusoidally at twice the frequency of the ac voltage. Hence the average power furnished to the capacitor is zero because the sinusoidal function is negative as much as it is positive. During half of the cycle, the capacitor is being charged and so energy is being stored in it. During the other half of the cycle, however, the capacitor is discharging and returning its stored energy to the power source. The net effect is the following:

In an ac circuit, the average power consumed by a perfect capacitor is zero.

Let us now examine how a capacitor behaves when the frequency of the ac source changes. If the frequency is nearly zero — one cycle per day, say — the source acts almost like a dc voltage. The current in the circuit is essentially zero (except at the first instant) because the charge on the capacitor is essentially constant. Hence the capacitor almost completely blocks the current at low frequencies. At high frequencies (such as $10^6$ Hz), however, the charge on the capacitor passes through the wire on the order of a million times each second. This means that the current in the wire is large at high frequencies. Hence the capacitor appears to block the current very little at high frequencies. We say that

The impeding effect of a capacitor is large at low frequencies and small at high frequencies.

We designate the ability of a capacitor to impede the flow of charge by the term **capacitive reactance,** which we designate by $X_C$. It is related to the rms current and voltage in the circuit of Fig. 20.4 through an Ohm's law-type relation:

$$V = IX_C \tag{20.1}$$

where $X_C$ replaces $R$ in Ohm's law. Using calculus, it is easy to show that

$$X_C = \frac{1}{2\pi fC}$$

where the units of $X_C$ are the same as for $R$, namely, ohms. Notice that, as our qualitative discussion implied, the impeding effect of the capacitor as expressed by $X_C$ is large when $f$ is small and small when $f$ is large.

---

**Example 20.2**

Suppose that an ac voltmeter across the voltage source in Fig. 20.4 reads 80 V and that $C = 0.40 \mu F$. Find the rms current in the circuit if the frequency of the ac voltage is (a) 20 Hz and (b) $2 \times 10^6$ Hz.

**Reasoning** We use $V = IX_C$, where $V = 80$ V and $I$ is the rms current:

$$X_C = \frac{1}{2\pi fC} = \frac{1}{2\pi f(4 \times 10^{-7} \text{ F})} = \frac{3.98 \times 10^5}{f} \, \Omega/s$$

This gives $X_C = 1.99 \times 10^4 \, \Omega$ at 20 Hz and $X_C = 0.199 \, \Omega$ at 2 MHz. Therefore we find (a) for $f = 20$ Hz,

$$I = \frac{V}{X_C} = \frac{80 \text{ V}}{1.99 \times 10^4 \, \Omega} = 4.02 \times 10^{-3} \text{ A}$$

and (b) for $f = 2$ MHz,

$$I = \frac{80 \text{ V}}{0.199 \ \Omega} = 402 \text{ A}$$

Notice that, as we have said before, at high frequencies a capacitor impedes the current much less than at low frequencies. ∎

*Exercise* What is the equation for the voltage furnished by the source in this case? *Answer: v = 113 sin 2πft V* ∎

## 20.5 Inductance circuits

The behavior of the simple self-inductance circuit in Fig. 20.5a can be analyzed in a manner similar to that used for the capacitance circuit. First we notice that the voltage difference between points *a* and *b* is equal to the source voltage $v_o \sin 2\pi ft$.

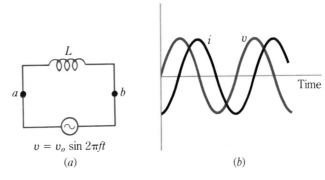

**FIGURE 20.5**

The voltage across the inductor leads the current through it by 90°, or $\frac{1}{4}$ cycle. Notice the symbol used for the inductor.

However, it is also equal to the voltage induced in the inductor by the changing current and flux in the circuit. In Chap. 19, we found this voltage to be $L \, \Delta i / \Delta t$. Equating these two voltages yields

$$v_o \sin 2\pi ft = L \frac{\Delta i}{\Delta t}$$

Since $L$ is a constant, we see at once that the source voltage is proportional to the rate of change of current in the circuit. Moreover, because the rate of change of current is the slope of the graph of current versus time, we have a way of finding one curve if the other is known. The voltage and current curves in the inductance circuit are shown in Fig. 20.5b. Notice that the voltage graph is indeed proportional to the value of the slope of the current graph.

Here, too, the current and voltage are 90° (or $\frac{1}{4}$ cycle) out of phase. In this case, though, the voltage is 90° ahead of the current. We say that the voltage leads the current by 90° in this case.

Once again, we can use the same reasoning as in Sec. 20.4 to show that the inductor consumes no energy on the average. Although the source stores energy in the inductor during part of the cycle, the inductor gives it back to the source in a later portion of the cycle. We showed in Chap. 19 that the energy stored in an inductor is $\frac{1}{2}Li^2$. You would do well to examine Fig. 20.5b and ascertain the part of the cycle during which the source is losing energy and the part during which energy is being returned to the source.

The general behavior of the inductance circuit as the source frequency is changed is also of interest. We know, of course, that the inductor always tries to counterbalance, or impede, the change in current. In fact, the induced emf in it is $L\,\Delta i/\Delta t$ and is therefore proportional to the rate of change of current. As a result, when the current is changing very slowly, the inductor does not have much effect. However, at very high frequencies, when the current is trying to change rapidly, the impeding effect of the inductor is very large.

The impeding effect of an inductor is large at high frequencies and small at low frequencies.

We represent the impeding effect of an inductor by the **inductive reactance** $X_L$, which is related to the rms current through the inductor and to the voltage across the inductor by

$$V = IX_L \tag{20.2}$$

where the inductive reactance $X_L = 2\pi fL$. As we expect, the reactance of the inductor is large at high frequencies and small at low frequencies. It is measured in ohms.

Notice that capacitors and inductors behave oppositely as a function of frequency. The current-impeding effect of capacitors is large at low frequencies and small at high frequencies. The reverse is true for inductors. Of course, the impedance effect of a resistor is independent of frequency.

---

**Example 20.3**

Suppose the inductor in Fig. 20.5$a$ has a value of 15 mH. The source voltage, as read by an ac meter, is 40 V, and its frequency is 60 Hz. Find the current through the inductor. Repeat for a frequency of $6 \times 10^5$ Hz.

**Reasoning**   We make use of the Ohm's law form $V = IX_L$. In the 60-Hz case, $V = 40$ V and

$$X_L = 2\pi fL = 5.65\ \Omega$$

Therefore, $I = 40$ V$/5.65\ \Omega = 7.08$ A.

At a frequency of $6 \times 10^5$ Hz, the value for $X_L$ becomes $5.7 \times 10^4\ \Omega$. Then we find that $I = 7.08 \times 10^{-4}$ A. Notice how very much larger the inductor's impeding effect is at high frequencies than at low frequencies.   ∎

---

## 20.6 Combined *LRC* circuits

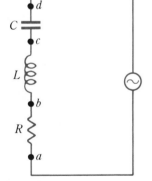

FIGURE 20.6
A series *LRC* circuit.

Let us consider next the series *LRC* circuit in Fig. 20.6. As in any series circuit, the current through each element in it is the same; let us say that it is $i = i_o \sin 2\pi ft$. A graph of this current is shown as the top curve in Fig. 20.7. For this current, the graph of the voltage $v_R$ across the resistor is also a sine curve, as shown by the second curve. (Recall that $v_R$ is in phase with the current.) However, $v_C$ and $v_L$, the voltages across the capacitor and the inductor, are 90° out of phase with the current. Their graphs are also shown in Fig. 20.7.

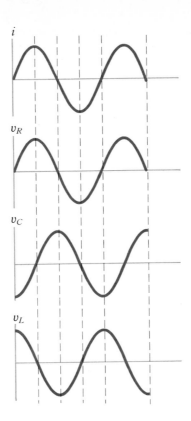

**FIGURE 20.7**

Only $v_R$ is in phase with $i$. Both $v_L$ and $v_C$ are $\frac{1}{4}$ cycle (90°) out of phase with $i$.

Notice from these graphs that $v_L$ and $v_C$ are always of opposite sign. Hence they tend to cancel each other. Suppose an ac voltmeter reads $V_L$ across the inductor and $V_C$ across the capacitor. If $V_C = V_L$, then the amplitude of $v_C$ and $v_L$ are the same and $v_C$ exactly cancels $v_L$. In this case, a voltmeter connected between points $b$ and $d$ in Fig. 20.6 would read zero, not $V_C + V_L$! Thus we see that ac voltmeter readings do not add to give proper voltage differences. Even though the instantaneous voltages add directly, the rms voltages read by ac meters are always positive and cannot show the cancellation effects that may be present.

Mathematical analysis of the $LRC$ circuit in Fig. 20.6 shows that the circuit obeys an Ohm's law relation of the form

$$V = IZ \qquad \text{with} \qquad Z = \sqrt{R^2 + (X_L - X_C)^2} \tag{20.3}$$

where $Z$ is called the **impedance** of the circuit and is measured in ohms. Notice that $X_L$ and $X_C$ enter the equation as $X_L - X_C$. As our qualitative discussion implied, the effects of the inductor and the capacitor tend to cancel. Indeed, when $X_L = X_C$, they together offer no contribution to the impedance.

Two interesting relations can be obtained from the square of the expression for $Z$. We have

$$Z^2 = R^2 + (X_L - X_C)^2 \tag{20.4}$$

and, after multiplying the whole equation by $I^2$,

$$I^2Z^2 = I^2R^2 + (IX_L - IX_C)^2$$

Recognizing that $IZ$ is the potential difference $V$ across the voltage source and that $IR = V_R$, $IX_L = V_L$, and $IX_C = V_C$, we can write this as

$$V^2 = V_R^2 + (V_L - V_C)^2 \qquad (20.5)$$

Both Eqs. 20.4 and 20.5 are reminiscent of the pythagorean theorem for a right triangle:

$$(\text{Hypotenuse})^2 = (\text{side 1})^2 + (\text{side 2})^2$$

Therefore we can picture the expressions for $Z^2$ and $V^2$ by right triangles, as shown in Fig. 20.8. Equation 20.4 follows directly from Fig. 20.8a, and Eq. 20.5 is the pythagorean-theorem expression for the right triangle in b.

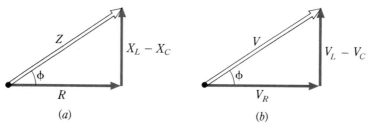

(a)            (b)

**FIGURE 20.8**
The pythagorean theorem gives the relations between the impedances as well as the voltage interrelations. What is the meaning of $\phi$?

The representations in Fig. 20.8 have still more meaning. Although the power loss in the series $LRC$ circuit occurs only in the resistance and is given by $I^2R$, there is another useful relation for the power loss:

$$\text{Power loss} = VI \cos \phi \qquad (20.6)$$

where $\phi$ is the angle shown in Fig. 20.8 and $\cos \phi$ is called the *power factor*. We see from the figure that

$$\tan \phi = \frac{X_L - X_C}{R}$$

It turns out that the angle $\phi$ is the phase difference between the current and the voltage in the $LRC$ circuit. We can easily check this fact and Eq. 20.6 in three limiting cases. Suppose that the circuit contains neither inductance nor capacitance, so that $X_L = X_C = 0$. Then, from Fig. 20.8, $\phi = 0$. (This is the correct value for $\phi$ because the current is in phase with the voltage in a pure resistance circuit.) Equation 20.6 then gives

$$\text{Power loss} = VI \cos 0 = VI$$

This answer agrees with our usual expression for the power loss, $I^2R$, as we can see by noticing that the potential difference across the resistance is $V$ in this case and that $V = IR$. Substitution in the above expression for the power loss gives, in this special case,

Power loss $= VI = (IR)(I) = I^2R$

As another check, when $R = X_L = 0$, there is no power loss because power losses occur only in resistive elements. We notice in this case that $\phi = 90°$, and so Eq. 20.6 gives

Power loss $= VI\cos 90° = 0$

which agrees with what we have just said. We leave it for you to show that Eq. 20.6 gives a reasonable result when $R = X_C = 0$.

---

**Example 20.4**

A power source ($V = 80$ V, $f = 2000$ Hz) is connected in series across a 300-$\Omega$ resistor and a 0.60-$\mu$F capacitor. Find ($a$) the current in the circuit, ($b$) the voltmeter readings across the resistor, ($c$) the reading across the capacitor, and ($d$) the power loss in the circuit.

*Reasoning*   We have

$$X_L = 0$$

$$X_C = \frac{1}{2\pi fC} = 133\ \Omega$$

$$Z = \sqrt{R^2 + X_C^2} = 328\ \Omega$$

($a$) $I = \dfrac{V}{Z} = \dfrac{80\ \text{V}}{328\ \Omega} = 0.244$ A

($b$) $V_R = IR = 73.2$ V

($c$) $V_C = IX_C = 32.6$ V

($d$) Power $= I^2R = 17.9$ W

Alternatively, since

$$\tan\phi = \frac{X_L - X_C}{R} = \frac{-133\ \Omega}{300\ \Omega}$$

$$\phi = -23.9°$$

we have again

Power loss $= VI\cos\phi = (80\text{ V})(0.244\text{ A})(\cos 23.9°) = 17.9$ W ■

---

*Exercise*   Show that these voltages obey the voltage triangle in Fig. 20.8 and that the triangle gives the proper power factor.   ■

---

**Example 20.5**

Suppose the voltage source in Fig. 20.6 has an rms value of 50 V and a frequency of 600 Hz. Suppose further that $R = 20\ \Omega$, $C = 10.0\ \mu$F, and $L = 4.0$ mH. Find ($a$) the current in the circuit and ($b$) the voltmeter reading across $R$, $C$, and $L$ individually.

***Reasoning*** We use $V = IZ$. Let us therefore find $X_L$ and $X_C$ at 600 Hz:

$$X_L = 2\pi fL = 15.1\ \Omega$$

$$X_C = \frac{1}{2\pi fC} = 26.5\ \Omega$$

Then we find that

$$Z = \sqrt{(20)^2 + (15.1 - 26.5)^2} = 23.0\ \Omega$$

Now, using $I = V/Z$, we find $I = 2.17$ A.

To find the voltage drop across $R$, we use $V_R = IR$ and note that $I = 2.17$ A. Therefore,

$$V_R = 43.4\ \text{V}$$

The voltage drop across the inductance is given by $V_L = IX_L$ to be

$$V_L = (2.17\ \text{A})(15.1\ \Omega) = 32.8\ \text{V}$$

Similarly, we see that

$$V_C = IX_C = 57.5\ \text{V}$$

Notice that the potential difference across the capacitor is larger than the source voltage. This points out once again that *in ac circuits, the sum of the voltmeter readings around a closed circuit is not zero; voltages do not add directly if rms voltage readings are used.* This fact is a result of the average character of the rms readings. They do not represent the instantaneous voltages, which can be either positive or negative. The instantaneous voltages do add directly. The rms voltages, however, are always positive by definition. Clearly, they cannot add to give zero. Kirchhoff's loop rule does not apply to them. ∎

## 20.7 Electrical resonance

Ac circuits that contain both capacitance and inductance show an important resonance phenomenon. To illustrate this fact, consider the series circuit shown in Fig. 20.9. We know that the current in this circuit, which has no resistance, is

$$I = \frac{V}{Z} = \frac{V}{X_L - X_C}$$

**FIGURE 20.9**

As the source frequency is changed, $X_L$ and $X_C$ change as shown in Fig. 20.10. The current in the circuit varies as shown in Fig. 20.11.

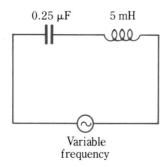

0.25 μF    5 mH

Variable
frequency

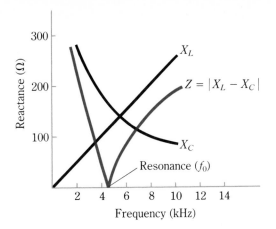

**FIGURE 20.10**

Both $X_L$ and $X_C$, as well as $Z$, for the circuit in Fig. 20.9 vary with source frequency as shown here.

Notice that when $X_L = X_C$, the current in the circuit should become infinite.

It is easy to obtain the condition $X_L - X_C = 0$ because $X_L$ increases with frequency while $X_C$ decreases with frequency. Figure 20.10 shows how these quantities vary for the $C$ and $L$ values given in the circuit of Fig. 20.9. We see that the impedance becomes zero at $f = 4500$ Hz in this case. This frequency, the frequency at which $X_L = X_C$, is called the **resonance frequency** of the circuit, and we denote it by $f_0$. Since $X_L = 2\pi f L$ and $X_C = 1/2\pi f C$, we have at resonance that

$$2\pi f_0 L = \frac{1}{2\pi f_0 C}$$

from which the resonance frequency is

$$f_0 = \frac{1}{2\pi} \sqrt{\frac{1}{LC}} \tag{20.7}$$

Figure 20.11 shows how the current in the circuit of Fig. 20.9 varies as the oscillator frequency is changed. (Of course, the voltage of the oscillator must be kept the same for all frequencies.) As we see, the current peaks sharply at the resonance frequency. In practical circuits, the peak is finite rather than infinite because all

**FIGURE 20.11**

As the frequency of the source in Fig. 20.9 is varied, the current in the circuit behaves as shown here.

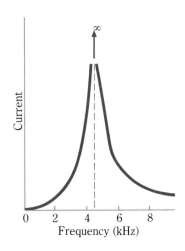

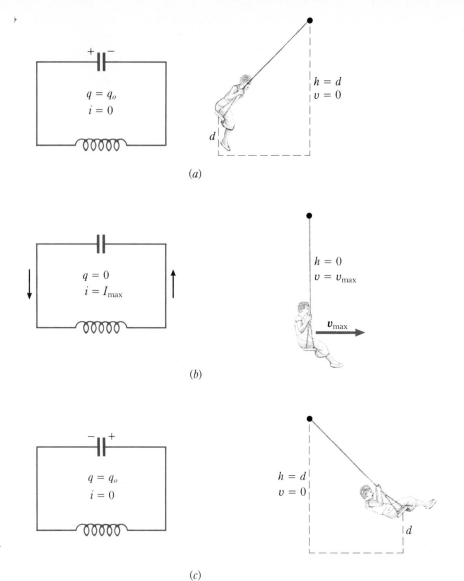

**FIGURE 20.12**

Just as the energy of the swing continually oscillates between potential and kinetic, the energy of the circuit is alternately stored in the capacitor and the inductor.

wires have some resistance. Even so, the effect is very dramatic and has important applications, as we shall see in the next chapter.

We can understand electrical resonance better if we recognize that it is much like mechanical resonance. You know that mechanical systems often have a natural frequency at which they vibrate. If pushed with this frequency, they will vibrate widely; in other words, they resonate. A simple $LC$ circuit also has a natural frequency of vibration. Let us explore this analogy between resonance in electrical and mechanical systems. Consider the $LC$ circuit and the child on a swing shown in Fig. 20.12. Suppose that at the starting instant, the current in the circuit is zero and the swing is at its highest position. The charge on the capacitor is $q_o$, and an energy $\frac{1}{2}(q_o^2/C)$ is stored in it. By analogy, the swing possesses gravitational potential energy.

We know that in the electrical system, the capacitor will begin to discharge through the inductor. The current will rise rather slowly because the inductor opposes any change in current. Similarly, the swing will begin to pick up speed as its inertia is overcome by the accelerating forces acting on it. Both the swing and the capacitor lose their potential energy. Once the swing has reached the bottom of its path, all its potential energy has been changed to kinetic energy. Similarly with the circuit: once the capacitor has lost all its charge, the current in the circuit has its largest value and the original energy is now stored in the inductor. Its value is $Li^2/2$. This situation is shown in Fig. 20.12b.

Of course, the swing does not stop at the bottom of the path. The system's inertia keeps it moving until it comes to rest in the position shown in Fig. 20.12c. Now its energy is all potential energy once again. Much the same thing happens in the electric circuit. The inductance, having inertia of a sort, opposes any change in current, and so the current does not stop at once. By the time the current finally stops, the capacitor is fully charged again, as in c. These processes repeat over and over.

Clearly, then, the electric circuit undergoes an energy interchange much like the child and swing. The swing's energy alternates between potential and kinetic, and the energy in the circuit is alternately stored in the capacitor and the inductor. Both systems would oscillate back and forth forever if there were no energy losses. In the case of the swing, friction losses eventually damp out the oscillation. In the electrical case, resistive effects cause some of the energy to be lost and so the oscillation slowly damps down in amplitude.

The analogy can be extended further. Both the swing and the circuit possess natural resonance frequencies for their motion. The swing system constitutes a pendulum; we computed its natural frequency of vibration in Sec. 13.6. The natural resonance frequency of the circuit is the resonant frequency computed in Eq. 20.7.

If we wish to cause the child to swing very high, we must push on the swing at just the proper time and with the same frequency as the resonant frequency of the swing. We have seen that a very large current could be built up in the $LC$ circuit if the oscillator "pushed" on the circuit at its resonant frequency. Hence, even the resonance behavior of the two systems is quite similar. It is shown in the next chapter that the $LC$ resonant circuit discussed here forms an integral part of any radio or television receiver.

## 20.8 Rectification of currents

Alternating currents are not suitable for many applications. For example, they cannot be used to charge batteries or to operate many components in electronic devices. Therefore, it is important to have methods for changing alternating current to direct current. Devices that do this are called *rectifiers*. There are two basic types of rectifiers: vacuum-tube diodes and solid-state diodes.

Figure 20.13 is a sketch of a vacuum diode and the associated circuit. Inside the vacuum chamber ("tube") are two elements: a metal plate and a thin filament, much like the filament of an incandescent lamp. The filament-voltage source $V_f$ simply sends a current through the filament and heats it white-hot. At this temperature, electrons boil out of the filament and enter the vacuum region. This process is called *thermionic emission*. If the plate-voltage source $V_P$ has its polarity as shown, the plate is positive. The electrons emitted by the filament are attracted to the plate, and so there is a current through the vacuum. Thus, for the polarity of $V_P$ shown, the tube carries current.

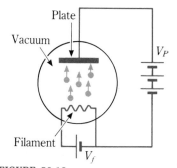

**FIGURE 20.13**

Electrons boiled from the white-hot filament travel to the cold plate. What happens if $V_P$ is reversed?

Suppose the polarity of $V_P$ is reversed, however. Then the plate is negative and repels the electrons emitted by the filament. The electrons can no longer reach the plate, and so the current through the tube is zero. Therefore we see that a vacuum diode conducts current if its plate is positive but does not allow current to pass if its plate is negative.

Until about 1950, vacuum tubes were much used, but since the invention of solid-state electronic devices, they have been largely supplanted. A device called a solid-state diode has replaced the vacuum-tube diode except in a few applications. It is sufficient for our purposes if we know that the symbol for a solid-state diode is ——▶⊢—— and that it has the following property: it conducts current in the direction of the arrow, but not in the reverse direction.

We can see how either type of diode rectifies current and voltage by referring to the circuit (called a half-wave rectifier) in Fig. 20.14. The sinusoidal ac input voltage to be rectified enters at the left. Notice that the diode in the circuit allows charge to

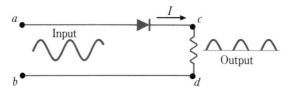

FIGURE 20.14

A half-wave rectifier.

flow in one direction only. As a result, there is a current only when point $a$ is positive relative to point $b$. This happens during only half the cycle. So the output voltage from $c$ to $d$ is positive during one half of the cycle and zero during the other half, as indicated by the output graph. Thus, when it is used this way, a diode is able to convert an ac voltage to one that does not alternate, in other words, to a dc voltage. Unfortunately, the output voltage is far from steady in this case.

The output of the half-wave rectifier can be smoothed somewhat by the use of a "filter" capacitor, as shown in Fig. 20.15. The filter capacitor $C$ is placed across the output. If no current were being drawn from the output, the capacitor would become

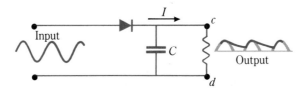

FIGURE 20.15

A filtered half-wave rectifier.

fully charged and would maintain a constant dc voltage at the output. In practice, the output acts like a resistance of value $R$. If $R$ is large enough, the time constant $RC$ is larger than the period of the voltage wave. Then the capacitor discharges slightly during each voltage pulse and the output is nearly steady. (We specify the "steadiness" by what is called the "ripple," the ratio of the voltage variation to the maximum voltage during the cycle.)

Another way to obtain a smoother output is shown in Fig. 20.16. This is a full-wave rectifier. The center-tap transformer (one that has a connection to the midpoint of the secondary, as shown) in essence provides two voltage sources that are one-half cycle out of phase. At the instant shown, the top end of the secondary is positive and the bottom is negative; the direction of the current is as indicated, from $c$ to $d$. One-half cycle later, the voltage reverses and the lower diode then conducts current from $c$ to $d$. As a result, the direction of the current is from $c$ to $d$ during both halves of the cycle and the output voltage is as shown.

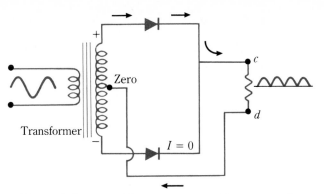

**FIGURE 20.16**
A full-wave rectifier.

None of these rectifier systems provides a smooth enough output voltage for many electronic devices. However, with additional filtering devices using capacitors and inductors, the ripple in the output voltage can be made very small. As a result, the steady dc voltages that were once available only from batteries are now most often furnished by rectifier systems.

## Minimum learning goals

Upon completion of this chapter, you should be able to

1 Define (*a*) *RC* time constant, (*b*) ac versus dc voltage or current, (*c*) effective and rms values, (*d*) capacitive reactance, (*e*) inductive reactance, (*f*) impedance, (*g*) power factor, (*h*) resonance in *LC* circuit, (*i*) rectifier, (*j*) thermionic emission.

2 Sketch the current and charge curves for an *RC* circuit during charging. Define the time constant for the circuit and relate it to the curves. Explain the significance of the time constant for capacitor discharge through a resistor.

3 Sketch a typical ac voltage or current curve, showing the peak, average, and rms values. Relate the rms value to the peak value in a quantitative way.

4 State the Ohm's law form that applies to an ac voltage impressed on a resistor. Sketch the current and voltage curves on the same graph. Compute the average power loss in the resistor when sufficient data are given.

5 Explain why the impeding effect of a capacitor should

be higher at low frequencies than at high frequencies. Use $V = IX_C$ in simple situations.

6 Sketch the current and voltage curves for a capacitor connected across an ac power source. State the average power loss in the capacitor.

7 Explain why the impeding effect of an inductor should be larger at high frequencies than at low frequencies. Use $V = IX_L$ in simple situations.

8 Sketch the current and voltage curves for an inductor connected across an ac source. State the average power loss in the inductor.

9 Use the relation $V = IZ$ for simple problems involving series *RCL* circuits.

10 Use $V = IZ$ to explain why a resonance frequency exists for an *LC* circuit. Show how to find the resonance frequency.

11 Draw a half-wave rectifier circuit and explain how it operates.

## Questions and guesstimates

1 You are given a 2-$\mu$F capacitor, a dry cell, and an extremely sensitive and versatile current-measuring device. How could you use these to measure a resistance that is thought to be about $10^8$ $\Omega$? Could you make the measurement using an ordinary voltmeter in place of the current meter?

2 In some places, low-frequency ac voltage (considerably less than 60 Hz) is used. Electric lights operated on this voltage flicker rapidly. Explain the cause of this flickering.

3 For which of the following uses would dc and ac volt-

age be equally acceptable: incandescent light, electric stove, electrolysis, television set, fluorescent light, neon-sign transformer, battery charger, toaster, electric clock?

**4** Draw an analogy between the vibration of a mass on a spring and the oscillation of an $LC$ circuit. What quantities in the mechanical system correspond to $L$ and $C$ in the electrical system? Explain.

**5** Compare the equation for the resonance frequency of a mass vibrating at the end of a spring with the resonance equation for an $LC$ circuit. What analogy can you draw between them?

**6** A dc voltmeter is connected across the terminals of a variable-frequency oscillator. How does the meter behave as the frequency of the oscillating voltage is slowly increased from 0.01 to 100 Hz? Explain.

**7** In an $RLC$ series circuit, when, if ever, is the current through the circuit in phase with the source voltage?

**8** The following statement was published in a daily newspaper. "A warning that home electrical appliances can cause fatal injuries has been sounded by City Health Director J. R. Smith. His comments followed the death of an 18-year-old boy who accidentally electrocuted himself by inserting a fork in a toaster. Dr. Smith pointed out that even adults can be killed by such electrical shocks. Ordinary house current is 110 V, but the voltage is increased if the current is grounded, he said." What is wrong with the last sentence, and how should it have been worded?

**9** The devices shown in Fig. P20.1 are called filters. When an ac voltage is put into the device, the output ac voltage depends on the frequency of the oscillating voltage. One of these devices lets the input voltage pass through undisturbed if the oscillation frequency is high. The other passes only low-frequency voltages. Explain which is which.

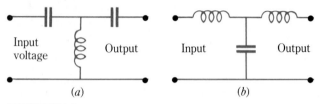

(a)                                    (b)

**FIGURE P20.1**

**10** The circuit in Fig. P20.2 is a full-wave rectifier (often referred to as a bridge-type rectifier). Explain why the current is rectified and why charge can flow during both halves of the cycle.

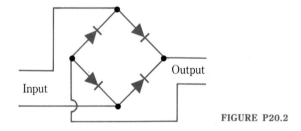

**FIGURE P20.2**

## Problems

*Unless otherwise specified, consider all meters to be perfect.*

**1** How large a resistor must be placed in series with a 0.50-$\mu$F capacitor if the time constant of the circuit is to be 3.0 s?

**2** About how long does it take for the charging current to drop to one-third its initial value when a 2.0-$\mu$F capacitor is being charged through a $10^8$-$\Omega$ resistor by a 9.0-V battery?

■ **3** A series circuit consists of an uncharged 3.0-$\mu$F capacitor, a $5 \times 10^6$-$\Omega$ resistor, a 12.0-V battery, and a switch. What is the current in the circuit and the charge on the capacitor (*a*) when the switch is first closed and (*b*) 1 time constant after the switch is closed?

■ **4** A series circuit consists of a 5.0-$\mu$F capacitor charged to 9.0 V, a switch, and a $10^7$-$\Omega$ resistor. What is the current in the circuit and the voltage across the capacitor (*a*) when the switch is first closed and (*b*) 1 time constant after the switch is closed?

■ **5** A series circuit consists of a 6.0-V battery, a $2 \times 10^6$-$\Omega$ resistor, a 4.0-$\mu$F capacitor, and an open switch. The capacitor is initially uncharged. The switch is now closed. (*a*) What is the time constant of the circuit? (*b*) About how long does it take for the capacitor to become two-thirds charged? (*c*) How much charge will flow onto the capacitor in the time calculated in part *b*? (*d*) About what is the average current into the capacitor during this interval?

■ **6** Suppose you measure the resistance of your body between your two hands with an ohmmeter and find it to be 62,000 $\Omega$. A 12.0-$\mu$F capacitor has been charged to 9.0 V and disconnected. You now grasp the two terminals of the capacitor with your two hands. (*a*) What is the time constant of the circuit involving your body and the capacitor? (*b*) What is the approximate potential difference across the capacitor after 0.75 s? (*c*) What is the charge on the capacitor when the potential across it is 9 V? (*d*) About what is the average current through your body in the 0.75 s?

**7** An ac ammeter in series with an incandescent bulb reads 0.25 A, and an ac voltmeter across the bulb reads

110 V. What are the maximum (peak) current through the bulb and the maximum voltage across it? How much average power does the bulb use?

**8** The current through an incandescent bulb is read to be 0.55 A by an ac ammeter, and the voltage to it is measured to be 115 V by an ac voltmeter. What are the peak current and voltage furnished to the bulb? How much power is used by the bulb?

**9** How much current does a toaster stamped 860 W/120 V draw from a 120-V ac power line? What is the resistance of the toaster during normal operation? How many calories of heat does it generate in 2 min?

**10** An electric heater is rated at 1400 W/120 V. How much current does the heater draw from a 120-V ac power line? How many calories of heat does it generate in 3.0 min?

▪ **11** There is a current given by the relation $i = 5 \sin 240t$ A through a 30-$\Omega$ resistor. How much power does this current dissipate in the resistor?

▪ **12** A voltage $v = 40 \cos 360t$ V is impressed across a 30-$\Omega$ resistor. How much power is dissipated in the resistor?

**13** How much current (rms) does a 3.0-$\mu$F capacitor draw from a 120-V/60-Hz source connected directly across it? Repeat for a 120-V/60,000-Hz source.

**14** A 2.5-$\mu$F capacitor is connected directly across a 30-V/200-Hz source. How much current (rms) does it draw from the source? Repeat for a source frequency of $2 \times 10^5$ Hz.

▪ **15** A 7.0-$\mu$F capacitor is connected directly across a 120-V/60-Hz power source. (a) What is the average value of the power consumed by the capacitor? (b) What is the rms current to the capacitor? (c) What is the maximum charge on the capacitor?

▪ **16** By what factor does the current to a capacitor change as the frequency of the source voltage across which it is connected is increased by a factor of 10,000? Assume that the circuit has no resistance and that the magnitude of the source voltage does not change.

**17** Find the reactance of a 3.0-mH inductance coil at a frequency of (a) 60 Hz and (b) $6 \times 10^5$ Hz.

**18** If an inductance coil is to have a reactance of 30 $\Omega$ for a frequency of 2000 Hz, what must its inductance be? What is its reactance at 3.0 MHz?

▪ **19** An ac voltage source is connected directly across a resistanceless 30-mH inductance coil. It causes a current of 0.50 A in the coil when the voltage is 6.0 V. (a) What is the frequency of the source? (b) If the frequency is tripled and the voltage kept steady at 6.0 V, what is the current in the coil?

▪ **20** An ac voltage source is connected directly across a resistanceless 0.70-mH inductance coil. How large must the voltage be to give a current of 1.20 A if the frequency is (a) 60 Hz and (b) $6 \times 10^5$ Hz?

**21** A 100-V/60-Hz source is connected across a 20-$\Omega$ resistor. (a) Find the current drawn from the voltage source. (b) Repeat for 6000 Hz. (c) How much power is dissipated in each case?

▪ **22** A 60-Hz voltage source is connected across a 2.00-$\mu$F capacitor. What value of inductor connected across this same voltage source would draw the same current?

▪ **23** A 20-$\Omega$ resistor in series with a 5.0-$\mu$F capacitor is connected across a voltage source. The ac voltage across the resistor is the same as that across the capacitor. What is the frequency of the source?

▪ **24** A 30-$\Omega$ resistor in series with a 2.0-mH resistanceless inductor is connected across a variable-frequency voltage source. At what frequency is the ac voltage across the resistor the same as that across the inductor?

▪ **25** A 2.0-mH inductor that has a resistance of 25.0 $\Omega$ is connected directly across a 40-V/5000-Hz power source. Find the current in the circuit and the power drawn from the source.

▪ **26** A 3.0-$\mu$F capacitor in series with a 500-$\Omega$ resistor is connected across a 25-V/120-Hz power source. Find the current in the circuit and the power drawn from the source.

▪ **27** A series circuit consists of a 40-V/2000-Hz power source, a 700-$\Omega$ resistor, and an unknown capacitor $C$. The voltage drop across the resistor is 30 V. What are the current in the circuit and the value of the capacitor?

▪ **28** A series circuit consists of a 60-V/3000-Hz power source, a 500-$\Omega$ resistor, and an unknown inductor $L$. The voltage drop across the resistor is 20 V. What are the current in the circuit and the value of the inductor?

**29** (a) How large a capacitor must be connected in series with a 0.20-H inductor if they are to resonate with a frequency of 60 Hz? (b) How large an inductor is needed to resonate at this same frequency with a 3-$\mu$F capacitor?

**30** When connected in series with a 3.0-$\mu$F capacitor, an inductor resonates fairly sharply to a frequency of 650 Hz. What is the inductance?

▪ **31** A capacitor and an inductor are connected in series across a 120-V/60-Hz source. The inductor has an inductance of 0.42 H, and its resistance is 800 $\Omega$. The capacitor is 2.0 $\mu$F. (a) Find the current in the circuit. (b) Repeat for a frequency of 6000 Hz.

▪ **32** How large an inductor must be connected in series with a 5-$\mu$F capacitor, a 20-$\Omega$ resistor, and a 100-V/60-Hz source if the current is to be 5 A?

■ **33** The following elements are connected in series across a 100-V/60-Hz source: $R = 30.0\ \Omega$, $C = 2.50\ \mu F$, $L = 2.50\ H$. (*a*) What current is drawn from the source when the elements are connected in series across it? (*b*) How much power is dissipated by the circuit? (*c*) Compute the power factor.

■ **34** The following elements are connected in series across a 40-V/2000-Hz power source: $R = 80\ \Omega$, $C = 1.0\ \mu F$, $L = 9.0\ mH$. Find (*a*) the current in the circuit, (*b*) the voltage drop across $C$, (*c*) the voltage drop across the $LC$ combination, and (*d*) the power loss in the circuit.

■ **35** An inductance coil wound on a plastic rod draws 0.60 A from a 12-V battery and 3.0 A from a 120-V/60-Hz source. Find the inductance of the coil and the power it draws from the ac source.

■ **36** When connected across a 20-V battery, an inductance coil draws 0.50 A. The same coil draws 2.0 A from a 90-V/60-Hz ac source. What is the inductance of the coil, and how much power does it draw from the ac source?

■ **37** How much power does a 40-mH inductor that has a resistance of 8.0 $\Omega$ dissipate when it is connected across a 110-V/60-Hz source?

■ **38** A series $RLC$ circuit in which $R = 40\ \Omega$ is connected across a 90-V/200-Hz power source. The ac voltage is the same across each of the elements. Find the voltage reading across the pure inductor. Find the values of $L$ and $C$.

■ **39** A half-wave rectifier is connected to a 30,000-$\Omega$ resistor as its load. It is rectifying 60-Hz ac voltage. (*a*) What magnitude of filter capacitor must be used (as in Fig. 20.15) if the time constant of the filter-load system is to be 20 times as large as the time during which the rectifier is not passing current? (*b*) Why is it difficult to obtain a smooth dc voltage from a 60-Hz rectifier if large currents are to be drawn from it?

■ **40** If the output voltage in Fig. 20.15 has a maximum value of 11.0 V and a minimum value of 9.8 V, what is the value of the ripple for the rectifier system?

# 21 Electromagnetic waves

**T**HE nature of many types of waves is nearly obvious to everyone. A wave on a string consists of the vibration of the string; waves on a water surface are the result of the motions of that surface; sound waves are disturbances in the air. But the nature of other waves is not so apparent. What vibrates in a radio wave? If light behaves as a wave, and it does, what is the nature of its vibrations? How are x-ray waves and heat radiation (a wave disturbance) carried through space? All these latter waves are electromagnetic waves. We learn about them in this chapter.

## 21.1 The generation of em waves

To begin our study of electromagnetic waves (em waves), let us consider the situation shown in Fig. 21.1. We see there a dipole (Sec. 16.8) consisting of two oppositely charged balls and the electric field the dipole generates. When the voltage source is a battery, the field is an electrostatic one and does not change. Suppose, however, that the battery is replaced by an ac voltage source. Then the charges on the balls will vary sinusoidally. The charge on the top ball will vary as

$$q = q_o \cos 2\pi ft$$

The equal-magnitude charge on the bottom ball is opposite in sign and will vary in

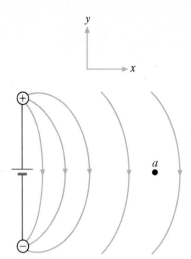

the same way. Thus the dipole continually reverses polarity, undergoing *f* cycles per second. What does this imply for the electric field outside the dipole?

Close to the dipole — at point *a*, for example — the field reverses in step with the charge reversal. Hence the field at *a* oscillates in the *y* direction and varies in a sinusoidal fashion with the same frequency as the source. Its equation is

$$E_y = -E_{oy} \cos 2\pi ft$$

What about the electric field far from the source, however? How does it behave?

We can think of the electric field as being the disturbance sent out by the dipole source much the same way a wave on a string is the disturbance sent down the string by an oscillating source. At a certain instant, the field sent out along the *x* axis is as shown in Fig. 21.2. The field shows the history of the charge on the dipole. The downward-directed fields were sent out when the top of the dipole was positive; the upward-directed fields were sent out one-half cycle later, when the top of the dipole was negative.

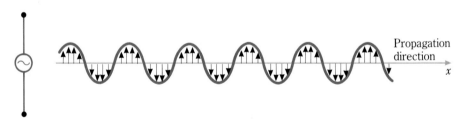

Propagation
direction

*x*

**FIGURE 21.2**

The alternating charges on the dipole antenna send an
electric-field disturbance out into space.

In the case of a radio station, the dipole (or antenna) is often simply a long wire. If you visit a radio transmitting site, you will see the antenna as a long wire stretched between two towers or as a vertical wire held by a single tower. Charges are placed on the antenna by an ac voltage from a transformer system. The electric field wave

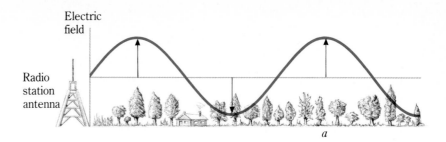

**FIGURE 21.3**

The electric field wave from the antenna blankets an area even quite distant from the station.

sent out by the antenna blankets the earth around it, as in Fig. 21.3. At a point such as *a* in the path of the wave, the electric field reverses periodically as the wave passes. The frequency of the oscillating electric field at *a* is the same as the frequency of the source.

Thus we see that an electric field wave is sent out by the oscillating dipole, the transmitting antenna. We should notice that, like all waves, the electric field wave obeys the following relation between frequency *f* and wavelength $\lambda$:

$$\lambda = \frac{v}{f} \tag{21.1}$$

where *v* is the speed at which the wave travels out through space. Further, we notice that the quantity that vibrates, namely the electric field vector, is always perpendicular to the direction of propagation. Hence the electric field wave is a transverse wave (Sec. 13.11).

It is easy to see that a radio station's antenna necessarily generates a magnetic field wave as it generates an electric field wave. To see this, refer to Fig. 21.4. At the radio station, charges are sent up and down the antenna in Fig. 21.4*a* to produce the alternating charges we have been discussing. This charge movement constitutes an alternating current in the antenna, and because a magnetic field circles a current, an oscillating magnetic field is produced, as shown in Fig. 21.4*b*. As with the

**FIGURE 21.4**

(*a*) As charge rushes up and down the antenna, (*b*) a magnetic field wave is sent out as shown.

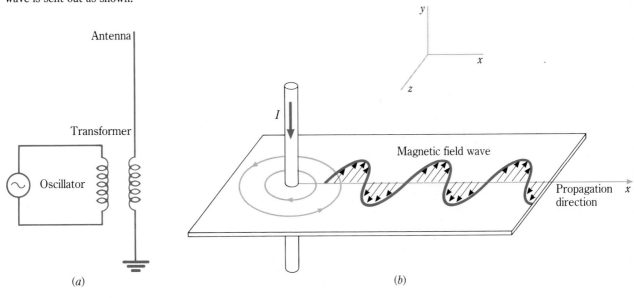

Antenna

Transformer

Oscillator

(*a*)

*y*

*x*

*z*

*I*

Magnetic field wave

Propagation direction    *x*

(*b*)

oscillating electric field, the magnetic field travels out along the $x$ axis as a transverse wave. Because the direction of the current oscillates, so too does that of the magnetic field.

Notice, however, that the magnetic field is in the $z$ direction, while the electric field is in the $y$ direction. As shown in Fig. 21.5, the magnetic field is perpendicular to both the electric field and the direction of propagation. The two waves are drawn in phase (that is, they reach their maxima together). That this is true is not obvious; it is the result of detailed computations.

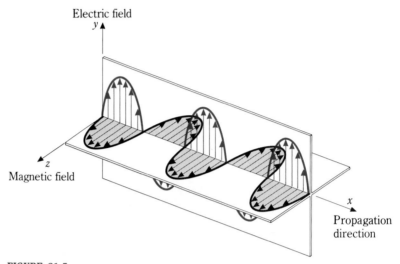

**FIGURE 21.5**

In an em wave, the magnetic field wave is perpendicular to both the electric field wave and the direction of propagation.

As we see, em waves are transverse waves and are much like waves on a string and other transverse waves. However, em waves consist of oscillating electric and magnetic fields, not of material particles. As such, they can travel through empty space (vacuum). And, as we shall see, they carry energy along their direction of propagation. Later we shall show that em waves travel through vacuum with the speed of light, which we designate by $c$. You will recall that, in the SI, the speed of light in vacuum is defined to be $c = 2.998 \times 10^8$ m/s.

There is one other feature of em wave generation that we should point out. Notice that the charges that oscillate up and down the antenna are accelerating. It turns out that whenever a charge undergoes acceleration, it emits em radiation; the larger the acceleration (or deceleration) of the charge, the more energy it emits as em radiation. Thus, if a fast-moving charged particle undergoes an impact, it will emit a burst of em radiation as it suddenly stops.

### *Example 21.1*
The oldest radio station in the United States is KDKA in Pittsburgh, which went on the air in 1920. It operates at a frequency of $1.02 \times 10^6$ Hz. What is the wavelength of its em wave? Assume the speed of em waves to be $3 \times 10^8$ m/s.

***Reasoning*** We know that $\lambda = v/f$ for any wave. In our case, $v = 3 \times 10^8$ m/s and $f = 1.02 \times 10^6$ Hz. Substitution gives $\lambda = 294$ m. ∎

***Exercise*** Radar waves (microwaves) have wavelengths of several centimeters. What is the frequency of an em wave that has a wavelength of 20 cm? ***Answer: 1.50 × 10⁹ Hz*** ∎

## 21.2 Types of electromagnetic waves

As we discuss at greater length later, radio-type waves were foreseen by a 34-year-old Scottish physicist, James Clerk Maxwell, in 1865, many years before the first radio was invented. Maxwell used the then-known facts about electricity to show that em radiation should exist. Furthermore, he was able to prove that these waves should have a speed of $3 \times 10^8$ m/s in vacuum. This was an astonishing prediction because the speed he found was a well-known speed, the speed of light in vacuum, $c$. Thus Maxwell was led to surmise that light waves are one form of em waves. Today we know that there are a variety of em waves that cover a wide range of wavelengths; we refer to this as the em wave *spectrum*.

The basic difference between the various types of em waves are the result of their different wavelengths. Since all electromagnetic radiation travels through vacuum with the speed of light, the relation $\lambda = v/f$ becomes $\lambda = c/f$ for electromagnetic radiation. Hence a difference in $\lambda$ implies a difference in the frequency of the radiation. The various types of electromagnetic radiation are shown in Fig. 21.6. Examine this chart carefully to become familiar with the wavelength ranges involved. The methods used to measure these wavelengths are discussed in Chap. 24. Let us now discuss briefly the nature of each type of radiation.

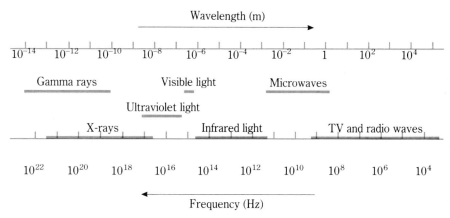

**FIGURE 21.6**
Types of electromagnetic radiation. The bars indicate the approximate wavelength range of each type of radiation.

### Radio waves

We have already discussed radio waves in some detail. Their wavelengths range from 1 m or so to more than about $3 \times 10^6$ m for the waves sent out by ac power lines. If one wished to obtain a wave with $\lambda = 10^8$ km, which is the distance from the earth to the sun, how frequently should one reverse the charges on the antenna?

### Microwaves

Microwaves are short-wavelength radio waves. They are sometimes called *radar waves*. The shortest wavelength given in Fig. 2.16 for microwaves ($10^{-3}$ m) represents the lower limit of wavelengths that can be generated electronically at present. Notice that, at a frequency of $10^{12}$ Hz, light can travel only 0.03 cm during one oscillation. Since material particles and energy cannot travel faster than the speed of light, only an antenna shorter than 0.03 cm can be charged during this short time. This should indicate why very-short-wavelength waves are difficult to produce electronically.

### Infrared waves

Infrared waves have wavelengths between those of visible light, $7 \times 10^{-7}$ m, and microwaves. Infrared radiation is readily absorbed by most materials. The energy contained in the waves is also absorbed, of course, and appears as thermal energy. For this reason, infrared radiation is also called *heat radiation*. The earth receives from the sun a large amount of infrared radiation as well as light. Warm objects of all types radiate infrared rays.

### Light waves

The wavelengths of the visible portion of electromagnetic radiation extend only from about $4 \times 10^{-7}$ to $7 \times 10^{-7}$ m, and we call this wavelength range *light*. We classify various wavelength regions in this range by the names of colors. The sensitivity of the normal human eye to wavelengths in this region is shown in Fig. 21.7. See also Color Plate II, which shows the light spectrum in color. You should learn the approximate wavelengths of the various colors. The "antenna" that generates light waves is charge accelerating within an atom.

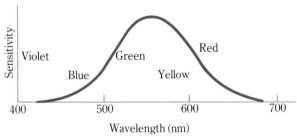

**FIGURE 21.7**
Sensitivity curve for the eye. The human eye is most sensitive to greenish-yellow light.

### Ultraviolet waves

Ultraviolet waves are radiation with wavelengths shorter than visible violet light but still longer than about 10 nm. At the shorter wavelengths, they are not distinct from x-rays.

### X-ray waves

X-rays are electromagnetic radiation with $\lambda \lesssim 10$ nm. Usually, this classification is reserved for the radiation given off by electrons in atoms that have been bombarded. This process is discussed in more detail in Chap. 26.

### Gamma-ray waves

Gamma rays ($\gamma$-rays) are electromagnetic radiation given off primarily by nuclei and in nuclear reactions. They differ from x-rays only in their manner of production. We discuss gamma rays more fully in our study of nuclear physics in Chap. 27.

Notice that the spectrum of electromagnetic radiation encompasses waves with wavelengths extending from longer than $10^6$ m to shorter than $10^{-15}$ m. Even though these waves are all electromagnetic waves, they differ considerably in their mode of interaction with matter. Much of the remainder of this book is concerned with various aspects of this subject.

## 21.3 Reception of radio waves

Television sets and radios are very sensitive electronic devices designed to detect electromagnetic waves with wavelengths in the radio range. Although we shall not discuss in detail the construction of such devices, let us see how they detect and tune to radio waves.

There are two basic fields present in an em wave, the electric and the magnetic field, and the wave can be detected by means of either. To detect the electric field part of the wave, we need only place a long wire (called the *receiving antenna*) in the path of the wave. Referring to Fig. 21.8*a*, we see that the electric field induces charges to oscillate in the wire of the antenna. When $\mathbf{E}_y$ is positive, the top of the

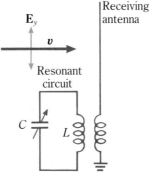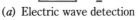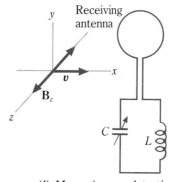

**FIGURE 21.8**

Two methods for detecting radio waves: (*a*) electric wave detection; (*b*) magnetic wave detection.

(*a*) Electric wave detection

(*b*) Magnetic wave detection

antenna is positive. An instant later, the antenna's polarity reverses as the electric field vector due to the wave reverses. This repeated action causes charge to flow up and down the wire of the antenna. As it does so, the current induces a voltage in the *LC* circuit that is coupled to it by means of a mutual inductance. (Although the figure shows the same coil for both the mutual inductance and the self-inductance, this need not be the case.) If the *LC* circuit is properly tuned, the induced voltage will cause resonance in it. Let us clarify this point. Each radio and television station is assigned its own individual frequency and sends out waves of that frequency only. Since the receiving antenna is buffeted by waves from many stations, some means must be found to select only the wave from the station one wishes to receive. In an ordinary radio, the selection is done by adjusting *C* within the resonant circuit until its resonant frquency is the same as the station's frequency. When this selection has been made, the circuit responds strongly to the station in question while effectively ignoring all others. The current and voltage variations in the resonant circuit are then amplified and modified within the radio or television to produce the sound and picture.

We can also detect the electromagnetic wave by means of its oscillating magnetic field. Since this field is varying rapidly, the wave induces an emf in a loop such as the one in Fig. 2.18*b*.* Notice that the loop must be properly oriented so that the flux lines of the magnetic field go through the loop. (It is for this reason that small radios that use this type of antenna show markedly different reception in different orientations.) The induced voltage in the loop antenna is impressed on an *LC* circuit. Tuning is accomplished as described above.

One might well ask why not all waves, light and x-rays included, can be detected by radio-type devices. The reason is quite simple. Very-high-frequency waves require *LC* resonant circuits that are thus far impossible to build. You will recall that the resonant frequency of a circuit is given by $1/2\pi\sqrt{LC}$. To make this frequency very high, both *L* and *C* must be very small. In the cases of infrared waves, light waves, and x-rays, two tiny wires lying side by side already have *L* and *C* values that are too large. We shall see in later chapters that a circuit of atomic size is needed to detect these waves. Indeed, we shall learn that individual atoms and molecules, in effect, become the resonant circuit for detecting very-high-frequency electromagnetic waves.

---

### 21.4 The speed of em waves (optional section)

Now that we understand many of the qualitative features of em waves, let us obtain an expression for their speed. We use a method that depends on a fact first pointed out clearly by Einstein in his theory of relativity, a fact that we discuss at greater length in Chap. 25: *only relative velocities can be determined.* An object can be said to be at rest relative to another object but not at rest in any absolute sense.

For example, as you read this, you are probably at rest relative to the earth, but the earth is in motion relative to the sun, and so too are you. Moreover, the sun is in motion in our galaxy, the Milky Way, and our galaxy is in motion relative to other galaxies in the universe. It makes sense to say that something is at rest relative to something else, but we cannot say which of two objects is at rest in any noncomparative way.

With this fact in mind, let us reconsider the force experienced by a charge $q$ moving with speed $v$ perpendicular to a magnetic field of magnitude $B_\perp$. We found in Chap. 18 that the force experienced by the charge has a magnitude

$$F = qvB_\perp$$

---

* In practice, the loop is a coil wound on a ferromagnetic rod.

However, who is to say that the charge is not at rest and the field moving instead? After all, only relative motion is observable. Therefore our experiment can be interpreted in the following alternative way: a field **B** moving with speed $v$ perpendicular to the field lines past a charge $q$ exerts a force $F = qvB_\perp$ on the charge. Because the force per unit charge $F/q$ is defined to be the magnitude of the electric field $E$, however, we can restate this as follows:

A magnetic field **B** moving with speed $v$ perpendicular to the field lines generates an electric field of magnitude $E = vB$ in the region through which it passes.

To illustrate this, consider the situation in Fig. 21.9. The magnet poles are moving with speed $v$ as shown. They carry the magnetic field lines with them, and so we have a moving magnetic field **B** in this region. Our previously stated result tells us that at a point such as $a$, through which the field lines are moving, an electric field **E** exists whose value is $E = vB$. You should be able to show that the direction of **E** is into the page.*

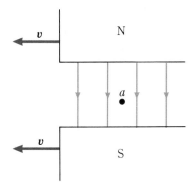

**FIGURE 21.9**

The magnetic field **B** (shown by the colored vertical lines) moves with the magnet poles past point $a$ with speed $v$. It generates an electric field $E = Bv$ directed into the page.

It seems from this line of reasoning that the magnetic field moving out from a radio antenna should generate an electric field in the region through which the magnetic field moves. At a given point, the electric field should be related to the speed $v$ of the magnetic field wave and to its magnitude $B$ at that point and instant through $E = vB$. The question now arises as to whether a moving *electric field* can generate a *magnetic field*. We now investigate that question, and we find that our investigation leads us to a very important result.

Consider the very long, uniformly charged wire shown in Fig. 21.10. The wire is moving to the right with speed $v$, and so the electric field lines from it are flying past

---

* *Hint:* Place a positive charge at $a$ and remember that the motion is relative.

**FIGURE 21.10**

The moving charged rod carries the electric field lines past point $P$.

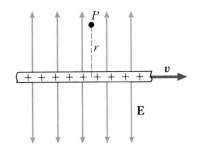

point $P$ with speed $v$. We know that the moving charged wire constitutes a current along the line of the wire. The magnitude of the current can be found by noticing how much charge passes $P$ each second.

Suppose the wire has a charge $\rho$ per unit length. Since a length $vt$ of the wire flies past $P$ in time $t$, we have

$$\text{Current} = \frac{\text{charge passing } P}{\text{time taken}} = \frac{\rho v t}{t} = \rho v$$

We see that the moving charged wire constitutes a current of magnitude $\rho v$.

A current produces a magnetic field, however, and so the moving charged wire is surrounded by a magnetic field. (You should be able to show that it circles the wire and is out of the page above the wire.) We learned in Chap. 18 that the magnetic field due to a current $I$ in a long, straight wire is $B = \mu_o I / 2\pi r$. If we apply this to the present case, we find the magnetic field at point $P$ to be

$$B = \frac{\mu_o \rho v}{2\pi r} \tag{21.2}$$

We wish to relate this to the electric field outside the wire at $P$.

The electric field outside a charged, long, straight wire is easily found using Coulomb's law and calculus. We simply state the result here. The electric field at point $P$ in Fig. 21.10 is

$$E = \frac{\rho}{2\pi \varepsilon_o r} \tag{21.3}$$

where $\varepsilon_o$ is the permittivity of free space, $8.85 \times 10^{-12}$ C$^2$/N $\cdot$ m$^2$.

Let us now eliminate $\rho$ between Eqs. 21.2 and 21.3. The result is

$$B = \varepsilon_o \mu_o v E \tag{21.4}$$

This is to be compared with the relation

$$B = \frac{1}{v} E \tag{21.5}$$

which we obtained for a moving magnetic field.

Although this is a very special situation in which a moving charged wire generates a magnetic field, it is typical. Moving charges generate a magnetic field, but the moving charges have associated with them an electric field that travels along with them. The magnetic field generated by the motion of the charges can equally well be attributed to the motion of the electric field. We therefore are led to conclude that

An electric field $\mathbf{E}$ moving with speed $v$ perpendicular to the field lines generates a magnetic field of magnitude $B = \varepsilon_o \mu_o v E$ in the region through which it passes.

Let us now return to Fig. 21.5, in which we see the electric and magnetic fields generated by an antenna. The fields are rushing out along the line of propagation with speed $v$. Consider the magnetic field as it flies past a point in space. It generates an electric field at that point. Similarly, the electric field from the antenna also flies past the same point and generates a magnetic field there.

If you spend a little time considering the situation in Fig. 21.5, you can see that the electric field shown there is in the same direction as the electric field generated by the moving magnetic field. In addition, the magnetic field shown is in the same direction as the magnetic field generated by the moving electric field. We are therefore tempted to say that the electric and magnetic fields in an em wave regenerate themselves as the wave travels out through space. Let us make this supposition and see where it leads us.

Suppose the electric and magnetic fields of an em wave generate each other as the wave moves with speed $v$ through space. Then both Eq. 21.4 and Eq. 21.5 must apply to the wave. If this is so, $B$ and $E$ must be related in the same way in the two equations, and so the proportionality constants between $E$ and $B$ must be the same. Hence

$$\varepsilon_o \mu_o v = \frac{1}{v}$$

Solving this relation for $v$, the speed of the em wave in vacuum, we find

$$v = \frac{1}{\sqrt{\mu_o \varepsilon_o}} \tag{21.6}$$

Since $\mu_o = 4\pi \times 10^{-7}$ and $\varepsilon_o = 8.85 \times 10^{-12}$ in SI units, we obtain the result

$$v = 2.998 \times 10^8 \text{ m/s}$$

This is the well-known measured speed of light in vacuum $c$. Hence we conclude that

All em waves travel through vacuum with the speed $c = 2.998 \times 10^8$ m/s, and light is one form of em wave.

Further, Eq. 21.5 gives us the relation between $B$ and $E$ in an em wave traveling through vacuum:

$$E = cB \tag{21.7}$$

## 21.5 Maxwell's equations

James Clerk Maxwell, using a much more rigorous approach than we used in the previous section, was the first to show that

**1** Electromagnetic waves, such as those we discussed qualitatively in Secs. 21.1 and 21.2, should exist. In them, $E$ and $B$ are in phase and $E = cB$.

**2** The speed of all electromagnetic waves in vacuum is $2.998 \times 10^8$ m/s, which is the speed of light in vacuum.

In addition, he provided convincing evidence that light is a form of electromagnetic radiation.

Maxwell derived these results in a rigorous, mathematical fashion by summarizing the experimental data available at the time (1865) in terms of four equations and then slightly modifying them as discussed below. The experimental results he described in equation form are

## James Clerk Maxwell (1831–1879)

Though born to wealth in Scotland, Maxwell had a somewhat spartan childhood. He was an undistinguished student until the age of 13, when his

extraordinary intellectual abilities suddenly became apparent. He developed into a shy, quietly humorous man, sociable with friends but often reticent and withdrawn among strangers. After studying at Edinburgh and Cambridge, he held professorships at Aberdeen and London. When his father died, he retired to his family's estate in Scotland, where he researched and wrote his famous treatise on magnetism and electricity. In 1871, he became the first Cavendish Professor of Experimental Physics at Cambridge, and he proceeded to design the famed Cavendish Laboratory. Though he is best known for his profoundly original theoretical work concerning the behavior of gases and electromagnetic waves, his experimental and mathematical talents were extraordinary as well. His contributions to science have proved as original and important as those of Newton and Einstein.

**1** Coulomb's law is an inverse-square law, and electric field lines can be used to represent forces expressed by it. Electric field lines begin at positive charges and end at negative charges.

**2** Magnetic field lines circle back on themselves, having no beginning and no end. This reflects the fact that a north pole is always accompanied by an equal-magnitude south pole.

**3** A varying magnetic field induces an emf, which is equivalent to stating that it induces an electric field. This is simply Faraday's law.

**4** Moving charges generate magnetic fields. The equation that describes this behavior was discovered by Ampère and is called *Ampère's law.*

Maxwell's four equations (which are universally referred to as *Maxwell's equations*) express in mathematical terms these four statements of fact, with one addition, which we shall now mention.

Upon examining his equations in detail, Maxwell noticed that electric and magnetic fields played analogous roles. The differences that existed, except one, could be explained by recognizing that free individual charges exist but free individual magnetic poles do not; a south pole of a coil or magnet is always accompanied by a north pole. The one difference for which no excuse could be found was the following: changing magnetic fields induce electric fields according to Faraday's law, but no evidence existed for the analogous effect — that changing electric fields induce magnetic fields.

Despite the lack of supporting experimental evidence, *Maxwell postulated that changing electric fields generate magnetic fields.* He showed that if his postulate were true, a new term must be added to Ampère's law, the equation that states how currents are related to the magnetic fields they generate. This term turned out to be negligibly small when he applied the equation to the experiments possible at that time. As a result, he could neither prove nor disprove the validity of the term.

In an attempt to justify the term, Maxwell applied his equations to various phenomena, focusing his attention on what happens when charges oscillate rapidly back and forth so as to produce a swiftly changing electric (and magnetic) field. Under those conditions, the postulated generation of magnetic fields by changing electric fields should be most obvious.

Maxwell found that the application of his equations to the field generated by oscillating charges leads to the prediction that em waves can exist. He found their speed to be *c*, the velocity of light in vacuum. Thus he established the true nature of light: light waves are a form of em wave.

It was not until 1887, eight years after Maxwell's death, that em waves were generated and detected by electrical means. This feat was performed by Heinrich Hertz. Using an *LC* circuit, he generated a spark that jumped back and forth across a gap in a loop of wire. The wave thus generated traveled from this "antenna" to a second loop of wire, the "receiver," and induced an emf in it. Subsequent experiments have fully confirmed Maxwell's supposition that a term was missing from Ampère's law. Today we recognize that Maxwell's equations are as fundamental and important to electromagnetism as Newton's laws are to mechanics. Maxwell's equations form the basis for all theoretical work in electromagnetism.

---

**Example 21.2**
An em wave passes through a certain point in space. The electric field there is $E = E_o \sin 2\pi ft$, with $E_o = 0.0042$ V/m. Find the magnetic field there.

**Reasoning**  We know that the electric and magnetic fields vibrate in phase (though perpendicular to each other), and so

$$B = B_o \sin 2\pi ft$$

To find $B_o$, we use the fact that $E = cB$:

$$B_o = \frac{E_o}{c} = \frac{4.2 \times 10^{-3} \text{ V/m}}{3 \times 10^8 \text{ m/s}} = 1.4 \times 10^{-11} \text{ T}$$

Notice how very small the magnetic field is in an em wave.  ▪

---

**Exercise**  If a field this large is perpendicular to the 25-cm² area of a single loop, and if it decreases to zero in $10^{-9}$ s, how large an emf does it induce in the loop? *Answer: $3.5 \times 10^{-5}$ V*  ▪

---

### 21.6 The energy carried by em waves

We have seen that em waves consist of moving electric and magnetic fields. Because these fields contain energy, the waves carry energy through space. For example, em waves from the sun warm the earth and provide the energy for plant growth.

# Magnetic monopoles

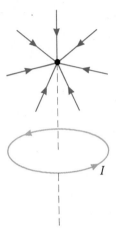

Magnetic poles always come in pairs, or so we have learned thus far. A current loop has both a north and a south pole; so does a bar magnet. Even particles such as the electron and proton act like tiny bar magnets with two poles, one north and the other south. It is this duet between north and south poles that allows us to picture all magnetic fields as the result of current loops.

There is a potential difficulty with this, however. In a theory he presented in 1931, the physicist Paul Dirac predicted that a particle that acts as a single north or south pole should exist. It has been named a *monopole*. Although many features of Dirac's theory proved to be correct, his prediction of monopoles was viewed with considerable skepticism and was largely ignored. New interest in his prediction was aroused when recent theories of the origin of the universe showed that, perhaps, monopoles were formed in the early stages of the development of the cosmos. Theory predicts that some of these monopoles should still be shooting through space. As a result of this new theoretical prediction, experiments have been designed to detect monopoles.

In a typical experiment, a current is established and monitored in a superconducting loop of wire. If an object that has both a north and a south pole passes through the loop, the current will change momentarily but will return to its original value. As the north pole, say, approaches and passes through the loop, it

induces an emf in the loop. However, the accompanying south pole induces an opposite emf when it passes through. The effects of the two emf's exactly cancel, and so the final current in the loop is the same as the initial current. The current in the superconducting loop should remain constant for years.

Suppose, however, that the south monopole shown in the accompanying figure falls through the loop. As it approaches the loop, a clockwise emf is induced in the loop, and this momentarily decreases the current in the loop. After it has passed through the loop, the falling monopole still induces a clockwise emf. (Use Lenz's law to show that this is true.) Hence the effect of the monopole's passage through the loop, unlike that of the magnet, is to change the current permanently, to decrease it in this case. We therefore have a method for detecting monopoles.

The first tentative detection of such an event occurred at Stanford University in 1982. Other investigators have been waiting for a similar event, and one was detected in 1985 at Imperial College in London. Many more such events must be detected before we can be sure that the effects observed were really caused by monopoles and were not just false alarms. By the time you read this, perhaps the question of the possible existence of monopoles will have been settled. If monopoles do exist, the ramifications could be astounding.

Those from a distant television station carry the energy that brings the picture and sound to your television. Let us now compute how much energy is carried to a surface by an em wave incident on the surface.

You will recall from Sec. 16.11 that the energy stored in unit volume of an electric field of magnitude $E$ is $\frac{1}{2}\varepsilon_o E^2$. Similarly, we showed in Sec. 19.7 that the energy stored in unit volume of a magnetic field of magnitude $B$ is $B^2/2\mu_o$. With these facts in mind, let us look at the beam of em radiation in Fig. 21.11; it carries energy through the plane indicated. The beam has an end area $A$ and travels to the right with the

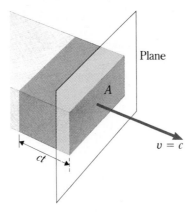

Plane

$A$

$v = c$

$ct$

**FIGURE 21.11**

A volume $Act$ of the beam goes through the plane in time $t$.

speed of light $c$. Because the beam travels a distance $ct$ during time $t$, a length $ct$ moves through the plane indicated in this time. Hence the volume of the beam that travels through the plane in time $t$ is $Act$. This is indicated by shading in the figure.

Let us take $t$ short enough for $ct$ to be much smaller than the wavelength of the beam's radiation. Then both $\mathbf{E}$ and $\mathbf{B}$ are substantially constant throughout the shaded volume. Therefore we can write that the energy carried through the plane by the beam volume $Act$ is

$$\begin{pmatrix} \text{Energy in} \\ \text{volume } Act \end{pmatrix} = \begin{pmatrix} \text{magnetic field} \\ \text{energy density} \end{pmatrix} (\text{volume}) + \begin{pmatrix} \text{electric field} \\ \text{energy density} \end{pmatrix} (\text{volume})$$

$$\text{Energy in } Act = \frac{B^2}{2\mu_o} Act + \tfrac{1}{2}\varepsilon_o E^2 Act$$

To find the energy carried through unit area of the plane in unit time, we need only divide by $t$ and the area $A$ of the beam. We then have for the energy transported through unit area in unit time

$$\begin{pmatrix} \text{Energy per unit} \\ \text{area per second} \end{pmatrix} = \frac{c}{2}\left(\frac{B^2}{\mu_o} + \varepsilon_o E^2\right)$$

We call this the *intensity $I$* of the wave. Because $B^2 = E^2/c^2 = E^2\varepsilon_o\mu_o$, this equation can be written in the form

$$I = \begin{pmatrix} \text{energy per unit} \\ \text{area per second} \end{pmatrix} = \tfrac{1}{2}c\varepsilon_o(E^2 + E^2) = c\varepsilon_o E^2$$

We can see in this last equation that the magnetic and electric field terms are equal in magnitude. Hence we conclude that

The electric field and the magnetic field of an em wave transport equal energies.

The average energy transported by the em beam is not equal to the value given above because $E$ and $B$ vary sinusoidally in a wave. We therefore need the average value of $E^2$ instead of the value that applies to the shaded region of Fig. 21.11. We learned in our study of alternating currents, however, that the average of $i^2$, for example, is $i_o^2/2$, half the amplitude squared. Because $E$ is also a sinusoidal quantity, the average value of $E^2$ is simply $E_o^2/2$, where $E_o$ is the amplitude of the electric field wave.

In view of this discussion, we can write the intensity of the wave as

$$I = \left(\begin{array}{c}\text{power per} \\ \text{unit area}\end{array}\right) = \left(\begin{array}{c}\text{average energy transported} \\ \text{per unit time per unit area}\end{array}\right) = \tfrac{1}{2}c\varepsilon_o E_o^2 \tag{21.8a}$$

Or, if we choose, we can write this in terms of $B_o$, the amplitude of the magnetic field wave. We recall that $E = cB$, and so we have

$$I = \left(\begin{array}{c}\text{power per} \\ \text{unit area}\end{array}\right) = \tfrac{1}{2}c\varepsilon_o c^2 B_o^2 = \frac{cB_o^2}{2\mu_o} \tag{21.8b}$$

where use has been made of $c = 1/\sqrt{\varepsilon_o\mu_o}$. We therefore conclude that

The power transported through unit area by an em wave incident perpendicular to the area is $\tfrac{1}{2}c\varepsilon_o E_o^2 = cB_o^2/2\mu_o$. This is called the intensity of the wave.

## 21.7 The inverse-square law for radiation

As was pointed out in the preceding section, the intensity of a radiation beam is defined in the following way. We imagine an area $A$ placed perpendicular to the beam, as in Fig. 21.12. Since the beam (let us say it is a beam of light) carries energy in the direction of propagation (in this case, to the right), a certain quantity of energy passes through the area in unit time. We define the intensity of the light $I$ by the relation

$$I = \frac{\text{energy}}{\text{area} \times \text{time}} \tag{21.9}$$

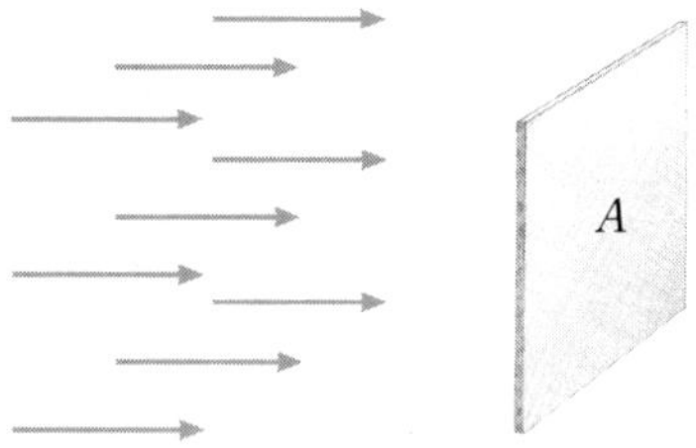

FIGURE 21.12

The intensity of a beam is the energy passing through unit area per second, provided the beam is perpendicular to the area.

Because energy $\div$ time is power, the SI unit for intensity is watts per square meter. Usually, of course, the beam is not uniform over a large area. Then we find the intensity at a point by letting $A$ become extremely small so that the beam does not change appreciably over it.

Let us now discuss the energy emitted by a small source of light, such as the one shown in Fig. 21.13. We assume the source to be small enough that it acts as a *point*

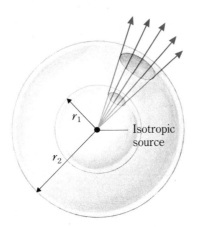

**FIGURE 21.13**

If the intensity at the surface of the sphere is $I_1$ and the radius of the sphere is $r_1$, how much power flows from the sphere?

*source;* further, we assume it to be an *isotropic source,* that is, one that emits light equally in all directions. To describe the energy that flows out from the source, let us imagine a spherical surface of radius $r_1$ concentric to it. We use the symbol $I_1$ to designate the intensity of the light from the source at the position of the surface. Because the surface area of the sphere is $4\pi r_1^2$ and because the intensity $I_1$ is uniform over the entire surface area at $r_1$, we have that

$$\left(\begin{array}{c}\text{Energy coming out of sphere} \\ \text{of radius } r_1 \text{ in unit time}\end{array}\right) = 4\pi r_1^2 I_1$$

Imagine a second concentric spherical surface of radius $r_2 > r_1$. If the light intensity due to the source is $I_2$ at the position of this surface, then the energy that flows out through it is

$$\left(\begin{array}{c}\text{Energy coming out of} \\ \text{sphere 2 in unit time}\end{array}\right) = 4\pi r_2^2 I_2$$

If radiation is not being absorbed in this region of space, the energy coming through the two concentric spherical surfaces must be the same. Thus

$$4\pi r_1^2 I_1 = 4\pi r_2^2 I_2$$

from which

$$\frac{I_1}{I_2} = \frac{r_2^2}{r_1^2} \tag{21.10}$$

This is the *inverse-square law for the radiation of energy from a point source.* It states that the light intensity from such a source decreases as the inverse square of the distance from the source. For example, if we triple the distance from the source, the light intensity decreases by a factor of 9.

### Example 21.3

Suppose a 1000-W radio station sends its power uniformly in all directions from its antenna. (An actual station beams its signal.) Find $E_o$ and $B_o$ in its wave 10 km from the antenna.

*Reasoning*   The power sent out by the station is spread over the surface of a sphere with radius $R$ centered on the antenna. At 10 km, the power per unit area of the station's beam (its intensity) is

$$\frac{\text{Power}}{\text{Area}} = \frac{1000\ \text{W}}{4\pi R^2} = \frac{1000\ \text{W}}{4\pi(10{,}000\ \text{m})^2} = 7.96 \times 10^{-7}\ \text{W/m}^2$$

We equate this to the value found in Eq. 21.8:

$$\tfrac{1}{2}c\varepsilon_o E_o^2 = 8.0 \times 10^{-7}\ \text{W/m}^2$$

Using $c = 3 \times 10^8$ m/s and $\varepsilon_o = 8.85 \times 10^{-12}$ C$^2$/N $\cdot$ m$^2$ gives $E_o = 0.024$ N/C. Notice how very small the electric field is.

To find the magnetic field, we recall that $E_o = cB_o$ and find $B_o = E_o/c = 8.2 \times 10^{-11}$ T. Do not become confused by the fact that the numerical value for $B$ is much smaller than that for $E$; the **E** and **B** fields transport equal energies. ∎

*Exercise*   Find the intensity of the radio wave 60 km from the station.   *Answer:* $2.21 \times 10^{-8}\ W/m^2$ ∎

---

### 21.8 Definition of electrical units

In Maxwell's time, the electrical units were defined differently from the way they are defined today. We now base our definition of these units directly on his discovery concerning the speed of em waves. Of course, our new definitions are chosen in such a way as to preserve as closely as possible the numerical results we are accustomed to from previous definitions.

We start with the value for the speed of light in vacuum. You will recall that $c$ is defined to have an exact value 299,792,458 m/s. (You will recall further that this defined value is then used to define the meter.) In this chapter, we learned that $c = 1/\sqrt{\varepsilon_o \mu_o}$, and so we have $\varepsilon_o \mu_o = c^{-2}$, from which

$$\varepsilon_o \mu_o = (299{,}792{,}458\ \text{m/s})^{-2}$$

The permeability of free space $\mu_o$ is defined to be exactly $4\pi \times 10^{-7}$ N $\cdot$ s$^2$/C$^2$. Using this value in the expression for $\varepsilon_o \mu_o$, we have the following value for $\varepsilon_o$, the permittivity of free space:

$$\varepsilon_o = 8.85 \times 10^{-12}\ \text{C}^2/\text{N} \cdot \text{m}^2$$

As pointed out in Sec. 18.10, the ampere is defined in terms of the force that one long, straight wire exerts on another parallel wire that carries an equal current. If the wires are a distance $d$ apart and the force per unit length is $F/L$, then the current in amperes is defined as

$$I = \sqrt{\frac{2\pi d}{\mu_o}\left(\frac{F}{L}\right)}$$

The coulomb is defined to be the charge carried through a cross section of a wire in one second when the current in the wire is one ampere. As a result, the definition of the coulomb is based directly on the measurement used to define the ampere.

To define the unit of the magnetic field vector, we make use of the relation

$$F = B_\perp IL$$

If a 1-m length of wire carries a current of 1 A perpendicular to a magnetic field and the force on that wire is 1 N, then the value of $B$ is, by definition, 1 T (or Wb/m$^2$), which is equal to 1 N $\cdot$ s/C $\cdot$ m.

The other quantities used in electricity have already been defined in terms of the quantities defined above, together with force, length, and time units. We shall not repeat them all here. However, we should point out that we have succeeded in defining all the electrical units in terms of definite experiments involving the measurement of force, length, and time. As a result, anyone who is able to duplicate our units for these three basic quantities will be able to duplicate our electrical units as well.

## Minimum learning goals

When you finish this chapter, you should be able to

1 Define (*a*) em wave, (*b*) radio wave, (*c*) radar or microwave, (*d*) infrared radiation, (*e*) light, (*f*) ultraviolet radiation, (*g*) x-rays, (*h*) gamma rays, (*i*) em wave spectrum, (*j*) Maxwell's equations, (*k*) intensity of an em wave.

2 Explain qualitatively how em waves are sent out from a radio station antenna.

3 Sketch the electric and magnetic fields in an em wave.

4 Describe two ways in which radio waves can be detected by a radio. Explain the function of the $LC$ circuit in the radio and how it is used to select a station.

5 Give the speed of em waves in vacuum.

6 Use $\lambda = v/f$ in simple situations.

7 Arrange a list of em wave types in order of decreasing wavelength. State the type of wave to which a given wavelength belongs.

8 When the value of $E$ or $B$ for an em wave is given, find the intensity of the wave.

9 Outline the method for defining $\mu_o$, $\varepsilon_o$, ampere, and coulomb.

10 Apply the inverse-square law for radiation to simple situations.

## Questions and guesstimates

1 Some radio stations have their transmitting antenna vertical, while others have theirs horizontal. Describe and compare the em waves generated by these two types of antennas. In particular, how are **B** and **E** directed relative to the earth's surface?

2 If you open up a transistor radio, you can see how its coil antenna is mounted. How could you use the radio to tell whether a distant station's antenna is vertical or horizontal?

3 Electromagnetic waves from most of the radio stations in the world are passing through the region around you. How does a radio or television set select the particular station you want to listen to? When you turn the dial on a radio, what is happening inside to select the various stations?

4 There are two types of radio and television receiving antennas in use. One picks up the electric part of the em wave, and the other picks up the magnetic part. Examine a

pocket transistor radio or a table radio and see which method is used. Is it possible to use both?

**5** From time to time, in the movies or on television, one sees the good guys trying to locate a clandestine radio transmitter by driving through the neighborhood with a device that has a slowly rotating coil on top. Explain how the device works.

**6** It is claimed that, in the vicinity of a very powerful radio transmitting antenna, one can at times see sparks jumping along a wire fence. What do you think of this claim?

**7** In microwave ovens, foods and utensils are subjected to very-high-frequency radar (em) waves. If a spoon is left in such an oven, it becomes very hot. What heats it? Can you explain the heating action in terms of the electric part of the wave? The magnetic? How are nonmetallic substances heated in the oven? Will a glass dish heat up in such an oven?

**8** There is some doubt about the safety of human exposure to intense radio waves and microwaves. Why would one expect the danger to depend on the frequency of the waves? Which would you expect to present more danger (if any), radio waves or microwaves?

**9** Refer to Fig. 21.9. Find the direction of the electric field at $a$ induced by the moving magnetic field.

**10** Refer to Fig. 21.10. Find the direction of the magnetic field at $P$ induced by the moving electric field.

**11** Are the direction and phase of the magnetic part of the wave in Fig. 21.5 drawn properly if the magnetic field is generated by the moving electric field? Repeat for the electric field generated by the moving magnetic field.

**12** Estimate the wavelength of the em wave generated by the vibration of a positively charged ball suspended as a pendulum. Compare this wavelength with the diameter of the earth, 12,700 km.

---

## Problems

**1** What is the wavelength of the em waves radiated by 60-Hz power?

**2** The pendulum of a grandfather clock is positively charged. What is the wavelength of the em waves emitted as the pendulum swings back and forth?

**3** When you are tuned to a radio station that is 161 km away, (a) how long does it take for an em signal from the station to reach you? (b) If the station operates at 1000 kHz, how many wavelengths away is it?

**4** One way to provide heat to muscles and other portions of the body is by *diathermy*. Radar waves are sent into the body much as a microwave oven sends waves into the material to be heated. The oscillating electric field of the wave causes dipolar molecules and ions in the body to move back and forth, thereby generating friction-type heat. The oscillating magnetic field of the wave induces emf's, which cause joule heating. Standard diathermy frequencies are 900 and 2560 MHz. (a) What are the wavelengths of em radiation in air that arise from these frequencies? (b) Can you see any problem in using diathermy on the face of someone whose teeth have metal fillings.

**5** An explosion occurs 3.0 km from an observer. How long after the observer sees the explosion will the sound from it be heard? (Take the speed of sound to be 340 m/s.)

**6** Galileo tried to measure the speed of light by stationing two people with lamps a distance $D$ apart. Person $A$ was to expose his lamp to person $B$; when $B$ saw the light from lamp $A$, he was to expose his lamp to $A$. The time

between $A$'s exposing his lamp and his seeing the light from lamp $B$ was to be measured. How large would $D$ have to be for this time to be 3.0 s?

**7** A radio is to be tuned to station WJR in Detroit, which operates at 760 kHz. If the radio's tuning circuit has an inductance of 0.030 mH, to what value must the capacitance be set to tune in this station?

**8** Television channel 5 has a frequency of about 80 MHz. If the inductance of a television's tuning circuit is 5.0 $\mu$H, what value must the tuning capacitor have for reception of channel 5?

**9** A certain bar magnet has $B = 0.65$ T at its end. The magnet is given a speed of 8.0 m/s, so that it moves perpendicular to its length. (a) As its end passes close to a certain point, how large an electric field is induced at the point? (b) Is this an easily observed electric field?

■ **10** Refer to Fig. 21.9. Assume that $v = 7.0$ m/s and that $B$ between the pole pieces is 0.40 T. (a) How large is the electric field at point $a$ at the instant shown? (b) Would this be easily observed? (c) Show that the direction of **E** at point $a$ is into the page.

■ **11** Refer to Fig. 21.10. Suppose the electric field at point $P$ due to the charge on the wire is $3 \times 10^4$ V/m and the speed $v = 7.0$ m/s. (a) What value does the induced magnetic field have at $P$? (b) Is this value large enough that the magnitude of the magnetic field could be measured easily? (c) What is the direction of the magnetic field at $P$?

■ **12** The electric field between the plates of an air-filled parallel-plate capacitor is $4 \times 10^4$ V/m. Suppose the capacitor is moving parallel to the plates at 6.0 m/s. (*a*) What is $B$ at a point through which the electric field is moving? (*b*) Would this be a strong enough magnetic field to measure easily?

**13** The amplitude of the electric field wave from a radio station should be greater than about $10^{-5}$ V/m if a radio is to respond well to it. What is the amplitude of the magnetic field wave when $E_o = 10^{-5}$ V/m?

■ **14** If the amplitude of the magnetic field wave in an em wave is to be 0.80 T, what must the amplitude of the electric field wave be? What would such a wave probably do to the air through which it is passing?

■ **15** The electric field wave of a radio wave has an amplitude of 0.80 mV/m at a certain point. What maximum voltage does the wave induce between the ends of a wire 30 cm long at that point? How must the wire be oriented to achieve this voltage? What is the amplitude of the accompanying magnetic field wave?

■ **16** An em wave has $E = 6.0 \times 10^{-4} \sin (8 \times 10^9 t)$ V/m. Write an equation for the magnetic field wave. What is the frequency of the wave? The wavelength?

■ **17** A certain em wave has $B = 3 \times 10^{-11} \sin (6 \times 10^9 t)$ T. (*a*) Find the frequency of the wave. (*b*) What is the change in $B$ as $t$ goes from zero to $\tau/4$, where $\tau$ is the period of the wave? (*c*) Find the average emf induced during this time interval in a loop of wire ($A = 8.0$ cm²) placed perpendicular to the magnetic field lines.

■ **18** A certain em wave has $E = 4.0 \times 10^{-3} \cos (1.5 \times 10^9 t)$ V/m. (*a*) Find the frequency of the wave. (*b*) What is the equation for the magnetic field in the wave? (*c*) Find the maximum emf induced in a metal rod 30 cm long held parallel to the electric field lines.

**19** In Bohr's picture of the hydrogen atom, the negative electron circles the positive nucleus with a frequency of $3.6 \times 10^{15}$ Hz. If this system obeys the laws studied in this chapter, it should act like an antenna radiating waves of this frequency. (*a*) Find the wavelength of the waves. (*b*) In which portion of the em spectrum should they be found?

**20** Many dipolar gas molecules (such as carbon dioxide and hydrogen chloride) act as two oppositely charged balls connected by a spring. For example, the HCl molecule has a natural resonant frequency for stretching vibrations of about $8 \times 10^{13}$ Hz. (*a*) Find the wavelength of em radiation that these vibrations should cause the molecule to

emit. (*b*) In which portion of the em spectrum are these waves found?

■ **21** A laser used in student laboratory experiments has a cylindrical beam that has a cross-sectional area of 3.0 mm². The power in the beam is 0.20 mW. Assuming the beam to consist of a single sinusoidal wave (not true), find the values of $E_o$ and $B_o$ in the beam.

■ **22** A searchlight sends out 5000 W of em radiation by means of a beam that has a cross-sectional area of 0.40 m². Assuming the beam to consist of a single sinusoidal wave (not true), find the values of $E_o$ and $B_o$ in the beam.

■ **23** The intensity of the wave from a distant 1.20-MHz radio station is $3.0 \times 10^{-11}$ W/m². Write the equations for the magnitudes of the electric and magnetic field waves in this region.

■ **24** A certain laser beam has a cross-sectional area of 2.0 mm² and a power of 0.80 mW. Find the intensity of the beam as well as $E_o$ and $B_o$ in it. Assume that the beam consists of a single sinusoidal wave.

■ **25** A certain 0.90-MHz radio transmitter sends out 50 W of radiation. Make the (poor) assumption that the radiation is uniform on a sphere with the transmitter at its center. (*a*) What is the intensity of the wave at 15 km from the source? (*b*) What are the amplitudes of the electric and magnetic field waves at this distance?

■ **26** A small light bulb hangs in the middle of a vacant lot. By what factor does the light intensity from the bulb decrease as one moves from a point that is 3.0 m from the bulb to a point that is 7.0 m from it?

■ **27** What is the approximate light intensity striking a table top that is 2 m from a 250-W bulb that is 8 percent efficient in generating light (that is, 8 percent of the power goes into light)? State any approximation you make and discuss its validity.

■ **28** The energy per second per square centimeter received at the earth from the sun is 0.134 J/cm² · s. (*a*) What is the intensity of the sun's radiation at the earth? (*b*) How much power does the sun radiate? The earth is $1.50 \times 10^{11}$ m from the sun, and the sun's radius is about $7 \times 10^8$ m.

■ **29** A very long, straight fluorescent light has an intensity $I_1$ at a radius $r_1$ and an intensity $I_2$ at $r_2$. If the two points are far from the ends and close to the bulb, show that $I_1/I_2 = r_2/r_1$.

# 22 The properties of light

IN this and the next two chapters, we are concerned primarily with a very small portion of the electromagnetic spectrum. This portion is composed of the small range of wavelengths to which the eye is sensitive, the wavelengths referred to as **light.** However, even though our primary concern is with light, much of what we learn is applicable to all em radiation.

## 22.1 The concept of light

Even in ancient times, the properties of light were a source of wonder and a stimulus to experimentation. The nature of light has always been a subject of great speculation. In Newton's time, nearly all scientists conducted scientific investigations into the properties and nature of light. Newton himself derived a great deal of his fame from his experiments with light.

In spite of this wide interest in light, its very nature remained in dispute until the first decade of the present century. During Newton's time, and for many years after, there was disagreement as to whether a light beam is a stream of "corpuscles" (particles) or a wave of some sort. Newton was a strong proponent of the corpuscular theory, and because of his prestige, many others were inclined to this view as well. In 1670 Christian Huygens, a contemporary of Newton's, was able to explain many of the properties of light by considering it to be wavelike. Both these ideas concerning the nature of light had their supporters.

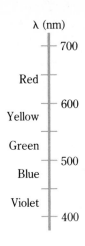

λ (nm)

700

Red

600

Yellow

Green

500

Blue

Violet

400

**FIGURE 22.1**

The correspondence between wavelengths and colors shown here is only approximate. Colors such as blue-green and orange occupy the intermediate regions. (See Color Plate II.)

It was not until 1803, when Thomas Young (and a little later Augustin Fresnel) presented evidence that light beams can interfere with one another much as sound waves do, that the wave theory of light became almost universally accepted. At about this time, the speed of light in water was measured. The observed speed was less in water than in air. This contradicted the corpuscular theory and supported the wave theory. Hence, by 1865, when Maxwell found theoretically that em waves should travel with the speed of light, the idea of light waves was fairly well accepted.

One would think, then, that by 1900 the nature of light was reasonably well understood. At that time, however, people still knew very little about the emission of light by atoms. It was not until about 1913 that Bohr gave the first reasonably correct interpretation of the mechanism of light emission. His concepts were greatly modified, and it was not until about 1930 that the emission of light could be said to be well understood. In addition, Einstein showed in 1905 that at least one property of light, the photoelectric effect, which is discussed in Chap. 25, is best explained by considering light to act as quanta, or particles. This concept has been expanded through the years until we now consider light to possess a sort of dual personality, part wavelike and part particlelike. More is said about these and other developments in later chapters.

It is clear that the subject of light has a long and varied scientific history. We expect that in years to come our understanding of the nature of light will continue to grow. For the next few chapters, however, it is sufficient to concentrate on those aspects of light that are evident from its em (wave) character. The characteristics of light that involve its particle nature and its atomic origin are discussed later in this book.

The wavelengths of visible light waves can be measured by methods discussed in Chap. 24. These wavelengths turn out to lie in the range 400 to 700 nm. The positions of the various colors on the wavelength scale are shown in Fig. 22.1 and in Color Plate II. When the complete spectrum of light shown in the upper part of the color plate enters the eye, as from a white-hot object, the light appears white and is called **white light.** (Objects that appear white in reflected light, such as this page, simply reflect the white light to our eyes.)

*Light waves are electromagnetic in nature, and thus consist of an electric field perpendicular to and in phase with a magnetic field,* as discussed in the preceding chapter. The electric field for a wave propagating in one direction in space is shown in Fig. 22.2. It is assumed that the wave is moving to the right along the *x* axis. Notice that the vibrating electric field is perpendicular to the *x* axis. Hence, *light waves are transverse waves* since the vibration is perpendicular to the direction of propagation. As such, they have many properties in common with waves on a string or waves on the surface of water since these too are transverse waves.

**FIGURE 22.2**

The electric field of an electromagnetic wave vibrates perpendicular to the direction of propagation. Hence the wave is transverse.

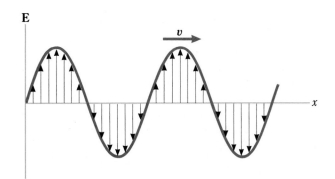

## 22.2 The speed of light

You will recall from Sec. 4.5 that, in the SI, the speed of light in vacuum is defined to be exactly $c = 299,792,458$ m/s, which we usually round off to $3.0 \times 10^8$ m/s. This definition was chosen so as to agree with the measured speed of light in terms of the meter, defined in Sec. 4.5. Let us see how the speed of light was accurately determined. The method we describe was used by Albert A. Michelson (1852–1931) to measure the time taken for light to travel between two mountaintops in California. The same method had been used over much smaller distances to measure the speed of light in materials other than air.

The method uses the apparatus shown in simplified form in Fig. 22.3. A beam of light from the source is reflected from one side of a cube that has mirrored surfaces on four sides. The beam is then reflected from mirror $M$ back to the cube, where it is reflected again as shown. If the cube is at just the right position, the beam will enter an observer's eye in the position indicated. (Actually the observer looks through a telescope system placed where the eye is shown and sees a reflection of the source.)

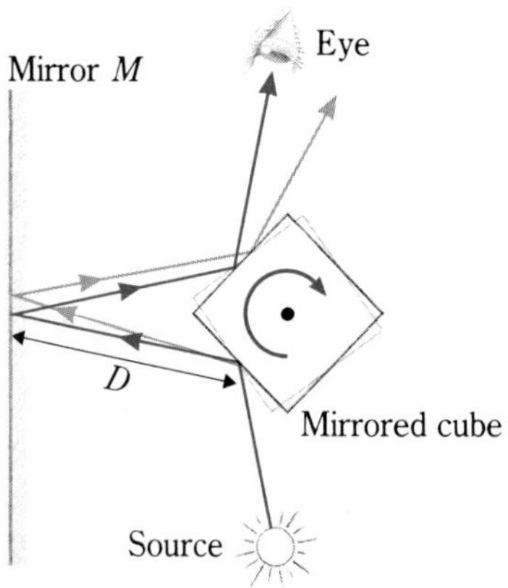

Mirror $M$  Eye

$D$

Mirrored cube

Source

Suppose, however, that the cube is rotating about an axis through its center, perpendicular to the page. When the cube is in the position indicated by the heavy lines in Fig. 22.3, the beam is reflected to the mirror as shown. By the time the beam returns to the cube from the mirror, however, the cube will have rotated, perhaps to the position represented by the lighter lines, and so the beam will not be reflected into the observer's eye. If the beam is to be reflected to the eye, the cube must rotate through 0.25 rev during the time the beam takes to travel to the mirror $M$ and back, for only then will the cube again be in a position given by the heavy lines in Fig. 22.3 and reflect the beam to the eye.

The measurement technique is to vary the speed of rotation of the cube until the reflected beam enters the eye. At that speed of rotation, we know that the time taken for one-fourth of a cube rotation is equal to the time the light takes to travel a distance $2D$. It is necessary to know only the speed of rotation of the cube and $D$ in order to compute the speed of the light. The result found is the same as that stated above.

Light travels fastest through vacuum. Its speed in other materials is always less than $c$. Moreover, its speed in materials other than vacuum depends on the wavelength of the light as well as on the material. We list in Table 22.1 the speed of light in various materials. (The wavelength used for the table is $\lambda = 589$ nm. This is the wavelength of the yellow light given off by a sodium-vapor lamp.)

### FIGURE 22.3

If the mirror is rotating at just the proper speed, the beam will be reflected into the eye of the observer. In practice, the distance $D$ is much larger than shown.

### ■ TABLE 22.1
**Speed of light at 589 nm**

| Material | Speed ($10^8$ m/s) |
| --- | --- |
| Vacuum | 2.99792 |
| Air | 2.9970 |
| Water | 2.25 |
| Ethanol | 2.20 |
| Benzene | 2.00 |
| Crown glass | 1.97 |
| Polystyrene | 1.89 |
| Flint glass | 1.81 |
| Diamond | 1.24 |

**The reflection of light**

When a stone is dropped into a large, still pond of water, a set of circular waves moves out from the point where the stone hit the water. We are all familiar with this situation, but let us look at it in some detail. The circular waves, or **wavefronts,** are shown in Fig. 22.4. They travel outward from the center in the directions shown by the arrows. These arrows in the direction of wavefront travel are called **rays.** Notice that they are always perpendicular to the wavefronts. Hence we can specify the motion of a wave either by use of rays or by drawing the wavefronts. Both methods are of value.

**FIGURE 22.4**

The rays are perpendicular to the wavefronts and show the direction of propagation of the wave.

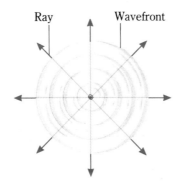

Ray    Wavefront

In Fig. 22.5, we see what happens when the wavefront is some distance away from the wave source. (It is 1 m away in this figure.) Notice that the rays are nearly parallel to each other; we refer to the limiting case of a far distant light source as *parallel light.* Also notice that the wavefronts are nearly straight lines. In three dimensions, the wavefronts are nearly planar. Hence for a far distant source, we refer to such waves as *plane waves.* Of course, the terms *parallel light* and *plane waves* are synonyms. It is often convenient to use plane waves in computations. When the source is far removed or when a suitable lens is used, the waves are very nearly of this sort.

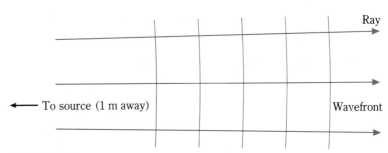

Ray

To source (1 m away)

Wavefront

**FIGURE 22.5**

Rays from a distant source are nearly parallel. Notice that the wavefronts are nearly flat. For an infinitely far away source, these are plane waves and the rays are parallel.

Suppose that a series of plane water waves is incident on a flat wall, as in Fig. 22.6a. The velocity of an incoming wave can be resolved into two components, $\mathbf{v}_\perp$ perpendicular to the wall and $\mathbf{v}_\parallel$ parallel to the wall. Upon striking the wall, $v_\perp$ is

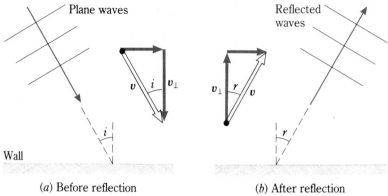

(a) Before reflection          (b) After reflection

**FIGURE 22.6**

The incident wave is reflected in such a way that· the angle of incidence $i$ equals the angle of reflection $r$.

reversed in direction, as shown in $b$. As a result, the wave is reflected backward from the surface. Let us now find how the angle of incidence $i$, shown in $a$, is related to the angle of reflection $r$, shown in $b$.

As you can see from Fig. 22.6a,

$$\cos i = \frac{v_\perp}{v}$$

In Fig. 22.6b, we see that

$$\cos r = \frac{v_\perp}{v}$$

Therefore, because the cosines are equal, the angle of incidence equals the angle of reflection.

The fact that a water wave is reflected in such a way that the angle of incidence equals the angle of reflection is of general validity. We could use the same reasoning to show that light waves are also reflected in this way. Notice that the only basic assumption made was that, upon reflection, the velocity component perpendicular to the surface was reversed and the parallel component was unchanged. Our result is true for any type of wave for which this assumption is true. Measurements on light and other forms of electromagnetic radiation confirm our deduction. We may therefore formulate the following rule:

The angle of incidence equals the angle of reflection.

The type of reflection shown in Fig. 22.7a, where the reflecting surface is perfectly smooth, is called *specular reflection*. Specular reflection occurs at a mirrorlike surface. Rougher surfaces, such as paper or painted walls, give rise to *diffuse reflection*, shown in Fig. 22.7b. For such surfaces, even though the law of reflection applies to individual rays in a light beam, the nonsmooth surface causes the rays to be reflected at various angles from the average plane of the surface.

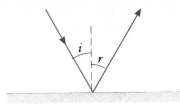

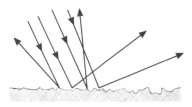

(a) Specular reflection

(b) Diffuse reflection

**FIGURE 22.7**

The angle of incidence equals the angle of reflection for each ray. A flat surface reflects all rays parallel to each other, thus giving rise to specular reflection. A rough surface diffuses the rays upon reflection, thus resulting in diffuse reflection.

## 22.4 Plane mirrors

Let us now apply what we have learned about reflection to the important topic of image formation by mirrors. First we consider how a plane mirror (one that is flat) forms an image.

Every day you look at yourself in a plane mirror and see an image of your face in front of you. If you stop to examine exactly what you are seeing, you perceive the image of your face as being behind the surface of the mirror. In fact, the image appears to be about as far behind the mirror as your face is in front of the mirror. Let us now examine such a reflection in order to understand clearly why the image is seen as it is.

Suppose you place an object in front of a mirror, as in Fig. 22.8a. You wish to find where your eye will perceive the image of the object as being. When you see the tip of the object, your eye sees a ray of light that was either emitted or reflected by the object tip. When you see the image of the object tip in the mirror, you are seeing light from the object tip that was reflected by the mirror, as in Fig. 22.8a. It is apparent from this same figure that you see the object as being off in the general direction from which the ray came after it was reflected by the mirror. Hence you know that the image of the object tip is somewhere along the line *EM* or its extension.

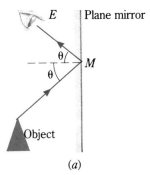

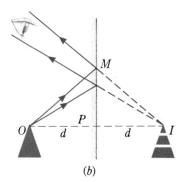

**FIGURE 22.8**

The image *I* formed by the plane mirror is as far behind the mirror as the object *O* is in front of it.

(a)

(b)

To see exactly where the image is, refer to Fig. 22.8b. The two reflected rays shown appear to the eye to come from the single point *I*. It is at this point that the image of the object tip appears to be. In fact, the same reasoning may be used to show that the total image of the object appears as shown at *I*. From the geometry of the triangles *OMP* and *IMP*, it is clear that, if the object is a distance *d* in front of the mirror, the image is the same distance *d* behind the mirror.

This type of image, one through which the observed rays do not actually pass, is called a **virtual** or an **imaginary image.** In other words, the rays reaching the eye do not really come from the point at which we see the image. There is no possibility whatsoever that an image of the object would appear on a sheet of paper placed at *I* behind the mirror. The mind merely imagines that the light comes from *I.* It is always true, of course, that the image of an object seen by reflection in a plane mirror is a virtual image. The image is always exactly as far behind the mirror as the object is in front of it.

## 22.5 The focus of a concave spherical mirror

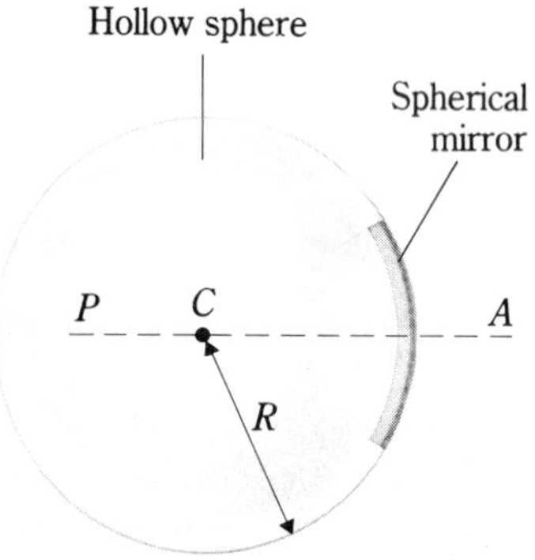

Plane mirrors are used by all of us, but spherical mirrors are not quite so common. However, makeup and shaving mirrors are portions of spheres, as are surveillance mirrors in stores. A spherical mirror is actually a portion of the surface of a hollow sphere, as shown in Fig. 22.9. The line *PA* is the *principal axis* of the mirror. If light is reflected from the inside surface of the sphere, as in Fig. 22.10*a*, the mirror is called a *concave* mirror. If light is reflected from the outside surface of the sphere, as in Fig. 22.10*b*, the mirror is called a *convex* mirror.

In drawing Fig. 22.10, we assumed that the light comes from a distant source so that the rays are parallel and the wavefronts are planes. As indicated in *a*, parallel rays that travel along the principal axis of a concave mirror are all reflected to a point *F*. This is approximately correct, as we shall soon show. The point to which the light from a distant object is reflected by a concave mirror is called the **focus** (or the **focal point**) of the concave mirror. If we were to reflect sunlight from a concave mirror, the light would be reflected into a tiny spot (an image of the sun) very near the focus of the mirror. This follows because the light waves from a source as distant as the sun are essentially planar by the time they reach the earth. Figure 22.10*b* shows what happens to parallel rays reflected from a convex mirror. The reflected rays appear to come from a point *F* behind the mirror. This point is the focus (or focal point) for a convex mirror.

**FIGURE 22.9**

A spherical mirror is a portion of a hollow sphere. The radius of the mirror is *R*, its center of curvature is point *C*, and its principal axis is the line *PA*. The principal axis goes through the center of curvature and the center point of the mirror's surface.

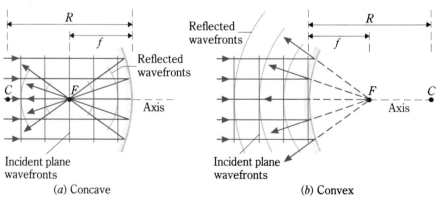

(*a*) Concave

(*b*) Convex

**FIGURE 22.10**

The reflected light from a concave mirror is converged to the focus *F*, while the light reflected from a convex mirror appears to diverge from the focus.

It is a simple matter to demonstrate how three special rays of the infinite number of rays from a source are reflected by a concave mirror. Once this is known, the position of an image formed by such a mirror becomes quite easy to ascertain. We now consider how a concave mirror reflects these three important rays.

First consider the ray shown in Fig. 22.11. It is parallel to the principal axis of the mirror and strikes the mirror at $A$. Let us draw a radius $CA$ from the center of the sphere of which the mirror is a part. Then, since $CA$ is a radius of the sphere, it is perpendicular to the mirror surface at $A$. The reflected ray makes the same angle to the perpendicular at $A$ as the incident ray does, since the angle of incidence at any smooth reflecting surface equals the angle of reflection. Moreover, since the incoming ray is parallel to the axis $CB$, the angle at $C$ must also be $i$.

**FIGURE 22.11**

A ray parallel to the principal axis of a concave mirror is reflected through the focal point.

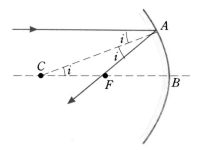

The triangle $CFA$ is isosceles, and $CF$ must therefore be equal to $FA$. If the angle $i$ is quite small, however, the length $FA$ is nearly the same as $FB$. Hence any ray parallel to the principal axis and not too far from it is reflected through a point $F$ that is halfway between the mirror surface $B$ and the center of the sphere of which the mirror is a part. In other words, if the mirror has a radius of curvature $R$, rays parallel to its axis are reflected through the point $F$, which is a distance $FB = R/2$ from the mirror. This point, the point to which parallel rays are focused, is the focus of the mirror, and the length $FB$ is called the **focal length** $f$ of the mirror: $f = R/2$.

It is not strictly true that all rays parallel to the principal axis are focused exactly at the focal point. An approximation was made in arriving at this result. If the length $AB$ in Fig. 22.11 is only a small fraction of the sphere's diameter, the approximation is fairly good. However, you should draw the case of a ray reflected from a point far away from the principal axis. It is not reflected through the focus. For this reason, the dimensions of a spherical mirror should always be much smaller than the diameter of the sphere from which it is made. We call this defect in the focusing action of spherical mirrors *spherical aberration*. Parabolic mirrors do not have this disadvantage, but they are more expensive to make.

## 22.6 | Three reflected rays and image formation

We now know that a ray that is parallel and near to the principal axis of a spherical mirror may be considered to reflect through the focal point. There are two other rays that are also easily traced. All three of these rays are shown in Fig. 22.12.

Ray 1 has already been discussed. Ray 2 comes through the focus and is reflected parallel to the axis. It is just the reverse of ray 1, and the geometric arguments used for ray 1 apply to it as well. Ray 3 passes through the center of curvature of the mirror before it strikes the mirror. Since it is traveling along a radius of the mirror

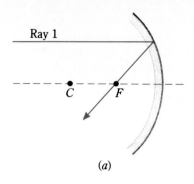

(a)

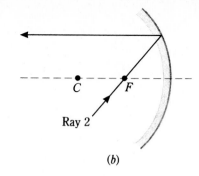

(b)

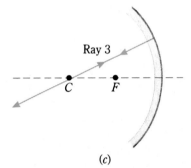

(c)

FIGURE 22.12

The three rays are easily drawn with a straightedge provided the center *C* and focus *F* of the mirror are known.

surface, it hits the mirror perpendicularly. Hence it is reflected straight back on itself. We therefore have the following rules *for concave mirrors:*

A ray parallel to the mirror axis is reflected through the focus.

A ray through the focus is reflected parallel to the mirror axis.

A ray through the center of curvature is reflected through the center of curvature.

These rules can be used to locate the position of images.

Suppose we wish to find the image of the object *O* formed by the mirror shown in Fig. 22.13. Let us say that the object is a light bulb. The bulb emits light in all directions, but we know how to treat only three of the millions of rays we could draw coming from it. These three rays travel as shown in Fig. 22.13. They are exactly the rays described by our rules, and you should trace each to see that it is properly drawn. Once the positions of *C* and *F* are known, only a straightedge is needed to draw these rays.

If you place your eye as shown in Fig. 22.13, the three rays appear to come from point *I.* In other words, you see an image of the light bulb at *I.* Moreover, since the rays actually do converge on point *I* and pass through it, a sheet of paper placed at *I* will show a lighted picture of the original bulb. This is a **real image:** *at a real image, the light actually passes through the image and reproduces the object.* Notice how this differs from the imaginary, or virtual, image found for the plane mirror.

Suppose the bulb is on the tip of the post represented in Fig. 22.13. Rays of light emanate from the post as well as from the bulb. (For example, light from the bulb might strike the post and be reflected from it. This light would appear to be emitted from the post, and it would act as if it were emitted rather than reflected light.) We

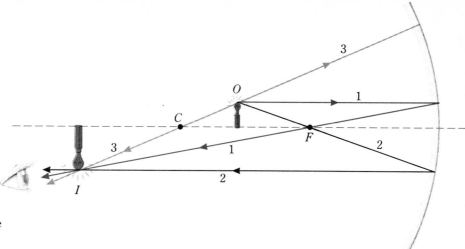

**FIGURE 22.13**

A real image *I* is formed from the object *O*. Trace the three rays from the object.

could treat each little portion of the post as a new light source and find its image. You should trace a few rays to show yourself that the image of the post lies along the image shown at *I* and that for this reason the image is upside down. Usually, once one point on the image has been found, the image position is obvious.

Consider the situation in Fig. 22.14, in which the distance between the object and the mirror is less than the focal length. Once again, we draw our three rays. Now, however, ray 2 does not go through the focus on its way to the mirror since the object is inside the focal point. However, it still appears to come from the focus and is reflected parallel to the axis, as always. Notice that all three rays appear to come from the image *I*. Hence, the eye sees the image as being behind the mirror in this case. Notice that the image is virtual (imaginary), erect (right side up), and magnified (larger than the object).

**FIGURE 22.14**

The three rays appear to come from the virtual image *I*. Notice especially rays 2 and 3 so that you can draw them in other situations.

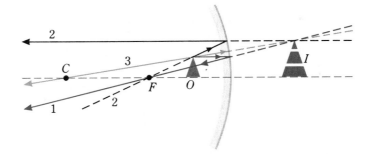

## 22.7 The mirror equation

To derive a mathematical equation that describes the location of an image, let us refer to Fig. 22.15. The distance *p* from the object to the mirror is called the **object distance.** The height of the object is called *O*. The height of the image is called *I*, and its distance from the mirror, the **image distance,** is called *i*. Notice that the distance *BF* from the mirror to the focal point is the focal length *f* of the mirror. The ray *ABE* in part *a* is not one of our usual three. However, it is reflected in such a

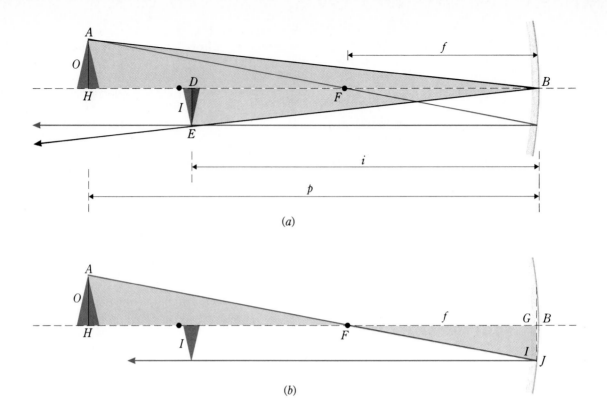

(a)

(b)

**FIGURE 22.15**

(*a*) Triangle *ABH* is similar to triangle *EBD*. (*b*) Triangle *AFH* is similar to triangle *JFG*. In the text, we assume the curvature of the mirror to be so small that distance *GB* is negligible.

way that angle *ABH* equals angle *DBE*. For this reason, the shaded triangles *ABH* and *DBED* in *a* are similar. Taking the ratios of corresponding sides gives

$$\frac{O}{I} = \frac{p}{i}$$

In Fig. 22.15*b*, the shaded triangles are also similar. Hence

$$\frac{O}{I} = \frac{HF}{FG}$$

But *HF* is just $p - f$, and *FG* is nearly *f*. (They differ by the distance *GB*.) To this approximation, we have

$$\frac{O}{I} = \frac{p - f}{f}$$

Equating this to the expression found in part *a* gives

$$\frac{p}{i} = \frac{p - f}{f}$$

After dividing this equation by *p* and rearranging, we have

$$\frac{1}{p} + \frac{1}{i} = \frac{1}{f} \qquad (22.1)$$

where $f = R/2$ and $R$ is the radius of curvature of the mirror.

Equation 22.1 is the **mirror equation.** It allows us to compute the distance $i$ of the image from the mirror surface, provided the object distance from the mirror surface $p$ and the focal length $f$ are known. To compute the relative heights of the object and the image, we note that $O/I = p/i$, as found previously. The **magnification** produced by the mirror is defined to be the ratio of the image height to the object height:

$$\text{Magnification} = \frac{I}{O} = \frac{i}{p} \qquad\qquad (22.2)$$

---

### Example 22.1

An object 2 cm high is placed 30 cm from a concave mirror with a radius of curvature of 10 cm. Find the position and size of the image.

*Reasoning*   A somewhat similar situation is shown in Fig. 22.15. The mirror relation applies, with $p = 30$ cm and $f = \frac{10}{2} = 5$ cm. Hence, with all distances in centimeters,

$$\frac{1}{30} + \frac{1}{i} = \frac{1}{5}$$

$$\frac{1}{i} = \frac{6}{30} - \frac{1}{30} = \frac{5}{30}$$

$$i = 6.00 \text{ cm}$$

As shown in Fig. 22.15, the image is on the same side of the mirror as the object, that is, on the side from which the incident light comes, the front side of the mirror. Because the light actually passes through and forms the image, the image is real. Its size is found from Eq. 22.2 to be, in centimeters,

$$\frac{I}{2} = \frac{6}{30}$$

$$I = \tfrac{2}{5} \text{ cm high}$$

It is always wise to check the algebraic solution by drawing the appropriate ray diagram.  ■

---

### Example 22.2

An object is placed 5.0 cm in front of a concave mirror with a 10-cm focal length. Find the location of the image.

*Reasoning*   If we refer to Fig. 22.14, we see that this example should give an image behind the mirror. Our answer should make this evident. Using the mirror equation, taking all distances in centimeters, we have

$$\frac{1}{5} + \frac{1}{i} = \frac{1}{10}$$

$$\frac{1}{i} = \frac{1}{10} - \frac{2}{10} = -\frac{1}{10}$$

$$i = -10.0 \text{ cm}$$

Notice that $i$ is negative. Hence, when an image is behind the mirror, the image distance is negative. This is not surprising in view of the fact that $i$ is taken as positive for the reverse case. An image behind the mirror is, of course, virtual. It does not appear on a sheet of paper placed there. ▪

*Exercise* What is the magnification, $I/O$? *Answer: 2.0* ▪

## 22.8 Convex mirrors

A convex spherical mirror is a portion of a sphere that reflects rays from outside the sphere, as shown in Fig. 22.10b. We saw there how parallel rays are reflected from such a mirror. They appear to diverge from a point behind the mirror. *Parallel rays incident on a convex mirror are reflected as though they came from the focal point.* To prove this, we proceed in much the same way as for the concave mirror.

Referring to Fig. 22.16, we see from the law of reflection and the geometry involved that several angles are equal. The triangle $AFC$ is isosceles, so that $AF = FC$. If the length $AB$ is small compared with the radius of curvature of the mirror, then $AF$ is nearly equal to $BF$. Consequently, $BF$ nearly equals $FC$, and so here too the focal point can be considered to be midway between the mirror and its center of curvature.

**FIGURE 22.16**

A ray parallel to the axis is reflected as though it came from the focal point.

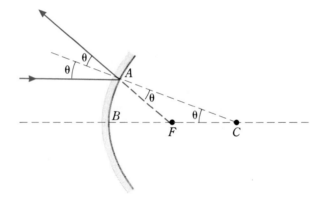

We are therefore able to write rules for drawing our three rays with a *convex mirror:*

A ray parallel to the axis is reflected as though it came from the focal point.

A ray heading toward the focal point is reflected parallel to the axis.

A ray heading toward the center of curvature is reflected back on itself.

These three rays are illustrated in Fig. 22.17. You should trace them to see that they conform to these rules. Notice that all three reflected rays appear to come from the image $I$ behind the mirror. As you see, the image is virtual, upright, and diminished in size.

The algebraic relation used in locating the image for a convex mirror can be obtained by reference to Fig. 22.18. You should be able to show that triangle $ABH$ is similar to $EBD$ in part $a$ and that triangle $JFG$ is similar to triangle $EFD$ in $b$. This

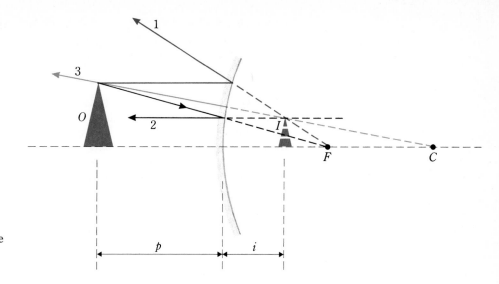

**FIGURE 22.17**
You should be able to draw the
three rays for any situation
involving a convex mirror.

being true, the following equations are found, as in the case of the concave mirror:

$$\frac{O}{I} = \frac{p}{i} \quad \text{and} \quad \frac{O}{I} = \frac{f}{f - i}$$

In writing these equations, the distance $BG$ was considered negligibly small.

**FIGURE 22.18**
Triangles $HBA$ and $DBE$ are
similar, as are triangles $DFE$ and
$GFJ$. We make the assumption that
the length $FG$ is essentially the
same as $FB$.

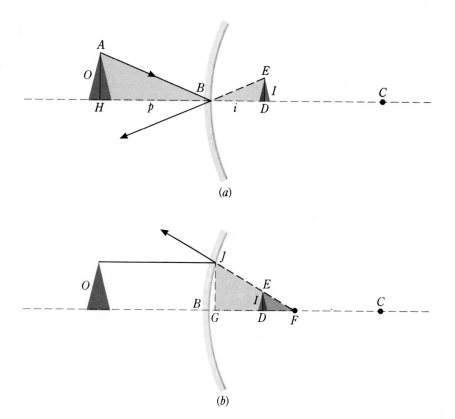

Equating the two expressions, inverting, dividing by $i$, and rearranging yield

$$\frac{1}{p} - \frac{1}{i} = -\frac{1}{f}$$

Notice that, except for signs, this equation is the same as Eq. 22.1 for the concave mirror. The difference in signs alerts us to the fact that the image in this case is behind the mirror rather than in front of it. Also, the negative focal-length term is the result of the mirror's being convex rather than concave.

Rather than remembering two mirror equations, we can set up rules that allow us to use Eq. 22.1 even for convex mirrors. If we agree to always call image distances behind a mirror, that is, virtual-image distances, negative, we can omit the negative sign from the $i$ term in the convex-mirror equation. Moreover, if we always say that the focal length of a convex mirror is negative, we can omit the other negative sign as well. Hence we can write for *all* mirrors

$$\frac{1}{p} + \frac{1}{i} = \frac{1}{f} \qquad \text{mirrors} \tag{22.1}$$

where we agree that

**1** Object distances are positive if the object is in front of the mirror and negative otherwise.

**2** Image distances are positive if the image is in front of the mirror and negative otherwise.

**3** The focal length is positive for a concave mirror and negative for a convex mirror.

In addition, we can use Eq. 22.2 for the magnification in either case without use of any negative signs whatsoever.

---

### Example 22.3

A convex mirror with a 100-cm radius of curvature is used to reflect the light from an object placed 75 cm in front of the mirror. Find the location of the image and its relative size.

*Reasoning*  This situation is approximately that shown in Fig. 22.18. Since the mirror is convex, $f = -R/2 = -50$ cm. Using the mirror equation and taking all distances in centimeters, we have

$$\frac{1}{75} + \frac{1}{i} = -\frac{1}{50}$$

$$i = -30.0 \text{ cm}$$

The negative sign tells us that the image is behind the mirror, and it is, of course, virtual. Its size relative to the height of the object is, from Eq. 22.2,

$$\frac{I}{O} = \frac{i}{p} = \frac{30}{75} = 0.400 \quad \blacksquare$$

---

*Exercise*  Check this solution graphically to see that no mistake has been made.  ■

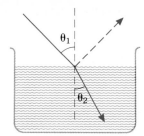

**FIGURE 22.19**

When a ray of light passes from air to water, the ray is refracted toward the normal to the surface.

## 22.9 The refraction of light: Snell's law

When a beam of light enters water from air, its path bends as shown in Fig. 22.19. This change in the direction of a ray as it passes from one material to another is called **refraction**. The angle $\theta_1$ is, of course, the angle of incidence, and angle $\theta_2$ is called the **angle of refraction**. Some of the light beam hitting the water surface is also reflected, as shown by the dashed ray in Fig. 22.19, but we ignore this reflection in this section.

In order to find a relation between the angle of incidence $\theta_1$ and the angle of refraction $\theta_2$, it is convenient to consider the motion of the wavefronts in a plane wave, as shown in Fig. 22.20. We assume that the speed of the wave is $v_1$ in material 1 and $v_2$ in material 2, with $v_1$ greater than $v_2$. The wavefronts have a bend in them at the interface between the two materials because the wave travels more slowly in material 2 than in material 1.

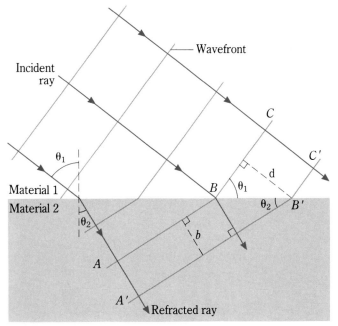

**FIGURE 22.20**

Because the wave travels more slowly in the lower material than in the upper one, distance $AA'$ is smaller than distance $CC'$.

Suppose it takes a time $t$ for wavefront $ABC$ to travel to the position $A'B'C'$. The distance $b$ the wavefront moves in material 2 in time $t$ is $b = v_2 t$, and the distance the front moves in material 1 is $d = v_1 t$. If we divide $d$ by $b$, we find that

$$\frac{d}{b} = \frac{v_1}{v_2}$$

Moreover, we see in the figure that

$$\frac{d}{BB'} = \sin \theta_1 \quad \text{and} \quad \frac{b}{BB'} = \sin \theta_2$$

which, after division of one by the other, gives the relation

$$\frac{d}{b} = \frac{\sin \theta_1}{\sin \theta_2}$$

Since $d/b = v_1/v_2$, the relation becomes

$$\frac{\sin \theta_1}{\sin \theta_2} = \frac{v_1}{v_2} \tag{22.3}$$

Although this equation describing the refraction phenomenon is helpful, a more useful form can be obtained if we define a quantity called the **index of refraction** $n$ of a material:

$$n = \frac{\text{speed of light in vacuum}}{\text{speed of light in material}} = \frac{c}{v} \tag{22.4}$$

Because light has its highest speed in vacuum, the index of refraction is always unity or larger.* Typical values for $n$ are listed in Table 22.2. As you see there, the refractive index for air is very close to unity, while the index for diamond is large, 2.42. Of course, the index of refraction for vacuum is exactly 1. The index of refraction varies slightly with wavelength, as we shall see later; it is larger for blue light than for red light.

■ **TABLE 22.2   Refractive indices at 589 nm**

| Material | $c/v = n$ | Material | $c/v = n$ |
|---|---|---|---|
| Air* | 1.0003 | Crown glass | 1.52 |
| Water | 1.33 | Sodium chloride | 1.53 |
| Ethanol | 1.36 | Polystyrene | 1.59 |
| Acetone | 1.36 | Carbon disulfide | 1.63 |
| Fused quartz | 1.46 | Flint glass | 1.66 |
| Benzene | 1.50 | Methylene iodide | 1.74 |
| Lucite or Plexiglas | 1.51 | Diamond | 2.42 |

* Standard temperature and pressure.

Because $v = c/n$ from the definition of refractive index, we can rewrite Eq. 22.3 as

$$\frac{\sin \theta_1}{\sin \theta_2} = \frac{c/n_1}{c/n_2} = \frac{n_2}{n_1}$$

which can be rewritten as

---

* We are speaking here of the group velocity, the velocity with which light transports energy.

$$n_1 \sin \theta_1 = n_2 \sin \theta_2 \qquad (22.5)$$

which we refer to as **Snell's law**.

We can see that if $n_2$ is greater than $n_1$, then $\sin \theta_1$ is larger than $\sin \theta_2$. In this case, $\theta_1$ is larger than $\theta_2$. This is the instance illustrated in Fig. 22.21a and is the most common one. Sometimes, however, we are interested in the reverse case, where $n_2$ is smaller than $n_1$. This is applicable to a beam of light going from glass to air, for example. Under these circumstances, Eq. 22.5 predicts that $\theta_2$ is larger than $\theta_1$, as shown in Fig. 22.21b.

If $n_2 > n_1$, the beam bends toward the normal; if $n_2 < n_1$, the beam bends away from the normal.

**FIGURE 22.21**

(a) If $n_2 > n_1$, the beam bends toward the normal. (b) If $n_2 < n_1$, the reverse is true.

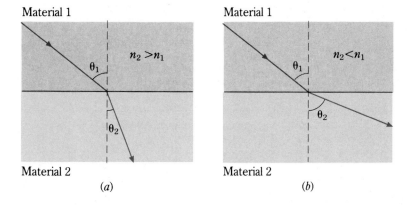

Material 1     $n_2 > n_1$     $\theta_1$     $\theta_2$     Material 2     (a)

Material 1     $n_2 < n_1$     $\theta_1$     $\theta_2$     Material 2     (b)

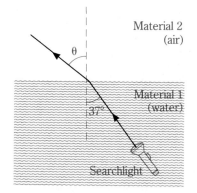

**FIGURE 22.22**

The underwater searchlight sends out a beam that bends away from the normal as it passes into the air.

### Example 22.4

A diver beneath the surface of the ocean shines a bright searchlight up at an angle of 37° to the vertical. At what angle does the light emerge into the air?

*Reasoning*  The situation is shown in Fig. 22.22. Notice that material 1 is water and material 2 is air. Applying Snell's law and using $n_1 = 1.33$ and $n_2 = 1.00$ (from Table 22.2), we have

$$1.33 \sin 37° = 1.00 \sin \theta$$

$$\sin \theta = 0.80$$

$$\theta = 53° \quad \blacksquare$$

*Exercise*  Find the refraction angle in water for light entering the water from air at an angle of incidence of 53°. *Answer: 37°*  ∎

### Example 22.5

Light is incident upon the surface of the water in a level, flat-bottomed glass dish as shown in Fig. 22.23 on the next page. At what angle does the light emerge from the bottom of the dish?

*Reasoning*  Up to this point, we have considered only what happens to light that moves from medium 1 into medium 2 and then stays in medium 2. Here, however, we have a situation in which the light ends up in the medium in which it started out. The result is quite different, as we shall see.

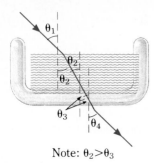

Note: $\theta_2 > \theta_3$

**FIGURE 22.23**

The direction of travel of a light beam is not altered by a parallel plate of transparent material.

At the air-water interface we have

$$1.00 \sin \theta_1 = n_w \sin \theta_2$$

At the water-glass interface, we have

$$n_w \sin \theta_2 = n_g \sin \theta_3$$

Since quantities equal to the same thing are equal to each other,

$$1.00 \sin \theta_1 = n_g \sin \theta_3$$

Notice that this equation is exactly the relation we would find if the water were not present and the light went directly into the glass from the air.

Proceeding, we have at the glass-air interface

$$n_g \sin \theta_3 = 1.00 \sin \theta_4$$

Combining this with the previous equation gives

$$\sin \theta_1 = \sin \theta_4 \qquad \text{from which } \theta_1 = \theta_4$$

This important result shows that *a uniform layer of transparent material does not change the direction of a beam of light* that first enters and then exits from the layer. The beam is usually slightly displaced sideways, however. (Why?) ▪

## 22.10 Total internal reflection

Aside from their romantic value, diamonds owe a great deal of their beauty to the phenomenon of total internal reflection. It is this property that causes diamonds to sparkle in all directions. What happens is that a beam of light becomes trapped within the diamond. When it finally does emerge, it can be emitted in any one of many directions. Hence the diamond gives off light (or sparkles) in random directions. There are many other cases in which total internal reflection is of importance. We now investigate this type of behavior of light in some detail.

Consider a light source $O$ below the surface of a lake, as in Fig. 22.24$a$. As ray $B$ passes from water to air, it is refracted away from the normal to the water surface; hence $\theta_2 > \theta_1$. Of course, some reflection also occurs at the surface, and $B'$ is the reflected portion of the ray. Ray $C$, which has a larger angle of incidence than $B$, is

**FIGURE 22.24**

When $\theta_1$ is greater than the critical angle $\theta_c$, the beam is totally internally reflected.

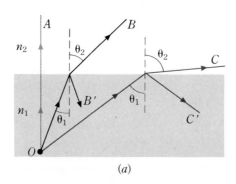

$(a)$

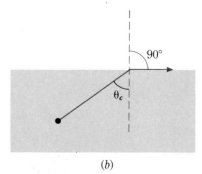

$(b)$

522    22 The properties of light

refracted nearly parallel to the water surface. In this case, the reflected ray $C'$ is more intense than the refracted ray.

Consider the case shown in Fig. 22.24$b$, where the angle of refraction is exactly 90° and the angle of incidence is $\theta_c$, called the *critical angle*. At incidence angles greater than $\theta_c$, the beam cannot leave the water; it is all reflected and there is no refracted beam. This situation is known as *total internal reflection*. Note that it can occur only if the light is moving from a material of high index of refraction to one of lower index of refraction. We can find the critical angle from Snell's law because $\theta_2 = 90°$ when $\theta_1 = \theta_c$:

$$n_1 \sin \theta_c = n_2 \sin 90°$$

$$\sin \theta_c = \frac{n_2}{n_1}$$

Typical critical angles are 49° for water, 42° for glass, and 24° for diamond. Since diamond has such a high index of refraction, the critical angle is quite small. Hence the beam of light within a diamond must hit the surface nearly straight on if it is to emerge into the air. The jeweler cuts a diamond in such a way that once a beam of light gets inside, the chance that it will strike a surface at an angle of 24° or less is quite small. As a result, the trapped beam reflects many times within the diamond before it is able to escape.

Total internal reflection makes it possible to "pipe" light around corners. Light that enters one end of a gently curved glass rod is totally internally reflected around the curve, as shown in Fig. 22.25$a$. When a bundle of such curved rods (which are called **optical fibers**) is used, the composite picture of an object can be piped from one place to another. Such a device is called a light pipe (Fig. 22.25$b$).

In recent years, optical fibers have been used in telecommunications. In this

**FIGURE 22.25**

($a$) Light is caused to follow a glass fiber by total internal reflection. ($b$) A glass-fiber gastroscope attached to a camera. A light source (outside the picture at the left) supplies light to the fiber bundle at the bottom. This light pipe is inserted through the throat into the stomach. Light reflected from the stomach wall is reflected up through the central fibers of the bundle and forms an image on the film of the camera. Often the camera is not used and the light is observed directly by eye. (*American Optical Corp., Fiber Optics Div.*)

($a$)

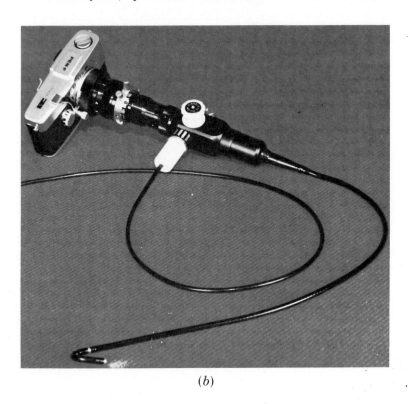

($b$)

application, modulated laser beams are used to carry the signals in place of the electric currents and radio waves of the conventional method used by telephone companies. Such an application has been made feasible by the development of fibers with extremely low energy losses. Because the frequency of light waves is much higher than that of conventional electric currents and radio waves, much more information can be transmitted per unit time on an optical beam in a fiber than through a conventional wire or on a comparable radio wave beam.

## 22.11 Lenses

The phenomenon of refraction finds its most useful application in lenses. A properly constructed lens is capable of focusing a beam of parallel light into a small region at a focal point. The mechanism by which this is done is illustrated in Fig. 22.26. We recall that a light wave travels more slowly in glass than in air. Hence, the central portion of the incident plane wave in Fig. 22.26a falls behind the outer portions because it travels a greater distance in the glass. The emergent wave is therefore curved as shown. Since the rays, that is, the direction of travel of the light, are perpendicular to the wavefronts, the light is converged toward point $F$ and this type of lens is called a **converging lens.** The central line through $F$ is called the *axis* of the lens.

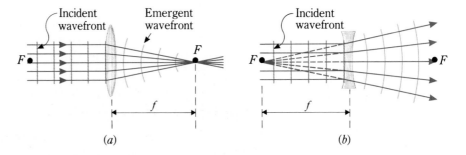

**FIGURE 22.26**
(*a*) Parallel rays are converged to the focal point by the converging lens. (*b*) They are diverged and appear to come from the focal point in a diverging lens.

Although we do not prove it here, the various rays converge to a single point, as shown, if the surfaces of the lens are portions of spheres. However, this is only an approximation, and it becomes a poor one if the lens surfaces are highly curved — in other words, if they constitute more than a very small portion of a sphere. Moreover, we assume the lens to be relatively thin. In these circumstances, the rays parallel to the axis are converged nearly to a point. *The point to which parallel rays are converged by a converging lens is called the* **focal point** *of the lens.*

There is a simple way to determine the position of the focal point of a converging lens. If sunlight is passed through it, an image of the sun is formed. Since the sun is very far away, the wavefronts from it are essentially planar and the rays are parallel. Hence, the place where the image of the sun is formed is the focal point of the lens.

A different type of lens is shown in Fig. 22.26b. Notice that, since this lens is thinner in the middle than on the edges, the outer portion of the wave falls behind the central portion. Now the emergent wave is spherical but diverging from the lens and this type of lens is called a **diverging lens.** Under the same restrictions stated for the converging lens, this diverging wave appears to come from point $F$. *This point, from which the original parallel beam appears to diverge, is called the* **focal point** *of the diverging lens.*

We see then that there are two general types of lens. Converging lenses, which are thickest in the middle, cause a parallel beam of light to converge to the focal point. Diverging lenses, which are thickest near the edges, cause a parallel beam of light to spread out (or diverge) as though from a focal point. Even if the light enters from the opposite side, these lenses still behave in this way. Hence each lens has two focal points, one on either side of it, and both are at the same distance from its center. This distance from the center of the lens to the focal point is called the **focal length** $f$ of the lens. Shapes of common lenses are shown in Fig. 22.27. Notice that the converging lenses are all thickest at the center.

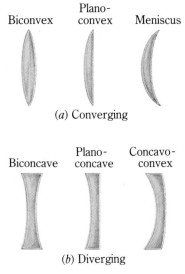

Biconvex    Plano-convex    Meniscus

(a) Converging

Biconcave    Plano-concave    Concavo-convex

(b) Diverging

If a glass lens is immersed in a liquid that has an index of refraction equal to that of the glass, a light wave travels with the same speed in the glass and in the liquid. The wave is not disturbed by passing through the lens, and the lens does not cause it to focus. Thus, the focal length of a lens depends on the medium in which it is used. Most commonly, this medium is air. However, this is not always the case. The focal length of a lens immersed in a medium other than air is

$$f_{medium} = \frac{f_{air} n_m (n_g - 1)}{n_g - n_m}$$

where $n_m$ is the refractive index of the medium and $n_g$ is the index of refraction of the glass. You should be able to show that this relation is reasonable for the limiting cases of $n_m \rightarrow n_g$ and $n_g \rightarrow 1.00$.

## 22.12 Ray diagrams for thin lenses

We just saw that a ray traveling parallel to a lens axis is bent by the lens. It is converged to the focus by a converging lens and diverged from the focus by a diverging lens. These rays are shown as ray 1 in Fig. 22.28.

The second ray comes from the object and goes through the focal point (or tries to in the diverging lens) before striking the lens, which converges or diverges it

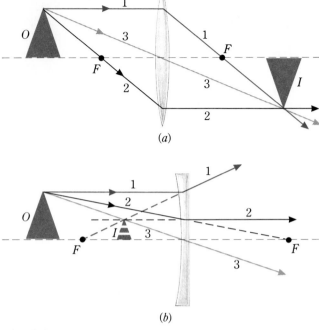

(a)

(b)

**FIGURE 22.28**

Graphical location of the images formed by thin lenses
consists of drawing the three rays.

parallel to the axis. Notice that this ray is exactly the same as ray 1 but traveling in
the reverse direction.

The third ray comes from the source and goes straight through the center of the
lens without deflection. It is easy to see why it behaves this way by referring to Fig.
22.29. Notice that the ray enters and leaves the lens at surfaces that are parallel to
each other. Hence, the ray behaves as though it had gone through a flat plate of
glass. Recall from Fig. 22.23 that a ray of light is not deviated in direction by a flat
plate that has parallel faces. Therefore, the ray that passes through the lens center
proceeds undeviated. It is displaced slightly, but this effect is negligible for a thin
lens.

We are now able to draw three of the many rays that pass through a lens. Any two
of these allow us to locate the image of an object. Two examples of this construction
are given in Fig. 22.28. Notice that the image in part *a* is real since the three rays
actually converge at it. A screen placed there would catch the light and show an
image of the object. In *b*, however, the image is virtual since the three rays merely
appear to come from the image position. A screen placed at that position would not
show an image because the three rays actually do not meet there.

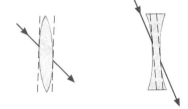

**FIGURE 22.29**

The ray passing through the
center of the lens essentially
passes through a flat plate (defined
by the broken lines) and is
therefore not deviated. A small
displacement of the ray occurs, but
this is not shown in the figure.
Why is it negligible for a thin lens?

---

### Example 22.6

A converging lens of focal length 10.0 cm is used to form an image of an object placed
5.0 cm in front of it. Draw a ray diagram to locate the image.

*Reasoning*   The appropriate ray diagram is shown in Fig. 22.30. We notice that
the eye will assume that the three rays come from the image position indicated. As
we see, the image is virtual, erect, and enlarged.   ∎

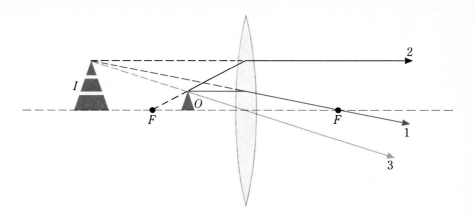

FIGURE 22.30

Virtual images are formed by convex lenses when the object is inside the focal point.

---

### Example 22.7

A diverging lens of focal length 10.0 cm is used to form an image of an object placed 5.0 cm in front of it. Find the image position by means of a ray diagram.

***Reasoning*** The appropriate ray diagram is shown in Fig. 22.31. Here too the image is virtual. It is erect and diminished in size. ∎

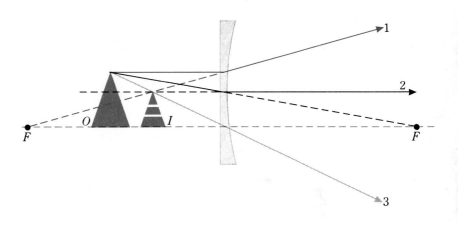

FIGURE 22.31

Is the image formed by a diverging lens always virtual if the object is real? What other shape could a diverging lens have? Why doesn't it make a difference?

## 22.13 The thin-lens formula

Consider the image formed by the converging lens in Fig. 22.32. In part *a*, triangles *ABH* and *EBD* are similar, and so we can write

$$\frac{I}{O} = \frac{i}{p}$$

From the two similar triangles *JFB* and *EDF* in *b*, we have

$$\frac{I}{O} = \frac{i-f}{f}$$

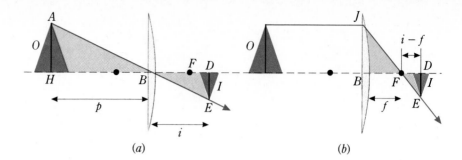

**FIGURE 22.32**

Triangles *ABH* and *EBD* are
similar, as are triangles *JFB* and
*EDF*.

Equating these two expressions and simplifying yield

$$\frac{1}{p} + \frac{1}{i} = \frac{1}{f}$$
(22.1)

This relation is exactly the same as the mirror equation (Eq. 22.1). We have obtained it by taking $p$ and $i$ as positive when the object and image are in their normal positions — that is, when the object is on the side from which the light comes and the image is on the side to which the light is going. Also, Eq. 22.2 for magnification applies to lenses since one of the above relations is identical to it.

We can derive the relation applicable to diverging lenses by referring to the sets of similar triangles in Fig. 22.33. We have

$$\frac{I}{O} = \frac{i}{p} \qquad \text{and} \qquad \frac{I}{O} = \frac{f-i}{f}$$

Equating and simplifying, we find that

$$\frac{1}{p} - \frac{1}{i} = -\frac{1}{f}$$

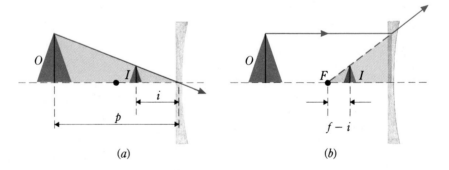

**FIGURE 22.33**

Consideration of the similar
triangles leads to the thin-lens
equation.

We want to eliminate the negative signs from this equation so as to make it the same as Eq. 22.1. To do this, we agree that diverging lenses have negative focal lengths, and so the negative sign preceding the $1/f$ term can be omitted. Moreover, if we agree to call image distances negative when the image lies on the side of the lens from which the light comes, then the negative sign on the $1/i$ term can be omitted. In that case, Eq. 22.1 applies to all lenses and mirrors. Notice also that Eq. 22.2 for magnification also applies to all lenses.

Let us briefly review our sign conventions in using Eq. 22.1,

$$\frac{1}{p} + \frac{1}{i} = \frac{1}{f} \qquad\qquad (22.1)$$

**1** The object distance is positive if the object is on the side of the lens or mirror from which the light is coming.

**2** The image distance is positive:

    a In the case of mirrors, if the image is on the side of the mirror where the light is, that is, in front of the mirror.

    b In the case of lenses, if the image is on the side of the lens to which the transmitted light goes.

**3** The focal length of converging mirrors and lenses is positive; the focal length of diverging mirrors and lenses is negative.

---

### Example 22.8

A diverging lens with a 20-cm focal length forms an image of a 3.0-cm-high object placed 40 cm from the lens. Find the image position and size.

*Reasoning*   From Eq. 22.1 we have, using all lengths in centimeters,

$$\frac{1}{40} + \frac{1}{i} = -\frac{1}{20}$$

$$i = -\tfrac{40}{3} \text{ cm}$$

The image is on the side from which the light is coming. It is imaginary, of course, as a ray diagram easily shows. To find the image size, we use Eq. 22.2:

$$\frac{I}{O} = \frac{i}{p}$$

$$I = \frac{(3)(\frac{40}{3})}{40} = 1 \text{ cm}$$

Notice that we do not carry the signs along when we find the image size.  ■

---

## 22.14 Combinations of lenses

Most optical instruments contain more than one lens. These systems are easy to deal with if we proceed in a systematic fashion. Let us consider the final image formed by the two lenses in Fig. 22.34*a*. The object is 20 cm from the first lens, which is 30 cm from the second lens. Both lenses are converging. To solve this problem by a ray diagram as well as by formula, we proceed as in Fig. 22.34*b* and *c*.

First, let us ignore the second lens completely and find the image cast by the first lens. The ray diagram locates this image as $I_1$ in part *b*. By formula we have, using centimeters,

$$\frac{1}{20} + \frac{1}{i_1} = \frac{1}{10}$$

$$i_1 = 20 \text{ cm}$$

where $i_1$ is as shown in the figure.

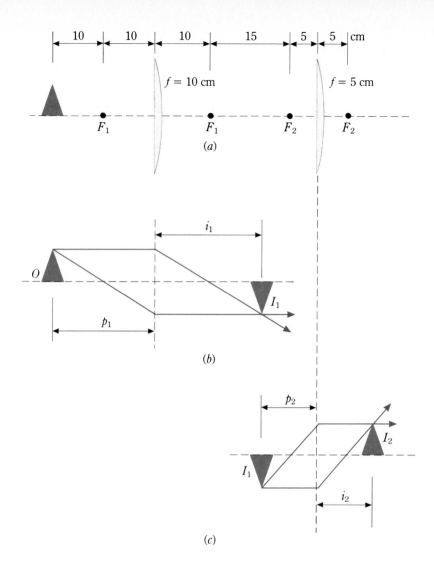

FIGURE 22.34

In order to find the image cast by a lens combination, we consider each lens in turn.

We now use the image formed by the first lens as the object for the second lens. This is possible because an eye or any other instrument placed to the right of $I_1$ will see this image as though there is really an object at that place. Ignoring the first lens completely and using $I_1$ as the object for the second lens, we draw the ray diagram in Fig. 22.34c. The final image is located at $I_2$, or, in equation form,

$$\frac{1}{p_2} + \frac{1}{i_2} = \frac{1}{f_2}$$

$$\frac{1}{30-20} + \frac{1}{i_2} = \frac{1}{5}$$

$$i_2 = 10 \text{ cm}$$

where $i_2$ is as shown. Clearly, the final image $I_2$ formed by the two lenses in combination is real and erect.

To find the size of the final image in terms of $O$, the height of the object, we apply Eq. 22.2 twice. For the situation of Fig. 22.34b, we have

$$\frac{I_1}{O} = \frac{20}{20}$$

$$I_1 = O$$

Now, using Fig. 22.34$c$ and remembering that the height of $I_1$ is the same as $O$ and that $I_1$ is really the object for the second lens, we have

$$\frac{I_2}{I_1} = \frac{i_2}{p_2}$$

$$I_2 = (O)(\tfrac{10}{10}) = O$$

Hence, we find the unusual situation where the image is the same height as the object. In general, this particular result is not found. However, the method we have used does have general applicability.

## Minimum learning goals

When you finish this chapter, you should be able to

**1** Define (*a*) light, (*b*) white light, (*c*) wavefront, (*d*) ray, (*e*) plane wave, (*f*) parallel light, (*g*) specular versus diffuse reflection, (*h*) virtual versus real image, (*i*) focal point and focal length, (*j*) index of refraction, (*k*) refraction and Snell's law, (*l*) total internal reflection, (*m*) critical angle, (*n*) concave, convex, diverging, and converging mirrors and lenses.

**2** Give the approximate wavelength limits of the visible spectrum and state the approximate color associated with a given wavelength.

**3** State the speed of light in vacuum. Compute $n$ for a material when the speed of light in it is given, and vice versa.

**4** Draw the appropriate rays for a given set of wavefronts, and vice versa. Explain why a distant source gives rise to parallel rays.

**5** Draw the reflected ray when the incident ray on a smooth surface is given.

**6** Use Snell's law in simple situations.

**7** Using a diagram, show why total internal reflection occurs only when $n_2 < n_1$. List a few uses of this phenomenon.

**8** Use ray diagrams to locate images for single mirrors and single lenses. Give the character of the image in each case.

**9** Use the lens and mirror equation to obtain $p$, $i$, or $f$ if two of the three are given or described to you. Relate $f$ to $R$ for a mirror. State the sign for $f$ in any given case.

**10** Find the size of an image when the object size is given, and vice versa.

**11** Tell whether a lens is diverging or converging in air when its shape is given.

**12** Explain how the focal point and focal length of a concave mirror and converging lens can be obtained by experiment.

## Questions and guesstimates

**1** Consider a concave mirror and an object at infinity. Where is the image formed? Is it erect or inverted? Is it larger or smaller than the object? Answer these questions as the object is slowly moved in toward the mirror. In particular, note the positions of the object at which any of the answers change.

**2** Repeat Question 1 for a convex mirror.

**3** Repeat Question 1 for a converging lens.

**4** Repeat Question 1 for a diverging lens.

**5** When you look down into a clear lake or vat of water, why does the water always appear shallower than it actually is?

**6** Using a wavefront diagram, explain why a lens can be

either converging or diverging depending on the material in which it is embedded.

**7** Can an empty water glass focus a beam of light? A full water glass? Is it possible to start a fire by accident if a bowl of water is set in a sunlit window?

**8** How can one use the mirror equation to find the position of the image of an object in a plane mirror?

**9** As light goes from air to glass, which of the following change: $f$, $\lambda$, $v$?

**10** Why can a smart fish in a calm lake see you on the bank by looking up at an angle of about 50° to the vertical?

**11** How do so-called one-way mirrors (or windows) operate?

**12** A "solar furnace" can be constructed by using a concave mirror to focus the sun's rays on a small region, the furnace region. How would you expect the temperature of the furnace to vary with the mirror area and focal length?

**13** A spherical air bubble in a piece of glass acts like a small lens. Explain. Is it converging or diverging?

**14** How can one determine the focal length of a converging lens? Of a diverging lens? Of a concave mirror? Of a convex mirror?

**15** Two plane mirrors are placed together so that they form a right angle. An object is then placed between them. How many images are formed? Repeat for an angle of 30° between the mirrors.

**16** About how much longer does it take for a pulse of light from the moon to reach the earth because of the presence of air rather than a vacuum above the earth?

**17** Newton believed that light consisted of a stream of particles and that these "light corpuscles" were strongly attracted by the water surface as light went from air to water. How would this lead to the observed effect of refraction?

**18** In various science museums (as well as in some unexpected places), a room is so designed that a person can whisper at one particular point in the room and be heard clearly at a certain distant point. How must the room be constructed to achieve this effect?

## Problems

**1** A laser beam from the earth is reflected back to the earth by a mirror on the moon, which is $3.84 \times 10^8$ m away. How long does it take for the beam to make the round trip?

**2** A radar beam is reflected by rain clouds that are 32 km away from the radar device. How long does it take the radar waves to make the round trip?

**3** Yellow light from a sodium arc lamp has a wavelength of 589 nm. When a beam of this light travels through water, what are its (*a*) speed, (*b*) wavelength, and (*c*) frequency?

**4** Blue light from a mercury arc has a wavelength of 436 nm. When a beam of this light travels through benzene, what are its (*a*) speed, (*b*) frequency, and (*c*) wavelength?

**5** Many cameras have a focusing mark that must be set to read the distance from the camera to the object being photographed. Suppose you wish to photograph yourself in a plane mirror. If you and the camera are 60 cm from the mirror, at what value should you set the distance indicator on the camera?

**6** An interior decorator wishes to mount a plane wall mirror in such a way that a person 160 cm tall will be able to see his or her full length in it. What is the shortest length of mirror possible for this use, and how high on the wall should it be mounted?

**7** A beam of light is reflected straight back on itself by a mirror that is perpendicular to the beam. The mirror is then turned so that its normal makes an angle of 15° to the beam. What is the new angle between the incident and reflected beams?

■ **8** If an object is placed between two parallel mirrors, an infinite number of images result. Suppose the mirrors are a distance $b$ apart and the object is midway between them. Find the distances of the first several images from the object.

■ **9** Two plane mirrors make an angle of 90° with each other, and a light ray is reflected first by one mirror and then by the other. Show that the direction of the final reflected ray is exactly opposite the direction of the original incident ray.

■ **10** Show that the result stated in Prob. 9 is a special case of the following: When a ray is reflected successively from two plane mirrors that make an angle $\theta$ with each other, the angle between the incoming and outgoing rays is $2\theta$.

**11** An object 0.50 cm high is placed 32 cm in front of a concave mirror that has a 16-cm radius of curvature. (*a*) Find the position and size of the image, and state whether it is real or virtual and erect or inverted. Repeat for object distances of (*b*) 16, (*c*) 12, and (*d*) 6 cm. Check your calculations with ray diagrams.

**12** A concave mirror that has a 20-cm radius of curvature forms an image of a 3.0-cm-high object that is placed 40 cm in front of the mirror. (*a*) Find the position and the size of the image. Is it real or virtual? Erect or inverted? Repeat for object distances of (*b*) 20, (*c*) 15, and (*d*) 8 cm. Check with a ray diagram.

■ **13** A concave mirror that has a 300-cm radius of curvature is used to form a real image of an object. (*a*) Where must the object be placed if the image distance is to equal the object distance? (*b*) Are the object and the image superimposed? (*c*) Compare the object and image sizes. Check with a ray diagram.

■ **14** (*a*) Given a concave mirror with a 50-cm focal length, where must an object be placed if the image is to be real and twice the size of the object? (*b*) Repeat for the case where the image is to be virtual.

**15** (*a*) Find the position, size, and nature (real or virtual, erect or inverted) of the image formed when a 2-cm-high object is placed 60 cm in front of a convex mirror that has a 30-cm radius of curvature. Repeat for object distances of (*b*) 25 and (*c*) 8 cm. Check with a ray diagram.

**16** (*a*) If an object is placed 40 cm in front of a convex mirror with a focal length of 20 cm, where is the image formed? What is the magnification? Is the image real or virtual? Repeat for object distances of (*b*) 20 and (*c*) 8 cm. Check with a ray diagram.

■ **17** A virtual image is formed by a convex mirror that has a 30-cm focal length. (*a*) Where must the object be placed if the image is to be half the size of the object? (*b*) Is it possible to obtain a virtual image that is larger than the object using this type of mirror?

■ **18** A virtual image one-fourth the size of the object is formed by a convex mirror that has a 50-cm focal length. (*a*) Where is the object? (*b*) Where is the image?

■ **19** (*a*) Where must an object be placed if the image formed by a convex mirror is to be one-third as far from the mirror as the object is? (*b*) How large is the magnification in this case?

■ **20** (*a*) Where must an object be placed if the image formed by a concave mirror is to be one-fourth as far from the mirror as the object is? (*b*) Is the image real or virtual?

**21** A beam of red light from a helium-neon laser ($\lambda = 633$ nm) enters a glass plate ($n = 1.56$) at an angle of $25°$ to the normal. (*a*) What is the speed of the beam in the glass? (*b*) What is its wavelength? (*c*) What angle does the beam make with the normal inside the glass?

**22** Green light with $\lambda = 546$ nm enters water in a dish at an angle of $70°$ to the normal. (*a*) What is the wavelength of the light in the water? (*b*) What angle does the beam make with the normal inside the water?

**23** Light enters a flat glass plate ($n = 1.56$) at an inci-

dence angle of $53°$. (*a*) What is the angle of refraction inside the glass? (*b*) After the beam leaves the plate, what is the angle between it and the beam incident on the plate?

**24** The index of refraction of glass is different for different wavelengths. Flint glass has an index of refraction of 1.650 for blue light ($\lambda = 430$ nm) and 1.615 for red light ($\lambda = 680$ nm). A beam of light consisting of these two colors is shone into flint glass at an incidence angle of $40°$. Find the angle between the two colored beams in the glass.

■ **25** At what angle of incidence should a beam strike a still pond if the angle between the reflected and refracted rays is to be $90°$?

■ **26** A beam of light has an incidence angle $\theta_1$ in a material of index of refraction $n_1$. It passes into a second material for which the refractive index is $n_2$, and the angle of refraction is $\theta_2 = 2\theta_1$. (*a*) Which is larger, $n_1$ or $n_2$? (*b*) Find the ratio $n_1/n_2$ in terms of $\theta_1$. (*c*) What is the ratio if $\theta_1$ is the critical angle for total internal reflection?

**27** At what angle to the vertical must a submerged fish look if it is to see an angler seated on the distant shore of a still pond?

**28** A light beam from a source 3.0 m below the surface of a calm pool of water is shone upward toward the surface. What is the maximum angle that the beam in the water can make with the vertical if part of the beam is to be able to escape into the air?

■ **29** A layer of water is placed on top of a glass plate that has $n = 1.560$. If a beam of light is to enter the water from the glass, what maximum angle can the beam in the glass make with the perpendicular to the glass?

■ **30** A solid glass hemisphere ($n = 1.56$) has its flat face perpendicular to a uniform parallel beam of light that illuminates the whole face. What fraction of the beam that enters the face is totally internally reflected as it strikes the curved surface for the first time?

■ **31** A converging lens that has a focal length of 40 cm and an index of refraction of 1.56 is immersed in water. (*a*) What is its focal length in the water? (*b*) Repeat for the lens immersed in methylene iodide.

■ **32** A converging lens ($n = 1.560$) has a focal length of 30 cm when immersed in benzene. What is its focal length when it is immersed in water?

**33** A converging lens with a 50-cm focal length forms an image of a 2.0-cm-high object. Find the position, size, and nature of the image for the following object distances: (*a*) 120 cm, (*b*) 70 cm, (*c*) 20 cm. Check your answers with ray diagrams.

**34** A converging lens with a 20-cm focal length is to form an image of a 2.0-cm-high object. Find the position, size, and nature of the image for the following object distances:

(*a*) 80 cm, (*b*) 40 cm, (*c*) 10 cm. Check your answers with ray diagrams.

**35** Find the position, size, and nature of the image formed by a diverging lens with a focal length of $-40$ cm if the 2.0-cm-high object is at a distance of (*a*) 90 cm, (*b*) 60 cm, (*c*) 20 cm. Check your answers with ray diagrams.

**36** An image of 3.0-cm-high object is formed by a diverging lens with $f=-20$ cm. Find the position, size, and nature of the image for the following object distances: (*a*) 80 cm, (*b*) 20 cm, (*c*) 8.0 cm. Check your answers with ray diagrams.

■ **37** A virtual image, that is half as far from the lens as the object, is to be formed by a diverging lens for which $f=-20$ cm. (*a*) Where should the object be placed? (*b*) What will the magnification be?

■ **38** (*a*) Where must an object be placed with respect to a converging lens if the image is to be half as large as the object? (*b*) Is the image real or virtual? Express your answer in terms of $f$.

■ **39** If a diverging lens is to be used to form an image that is one-third the size of the object, where must the object be placed?

■ **40** A single lens is to be used to form a virtual image of an object. The image is to be twice the size of the object. What kind of lens should be used, and where should the object be placed?

■ **41** An alternative form of the lens and mirror equations is

$$s_i s_o = f^2$$

where $s_o$ is the distance of the object from the focal point and $s_i$ is the distance of the image from the focal point. Derive this relation. Show $s_o$ and $s_i$ in a sketch for mirrors and for lenses.

■ **42** In certain cases, when an object and a screen are separated by a distance $D$, two positions of a converging lens relative to the object will give an image on the screen. Show that these two values are

$$x = \tfrac{1}{2}D\left(1 \pm \sqrt{1 - \frac{4f}{D}}\right)$$

where $x$ gives the positions of the lens. Under what conditions will no image be found?

■ **43** Two identical converging lenses with $f=40$ cm are 90 cm apart. (*a*) Find the final image position for an object placed 120 cm in front of the first lens. (*b*) How large is the magnification of the system? (*c*) Is the image real or virtual? Erect or inverted?

■ **44** An object is placed 12 cm in front of a converging lens ($f=8$ cm). At a distance of 40 cm beyond the first lens is a diverging lens with $f=-6$ cm. (*a*) Find the position and magnification of the final image. (*b*) Is the image real or virtual? (*c*) Erect or inverted?

■ **45** A converging lens with $f=20$ cm is followed by a diverging lens with $f=-30$ cm; their separation is 30 cm. If an object is placed 5 cm in front of the first lens, find (*a*) the position and (*b*) the magnification of the final image.

■ **46** An object is placed 30 cm in front of a converging lens with $f=20$ cm, which in turn is 50 cm in front of a plane mirror. Find all the images formed by this system and state whether each is real or virtual.

■ **47** A diverging lens with an 8-cm focal length is placed 16 cm to the left of a concave spherical mirror that has a radius of 20 cm. If an object is placed 8 cm to the left of the lens, find all the images formed by the system and state whether each is real or virtual.

# 23 Optical devices

**N**OW that we understand lenses and mirrors, we are in a position to discuss some common optical devices. We discuss the operation of such widely different devices as the human eye and the telescope, as well as the microscope and other important optical instruments. In so doing, not only do we obtain practice in the use of lenses and mirrors, but also we become more intelligent operators of the optical devices we are frequently called on to use.

## 23.1 The eye

The most familiar of all optical devices, the eye, is also one of the most complicated. Although the lens system of the eye is not very complex, the associated interpretive equipment is as complex as humans themselves. The exact way in which the image formed on the retina of the eye is transformed to our sensation of sight is a problem still challenging biophysicists. We restrict our discussion to the lens system and to the formation of the image on the retina.

A simplified diagram of the eye is shown in Fig. 23.1. As you probably already know, the cornea is a protective cover, the iris diaphragm controls the amount of light entering, and the retina is a sensitive surface that changes the image formed on it to electrical energy, which is then transmitted to the brain. A ray of light entering the eye is refracted at the cornea. Lesser refraction effects occur in the pupil and lens since the refractive indices of the cornea, pupil, lens, and fluid portions of the eye are quite similar.

FIGURE 23.1
Diagram of the human eye.

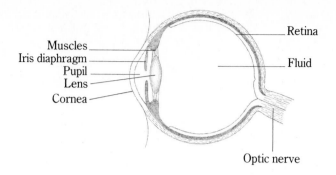

For the normal relaxed eye, these combined refraction effects form an image of distant objects on the retina. Hence, the focal length of the eye is about the distance from the retina to the lens. If you draw ray diagrams for a converging lens, you will find that the image distance increases as the object is brought closer to the lens. In the eye, however, the image distance must always be such that the image is formed at the retina. This coincidence of image and retina can be maintained if the lens is made more converging as the object is brought closer to it. The muscles of the eye alter the shape of the rather deformable lens so as to make it thicker (more converging) when viewing close objects. A person with normal eyesight cannot make the lens converging enough to see objects closer than about 25 cm, the normal **near point** of the eye.

Normally, one is able to relax the eye lens to the extent necessary to focus a distant object on the retina. A person who is unable to do this is said to be **nearsighted,** or **myopic.** The myopic eye is able to focus only objects that are less than a certain distance from the eye. This distance is called the **far point** of the eye. Since the eye remains too converging to allow proper focus for very distant objects, a myopic person must be fitted with spectacles that have a diverging lens in order to see far-away objects clearly. We illustrate this effect in Fig. 23.2*a*, where the dashed lines indicate the position to which a distant object would be focused without the corrective lens.

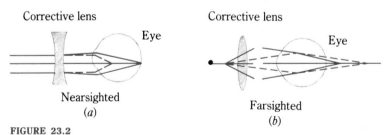

FIGURE 23.2
A diverging lens is used to correct myopia, whereas a converging lens is needed to correct hyperopia. The broken lines show the path the rays would follow without the spectacle lens.

People who have **hyperopia,** that is, **farsighted** people, are able to relax the eye to see distant objects but are unable to make the lens thick enough to focus nearby objects onto the retina. A farsighted person must wear a converging spectacle lens to help the eye bring nearby objects to focus on the retina. This situation is shown in Fig. 23.2*b*.

If the eye has very little ability to alter the shape of the lens, we say that the eye

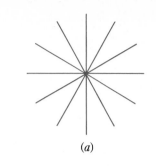

(a)

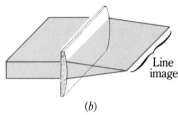

Line image

(b)

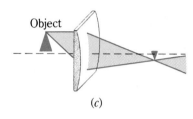

Object

(c)

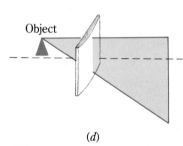

Object

(d)

**FIGURE 23.3**
Cylindrical lenses form line images rather than point images. Rays in various planes are focused differently, as shown in (c) and (d).

has lost its **accommodation.** Such an eye may be able to focus neither very distant nor very close objects. The use of bifocal spectacles allows one to look through diverging lenses when looking straight ahead and converging lenses when looking down. Some people have three types of lenses built into a single spectacle lens, a trifocal lens. It furnishes good visibility for objects that are at far, intermediate, and near distances.

Another common type of eye defect is **astigmatism.** When a person with this type of eye defect views the test pattern in Fig. 23.3a, some of the lines appear darker than others. This is caused by a nonspherical lens. As you know, a spherical lens focuses parallel rays at a single point, the focal point. A slightly misshapen lens, however, often acts as a spherical lens with a cylindrical lens superimposed on it.

Notice in Fig. 23.3b that the lens focuses parallel rays into a line image rather than a point image. Rays in a plane perpendicular to the cylinder axis, as in c, are focused, but rays in a plane parallel to the axis, as in d, are not focused at all. This causes the eye lens to focus the lines at different positions. When a vertical line is in proper focus, a horizontal line, for example, may not be in focus. Consequently, lines oriented at different angles appear different to a person who has this type of eye defect. To compensate for this, an eyeglass that exactly cancels the cylindrical part of the eye lens must be used.

---

***Example 23.1***
A farsighted man is able to read the newspaper only when it is held at least 75 cm from his eyes. What focal length must the lenses of his reading glasses have? Assume that the distance between the glasses and his eyes is negligible.

***Reasoning*** We want the man to be able to see print clearly when the reading material is 25 cm from his eyes, that is, at the normal near point. Hence, what is needed is a lens that gives a *virtual* image at $i = -75$ cm when the object distance is $p = 25$ cm. Using the lens equation, we get

$$\frac{1}{25} + \frac{1}{-75} = \frac{1}{f}$$

$$f = 37.5 \text{ cm}$$

Notice that a converging lens is needed. ■

---

***Exercise*** If your eyeglasses have $f = 60$ cm, what is your probable near point?
*Answer: 43 cm* ■

---

***Example 23.2***
What must the focal length of a corrective lens be for a woman whose far point is 50 cm?

***Reasoning*** The corrective lens must form an image of a distant object that is virtual and 50 cm from the eye. Therefore, $i = -50$ cm, $p \to \infty$, and the lens equation is

$$\frac{1}{\infty} + \frac{1}{-50} = \frac{1}{f}$$

$$f = -50 \text{ cm}$$

Notice that the required lens is diverging. ■

## 23.2  The simple camera

A camera (Fig. 23.4) operates very much like the human eye. It uses a lens to produce an image of an object on a film. The film serves the purpose of the retina in the eye — that is, the camera lens produces an image on the film in much the same way that the eye lens produces an image on the retina. The image is inverted on the film, and its size *I* is related to the object size *O* by the usual relation:

$$\frac{I}{O} = \frac{i}{p}$$

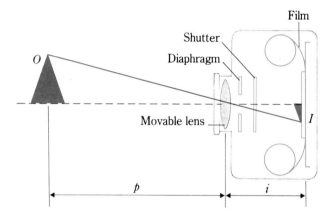

**FIGURE 23.4**

A simple camera. How is the image brought to focus on the film?

Unlike that of the eye, the lens of a camera cannot be made with variable focal length. Hence, to achieve good focus on the film, the lens must be moved back and forth as the distance to the object changes. Cameras that do not have movable lenses usually have only a very small hole in front of the lens; they operate much like a pinhole camera, which has no lens at all but merely uses a small pinhole to admit the light to the film. The operation of such a camera is treated as a question at the end of this chapter.

Expensive cameras possess very complex lens systems instead of a single lens. The complexity is necessary if a camera is to give sharp images with fast shutter speeds. It is clear why sharp images are advantageous, and fast shutter speeds allow one to take sharper pictures of swiftly moving objects. Any movement will blur the image somewhat, but the shorter the time the camera shutter is open, the less blurred the image will be. Since the shutter must be open long enough to allow sufficient light to hit the film, fast shutter speeds mean that the lens must be large so that a large amount of light can enter the camera.

As we saw in Sec. 22.11, only the central portion of a spherical lens can be used if a clear image is desired. This becomes even more important if a camera is to be used to take close-up pictures, since then the lens must be very convex. It is only by making a complicated combination lens that the focusing errors inherent in a single lens can be eliminated. We say that such a lens has been corrected for **spherical aberration.**

Another lens defect causes images to have colored edges. This is called **chromatic aberration.** It results from the fact that the speed of light in glass varies with wavelength. As a result, the index of refraction of the glass is not the same for all colors. Blue light is refracted more strongly by the lens than red light is. This causes the colors in a beam of ordinary light to separate, and the image is therefore colored.

To correct for this defect, two or more types of glass must be layered together to form the lens. Expensive lenses consist of several individual lenses cemented together. A lens that has been partly corrected for chromatic aberration is called an **achromatic lens.** However, it is impossible to free a lens of this defect completely. Indeed, lenses used in expensive optical systems, such as a very expensive camera or microscope, are very complex. They correct the system not only for spherical and chromatic aberration, but also for other lens defects. The design of precise optical instruments is very involved. You can read more about it in advanced texts concerned with geometric optics.

## 23.3 Lenses in close combination: diopter units

Perhaps you have had your eyes tested and noticed that the examiner often places several spectacle lenses, one in front of the other, in front of your eye at once. In order to make use of the observed best combination, he or she needs to know how to add the effects of thin lenses in close combination. The necessary simple formula can be derived readily, as we now show. We consider only the case in which the focal lengths of the lenses are much longer than the distances from one lens to the next.

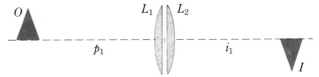

**FIGURE 23.5**

If the two lenses are very close together, their combined effect is to act as a single lens with a focal length $1/f = 1/f_1 + 1/f_2$.

Referring to Fig. 23.5, we find first the image formed by the first lens:

$$\frac{1}{p_1} + \frac{1}{i_1} = \frac{1}{f_1}$$

This image is then used as the object for the second lens. The object distance is $p_2 = -i_1$ if we ignore the small separation of the lenses. We use a negative sign because the object is on the side of the lens to which light is going, and our sign conventions require a negative object distance in such cases. Now, writing the lens equation for the second lens yields

$$\frac{1}{-i_1} + \frac{1}{i_2} = \frac{1}{f_2}$$

If we add these two equations, we have

$$\frac{1}{p_1} + \frac{1}{i_2} = \frac{1}{f_1} + \frac{1}{f_2}$$

$$\frac{1}{p_1} + \frac{1}{i_2} = \frac{1}{f}$$

where the combined focal length $f$ of the two lenses is

$$\frac{1}{f} = \frac{1}{f_1} + \frac{1}{f_2} \qquad \text{close combination} \qquad (23.1)$$

This is the focal length of two thin lenses in close combination.

In order to save the eye examiner the trouble of adding reciprocals, a unit using the reciprocal of the focal length is defined. *The* **power** *of a lens in* **diopters** *is the reciprocal of the focal length in meters.* For example, the power of a diverging lens with a 20-cm focal length is $1/(-0.20)$, or $-5$ diopters. The power of a lens combination for thin lenses in close contact is simply the algebraic sum of the individual powers.

---

### Example 23.3

Three lenses of focal lengths 20, $-30$, and 60 cm are placed in contact with each other. Find the focal length of the combination.

***Reasoning*** The powers of the lenses are 5, $-\frac{10}{3}$, and $\frac{10}{6}$ diopters. Their combined power is

$$\frac{30}{6} - \frac{20}{6} + \frac{10}{6} = \frac{20}{6} \text{ diopters}$$

Hence the combined focal length is $\frac{6}{20}$, or 0.30 m. It is interesting to notice that if the $\frac{10}{6}$-diopter lens had been negative, the three lenses would have substantially the same effect as a flat plate of glass. ∎

---

## 23.4 The magnifying glass

One of the simplest optical instruments is the magnifying glass. A typical use of this instrument is shown in Fig. 23.6. Since the magnifier is one of the basic constituents

**FIGURE 23.6**

Why is only the magnified text in focus for the camera that took this photo?

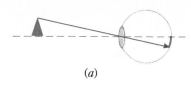

(a)

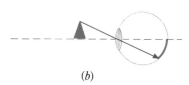

(b)

FIGURE 23.7

When an object is brought closer to the eye, the image on the retina becomes larger.

in many optical devices, we describe its characteristics. It is a converging lens, and its focal length is designated $f$. In practice, one uses a magnifier to form an enlarged image of an object on the retina of the eye.

We can understand this most easily by referring to Fig. 23.7. As you can see, the size of the image formed on the retina increases as the object is brought closer and closer to the eye. However, the human eye is unable to focus well on objects that are closer than the near point. We represent this distance by $\gamma$; it is about 25 cm for the normal eye.* If we use a converging lens in front of the eye, as in Fig. 23.8, we can view the virtual image it forms. Even though the object is inside the near point (and therefore too close to be seen clearly), the image is right at the near point. Because the object is much closer than the near point, its image on the retina is far larger than what it would be if the object were being viewed with the naked eye, and therefore more detail in the image can be seen.

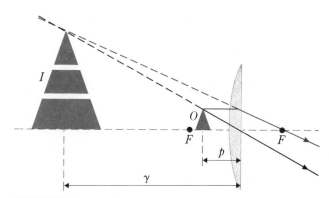

FIGURE 23.8

A magnifying glass allows one to place the object being examined far inside the eye's near point. Doing so enlarges the image on the retina.

Two methods are used to measure the magnifying effect in this case. The magnification we defined in Chap. 22 is

$$M = \frac{I}{O}$$

and we call this the *lateral* (or *linear*) *magnification*. This was shown to be equivalent to the ratio

$$\frac{\text{Image distance}}{\text{Object distance}} = \frac{i}{p}$$

As we see in Fig. 23.8, $i = \gamma$ when the image cast by the magnifier is at the near point. Thus

$$M = \frac{\gamma}{p} = 1 + \frac{\gamma}{f} \qquad (23.2)$$

* You can find the near-point distance $\gamma$ for your eyes by seeing how close you can hold a page and still read it easily.

where the lens equation has been used to replace $1/p$ by $(1/f) - (1/i)$ with $i = -\gamma$.

The second method for describing magnification is to use a quantity called the *angular magnification* (or *magnifying power*). We define it in reference to Fig. 23.9. Notice that when the object is viewed at the near point of the eye, as in Fig. 23.9*a*, it subtends an angle $\phi$ at the eye. However, when it is viewed through the magnifier, the object subtends an angle $\phi'$ at the eye. We define

$$\text{Angular magnification} = \frac{\phi'}{\phi} \tag{23.3}$$

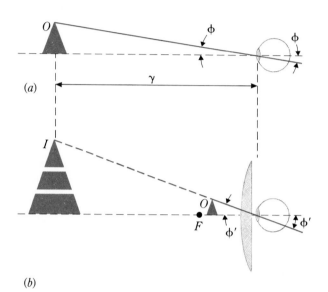

**FIGURE 23.9**
In both cases the eye is focused at the near point. (*a*) When the object is at the near point, the angle it subtends at the eye (and on the retina) is $\phi$. (*b*) When the object is far inside the near point, the much larger angle it subtends is $\phi'$. Because the image formed by the magnifying glass is at the near point, it can be seen clearly by the eye.

To obtain an expression for this in the present case, we note from Fig. 23.9 that

$$\tan \phi = \frac{O}{\gamma} \qquad \tan \phi' = \frac{I}{\gamma}$$

Then, since the angles usually encountered in such situations are small, the tangents are equal to their angles and we have

$$\text{Angular magnification} = \frac{I/\gamma}{O/\gamma} = \frac{I}{O} = \frac{\gamma}{p}$$

an expression identical to Eq. 23.2 for the linear magnification.

As we see, the two definitions give the same results under the present conditions. In practice, the image is often viewed at infinity rather than at a distance $\gamma$. Then $p = f$ and the magnification is simply $\gamma/f$. As you see, $M$ depends upon how the magnifier is used.

A typical simple magnifying glass might have a focal length of 5 or 10 cm. Since $\gamma \approx 25$ cm, such a magnifier would provide a magnification of between 2.5 and 5. In other words, if all other factors remained constant, such a lens would allow one to observe details with dimensions as small as one-fifth the size that would be possible with the naked eye. Usually, however, other factors must also be considered. Among these are blurring of the image due to spherical and chromatic aberrations of the lens. Even in the case of perfect lenses, there is a limit to the detail resolvable. We discuss the causes of this limit in the next chapter.

## 23.5 The compound microscope

In the compound microscope, a lens called the *objective lens* (or simply the *objective*) gives an enlarged, real image of an object. This image is then examined by a magnifier called the *eyepiece lens* (or *eyepiece*). The simplest microscope of this type is shown in Fig. 23.10. In Fig. 23.10b, the appropriate ray diagram has been drawn so that the image locations and characteristics can be clearly seen. To find the magnification obtained with this microscope, we notice that the objective lens causes a magnification

$$M_o = \frac{i_o}{p_o} = i_o \left( \frac{1}{f_o} - \frac{1}{i_o} \right)$$

**FIGURE 23.10**

In the compound microscope, the eyepiece is used as a magnifying glass to observe the image cast by the objective lens.

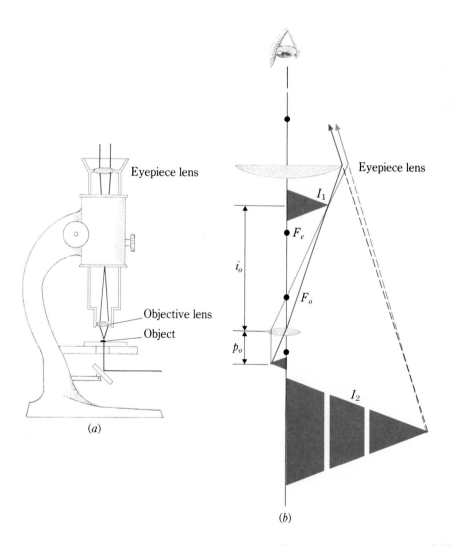

(a)

(b)

where $p_o$ is the distance from the object to the objective and $i_o$ is the distance from the objective to the image it forms. This gives

$$M_o = \frac{i_o}{f_o} - 1$$

where we have used the lens equation.

The eyepiece lens functions as a magnifier. For it, from Eq. 23.2,

$$M_e = 1 + \frac{\gamma}{f_e}$$

The total magnification is the product $M_e M_o$ (why not their sum?):

$$M = \left(\frac{i_o}{f_o} - 1\right)\left(\frac{\gamma}{f_e} + 1\right) \approx \frac{i_o \gamma}{f_o f_e}$$

Why are the approximations leading to the latter form usually justified? In practice, $i_o$ is about equal to the length of the microscope (about 18 cm) and $\gamma$ is about 25 cm.

As we see, $f_1$ and $f_2$ should be small for highest magnification. In order to achieve small $f_1$ and $f_2$, a good microscope uses quite complex combinations of lenses for the objective and eyepiece lenses. Great care must be taken in the lens design, or else various lens aberrations will so seriously distort and color the image as to make the instrument nearly worthless.

## 23.6 The astronomical telescope

Two basic problems confront astronomers when they design a telescope: (1) they would like the distant object to appear larger than when viewed with the naked eye and (2) they would like to increase the faint amount of light that reaches the eye from the object. To accomplish the latter, they use an objective lens that is very large so that it collects a great amount of light from the object. One telescope, at the Yerkes Observatory, has an objective lens that is 100 cm in diameter and has a 19-m focal length. Clearly, such a large lens will gather much more light than the very small opening in the eye. A diagram of such a telescope is shown in Fig. 23.11.

The objective lens forms an image $I_o$ of the distant object. This image falls very close to $f_o$, the focal point of the objective lens, since the light from the distant object is nearly parallel. The eyepiece lens acts, as usual, as a magnifying glass to look at

**FIGURE 23.11**

The two-lens telescope considerably increases the angle subtended at the eye, but the final image is inverted.

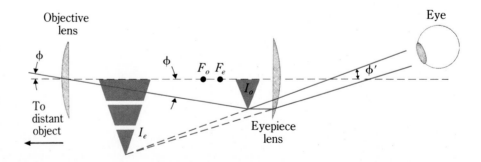

the image formed by the objective lens. Of course, the image $I_e$ formed by the eyepiece is virtual and about 25 cm from the eyepiece lens.

To find the magnifying power of this device, we notice that if the naked eye were to look at the distant object, the angle $\phi$ subtended would be the same as the angle subtended at the eyepiece. (Why?) If $\phi$ is small, so that it can be replaced by its tangent, we have, since the image formed by the objective is very close to the focal point,

$$\phi \approx \tan \phi \approx \frac{I_o}{f_o}$$

where $I_o$ is the height of the image formed by the objective and $f_o$ is the focal length of the objective. Also, since the focal point of the eyepiece is close to $I_o$, we have

$$\phi' \approx \tan \phi' \approx \frac{I_o}{f_e}$$

If we divide this equation by the preceding one, we find the magnifying power of the telescope to be

$$M = \frac{\phi'}{\phi} = \frac{f_o}{f_e}$$

We see that the focal length of the objective should be long and that of the eyepiece should be short.

Since it is very difficult to make perfect large lenses, many large astronomical telescopes use a concave mirror in place of an objective lens. One arrangement for such a reflecting telescope is shown in Fig. 23.12. (The size of the flat deflecting mirror is exaggerated.) A reflecting telescope is not affected by chromatic aberration except in the eyepiece. Moreover, spherical aberration in the mirror is eliminated by using parabolic mirrors. Since it is considerably easier to produce a good large mirror than a good large lens, it is clear why reflecting telescopes are sometimes preferred.

**FIGURE 23.12**

A schematic diagram of a reflecting telescope. The flat mirror is much smaller than shown.

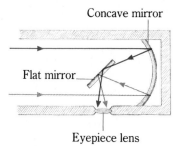

Concave mirror

Flat mirror

Eyepiece lens

## 23.7 The binocular

As you can see in Fig. 23.11, a telescope gives an inverted image of a distant object. Moreover, right and left are reversed in the image, as you can confirm with a ray diagram. Because of this, a simple telescope is not very suitable for viewing objects on the earth. Usually we use binoculars for this purpose.

# *Radio astronomy*

Until about 1950, optical telescopes were our only means of looking into the depths of space. With such devices, we were limited to those observations that could be made at wavelengths near the visible range. We know, however, that visible wavelengths are only a tiny fraction of the electromagnetic spectrum. As a result, with optical telescopes we are able to observe only objects that strongly emit light. These objects must of course be white-hot in order to appear luminous.

We are led to wonder whether we are not missing much that exists in outer space, since with optical telescopes we can see only the hot suns, the visibly glowing objects. If we were able to "see" much cooler bodies in space, perhaps we would find new objects, such as suns that have cooled or embryonic suns in the process of birth. Using the radar techniques developed during (and after) World War II, scientists designed and constructed telescopes that "see" the radio waves reaching the earth from space.

The large reflecting telescope shown here detects microwaves (short-wavelength radio waves) reaching it from the distant reaches of space. Because the telescope is designed to detect wavelengths in the 10- to 20-cm range, the telescope "mirror" need not be perfectly smooth. For reasons we shall begin to learn about in the next chapter, even a wire-mesh surface reflects these waves well. Hence the huge 64-m-diameter parabolic reflector is covered with wire mesh and looks not at all like a mirror surface. Even so, microwaves from space are focused by it to an electronic detector at its focal point. Provision is made for orienting the reflector so that the telescope can be aimed at any area of the heavens that is 30° above the horizon.

Telescopes such as this have enabled us to detect radio-wave sources nearly 10 billion light-years away from the earth — 20 times farther than is possible with present optical telescopes. Using them, scientists have discovered many previously unknown objects in the far reaches of space.

---

A diagram of a binocular is shown in Fig. 23.13*a*. Each half of the binocular consists of an objective lens, an eyepiece lens, and two prisms mounted at right angles to each other. The prisms are used to reflect the light by total internal reflection, as shown in part *b*. Notice how the reflection inverts rays 1 and 2. As a result, the prisms invert the image and reverse right and left, thereby counteracting the reversals caused by the objective lens. The final image formed by the binocular for viewing by the eye is therefore erect and preserves the normal right-to-left orientation.

## 23.8 The prism spectrometer

A prism, usually made of glass, is frequently used to separate light into its various colors. A beam of light is usually bent twice in a prism, once when it enters and once

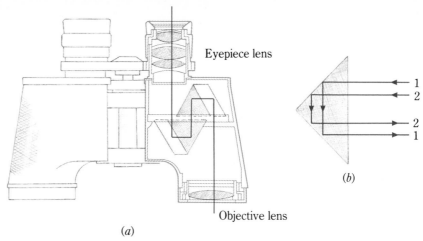

Eyepiece lens

1
2

2
1

(b)

Objective lens

(a)

**FIGURE 23.13**
The prism binocular.

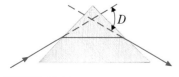

D

**FIGURE 23.14**
A prism deviates the light beam through the angle *D*.

when it leaves. We call the total angle through which the ray is bent the **angle of deviation.** It is shown as angle *D* in Fig. 23.14.

With the incident angle, the angles of the prism, and the refractive index of the glass known, it is possible to compute *D* by using Snell's law. The higher the index of refraction of the glass, the larger the deviation of the beam. This has important consequences, as we now see.

We mentioned in Sec. 22.9 that the speed of light in most materials varies with wavelength. This is equivalent to saying that the index of refraction of the material depends on the color of the light. For most materials, the index of refraction for violet light is larger than that for red light. Hence, violet light is bent more by a glass prism than red light is. Consequently, if a beam of white light enters a prism, as in Fig. 23.15 and Color Plate III, the light is dispersed into its colors. The ability to disperse light varies from material to material. For high dispersion, the index of refraction must change markedly with wavelength.

**FIGURE 23.15**

The angle of deviation by a prism is not the same for all wavelengths of light. Hence the prism disperses white light into its constituent colors.

White light

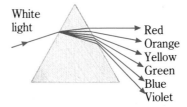

Red
Orange
Yellow
Green
Blue
Violet

Prisms are of great importance in science and industry since they allow us to separate a beam of light into its various wavelengths. Because each atom and molecule can be induced to emit its own characteristic wavelengths of em radiation, the wavelengths emitted by a substance help us to determine what the substance is. A device that uses a prism to separate a beam into the wavelengths that compose it is called a *spectroscope* or *spectrometer*. The simple prism spectrometer sketched in Fig. 23.16 is being used to analyze the wavelengths emitted by the light source. Let us suppose for this discussion that the source emits a single wavelength. (Sodium vapor lamps, the yellow lights often used on highways, emit essentially one visible wavelength, 589 nm.)

23.8 The prism spectrometer     **547**

FIGURE 23.16

An image of the slit is obtained on the photographic plate of the prism spectrometer. If the light had contained more than one wavelength, multiple images would have been found in the photograph.

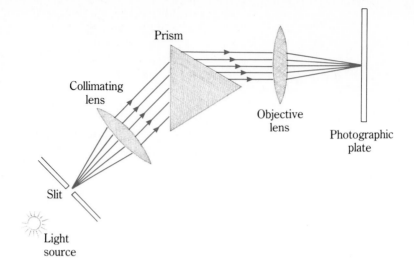

Light from the source enters the spectrometer through a narrow slit that is placed at the focal point of the *collimating lens*. Because the slit acts as an object placed at this focal point, the lens produces parallel light. Since they are of a single wavelength, the light rays are all deviated through the same angle by the prism and therefore emerge from it as parallel rays. As they pass through the *objective lens,* the parallel rays are brought to a focus at its focal point. There they produce an image of the object that produced them, namely, the slit. If a photographic plate or film is placed at the focal point of the objective, the image of the slit appears as a narrow line on the plate or film. For reasons we shall soon learn about, this image is called a **spectral line.** Often the plate is replaced by a telescope so that the image of the slit can be seen visually or, perhaps, monitored using a photocell.

Each type of light source emits its own characteristic wavelengths, and we learn about the inner workings of atoms and molecules from them (Chap. 26). If a mercury vapor lamp (the bluish lamps often used as yard lamps) is used as the light source for the spectrometer, several spectral lines appear on the photographic plate, as shown in Fig. 23.17. Each line represents a wavelength in the spectrum of light emitted by mercury atoms. The spectra emitted by several types of atom are shown in Color Plate II.

**FIGURE 23.17**

When a spectrometer is used to photograph a slit illuminated by a mercury arc, several images of the slit (or spectral lines) appear on the photograph.

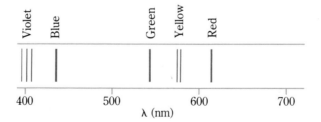

## 23.9 Polarized light

Many optical devices make use of the fact that light is a transverse vibration. As we shall see, this fact is important when light is transmitted through certain materials.

It is also a factor when light is reflected. Although our previous discussion was not concerned with this feature of light waves, it is fundamental to the behavior of light that concerns us in this section.

We saw in previous chapters that light is em radiation. It consists of waves such as the one shown in Fig. 23.18. The electric field vector is sinusoidal and perpendicular to the direction of propagation. If the wave is traveling along the $x$ axis, the electric field vibrates up and down at a given point in space as the wave passes by. There is a magnetic field wave perpendicular to the page and in step with the electric field. We call a wave such as this a **plane-polarized wave.** It derives its name from the fact that the electric field vector vibrates in only one plane, the plane of the page in this case.

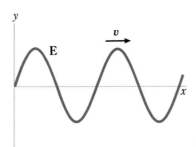

**FIGURE 23.18**

The electric-field vector vibrates in a single plane when a beam of light is plane-polarized.

Most light consists of many, many waves like the one in Fig. 23.18. If the direction of propagation is to the right, the electric field vectors must all vibrate perpendicular to the $x$ axis. However, they need not all vibrate in the plane of the page, and actually most of them do not. Let us stand at the end of the $x$ axis in Fig. 23.18 and look back along it toward the coordinate origin, in other words, with the wave traveling straight toward us. The great multitude of waves coming toward us give rise to many individual electric field vectors that are randomly oriented, as in Fig. 23.19a. If the waves were plane-polarized vertically, that is, in the plane of the page in Fig. 23.18, the approaching vectors would appear as shown in Fig. 23.19b. For a horizontally plane-polarized wave, the vectors would appear as in Fig. 23.19c.

**FIGURE 23.19**

If a narrow beam of light is coming straight out of the page, the electric-field vibration will be as shown for three types of beams.

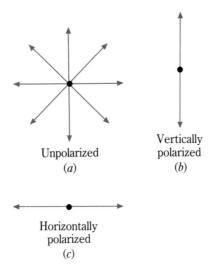

Unpolarized
(a)

Vertically polarized
(b)

Horizontally polarized
(c)

Unpolarized light can be conveniently plane-polarized using a polarizing sheet. This is a sheet of transparent plastic in which special needlelike crystals of iodoquinine sulfate have been embedded and oriented. The resulting sheet allows light to pass through it only if the electric field vector is vibrating in a specific direction. Hence, if unpolarized light is incident upon the sheet, the transmitted light will be plane-polarized and will consist of the sum of the electric field vector components parallel to the permitted direction. (Before the invention of the polarizing sheet by the Polaroid-Land Company, other methods were used, but because of their convenience and low cost, polarizing sheets have displaced these other methods except in certain very exacting situations.)

Any vector can be thought of as consisting of two perpendicular components. Hence, if the electric field is oriented as shown in Fig. 23.20a, it can be thought of as consisting of a vertical and horizontal component, as shown in b. If we pass light that is vibrating at the angle shown in Fig. 23.20 through a polarizing sheet whose transmission direction is vertical, the vertical component of the vibration will pass through and the horizontal component will be stopped.

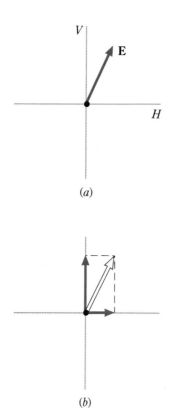

**FIGURE 23.20**

The electric-field vector can be split into $x$ and $y$ components.

Consider what happens when unpolarized light is passed through two polarizing sheets, as shown in Fig. 23.21. In part *a*, the polarizer (the first sheet) allows only the vertical vibrations to pass. These are also transmitted by the analyzer (the second sheet) since it too is vertical. In part *b*, however, the polarizer has been rotated through 90° and now allows only horizontal vibrations to pass. These are completely stopped by the vertically oriented analyzer. Therefore (almost) no light comes through the combination. In this latter case we say that the polarizer and analyzer are **crossed.**

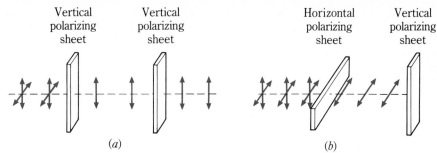

Vertical polarizing sheet    Vertical polarizing sheet      Horizontal polarizing sheet    Vertical polarizing sheet

(a)        (b)

**FIGURE 23.21**

(*a*) The unpolarized light is polarized by the first polarizing sheet (the polarizer). (*b*) The second polarizing sheet (the analyzer) and the polarizer are crossed, and the beam is completely stopped by the analyzer.

The polarization of light is used in many technical and scientific applications. For example, under a microscope, details can often be seen more clearly if they are examined between crossed polarizing sheets. Portions of the object that appear the same in ordinary light may differ considerably in their ability to change the polarization of the transmitted light. Hence, these otherwise unobservable details can easily be seen. When a transparent object is under high stress, it often rotates the plane of polarization of the transmitted light. As a result, a nonuniformly stressed object observed between crossed sheets shows alternate dark and bright bands, as in Fig. 23.22. Where the bands are closest together, the stress is most uneven. By examining plastic models of strained objects like that in Fig. 23.22, it is possible to tell exactly how the stress is distributed. This is of great importance in the design of various parts for machines.

**FIGURE 23.22**

A strained transparent object viewed through crossed Polaroids shows alternate dark and bright bands. The stress variation is greatest where the bands are closest together.

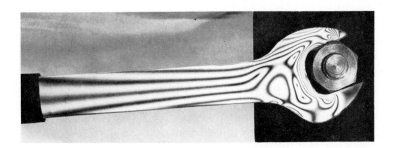

We seldom realize that the light coming from the blue sky is largely plane-polarized. Nor do we often remember that the glare of light from the surface of a lake or white concrete is considerably polarized. (This latter fact forms the basis for Polaroid sunglasses.) Since polarized light can be recognized only by the use of an analyzer, our eyes do not tell us of its presence. The whole subject of polarized light is an interesting one, and you may wish to pursue it further.*

---

\* A readable booklet on the subject in the Momentum Book Series is W. A. Shurcliff and S. S. Ballard, *Polarized Light,* Van Nostrand, Princeton, N.J., 1965.

# Minimum learning goals

When you finish this chapter, you should be able to

1 Define (*a*) myopia versus hyperopia, (*b*) astigmatism, (*c*) near point versus far point, (*d*) spherical aberration, (*e*) chromatic aberration, (*f*) diopter, (*g*) angular magnification and magnifying power, (*h*) spectral line, (*i*) plane-polarized wave, (*j*) crossed polarizing sheets.

2 Sketch the important features of the eye and explain the function of each.

3 Explain how eyeglasses can remedy myopia, hyperopia, and astigmatism. Work problems such as those posed in Examples 23.1 and 23.2.

4 Find the equivalent focal length for several thin lenses in close contact.

5 Explain the operation of the simple magnifier and compute its approximate magnification.

6 Show how the compound microscope operates by sketching its objective and eyepiece lenses and drawing a ray diagram for it.

7 Sketch the optical system for the astronomical telescope and locate the images it produces.

8 Explain how a prism spectrometer gives rise to line images. Describe how it separates colors and can be used to analyze a beam of light.

9 Explain how polarized light is produced and how it can be distinguished from unpolarized light.

# Questions and guesstimates

1 In Chap. 22, we learned that the image in a plane mirror is the same size as the object. Why then do we place our face very close to the mirror when we examine our bloodshot eyes?

2 Show that a real image of a woman formed by a converging lens is inverted but that she and her image still have the same right hand. Show that exactly the reverse is true for an image formed by a plane mirror.

3 Clearer images are obtained in optical instruments when only a small portion of the lens is used. In the case of the pinhole camera, no lens is needed. To see how this is possible, draw a small, bright object about 1 mm high 10 cm from a 1-cm opening in a large, opaque screen. Show how the bright spot cast by the object on a screen 5 cm behind the opening decreases in size as the opening is made smaller. Show that in the limit of a pinhole opening, two objects 1 cm apart that are both 10 cm from the opening give rise to well-defined images on the screen.

4 Show why a pinhole placed in front of a lens leads to a good image even when the image is not quite in focus. (See Question 3.)

5 A glass prism deviates a beam of blue light somewhat more than a beam of red light. Show by means of wavefronts how this leads us to conclude that red light travels faster in glass.

6 Which of the following, as normally used, form real images: (*a*) eye, (*b*) camera, (*c*) microscope, (*d*) telescope, (*e*) binocular, (*f*) slide projector, (*g*) plane mirror, (*h*) concave shaving mirror, (*i*) searchlight mirror?

7 Explain clearly why spectral lines are called lines.

8 One can buy a cheap microscope for use by children. Invariably, the images seen in such a microscope have colored edges. Why is this so?

9 Suppose that the inside of a box camera is filled with water and that the lens is made stronger so that the image still falls on the film surface. Will the pictures that the camera takes be changed in any way? Repeat for a box with only a pinhole and no lens.

10 Why are the shutter speed and lens speed of a camera important? What design factors influence these speeds?

11 What happens to the light energy that a polarizing sheet does not transmit when unpolarized light is incident on it? Can you think of any drawback this might pose to using a polarizing sheet?

12 How can one determine whether a beam of light is polarized? Whether it is composed of two beams, one polarized and the other not?

13 For a commercial camera in which the diameter of the lens opening, that is, the *aperture diameter,* is 5 mm, the proper exposure time for a scene is $\frac{1}{60}$ s. What is the proper exposure time for a pinhole camera using the same type of film if the diameter of the pinhole is 0.50 mm?

14 You have available a long, cylindrical, cardboard mailing tube and two lenses with focal lengths of 60 and 10 cm that can be fitted into the tube. Use these to design a toy telescope.

# Problems

1  A tree 3.0 m high is 14 m from a person. What is the height of the image on the person's retina? Assume that the eye lens is 1.50 cm from the retina.

2  If the length of an image on the retina is $b$ when the object is at the eye's near point (25 cm), what is it when the object is 2.0 m away?

■ 3  A boy is able to see the title on the outside of this book clearly when the book is no more than 110 cm away. What should the focal length of his eyeglasses be if he is to see distant objects clearly?

■ 4  A certain eyeglass prescription is listed as $f = -50$ cm. Describe the eye defect the lens is designed to correct.

■ 5  A student is able to see printing in a book clearly when the book is 40 cm away but not when it is closer. ($a$) Is she near- or farsighted? ($b$) What type of lens should she use to correct her sight, and what should its focal length be?

■ 6  A man is not able to see objects clearly unless they are closer than about 0.90 m. ($a$) Is he myopic or hyperopic? ($b$) What type of lens should he use to correct his sight, and what should its focal length be?

■ 7  A little boy wears thick, magnifying-glass-type eyeglasses. His older sister holds the eyeglasses in sunlight and obtains images of the sun. If each lens gives an image 35 cm from the lens, what are the boy's probable far point and near point without glasses?

■ 8  A teacher notices that a child in his class holds pages very close to her eyes when reading. The usual position for this child is 12.0 cm. ($a$) Is the child near- or farsighted? ($b$) What kind of lens should be used in the child's eyeglasses, and what should its focal length be?

9  If a certain single-lens camera has a lens-to-film distance of 8.0 cm and takes pictures that are $8 \times 6$ cm, how far from a painting that is $100 \times 100$ cm must the camera be placed if the image of the painting is to just fit on the photograph?

10  When the camera of Prob. 9 is used to photograph a tree from a distance of 25 m, the image on the film turns out to be 2.0 cm high. How tall is the tree?

■ 11  Show that the length of the image on the retina varies inversely with the distance of the object from the eye.

■ 12  A single-lens camera forms a clear image of a distant object when the lens is 8.0 cm from the film. ($a$) What is the focal length of the lens? ($b$) How far should the lens be moved to best focus an object that is 2.0 m away?

■ 13  A fixed-lens camera uses a lens with a focal length of 9.0 cm, and the film is 9.0 cm from the lens. An object 2.0 m from the camera is photographed. How far from the film is the image formed?

14  Find the focal length of the following thin lenses when they are placed in close combination: 8.0, $-5.0$, and 7.0 diopters.

15  Find the ($a$) power and ($b$) focal length of the following thin lenses when they are placed in close combination: focal lengths of 80, $-30$, and $-60$ cm.

16  What must the focal length of a thin lens be if the lens is to be placed in contact with a 2.0-diopter lens to form a combination lens that has a focal length of $-80$ cm?

17  A person whose near point is 20 cm views an object through a magnifying glass that has a focal length of 8.0 cm. What magnification does he obtain if the image is at ($a$) his near point and ($b$) infinity?

18  A magnifying glass with a 6.0-cm focal length is used by a nearsighted person in such a way that the final image is at her near point, 11.0 cm. What magnification does she obtain?

■ 19  An insect 0.20 mm long is viewed by the naked eye of a person whose near point is 25 cm. The person then views the same insect through a magnifying glass that has a focal length of 7.0 cm. What is the approximate ratio of the two image sizes on the retina?

■ 20  A person whose near point is 18 cm tries to use a diverging lens as a magnifier. ($a$) What must the focal length of the lens be if the person is to see a distinct image? ($b$) If $f$ is $-60$ cm for the lens, what maximum magnification can be achieved?

21  What is the approximate magnification of a simple microscope if the focal length of the objective lens is 2.0 cm and that of the eyepiece lens is 8.0 cm?

22  A simple microscope uses an eyepiece lens of focal length 5.0 cm and an objective lens of focal length 3.0 cm. What is the approximate magnification?

■ 23  A boy makes a simple microscope by cementing a lens that has a focal length of 5.0 cm to one end of a tube 15 cm long and a lens that has a focal length of 3.0 cm to the other end. ($a$) If he uses the 3.0-cm lens as the eyepiece, about how far in front of the objective must he place the specimen he is observing? ($b$) What is the approximate magnifying power of his microscope?

■ 24  Two marks 0.0200 mm apart are viewed in a microscope that has a magnifying power of 300. When viewed through the microscope, what angle (in degrees) do the marks subtend at the eye?

25 A telescope with an objective lens that is 15 cm in diameter requires a 3.0-min exposure to properly photograph a distant star. Assume that an identical telescope could be built but with a 25-cm diameter for its objective lens. What would the proper exposure time be using this second telescope?

26 A camera that has a lens-opening diameter (this is called the aperture) of 1.2 cm photographs a scene adequately when the exposure time is 0.015 s. If the aperture is decreased to 0.40 cm, what exposure time should be used?

27 A telescope at the Yerkes Observatory has an objective lens with a focal length of about 19 m. When this telescope is used to observe the moon, how many kilometers on the moon correspond to 1.0 cm on the image cast by the objective lens? The distance to the moon is $3.8 \times 10^8$ m.

28 A certain reflecting telescope uses as its objective a mirror that has a focal length of 70 cm. (a) How large an image of the moon does this mirror produce? (b) If the telescope uses an eyepiece that has a focal length of 4.0 cm, what is the magnifying power? (The distance to the moon = $3.8 \times 10^8$ m, and the moon's diameter is $3.5 \times 10^6$ m.)

29 You will notice in Fig. 23.11 that the telescope gives an inverted image. This is objectionable if one wishes to view an opera from a distant seat in an opera house. Instead, one can use an opera glass (that is, a *galilean telescope*), such as the one pictured in Fig. P23.1. For the one shown there, locate the position of the final image of a distant object. Is it real or virtual? Erect or inverted? What is the magnifying power of this telescope?

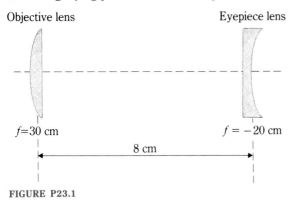

Objective lens           Eyepiece lens

$f = 30$ cm        $f = -20$ cm

8 cm

FIGURE P23.1

30 (a) By what factor is the light intensity in a telescope increased if the diameter of the objective lens is changed from 0.50 to 3.0 cm but all other dimensions remain constant? (b) By what factor is the light intensity increased if the objective diameter is increased from 0.50 to 3.0 cm and the focal length of the objective is tripled?

31 What is the magnifying power of a telescope that uses a 1.25-diopter objective and an eyepiece that has a magnifying power of 5?

32 What is the magnifying power of an astronomical telescope with a 1.50-diopter objective and an eyepiece that has a magnifying power of 3?

33 In a certain telescope, the distance between the eyepiece and the objective is 90 cm. The angular magnification of the telescope is 60. Find the focal lengths of the two lenses.

34 A girl has available two eyeglass lenses with focal lengths of $+90$ cm and $+30$ cm. She.wishes to place them in a cylindrical cardboard tube in such a way as to make a telescope that is as short as possible and yet has the largest possible value of angular magnification. (a) About how far apart should the lenses be? (b) About what will the magnification be?

35 The index of refraction of a certain glass is 1.4650 for $\lambda = 440$ nm and 1.4570 for $\lambda = 580$ nm. A beam of light consisting of only these two wavelengths enters a plate of this glass at an angle of 50°. Find the angle between the two beams inside the glass.

36 Show that if the apex (top) angle $A$ of a prism is very small (that is, if the prism is very thin) and a beam of light strikes the prism perpendicular to one of the faces, the deviation $D$ of the beam is given by $D = A(n - 1)$.

37 In Fig. 23.14, take the apex (top) angle of the prism to be 60°. If $n = 1.60$ and the angle of incidence is 53°, find (a) the angle at which the beam leaves the prism and (b) the angle of deviation $D$.

38 A certain piece of optical equipment contains a lens of focal length 40 cm made of glass for which $n = 1.60$. In order to cool the lens, that portion of the apparatus containing it is immersed in a rectangular vessel of water. In order for the apparatus to operate properly under water, a second lens made of the same type of glass must be placed in the water in contact with the original lens. What must the focal length of this second lens be?

# 24 Interference and diffraction

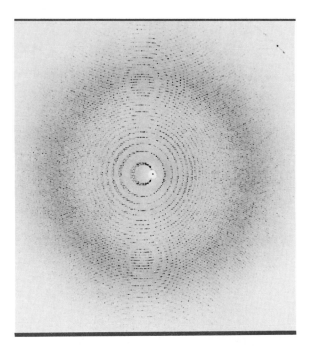

IN the preceding two chapters, we discussed the behavior of lenses and mirrors, using the concept of light rays. We did not need to know whether the light consisted of particles or waves for these discussions. This is not true of the topics treated in this chapter. We shall see that the wave nature of light gives rise to interference phenomena much like those we encountered in our study of wave motion and sound. The mere existence of these phenomena, as well as other effects discussed in this chapter, led to the final acceptance of the wave nature of light.

## 24.1 Huygens' principle and diffraction

Have you ever watched carefully as gentle water waves lap against a post or some other obstacle in their path? If you have, you have noticed that the waves seem to bend around the post instead of casting a clear shadow of it. A related situation is shown in Fig. 24.1, where we see plane water waves generated in a ripple tank. They strike a barrier that has a small hole in it. Notice how the waves pass through the hole and spread out to fill the whole region beyond the barrier.

This general type of behavior can be seen not only with water waves but also with sound waves and electromagnetic waves. It is a characteristic behavior of waves, and we give it a special name, diffraction:

Direction
of
wave motion
⟶

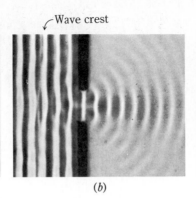

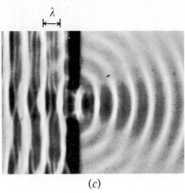

(a)

(b)

(c)

**FIGURE 24.1**
Plane water waves incident on a hole in a barrier.
Diffraction causes the waves to spread into the entire
region beyond the barrier. (*D.C. Heath/Education
Development Center*)

Waves are capable of bending around obstacles, a phenomenon called
**diffraction.** In other words, obstacles do not cast perfectly sharp shadows.

As we shall see, diffraction causes waves of all types to behave in seemingly peculiar
ways.
To explain the phenomenon of diffraction, Christian Huygens, a contemporary of
Newton's, postulated what is now known as **Huygens' principle**:

Each point on a wave crest acts as a new source of waves.

For example, in Fig. 24.1, the portion of the wave crest that strikes the tiny hole in
the barrier acts as a new wave source. As a result, waves spread outward from the
hole and fill the whole region beyond the barrier.
We see from this that obstacles in the path of waves cannot be expected to cast
clear shadows. This seems to disagree with what we know about light waves,
because objects in the path of light cast shadows that are easily observed. A clue to
the resolution of this discrepancy can be found in Fig. 24.1. Notice that in *a* the
wavelength λ is only about one-third as large as the width of the opening and the
waves spread only slightly into the region of shadow. When the wavelength is larger,
however, as in *c*, the waves spread into the shadow region much more. We shall
return to this phenomenon later.
During the centuries since Huygens proposed his principle, both experimental
work and theoretical work have established the basis for the principle and confirmed
its validity. We therefore assume, when convenient, that each point on a wave crest
acts as a new wave source. This simple artifice is useful in explaining the most
complex wave behavior, and you will see examples of its use as our discussion
proceeds.

## 24.2 Interference

An interesting experiment involving water waves is shown in Fig. 24.2. We see
there two oscillators that send two sets of identical water waves out along the water
surface. Notice what happens as the waves from the two sources interact with one

FIGURE 24.2

The two sources send out identical water waves that interfere constructively along the lines labeled *B* and destructively along those labeled *D*. (*D.C. Heath/Education Development Center*)

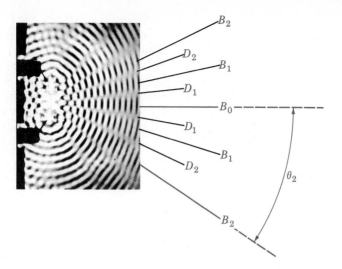

another. Along certain lines that radiate from the source (labeled *B*), they create very large wave crests; along other lines (labeled *D*), no wave crests are seen. Apparently the water waves from the two sources reinforce one another at certain points and cancel one another at other points. Let us now investigate such phenomena.

You will recall from our work with sound waves and waves on a string that identical waves can reinforce or cancel each other. To review this fact, consider the two waves *A* and *B* in Fig. 24.3. In part *a*, the two waves are in phase, crest on crest and trough on trough. When they are added, the resultant wave is twice as large as the original waves. The waves in *a* undergo *constructive* interference.

The situation in *b* is quite different, however. There wave *B* has been held back through one-half wavelength, $\lambda/2$, so that crest falls on trough for the two waves. The waves are $\lambda/2$, or $180°$, out of phase. Now when they are added, they exactly cancel each other, and so their sum wave is zero. The waves undergo destructive interference.

FIGURE 24.3

Identical waves can reinforce or cancel each other, depending on their relative phases.

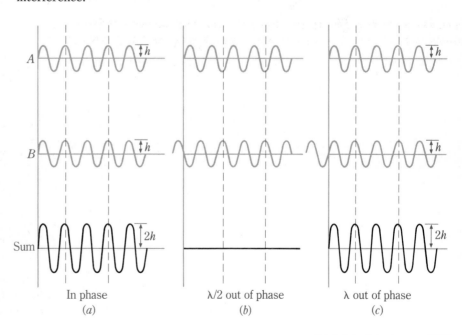

In *c*, we see what happens when wave *B* is $\lambda$ behind wave *A*. The two waves are now $\lambda$, or 360°, out of phase. Now crest falls on crest, and so the two waves add to give a resultant wave that is twice as large. The waves interfere constructively.

In general, we conclude (as we did in the case of sound waves) that *two identical waves interfere constructively if they are in phase with each other*. If one of the waves is retarded by a distance $\lambda$, $2\lambda$, $3\lambda$, and so on, relative to the other, they will still reinforce when combined because crest will still fall on crest. If the relative retardation is $\lambda/2$, $3(\lambda/2)$, $5(\lambda/2)$, and so on, then crest will fall on trough and the two waves will interfere destructively; they will cancel each other.

Let us now return to our discussion of the combined effect of two wave sources. We wish to find out why the waves from these sources reinforce in certain regions and cancel in others. This question is easily settled if we refer to Fig. 24.4. The two sources are sketched at points *A* and *B*. They send out identical waves in all directions. Consider first the waves they send in the horizontal direction, as shown in *a*. The waves are in phase, crest on crest and trough on trough. Hence waves that leave the source in this direction reinforce each other. This is why reinforcement occurs along line $B_0$ in Fig. 24.2.

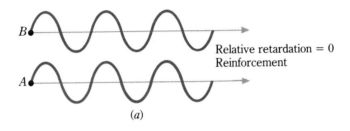

(a)

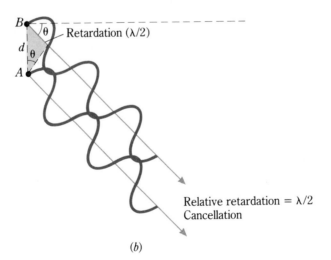

(b)

**FIGURE 24.4**

Reinforcement occurs at those angles for which the relative retardation is $\lambda$, $2\lambda$, $3\lambda$, and so on.

Consider next the waves that the sources send out in the direction shown in Fig. 24.4*b*. In this direction, the wave from source *B* is retarded by $\lambda/2$ relative to the wave from *A*. Now the crests of one wave fall on the troughs of the other. As a result, the waves leaving the sources in this direction cancel. That is why the water waves in Fig. 24.2 cancel along the two lines labeled $D_1$. (You should be able to explain why there are two lines labeled $D_1$ in Fig. 24.2.)

If you consider what would happen if the angle $\theta$ in Fig. 24.4 were increased still further, you see that wave $B$ would be held back still more relative to wave $A$. But if $\theta$ and thus the relative retardation were increased until the relative retardation was $\lambda$, then the two waves would reinforce each other again. That is what is happening along lines $B_1$ in Fig. 24.2. Along these two lines, the relative retardation is $\lambda$.

Reasoning in this way, you can convince yourself that the relative retardation is $3(\lambda/2)$ along cancellation lines $D_2$ and $2\lambda$ along reinforcement lines $B_2$. As we see, lines $B_0$, $B_1$, $B_2$, and similar lines represent lines along which the waves reinforce. Along them, the relative retardation between the waves from the two sources is 0, $\lambda$, $2\lambda$, and so on. The waves from the two sources therefore reinforce each other.

Let us now obtain a mathematical relation for the angles at which these reinforcing lines occur. To do this, examine the small shaded triangle in Fig. 24.4. Notice that the angle $\theta$ in it is equal to the angle $\theta$ that the rays make with the horizontal. We see at once that in the shaded triangle,

$$\text{Relative retardation} = d \sin \theta$$

where $d$ is the distance between the sources.

To find the angles at which reinforcement occurs, in other words, the angles to the points labeled $B$ in Fig. 24.2, we recall that, for reinforcement, the relative retardation must be 0, $\lambda$, $2\lambda$, $3\lambda$, or, in general, $n\lambda$, where $n$ is an integer. Therefore, if $\theta_n$ is the angle for which the relative retardation is $n\lambda$, we have

$$n\lambda = d \sin \theta_n \tag{24.1}$$

This equation tells us at what angles $\theta_n$ reinforcement occurs.

For example, along line $B_0$ in Fig. 24.2, we have $n = 0$ (because the waves are not retarded relative to each other), and so $\theta_0$ is

$$0 = d \sin \theta_0$$

$$\theta_0 = 0$$

Similarly, along line $B_2$ we have $n = 2$, and so

$$2\lambda = d \sin \theta_2$$

---

### Example 24.1

Suppose that the sources in Fig. 24.2 are 2.0 cm apart and the waves have a crest-to-crest distance (a wavelength) of 0.70 cm. What is the angle at which the reinforcement line $B_2$ occurs?

*Reasoning*  We are told that $d = 2.0$ cm and $\lambda = 0.70$ cm, and we are interested in the situation for which $n = 2$. Substituting in Eq. 24.1 gives

$$\sin \theta_2 = \frac{(2)(0.70 \text{ cm})}{2.0 \text{ cm}} = 0.70$$

from which $\theta_2 = 44°$. The lines $B_2$ in Fig. 24.2 would make $44°$ angles with the horizontal.  ■

---

*Exercise*  At what angle would line $B_1$ be found?  *Answer: 20.5°*  ■

## 24.3 Young's double-slit experiment

The experiment described in Sec. 24.2, on the interference of waves from two sources is not peculiar to water waves. You will recall from Sec. 14.7 that the prongs of a tuning fork can give rise to interference in sound waves. The explanation for this phenomenon is similar to the description of the interfering water waves, except that the waves are longitudinal sound waves rather than transverse water waves. It should be clear that any identical waves, transverse or longitudinal, are capable of exhibiting interference phenomena.

As we mentioned previously, Newton believed light to be corpuscular. He pictured it as a stream of particles shot out from light sources. Naturally, these particles must travel in straight lines, much as baseballs do. Although the Italian scientist Grimaldi had shown as early as 1660 that light can be diffracted, that is, bent around objects, Newton was able to devise an explanation of this fact in terms of his light corpuscles. His explanation was not very satisfying, but most people respected him highly and therefore accepted his pronouncements concerning the nature of light. It was not until after 1803 that the wavelike nature of light became widely accepted.

In 1803 and 1807, Thomas Young published the results of his experiments demonstrating the phenomenon of interference of light waves. He allowed a narrow beam of sunlight passing through a hole in a window shutter to fall on two narrow, parallel slits in a piece of cardboard, as illustrated schematically in Fig. 24.5. On a viewing screen placed to the right of the slits, he observed a pattern of alternating bright and dark regions called **fringes.** His observations of these fringes, and his inference that light is a wave phenomenon, allowed him to compute the wavelength of light for the first time. Let us see how he did it.

**FIGURE 24.5**

The two slits $S_1$ and $S_2$ act as sources for two coherent waves. In the case of light waves, the interference fringes are usually only a few millimeters apart. (Compare this experiment with Fig. 24.2 for water waves.)

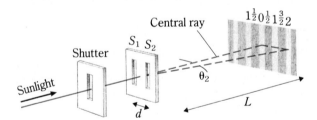

If you look back at Fig. 24.2, you will notice that a vertical wall placed at the right-hand edge of the figure would show a water-wave pattern. At the points labeled $B$, the water-wave crests are high, but where the lines labeled $D$ intersect the wall, the water is calm. This interference pattern on the wall corresponds to the fringes in Fig. 24.5. The bright fringes in Fig. 24.5 correspond to the positions labeled $B$ in the (imagined) water-wave interference pattern of Fig. 24.2. As you might suspect, the $D$ positions correspond to the dark positions in Young's double-slit pattern.

Using this correspondence with the water interference experiment, we can interpret Young's pattern as follows. The two slits act as two light sources that send out identical waves. Brightness occurs on the screen at the fringe labeled 0 because the waves traveling to this position reinforce each other; their relative retardation is zero.

At fringe 2, the two waves again reinforce. These fringes correspond to regions $B_2$ in Fig. 24.2, and so the relative retardation between the waves reaching there is

$2\lambda$, where $\lambda$ is the wavelength of the light waves. As we see, the situations in Figs. 24.2 and 24.5 are completely analogous. We can therefore apply Eq. 24.1 to Young's double-slit fringes and write

$$n\lambda = d \sin \theta_n \qquad (24.1)$$

In this expression, $n$ is the fringe number as labeled in Fig. 24.5. The slit separation is $d$, and the wavelength of the light is $\lambda$. Just as $\theta_2$ is the angle between the ray to the central fringe (0) and the ray to fringe 2, $\theta_n$ is the angle between the central ray and the ray to fringe $n$. Frequently we call $n$ the *order number* of the fringe. Using this terminology, we call the $\theta_2$ fringe the *second-order fringe*.

Young used Eq. 24.1 to compute the wavelength of light. Because he used sunlight, his light beam consisted of all the colors of the rainbow and so contained an infinite number of wavelengths. Each wavelength had its bright fringes at the angles given by Eq. 24.1. Consequently, his fringes consisted of a multitude of interlocking fringes, one for each wavelength in his light. Because of this, his fringes were colored; the portion of the fringe closest to the center was blue, while the outer part was red. (Why?)

In a typical experiment, consider the distance $L$ in Fig. 24.5 to be 120 cm. The slit separation $d = 0.025$ cm, and the distance from the center of the pattern to the approximate center of the fringe labeled 2 is 0.50 cm. To find $\theta_2$, we see from Fig. 24.5 that

$$\tan \theta_2 = \frac{\text{distance } 0 \rightarrow 2}{L} = \frac{0.50 \text{ cm}}{120 \text{ cm}} = 0.00417$$

from which $\theta_2 = 0.24°$.

Young made use of data such as these to compute the wavelength of the light near the center of a typical fringe. Substituting in Eq. 24.1, he obtained

$$\lambda = \frac{d}{n} \sin \theta_n = \frac{0.025 \times 10^{-2} \text{ m}}{2} \sin 0.24 = 5.2 \times 10^{-7} \text{ m}$$

He was therefore able to conclude that light had a wavelength of about 500 nm, with the wavelength of blue light being somewhat shorter than this and that of red light somewhat longer.

Today, of course, we are not restricted to sunlight in performing a double-slit experiment. Laser light, for example, is monochromatic (that is, of a single wavelength). Even though the laser beam consists of only a single wavelength, the interference fringes obtained in a double-slit experiment are still quite wide. A photograph of a typical fringe pattern using monochromatic light is shown in Fig. 24.6. The numbers are analogous to those in Fig. 24.5.

**FIGURE 24.6**

Interference fringes produced by a double-slit system using monochromatic (single-wavelength) light. (*After Jenkins and White*)

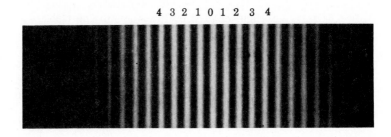

4 3 2 1 0 1 2 3 4

# Thomas Young 1773 – 1829

An English physicist and physician, Young was a child prodigy whose marvelous intellect served science throughout his life. By the age of 4, he had twice read the Bible from cover to cover. During his youth, he learned several languages, including not only Latin and Greek but also Hebrew, Arabic, Persian, and Turkish. He studied medicine, but his poor bedside manner hindered his progress in his chosen profession. Luckily, he inherited considerable wealth from an uncle and became financially independent. He was therefore free to follow his interest in science.

Young's earliest research involved the human eye. While still a medical student, he discovered the mechanism of the accommodation ability of the eye. The cause of astigmatism was another of his discoveries. His interest in the human eye led him to investigate the nature of light, as explained in this chapter. Other investigations culminated in the discovery of how the eye perceives color; he was also one of the first to realize that light waves are transverse.

In addition to his studies involving light, Young made notable discoveries in the areas of heat, elasticity, and surface tension. He wrote many articles for the *Encyclopedia Britannica* and contributed much to the scientific discussions of his day. Impelled by his wide-ranging interests, he became an expert on ancient Egyptian hieroglyphics and was among the first to decipher them.

## 24.4 Coherent waves and sources

You might suppose that we could obtain a pattern similar to that of the double slit using two tiny light bulbs in place of the two slits. The bulbs should act as two sources, and the waves from them should interfere. This would be a false assumption, however, for reasons that we point out now.

In describing interference effects, we have used waves that are identical in shape and wavelength. Moreover, we have always assumed that the waves maintain a fixed phase relation to each other. Two such waves are said to be *coherent*:

Coherent waves have the same form and wavelength and have a fixed phase relation to each other.

Sources of coherent waves are called *coherent sources.*

The light beams coming from two light bulbs are not coherent, however. The light emitted by the atoms in one bulb may have the same wavelength as that emitted by the atoms in the other, but the atoms in one act independently of those in the other and so the waves sent out by the atoms have random phases. In other words, the crests of one have no fixed relation to the crests of the other. Moreover, the wave emitted by a source is the combination of the multitude of waves emitted by its atoms. The shapes of the two waves from the two sources are very complex and quite unlike each other. We say that waves such as this are *noncoherent.* They cannot give rise to the steady interference effects we have been discussing.

Because two sources of light are almost always noncoherent, it is usually necessary to divide a single light beam into two parts to obtain an interference pattern. For example, in the double-slit experiment, the two slits are illuminated by the same beam, the same light wave, and this wave is divided into two distinct parts by the two slits. Because the two waves thus generated are part of the same wave, they are coherent and give rise to the interference effects discussed. We shall see in the next section another common way of obtaining two coherent waves.

## 24.5 Interference in thin films

One of the most common interference effects is the colored fringes we often see in soap and oil films (Color Plate III). When a soapy plate drains, colored reflections often occur from it. A similar effect occurs when light is reflected from a rain puddle that has an oil slick on it. Let us now analyze this important type of interference.

Consider what happens when a monochromatic beam of light is reflected from the wedge of glass in Fig. 24.7. Two reflected beams are obtained, one from the upper surface and one from the lower surface. Because they are part of the same beam, the

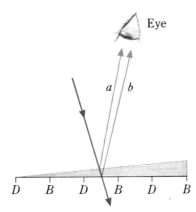

**FIGURE 24.7**

Interference occurs between the two reflected beams. Bright fringes occur at *B*, and dark ones occur at *D*. We neglect the small deflections due to refraction at the surfaces.

two reflected beams are coherent. They therefore give rise to interference of the type we have been discussing. Ray *b* is retarded, or delayed, relative to ray *a* because *b* has traveled farther. We know that the following occurs for various relative retardations:

Reinforcement (brightness) occurs for relative retardations of $\lambda$, $2\lambda$, $3\lambda$, and so on. Cancellation (darkness) occurs for relative retardations of $\lambda/2$, $\lambda + \lambda/2$, $2\lambda + \lambda/2$, and so on.

It is not surprising, then, that we see interference effects as we look at the wedge in Fig. 24.7. If we use monochromatic light, we see alternate regions of darkness (at positions *D*) and brightness (at *B*), which we have called fringes. At positions of darkness, the relative retardation is $n\lambda + \lambda/2$, where *n* is any integer. At positions of brightness, the relative retardation is $n\lambda$.

If white light (which contains all colors) is used in a situation such as this, the different wavelengths reinforce at different places. As a result, the fringes are highly colored, and we have the rainbow effect mentioned earlier for soap and oil films, which are often very thin and wedgelike. Which would you expect to be more closely spaced in such an interference pattern, the blue or the red fringes?

A famous experiment conducted by Newton that illustrates this phenomenon is shown in Fig. 24.8. A planoconvex lens (much less curved than the one shown) is placed on a flat glass plate and illuminated from above with monochromatic light. Rays reflected to the eye from the surfaces of the air wedge formed by the lens and the plate give rise to the fringe pattern shown in Fig. 24.8b, a pattern called *Newton's rings*. We recognize that the air wedge between the lens and the plate gives rise to the same general effect we diagramed in Fig. 24.7. Here, however, the fringes are circular because of the circular geometry of the wedge formed by the lens.

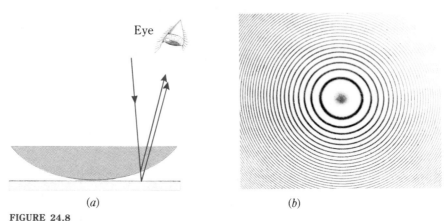

Eye

(a)                    (b)

**FIGURE 24.8**

The ray reflected from the lower side of the lens in (a) interferes with the ray reflected from the glass plate and gives rise to the Newton's rings shown in (b). Why is the center point dark? The angles in (a) are distorted so as to show the two rays clearly. (*Bausch & Lomb*)

One would think that the center point in the ring pattern should be bright since the separation there is nearly zero and both rays travel the same distance.* Thinking that there must be air between the lens and the plate, early workers tried in vain to push the plate closer to the lens to produce a bright center spot. The better the contact between the two surfaces, however, the clearer it became that the center point was dark, not bright. This is now known to be the result of the fact that the ray reflected from the lower surface suffers a 180° phase change upon reflection. It is, in effect, held back through $\lambda/2$ by the reflective process. It therefore interferes destructively with the other beam, since that beam undergoes no such phase change on reflection. There is much evidence, both experimental and theoretical, for this 180° phase change (or $\lambda/2$ retardation) on reflection, and we state that

Light reflected near normal incidence undergoes a 180° phase change if the refractive index of the reflecting material is larger than that of the surrounding material.

Reflection by a material of lower refractive index than the medium in which the rays are traveling causes no phase change.

---

* This could also be said about the left edge of the wedge in Fig. 24.7. Experiment shows, however, that it too is the location of a dark fringe.

Figure 24.7 also shows this effect. Notice there that a dark fringe occurs at the vertex of the wedge. In order to locate the positions of the other fringes, however, we must determine how much ray $b$ is retarded as a result of moving through the material of the wedge. We investigate this question in the next section.

## 24.6 Equivalent optical path length

Light waves travel fastest in vacuum; their speed there is $c$. It therefore takes longer for a light wave to travel through a length $L$ of glass than to travel through the same length of vacuum. In effect, the path length through the glass appears to be longer than $L$. We wish now to find the path length in vacuum that is equivalent to a path length $L$ in another material, such as the coating on the glass in Fig. 24.9.

The fundamental equation relating the wavelength, frequency, and speed of a wave is

$$\lambda = \frac{\text{speed}}{\text{frequency}}$$

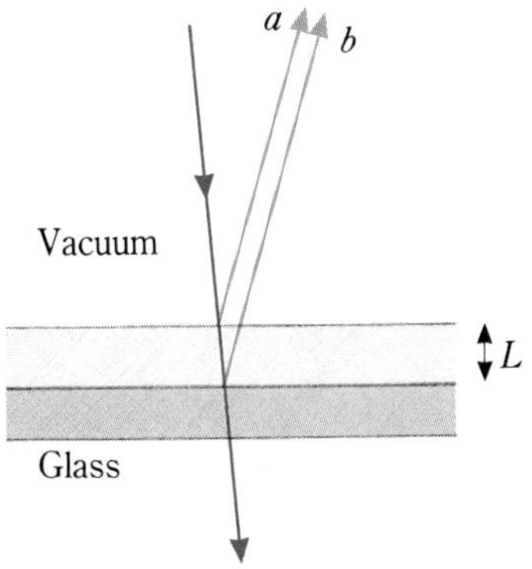

When a beam of light travels from one material into another, its speed changes but its frequency does not.* The beam in Fig. 24.9 travels from vacuum into a layer of material with refractive index $n$, in which the speed of light is $v$. If we designate the wavelength in vacuum by $\lambda$ and that in the layer by $\lambda_m$, we have

$$\lambda = \frac{c}{f} \qquad \text{and} \qquad \lambda_m = \frac{v}{f}$$

Dividing $\lambda$ by $\lambda_m$ gives

$$\frac{\lambda}{\lambda_m} = \frac{c}{v} = n$$

$$\lambda_m = \frac{\lambda}{n} \tag{24.2}$$

Because the index of refraction $n$ is greater than unity, the wavelength is shorter in the material than in vacuum.

Let us compute how many wavelengths thick the layer on the glass plate in Fig. 24.9 is. If we call the thickness of the layer $L$, we have

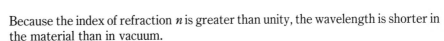

$$\text{Number of wavelengths in layer} = \frac{L}{\lambda_m} = \frac{L}{\lambda/n} = \frac{Ln}{\lambda}$$

In other words, a length $L$ of the material contains as many wavelengths as there are in a length $nL$ of vacuum. Therefore, in terms of wavelengths, *a length $L$ in a material of refractive index $n$ is equivalent to a path length $nL$ in vacuum.* We term the quantity $nL$ the **equivalent optical path length.**

---

* According to Huygens' principle, each wave crest striking the boundary acts as a new source of waves; the frequency of waves entering the boundary therefore equals the frequency of those leaving it.

---

### FIGURE 24.9

Vacuum

Glass

$L$

$a$ $b$

The coating on the glass plate has a thickness $L$ and a refractive index intermediate to that of vacuum and glass.

The equivalent optical path length tells us how much a beam is retarded by traversing a length $L$ of a material that has a refractive index $n$. In that length, there are the same number of wavelengths as there are in a length $nL$ of vacuum. Hence the retardation is $nL/\lambda$ wavelengths.

---

**Example 24.2**

In Fig. 24.8, by how much does the thickness of the air gap increase as you go from the location of one bright fringe to the next larger bright fringe? Assume the light used has $\lambda = 589$ nm.

*Reasoning*    As we go from one bright fringe to the next (that is, from reinforcement to reinforcement), one ray must be held back an additional wavelength relative to the other. Since the beam traverses the gap twice, going down and then up, the gap thickness increases by $\lambda/2$ as we go from one bright fringe to the next. Therefore,

$$\text{Gap thickness increase} = \lambda/2 = \frac{589 \text{ nm}}{2} = 295 \text{ nm} \quad \blacksquare$$

---

*Exercise*    Suppose the gap is filled with water. What is the gap increase between bright fringes?    *Answer: 221 nm* ▪

*Exercise*    How thick is the air gap at the seventh bright fringe?
*Answer: 1914 nm* ▪

*Exercise*    How thick is the air gap at the first bright fringe?
*Answer: 147.2 nm* ▪

---

**Example 24.3**

In order to reduce reflection, camera lenses are sometimes coated with a thin layer of magnesium fluoride ($n = 1.38$). How thick should such a coating be for minimum reflection at 550 nm?

*Reasoning*    We can use Fig. 24.9 and assume that the thickness $L$ is the magnesium fluoride layer coating the glass of the lens. If we assume normal incidence, ray $b$ travels $2L$ farther through the magnesium fluoride than ray $a$ does. Since both rays undergo a $180°$ phase change in reflection (why?), no relative retardation results from this reflection. The retardation between $a$ and $b$ is $2nL$. For no reflection, $a$ and $b$ must interfere destructively with each other. This occurs if

$2nL = \lambda/2, \lambda + \lambda/2$, and so on

For the thinnest possible coating, $2nL = \frac{1}{2}\lambda$, and so

$$L = \frac{\lambda/2}{2n} = \frac{(550 \text{ nm})/2}{2(1.38)} = 99.6 \text{ nm} \quad \blacksquare$$

---

## 24.7 The diffraction grating

Although Thomas Young used his double-slit experiment to measure the wavelength of light, the double-slit pattern (Fig. 24.6) is far too diffuse to give accurate results. It turns out that a large number of equally spaced slits gives a much sharper fringe system. For example, Fig. 24.10 shows the interference pattern for 20 parallel slits. Notice how very sharp the fringes are. To measure wavelengths with high precision, a large number of parallel, equally spaced slits is used. We call such a

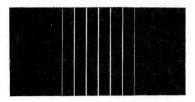

**FIGURE 24.10**

The interference pattern produced by 20 parallel, equally spaced slits. (*After Jenkins and White*)

device a **diffraction grating.** A typical grating might consist of 10,000 parallel slits, each separated from the next by a distance $d = 10^{-4}$ cm. Let us now discuss the behavior of such a grating.

A diffraction grating is typically used as in Fig. 24.11. The device shown there is similar to the prism spectrometer we discussed in connection with Fig. 23.16, but here the prism is replaced by a diffraction grating. The entrance slit is brightly lit by the light source whose wavelengths we wish to examine. Because the slit is at the

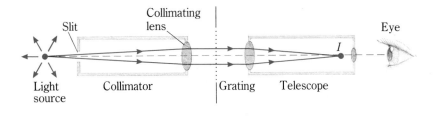

**FIGURE 24.11**

A schematic diagram of a grating spectrometer.

focal point of the collimating lens, parallel light exits from this lens and is incident perpendicularly on the grating. The light passing through the grating is observed with the telescope. As with the prism spectrometer, we see a sharp image of the slit when we view the undeviated beam.

Now suppose we swing the telescope to the side, as in Fig. 24.12. At most angles $\theta$, we see nothing. At certain angles, however, we see very sharp images of the slit,

**FIGURE 24.12**

When the telescope is rotated on the arc of a circle, an image of the slit is formed by interference at an angle $\theta$ to the straight-through beam.

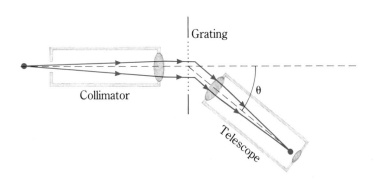

images we have called spectral lines. These images are the equivalent of the fringes in Fig. 24.10, but these images produced by the grating are even sharper than those produced by only 20 parallel slits. We now discuss how these images of the slit, the spectral lines, can be used to find the wavelength of the light entering the spectrometer.

Consider what happens when $\theta = 0$ in Fig. 24.12. We then see the straight-through beam, the beam shown in Fig. 24.13a. All the beams from all the slits travel the same distance, and so they reinforce. When the telescope is at $\theta = 0$, therefore, we see a clear, bright image of the slit. This image at $0°$ is called by various names: the *central maximum*, the *zeroth-order maximum*, and the *central image*. It has the same color as the source since it contains all wavelengths emitted by the source.

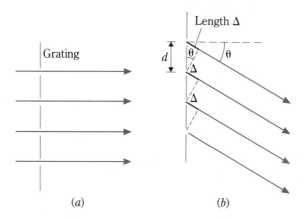

FIGURE 24.13

(a) The relative retardation of the straight-through rays is zero. (b) When the retardation $\Delta$ is a whole number of wavelengths long, the rays all reinforce one another. It is at these angles that the grating yields maxima.

If we now move the telescope to other positions, we find no light at most values of $\theta$. If the source emits only one wavelength, however (or perhaps only a few), then at certain angles we observe very sharp, highly colored images of the slit. To understand how these arise, look at Fig. 24.13b.

The rays of light from the various slits are retarded relative to each other, and the difference in path length for adjacent slits is $\Delta$. We know that if $\Delta$ equals $\lambda$ or $2\lambda$ or $3\lambda$ or in fact $n\lambda$, where $n$ is an integer, then all the rays reinforce each other. Hence if $\theta$ is just right for $\Delta$ to be equal to a whole number of wavelengths, then the light from all the slits will reinforce. We thus find that brightness is observed in the telescope when $\Delta = n\lambda$.

From each little triangle in Fig. 24.13b, we have that

$$\sin \theta = \frac{\Delta}{d}$$

where $d$ is the distance between slits, called the *grating spacing*. For brightness, we must have $\Delta = n\lambda$, and so we find bright images of the slit when $\theta$ is equal to values $\theta_n$ given by

$$n\lambda = d \sin \theta_n \qquad n = 1, 2, 3 \ldots \qquad (24.3)$$

This is called the grating equation.

To better understand the grating equation, suppose the light source used contains only two wavelengths: 500 nm and 600 nm. Suppose further that $d = 2 \times 10^{-6}$ m. Substituting these values in Eq. 24.3, we find the image positions listed in

**TABLE 24.1**
**Positions of spectral lines**

| $\lambda$ (nm) | $n$ | $\theta_n$ (degrees) |
|---|---|---|
| 500 and 600 | 0 | 0 |
| 500 | 1 | 14.5 |
| 600 | 1 | 17.5 |
| 500 | 2 | 30.0 |
| 600 | 2 | 36.9 |
| 500 | 3 | 48.6 |
| 600 | 3 | 64.2 |
| 500 | 4 | 90 |
| 600 | 4 | missing |

Table 24.1. These images are also shown in Fig. 24.14 along with the names we use to describe them. Because the fourth-order spectral lines occur at $\theta = 90°$ and larger, they cannot be observed. Also notice that the lines appear on both sides of the central maximum (the zeroth-order image).

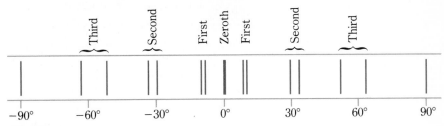

**FIGURE 24.14**
The first, second, and third spectral orders each contain two lines, one for the 500-nm light and one for the 600-nm light.

Since we can measure with high precision the angle $\theta_n$ at which the $n$th-order maximum occurs, it is necessary to know only the grating spacing $d$ in order to determine $\lambda$ accurately. For example, if the yellow light from a sodium-arc lamp is used in even a simple spectrometer, it is not difficult to see that the sodium light gives *two* slit images (or lines) at each order position. These lines are very close together and have wavelengths of 589.0 and 589.6 nm. The mere fact that one is able to see these two lines as distinct images provides some measure of the potential accuracy of such a device.

If mercury light is used in a grating spectrometer, several lines are seen at each order. As pointed out in Fig. 23.17, mercury light consists of a yellow line (579.0 nm), a greenish-yellow line (546.1 nm), a blue line (435.8 nm), and a violet line (404.7 nm). Other fainter lines are also visible. In later chapters, we shall see that other types of light sources also exist. While the light from a mercury arc consists of several easily observed discrete lines (called a *bright-line spectrum*), the light from an incandescent bulb contains all the colors, and so no sharp lines are observed. The incandescent source gives off what is called a *continuous spectrum* because a continuous band of color is seen when it is used in a spectrometer. These facts are shown clearly in Color Plate II.

---

**Example 24.4**
A particular grating has 10,000 lines per centimeter. At what angles does the 589.0-nm line appear?

**Reasoning**   The grating spacing $d$ is 0.0001 cm $= 10^{-6}$ m. Using the grating equation, we have

$$\sin \theta_1 = \frac{5.89 \times 10^{-7}}{10^{-6}} = 0.589$$

From a table of sines, $\theta_1 = 36°$, and so the first-order images are found at this angle on each side of the central maximum. For the second order, we have

$$\sin \theta_2 = 2\left(\frac{5.89 \times 10^{-7}}{10^{-6}}\right) = 1.178$$

Because it is impossible for the sine of an angle to be greater than unity, this and higher-order images do not exist. ■

## 24.8 Diffraction by a single slit

Until now, we have assumed that the slit width (the width of the open space) was negligible relative to the wavelength of light being used. If you look at Fig. 24.1, you will see that the diffraction for longer wavelengths is greater than the diffraction for shorter wavelengths. Thus diffraction appears to depend on the size of the wavelength relative to the width of the slit. We wish now to investigate the reasons for this effect and to describe more fully how a single slit diffracts light. The result we obtain has fundamental importance and places limits on our ability to make measurements.

To see the effect of the diffraction of light waves, we can send light through a single slit and record the transmitted light on a photographic film, as shown in Fig. 24.15. The central bright spot is considerably wider than the slit. Moreover, bright bands occur on each side of the central image, and these must result from some sort of interference effect. Let us now see what is involved in this situation.

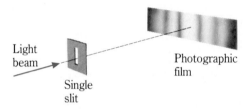

**FIGURE 24.15**
Single-slit diffraction pattern. (Not to scale.)

Consider a wave crest as it strikes the slit. According to Huygens' principle, each tiny point on the crest acts as a source of new waves. Thus light rays emanate from all points along the crest. Some rays travel straight forward, while others make an angle $\theta$ to the forward direction. As shown in Fig. 24.16$a$, the light rays that go straight through the slit are all in phase with one another. For this reason, the straight-through position is bright and gives rise to the central bright spot in Fig. 24.15. However, at an angle $\theta$ to the straight-through beam, rays from various parts of the slit travel different distances to the film. The most important situations are shown in Fig. 24.16$b$, $c$, and $d$.*

In $b$, ray $B$ from the middle of the slit is half a wavelength behind ray $A$. As a result, these two rays cancel each other. That is not all, however, because we see that the rays leaving the slit from positions just above $A$ and $B$ also cancel since they, too, have a path difference of $\lambda/2$. In fact, for each ray leaving the lower half of the slit there is leaving the upper half a corresponding ray that cancels it. Hence, at this angle $\theta$, no light reaches the film from the slit and we observe darkness. As you can see from the figure, this situation occurs when $\sin \theta = \lambda/b$, where $b$ is the slit width. Notice that if the slit width $b$ is equal to the wavelength of the light, the dark spot occurs at $\theta = 90°$. In other words, *if the slit width is decreased until it is as small*

---

* If the rays were exactly parallel, they could not meet; as a result, they would not interfere with each other. We consider here either of the following situations: ($a$) a lens focuses the parallel rays to a point or ($b$) slight nonparallelism causes the rays to meet at a point.

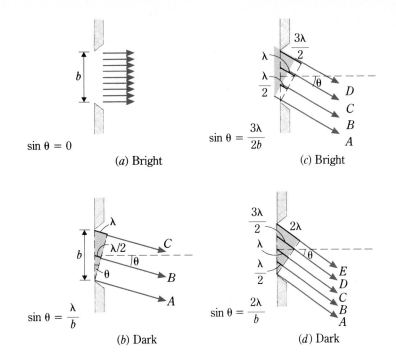

**FIGURE 24.16**
In analyzing the single-slit pattern qualitatively, we section the slit into portions whose rays differ by $\lambda/2$ in path length. Why?

as $\lambda$, the image of the slit spreads to become infinitely wide.

If $b$ is considerably larger than $\lambda$, as in Fig. 24.16, a side bright fringe occurs for the angle $\theta$ shown in part $c$. In this case, the rays from the bottom third of the slit cancel those from the center third while the top third is left uncanceled. Darkness is again achieved at the larger angle shown in part $d$. Here the slit can be thought of as being divided into fourths. The bottom one-fourth of the slit is canceled by the portion just above it. Similarly, the two upper sections also cancel. Hence darkness is observed at this angle.

The most important feature of the single-slit pattern, for our purposes, is the position of the first minimum next to the central maximum. Calling the angle between the central maximum and the first minimum $\theta_c$, we have found that

$$\sin \theta_c = \frac{\lambda}{b} \qquad (24.4)$$

This relation is used in the next section.

## 24.9 Diffraction and the limits of resolution

One of the most important consequences of diffraction is that it limits our ability to observe very fine details. We can appreciate this difficulty by referring to Fig. 24.17: two light sources sending light through a slit to a viewing screen. If the slit is quite small, the images cast on the screen are accompanied by nonnegligible diffraction fringes, as shown. These fringes are the result of the light's having passed through the slit, whose width we specify to be $b$.

You can begin to understand the difficulty this presents if you consider the following rough analogy, which we pursue further later. The pupil of the eye

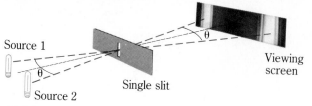

Source 1

Source 2

Single slit

θ

Viewing screen

**FIGURE 24.17**

In the situation diagrammed here, the two sources are well resolved on the screen because their diffraction patterns do not overlap seriously.

corresponds roughly to the slit, and two lines on an object being viewed by the eye are the two details the eye is looking at. The retina acts as a screen. Because the images on the retina are made fuzzy by the diffraction effects of the slit (the pupil), the eye is hindered in seeing fine detail in the object being viewed.

Returning to the situation in Fig. 24.17, we can see the images of the two objects on the screen as separate entities as long as the angle $\theta$ is not too small. Difficulty arises when $\theta$ is so small that the diffraction patterns overlap appreciably. They can no longer be seen as separate (that is, they can no longer be *resolved* ) when they are close enough together for the central maximum of one pattern to fall on the first minimum of the other pattern. In that situation, the case of *minimum resolution,* angle $\theta = \theta_c$, where $\theta_c$ is as defined in the preceding section.

Using the result of the preceding section, we can resolve the sources only if their angular separation $\theta$ is larger than $\theta_c$, where

$$\sin \theta_c = \frac{\lambda}{b} \tag{24.4}$$

As we expect, the smaller the slit width $b$, the farther apart the objects must be if they are to be resolved, for the interference pattern broadens with decreasing slit width.

Although our discussion has been in terms of objects viewed through a slit, a similar expression applies if the slit is replaced by a circular hole, such as the pupil of the eye or the objective lens of a microscope. For those situations,

The limiting angle $\theta_c$ for the resolution of two objects viewed through a circular aperture of diameter $D$ is

$$\sin \theta_c = 1.22 \frac{\lambda}{D} \tag{24.5}$$

Now let us see what sort of limit this diffraction effect places on our ability to view objects with a microscope.

We show in Fig. 24.18 the objective lens of a microscope and two details $S_1$ and $S_2$ of an object being viewed. The details are a distance $s$ apart, where $s$ is much smaller than shown, the objective lens has a diameter $D$, and the details are a distance $d$ from the lens. How close together can the details be and still be resolved?

According to Eq. 24.5, the details can just be resolved if the angle $\theta_c$ they subtend is

$$\sin \theta_c = 1.22 \frac{\lambda}{D}$$

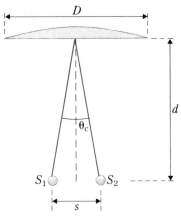

**FIGURE 24.18**
The two sources can just be resolved if $\theta = \theta_c$.

From Fig. 24.18, we see that

$$\sin(\theta_c/2) = \frac{s/2}{\sqrt{d^2 + (s/2)^2}} \cong \frac{s}{2d}$$

because $s$ is actually very much smaller than $d$.

For small angles, the angle in radians is equal to its sine. Since $\theta_c$ is usually very small, we can replace the sine by its argument in radians to give

$$\theta_c = \frac{s}{d}$$

Making this same approximation in Eq. 24.5, we have

$$\theta_c = 1.22\,\frac{\lambda}{D}$$

Equating these two expressions for $\theta_c$ yields, after a little arithmetic,

$$s = 1.22\left(\frac{d}{D}\right)\lambda$$

If we look at Fig. 24.18, we see that $d/D$ is the ratio of the object distance to the lens diameter. In all normal uses of microscopes, this ratio is approximately unity. As a result, to a rough approximation,

$$s \approx \lambda$$

In other words, *the smallest detail that can be seen in a microscope is about the same size as the wavelength of light being used.* This is a fundamental restriction imposed by diffraction; it cannot be circumvented by use of perfect lenses or ingenious microscope design.

As we see, diffraction effects cause images to be fuzzy. Another example of this

# Picturing the interior of the human body

When a dentist x-rays your teeth or a doctor uses x-rays to examine a bone fracture, a shadow diagram is the result. A typical dental x-ray is shown in Fig. *a*. Notice that it shows the shadow cast by the teeth as an x-ray beam passes through them to a film.

With the advent of extremely fast and compact computers, the x-ray method has become far more powerful. The method we describe here is a simplified outline of a CAT scan (for *c*omputer-*a*ssisted *t*omography). It provides a picture of a flat slice through the body. (The word tomography comes from the Greek *tomos* = slice and *graph* = picture.) This same general technique can be used with ultrasound and, indeed, with any penetrating radiation whose wavelength is small enough to make diffraction effects minimal.

Suppose an x-ray beam is sent through a person's body, as in *b*. Each little piece of the body through which it passes absorbs some of the beam. A little piece of mass $\Delta m$ in the path of the beam absorbs a fraction $k \Delta m$ of the beam, where $k$ represents the absorbing ability of the material. For x-rays, bone has a larger $k$ value than flesh, for example. Think of the path of the x-ray beam through the body as traversing $N$ little pieces of mass laid end to end. Hence, in passing through the body, the beam passes through $N$ absorbing masses. Therefore, the fraction of the beam absorbed before it reaches the detector is

$$\text{Fraction} = k_1 \Delta m_1 + k_2 \Delta m_2 + k_3 \Delta m_3 + \cdots + k_N \Delta m_N$$

which is simply the sum of the beam fractions that are absorbed by the $N$ tiny masses in its path. (This equation is only approximate, but it is sufficient for our purposes.)

We would like this measurement of the fraction of radiation absorbed to tell us about the little pieces of mass through which the beam passes. If it could tell us the values of $k_1$, $k_2$, . . . , $k_N$, then we would have a good idea of which pieces are flesh, which are bone, and so on. This is a single equation with $N$ unknowns,

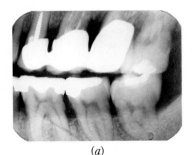

(a)

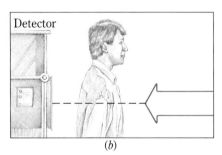

(b)

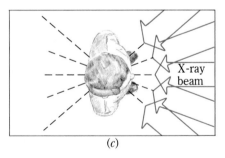

(c)

fact is shown in Fig. 24.19. The shadow of the washer shown there is surrounded by diffraction fringes, and the situation becomes even worse for a smaller object. In the case of objects whose size is comparable to the wavelength of light being used, details of the object are completely obscured by diffraction. We must therefore conclude that *it is impossible to obtain images of objects with detail comparable in*

however (all the $k$'s). As you know from algebra, we need $N$ independent equations to find $N$ unknowns. Let us now see how we can obtain more data and thus more equations.

Suppose the x-ray beam is rotated horizontally around the body, as shown in $c$. Each position gives a new measurement of the fraction absorbed. Since the beam passes through at least a few new little pieces of mass, the fraction-absorbed equation is new for each beam position. If we take measurements at (let us say) 1000 positions, we will have 1000 equations that can be solved simultaneously for 1000 $k$ values. Thus we can, in effect, split a horizontal slice of the body into 1000

tiny parts and compute the $k$ value for each. This computation requires a high-speed computer, but with its aid the computation can be made very quickly.

Since different types of flesh, as well as other materials, have different abilities to absorb radiation, they have different $k$ values. Hence a plot of the $k$ values on the slice through the body shows details of the body along the slice. The computer prints out the $k$ values as varying shades of gray (or in colors). As a result, the varying features of the body on this slice through it are shown clearly. A typical CAT scan printout for the head of a person is shown in Fig. $d$. The entire setup appears in Fig. $e$.

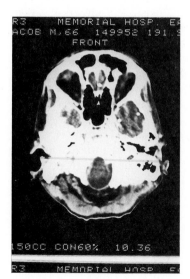

(d)

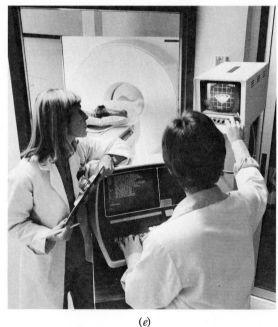

(e)

*size to the wavelength of radiation being used.* For this reason, even the best microscopes cannot discern details comparable to the wavelength of light or smaller. Although a precise statement of the ability of optical devices to resolve the details of very small objects is beyond the scope of this book, *a rough rule of thumb is that detail smaller than a few wavelengths of light cannot be seen.*

24.9 Diffraction and the limits of resolution    575

## 24.10 The diffraction of x-rays by crystals

One of the most important applications of interference and diffraction is in the study of the structure of crystals. Because the layers of atoms in crystals are less than 1 nm apart, it is necessary to use very-short-wavelength radiation to study them. As we learned in Chap. 21, these very-short-wavelength waves are called x-rays. Let us now see how x-rays give an interference pattern when they are incident on a crystal.

For the purposes of our discussion, suppose we are using a crystal of sodium chloride, which is common table salt. Its lattice is shown in Fig. 9.2, and we show it in cross section in Fig. 24.20*a*. The atoms of the crystal are uniformly spaced in planes and are a distance *d* apart. If a beam of x-rays is incident on the crystal as shown in Fig. 24.20*b*, beam *a*, reflected from the top layer of atoms, does not travel as far as beam *b*. If the excess distance traveled by beam *b*, $2d \sin \theta$, is a whole number of wavelengths, the beams reinforce. When that happens, all the layers reflect beams that reinforce one another. Hence, strong reflection occurs when

$$n\lambda = 2d \sin \theta \qquad (24.6)$$

$(a)$ $(b)$

FIGURE 24.20
The atoms in crystals lie in evenly spaced planes. A simple example is shown in (*a*). When the atomic planes reflect x-rays, as in (*b*), the reflected rays give rise to interference effects.

where $n$ is an integer. This is called the **Bragg equation** after W. H. Bragg and his son W. L. Bragg, who first made extensive use of it in 1913.

Notice that Eq. 24.6 is similar to the grating equation (24.3). However, it differs by a factor of 2. In addition, the angle $\theta$ is defined differently in the two cases.

Equation 24.6 is basic to many fundamental measurements. For example, consider what happens in Bragg reflection from a crystal of salt. Since the spacing of the atomic layers can be found from the density of the salt and the mass of its atoms, the distance $d$ is known. Since both $n$ and $\theta$ can be measured, the wavelength of the x-rays can be found by using Eq. 24.6. This is one way in which we have learned that the wavelengths of x-rays are in the range near 0.1 nm. Of course, if $\lambda$ is known, the distance $d$ can be measured. This is the basis for the field of x-ray crystallography, in which the structure of crystals is determined by using x-rays. Through measurements such as these, the structures of even such complex biological molecules as DNA have been determined.

X-ray techniques for crystal-structure determination are quite involved because there are many possible layers of planes in the crystal. A few are shown in Fig. 24.21. When one considers the crystal in three dimensions, the situation is even more complex. Depending on how the crystal is used, the photograph of the diffraction pattern may be a series of spots, a set of circular rings, parallel lines, and so on. Two types of photographs obtained are shown in Fig. 24.22. The photograph in *a* was taken by sending a beam of x-rays through a single crystal, and the one in *b* was obtained in a similar way using a polycrystalline material. Interpretation of such photographs is an involved science. However, most of our knowledge concerning the structure of crystals has been obtained from the analysis of photographs such as these.

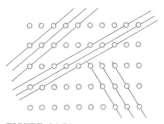

**FIGURE 24.21**

Many possible parallel-layer systems of atoms exist in a crystal. Three such systems are shown here.

**FIGURE 24.22**

The diffraction pattern produced by a beam of x-rays incident on (*a*) a single crystal and (*b*) a crystalline powder.

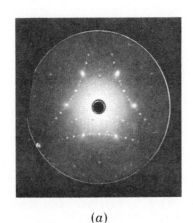

(*a*)

(*b*)

## Minimum learning goals

When you finish this chapter, you should be able to

1 Define (*a*) diffraction, (*b*) Huygens' principle, (*c*) order number of fringe or spectral line, (*d*) coherent waves, (*e*) Newton's rings, (*f*) equivalent optical path length, (*g*) diffraction grating, (*h*) Bragg equation.

2 Describe a water-wave experiment that illustrates the phenomenon of diffraction.

3 Show the phase relation of two identical waves if they are to interfere (*a*) constructively and (*b*) destructively.

4 Describe Young's experiment and how two coherent beams are obtained in it. Using a diagram, show why these two beams can interfere destructively and constructively at various points. Consider the diagram and justify the relation $n\lambda = d \sin \theta_n$ for the bright-fringe positions.

5 Use a double-slit interference pattern to determine $\lambda$ if sufficient data are given.

6 Explain how interference is produced by a thin film or wedge. Tell why the fringes formed in white light are colored. Compute the wedge-thickness difference between adjacent dark or bright fringes.

7 Explain how a diffraction grating is used to measure the wavelength of a spectral line.

8 Describe what happens to a beam of light transmitted through a slit as the slit is made very narrow. Pay particular attention to what happens when the slit width approaches $\lambda$. Explain the importance of this effect to our ability to observe details.

9 Explain the parameters in the Bragg relation. From a consideration of reflection from crystal planes, show how the relation arises.

## Questions and guesstimates

1 The two loudspeakers in Fig. P24.1 are connected to the same oscillator and therefore send out identical sound waves. Under what conditions would you be able to notice an interference effect as you walked along line *AB*? What if the loudspeakers were replaced by light bulbs?

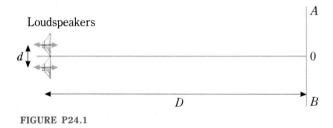

**FIGURE P24.1**

2 Two cars sit side by side in a large, vacant parking lot with their horns blowing. Would you expect to be able to notice interference effects from the two sound sources? What if the horns were replaced by two violins playing the same note?

3 A telephone pole casts a clear shadow in the light from a distant source. Why is no such effect noticed for the sound from a distant car horn?

4 Why is it impossible to obtain interference fringes in a double-slit experiment if the slit separation is less than the wavelength of the light being used?

5 Devise a Young's double-slit experiment for sound, using a single loudspeaker as a wave source.

6 Mercury light consists of several distinct wavelengths. Suppose that in a double-slit experiment, filters are placed over the slits so that $\lambda = 436$ nm (blue) light goes through one slit and $\lambda = 546$ nm (green) light goes through the other. Is it possible to see an interference pattern on the screen?

7 What change occurs in a Young's double-slit experiment when the whole apparatus is immersed in water rather than air? What change is observed in an arrangement of Newton's rings if the space between plate and lens is filled with water?

8 Very thin films are sometimes deposited on glass plates. The thickness of the film can be controlled by observing the change in color of white light reflected from the surface as the film's thickness is increased. Explain.

9 Why does a glass or metal surface that has a thin oil film on it often reflect a rainbow of color when white light is reflected from it?

10 Figure P24.2 shows the interference fringes that are observed when glass plates are placed on top of optically flat surfaces (called *optical flats*). Tell as much as you can about the surface of the two plates used here.

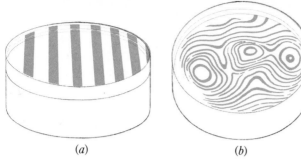

(a)              (b)

**FIGURE P24.2**

**11** Suppose two additional slits are added to the two slits in a Young's double-slit experiment, one on each side of the original two, so that there are four equally spaced slits. For a certain slit-to-screen distance, it is noticed that the center point of the fringe pattern is dark. Explain how this could occur.

**12** Explain the following statement: The difference in thickness between the position of two adjacent bright fringes in a thin-film interference pattern is zero or $\lambda/2n$, where $\lambda$ is the wavelength of light used and $n$ is the index of refraction of the film.

**13** Should a microscope have better resolving power when blue rather than red light is used? Explain.

**14** Suppose you are given a diffraction grating whose characteristics are unknown. How can it be used to determine the wavelength of an unknown spectral line?

**15** Press two pieces of flat glass (microscope slides are ideal) together in various ways and estimate from the interfering reflected light how close together the surfaces are. (You can see the interference pattern easily in any lighted room *provided* you get the plates close enough together.)

**16** Assuming that diffraction caused by the pupil of your eye is the limiting factor, about how far away from you is an oncoming car if its headlights are just resolved?

## Problems

**1** Two identical sound sources at the coordinate origin send in-phase waves for which $\lambda = 80$ cm toward a listener on the *x* axis at $x = 5.0$ m. One of the sources is now slowly moved in the negative *x* direction. What are its first three *x*-coordinate positions at which the listener hears (*a*) loud sound and (*b*) weak sound?

**2** Suppose the sources in Problem 1 are at the origin and send out in-phase waves of unknown wavelength. As one source is slowly moved to negative *x* values, the listener notices that loudness is heard at several points on the *x* axis and that the distance between adjacent points is 25 cm. What is the wavelength of the sound?

**3** Figure P24.1 shows two identical sound sources that vibrate in phase and send out waves for which $\lambda = 18$ cm. Maxima and minima of sound are heard as a receiver is moved along line *AB*. What is the path difference from the two sources at (*a*) the first maximum to one side of 0 and (*b*) the second minimum to one side of 0?

■ **4** The identical sound sources in Fig. P24.1 send out in-phase waves. An observer at *A* notices that loud sound occurs when the line from the sources to *A* makes an angle of 27° with the line from the sources to 0. If $d = 40$ cm, what possible wavelengths does the sound have? Assume $D \gg d$.

■ **5** Two identical sound sources vibrate in phase and send waves for which $\lambda = 50$ cm toward each other along the *x* axis. One source is at $x = 0$, and the other is at $x = 4.0$ m. At what points on the *x* axis between them is the combined sound (*a*) maximum and (*b*) minimum?

■ **6** Two identical variable-frequency sound sources vibrate in phase and send sound toward each other along the *x* axis. One source is at $x = 0$, and the other is at $x = 4.0$ m. A sound detector is placed at $x = 2.5$ m. The equal frequencies of the sources are increased from zero together. The sound intensity at the detector decreases with increasing frequency until it reaches a minimum and then begins to increase again. What is the wavelength of the sound at the minimum?

**7** Light with $\lambda = 436$ nm is used in a Young's double-slit experiment, and the first-order maximum is at 3.5°. (*a*) What is the slit separation? (*b*) At what angle does the second-order maximum occur?

**8** The slits in a Young's double-slit experiment are 0.013 cm apart, and light for which $\lambda = 633$ nm is used. (*a*) At what angle does the third-order maximum occur? (*b*) The fifth-order maximum?

■ **9** The sound sources in Fig. P24.1 send out identical, in-phase waves for which $\lambda = 50$ cm. If $d = 8.0$ m and $D = 25$ m, how far along *AB* from 0 are (*a*) the first-order maximum and (*b*) the first-order minimum?

■ **10** In a double-slit experiment, the slit separation is 0.25 cm and the slit-to-screen distance is 100 cm. Call the position of the central bright fringe zero, and locate the positions of the first three (*a*) maxima and (*b*) minima on both sides of the central maximum. The wavelength of the light through the slit is 450 nm.

■ **11** What slit separation in a double-slit experiment gives a second-order maximum 0.50 cm from the central bright spot? The screen-to-slit distance is 2.0 m, and $\lambda = 500$ nm.

■ **12** When mercury light ($\lambda = 436$ nm) is used in a double-slit experiment, the first-order maximum is at an angle of $3.0 \times 10^{-4}$ rad. When this light is replaced by a source of unknown wavelength, the second maximum occurs at $4.0 \times 10^{-4}$ rad. (*a*) What is the wavelength of the second light? (*b*) In what region of the spectrum is it found?

■ **13** Two flat glass plates form a very thin air wedge between them. When the combination is viewed with light for which $\lambda = 550$ nm, a dark fringe exists at the line of contact. What is the thickness of the air wedge at (*a*) the first bright fringe and (*b*) the third bright fringe?

■ **14** When blue light (589 nm) is reflected from an air wedge formed by two flat plates of glass, the bright fringes are 0.50 cm apart. (*a*) How thick is the air wedge 4.0 cm from the line of contact of the plates? Assume the wedge is viewed at normal incidence. (*b*) Repeat for the wedge filled with water rather than air.

■ **15** In Fig. 24.8*b*, suppose that the light used to produce Newton's rings is the mercury green line (546 nm) and that the radius of the ninth dark ring is 1.50 cm. (*a*) How large is the air gap at this position? (*b*) If the gap is now filled with water, how big is the gap at the new position of this ninth ring? The center point of the pattern is dark.

■ **16** Two parallel glass plates are originally in contact and viewed from directly above with 540-nm light (green) reflected nearly perpendicularly by the surfaces. As the plates are slowly separated from each other, darkness is observed at certain separation distances. (*a*) What are the values of the first four separation distances? *Hint:* Darkness is observed when the plate separation is zero. (*b*) Repeat for the gap between the plates filled with water.

■ **17** A wedge-shaped sliver of glass has an index of refraction of 1.56. When it is viewed from directly above with 436-nm light, the sharp edge of the wedge is dark. What is the thickness of the wedge at the fifth bright fringe?

■ **18** An oil slick ($n = 1.42$) on a water puddle shows interference fringes. What are the possible values for the difference in thickness of the oil slick at adjacent red fringes? Assume $\lambda = 600$ nm.

■ **19** A Young's double-slit experiment uses slits 0.20 mm apart. The entire apparatus is under water. At what angles do the first two interference maxima occur if 540-nm light is used?

■ **20** Light for which $\lambda = 590$ nm falls on two slits (separation unknown), together with light of unknown wavelength. The fourth-order maximum of the 590-nm light falls at the same position as the fifth-order maximum of the light of unknown wavelength. (*a*) What is the wavelength

in air of this light? (*b*) Repeat for the entire system in water.

■ **21** A double-slit system immersed in water is illuminated by 633-nm light. An interference pattern is formed on a screen 2.0 m away in the same tank of water. What is the distance on the screen from the central maximum to the first-order maximum if the slit separation is 0.040 cm?

■ **22** A metal mirror has a thin layer of plastic ($n = 1.50$) coating its surface. By observation of how various wavelengths of light are reflected from the mirror, it is found that the reflected intensity is a minimum for $\lambda = 540$ nm. Find two possible thicknesses for the coating. (*Hint:* In effect, $n \to \infty$ for a metal.)

**23** To calibrate a diffraction grating, a student sends red light from a helium-neon laser (632.8 nm) through it. The first-order maximum occurs at an angle of 28°. (*a*) What is the grating spacing? (*b*) At what angle does the second-order maximum occur?

**24** Sodium-arc yellow light is a doublet composed of two wavelengths, 588.995 and 589.592 nm. Compute the angular separation between these two lines in the first-order spectrum for a grating with 4000 slits per centimeter. Repeat for the second-order spectrum.

**25** A certain grating has 5000 lines per centimeter. What is the angular separation between the blue (435.8 nm) and green (546.1 nm) mercury lines in (*a*) the first-order spectrum and (*b*) the second-order spectrum?

**26** A certain diffraction grating has 6000 lines per centimeter. At what angle does the second-order spectrum of the sodium yellow line occur ($\lambda = 589.0$ nm)?

**27** For a particular grating, the second-order mercury blue line (435.8 nm) lies exactly at 37°. At what angle is the first-order yellow line (579.0 nm) found?

■ **28** A diffraction grating having 5000 lines per centimeter is used in a large tank of water. What are the four smallest angles (in the water) at which the blue mercury line (435.8 nm) appears?

■ **29** Modern grating spectrometers often use a reflection grating. This is a mirror on the surface of which are a series of reflecting lines (equivalent to the slits). This situation is shown in Fig. P24.3, where the distance between the centers of the reflecting lines is *d*. Find the grating equation for this device, that is, the angles $\theta$ at which interference maxima occur.

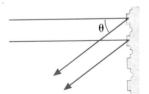

FIGURE P24.3

**30** (*a*) Is it possible to design a grating in such a way that the first-order 600-nm red line will lie on top of the second-order 400-nm violet line? (*b*) If so, how? (*c*) If not, could it be done for any other combination of orders? (*d*) If so, how?

**31** Steel sheds often have a corrugated metal surface with corrugation repeating every 10 cm or so. Under appropriate conditions, this type of wall can act as a reflecting diffraction grating for sound waves (see Prob. 29). What value of $\lambda$ for waves at normal incidence will give rise to a first-order maximum at an angle of $37°$ to the normal?

**32** Find the angular width of the central maximum (that is, the angle between the two side minima) for a single slit that is 0.040 cm wide and illuminated with 600-nm light.

**33** A single slit is illuminated with 540-nm light, and its first diffraction minimum occurs at an angle of $2.0°$ from the center of the diffraction pattern. What is the width of the slit?

**34** To determine the width of a very narrow slit, a student sends red light from a helium-neon laser ($\lambda = 633$ nm) through it in a single-slit diffraction experiment. Since she (rightly) does not wish to view the laser light directly, the diffracted beam is allowed to strike a screen 5.0 m from the slit. She measures the distance from the central maximum to the first minimum to be 2.0 cm. How wide is the slit?

**35** A man is looking at the headlights of a distant truck. If his eye opening has a diameter of 0.25 cm, how far away is the truck when the two headlights are just resolved? Assume that the limiting factor is diffraction caused by the eye opening, that the wavelength of the light is 500 nm, and that the light separation is 150 cm. What can you conclude from your result?

**36** An image of a photographic slide is projected onto a screen 3.0 m away by a lens with a diameter of 3.0 cm. The lens is 12 cm from the slide. Assume the lens is perfect, so that diffraction is the only limit on its imaging ability. How close together can two tiny spots on the slide be if they are to be resolved on the screen? Assume $\lambda = 500$ nm. How close together on the screen are these spots?

**37** Strong reflection of 0.148-nm x-rays occurs from a crystal when the incident beam makes an angle of $50°$ with the beam that is reflected backward. What are the possible lattice plane separations that give rise to this reflection?

**38** When a beam of 0.45-nm x-rays is incident on a crystal, a strong reflection occurs backward along the incident beam but at an angle of $30°$ to it. What possible lattice spacings could give rise to this reflection?

# 25 Three revolutionary concepts

**B**Y 1900, many scientists felt that most of the great discoveries in physics had been made. To be sure, a few vexing problems remained, but it appeared that nearly all the fundamental physical laws had been found. As we shall see in this chapter, such a view was completely incorrect. Vast areas of nature's physical behavior were still unknown at that time.

As we look through the history of science, we see that each truly great scientific advance is associated with the name of a single person. Galileo is recognized as the leader in our understanding of how objects undergo translational motion. Newton's name is enshrined in his three laws of motion and in the law of gravitation. Faraday pioneered the way to an understanding of magnetism, and Maxwell unified all electricity with his four fundamental equations. These and many other similar examples attest to the fact that the intellect of a single individual has the power to illuminate large areas of science for us all.

This is not to say that these individuals made their discoveries in isolation. Quite the contrary. Historians of science show clearly that each of these discoveries was the culmination of years of work by many others. Indeed, Newton once wrote, "If I have seen further than other men, it is because I stood on the shoulders of giants." Even so, other people stood on the shoulders of these same giants and saw nothing. While we must pay due respect to their predecessors, the insight and genius of these great scientists should not be underestimated. We should not stand in such awe of our scientific ancestors that we underestimate our own capabilities, however. The discoveries we discuss in this chapter and in those that follow often came from unexpected sources.

# PART I: RELATIVITY

## 25.1 The postulates of relativity

Over the centuries, multitudes of experiments have been carried out to learn the laws of nature. In 1905, Albert Einstein became convinced that the experimental data force us to accept two seemingly innocuous facts of nature:

**1** The speed of light in vacuum is always measured to be the same ($c = 2.998 \times 10^8$ m/s), no matter how fast the light source or observer may be moving.

**2** Absolute speeds cannot be measured. Only speeds relative to some other object can be determined.

Assuming the truth of these statements, Einstein was able to show that many unexpected facets of the world about us were yet to be discovered. His line of reasoning is known as the *theory of relativity,** and the two statements of apparent fact given above are the basic postulates of his theory.

It is not possible to prove these postulates directly. They are the consensus of all the experimental facts known. We consider it possible, though unlikely, that some experiment will someday disprove one of them, but they are as of now supported by many unsuccessful attempts to disprove them. Moreover, as we shall see, they lead to astounding conclusions that have been well verified by experiment.

The second postulate needs some explanation, perhaps. It is easy to measure the relative speeds of objects. A car's speedometer tells us at once how fast the car is moving relative to the roadway, but this is not an absolute speed. The earth is moving because of both its rotation on its axis and its motion around the sun. Since we know both these speeds, we could, if required, find the car's motion relative to the sun.

The sun itself is moving in our galaxy, the Milky Way, however, and the center of the galaxy is in motion relative to more distant stars. There seems to be no way to define a definite, absolute speed of an object since everything appears to be moving. We can state only how fast one object is moving relative to another.

There is another way to state the second postulate, a way that gives us an inkling of its fundamental importance. This alternate statement is usually made in terms of reference frames. *A **reference frame** is any coordinate system relative to which measurements are taken.* For example, the position of a sofa, table, and chairs can be described relative to the walls of a room. The room is then the reference frame used. Or, perhaps a fly is sitting on a window in a moving car. We can describe the fly's position in the car, using the walls as a reference frame. Alternatively, we can describe the position of a spaceship relative to the positions of the distant stars. A coordinate system based on these stars is then the reference frame.

The second postulate can be stated in terms of reference frames in the following way:

**2′** The basic laws of nature are the same in all reference frames moving with constant velocity relative to each other.

---

\* We discuss here Einstein's *special theory*. It applies only to measurements on objects that are not accelerating. In 1916, Einstein extended the theory to objects that are accelerating in his *general theory*.

Often this statement is shortened by using the term inertial reference frame. *An* **inertial reference frame** *is a coordinate system in which the law of inertia applies:* a body at rest remains at rest unless an unbalanced force on it causes it to be accelerated. The other laws of nature also apply in such a system. To a very good approximation, all reference systems moving with constant velocity relative to the distant stars are inertial frames.

*2″* The basic laws of nature are the same in all inertial reference frames.

You can understand the relation between these two alternative ways of stating the second postulate by considering the following. When we say that only relative speeds can be measured, we are assuming a lack of bias in reference frames. For example, a spaceship may be heading for the moon at a speed of $10^5$ km/day relative to the moon, but it is also true that the moon is heading toward the ship at $10^5$ km/day relative to the ship. The fact that one is moving relative to the other is easily ascertained, but the statements are equivalent to each other, and neither object can be said to be at rest in an absolute sense.

Suppose, though, that some law of nature depended on the speed of the reference frame. The people in the spaceship could use such a law to determine their speed. People on the moon could do likewise. The two measured speeds would be different. As a result, people would be capable of measuring more than just their relative speeds. In fact, the law could be used to set up an absolute ranking of speeds. This would contradict the second postulate, however, which we, along with Einstein, assume to be correct. We therefore conclude that all nature's laws must be the same in all inertial reference frames.

## 25.2 The speed of light as a limiting speed

By use of Einstein's two postulates, we can prove by logic alone that

No material object can be accelerated to speeds in excess of the speed of light in vacuum.

The validity of this statement is easily demonstrated in the following simple way. We prove it by the technique called *reductio ad absurdum,* in which we disprove a proposition (in this case, that an object can travel faster than $c$) by showing that it leads to a known false result (in this case, that an observer will measure a value different from $c$ for the speed of light).

Suppose we have two nonaccelerating stations in space, shown as $A$ and $B$ in Fig. 25.1. They act as inertial reference frames. Observers at $A$ and $B$ have instructed the spaceship operator to follow a straight-line path between $A$ and $B$. The ship is to travel at its top constant speed and is to send a light pulse from the front of the ship toward $B$ as it passes $A$. Of course, $A$ and $B$, working in partnership, can determine the speed of the spaceship by timing its flight from $A$ to $B$. Let us now make the false assumption that they find the speed of the ship to be $2c$.

The spaceship sent out a pulse of light as it passed $A$, and since the laws of nature must apply to all three inertial observers ($A$, $B$, and the person in the ship), the light pulse must behave in a normal way for each of them. Remember, the observer in the ship cannot tell whether or not the ship is moving, except in a relative sense. Therefore, for this observer, the light pulse must precede the ship and must reach $B$

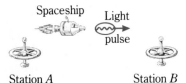

Spaceship  Light pulse

Station $A$    Station $B$

**FIGURE 25.1**

What is the maximum speed with which the ship can pass between the two space stations?

before the ship does. Therefore $A$ and $B$, working together, would find that the light pulse is moving faster than the ship. But they measure the ship as moving with speed $2c$, and so they find that the speed of the light pulse is greater than $2c$. But this is an impossible result, since it contradicts the known fact that all observers will obtain $c$ for the speed of light. We therefore conclude that our original assumption was false; the spaceship could not have been moving between $A$ and $B$ with a speed of $2c$.

This experiment will always lead to this contradiction as long as we insist that the speed of the ship exceed $c$. We therefore conclude that the spaceship cannot exceed the measured speed of light $c$. Indeed, we can enlarge this line of reasoning to include all material objects and signals that carry energy. As a result we can state:

Nothing that carries energy can be accelerated to the speed of light $c$.

As we proceed, we see that this result of Einstein's theory repeatedly has been tested carefully and has been found correct in every test.

## 25.3 Simultaneity

Ordinarily we expect that two observers will agree as to whether or not two events occur at the same time. Einstein showed, however, that under certain circumstances the expected result does not correspond to reality. The basic postulates of relativity force us to conclude that events that are simultaneous in one inertial reference frame may not be simultaneous in another. To show this simply, we again resort to a thought experiment. The progress of a light pulse, as noted by two inertial observers, forms the basis for our experiment.

Suppose that a boxcar is traveling to the right at a very high constant velocity, as in Fig. 25.2$a$. At the exact center of the car is a high-speed flashbulb that will send out light pulses to the right and left when it flashes. The boxcar is fitted with photocells at each end, so that a man in the boxcar can detect when the light pulses strike the ends of the car. By some ingenious device, a woman at rest on the earth is also able to measure the progress of the two pulses. Because both observers are in inertial reference frames (one is the moving boxcar, the other is the earth), each must see the light pulses behave "normally" in his or her reference frame. Of course, "normal" for the woman on earth is that the light pulses travel with speed $c$ to the right and left from the flashbulb. "Normal" for the man in the car is that the two light pulses strike the detectors at opposite ends of his car simultaneously.

Consider first the man in the car. To him, the experiment is very simple. The flashbulb is at rest relative to him in the center of the car. When the bulb flashes, two pulses travel the equal distances to the two ends of the car in equal times. (Remember, for him the experiment must be the same whether or not the car is moving, because postulate 2 implies identical results for any inertial reference frame.) Hence the light pulses hit the two ends of the car simultaneously.

Now consider how the woman who is stationary on the earth sees the experiment. Her measurements show that the experiment proceeds "normally" (for her), and so the situation progresses as in Fig. 25.2$b$ and $c$. Notice that the pulses travel equal distances to the right and left in equal time. But since the boxcar is moving to the right, the distance to the left end is shortened. As a result, the observer who is stationary on the earth measures the pulse on the left as striking the end of the boxcar before the other pulse strikes the opposite end. According to her, the light pulses do not hit the two ends of the car simultaneously.

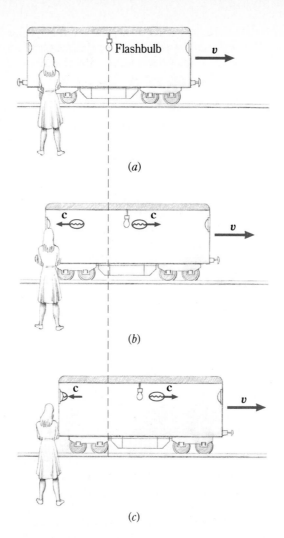

(a)

(b)

**FIGURE 25.2**
Unlike the inertial observer in the moving frame, the stationary observer on earth does not see the light pulses strike the ends of the car simultaneously.

(c)

We must therefore conclude that time is not a simple quantity because

Events that are simultaneous in one inertial system may not be simultaneous in another.

Further considerations show that this situation exists only if the two events occur at different locations. In this case, one event took place at one end of the car and the other was at the opposite end.

## 25.4 Moving clocks run too slowly

As you might suspect from the results of the previous section, time is not a simple quantity. Einstein pointed this out when he showed that a clock ticks out time differently for an observer who holds the clock and for one moving past it. We demonstrate this effect in a thought experiment using a very special clock, but it was proved to be true in general by Einstein.

Consider the clock held by the woman in Fig. 25.3. It consists of a pulse of light reflecting between two mirrors in a cylindrical vacuum tube. Each time the light pulse strikes the lower mirror, it clicks out a unit of time that we shall call a "click." If the tube is 1.5 m long, the woman can compute easily that

$$1 \text{ click} = \frac{2d}{c} = \frac{3.0 \text{ m}}{3.0 \times 10^8 \text{ m/s}} = 10^{-8} \text{ s}$$

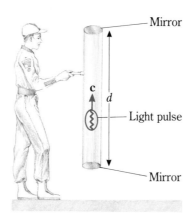

**FIGURE 25.3**

The light clock registers one click each time the light pulse is reflected from the lower mirror.

Suppose that several copies of this clock are made and that one is being used by a man in a spaceship. The woman with her identical clock looks out the window of her laboratory (which is in another spaceship) and sees the man shoot past her with speed $v$. She is pleased to see that he is using a clock similar to hers and contacts him by radio. He tells her that the clock is functioning well and is ticking out time as usual, one click each $2d/c$ seconds.

After thinking about it a bit, the woman discovers that there is something very peculiar about this. She concludes that the man's clock must be ticking out time more slowly than hers. We can understand her reasoning as follows.

Since the man's clock is operating properly for him, she knows it must be operating as in Fig. 25.4. We see there the clock in its position at two consecutive clicks. The woman knows that the light pulse moves along the path indicated. Although the man sees the pulse move straight up and down in the clock, the woman

**FIGURE 25.4**

The light pulse in the moving clock must travel a distance larger than $2d$ during one click interval. The light-pulse path length is $2\sqrt{d^2 + (\frac{1}{2}vt_w)^2}$.

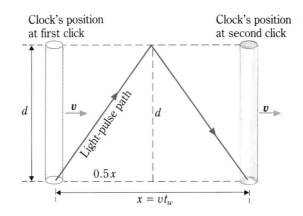

## Albert Einstein (1879–1955)

As a student, Einstein was unhappy in the rigid, militaristic school system of his native Germany. His exceptional abilities and interest in physics flourished after he moved to Zurich to complete his undergraduate studies. Despite excellent references, he could not get a teaching job and found work in the patent office instead, continuing his studies in theoretical physics in his spare time. At the age of 26, he published papers on Brownian motion, the photoelectric effect, and relativity. Any one of these would have merited a Nobel prize (he won it for his paper on the photoelectric effect). These famous studies were but the beginning of a long and brilliant career. He fled the persecutions of Hitler's Germany, and in 1940 he became a U.S. citizen. Partly through his efforts, the United States assembled many of the world's best scientists to develop atomic bombs

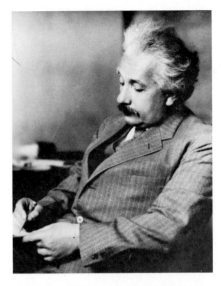

before Nazi Germany could do so. In addition to being the most famous modern scientist, Einstein was noted for his warm and compassionate personality and his deep interest in peace and social justice.

knows that the pulse moves to the right as well, because of the movement of the clock to the right.* The woman computes the time between clicks on the man's clock as follows.

According to the woman, the pulse moves a distance given by the colored line in the figure. From the pythagorean theorem and the dimensions given in the figure, we see that

$$\text{Pulse path length} = 2\sqrt{d^2 + (\tfrac{1}{2}x)^2}$$

The woman knows that the man's clock is traveling past her with speed $v$. Further, according to her clock, it will take the man's clock a time $t_w$ to move from one position to the other. Therefore, she knows that $x = vt_w$. As a result, according to the woman,

$$\text{Pulse path length} = 2\sqrt{d^2 + (\tfrac{1}{2}vt_w)^2}$$

Further, she knows that a light pulse always travels through vacuum with speed $c$. According to her, then, the time taken for the change in position shown in the figure should be

---

* You might ask, Who is correct? They both are, as we shall soon see. Both are describing the behavior correctly as measured in their own reference frames.

$$t_w = \frac{\text{pulse path length}}{c} = \frac{2\sqrt{d^2 + (\frac{1}{2}vt_w)^2}}{c}$$

We can solve for $t_w$ in this equation and find (after squaring both sides, rearranging, and taking square roots)

$$t_w = \frac{2d/c}{\sqrt{1 - (v/c)^2}}$$

But we recognize $2d/c$ as the time that the man insists it takes for his clock to make one click. We therefore have the following result:

$$\frac{\text{Time interval on}}{\text{stationary clock}} = \left[\frac{1}{\sqrt{1 - (v/c)^2}}\right] \left(\frac{\text{time interval on}}{\text{moving clock}}\right)$$

The quantity $\sqrt{1 - (v/c)^2}$ is called the **relativistic factor.**

As an example, suppose that the man is moving past the woman at a speed of $0.75c$. Then $\sqrt{1 - (v/c)^2}$ has a value 0.66, and the inverse of this is 1.51. Under these conditions the woman's clock will tick out 1.51 clicks during the time she knows the man's clock takes to tick out 1 click. As we see, the moving clock ticks out time more slowly than the stationary clock.

A clock moving with speed $v$ ticks out a time of $\sqrt{1 - (v/c)^2}$ second during 1 s on a stationary clock.

After arriving at this unexpected result, the woman contacts the man by radio and informs him that she has discovered that moving clocks tick out time too slowly. Before she can give him the details, he states that he has been thinking along the same lines. He has discovered that her clock, which was moving past him with speed $v$, was ticking out time too slowly. Then they both recall that only relative motion has meaning. Neither clock is special.

Any clock that is moving relative to an observer will appear to tick out time more slowly than a clock that is stationary with respect to the observer.

We call this effect **time dilation,** since time is stretched out, so to speak, for moving clocks.

This astonishing result applies to all timing mechanisms, no matter how complex. If the man had been using the growth rate of a fungus as a clock, the woman would have found the fungus growth rate to be slowed by its motion. Even aging of the human body will be slowed by motion at high speed, as we see in one of the following examples.

There is one point we should always remember, however. A good clock always behaves normally to a person at rest relative to it. Observers moving past the clock may claim that it ticks out time too slowly. In spite of this, the clock still ticks out time properly as viewed by an observer stationary relative to it. The time ticked out by a clock that is stationary relative to the observer is called the **proper time.**

---

### Example 25.1

One striking example of time dilation is obtained by measuring how long unstable particles "live." For example, a particle called the pion lives on the average only about $2.6 \times 10^{-8}$ s when at rest in the laboratory. It then changes to another form. How long would such a particle live when shooting through the laboratory at $0.95c$?

**Reasoning**   In the second case, the pion is moving with speed $0.95c$ relative to the observers in the laboratory. Experiments should show that the internal clock of the pion, which controls how long it lives, should be slowed because of its motion. A time of $2.6 \times 10^{-8}$ s read by the moving clock should be as follows when timed by the laboratory clock:

$$\text{Life according to lab clock} = \frac{2.6 \times 10^{-8} \text{ s}}{\sqrt{1 - (0.95)^2}}$$

which turns out to be $8.3 \times 10^{-8}$ s. As we see, the moving pion should live about 3 times as long as a stationary one. This experiment and variations of it have been carried out. The results found by experiment agree with the computed results.   ∎

**Exercise**   How fast must the pion be moving if it is to "live" $10^{-7}$ s?   *Answer: v/c = 0.966*   ∎

### Example 25.2
The star closest to our solar system is Alpha Centauri, which is $4.3 \times 10^{16}$ m away. Since light moves with a speed of $3 \times 10^8$ m/s, it would take a pulse of light $1.43 \times 10^8$ s, or 4.5 yr, to reach there from the earth. (We say that the distance to the star is 4.5 light-years.) How long would it take, according to earth clocks, for a spaceship to make the round trip if its speed is $0.9990c$? According to clocks on the spaceship, how long would it take?

**Reasoning**   To a good approximation, we can take the spaceship speed as $c$ for this computation, and so the round trip would require 9.0 yr according to earth clocks.

The spaceship clocks will appear to run too slow by the relativistic factor

$$\sqrt{1 - (0.999)^2} \approx 0.045$$

Therefore the spaceship clocks will read $(0.045)(9.0) = 0.40$ yr instead of the 9.0 yr read from the earth clocks. As a result, the journey would seem to take only about 5 months according to the crew of the spaceship—far more tolerable than the 9.0 yr that people on earth would record.

Incidentally, the twin of one of the crew who was left behind on the earth would age 9.0 yr during the voyage. The twin in the spaceship, however, would age only 5 months. This phenomenon, the **twin paradox,** has been discussed at length by scientists. They generally agree that this result is valid and that the two twins actually will age differently.*   ∎

### Example 25.3
Graph the relativistic factor $\sqrt{1 - (v/c)^2}$ as a function of $v/c$. Explain why we do not observe relativistic time dilation in everyday phenomena.

---

* To test this effect with actual clocks, an extremely precise clock transported around the earth by plane has been compared with a stationary "twin." The expected result was found. For a discussion of the experiment, see J. Hafele, *Physics Teacher,* **9**:416 (1971).

*Reasoning* The appropriate graph is shown in Fig. 25.5. Notice that the relativistic factor departs from unity only at extremely high speeds. For most purposes, speeds below $0.10c$, where the relativistic factor is 0.995, are too small to show

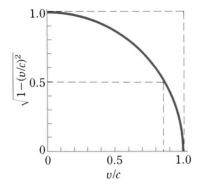

FIGURE 25.5

The relativistic factor differs appreciably from unity only at speeds that approach the speed of light.

appreciable relativistic effects. In everyday life, our clocks never come anywhere close to such high speeds. However, as we learned in our study of electricity, the electrons in a beam such as that in a television tube are easily accelerated to relativistic speeds. ■

*Exercise* How fast must a particle be moving to have a relativistic factor of 0.0100? *Answer: v/c = 0.999950* ■

## 25.5 Relativistic length contraction

The time-dilation effect implies a peculiar effect involving measured lengths. To see what this effect is, consider once again the man and woman of the previous section. Let us say that the woman is on the earth while the man is traveling with speed $v$ along a straight line from earth to the nearest star, Alpha Centauri. Astronomers based on earth tell us that the star is $d = 4.3 \times 10^{16}$ m away from the earth. Because relative speeds can be measured easily, the man and woman agree that their speed relative to each other is $v$ as the man in the spaceship shoots from the earth to the star. The woman is at rest in a reference frame in which the earth and the star are also at rest.* She sees the man shooting past her at speed $v$.

The man, on the other hand, is at rest relative to his spaceship, and he takes the ship itself to be his reference frame. Relative to the ship, both the earth and the star are moving with speed $v$. Let us now examine the man's flight from the earth to the star from the woman's vantage point.

The woman knows that the distance from the earth to the star, both of which are at rest relative to her reference frame, is $d_e = 4.3 \times 10^{16}$ m, where the subscript $e$ stands for "earth." Using $x = vt$, she computes that the time registered by her earth clock for the man's trip to the star will be

$$t_e = \text{earth time} = \frac{d_e}{v}$$

---

* We neglect the comparatively small relative motion of the earth and the star.

Indeed, when the ship turns around at the star and returns to earth, the total time the ship has been in flight is $2t_e = 2d_e/v$.

However, the man's computation will be somewhat different. Using spaceship clocks, he times his flight from the earth to the star and finds that it takes a time $t_s$. He can then compute the distance to the star by use of $x = vt$ and obtain

$$d_s = vt_s$$

where the subscript $s$ refers to measurements in a reference frame at rest relative to the spaceship. A similar computation for the return trip tells him that his total flight covered a distance $2d_s$ in $2t_s$.

We therefore have the following equations that are undeniably correct for the two observers who formulated them:

$$2d_e = v(2t_e)$$

and

$$2d_s = v(2t_s)$$

But we know that the time-dilation effect influences the spaceship clock in such a way that when it is compared with the earth clock after the spaceship returns to earth, we have

$$t_s = \sqrt{1 - \left(\frac{v}{c}\right)^2}\, t_e$$

The spaceship clock ticked out time more slowly than the clock on earth.

Substituting this value for $t_s$ in the expression for $d_s$ yields

$$d_s = v\sqrt{1 - \left(\frac{v}{c}\right)^2}\, t_e$$

However, $d_e = vt_e$, and so $t_e = d_e/v$. Using this value for $t_e$ gives

$$d_s = \sqrt{1 - \left(\frac{v}{c}\right)^2}\, d_e$$

In other words, the distance from the earth to the star measured by the man in the spaceship is smaller than the distance measured by astronomers on earth. Apparently, if you are in motion relative to two points that are a fixed distance apart, the distance between the two points appears shorter than if you were at rest relative to them. The ratio of the two distances is the relativistic factor, $\sqrt{1 - (v/c)^2}$.

Einstein found this to be a general result. We can summarize it as follows:

If an object and an observer are in relative motion with speed $v$, then the observer will measure the object as having contracted along the line of motion. The contraction factor is $\sqrt{1 - (v/c)^2}$.

Notice that the contraction occurs only along the line of motion. No such contraction is observed perpendicular to the direction of motion. The length of an object measured by an observer at rest relative to it is called the **proper length.**

## Example 25.4

A woman traveling at high speed in a spaceship holds a meterstick in her hand. What does she notice about the length of the stick as she rotates it from a position that is parallel to the line of motion to a position that is perpendicular?

*Reasoning*   She notices no change in the stick's length. The length-contraction effect concerns objects moving at high speed relative to the observer. The meterstick is at rest relative to the woman.   ■

## 25.6 The relativistic mass-energy relation

The postulates of relativity tell us that no object can be accelerated to speeds in excess of the speed of light. It can be seen at once that this conflicts with prerelativity ideas. For example, consider an object of mass $m$ being accelerated by a constant force $F$. We would usually state that the acceleration is $a = F/m$. As a result, the object's final velocity after time $t$ is simply

$$v = v_o + at = v_o + \frac{F}{m} t$$

*This relation must be wrong.* It predicts that the velocity can increase without limit. As long as the force keeps acting, the velocity will keep on increasing. This contradicts the conclusion that the object's speed cannot exceed $c$. Something is obviously wrong with our nonrelativistic ideas for objects moving at very high speeds.

A consideration of the postulates of relativity leads to the source of the difficulty. Using reasoning too lengthy for us to give here, Einstein was able to show that his postulates, together with the momentum-conservation law, yield the fact that the mass of an object varies with the speed of the object. At rest, the object has a mass $m_o$, its **rest mass;** at high speeds, however, the object's mass is larger.

An object of rest mass $m_o$ has an apparent mass $m = m_o/\sqrt{1 - (v/c)^2}$ when moving with speed $v$ past an observer.

This variation of mass with speed is shown in Fig. 25.6. As we see, the apparent mass $m$ is close to the rest mass $m_o$ as long as $v/c$ is less than a few tenths. But when $v$ becomes close to the speed of light, so that $v/c$ approaches 1, the apparent mass becomes much larger than $m_o$. In terms of the equation for the apparent mass,

**FIGURE 25.6**

The apparent mass of a moving object approaches infinity as the object's speed approaches the speed of light.

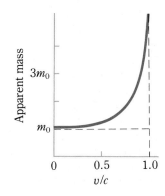

$$m = \frac{m_o}{\sqrt{1 - (v/c)^2}} \qquad (25.1)$$

We notice that the quantity $v/c \to 1$ as $v$ approaches $c$. Therefore, for $v \to c$,

$$m = \frac{m_o}{\sqrt{1 - (v/c)^2}} \to \frac{m_o}{\sqrt{1 - 1}} \to \frac{m_o}{0} \to \infty$$

The apparent mass becomes infinite if the object's speed reaches the speed of light.

The variation of mass with speed can be used to justify the fact that no object can be accelerated to a speed in excess of the speed of light. An infinite mass would require an infinite force to accelerate it. Because infinite forces are not available, it is apparent that an object with a speed $v \to c$ cannot be accelerated to the speed of light, a speed at which its mass would be infinite.

The force that acts to accelerate an object gives energy to the object. At low speeds, we know that the work done by the applied force equals the increase in the kinetic energy of the object, provided changes in potential energy and friction work are negligible. This is still true at speeds close to $c$, but the kinetic energy of an object is no longer given by $\frac{1}{2}m_o v^2$. Nor is it, as one might guess, $\frac{1}{2}mv^2$. Instead it is found that the kinetic energy of an object is given by

$$KE = (m - m_o)(c^2) \qquad (25.2)$$

When $v \ll c$, Eq. 25.2 reduces* to $KE = \frac{1}{2}m_o v^2$. We see, then, that the correct expression for kinetic energy is given by Eq. 25.2. The expression $KE = \frac{1}{2}m_o v^2$ is very accurate for speeds low enough to ensure that $m \approx m_o$. But when $v \to c$, the value $(m - m_o)(c^2)$ must be used for kinetic energy.

It is possible to show that a relation similar to Eq. 25.2 applies to all types of energy. Einstein showed that *for any change in the energy of an object, there is a corresponding change in mass*. The result is that for a change in energy $\Delta E$, the object's mass changes by an amount $\Delta m$, given by

$$\Delta E = \Delta m\, c^2 \qquad (25.3)$$

(This is often written as $E = mc^2$.) As an example, if you increase the potential energy of an object, its mass increases in accordance with Eq. 25.3. In fact, this equation predicts that mass can be created by providing energy (we see examples of this later). More spectacularly, mass can be destroyed to provide energy. In either case, the change in mass $\Delta m$ is equivalent to a change in energy of $\Delta m\, c^2$. Perhaps you already know that the nuclear energy of a reactor or nuclear bomb results from the fact that an amount of mass $\Delta m$ is destroyed to produce an energy $\Delta m\, c^2$.

---

* To show this, make use of the mathematical fact that, for $x \ll 1$, the quantity $1/\sqrt{1 - x} \cong 1 + \frac{1}{2}x$. Then, if we call $(v/c)^2$ the quantity $x$, in the case when $(v/c)^2 \ll 1$,

$$m = \frac{m_o}{\sqrt{1 - (v/c)^2}} = m_o\left(1 + \frac{1}{2}\frac{v^2}{c^2}\right)$$

Therefore Eq. 25.2 becomes

$$KE = mc^2 - m_o c^2 = m_o c^2 + \frac{m_o}{2}\frac{v^2}{c^2}c^2 - m_o c^2 = \frac{1}{2}m_o v^2$$

### Example 25.5

The available chemical energy in a 100-g apple is about 100 kcal (the nutritionists leave off the prefix *kilo* and call them calories). We learned in our study of heat that 1 cal is 4.184 J of energy, and so an apple contains about 420 kJ of available energy. Compare this with the energy one could obtain by changing all the mass to energy.

*Reasoning*  According to the mass-energy relation,

$$\text{Energy} = \Delta m\, c^2$$

In this case $\Delta m = 0.10$ kg and $c = 3 \times 10^8$ m/s, giving

$$\text{Energy} = 9 \times 10^{15} \text{ J}$$

We see from this that when we eat an apple, we obtain only a very small fraction ($5 \times 10^{-11}$) of its total energy.  ■

---

### Example 25.6

The light given off by a television tube comes from electrons shooting down the tube and hitting a fluorescent screen at its end. Their speeds are of the order of one-third the speed of light. What is the apparent mass of such a high-speed electron ($m_o = 9.1 \times 10^{-31}$ kg)?

*Reasoning*  The electrons are moving with speed $v = c/3$ relative to the person watching the television set. As a result,

$$m = \frac{m_o}{\sqrt{1 - (\tfrac{1}{3})^2}} = \frac{m_o}{\sqrt{0.89}} = 1.06\,m_o$$

and so $m = 9.6 \times 10^{-31}$ kg. Even at this very high speed the electron mass has increased by only 6 percent.  ■

---

*Exercise*  How fast must a particle be moving if $m/m_o = 800$ for it?  *Answer: $v/c = 0.99999920$*  ■

## PART II: PHOTONS

---

### 25.7 Planck's discovery

In 1900, five years before Einstein proposed his theory of relativity, Max Planck (1858–1947) made a discovery that seemed less than earth-shaking at the time but that we now recognize as the first of a Pandora's box of surprises. Planck, along with others, had been trying to interpret the radiation given off by hot, nonreflecting objects, so-called *blackbodies* (Sec. 11.11). Careful measurements of the intensity of light (as well as of infrared and ultraviolet radiation) given off by red-hot objects indicated that the intensity varies with wavelength as shown in Fig. 25.7. As we see, only a small fraction of the emitted radiation has wavelengths in the visible range. Most is in the infrared (or heat) wavelength range. Furthermore, the curves show that the radiation maximum shifts from the infrared to the visible as the tempera-

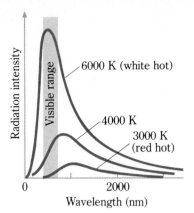

**FIGURE 25.7**

Blackbody radiation. For compara-
tive purposes, the temperatures
correspond as follows: 6000 K
(sun's surface), 4000 K (carbon
arc), 3000 K (very hot tungsten
lamp).

ture is increased.* This agrees with our experience that a white-hot body is hotter
than a red-hot one.

In order to interpret these curves, we are led to ask what sort of transmitting
antenna could be sending out em radiation from the hot object. Since the wavelengths involved are very short, the frequency of the vibrating charges must be very
large. For example, at a wavelength of 1000 nm we have

$$\text{Frequency} = f = \frac{c}{\lambda} = \frac{3 \times 10^8 \text{ m/s}}{10^{-6} \text{ m}} = 3 \times 10^{14} \text{ Hz}$$

Notice how very high this frequency is. Only in atomic-size antennas can charges be
oscillated this fast. As a result, we expect that the em radiation is being emitted by
the vibrating charges within the atoms and molecules that compose the hot object.

There are many models that we could postulate for these atomic or molecular
oscillators. For example, if the object were composed of diatomic polar molecules,
the vibrating molecule would be represented as shown in Fig. 25.8. The two atoms
are held together by a springlike force, and since the molecule is polar, its two atoms
carry equal and opposite charges. As the atoms vibrate back and forth, they act as
vibrating charges on an antenna and therefore emit em radiation of frequency $f_o$,
where $f_o$ is the natural frequency of vibration of the molecular spring system. At
least, this is the way Planck and his contemporaries reasoned.

**FIGURE 25.8**

Before 1900, it was thought that
the dipolar molecule acts like a
radio antenna and sends out em
waves as it vibrates.

It turns out, however, that all theories of radiation based on this model failed to
describe the radiation from hot objects accurately. The theories were capable of
duplicating the curves of Fig. 25.7 at long wavelengths but gave completely incorrect predictions at short wavelengths. It was Max Planck who discovered how the
theory could be modified to agree with experiment. His modification is easily
understood but difficult to justify. In fact, his only justification for it was that it gives

---

* The position of the maximum in the curve is given by Wien's law: $\lambda_m T = 2.90 \times 10^{-3}$
m · K, where $T$ is the kelvin temperature and $\lambda_m$ is the wavelength at which the maximum
occurs.

the correct answer. Let us now see what he had to assume to get agreement between theory and experiment.

As we know, the amplitude of vibration of a system such as that in Fig. 25.8 depends on the energy of the system. This is true for a mass on a spring, a pendulum, and all other oscillators. In all cases, the more energy the system has, the bigger the amplitude of the vibration. Although the frequency of vibration is always $f_o$, the natural vibration frequency, the amplitude increases as the energy increases. Planck found that he had to assume that a system could not vibrate with just any energy:

An oscillator of natural frequency $f_o$ can vibrate only with energies $hf_o$, $2hf_o$, $3hf_o$, . . . , and with no others.*

The quantity $h$ is simply a constant (it is not the amplitude of vibration!) that must have the value

$$h = 6.626 \times 10^{-34} \text{ J} \cdot \text{s}$$

if agreement with experiment is to be achieved. It is now called **Planck's constant.**

Planck's assumption is truly astonishing. It means that a vibrator can oscillate only with certain amplitudes and with no others. For example, since the total energy of a pendulum is $mgH$, where $H$ is the height to which the bob swings, Planck says that $mgH$ can be only $hf_o$, $2hf_o$, and so on, and nothing in between. To see what this means, let us consider a pendulum that has a natural frequency $f_o = 1$ Hz and a bob of mass 100 g. Then the heights to which this pendulum could swing would be

$$H_1 = \frac{hf_o}{mg} = \frac{(6.6 \times 10^{-34} \text{ J} \cdot \text{s})(1 \text{ s}^{-1})}{(0.10 \text{ kg})(9.8 \text{ m/s}^2)} = 6.7 \times 10^{-34} \text{ m}$$

or

$$H_2 = 2H_1 = 13 \times 10^{-34} \text{ m}$$

or

$$H_3 = 3H_1 = 20 \times 10^{-34} \text{ m}$$

and so on. No intermediate maximum vibration heights are possible. It could not be induced to vibrate to a height of $16 \times 10^{-34}$ m, for example.

Notice that the difference between successive heights of vibration as predicted by Planck is only about $10^{-33}$ m. This is far too small a difference to measure. As a result, we can never tell whether Planck is correct by observing the vibration of a pendulum. The gaps between the allowed energies are too small to be measured. This turns out to be true for all common vibrating systems. Hence, *we can neither prove nor disprove Planck's assumption using large-scale (laboratory-size) vibrating systems.*

Planck was therefore faced with a disturbing situation. He could obtain a suitable theory for the radiation from a hot object provided that he was willing to make the

---

* Zero energy is not allowed because it leads to conflict with the uncertainty principle, mentioned later.

## Max Planck (1858–1947)

As a young German student, Planck came close to choosing music rather than physics as his major interest. His long scientific career was devoted almost entirely to the fields of heat and thermodynamics. He discovered the quantized nature of energy while he was a professor in Berlin, publishing his findings in a series of papers between 1897 and 1901. He developed a formula that agreed with his experimental data on heat radiation, and then he had to introduce the concept of energy gaps to explain his results. At first, Planck thought the success of his concept might be nothing more than a happy mathematical accident. As others made further important progress based on the concept, his doubts diminished, but he continued to look in vain for ways to reconcile quantum physics with the classical newtonian view. By 1918, when he received the Nobel prize, his preeminence as a scientist and teacher was well established. In 1930, he became head of the Kaiser Wilhelm Society, the most prestigious scientific organization in Germany, but was forced to resign in 1937 because of his opposition to Hitler's persecutions. His son was executed in 1944 after an attempt on Hitler's life. After the war, when he was nearly 90, Planck was reappointed head of the society, which was renamed the Max Planck Society in his honor.

assumption outlined above. The experimental test of it for other vibrating systems appeared to be impossible. Therefore, at the time, it was viewed by both Planck and his contemporaries as a rather curious result but one of doubtful validity. We shall see, however, that it appears to be correct and of extreme importance.

### 25.8 Einstein's use of Planck's concept

Five years after Planck's discovery, another natural phenomenon was shown to involve Planck's constant $h$. This discovery was made by Albert Einstein in the same year (1905) he proposed his theory of relativity. As we shall see, he was able to explain in a detailed way the results of an experiment first performed by Heinrich Hertz. In so doing, he postulated that light had corpuscular as well as wave properties. Einstein's postulate, later verified, has become an integral part of modern physics.

It was discovered in 1887 by Hertz (who also produced and detected the first radio waves) that light could dislodge electrons from a metal plate. We now know this to be a general phenomenon: *short-wavelength em energy incident on a solid can cause the solid to emit electrons. This is called the* **photoelectric effect,** *and the emitted electrons are called* **photoelectrons.**

An experiment for observing the photoelectric effect is shown in Fig. 25.9. A metal plate is sealed in a vacuum tube, together with a small wire called the collector. (Such an arrangement is called a photocell.) These elements are connected in a battery and galvanometer circuit, as shown. When the tube is covered so that no light enters it, the current through the galvanometer is zero, since the

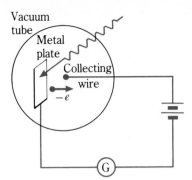

**FIGURE 25.9**

When light strikes the plate, electrons are emitted from it.

portion of the circuit from plate to collector inside the bulb lacks a connection. The vacuum space has essentially infinite resistance.

If short-wavelength light is incident on the plate, the galvanometer needle deflects. The direction of the current shows that electrons are leaving the plate and traveling through the vacuum tube to the collector. One might guess that the light heats up the plate and that when it becomes hot, electrons with high thermal energy escape from it. This is not the case, however. Careful experiments have shown that no matter how feeble the light and no matter how massive the metal plate, a stream of electrons is emitted from the plate the instant the light reaches it. No heating is required.

It is further observed that, for a given light source, the number of electrons emitted from the plate is proportional to the intensity of the light. If the battery voltage is large enough to attract all the emitted electrons to the collector, then the current in the galvanometer is directly proportional to the light intensity. (It is for this reason that a photoelectric cell is used to measure light intensity.)

A more startling feature is shown in Fig. 25.10. Suppose that the wavelength of a light beam can be varied while its intensity (its energy per unit area per second) is kept constant. Then the current in the circuit of Fig. 25.9 can be monitored as the variable-wavelength beam is incident on the plate of the photocell. It is found that the current varies with wavelength in the way shown in Fig. 25.10. Other plate materials yield similar curves but with different values for $\lambda_o$, the wavelength at which the current in the circuit becomes zero.

**FIGURE 25.10**

The current in the circuit of Fig. 25.9 varies with wavelength as shown for the metals sodium, potassium, and cesium. What is the meaning of the $\lambda_o$ value indicated in each case?

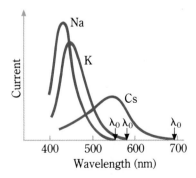

The most startling feature of these curves is that no electrons are emitted if the wavelength of the light is larger than $\lambda_o$. This wavelength is called the **photoelectric-threshold wavelength.** No matter how intense the light, no electrons are emitted if its wavelength is even slightly longer than $\lambda_o$. No matter how weak the light, if the light has a wavelength shorter than $\lambda_o$, electrons are emitted essentially as soon as the light is turned on. The particular value of $\lambda_o$, the critical wavelength for electron emission, depends on the material from which the plate is made.

Another experiment involving the circuit of Fig. 25.9 yields further important data. In this experiment, a beam of known wavelength and intensity is directed at the plate. The energy of the fastest electron emitted by the plate is then measured. To carry out that measurement, the battery of Fig. 25.9 is replaced by a variable-voltage source *of reversed polarity*. Because the collector is now negative instead of positive, it repels the photoelectrons. The current in the circuit drops to zero when this reverse voltage is made large enough. At the voltage $V_o$ (the *stopping potential*) that results in zero current, the work done by the fastest photoelectron as it travels from the plate to the collector is $V_o e$ because it moves through a voltage difference $V_o$. This work, however, must equal the kinetic energy of the most energetic photoelectron. We can therefore determine the maximum kinetic energy of the photoelectrons by measuring the stopping potential $V_o$. They are related through

$$(KE)_{max} = V_o e$$

If $V_o$, and therefore $(KE)_{max}$, is measured for beams of light of various wavelengths that are incident on the plate, an interesting result is found. When $(KE)_{max}$ is plotted against $1/\lambda$, the result is a straight-line relation, as shown in Fig. 25.11. Moreover, the value of $\lambda$ at which $(KE)_{max}$ becomes zero is the threshold wave-

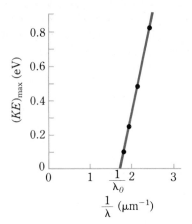

FIGURE 25.11

The photoelectron energy varies inversely as the wavelength. This particular graph is for sodium.

length, $\lambda_o$. The equation of a straight line, $y = mx + b$, becomes in this case

$$(KE)_{max} = \frac{A}{\lambda} - B \tag{25.4}$$

where $1/\lambda$ replaces $x$ and the intercept $b$ is $-B$. The constant $B$ varies from substance to substance, but $A$, the slope of the line, is the same for all materials and has a value of $2.0 \times 10^{-25}$ J · m.

Many attempts have been made to explain all these observations in terms of the wave nature of light. None has been successful. Two basic difficulties are encountered by any wave interpretation:

**1** How can one conceive of waves giving rise to a threshold wavelength? Light with $\lambda$ just slightly less than $\lambda_o$ does not differ appreciably from light with $\lambda$ just slightly greater than $\lambda_o$. Yet wavelengths slightly shorter than $\lambda_o$ cause electrons to be emitted, whereas those that are just slightly longer do not.

**2** How can even the weakest possible beam of light cause electrons to be emitted as soon as the light is turned on? The light energy seems to localize on one electron instantaneously and cause it to break free from the solid.

Thus it appeared that a new approach was needed to explain the photoelectric effect. This bold, imaginative step was taken by Einstein, who seized on Planck's ideas of special oscillator energies. Planck, it will be recalled, postulated that an oscillator with natural frequency $f_o$ could take on only certain discrete energies, namely, $hf_o$, $2hf_o$, . . . , where $h = 6.626 \times 10^{-34}$ J · s. We say that the oscillator's energies are *quantized* and that the possible energies differ by an *energy quantum* $hf_o$.

As we have seen, the em radiation (including light) emitted by a hot object was considered to be emitted by atomic and molecular oscillators composing the object. Einstein reasoned that if these atomic oscillators were to emit radiation in the way Planck visualized, the energy must be emitted in little bursts or packets. For example, since em radiation carries energy, an oscillator that is emitting light must be sending out energy. However, since an oscillator can have only certain discrete energies, it cannot throw out energy continuously. It must throw out the energy in bursts of magnitude $hf_o$ because this is the spacing between the allowed energies of the oscillator.

To be specific, suppose an oscillator has energy $37hf_o$. If it loses energy by sending out radiation, its energy can change to $36hf_o$ but not to anything in between these two values, since the oscillator's energies are quantized. But in so doing, the oscillator must have thrown out a pulse of light or other radiation with energy $hf_o$. We call such a pulse of em energy a **light quantum** or **photon.** Hence we see that there is some justification for thinking that a beam of light consists of a series of energy packets called photons, each of energy $hf_o$.

Einstein therefore postulated the following character for light:

A beam of light with wavelength $\lambda$ (and frequency $f = c/\lambda$) consists of a stream of photons. Each photon carries energy $hf$.

We see later how the photon energy is related to the structure of atoms and molecules. Let us now apply Einstein's model for a light beam to the photoelectric effect.

If light does consist of little particles of energy, these quanta, or photons, will collide with individual electrons as the light beam strikes a substance. When the energy of the photon is greater than the energy needed to tear an electron loose from the substance, electrons are emitted the instant the light is turned on. When the energy of a light quantum is less than that value, no electrons are emitted no matter how intense the light. (The chance of two photons hitting the same electron simultaneously is practically zero.) We see at once that the energy needed to tear an electron out of the plate is exactly equal to the energy of a light quantum with the threshold wavelength. Hence the minimum work needed to tear an electron loose from a solid is given by

$$\text{Minimum work} \equiv \phi = \frac{hc}{\lambda_o} = hf_o$$

where this minimum work is represented by $\phi$ and is called the **work function** of the material.

In the event that the light quantum has more energy than this, that is, if $\lambda$ is smaller than $\lambda_o$, not only can an electron be knocked out of the plate, but also it can have kinetic energy to spare. That is, the energy $hc/\lambda$ of the photon is lost partly in doing work $\phi$, or tearing the electron loose, and the remainder of the energy appears as the kinetic energy of the electron. We may therefore write, for nonrelativistic energies,

$$(\tfrac{1}{2}mv^2)_{\text{max}} = \frac{hc}{\lambda} - \phi \tag{25.5}$$

This is called the **photoelectric equation.**

Most of the emitted photoelectrons have less kinetic energy than the $(\tfrac{1}{2}mv^2)_{\text{max}}$ given by Eq. 25.5 because they undergo collisions before they escape from the material. Thus $\tfrac{1}{2}mv^2$ in Eq. 25.5 is the same as $(KE)_{\text{max}}$ in Eq. 25.4. Comparison of Eq. 25.5 with Eq. 25.4 tells us that $A$ in Eq. 25.4 should be $hc$. Experiment shows that the numerical value of $A$ is indeed equal to $hc$. As a final confirmation of Eq. 25.5, the work function $\phi$ determined by equating it to the experimental value of $B$ in Eq. 25.4 is the same as the work function determined by entirely different experiments.

Thus we can conclude that photoelectrons are ejected from a material if an incident photon has enough energy to eject one. The photon energy is $hf$, which is the same as $hc/\lambda$. A photon with the threshold wavelength $\lambda_o$ has an energy $hc/\lambda_o$, and this energy is equal to $\phi$, the work function. Such a photon is just barely capable of ejecting photoelectrons. Photons of wavelength shorter than $\lambda_o$ have more than enough energy to eject photoelectrons, and the excess energy appears as the kinetic energy of the photoelectron.

---

### Example 25.7

What is the energy of a photon in a beam of infrared radiation whose wavelength is 1240 nm?

*Reasoning*   We have that

$$\text{Photon energy} = \frac{hc}{\lambda} = \frac{(6.626 \times 10^{-34} \text{ J} \cdot \text{s})(3.00 \times 10^8 \text{ m/s})}{1240 \times 10^{-9} \text{ m}}$$

$$= 1.602 \times 10^{-19} \text{ J} = 1.00 \text{ eV}$$

where use has been made of the conversion factor $1 \text{ eV} = 1.602 \times 10^{-19}$ J, as pointed out in Sec. 16.6.

It is convenient to remember this result: *The photons in 1240-nm em radiation have an energy of 1 eV.* For example, light of wavelength 1240/4 nm has photon energies of $4 \times 1$ eV.   ∎

---

### Example 25.8

Find the energy of a photon in each of the following: (*a*) radio waves for which $\lambda = 100$ m; (*b*) green light with $\lambda = 550$ nm; (*c*) x-rays with $\lambda = 0.200$ nm.

*Reasoning*   Using the result of the previous example,

$$(a) \text{ 100-m photon energy} = \frac{1240 \times 10^{-9} \text{ m}}{100 \text{ m}} \times 1 \text{ eV} = 1.24 \times 10^{-8} \text{ eV}$$

(b) 550-nm photon energy $= \dfrac{1240 \text{ nm}}{550 \text{ nm}} \times 1 \text{ eV} = 2.25 \text{ eV}$

(c) 0.200-nm photon energy $= \dfrac{1240 \text{ nm}}{0.200 \text{ nm}} \times 1 \text{ eV} = 6200 \text{ eV}$

Notice what high energies x-ray photons have.  ■

**Exercise**   A laser beam ($\lambda = 633$ nm) has a power of 2.0 mW; that is, it carries an energy of 2.0 mJ past a given point each second. How many photons pass a point in the path of the beam each second?   *Answer: 2.45 × 10¹⁶*  ■

**Example 25.9**

When light of wavelength $5 \times 10^{-5}$ cm is incident on a particular surface, the stopping potential is 0.60 V. What is the value of the work function for this material?

**Reasoning**   We make use of the photoelectric equation (25.5):

$$\phi = \frac{hc}{\lambda} - (\tfrac{1}{2}mv^2)_{\text{max}}$$

But we had $(\text{KE})_{\text{max}} = V_o e$, where $V_o$ is the stopping potential. And so

$$\phi = \frac{hc}{\lambda} - V_o e = \frac{(6.63 \times 10^{-34} \text{ J} \cdot \text{s})(3 \times 10^8 \text{ m/s})}{5 \times 10^{-7} \text{ m}} - (0.60 \text{ V})(1.60 \times 10^{-19} \text{ C})$$

from which $\phi = 3.02 \times 10^{-19}$ J $= 1.88$ eV. Most metals have a work function several times larger than this value. However, various oxides and more complex compounds have work functions in this range.  ■

**Exercise**   What is the threshold wavelength of this material?   *Answer: 660 nm*  ■

## 25.9 The Compton effect

Since both light and x-rays are em waves, the photon concept should apply to x-rays as well. Direct evidence for the x-ray photon was first provided by A. H. Compton in 1923. He noticed that when a monochromatic, or single-wavelength, beam of x-rays was shone on a block made of graphite, two kinds of x-rays were scattered from the block. See Fig. 25.12. One kind had the same wavelength as the incoming radiation, and the other kind had a longer wavelength than the incident rays. The unchanged-wavelength portion of the beam can be pictured as arising in the following way: the oscillating electric field in the incident beam causes the charges in the atoms to oscillate with the same frequency as the wave. These oscillating charges act as antennas, radiating waves of the same frequency and wavelength. Hence the scattered x-rays are reradiated waves from the oscillating atomic charges.

As we stated, in addition to this beam of scattered x-rays, there is another type of scattered x-ray, one that has a slightly longer wavelength. The exact wavelength of these anomalous x-rays varies in a precise and relatively simple way, depending on

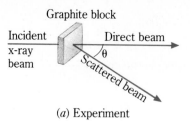

Graphite block

Incident
x-ray
beam

Direct beam

θ

Scattered beam

(a) Experiment

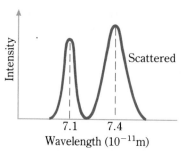

Intensity

Scattered

7.1    7.4

Wavelength ($10^{-11}$m)

(b) Spectrum of scattered beam

**FIGURE 25.12**

The Compton effect. When x-rays ($\lambda = 0.071$ nm in this case) are scattered, the scattered beam has two components. One has the same wavelength as the original beam; the other has a longer wavelength.

the angle $\theta$ at which they are scattered. No explanation for their existence appeared possible using simply the wave picture of x-rays.

A simple explanation for this phenomenon was simultaneously and independently presented by Compton and Peter Debye. They considered that the x-ray beam consists of photons, each of energy $hf$, and that a photon collides with an electron much as two balls collide, as shown in Fig. 25.13. The photon gives up some of its energy to the electron and bounces off, as shown in part $b$. Since its energy is less after the collision, its wavelength must be longer because energy equals $hc/\lambda$.

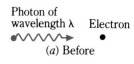

Photon of
wavelength λ    Electron

(a) Before

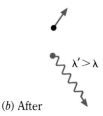

$\lambda' > \lambda$

(b) After

**FIGURE 25.13**

In the Compton effect, a photon collides with an electron. Both energy and momentum are conserved in the collision.

They then treated the problem mathematically exactly as we would treat the problem of the elastic collision of two balls — that is, both energy and momentum must be conserved. Thus,

Energy before collision = energy after collision

Momentum before collision = momentum after collision

These equations, then, could be solved for the photon's energy loss as a function of the angle at which it was scattered. The result could be used to compute the wavelength of the scattered photon by using the fact that a photon's energy is $hc/\lambda$. When this was done, the computed wavelengths of the scattered photons were found to coincide exactly with the measured values. Here again was a striking confirmation of the particle properties of em waves.

## 25.10 The momentum of the photon

Since a photon has energy, we would expect it to have momentum and mass as well. We can see, however, that the rest mass of a photon must be zero. Since it travels with speed $c$ in vacuum, we have

$$m = \frac{m_o}{\sqrt{1 - (v/c)^2}} = \frac{m_o}{\sqrt{1 - 1}} = \frac{m_o}{0}$$

If $m_o$ were anything but zero, the photon would have infinite mass. Since $E = mc^2$, however, infinite mass implies infinite photon energy, and we know this to be untrue. Therefore, we must conclude that *the rest mass of the photon is zero*.

To find the momentum of the photon, we use the two relations for photon energy:

Photon energy $= (m - m_o)(c^2) = mc^2$

Photon energy $= hf$

If we equate these two expressions, we can solve for the momentum of the photon, which is $mv = mc$ in this instance. Doing so, we find

$$\text{Photon momentum} = p = \frac{hf}{c} = \frac{h}{\lambda} \qquad (25.6)$$

This is the value used by Compton and Debye in their theories of the Compton effect.

We therefore have a method for assigning both energy and momentum to the photons of electromagnetic waves. Although these waves appear to possess a particlelike nature under certain circumstances, the *photons owe all their mass to their kinetic energy. They have no rest mass* and always travel with speed $c$ in vacuum.

# PART III: QUANTUM MECHANICS

## 25.11 The de Broglie wavelength

As we have seen, em radiation has a dual nature. It has a wavelike character that causes it to show interference and diffraction effects. It also has particlelike behavior, as shown by its photon properties. Given this duality, it is natural to speculate that the electron, and perhaps other particles, may have wave properties.

Louis de Broglie was the first to seriously propose the dual nature of the electron. He was led to his proposal by a suggestive theory for the hydrogen atom presented earlier by Niels Bohr. De Broglie found (in 1923) that he could rationalize one of Bohr's major assumptions by assuming the electron to have wave properties. Rather than follow through the historical chain of events, we will proceed directly to de Broglie's result.

The photon has a momentum given by Eq. 25.6 to be

$$p_{\text{photon}} = \frac{hf}{c} = \frac{h}{\lambda} \quad \text{and} \quad \lambda = \frac{h}{p_{\text{photon}}}$$

By analogy, if a particle such as an electron has wave properties, perhaps its associated wavelength and momentum could also be related by a similar equation. De Broglie postulated that particles have wave properties and that their wavelength is

$$\text{de Broglie wavelength} = \lambda = \frac{h}{p} \tag{25.7}$$

where $h$ is Planck's constant and $p$ is the momentum of the particle in question.

Direct experimental confirmation of de Broglie's supposition was obtained somewhat by accident in 1927 by C. J. Davisson and L. H. Germer. They were investigating the scattering of a beam of electrons by a metal crystal (nickel). Their apparatus, which was enclosed in a vacuum chamber, is sketched schematically in Fig. 25.14. A beam of electrons is given a known energy by accelerating the electrons through the potential difference $V$. As shown in the figure, measurements were made of the number of electrons scattered by a nickel crystal upon which the beam was incident. The unexpected result was that the electrons reflected very strongly at certain special angles and not at others. These results were reported as unexplained by Davisson and Germer.

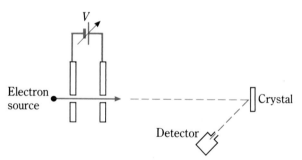

FIGURE 25.14
Davisson and Germer measured the numbers of electrons reflected from the crystal at various angles.

It was then suggested to the two investigators that perhaps this was evidence of de Broglie's radical ideas. They therefore undertook further measurements with properly oriented crystals to see whether one could apply Bragg's law (for the reflection of x-rays) to explain the data. Recall from Chap. 24 that interference of waves reflected from various planes within a crystal can be described by Bragg's law: *If the spacing between crystal planes is* d *and if the waves have wavelength* λ, *then strong reflection should occur at angles* θ *given by*

$$n\lambda = 2d \sin \theta \qquad n = 1, 2, \ldots$$

where $\theta$ is the angle between the reflected beam and the reflecting crystal plane.

Since Davisson and Germer knew the value of $d$ and the positions of strong reflection $\theta$ of the electrons, they could compute $\lambda$. In addition, the momentum of the electrons could be found, since

$$\tfrac{1}{2}mv^2 = Ve$$
$$p = mv = \sqrt{2Vme}$$

where $V$ is the potential difference through which the beam is accelerated. The de Broglie wavelength is

$$\lambda = \frac{h}{p}$$

and so it too could be obtained. Davisson and Germer found that these two wavelengths were identical. In other words, the electrons are reflected in the same way as their de Broglie waves should be reflected. This is very direct evidence for de Broglie's idea that electrons have wave properties.

Over the years, it has been found that neutrons, protons, atoms, and molecules, as well as other particles, show the same wave effects that can be obtained with electrons. We are therefore compelled to believe that particles, when moving through space, behave like waves of wavelength $h/p$, where $h$ is Planck's constant and $p$ is the momentum of the particle in question. Why this behavior had not previously been noted for macroscopic particles is discussed in Example 25.11.

---

### Example 25.10

An electron in a television tube may have a speed of $5 \times 10^7$ m/s. If we neglect relativistic effects, what is the de Broglie wavelength associated with this electron?

*Reasoning*   Substituting in Eq. 25.7, we find

$$\lambda = 0.145 \times 10^{-10} \text{ m}$$

Apparently, the wavelength associated with an electron is in the x-ray range of lengths. (We do not mean to imply that de Broglie waves are related to electromagnetic waves. They most certainly are not electromagnetic in nature. More will be said about this in the following section.)   ∎

---

### Example 25.11

Describe the diffraction pattern that would be obtained by shooting a bullet ($m = 0.10$ g, $v = 200$ m/s) through a slit 0.20 cm wide.

*Reasoning*   The wavelength of the de Broglie wave associated with the bullet is

$$\lambda = \frac{h}{p} = \frac{6.60 \times 10^{-34}}{(10^{-4})(2 \times 10^2)} = 3.30 \times 10^{-32} \text{ m}$$

From the above determination and the knowledge that interference and diffraction effects become large only if $\lambda$ is comparable with the slit width or separation (see Sec. 24.9), we can conclude that interference effects are negligible. To show this clearly, however, let us find the angle $\theta$ between the straight-through beam and the first diffraction minimum. This minimum occurs when (Eq. 24.4)

$$\sin \theta = \frac{\lambda}{\text{slit width}} = 1.6 \times 10^{-29}$$

In other words, the diffraction angles will be so small that all the particles will travel essentially straight through the slit. Straight-line motion results, and the wave effects are unobservable. This situation always occurs for macroscopic experiments, and it is for this reason that the de Broglie wave effects are unobservable in the motion of macroscopic particles. ■

## 25.12 Wave mechanics versus classical mechanics

The discovery of the wave properties of particles has serious implications both for the interpretation of particle motion and for mechanics in general. We must investigate under what circumstances the wave nature of particles is important. In carrying out our investigation, we can rely on our previously acquired knowledge about wave behavior, but first let us discuss what de Broglie waves tell us about particles.

In Fig. 25.15a we see the x-ray diffraction pattern for a beam of x-rays passing through aluminum foil. We discussed this phenomenon, x-ray diffraction, in Sec. 24.10. Figure 25.15b shows the pattern formed when electrons are shot through the same foil. The speed of the electrons is such that their de Broglie wavelength is the same as the em wavelength of the x-rays. As you see, the diffraction rings for the electrons are just like those for the x-rays.

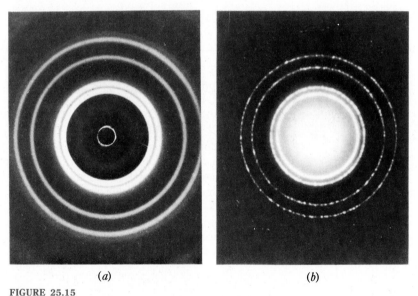

(a)                                                     (b)

FIGURE 25.15
Diffraction pattern produced by a beam of (a) x-rays and
(b) electrons incident on an aluminum foil target.
(*Education Development Center, Newton, Mass.*)

If we think of the x-ray beam as a stream of photons, then the bright diffraction rings show where the photons strike the screen (or photographic film in this case). Since we know the wavelength of the x-rays and the crystal lattice structure of aluminum, we can calculate how the x-rays should be diffracted. Thus we can calculate the diffraction pattern in Fig. 25.15a. Indeed, experiments have now been done to collect data such as those in Fig. 25.15a by measuring individual photons one by one. We cannot predict where in the pattern a given photon will strike, but it is more probable that a photon will strike near a maximum than near a minimum. And never does a photon strike the pattern at the precise location of the minima.

We can summarize as follows: a calculation using the wave picture of x-rays tells us where the photons go. When we calculated interference and diffraction effects for em waves, we actually were calculating where the photons strike the screen. Most of the photons in the beam strike near the maxima; very few strike near the minima.

The similarity between the x-ray and electron diffraction patterns indicates that a similar situation exists for de Broglie waves. These waves allow us to calculate where the electrons strike the screen. Using the de Broglie wavelength appropriate to the electrons, we can predict their interference and diffraction patterns just as we have done for other types of waves. The electrons have a high probability of striking near the maxima and zero probability of striking the minima.

The way in which the relationship between em waves and photons parallels the relationship between de Broglie waves and particles is of great importance. It helps us to understand when the wave properties of particles invalidate the laws of classical mechanics. Consider the two experimental situations in Fig. 25.16. There we see what happens when a wave disturbance such as a beam of light passes through two slits. We know that if the two slits are the result of two missing planks in a board fence, then the situation will be like that in Fig. 25.16a. Two distinct beams of light will be formed, and a sharp shadow will be seen. Under these conditions, where the slit dimensions are far larger than the wavelength, light travels in straight lines and clear shadows are cast; interference and diffraction effects are negligible.

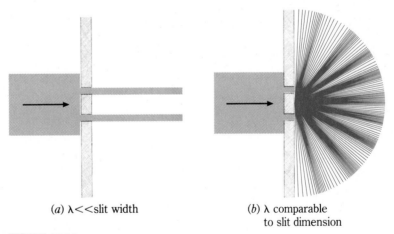

(a) λ<<slit width

(b) λ comparable to slit dimension

**FIGURE 25.16**

(a) When the wavelength associated with a particle is much smaller than the slit width, clear images of the slits are formed. (b) When λ is comparable to the slit width, however, typical wave interference phenomena are observed.

If the slit dimensions are comparable to the wavelength of the light, however, then the conditions in Fig. 25.16b will be found. The light no longer follows a straight-line path. Interference and diffraction effects are clearly visible. We thus conclude the following:

Wave effects become noticeable when the dimensions involved in the experiment are comparable to or smaller than the wavelength.

Because of the close similarity among all waves, including de Broglie waves, we can conclude that the wave nature of particles will be noticeable only when the particle wavelength is comparable to or larger than the dimensions involved in the experiment. We saw in Example 25.11 that bullets and other macroscopic objects have wavelengths far smaller than ordinary dimensions. Therefore we cannot detect their wave nature. But because the de Broglie wavelength is

$$\lambda = \frac{h}{mv}$$

particles of very small mass (such as electrons) have much larger wavelengths. Their wavelengths may be comparable to experimental dimensions, and so their wave nature may be discernible.

Returning, then, to our original question of when classical mechanics will fail, we can state the following:

Classical mechanics becomes invalid when the particle wavelength is comparable to or larger than the smallest dimension of the experiment.

This is likely to happen only when we are dealing with particles that are atomic in size or smaller. In particular, the wave effects of electrons within atoms are easily noticed. Under these conditions we must replace classical mechanics by *wave mechanics,* the mechanical description that includes the wave nature of particles. For reasons you will soon encounter, wave mechanics is often referred to as *quantum mechanics.*

Soon after de Broglie suggested the wave nature of particles, Erwin Schrödinger developed an equation to describe the behavior of particles in terms of their wave nature. **Schrödinger's equation,** which is analogous to the equation used to describe the behavior of electromagnetic waves, forms the basis of wave, or quantum, mechanics. It is now believed that, with slight modification, Schrödinger's equation will predict the observed behavior of particles under all nonrelativistic conditions. Therefore we should perhaps discard all the mechanics we learned in this text and start once again from this new basic equation. That would be foolhardy, however, because Schrödinger's equation is exceptionally difficult to use to obtain practical answers except in the simplest problems. Physicists therefore retain newtonian and classical concepts and use them to solve most problems. They worry about relativistic effects only when particle speeds approach the speed of light or when very accurate results are required. They replace newtonian mechanics by quantum mechanics only when they are dealing with dimensions comparable to particle wavelengths. In the latter case, despite the difficulties involved, quantum mechanics must be used in order to obtain reliable answers. We see in the next chapter that quantum mechanics must be used in the discussion of the internal workings of atoms.

## 25.13 Resonance in de Broglie waves: stationary states

When we dealt with other types of waves, such as waves on a string and sound waves in tubes, we found that wave resonance was of great importance. This is also true for

This energy is large enough for the energy gaps to be measurable. We therefore conclude that the wave nature of particles and the quantized character of their energies will be discernible in atomic-size systems.

Let us now return to the fact that the resonance forms in Fig. 25.17 have nodes. Wave mechanics tells us that the particle will never be found at the nodal positions. This contradicts classical physics, which tells us the particle is equally likely to be found anywhere in the tube. We can reconcile the classical and wave mechanic results if we notice that the nodes are $\frac{1}{2}\lambda$ apart. For a dust particle in a tube 50 cm long, we have

$$\lambda_n = \frac{2L}{n} = \frac{1}{n}$$

The energy of such a particle was found to be $2 \times 10^{-52} \ n^2$ J, however, whereas thermal energies are about $10^{-21}$ J. Hence a dust particle with thermal energy would be in the $n$th stationary state, which is given by

$$(2 \times 10^{-52})(n^2) = 10^{-21}$$

from which $n = 2 \times 10^{15}$. However, the distance between nodes is $\lambda_n/2$, with $\lambda_n = 1/n$ m in our case. Therefore the distance between nodes is about $10^{-15}$ m, which is far too small a distance to measure. Here, too, we find that our ordinary experience cannot be used to test the wave mechanical result. Even so, the wave approach shows clearly the quantized nature of energy and position. It is for this reason that wave mechanics is often referred to as quantum mechanics.

### Case 2: A harmonic oscillator

A mass $m$ vibrating under a Hooke's law spring force, as in Fig. 25.18, is called a *harmonic oscillator*. To a first approximation, the vibrating atoms in molecules are harmonic oscillators. In many ways, the harmonic oscillator is like the particle in a tube that we have just discussed, but the problem is complicated by the fact that the system has a variable potential energy as the spring is distorted. Even so, the resonance motion of the system can be found by solving Schrödinger's equation. The end result is not too different from that for the particle in a tube. In particular, the energy is quantized, with values

$$E_n = (n + \tfrac{1}{2})\left(\frac{h}{2\pi}\right)\sqrt{\frac{k}{m}} \qquad n = 0, 1, 2, \ldots$$

where $k$ is the spring constant.

We can place this result in an interesting form if we recall that the resonance frequency $f_o$ for a mass at the end of a spring is

$$f_o = \frac{1}{2\pi}\sqrt{\frac{k}{m}}$$

Substitution of this value in the expression for $E_n$ gives

$$E_n = (n + \tfrac{1}{2})(hf_o) \qquad n = 0, 1, 2, \ldots \tag{25.9}$$

Therefore the energies of a Hooke's law oscillator are quantized; the gaps between allowed energies are $hf_o$.

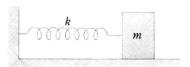

**FIGURE 25.18**

A harmonic oscillator consisting of a mass $m$ at the end of a spring whose constant is $k$.

This astounding result is simply the property that Planck had to attribute to oscillators in order to explain blackbody radiation. Some 25 years after Planck guessed that this should be the case, the use of de Broglie wave concepts showed why it must be true. We saw in Sec. 25.7 that Planck's supposition could not be tested for laboratory-size oscillators. Now we see that his unsubstantiated guess is buttressed by the many other successes of quantum theory. Further substantiation for wave mechanics will be found in the next chapter.

## 25.14 The uncertainty principle

Since the discovery of the wavelike nature of the electron, many experiments have been carried out to see whether other particles also exhibit this behavior. As we saw in Example 25.11, it is impossible to devise an experiment to observe the wave nature of large objects. However, particles of atomic size can be investigated relatively easily for wavelike effects. No exception to de Broglie's wavelength equation has ever been found. In fact, the use of electrons and neutrons as well as x-rays in diffraction experiments designed to investigate crystal structure is now commonplace.

The wave nature of all particles leads to a great philosophical principle. Prior to this discovery, philosophers had often argued about whether the fate of the universe was completely determined. Could we, in principle at least, determine the position, speed, and energy of all the particles in the universe and then predict the course of all future events? It appears that the wave nature of all particles requires us to give a negative answer to this question. This fact is embodied in the *Heisenberg uncertainty principle,* which we now examine.

Let us consider how we would locate the position, speed, and kinetic energy of an object or a particle. In order to locate the particle, we must either touch it with another particle or look at it in a beam of light. Let us make the light beam as weak as possible so that its momentum will not disturb the object at which we are looking. To that end, we look at the object by using a single photon. Or, if we choose, we touch the object with a single, extremely small particle. We call the photon or particle that we use to investigate the object the *probe particle.*

To minimize the disturbance the probe particle will cause, we use as low an energy as we possibly can. There is, however, a lower limit on this energy, because the wavelength of the probe particle must be smaller than the object we are looking at. Otherwise, as we saw in Chap. 24, interference and diffraction effects will cause the waves associated with the probe to cast extremely blurred images of the object. In particular, we saw in Chap. 24 that the finest detail we can see using waves (either light or particle waves) is detail of the same size as the wavelength. Hence, the position of the object at which we are looking may be in error by an amount

$$\Delta x \approx \lambda$$

Moreover, the momentum of the probe particle (whether photon or material) is given by $p = h/\lambda$. When it touches the object we are looking at, some of this momentum is transferred to the object, and this disturbance may alter the momentum of the object. The uncertainty in the momentum of the object $\Delta p$ may be as large as the probe particle's momentum $p$. Therefore,

$$\Delta p \approx \frac{h}{\lambda}$$

If we multiply the expressions for $\Delta p$ and $\Delta x$, we find

$$\Delta p \, \Delta x \approx h$$

In other words, when we use the most precise experiment imaginable to locate the position of an object and measure its momentum simultaneously, the product of the intrinsic uncertainties in these two measurements will be approximately as large as Planck's constant $h$. This appears to be a perfectly general relation, and it is one form of Heisenberg's uncertainty principle.

A second form of the uncertainty principle can be obtained through similar reasoning. As the probe particle passes by the object we are looking at, the position of the object is uncertain to within a distance of about $\lambda$, as we have just seen. Of course, $\lambda$ is the wavelength associated with the probe particle. If the speed of the probe particle is $v$, the time taken for the particle to pass through this distance of uncertainty is $\lambda/v$. Therefore, the exact amount of time during which the object is at a particular position is uncertain by an amount

$$\Delta t \approx \frac{\lambda}{v}$$

In addition, the energy of the probe particle is partly lost to the object under observation when the two come into contact. As a result, the uncertainty in the energy of the object is of the order of the probe particle's energy. Therefore,

$$\Delta E \approx \frac{hv}{\lambda}$$

Multiplying these two expressions yields

$$\Delta E \, \Delta t \approx h$$

We have arrived, therefore, at two uncertainty relations, one involving momentum and the other involving energy. These relations were first proposed by Werner Heisenberg in 1927. Let us now restate them in a more exact form.

**Heisenberg Uncertainty Principle** *In a simultaneous measurement of coordinate x and momentum p of a particle,*

$$\Delta x \, \Delta p \geq \frac{h}{4\pi} \tag{25.10}$$

*where $\Delta x$ and $\Delta p$ are the uncertainties in x and p. Similarly, if the energy E of a particle at time t is measured, then the uncertainties $\Delta E$ and $\Delta t$ are such that*

$$\Delta E \, \Delta t \geq \frac{h}{4\pi} \tag{25.11}$$

It is impossible, then, even in principle, to know everything about an object. There will always be uncertainty about its exact energy at a given time and its exact momentum at a given place. This is one of the fundamental results inherent in the concepts of light quanta and particle waves. Clearly, a new formalism is needed to describe atomic particles and light quanta in situations in which these effects are important. The methods of quantum or wave mechanics must be used to handle these phenomena.

# Minimum learning goals

When you finish this chapter, you should be able to

1 Define (*a*) reference frame, (*b*) inertial reference frame, (*c*) time dilation, (*d*) relativistic factor, (*e*) length contraction, (*f*) rest mass, (*g*) mass-energy relation, (*h*) light quantum, (*i*) photon, (*j*) photoelectric effect, (*k*) work function, (*l*) threshold wavelength, (*m*) Compton effect, (*n*) de Broglie wavelength, (*o*) Davisson-Germer experiment, (*p*) quantum (wave) mechanics, (*q*) stationary state, (*r*) quantized energy, (*s*) uncertainty principle.

2 State the two basic postulates of relativity.

3 State the conclusions that relativity gives rise to with respect to the following: maximum speed of objects, simultaneous events, time dilation, length contraction, mass variation with speed, kinetic energy, and mass-energy conversion. Compute the answers to simple problems involving these conclusions.

4 Compute the allowed energies (according to Planck) for an oscillator with a known natural frequency provided Planck's constant is given. Explain why the energy of a pendulum appears to be continuous.

5 Sketch a graph of radiation intensity versus $\lambda$ for a hot object. Show how the graph changes with temperature.

6 Describe the photoelectric effect and point out what is meant by the photoelectric threshold. State the energy of a photon in terms of its wavelength. Explain how the photon concept applies to the photoelectric effect. Compute the threshold wavelength from a knowledge of the work function. Use the photoelectric equation in simple situations.

7 Describe the Compton effect and explain how it can be interpreted in terms of photon scattering.

8 State the rest mass and momentum of the photon.

9 Give the de Broglie wavelength of a particle of known mass moving with a known speed. Explain why the wave properties of electrons are easily noticed, while those of a baseball are not noticeable.

10 Describe the stationary states of a particle in a tube. Detail the novel predictions of the wave theory with regard to position and energy. Explain why these predictions do not violate common experience.

11 Explain under what conditions classical newtonian mechanics must be replaced by quantum mechanics. Reasoning from the interference effects observed for light, show why newtonian mechanics breaks down under these conditions.

# Questions and guesstimates

1 Suppose you are in a spaceship traveling away from earth at a speed of $0.90c$. A laser beam is directed at the ship from the earth. If you measure the speed of the laser light relative to your ship, what will be the light's speed?

2 Suppose an astronaut has perfect pitch and so can recognize at once that a particular tuning fork gives off a sound of middle C when struck. What would she hear if she listened to the tuning fork inside her spaceship while traveling through space at a speed of $0.9c$?

3 Most human beings live less than 100 yr. Since the maximum velocity one can acquire relative to the earth is $c$, the speed of light, a person from earth can travel no farther than 100 light-years into space in 100 yr. Does this necessarily mean that no person from earth will ever be able to travel farther from earth than 100 light-years? (One light-year is the distance light travels in one year, or $9.46 \times 10^{15}$ m.)

4 Suppose that the speed of light were only 20 m/s, and all the relativistic results applied when this speed was used for $c$. Discuss how our lives would be changed.

5 It should be clear from this chapter that the statement "matter can be neither created nor destroyed" is false. What can one say instead?

6 Discuss how our world would be affected if nature were suddenly to change in such a way that Planck's constant became $10^{32}$ times larger than it is. Consider the situation from two different aspects: (*a*) quantization of energy of oscillators, and (*b*) the uncertainty principle.

7 How does the photon picture of light explain the following features of the photoelectric effect: (*a*) the critical wavelength; (*b*) the stopping potential being inversely proportional to wavelength?

8 How can the work function of a metal be measured? Planck's constant?

9 Make a list of experiments in which light behaves as a wave and a list of experiments in which its quantum char-

acter is important. Is there any experiment in your list that can be explained from both standpoints?

**10** When light shines on a reflecting surface in vacuum, a pressure is exerted on the surface by the light. Explain. Would the pressure be different if the surface were black, so that it absorbed the light?

**11** If all the mass energy of a fuel could be utilized, about how many kilograms of fuel would be needed to furnish the energy required by a city of 300,000 people for 1 yr?

**12** Estimate the power change for a local radio-station antenna system as it changes from one quantized oscillation energy state to an adjacent state. What energy photons does the station radiate? What wavelength photons? What frequency?

**13** Ultraviolet light causes sunburn, whereas visible light does not. Explain why. Some people insist that they sunburn most easily when their skin is wet. Do you see any reason for this?

---

## Problems

**1** An airplane is flying at a speed of 300 m/s parallel to the earth's surface when a screw falls from the ceiling. Relative to the point on the floor directly below its original position, where does the screw land? The ceiling is 2.7 m above the floor.

**2** (a) Suppose that you are in an elevator that is rising at a constant speed of 3.0 m/s. You drop a penny from your hand, which is 1.50 m above the floor. How long does it take for the penny to reach the floor? (b) Repeat if the elevator is standing still.

**3** An astronaut uses his very accurate wrist watch to measure his pulse as he shoots through space at a speed of $0.80c$. The result he obtains is 55 beats/min. His pulse is also being recorded by telemetry to a clock on earth. What is his pulse according to the earth clock?

**4** An apprentice astronaut has been given special dispensation to take her physics test, which is normally 50 min long, while she is on a space cruise moving at a speed of $0.95c$ relative to the earth. How long should she be allowed for the test by a monitor (a) in the ship and (b) on earth?

**5** A certain unstable substance disintegrates in such a way that half is lost in a time of 820 days. When this substance is in a spaceship traveling past the earth at a speed of $0.94c$, how long will it take for half to disintegrate according to (a) observers in the spaceship and (b) observers on earth?

**6** Suppose superior beings on a planet near Alpha Centauri, which is $4.3 \times 10^{16}$ m away, send a spaceship to us at a speed of $0.9990c$. It is contaminated by a pair of microbes that reproduce on earth in such a way that the population doubles every $6.4 \times 10^5$ s. How many microbes will be on board when the spaceship hits the earth?

**7** A certain particle has a lifetime of $2.00 \times 10^{-8}$ s when at rest. How fast must a beam of these particles be moving if they are to travel a distance of 18 m in the laboratory before changing form?

**8** It is found that a newly created beam of particles travels 3.5 m in the laboratory before the particles change form. Their speed in the laboratory is $0.9980c$. What is the lifetime of these particles when they are at rest relative to the laboratory?

**9** A 30-year-old astronaut marries a 20-year-old woman just before setting out on a space voyage. When he returns to earth, she is 35 and he is 32. How long was he gone according to earth clocks, and what was his average speed during the trip?

**10** When the length of a spaceship at rest on earth is measured by observers on earth, it is found to be 36.38 m. What will observers on earth measure the ship's length to be as the ship flies past the earth with a speed of (a) $0.20c$ and (b) $0.9920c$?

**11** The roads in rural Iowa are directed mostly north-south (N-S) or east-west (E-W) and are 1.00 mi apart. (a) How fast must a westward-directed plane be moving if passengers on it are to see the N-S roads half as far apart as the E-W roads? (b) An Iowan looking up at the plane as it flies overhead measures its length as 18 m. What is the plane's length when it is at rest at the airport? (c) A passenger uses her elaborate watch to time the flight from one road to the next. What value does she find from her watch? (d) An Iowan, with the aid of a friend 1 mi to the west, measures how long it takes for the plane to move from one road to the next. What value do they obtain?

**12** A queer particle is shaped like a cube with edge length $b$. If it is set into motion with a high speed $v$ in a direction parallel to one of its edges, what will be its volume if it is measured while it hurtles through the laboratory?

**13** The insignia on a particular spaceship is a square with a dot in its center. How fast must the ship be moving past the earth for people on earth to measure the square and perceive it as being the insignia of space invaders who use a rectangle with sides in the ratio 1:2?

**14** The straight-line distance between the earth and the star Alpha Centauri is about $4.3 \times 10^{16}$ m. Suppose a spaceship could be sent to the star with a speed of $2.5 \times 10^8$ m/s. (*a*) How long will the trip take according to earth clocks? (*b*) How long a time will the spaceship's clocks record for this journey? (*c*) What measurement will the spaceship's occupants get for the earth-to-star distance? (*d*) What speed will the spaceship's occupants compute from the results of (*b*) and (*c*)?

**15** The rest mass of an electron is $m_o = 9.11 \times 10^{-31}$ kg. Find $m/m_o$ for it when its speed is (*a*) $3 \times 10^5$ m/s, (*b*) $3 \times 10^7$ m/s, (*c*) $2.0 \times 10^8$ m/s, (*d*) $2.9 \times 10^8$ m/s.

**16** How fast must a particle be going (in relation to $c$) if its $m = 1000 m_o$?

**17** In the radioactive decay of a radium nucleus, the nucleus emits an $\alpha$ particle (that is, a helium nucleus). The kinetic energy of the resultant particles is about 4.9 MeV. How much rest mass is converted to kinetic energy in the decay process? For comparison purposes, the mass of the original radium nucleus is about $3.8 \times 10^{-25}$ kg.

**18** Find the kinetic energy of an electron that is moving with a speed of $0.95c$ ($m_e = 9.1 \times 10^{-31}$ kg). Express your answer in both joules and electronvolts.

■ **19** In a television tube, the electrons in the electron beam are accelerated through a potential difference of about 20,000 V and thereby acquire an energy of 20,000 eV. Find the ratio of $m/m_o$ for these fast-moving electrons.

■ **20** In modern nuclear accelerators, particles are sometimes accelerated to energies of billions of electronvolts. (*a*) What is the mass of a proton of $3 \times 10^9$ eV? (*b*) How fast is it moving ($m_o = 1.67 \times 10^{-27}$ kg)?

**21** To melt 1.00 g of ice at $0°C$ requires an energy of 80 cal. By what percentage does the mass of the ice increase because of the energy added to melt it?

**22** Chemists sometimes say that "the mass of the reactants equals the mass of the products" in a chemical reaction. When 2 g of hydrogen is burned with 16 g of oxygen to form 18 g of water, the reaction gives off about 60,000 cal of heat energy. How much mass is lost in the process?

■ **23** At low speeds, if a man moving with a speed $v$ relative to the earth shoots a projectile out along his line of motion with a speed $u$ relative to himself, then the speed of the projectile relative to the earth will be simply $v + u$. This cannot be correct at speeds near $c$, since speeds in excess of $c$ would be predicted. (For example, if $v = 0.7c$ and $u = 0.7c$, the speed relative to the earth would be $1.4c$, an impossibility.) Einstein showed the relative speed to be given by the formula

$$\frac{v + u}{1 + vu/c^2}$$

If a spaceship with a speed of $0.8c$ past the earth shoots a projectile out from itself in its line of motion with a speed of $0.9c$, what is the speed of the projectile relative to the earth?

■ **24** Repeat Prob. 23, replacing the projectile by a light pulse. Before you work the problem mathematically, can you give the answer from a consideration of what you already know?

**25** A 60-g mass hangs at the end of a spring whose spring constant is 0.030 N/m. (*a*) What is the natural vibration frequency for this system? (*b*) What is the energy gap between allowed energies for this oscillator? Give your answer in both joules and electronvolts.

**26** The hydrogen chloride molecule acts in many ways like two balls joined by a spring. It has a back-and-forth vibration (stretching and compressing of the spring) with a natural frequency of $8.5 \times 10^{13}$ Hz. What is the energy gap between allowed energies for this oscillator? Express your answer in both joules and electronvolts.

**27** The helium-neon laser emits red light with $\lambda = 6.328 \times 10^{-7}$ m. What is the energy of a photon in such a beam? Express your answer in both joules and electronvolts.

**28** The critical wavelength for photoelectric emission from a certain substance is 400 nm. What is the work function (in electronvolts) of this material?

**29** Find the work function of a material that has a threshold wavelength of 430 nm. Express your answer in electronvolts.

**30** (*a*) Find the threshold wavelength for a material (gold) that has a work function of 3.8 eV. (*b*) What region of the spectrum is this in?

■ **31** (*a*) Find the energy of the em wave quanta sent out by a radio station operating at $1.20 \times 10^6$ Hz. Express the answer in electronvolts. (*b*) What would be the speed of an electron that has this amount of kinetic energy?

**32** The average thermal translational kinetic energy of a particle is $\frac{3}{2}kT$. (*a*) What photon wavelength is equivalent to this average thermal energy at $27°C$? (*b*) What type of radiation is this?

■ **33** Light with a wavelength of 400 nm is shone on a surface that has a work function of 1.8 eV. Find the speed of the fastest photoelectron emitted from the surface.

■ **34** When 400-nm light is shone on a certain substance, the stopping potential measured is 0.30 V. What is the work function of the substance?

■ **35** The energy needed to break apart the atoms in the CN molecule (its dissociation energy) is 7.61 eV. What is the maximum wavelength of radiation that would be capable of causing this molecule to dissociate? In what region of the spectrum is this?

**36** The energy of the carbon-carbon bond in organic molecules is about 80 kcal/mol. If all the energy of a photon could be utilized in breaking this bond, photons of what wavelength could just accomplish it? *Hint:* 80 kcal/mol = 80,000/(6.02 × 10²³) cal per bond.

**37** An energy of 13.6 eV is needed to tear loose the electron from a hydrogen atom, that is, to ionize the atom. If this is to be done by striking the atom with a photon, what is the longest-wavelength photon that can accomplish it? (Assume all the photon energy to be effective.)

**38** A helium-neon laser for laboratory use might have a beam power of 0.5 mW. Assume that the beam is completely absorbed in a 2-g silver coin. (*a*) How many silver atoms are there in the coin? (*b*) Assume that there is one free electron per atom and that all the energy is taken up by these electrons. For how long would the beam have to strike the coin to give each free electron an average energy of 4.6 eV, the work function for silver?

**39** When light with $\lambda = 500$ nm is incident on a surface, the photoelectric stopping potential is found to be 0.15 V. What is the work function of the surface in electronvolts?

**40** Ultraviolet radiation with $\lambda = 300$ nm strikes a surface that has a work function of 3.7 eV. What will be the photoelectric stopping potential in this case?

**41** (*a*) What is the momentum of a 20-eV photon? (*b*) How does this compare with the momentum of a 20-eV electron?

**42** How large an impulse does a 500-nm photon exert on a surface if the photon is (*a*) absorbed by the surface and (*b*) reflected straight back by the surface?

**43** When a photon of wavelength $\lambda$ strikes a free, stationary particle of mass $m$, it is scattered at an angle $\theta$ to the original line of motion. The wavelength of the scattered photon is given by the *Compton equation:*

$$\lambda' - \lambda = \left(\frac{h}{mc}\right)(1 - \cos\theta)$$

Evaluate $(\lambda' - \lambda)/\lambda$ for a photon striking a free electron and rebounding straight backwards if (*a*) $\lambda = 500$ nm and (*b*) $\lambda = 0.50$ nm.

**44** A photon with $\lambda = 0.050$ nm strikes a free electron head-on and is scattered straight backward. If the electron is initially at rest, what is its speed after the collision?

**45** A laser beam has a wavelength of 633 nm. Its beam power is 0.50 mW, and its cross-sectional area is 4.0 mm². (*a*) How many photons strike a surface perpendicular to the beam each second? (*b*) What is the momentum of each photon? (*c*) What force does the beam exert on the surface if it is totally absorbed by the surface? (*d*) If it is totally reflected?

**46** A proton is accelerated through a potential difference of 1000 V. What is its de Broglie wavelength?

**47** A helium nucleus ($m = 4 \times 1.67 \times 10^{-27}$ kg, $q = 2e$) is accelerated through 1000 V. What is its de Broglie wavelength?

**48** The average kinetic energy of a free electron in a metal is $3kT/2$ at high temperatures. (*a*) At what temperature would the electron's average de Broglie wavelength be 0.5 nm? (*b*) Repeat for a helium atom that has an energy $3kT/2$ and a mass $4 \times 1.67 \times 10^{-27}$ kg.

**49** The nuclei of atoms have radii of the order of $10^{-15}$ m. Consider a hypothetical situation of a proton confined to a narrow tube with length $2 \times 10^{-15}$ m. What will be the the de Broglie wavelengths that will resonate in the tube? To what momentum does the longest wavelength correspond? If relativistic effects are assumed to be negligible, to what energy (in electronvolts) does this correspond?

**50** Consider a beam of electrons shot toward a crystal, as shown in Fig. P25.1. The crystal spacing is $b$, as indicated. For what de Broglie wavelengths will the electron beam be strongly reflected straight back on itself? For what electron kinetic energies? It is found by experiment that electrons with these energies are unable to move through such a crystal in the direction shown. Evaluate the energies in electronvolts for $b = 2 \times 10^{-10}$ m.

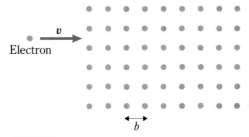

**FIGURE P25.1**

**51** A beam of electrons is shot at a crystal, as shown in Fig. P25.1. What must the de Broglie wavelengths of the electron beam be if the beam is not to reflect straight back on itself? What must the electron energies be? Evaluate the energies in electronvolts for $b = 2 \times 10^{-10}$ m.

**52** The average translational kinetic energy due to thermal motion of a particle of mass $m$ is $3kT/2$. This corresponds to an average momentum of about $\sqrt{3kTm}$. At any instant, a ball sitting on a table might be moving with about this momentum as a result of the thermal impacts of the molecules touching it. Assuming the uncertainty in the momentum of a ball to be this large, about what is the minimum uncertainty in the position of the ball? (Take $m = 10$ g and $T = 300$ K.) Repeat for a molecule of mass $10^{-20}$ g.

■ **53** The radius of a typical atomic nucleus is about $5 \times 10^{-15}$ m. Assuming the position uncertainty of a proton in the nucleus to be $5 \times 10^{-15}$ m, what will be the smallest uncertainty in the proton's momentum? In its energy in electronvolts?

■ **54** If Planck's constant were 660 J $\cdot$ s instead of $6.63 \times 10^{-34}$ J $\cdot$ s, our world would be much more complicated. In that case, how large would the de Broglie wavelength of a 100-kg football player running at 5 m/s be? About what would be the least uncertainty of his location according to an opposing player?

# 26 Energy levels and spectra

THE five years from 1923 to 1928 were of exceptional importance in physics. The discovery of the wave properties of particles in 1923 cleared the way for an understanding of the behavior of electrons in atoms. By 1928, through the use of the Schrödinger representation of wave mechanics, atomic energy-level structures and the way atoms emit and absorb light were no longer a mystery. In this chapter we learn how the wave picture explains the internal workings of atoms.

## 26.1 The modern history of atoms

Although there had been much prior speculation about atoms, it was not until 1911 that the nuclear model of the atom was validated. In that year, Ernest Rutherford and his associates carried out the experiment sketched in Fig. 26.1. He used particles shot out from the radioactive element radium as projectiles. These particles, called alpha ($\alpha$) particles, are now known to be simply the nuclei of helium atoms. A beam of such particles was shot at a thin film of gold known to be only a few hundred atoms thick.

Rutherford expected the result shown in part $a$: like bullets passing through cardboard, the particles would be slowed by the atoms and perhaps deflected slightly. Instead, the result was as shown in part $b$: although most of the particles were not deflected by the film a very few were strongly deflected, as though they

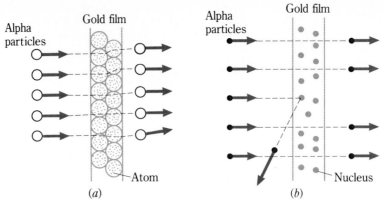

FIGURE 26.1

Rutherford shot $\alpha$ particles through thin films of gold. (*a*) The original concept of what the situation was. (*b*) The concept required to explain the experimental results.

had struck a massive object. From these observations, and the detailed analysis of them, Rutherford developed our modern concept of the atom, the so-called **nuclear atom.**

At the center of the atom is a tiny object called the *nucleus;* it is about $10^{-15}$ m in radius and contains about 99.9 percent of the atom's mass. The nucleus carries a positive charge $Ze$, where $e$ is the charge quantum and $Z$ is the atomic number of the element in question. ($Z = 1$ for H, 2 for He, 3 for Li, and so on.) The radius of the atom is of the order of 40,000 times that of the nucleus. Hence the nucleus is a tiny speck at the center of the atom. Floating in the (relatively) vast reaches of the atom outside the nucleus are the $Z$ atomic electrons. They carry a combined charge $-Ze$, and so the atom is electrically neutral. Today we know that the electron is essentially a point charge and has a radius close to, if not equal to, zero. As we see, the atom is indeed mostly empty space.

The simplest of all atoms is hydrogen. It consists of a proton as nucleus and a single electron. The model for it shown in Fig. 26.2 is in agreement with Rutherford's results. The electron orbits the nucleus, with the coulomb attraction exerted on it by the nucleus furnishing the required centripetal force. Such a model, however, should act like an em wave antenna because it is much like an oscillating dipole. If it did, an atom would "run down" as it lost energy in radiation, and its electron would spiral into the nucleus. However, hydrogen atoms don't behave this way. Usually hydrogen atoms do not radiate, and they never seem to collapse. Therefore this model for hydrogen must be in error in some way.

Hydrogen atoms can be induced to emit radiation, however. Long before 1900, it had been shown that gases and even vaporized solids can be made to emit light (that is, their atoms can be *excited*) by sending a spark or high-voltage discharge through them. (For example, neon gas in the familiar neon sign emits light when a gas discharge is set up in it by high-voltage electrodes at the tube ends.) The wavelengths of light given off by these hot gases, or their spectrum, can be investigated by use of a spectrometer, as discussed in Sec. 24.6 and shown in Color Plate II.

The spectral lines emitted by many atoms had been measured in detail even before 1900. Not knowing the structure of atoms, however, scientists were unable to give a meaningful interpretation of these spectra. Hydrogen atoms (but not molecules) give the simplest of all spectra; the hydrogen emission spectrum consists of the series of spectral lines in Fig. 26.3. (Recall from Sec. 24.6 that a spectral line is actually an image of the slit of the spectrometer. Each wavelength gives a

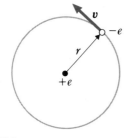

FIGURE 26.2

Classical physics pictured the electron as moving in a circular orbit around the hydrogen nucleus.

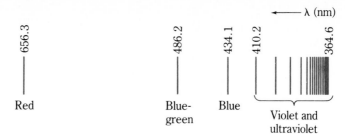

FIGURE 26.3

The Balmer series of lines for hydrogen.

separate image.) The lines in the near ultraviolet were visible only in photographs, of course, because the eye cannot see ultraviolet waves.

Notice that the lines at shorter wavelengths are closer and closer together. However, there are no lines of wavelength shorter than $\lambda = 364.6$ nm (in this region), and this shortest wavelength of the series is called the **series limit.** According to the theory we shall present shortly, there should be an infinite number of lines in this series. About 40 have actually been resolved. The remainder are too close together to be seen distinctly.

Since these spectral lines seem to have a definite sort of order, it is perhaps natural to try to fit their wavelengths to an empirical formula. This was first done by Balmer in about 1885, and this series is now known as the Balmer series. He found that the wavelengths of the lines could be expressed by the following remarkably simple formula:

$$\text{Balmer:} \quad \frac{1}{\lambda} = R\left(\frac{1}{2^2} - \frac{1}{n^2}\right) \tag{26.1}$$

where $R$ is a constant of value $1.0974 \times 10^7$ m$^{-1}$ and $n = 3$, 4, 5, and so on. Of course, $2^2$ is simply 4. If $n$ is set equal to 3, $\lambda$ turns out to be 656 nm. This is also the first line in the Balmer series, shown in Fig. 26.3. For $n = 4$, $\lambda$ is given as 486.2 nm, and so on. The integers from 3 to infinity, when placed in Eq. 26.1, yield the lines of the Balmer series. When $n$ is set equal to infinity, the formula yields the series limit, 364.6 nm. The empirical constant $R$ is called the **Rydberg constant** in honor of the man who accurately determined its value.

Later, it was found that hydrogen atoms emit wavelengths other than those found in the Balmer series. The Lyman series occurs in the far ultraviolet, and the Paschen series is in the infrared. These series are illustrated in Fig. 26.4. Other series still

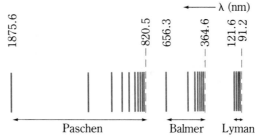

FIGURE 26.4

The three shortest-wavelength spectral series of lines given off by hydrogen atoms.

farther in the infrared have also been found. Amazingly enough, these series also follow formulas very much like Balmer's. It is found that

$$\text{Lyman: } \frac{1}{\lambda} = R\left(\frac{1}{1^2} - \frac{1}{n^2}\right) \qquad n = 2, 3, \ldots$$

$$\text{Balmer: } \frac{1}{\lambda} = R\left(\frac{1}{2^2} - \frac{1}{n^2}\right) \qquad n = 3, 4, \ldots$$

$$\text{Paschen: } \frac{1}{\lambda} = R\left(\frac{1}{3^2} - \frac{1}{n^2}\right) \qquad n = 4, 5, \ldots$$

and so on, with $R = 1.0974 \times 10^7 \text{ m}^{-1}$.

It is apparently more than mere coincidence that such simple formulas should apply to a phenomenon as complicated as light emission. Clearly some great simplicity in atomic behavior must be responsible for this remarkable set of relations.

In 1912 Niels Bohr, a student from Denmark who was spending a postdoctoral year in Rutherford's laboratory in England, devised the first reasonable interpretation of the hydrogen spectrum. He started with the classical model in Fig. 26.2. To circumvent the difficulty associated with the fact that this model predicts antenna-like radiation, he simply accepted as fact that, in certain stable orbits, the atom does not radiate. Why this should be was not clear, but with this assumption he was able to show how the observed hydrogen spectral lines originate.

Bohr's theory, although important in its time as an inspiration and guide for later workers, has now been supplanted. Its greatest drawback was the fact that his bold assumption that stable orbits exist was not buttressed by an explanation of why they exist. Such an explanation became possible in 1923, when de Broglie found out that the electron has wave properties. We shall therefore jump forward in history and describe an early model of the hydrogen atom that makes use of the wave nature of the electron. We call it the *semiclassical* (SC) *theory* of the atom. Although a proper treatment of the atom using quantum mechanics has supplanted the SC theory, we shall find that the SC theory prepares us to understand the presently accepted model.

## 26.2 The semiclassical hydrogen atom

Let us assume that the hydrogen atom consists of an electron of mass $m$ orbiting the nucleus, as in Fig. 26.5. (So that we may apply our calculations to other atoms with $Z$ different from unity, we take the nuclear charge to be $Ze$. For hydrogen, $Z = 1$.) We know that the electron has wave properties and that its de Broglie wavelength is $\lambda = h/mv$, where $v$ is the speed of the electron in orbit. However, the electron will not be in a stable state, a stationary state, unless its de Broglie wave resonates on

FIGURE 26.5

The electron is assumed to travel in a circle around the nucleus. The centripetal force is furnished by the coulomb attraction to the nucleus.

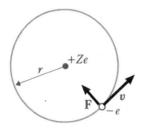

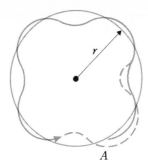

**FIGURE 26.6**

If the orbit length $2\pi r$ is an integral number of wavelengths, the wave will reinforce itself when it returns to the starting point $A$. In the case shown, $2\pi r = 4\lambda$.

the orbit. For resonance, the orbit length, $2\pi r$, must be a whole number of wavelengths long.

For an example of the resonance of an electron's de Broglie wave on a circular orbit, refer to Fig. 26.6. Start at point $A$ and trace the wave around the orbit. As you see, the orbit is exactly 4 wavelengths long. Therefore, when the wave returns to point $A$, it reinforces the original wave. Hence, as the wave winds itself around the orbit again and again, crest will fall on crest and trough on trough; this is the condition for a stationary state and resonance. Thus the resonance condition is, for an orbit that is $n$ de Broglie wavelengths long,

$$n\lambda_{\text{electron}} = 2\pi r_n \qquad n = 1, 2, 3, \ldots \qquad (26.2)$$

Wave mechanical considerations too involved for us to present here show that an electron in a stationary state does not radiate energy. Hence an orbit that satisfies Eq. 26.2 is stable.

Because $\lambda_{\text{electron}} = h/mv$, we can rewrite Eq. 26.2 and solve for the angular momentum $r_n mv_n$ of the electron in the $n$th orbit:

$$r_n mv_n = n\left(\frac{h}{2\pi}\right) \qquad (26.3)$$

This equation for the angular momentum is exactly the same criterion that Bohr found he had to use to select his stable orbits. He, however, could give no physical justification for it. Now we see why it must be true; it is the condition for electron resonance within the atom. Unfortunately, neither $v_n$ nor $r_n$ in Eq. 26.3 is known; we need yet another equation to evaluate these two quantities for the electronic orbits. Bohr showed how this can be done.

We can find a second equation by noticing that the coulomb electrostatic force supplies the centripetal force that holds the electron in orbit. Therefore, if we assume that the massive nucleus remains at rest, we can write for the orbiting electron

Centripetal force = coulomb force

$$\frac{mv_n^2}{r_n} = k\frac{(Ze)(e)}{r_n^2} \qquad (26.4)$$

where $k = 8.99 \times 10^9$ N $\cdot$ m²/C².

We can solve Eqs. 26.3 and 26.4 simultaneously for the electron's speed $v_n$ and the orbital radius $r_n$. The results are, for $n = 1, 2, 3, \ldots$ ,

$$r_n = n^2 r_1 \qquad \text{and} \qquad v_n = \frac{h}{2\pi n m r_1} \qquad (26.5)$$

where $r_1$ is the radius of the smallest possible orbit, given by

$$r_1 = \frac{h^2}{4\pi^2 Ze^2 mk} \qquad (26.6)$$

For hydrogen, $Z = 1$ and $r_1 = 0.53 \times 10^{-10}$ m. This is called the *Bohr radius*, since it is the radius that Bohr predicted for the unexcited hydrogen atom. Further, Bohr predicted the stable orbits whose radii are given by Eq. 26.5, and these orbits are called the *Bohr orbits*. Experiments show that unexcited hydrogen atoms do indeed have the 0.053-nm radius predicted by theory. In the next two sections, we shall see how the theory explains the observed emission spectrum of hydrogen.

## 26.3 Hydrogen energy levels

We have found that the hydrogen atom should have certain stationary states in which it is stable. The theory we have outlined finds that the stable states consist of circular orbits whose radii, for hydrogen, are

$$r_n = n^2(0.53 \times 10^{-10} \text{ m}) \qquad n = 1, 2, \ldots$$

A sketch showing the first few stable orbits is given in Fig. 26.7. Let us now see what energy the atom has in each of these states.

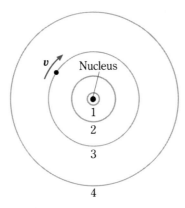

**FIGURE 26.7**

The electron can orbit the nucleus in a series of stable orbits.

Each stationary state we have found for the atom must have its own characteristic energy. The energy of the atom consists of two parts. One part is the kinetic energy of the electron as it moves in its orbit. This energy is, for the $n$th stationary state,

$$(KE)_n = \tfrac{1}{2} m v_n^2$$

where relativistic effects are neglected.* By use of Eq. 26.4, this becomes

$$(KE)_n = \frac{Ze^2 k}{2r_n} \qquad\qquad (26.7)$$

In addition to its kinetic energy, the electron has electrical potential energy. The electrical potential energy of the electron is negative. This is so because we define the potential energy of two charges to be zero when the charges are infinitely far apart. As the electron comes closer to the nucleus, however, it is going "downhill" in potential energy because the nucleus attracts it. It therefore moves to potential energies that are less than zero, or to negative potential energy. (The situation is analogous to the situation in gravitational potential energy where we define zero potential energy as being at the ceiling of a room. Then everything in the room has

---

* In Prob. 5 you will see that for hydrogen this neglect is not serious.

negative potential energy.) In the present electrical case, the potential energy of an electron that is a distance $r_n$ from a positive charge $Ze$ is

$$(PE)_n = \frac{-Ze^2k}{r_n} \tag{26.8}$$

Adding this to the kinetic energy of the electron in the $n$th orbit (Eq. 26.7) gives the total energy of the atom in the $n$th stationary state:

$$E_n = \frac{-Ze^2k}{2r_n} \tag{26.9}$$

Notice that the energy of the atom is negative and becomes more negative as $r_n$ becomes smaller (in other words, as the electron gets closer to the nucleus). This is the result of the fact that the electron loses potential energy as the coulomb force pulls it closer to the nucleus.

We can put Eq. 26.9 in a more convenient form by using Eq. 26.6 to replace $r_n$:

$$E_n = -\left(\frac{1}{n^2}\right)\left(\frac{2\pi^2 Z^2 e^4 k^2 m}{h^2}\right) \tag{26.10}$$

Evaluation of the constants in this expression yields, for $Z = 1$,

$$E_n = -\frac{2.18 \times 10^{-18}\,\text{J}}{n^2} = -\frac{13.6}{n^2}\,\text{eV} \tag{26.11}$$

Do not be puzzled by the fact that $E_n$ is negative. This simply reflects the fact that the zero point for energy was taken as $r = \infty$.

Remember that each value of $n$ corresponds to a given stationary state of the atom. In terms of our semiclassical model, $n = 1$ corresponds to the electron orbiting in its smallest possible orbit, $r_1$. The atom's energy in this state, called the **ground state**, is $E_1 = -13.6$ eV. Because systems that are left to themselves fall to their lowest possible energy (stones roll downhill), hydrogen atoms are usually found in the $n = 1$ state. For $n = 2$, the next higher energy state, the orbit radius is (from Eq. 26.5) $4r_1$. The energy of the atom is then

$$E_2 = -\frac{13.6}{2^2}\,\text{eV} = -3.4\,\text{eV}$$

Notice that $E_2$ is greater than $E_1$. The atom has a higher energy in state 2 than in state 1. In general,

$$r_n = n^2 r_1 \qquad \text{and} \qquad E_n = \frac{E_1}{n^2} = -\frac{13.6}{n^2}\,\text{eV} \tag{26.12}$$

As we see, like the energies of a particle in a tube, the energies of an electron in an atom are quantized.

It is convenient to represent the energies of quantized systems (such as atoms) on what is called an **energy-level diagram.** The energy-level diagram for hydrogen is shown in Fig. 26.8. It is a vertical energy scale with horizontal lines drawn at the energies of the stationary states of the atom. For example, the $n = 1$ line is drawn at $E_1 = -13.6$ eV. The $n = 2$ line appears at

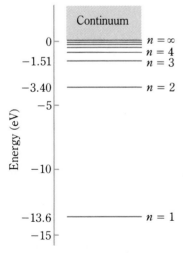

**FIGURE 26.8**

The energy-level diagram for hydrogen. There are an infinite number of levels between the levels where $n = 4$ and $n = \infty$.

## Niels Bohr (1885 – 1962)

Bohr received his Ph.D. from the University of Copenhagen in 1911 and went to Manchester the following spring to work in Rutherford's laboratory. He brought with him a good understanding of the quantum ideas of Planck and Einstein, and he proceeded to apply them to Rutherford's nuclear atom. His daring assumption that electron orbits are quantized revolutionized atomic theory, although most physicists accepted it only slowly. In 1913, Bohr returned to Copenhagen to become the first director of the Institute for Theoretical Physics. This "Copenhagen school" soon became a meeting place for theoretical physicists from all over the world; it was particularly influential in forming our present interpretation of quantum mechanics. During World War II, he escaped from German-occupied Denmark and took part in the U.S. effort to develop the fission reactor and the nuclear bomb. He was greatly concerned about the need to control the enormous military and

political power of the atomic bomb. Following the war, he devoted himself wholeheartedly to the development of peaceful uses of atomic power and to attemps to control the threat posed by nuclear weapons.

$$E_2 = -\frac{13.6}{2^2} = -3.4 \text{ eV}$$

Similarly, $E_3 = -1.51$ eV, and so on.

We are able to show only the first several levels because at higher values of $n$ the levels are too close together to draw. This is obvious from the fact that all the levels from $n = 3$ to $n = \infty$ must fall within the small gap between $-1.51$ eV and zero. Since the radius of the orbit increases rapidly with $n$, the electron is free from the nucleus when $n = \infty$; under that condition, the atom is ionized.

You will notice on the energy-level diagram a region labeled *continuum* at energies above zero. At the level at which $n = \infty$, the electron is free from the atom and is at rest. Higher energies represent the translational kinetic energy of the free electron. This energy is not quantized, and so all energies above $E = 0$ are allowed.

### Example 26.1
What is the ionization energy for the hydrogen atom?

***Reasoning*** In order to ionize an atom, the electron must be pulled loose from it. Like other physical systems, the unexcited atom is in its lowest possible energy state, and $n = 1$ for this state. The atom's energy there is $-13.6$ eV. When the electron is pulled loose from the atom, however, the atom is raised to the state at which $n = \infty$, in which $r = \infty$. Its energy there is 0 eV. We thus see that to ionize the atom, we must raise it from the $-13.6$-eV level to the 0-eV level, an energy difference of 13.6 eV. Therefore the ionization energy for the hydrogen atom is 13.6 eV. ∎

## 26.4 Light emission from hydrogen

Hydrogen atoms are usually in their lowest energy state, where $n = 1$. In this state, they are said to be unexcited. However, if you bombard hydrogen with particles such as electrons or protons, collisions can excite the atom. In other words, a collision may give the atom enough energy to change it from the stationary state at which $n = 1$ to a higher stationary state.

From the energy-level diagram in Fig. 26.8, we see that the difference in energy between the $n = 1$ and $n = 2$ states is

$$E = E_2 - E_1 = 13.6 - 3.4 = 10.2 \text{ eV}$$

Therefore the bombarding particle must provide an energy of at least 10.2 eV to excite the atom from the $n = 1$ state to the $n = 2$ state. Similarly, to excite the atom from the $n = 1$ state to the $n = 3$ state requires an energy

$$E = E_3 - E_1 = 13.6 - 1.51 = 12.1 \text{ eV}$$

In order to ionize the atom, the bombarding particles must furnish an energy of at least 13.6 eV. Why?

In one of the most common ways for exciting gas atoms (Fig. 26.9), a high voltage is applied to a low-pressure gas. The few free electrons and ions that are always

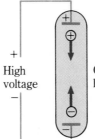

**FIGURE 26.9**

The high voltage across the discharge tube causes free electrons and ions within it to accelerate. If the voltage is high enough, these moving charges will ionize other atoms by collision.

present (because of natural radioactivity and cosmic rays — see Chaps. 27 and 28) are accelerated by the voltage, collide with the gas atoms, and generate an avalanche of charged particles by collision. As a result, the gas within the tube, called a *discharge tube,* contains a large number of ionized and highly excited atoms. Neon signs are typical examples of discharge tubes. As you may know, they give off characteristic colors of light. Let us now see why a hydrogen gas discharge tube should emit light.

Like all physical systems, atoms tend to fall to the lowest energy state possible. Excited hydrogen atoms spontaneously lose energy and fall to lower-energy states. For example, an excited atom in the $n = 3$ state may fall to the $n = 2$ state. In doing so, it must somehow lose the difference in energy between these two states, namely, $3.4 - 1.5 = 1.9$ eV. It is possible for the atom to lose this energy through collisions with other atoms. Much of the energy lost in this way eventually appears as thermal energy. However, there is another important means by which the atom can rid iself of excess energy: it can emit a photon.

Suppose a hydrogen atom emits a photon as its electron falls from the $n = j$ level to the $n = i$ level. The difference in energy between the two levels is $E_j - E_i$, and the emitted photon must have this energy. The energy of a photon is $hc/\lambda$, however, and so we have

Photon energy $= E_j - E_i$

or

$$\frac{hc}{\lambda} = E_j - E_i$$

If we use Eq. 26.10 to replace $E_i$ and $E_j$ we find

$$\frac{1}{\lambda} = \frac{2\pi^2 Z^2 e^4 k^2\, m}{h^3 c} \left( \frac{1}{i^2} - \frac{1}{j^2} \right) \tag{26.13}$$

which becomes, in the case of $Z = 1$,

$$\frac{1}{\lambda} = 1.0974 \times 10^7 \left( \frac{1}{i^2} - \frac{1}{j^2} \right) \text{ m}^{-1}$$

The theoretical constant in this equation turns out to be equal to the Rydberg constant $R$, which we discussed in Sec. 26.1. Writing the equation in terms of it, we find

$$\frac{1}{\lambda} = R \left( \frac{1}{i^2} - \frac{1}{j^2} \right) \tag{26.14}$$

If you compare this with the experimental wavelength formulas given in Sec. 26.1, you will see that the Lyman series of spectral lines is obtained by setting $i = 1$. Therefore, the Lyman series of lines must be emitted as the electron falls from outer orbits to the $n = 1$ (or innermost) orbit.

Suppose, for example, that a collision has knocked the electron into the $n = 3$ orbit, as shown in Fig. 26.10. If the electron falls back to the $n = 1$ orbit, a photon

FIGURE 26.10

A hydrogen atom in the ground state is excited to the $n = 3$ state. It emits a photon as it falls back to the $n = 1$ state. (Orbits not to scale.)

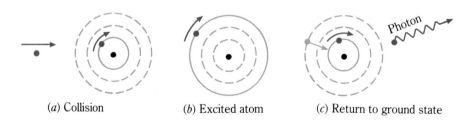

(a) Collision      (b) Excited atom      (c) Return to ground state

630    26 Energy levels and spectra

will be emitted to carry away the lost energy. From Eq. 26.14, we would have

$$\frac{1}{\lambda} = R\left(\frac{1}{1^2} - \frac{1}{3^2}\right)$$

which turns out to be the second line in the Lyman series. Indeed, we can obtain the whole Lyman series if in Eq. 26.14 we let $i = 1$ and $j = 2, 3, 4, \ldots$ ; *the Lyman series of lines is emitted as the electron falls from outer orbits to the $n = 1$ orbit.*

Similarly, if the electron falls from outer orbits to the $n = 2$ orbit, then we obtain a series of wavelengths given by

$$\frac{1}{\lambda} = R\left(\frac{1}{2^2} - \frac{1}{j^2}\right) \qquad j = 3, 4, \ldots$$

which is the Balmer series. Thus *the Balmer series of lines are emitted as the electron falls to the $n = 2$ orbit.* As you might expect, the Paschen series arises from transitions to the $n = 3$ orbit. These facts are summarized in Fig. 26.11, where only a few possible transitions are shown.

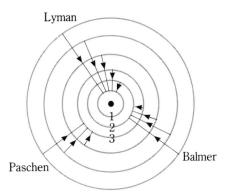

**FIGURE 26.11**

Origin of the various spectral series for hydrogen. (Orbits not to scale.)

The energy difference between the levels decreases rapidly as we go to higher and higher orbits. Hence, nearly as much energy is emitted when the electron falls from orbit 10 to 2 as when it falls from orbit 100 to 2. This means that the lines in the Balmer series become very closely spaced as we go to the wavelengths emitted by transitions from the outermost orbits to orbit 2. Of course, the most energy is emitted if the electron falls from outside the atom ($n = \infty$) to orbit 2. This results in the emission of the series-limit wavelength.

We can further clarify the origin of these spectral series by referring to the energy-level diagram for the atom in Fig. 26.8. The diagram is redrawn in Fig. 26.12, with vertical arrows showing possible electronic transitions. There is a way to see at a glance how the wavelengths of the emitted lines vary. The energy of a transition is proportional to the length of its transition arrow. Hence the Lyman-series arrows (not all of which are shown) are longer than those of the Balmer series, telling us at once that the Lyman-series wavelengths will be shorter. We can also easily see from this diagram that the spectral lines in a series that correspond to transitions from the higher $n$ values will lie very close together, since these energy levels have nearly the same values.

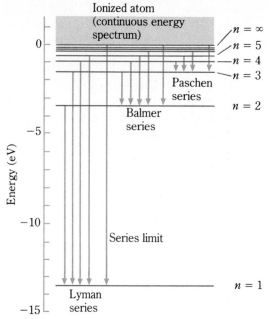

**FIGURE 26.12**

Origin of the various hydrogen spectral series.

---

**Example 26.2**

Find the wavelength of the fourth line in the Paschen series.

**Reasoning**   We know that the Paschen series results from transitions to the $n = 3$ state (Fig. 26.12). The fourth line occurs when the atom falls from the $n = 7$ state. Therefore, from Eq. 26.14,

$$\frac{1}{\lambda} = R\left(\frac{1}{3^2} - \frac{1}{7^2}\right)$$

Using $R = 1.0974 \times 10^7$ m$^{-1}$ gives $\lambda = 1005$ nm, a wavelength in the near-infra-red.  ∎

---

**Exercise**   What is the wavelength of the second line of the Paschen series? *Answer: 1281 nm*  ∎

---

**Example 26.3**

Singly ionized helium is a helium atom that has lost one of its two electrons. (*a*) Draw the energy-level diagram for this ion. (*b*) Find the first line of its Balmer-type series.

**Reasoning**   (*a*) The singly ionized helium atom is much like a hydrogen atom, except that the charge on the nucleus is $+2e$, and so $Z = 2$. From Eqs. 26.10 and 26.11, we have, after setting $Z = 2$,

$$E_n = -\frac{54.4}{n^2}\text{ eV}$$

This yields $E_1 = -54.4$ eV, $E_2 = -13.6$ eV, $E_3 = -6.04$ eV, and so on. These values are shown in the energy-level diagram of Fig. 26.13. (*b*) The first line in the Balmer series arises from the transition from $n = 3$ to $n = 2$. From Fig. 26.13, we

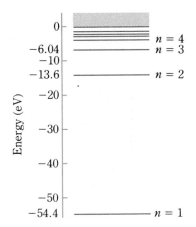

FIGURE 26.13

The energy-level diagram for singly ionized helium atoms.

have $E_3 = -6.04$ eV and $E_2 = -13.6$ eV. Therefore, the energy of the emitted photon is

Photon energy $= 13.6 - 6.04 = 7.6$ eV

Recalling from Example 25.7 that 1 eV corresponds to a wavelength of 1240 nm and that wavelength varies inversely with energy, we find

$$\lambda = \left(\frac{1 \text{ eV}}{7.6 \text{ eV}}\right)(1240 \text{ nm}) = 163 \text{ nm}$$

which is in the far ultraviolet. ∎

*Exercise* Find the series limit for the Paschen series of the helium atom. *Answer: 205 nm* ∎

## 26.5 The absorption spectrum of hydrogen

Not only do atoms emit light, they also absorb light. To learn about light absorption, let us now examine what happens in the experiment sketched in Fig. 26.14*a*. A beam of light passes through a tube filled with hydrogen atoms. The incident beam contains a continuous spectrum (a continuous range of wavelengths), as shown in Fig. 26.14*b*. However, certain discrete wavelengths are missing from the transmitted beam. Consequently, if the transmitted beam is examined with a spectrograph, the spectrum in Fig. 26.14*c* is obtained. We wish to find out which wavelengths the hydrogen atoms absorb from the beam.

To do so, we examine what happens as photons within the incident beam collide with the atoms. Normally the hydrogen atoms are in their ground state, the state in which $n = 1$. When a photon strikes an atom, the photon loses either all its energy

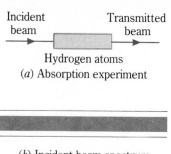

Incident        Transmitted
beam               beam

Hydrogen atoms
(a) Absorption experiment

(b) Incident beam spectrum

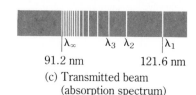

$\lambda_\infty$     $\lambda_3$ $\lambda_2$   $\lambda_1$

91.2 nm                    121.6 nm

(c) Transmitted beam
(absorption spectrum)

**FIGURE 26.14**

Hydrogen atoms absorb only
certain wavelengths. What are they?

or none of it.* The primary factor determining which of these two possibilities will
prevail is the following: when the energy of the colliding photon is exactly equal to
the energy difference between level $n = 1$ and some other level, the photon can be
absorbed. Otherwise, it must continue with its original energy.

The reason for this is quite simple. Since the electron in a hydrogen atom can
exist in only one of the discrete energy levels, it can take on only increments of
energy that will transfer it from one level to another. For example, if the atom is in
the level $n = 1$, as it normally is, it can absorb only energies equal to the energy
differences between the level $n = 1$ and the higher levels. As we see in Fig. 26.12,
these transitions correspond to energies that (in emission) give the Lyman series of
lines. Therefore, photons with wavelength equal to that of the first line of the Lyman
series (121.6 nm) will have enough energy to excite the atom from the state $n = 1$
to $n = 2$. They can therefore be absorbed by the atom.

Similarly, photons with wavelength equivalent to any of the lines in the Lyman
series can be absorbed by hydrogen atoms in the ground state. No other interme-
diate-wavelength photons can be absorbed, since their energies will not correspond
to an allowed transition for the electron. However, photons with wavelengths less
than the Lyman series limit, 91.2 nm, can be absorbed. These photons have enough
energy to excite the electron into the continuum, the region of continuous energy
levels. Photons with this much energy tear the electron completely loose from the
atom (that is, ionize it) and give it additional kinetic energy. This type of photon
absorption process is similar to photoelectric emission of electrons from a solid, and
is referred to as the *atomic photoelectric effect.*

From what has been said, we can state what will happen when a continuous
spectrum of radiation is passed through a gas of atomic hydrogen. Most of the
wavelengths will not be absorbed, since their photons do not have the proper
energies to excite the atom to an allowed energy state. However, wavelengths that
correspond to lines in the Lyman series will be absorbed, since the corresponding
photons have the proper energy to excite the atom to an allowed energy state. We
call such an absorption spectrum a *line absorption spectrum.* Wavelengths shorter

* We ignore the Compton effect (Sec. 25.9) in this discussion because it is negligible in
comparison with the effect we are discussing.

than the Lyman series limit will be absorbed, since these photons will ionize the atom and carry the electron into the continuum. The absorption in this wavelength region, not shown in Fig. 26.14, is called a *continuous absorption spectrum* because a whole range of wavelengths is absorbed.

Finally, we should note that absorption lines that correspond to the Balmer series lines do not exist, except perhaps extremely weakly. The reason for this is as follows. We know the Balmer series corresponds to transitions between the $n = 2$ and higher levels. Since very few electrons are normally in the state $n = 2$, only a very few atoms are capable of having an electron knocked from the state at which $n = 2$ to higher states. Therefore photons that correspond to these energies will not be strongly absorbed. Of course, in highly excited hydrogen gas, the situation becomes more favorable for detecting absorption at the Balmer line wavelengths. Why?

---

### Example 26.4

What is the longest wavelength of light capable of ionizing a hydrogen atom?

**Reasoning** The desired wavelength has enough energy to tear the electron loose from the atom, that is, to raise it to the $n \to \infty$ energy level. This same energy is emitted when the electron falls from orbit $\infty$ to orbit 1. Clearly, the wavelength is given by Eq. 26.14 with $i = 1$, $j = \infty$, and the constant $R = 1.097 \times 10^7$ m$^{-1}$,

$$\frac{1}{\lambda} = 1.097 \times 10^7 \left( \frac{1}{1} - \frac{1}{\infty} \right) \text{ m}^{-1}$$

$$\lambda = 0.912 \times 10^{-7} \text{ m} = 91.2 \text{ nm}$$

This, of course, is also the series limit for the Lyman series.

This answer could have been obtained more quickly by noting from Fig. 26.8 or Eq. 26.11 that $E_\infty - E_1 = 13.6$ eV. From the fact that a 1-eV photon has $\lambda = 1240$ nm, and that

$$\lambda = \left( \frac{1 \text{ eV}}{13.6 \text{ eV}} \right) (1240 \text{ nm}) = 91.2 \text{ nm} \quad \blacksquare$$

---

### 26.6 The wave theory of the atom

As we have seen, the Bohr theory of the hydrogen atom predicts the correct energy levels for the atom. It also explains the spectrum that hydrogen atoms emit and absorb. Using the wave properties of the electron, we have been able to justify Bohr's assumption that the electron exists only in certain stable states. Bohr assumed at the outset that these stable states consist of circular orbits around the nucleus. A better approach would be to start with Schrödinger's equation (Sec. 25.12) for the behavior of de Broglie waves and determine its resonance solutions for an electron in the coulomb potential of the nucleus.

Recall from Sec. 25.13 that the wave resonances of a particle in a tube told us where the particle is likely (and unlikely) to be found. Each resonance form was characterized by a quantum number, which was an integer between 1 and $\infty$. It turns out that resonances in three dimensions require three quantum numbers to specify their form. Thus we should expect the hydrogen atom's resonance forms to be characterized by three quantum numbers, not the single one we used in connection with the Bohr theory. Even so, the resonance forms should tell us where the electron is likely to be found when the atom is in a given resonance state. Let us now discuss the results one obtains for the hydrogen atom when the Schrödinger equation is used to find its resonance states.

The wave theory of the hydrogen atom gives the same energy levels found previously. We have

$$E_n = -\frac{13.6}{n^2} \text{ eV}$$

This ensures that the wave theory (or quantum mechanical) result will predict the observed hydrogen spectrum. Each energy state is characterized by the quantum number $n$, which we call the **principal quantum number.**

The resonance form found for the $n = 1$ state differs substantially from the circular orbit Bohr postulated for the $n = 1$ state. It turns out that the electron is likely to be found in the fuzzy part of a fuzzy circular shell centered on the nucleus. Figure 26.15 shows a cross section of the fuzzy shell, and the electron is most likely

**FIGURE 26.15**

In the ground state, the electron is most likely to be found in a fuzzy spherical shell centered on the nucleus. Shown is a cross section through the shell.

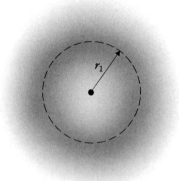

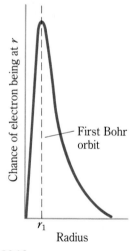

Chance of electron being at $r$

First Bohr orbit

$r_1$

Radius

**FIGURE 26.16**

Wave theory predicts the relative probabilities shown that the electron will be found at various radii from the atom's center.

to be found where the shading is heaviest. Although it is most likely to be found at a radius $r_1$ from the nucleus, the electron may exist anywhere in the shaded region of the spherical shell. Be certain that you understand how this result differs from Bohr's concept of a single circular orbit. The wave theory replaces the circle with a spherical shell and, in addition, does not restrict the electron to a definite radius; the shell is very fuzzy. This is shown graphically in Fig. 26.16.

The resonance form predicted by the wave theory for the $n = 2$ state is much more complicated than that for the $n = 1$ state. It turns out there are three resonances that have the $n = 2$ energy. We can picture these resonances, and those for still larger values of $n$, by means of diagrams that attempt to show possible electron positions in space. They show the chance of finding the electron at various positions within the atom. Figure 26.17a shows such a diagram for the $n = 1$ resonance. To get the three-dimensional picture of the physical situation, the pattern shown should be rotated about a vertical axis through the center; the intensity of the pattern in space gives the chance of finding the electron there. The three $n = 2$ resonances of the atom are shown in b, c, and d. Although the resonance in c approaches an orbital shape, the other two do not conform at all to Bohr orbits.

When we go to $n = 3$, there again turn out to be several resonances possible. As we saw, for $n = 1$ there was one resonance form. For $n = 2$ there are three resonance forms. For $n = 3$, there are six different resonance patterns. At larger

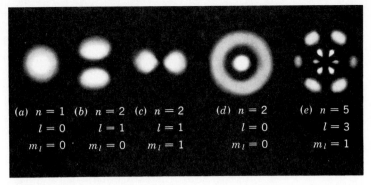

**FIGURE 26.17**

To obtain the electron distribution in space, the figures
must be rotated about a vertical axis through the center
of each. All the distributions are constant on any
horizontal circle with the axis as center.

$n$'s, some of the resonance patterns become quite complicated; one of the resonances for $n = 5$ is shown in Fig. 26.17$e$.

We see from all this that the Bohr theory is indeed a gross oversimplification of
the electron behavior in the hydrogen atom. In particular, Bohr's concept of fixed
orbits is untenable. The energy levels of the atom are predicted correctly by the
Bohr theory, however, and the principal quantum number $n$ that Bohr introduced
has great importance. Although we should always keep the limitations of the Bohr
model in mind, it offers a framework for a systematic description of atoms, and we
make frequent reference to it.

## 26.7 Quantum numbers and the Pauli exclusion principle

As we saw in the previous sections, the hydrogen atom and its electron can exist in
certain discrete energy levels characterized by an integer $n$. In particular, we have
seen that the energy levels are given by the equation

$$E_n = \frac{-13.6Z^2}{n^2} \text{ eV}$$

where $Z = 1$ for hydrogen. The integer $n$ ranges from 1 to infinity as the atom
assumes its various allowed energies. Although we arrived at this result by use of
Bohr's model, the wave picture, based on the solution of the Schrödinger equation,
leads to the same result. Hence it is seen that $n$ is a fundamental parameter needed
to describe the state of a hydrogen atom. As mentioned earlier, it is called the
*principal quantum number.* It characterizes the energy level in which the electron
is to be found. Bohr pictured each value of $n$ as being associated with a particular
orbit for the electron, but this proves untenable, as pointed out in the preceding
section. Nevertheless, it is common to say that each value of $n$ corresponds to a
particular **shell** (rather than orbit) about the nucleus. For example, when the atom
is in the $n = 3$ energy level, it is customary to say that the electron is in the $n = 3$
shell.

We also saw in the preceding section that more than one wave resonance form is possible for the same value of the principal quantum number $n$. The wave theory shows that two other quantum numbers must be specified in order to designate a particular wave resonance within the atom. One of these, the **orbital quantum number,** is related to the angular momentum of the Bohr electron in its resonance orbit. It is represented by the letter $l$ and can assume integer values from 0 to $n - 1$. For example, when $n = 1$, the possible values for $l$ are limited to a single value, namely $l = 0$. When $n = 2$, it is apparent that $l$ can take on the values 0 and 1, since $n - 1 = 1$ in this case. Notice that $l$ is always less than $n$.

The third quantum number, the **magnetic quantum number,** can assume the values $0, \pm 1, \pm 2, \ldots, \pm l$. It is represented by $m_l$. When $n = 4$, for example, the largest possible value for $l$ is 3, and hence $m_l$ can take on the values $-3, -2, -1, 0, +1, 2, 3$. This means that when the atom is in the $n = 4$ energy level, seven different resonance forms are possible that have $l = 3$. In addition, there are five resonance forms with $l = 2$, three resonance forms with $l = 1$, and one resonance form with $l = 0$. Hence the atom can exist in $7 + 5 + 3 + 1 = 16$ different resonances, each with the same energy, namely, the energy of the $n = 4$ level.

Finally, a quantum condition exists for the electron itself. As we mentioned before, the electron acts as a small magnet because of its spin about its axis. This magnet can take up only two orientations relative to an external magnetic field in which the atom may find itself. It can align either parallel or antiparallel to the field line direction. We characterize this by assigning a **spin quantum number,** designated $m_s = \pm \frac{1}{2}$; the two signs represent the aligned and antialigned positions.

We see therefore that four quantum numbers are needed to describe the state of an electron in an atom:

Principal: $n = 1, 2, \ldots$

Orbital: $l = 0, 1, \ldots, n - 1$

Magnetic: $m_l = 0, \pm 1, \ldots, \pm l$

Spin: $m_s = \pm \frac{1}{2}$

Or, in summary,

$$0 \le |m_l| \le l \le n - 1 < n \le \infty$$

and

$$m_s = \pm \frac{1}{2}$$

For a given set of $n$, $l$, and $m_l$ values, there is a very definite wave resonance pattern for the electron wave within the atom. To specify the electron completely, however, we must know whether the electron spin is aligned with ($m_s = \frac{1}{2}$) or opposite to ($m_s = -\frac{1}{2}$) the magnetic field. We call each combination of the quantum numbers $n$, $l$, $m_l$, and $m_s$ an electronic **state** of the atom. We shall now see that an extremely important law of nature applies to the behavior of electrons in the available states.

The importance of designating these states as we have done was first appreciated fully by Wolfgang Pauli in 1925. He wished to extend these concepts to atoms other than hydrogen. In order to properly assign states to the various electrons in multi-electron atoms, he arrived at the following conclusion, which is known as the *Pauli exclusion principle:*

No two electrons in an atom can have the same four quantum numbers; that is, no two electrons can exist in the same state.

This principle is basic to an understanding of the electronic structure of atoms, as we see in the next section.

## 26.8 The periodic table

Until now we have been primarily concerned with an atom that has only one electron. This might be hydrogen, singly ionized helium, doubly ionized lithium, and so on. We are now in a position to discuss how the additional electrons are arranged in the multielectron atoms found in nature and listed in the periodic table. To do this, we once again use the concept of electron shells about the nucleus; each value of $n$ has a shell associated with it. Moreover, we assume that the same resonances found for the single-electron atom can be carried over qualitatively to more complex atoms. That is, we use electronic states specified by the $n$, $l$, $m_l$, and $m_s$ quantum numbers described in the previous section.

The question we must now answer is: How do the electrons arrange themselves in the various atomic states when there is more than one electron in an atom? For example, there are six electrons in each carbon atom. In which energy levels and electronic states are they to be found? We can answer this question by using the following three rules, which we already discussed:

**1** A neutral atom has a number of electrons equal to its atomic number $Z$.

**2** In an unexcited atom, the electrons are in the lowest possible energy states.

**3** No two electrons in an atom can have the same four quantum numbers (the exclusion principle).

Let us now use these rules to determine the electronic structure of the unexcited atoms in the periodic table.

### Hydrogen (Z = 1)

Its single electron will be in the $n = 1$ level. This is the lowest possible energy level, and no violation of the exclusion principle occurs.

### Helium (Z = 2)

Its two electrons can both exist in the $n = 1$ level since they can have the following nonidentical quantum numbers:

| Electron | $n$ | $l$ | $m_l$ | $m_s$ |
|----------|-----|-----|-------|-------|
| 1 | 1 | 0 | 0 | $\frac{1}{2}$ |
| 2 | 1 | 0 | 0 | $-\frac{1}{2}$ |

However, since these are the only combinations of quantum numbers possible for $n = 1$, a third electron cannot enter this shell. The shell is filled.

### Lithium (Z = 3)

This atom has three electrons, and so the third must go into the $n = 2$ shell. We have

| Electron | $n$ | $l$ | $m_l$ | $m_s$ |
|----------|-----|-----|-------|-------|
| 1 | 1 | 0 | 0 | $\frac{1}{2}$ |
| 2 | 1 | 0 | 0 | $-\frac{1}{2}$ |
| 3 | 2 | 0 | 0 | $\frac{1}{2}$ |

Since this third electron is in the second energy level, it is much more easily removed from the atom than the first two are. Hence lithium loses one electron in chemical reactions and is univalent.

### Larger Z Atoms

Obviously there are quite a few possible combinations for the quantum numbers when $n = 2$. If you count them, you will find there are eight:

| $n$ | $l$ | $m_l$ | $m_s$ |
|-----|-----|-------|-------|
| 2 | 0 | 0 | $\pm\frac{1}{2}$ |
| 2 | 1 | 0 | $\pm\frac{1}{2}$ |
| 2 | 1 | $+1$ | $\pm\frac{1}{2}$ |
| 2 | 1 | $-1$ | $\pm\frac{1}{2}$ |

Therefore eight electrons can exist in the $n = 2$ shell. This means that the shell will not become closed until element $Z = 10$, neon, is reached. You probably know that this is an unreactive gas, unreactive because it has a closed shell. The next element, $Z = 11$, is sodium. This is univalent, since its extra electron is alone out in the $n = 3$ shell and is rather easily removed.

As one proceeds to the very-high-$Z$ elements in the table, the concept of shells becomes less useful. The trouble arises primarily because the separation between energy levels is relatively small at high $n$ values. In these cases the repulsions between the various electrons in the atom sometimes contribute energies large enough to cancel the influence of energy differences between shells. Despite this complication, the shell approach still proves useful for qualitative considerations.

## 26.9 X-rays and the spectra of multielectron atoms

As we have seen, the Pauli exclusion principle tells us how the electrons pack into an atom. When the atom is unexcited, the electrons fill the lowest energy levels possible, consistent with the exclusion principle. To a first approximation, Eq. 26.10 gives the energy of any electron in the $n$th state. Therefore the energy of an electron in a multielectron atom is the same as $Z^2$ times the energy of the electron in the same state in the hydrogen atom. This approximation breaks down for the outer electrons of the atom, however, because the interaction energies between these electrons are comparable to the energy differences between the Bohr energy levels. Therefore the Bohr energies cannot apply to these outer electrons.

However, the interaction energy between electrons is small relative to the energy differences between the $n = 1$ and $n = 2$ states. To see this, let us compute the energy values $E_n$ as given by the Bohr formula for an atom with many electrons.

Taking zinc as an example ($Z = 30$), the Bohr energies are

$$E_n = -\frac{13.6Z^2}{n^2} \text{ eV} = -\frac{12{,}240}{n^2} \text{ eV}$$

Notice the large numerical value of the energy. The situation is even more startling for gold ($Z = 79$):

$$E_n = -\frac{84{,}900}{n^2} \text{ eV}$$

As we see, the energies of states $E_1$ and $E_2$ in these atoms differ by tens of thousands of electronvolts. Compared with these inner-shell energies, the coulomb interaction energies between electrons are small. Hence the Bohr energies are nearly correct for electrons in the $n = 1$ and $n = 2$ shells of atoms that have high atomic numbers.

For an electron in an outer shell, however, the situation is quite different. First, the inner-shell electrons, being closer to the nucleus, appear (to first approximation) to cancel part of the nuclear charge. Hence the $n = 2$ electrons "see" a nuclear charge of about $(Z - 2)(e)$ rather than $Ze$; similarly, the $n = 3$ electrons "see" a nuclear charge of about $(Z - 10)(e)$ because of the two $n = 1$ electrons and the eight $n = 2$ electrons. We say that the inner electrons *screen* the nuclear charge from the outer electrons.

In addition to this effect, the outer-shell electrons are subject to the energies from electron-electron interaction involving all the other electrons of the atom. As mentioned, these energies are comparable to the small energy differences between outer shells. Hence the Bohr energy formula does not apply to them.

To cause an atom to emit radiation, some of its electrons must be excited to higher energies. Because the outer-shell electrons require only small amounts of energy to excite them to empty states, it is not difficult to obtain visible light from atoms of high $Z$. The atom is simply vaporized and used in a discharge tube much like the one we saw in Fig. 26.9. However, the spectral lines emitted by transitions between these outer-shell levels are extremely numerous and complex. For example, the calcium spectrum shown in Color Plate II shows only a small fraction of the strongest lines that calcium emits at visible wavelengths.

The situation is quite different for transitions involving inner-shell electrons. In the case of zinc, which we mentioned previously, nearly 12,000 eV of energy is needed if an $n = 1$ electron is to be thrown to a higher vacant state. Moreover, when an $n = 2$ electron falls back to the $n = 1$ state, it will release a photon whose energy is about 9000 eV. This corresponds to a wavelength of

$$\lambda = \left(\frac{1 \text{ eV}}{9000 \text{ eV}}\right)(1240 \text{ nm}) = 0.14 \text{ nm}$$

This wavelength is in the x-ray region. Thus we see that transitions between inner shells of a high-$Z$ atom give rise to x-rays. To generate x-rays, we need to excite the inner-shell electrons, and, as we have seen, this requires large amounts of energy.

A typical x-ray tube circuit used to generate x-rays is shown in Fig. 26.18. Electrons emitted from the hot filament are accelerated through potential differences of the order of $10^5$ V. When these high-energy electrons strike the high-$Z$ atoms in the target, electrons are knocked out of the inner shells of the atoms. As other electrons fall into the vacancies, x-ray photons are emitted. The x-rays so generated have wavelengths that are characteristic of the energy differences between the various shells within the atom. That is, the emitted photons carry an

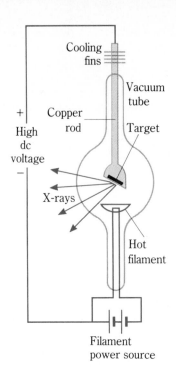

Cooling
fins

Vacuum
tube

+

Copper
rod

Target

High
dc
voltage

X-rays

−

Hot
filament

Filament
power source

**FIGURE 26.18**

Electrons emitted by the hot
filament bombard the target. The
target then emits x-rays.

energy equal to the difference between the energies of the two shells that act as the
starting point and the end point for the electron that falls into the vacancy. X-rays
emitted by this process are referred to as **characteristic x-rays**.

Another type of x-ray emitted from a target when it is bombarded by electrons is
referred to as **bremsstrahlung,** from the German "braking radiation." As the
name implies, these x-rays are emitted by the bombarding electrons as they are
suddenly slowed on impact with the target. We know that any accelerating charge
emits electromagnetic radiation (a charge oscillating on an antenna, for example).
Hence these impacting electrons also emit radiation as they are strongly decelerated by the target. Since the rate of deceleration is so large, the emitted radiation is
correspondingly of short wavelength, and so the bremsstrahlung is in the x-ray
region. However, unlike characteristic x-rays, the bremsstrahlung has a continuous
range of wavelengths. This reflects the fact that deceleration can occur in a nearly
infinite number of different ways, so that the energy released varies widely from one
impact to another.

Figure 26.19 is a graph of the radiation emitted from a molybdenum target when
it is bombarded by 35,000-eV electrons. The two sharp peaks are the characteristic

**FIGURE 26.19**

X-rays emitted from a molybdenum
target when it is bombarded by
35,000-eV electrons.

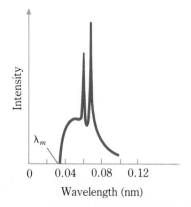

Intensity

$\lambda_m$

0    0.04   0.08   0.12

Wavelength (nm)

x-rays emitted as electrons fall to the $n = 1$ shell of this atom from its $n = 2$ and $n = 3$ shells. The shorter wavelength, of course, corresponds to the higher-energy transition, the $n = 3$ to $n = 1$ transition. Bremsstrahlung is the cause of the lower-intensity radiation spread over all wavelengths longer than $\lambda_m$. Since the energy of the electrons in the impacting beam was 35,000 eV, the emitted photons cannot have energies larger than this value. Using our conversion based on 1240 nm being equivalent to 1 eV, we find that 35,000 eV corresponds to $1240/35{,}000 \approx 0.035$ nm. As we see from Fig. 26.19, the highest-energy bremsstrahlung does indeed have this wavelength.

---

**Example 26.5**

From the data in Fig. 26.19, find the energy difference between the $n = 1$ and $n = 2$ levels in molybdenum.

**Reasoning** As we saw in our discussion of Fig. 26.19, the long-wavelength peak in that figure, 0.070 nm, results from the $n = 2$ to $n = 1$ transition. Therefore the photon of wavelength 0.07 nm carries away the energy lost by an electron as it falls from the $n = 2$ to the $n = 1$ shell. Since 1240 nm corresponds to 1 eV, 0.070 nm corresponds to an energy of 1240/0.070, or about 18,000 eV. Therefore the energy difference between these two shells in molybdenum atoms must be about 18,000 eV. ∎

---

**Exercise** A zinc target is bombarded by 13,000-eV electrons. What is the shortest-wavelength x-ray emitted by the target? What approximate wavelength corresponds to the $n = 3$ to $n = 1$ transition? *Answer: 0.095 nm; 0.114 nm* ∎

---

## 26.10 Laser light

A beam of ordinary light is the sum of a multitude of individual waves sent out by independent atoms. Although the waves in a monochromatic light beam all have the same wavelength, the waves sent out by the individual atoms are not in phase. Indeed, they do not maintain a fixed phase relation relative to each other. In other words, the waves are not coherent. It turns out that, if the amplitude of each wave is $A$, then the amplitude of the wave that results from the sum of $N$ such waves is $A\sqrt{N}$.

Suppose, however, that we could synchronize the atoms in a light source so that they emit light waves that are in phase with one another. Then all the waves would be in phase and would be coherent. The resultant wave due to $N$ in-phase coherent waves, each of amplitude $A$, is simply the direct sum of the waves, namely $AN$. If we compare this with the amplitude of $N$ incoherent waves, $A\sqrt{N}$, we see that the amplitudes are in the ratio of $AN \div A\sqrt{N}$. The intensity of a wave is proportional to its amplitude squared, however, and so we find that

$$\frac{\text{Intensity (coherent)}}{\text{Intensity (incoherent)}} = \left(\frac{AN}{A\sqrt{N}}\right)^2 = N$$

*A beam consisting of N waves is N times more intense if the waves are coherent than if they are incoherent.* Because a typical beam might consist of perhaps a million individual waves at a given point, a coherent beam might be a million times more intense than a similar incoherent beam.

# The discovery of x-rays

Many of the laws of physics have been discovered by interpreting the results of experiments that were carefully designed to determine these laws. This was the case in Galileo's discovery of the law of falling bodies. Newton's laws are also of this same general type. However, sometimes a curious, unexpected laboratory circumstance leads to the discovery of an important phenomenon. This was how Wilhelm Konrad Röntgen (1845–1923) discovered x-rays.

Experimenting with high-voltage discharges, he applied a potential difference of several thousand volts to electrodes in the two ends of a partly evacuated tube. Under such conditions, a glow much like that observed in neon signs occurs. However, if the pressure of the gas in the tube is reduced enough, the glow nearly ceases. While performing experiments with a highly evacuated discharge tube in his darkened laboratory in 1895, Röntgen observed

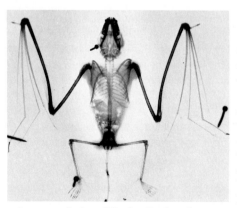

An x-ray of a bat. (*Hewlett-Packard*)

that a nearby fluorescent screen (much like that in present-day TV sets) was also glowing in the darkness. By moving the screen about the room, he was able to show that the light given off by the fluorescent screen resulted from something taking place in the discharge tube. Since a light-tight cover could be placed over the tube without greatly affecting the glow on the screen, it was clear that the fluorescent glow was caused by something other than the light given off by the tube. Röntgen named this unknown radiation coming from the tube and striking the screen x-rays.

He carried out many experiments with the rays from the tube and found that they were highly penetrating, even being able to pass through a book. Heavy metals and bone (among other materials) were not nearly as transparent to the rays as such materials as wood and paper. He was even able to cast a shadow with these rays. Shadows could be produced showing the bones of his wife's hand and the ring on her finger. Because of this ability to produce shadows of bones and metallic objects, x-rays found immediate widespread use in the scientific, medical, and industrial worlds. Unfortunately, x-rays damage human tissue, and overexposure can cause severe burns. Even when visible burns do not occur, other damage may be present. Many early workers in the field of x-rays suffered deteriorating health and even death because of their ignorance of the harmful properties of these little-understood rays.

It was not until the 1950s that a light source that emits coherent waves from its individual atoms was devised. This type of source is called a **laser**. (Laser is an acronym for *l*ight *a*mplification by *s*timulated *e*mission of *r*adiation.) It makes use of a fact pointed out in 1917 by Einstein: *excited atoms can be stimulated to emit a light wave by a wave of the same frequency incident on the atom; moreover, the*

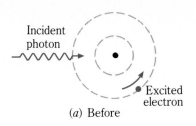

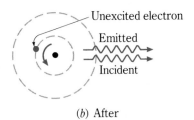

(a) Before

Incident photon

Excited electron

Unexcited electron

Emitted

Incident

(b) After

**FIGURE 26.20**

Stimulated emission produces waves that are in phase.

*emitted and incident waves will travel away in phase.* This process, called **stimulated emission,** is shown in Fig. 26.20. Let us see how the laser makes use of this phenomenon.

The laser is basically an electrical discharge tube like the discharge tube of a neon sign. In this case, however, the tube is straight and contains two specially chosen gases. (A helium-neon laser has 15 percent helium and 85 percent neon.) The atomic system made up of the helium and neon atoms contains three energy levels that are of particular interest to us. They are shown as $E_1$, $E_2$, and $E_3$ in Fig. 26.21. Before the discharge is activated, the atoms are not excited; level $E_1$ is occupied, but

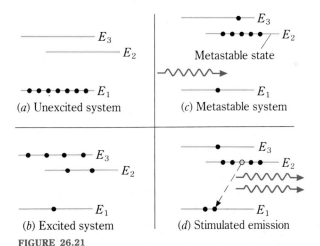

(a) Unexcited system

(b) Excited system

(c) Metastable system

Metastable state

(d) Stimulated emission

**FIGURE 26.21**
In a laser, a population inversion, a metastable state, and stimulated emission are required.

levels $E_2$ and $E_3$ are empty, as indicated in *a*. When a high voltage is impressed across the tube so as to cause an electrical discharge in it, electrons are excited to levels $E_3$ and $E_2$. The electrons in level $E_3$ fall to level $E_2$, which is of a very special sort. Level $E_2$ is called a **metastable state:**

A metastable state is one in which the electron is inordinately stable and from which the electron will fall only after a relatively long time.

As a result of its ability to retain electrons, level $E_2$ soon contains a large number of electrons. Indeed, it eventually has more electrons in it than does the $E_1$ level. We have achieved what is called a **population inversion:** level $E_1$, which normally has more electrons than $E_2$, now has fewer electrons.

Now suppose that a light wave whose photons have energy equal to $E_2 - E_1$ is incident on the gas. These photons can be absorbed and excite the few electrons in level $E_1$ to $E_2$. Also, in Fig. 26.21$d$, they can cause electrons to fall from $E_2$ to $E_1$, giving rise to stimulated emission of waves that are identical to the incident wave. Because of the population inversion, the stimulated emission overwhelms further absorption, and so the intensity of the emitted waves grows as they pass through the gas. The end result is a coherent beam traveling through the discharge tube.

In a helium-neon laser, the situation is much like that in Fig. 26.22. The ends of the discharge tube consist of accurately parallel plane mirrors. However, the mirror at the right is only lightly metallized so that it reflects perhaps only 99 percent of the

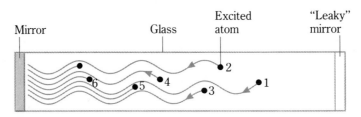

**FIGURE 26.22**
Schematic diagram showing how stimulated emission builds up a strong coherent wave in a laser tube.

light. Suppose that atom 1 in the figure is in a metastable state $E_2$, but eventually falls from state $E_2$ to $E_1$. In doing so, it emits a photon whose wavelength is 632.8 nm (red light). This emitted wave travels past atoms 2, 3, and so on, and stimulates them to emit similar photons. Then, in succession, a whole multitude of excited neon atoms emit identical photons, and these photons are all in phase, as indicated in Fig. 26.22. In a short time, the tube is filled with coherent waves moving back and forth between the two mirrors at the ends of the tube. A very intense, monochromatic, coherent beam is set up in the tube. A small fraction of the beam exits from the tube through the "leaky" mirror at one end.

Because all the waves issuing from the end of the laser tube are in phase, the beam is of high intensity. Its wavelength is sharply defined, 632.8 nm, because all the waves are identical. Not only is the beam intense and coherent, but also it is a very thin, straight beam that diverges very little. Any rays within the tube that diverged much from its axis were lost out the sides during their many trips back and forth. The fact that the beam does not diverge appreciably is of great practical importance. Unlike light from a bulb, the laser beam's energy does not fan out into space. Instead, it flows out into space through a thin cylinder and maintains its intensity over very long distances. For example, the laser whose beam is shown in Fig. 26.23 has a power of only 0.0005 W, but the light energy is confined to such a narrow cylindrical path that the beam far outshines the high-power bulbs in the foreground.

Although you are probably most accustomed to the helium-neon laser, which has an output of only a fraction of a milliwatt, there are many types of lasers available today. All make use of a metastable state to achieve a population inversion so that

FIGURE 26.23

Even though the laser at $L$ has an output of only 0.0005 W, it outshines the high-intensity lights on the San Francisco–Oakland Bay Bridge. (*Electro Optics Associates, Palo Alto, Calif.*)

stimulated emission can give rise to an in-phase coherent set of waves. They differ in wavelength, ranging from the far infrared to the long x-ray range. Their energies range from a small fraction of a milliwatt (in lasers used in laser disk audio systems) to millions of watts (in lasers being developed for weapons).

Today the laser has widespread use. In surgery, the needle-thin intense beam allows surgeons to destroy tissue at a well-defined point or to "weld" broken tissue, such as in the retina of the eye. The straightness of the beam makes it useful in surveying applications. Even the distances to points on the moon's surface are being monitored with laser signals. This same property of the laser beam is widely applied in industry and research to align tools and equipment. And the intense power of a laser beam allows it to burn through metal faster and cleaner than a drill; hence it is useful in many industrial and potential military applications.

Tiny solid-state lasers are now in widespread use. In compact disk (CD) audio systems, a laser beam replaces the phonograph needle. After it is reflected from the disk, the beam is converted to an electric signal that contains the information stored on the disk. Computer circuits analyze this signal and transform it to the voltage that drives the loudspeakers. It is interesting to notice that this, as well as many other common devices, were made possible by worldwide research programs in microelectronics and solid-state physics.

## Minimum learning goals

When you finish this chapter, you should be able to

**1** Define (*a*) nuclear atom; (*b*) spectral series; (*c*) series limit; (*d*) Rydberg constant; (*e*) Lyman, Balmer, and Paschen series; (*f*) Bohr orbits; (*g*) energy-level diagram; (*h*) ground state of atom; (*i*) ionization energy; (*j*) Pauli exclusion principle; (*k*) quantum number; (*l*) atomic shells; (*m*) characteristic x-rays versus bremsstrahlung; (*n*) laser; (*o*) coherent waves; (*p*) stimulated emission; (*q*) metastable state; (*r*) population inversion.

**2** Explain how Rutherford's experiment presented evidence for the concept of the nuclear atom.

**3** Give the approximate diameter of an atom.

**4** Sketch the lines of the Balmer series and write the Balmer formula. Compute the wavelength of a given line of the Balmer series when the Rydberg constant is given. Repeat for both the Lyman series and the Paschen series. State which series are partly in the visible region.

**5** Explain how the wave properties of the electron lead to the Bohr orbits and the Bohr energy levels.

**6** Give the general formula for the energy levels of the hydrogen atom in electronvolts. Sketch the energy-level diagram for hydrogen.

**7** Compute the wavelength emitted by the hydrogen atom in any specified transition. Show on an energy-level diagram how the Lyman, Balmer, and Paschen series arise.

**8** Explain why hydrogen atoms normally absorb the Lyman series wavelengths but not those of the Balmer series.

**9** Explain the meaning of an electron-distribution diagram, such as those in Figs. 26.15 and 26.17.

**10** State the Pauli exclusion principle and show why it predicts lithium to be univalent.

**11** Describe how x-rays are produced in an x-ray tube. Compute the shortest-wavelength x-rays emitted by a target impacted by electrons of a given high energy.

**12** State the important features of a laser beam with regard to coherency, phase, and shape. Point out how these features lead to specialized uses for lasers.

# Questions and guesstimates

**1** Why doesn't the hydrogen gas prepared by students in the laboratory glow and give off light?

**2** Suppose you are given a glass tube containing two electrodes that is sealed at both ends. The gas inside is either hydrogen or helium. How can you tell which it is without breaking the tube? If the gas is at high pressure, what difficulty might you have?

**3** When white light is passed through a vessel containing hydrogen gas, it is found that wavelengths of the Balmer series as well as those of the Lyman series are absorbed. We conclude from this that the gas is very hot. Why can we draw this conclusion? (This is actually the basis for one method of measuring the temperature of a hot gas.)

**4** Explain clearly why x-ray emission lines in the range of 0.1 nm are not observed from an x-ray tube when a low-atomic-number metal is used as the target in the tube.

**5** A steel company suspects that one of its competitors is adding a fraction of a percent of a rare-earth element to its (the competitor's) product. How can the element quickly be identified and its concentration determined?

**6** In the helium atom, the two electrons are in the same shell but avoid each other well enough for their interaction to be of only secondary importance. Estimate the ionization energy (in electronvolts) for helium, that is, the energy required to tear one electron loose. Also, estimate the energy needed to tear the second electron loose. Which of these two values is more reliable?

**7** The ionization energies for lithium, sodium, and potassium are 5.4, 5.1, and 4.3 eV, respectively, while those for helium, neon, and argon are 24.6, 21.6, and 15.8 eV, respectively. Explain in terms of atomic structure why these values are to be expected.

**8** Estimate how much energy a photon must have if it is to be capable of expelling an electron from the innermost shell of a gold atom.

**9** The diameter of a nucleus is about $10^{-15}$ m. Estimate the least momentum a proton must have if it is to be a part of the nucleus.

# Problems

**1** The radius of the gold nucleus is about $6 \times 10^{-15}$ m, while its atomic radius is about $1.5 \times 10^{-10}$ m. Suppose one wishes to draw a scale diagram of the gold atom using a dot whose diameter is 2.0 mm for the nucleus. How far away from the dot would be the outer edge of the atom?

**2** Ten thousand tiny projectiles are shot at random places into a 4000-cm² window whose pane is partly broken out. (*a*) Only 700 of the projectiles go through the window. How large is the area of the hole in the pane? (*b*) The window pane is now completely removed, and 300 tiny spheres suspended by tiny threads are placed in the window opening. Now 9400 of the 10,000 projectiles go straight through the window. About what is the cross-sectional area of each sphere? (*c*) To what in Rutherford's experiment do the spheres of part *b* correspond?

■ **3** Rutherford and his coworkers shot $\alpha$ particles ($q = +2e$) at gold atoms ($Z = 79$). Some of the particles had a kinetic energy of 4.8 MeV. (*a*) What is the potential energy (in terms of $r$) of an $\alpha$ particle a distance $r$ from the gold nucleus? (*b*) How close can Rutherford's particles come to the center of the gold nucleus? Assume that the gold nucleus remains essentially stationary, and neglect the effect of the (relatively distant) atomic electrons.

■ **4** The density of gold is 19.3 g/cm³, and its atomic mass is 197 kg/kmol. (*a*) What is the mass of a gold atom? (*b*) How many gold atoms are there in a 1-cm² area of gold film that is 0.020 mm thick? (*c*) The diameter of a gold nucleus is about $10^{-14}$ m. Assuming no overlap, how much of the 1.0-cm² total area do the gold nuclei cover? (*d*) If he had used a film of this thickness, about what fraction of the $\alpha$ particles would Rutherford have observed to be strongly deflected?

■ **5** Bohr assumed that the electron in the hydrogen atom was orbiting with nonrelativistic speed. What speed does the electron have in the $n = 1$ orbit, according to his theory? Compare it to $c$.

■ **6** In the singly ionized helium atom, a single electron orbits the nucleus, which has a charge $+2e$. What is the radius of the first Bohr orbit for this ion?

■ **7** Suppose that the semiclassical theory can be applied to the innermost electron in a lead atom ($Z = 82$) if the presence of all the other electrons is neglected. (This is really not too bad an approximation.) (*a*) Show that the energy needed to remove this electron from the atom is $(13.6)(82)^2$ eV. (*b*) What is the radius of the first Bohr orbit for this atom?

**8** Suppose an electron revolves about the hydrogen nucleus in a circular path of radius 0.050 nm. (*a*) What speed must the electron have if the coulomb force is to furnish the centripetal force? (*b*) What is the frequency of the electron in the orbit? (*c*) On the basis of classical theory, to what wavelength of radiation should this give rise?

**9** (*a*) What is the formula for the energy levels of doubly ionized lithium, written in the form $E_n = -(\text{const})/n^2$ electronvolts? (*b*) How much energy (in electronvolts) is needed to remove the last electron from doubly ionized lithium?

**10** Repeat Prob. 9 for a carbon atom that has been stripped of all but one of its electrons. For carbon, $Z = 6$.

**11** Assume that the angular momentum of the earth's rotation about the sun obeys the de Broglie wave resonance condition $n\lambda_{\text{earth}} = 2\pi r_n$. What would be the value of the quantum number $n$ in this case? (This is an example of the fact that macroscopic systems normally correspond to very large quantum numbers and behave classically.)

**12** Classically, a hydrogen atom with an orbital diameter of a few meters should act like a radio antenna and emit radiation with a frequency equal to the frequency of the electron in orbit. Wave theory must also predict this result since it applies to radio antennas as well as to atoms. Show that the orbital frequency of the electron is

$$f_{\text{orbital}} = \frac{me^4}{4\epsilon_0^2 h^3 n^3}$$

Now compute the frequency emitted by the hydrogen atom as it falls from state $n$ to state $n - 1$. Show that when $n \gg 1$, this frequency is the same as the orbital frequency.

**13** Compute the wavelength of (*a*) the fifth line of the Lyman series and (*b*) the sixth line of the Balmer series.

**14** Find the wavelength of the 12th line of the Balmer series and compare it with the wavelength of the 13th line.

**15** (*a*) Give the lowest four energy levels of doubly ionized lithium ($Z = 3$ for lithium). (*b*) Sketch the energy-level diagram for it.

**16** (*a*) For sodium, $Z = 11$. Suppose that 10 of its 11 electrons have been stripped away. Find the lowest four energy levels of the remaining atom. (*b*) Sketch the energy-level diagram for it.

**17** Electrons with energy of 11.6 eV are shot into a gas of hydrogen atoms. What wavelengths of radiation will be emitted strongly by the gas?

**18** Electrons with energy of 12.6 eV are shot into a gas of hydrogen atoms. What wavelengths of radiation will be emitted by the gas?

**19** What are the energies (in electronvolts) of the four lowest-energy photons that unexcited hydrogen gas atoms will absorb?

**20** Repeat Prob. 19 for absorption by singly ionized helium atoms.

**21** A beam of ultraviolet light with $\lambda = 70$ nm is incident on unexcited hydrogen atoms. When a photon strikes one of the atoms and ejects its electron, what is the kinetic energy of the electron once it is free of the atom? This is called the *atomic photoelectric effect*.

**22** A beam of 6.0-nm-wavelength x-rays shines on a gas of unexcited hydrogen atoms. It expels electrons (photoelectrons) from the hydrogen atoms. (*a*) What is the energy of the expelled photoelectrons? (*b*) What is their speed?

**23** The ionization energy of unexcited helium atoms is 24.6 eV. Suppose that 20-nm ultraviolet light is incident on such atoms. (*a*) What is the energy of the fastest electron ejected from the atoms by the ultraviolet light? (*b*) What is the speed of the electron?

**24** A room-temperature gas of hydrogen atoms is bombarded by a beam of electrons that has been accelerated through a potential difference of 13.1 V. What wavelengths of light will the gas emit as a result of the bombardment?

**25** As a very crude model, suppose that an atomic nucleus consists of noninteracting protons and neutrons traveling in circular paths within the nucleus. Since the radius of a typical large nucleus might be $5 \times 10^{-15}$ m, assume particles in the ground state to have orbit radii of $5 \times 10^{-15}$ m. What must be the de Broglie wavelength for a neutron that resonates in such an orbit in its ground state? What is the kinetic energy (in electronvolts) of the neutron? Neglect relativistic effect.

**26** The perimeter of the benzene molecule is a hexagon, with each side having a length of 0.140 nm. Since the molecule has three double bonds, it is not totally unreasonable to assume that one electron in the molecule can circulate freely around this perimeter much as though it were a free electron restricted to a hexagonal path. Using reasoning based on resonance and de Broglie wavelength, show that the energy levels for such an electron should be (to this approximation)

$$E_n = (7.1 \times 10^{17}) \left( \frac{n^2 h^2}{m} \right)$$

with all quantities in SI units. If the result of this computation is correct, at what wavelengths would you expect benzene to absorb light? Does this contradict the fact that benzene is a crystal-clear liquid?

**27** (*a*) Calculate the recoil speed of a hydrogen atom due to its emission of a photon with wavelength 656 nm, the first line of the Balmer series. (*b*) Find the ratio of this recoil energy of the atom to the difference in energy between the two states that give rise to the emission line.

■ **28** Eight years before Bohr's theory of the hydrogen atom was proposed, the *Ritz combination principle* was discovered. One example of it is as follows: the frequency of the second line in the Lyman series equals the sum of the frequency of the first line in the Lyman series and the frequency of the first line in the Balmer series. (*a*) Using semiclassical theory, show that this result should be true. (*b*) What would the principle predict for the third line of the Lyman series?

**29** How many electrons can exist in the $n = 3$ shell of a Bohr-type atom?

■ **30** How many electrons can exist in the $n = 4$ shell of a Bohr-type atom?

■ **31** An atomic *subshell* consists of those electrons in an atom that have the same $n$ and $l$ but different $m_l$ and $m_s$. How many electrons exist in the $n = 3$, $l = 2$ subshell of gold?

■ **32** How many electrons are needed to fill the $n = 3$, $l = 1$ subshell? (See Prob. 31 for the definition of subshell.)

■ **33** Suppose the electron did not have spin, so that its spin quantum number did not exist. What would be the first four elements in the periodic table that would show a valance of $+1$?

■ **34** Make a table showing the quantum numbers for the various electrons in the chlorine atom. ($Z$ for chlorine = 17.)

**35** Modern color television sets often have electron beams accelerated through more than 20,000 V. What are the shortest-wavelength x-rays generated by a 25,000-V beam as it hits the end of the television tube? (Some early television sets were not properly shielded and leaked appreciable amounts of x-rays outside the set.)

**36** To reach tumors deep within a person's body, so-called "hard" x-rays are used. These are generated using very high voltages. What is the shortest-wavelength x-ray generated by an x-ray tube operating at 180 kV?

■ **37** An ordinary x-ray tube might easily operate at 40,000

V with a current of 10 mA bombarding the plate. (*a*) How much heat is produced in the plate per second by this bombardment? (*b*) If the specific-heat capacity of the metal of the 150-g plate is 0.10 cal/g · C°, what temperature rise will occur in 1 min if no heat is lost?

■ **38** Some very-high-energy x-ray tubes have water circulated through the inside of the metal plate in order to cool it. Suppose that a certain tube is operating at 20 mA and 180,000 V. How much would the cooling water be heated if it circulated through the plate at 1.50 liters/min?

■ **39** (*a*) Estimate the minimum voltage needed across an x-ray tube whose target is made of titanium ($Z = 22$) if the $n = 1$ electron is to be excited? (*b*) Estimate the longest wavelength of the x-ray emitted as the atom undergoes an $n = 2$ to $n = 1$ transition.

■ **40** (*a*) From the data of Fig. 26.19, determine the energy difference between the $n = 2$ and $n = 3$ levels in molybdenum. (*b*) If you wished to construct the energy-level diagram for this atom, what further data would you need?

■ **41** An argon laser ($\lambda = 456.5$ nm) emits a pulse of light to "weld" the detached retina in a person's eye. If the pulse lasts $1 \times 10^{-8}$ s and contains an energy of $1.5 \times 10^{-3}$ J, what is the instantaneous power delivered to the weld point?

■ **42** The beam of a laser diverges slightly because of diffraction effects at the end of the laser tube. Assume that the beam of a helium-neon laser ($\lambda = 632.8$ nm) has a diameter of 3.0 mm as it leaves the laser. About how large will the beam diameter be when the beam strikes a target that is 200 m away? Assume that beam spreading is due solely to diffraction.

■ **43** Two different laser beams of the same type will be coherent if the lasers emit exactly the same wavelength. Even if the wavelengths are slightly different, the two beams will show an interference effect. When joined, they will give a resultant beam that fluctuates over time from brightness to darkness. This is similar to the phenomenon of beats of sound waves, discussed in Chap. 14. If one beam has a wavelength of exactly 600 nm, what must the wavelength of the other beam be to produce maximum brightness once each second? [You may want to use the fact that, for $x \ll 1$, $1/(1 \pm x) \cong 1 \mp x$.]

# 27 The atomic nucleus

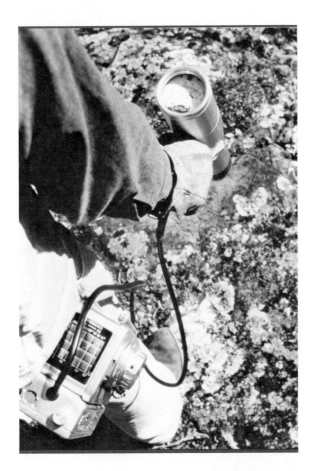

$\mathbf{A}$T the very center of each atom is a positively charged nucleus. Although it constitutes only about $10^{-13}$ percent of the atom's volume, the nucleus contains all but 0.1 percent of its mass. In this chapter, we examine the salient features of the nucleus, how it is constructed, and what influences its stability. We also discuss a few of the many applications of nuclear physics in our present-day world.

## 27.1 Atomic number and mass number

Rutherford's investigations, discussed in Chap. 26, have been extended in many ways as the years have gone by. Today we know that the nucleus is composed of protons (p) and neutrons (n). You will recall that the charge on the proton is $+e$ and that the neutron has no charge. Furthermore,

$$m_p = 1.673 \times 10^{-27} \text{ kg} = 1.007276 \text{ u}$$

and

$$m_n = 1.675 \times 10^{-27} \text{ kg} = 1.008665 \text{ u}$$

where the mass unit u is the **atomic mass unit** (sometimes written amu). We shall define this unit precisely in Sec. 27.2. For now we simply assert that

$$1\ u = 1.660566 \times 10^{-27}\ kg$$

Notice that the neutron and proton masses are nearly, but not exactly, the same. Like the electron, the proton and neutron have spin of $\frac{1}{2}$ and obey the Pauli exclusion principle. For future reference, the electron's mass is

$$m_e = 9.1094 \times 10^{-31}\ kg = 5.486 \times 10^{-4}\ u$$

As we saw in Chap. 26, the chemical behavior of an atom is determined by the electrons that exist in the relatively vast reaches of space outside the nucleus. All unexcited atoms of a given element have the same number of electrons and the same electronic structure. Each carbon atom has 6 electrons, for example, and each gold atom has 79 electrons. As stated in Chap. 26, we designate the number of electrons in a neutral atom by $Z$, the **atomic number** for that chemical species. Thus for carbon $Z = 6$ and for gold $Z = 79$. The atomic numbers of the elements are given in Appendix 1.

Because the atom is neutral, the nucleus of an element with atomic number $Z$ must have a charge of $+Ze$ to counterbalance the charge of the atom's $Z$ electrons. This charge is ascribed to $Z$ protons, each with a charge of $+e$, within the nucleus. But the mass of the nucleus (except in the case of hydrogen) is larger than the mass of the $Z$ protons. This additional mass is due to neutrons in the nucleus. We therefore conclude that, for an element of atomic number $Z$, the nucleus contains $Z$ protons together with additional neutrons. We designate the protons and neutrons in the nucleus as **nucleons.**

Because the nucleus is composed of protons and neutrons, and each of these particles has a mass close to 1 u, we would expect the nuclear mass to be nearly an integer when measured in atomic mass units. For example, the nucleus of the common helium atom contains two neutrons and two protons, and so its mass should be (and is) about 4 u. Similarly, the mass of the nucleus of the common carbon atom (which contains six neutrons and six protons) is 12 u. With this in mind, we assign a mass number $A$ to each nucleus; the **mass number** $A$ of an atom (and its nucleus) is equal to the number of nucleons (protons + neutrons) in the nucleus, and this is approximately equal to the mass (in atomic mass units u) of the nucleus. We call the number of neutrons in a nucleus the **neutron number** $N$. The number of protons, of course, is $Z$.

## 27.2 Nuclear masses; isotopes

The masses of nuclei have been measured to high precision. These measurements are carried out with *mass spectrometers;* a schematic diagram of one type is given in Fig. 27.1. As indicated, ions of the element under consideration are allowed to escape from the ion source. After being accelerated through the potential difference $V$, the ion beam is collimated by means of slits such as $S_2$. When they leave $S_2$, the ions are moving with speed $v$ and are deflected into a circular path by the magnetic field, as shown. The radius $r$ of this path is measured by noting the positions at which the ions strike a photographic plate or some other detector.

To find the working equation for the mass spectrometer, we note that the energy furnished by the accelerating potential is

$$Vq = \tfrac{1}{2} mv^2$$

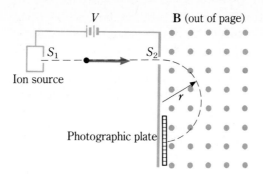

FIGURE 27.1

In the mass spectrometer, ions are deflected by a magnetic field.

where $q$ and $m$ are the charge and mass of the ion. When the ion enters the magnetic field, a centripetal force is furnished by the field, and so

$$\frac{mv^2}{r} = qvB$$

Solving these two equations simultaneously for $m$, we find

$$m = \frac{r^2 B^2 q}{2V} \qquad (27.1)$$

We can therefore compute the mass $m$ of the ion if $r$, $B$, $q$, and $V$ are known. To find the mass of the nucleus, we subtract from $m$ the mass of the electrons associated with the ion.

When the mass spectrometer is used to measure nuclear masses, an interesting effect is observed. Very frequently, an element will give rise to two or more different beams in the spectrometer. That is, particles will appear at the detector at two or more very well defined radii. From Eq. 27.1 we conclude that nuclei of the *same element* may have different masses.

As an illustration, when chemically pure chlorine is sent through the mass spectrometer, it appears to consist of two different types of nuclei:

Species 1     Mass = 34.97 u     Relative percentage = 75.4

Species 2     Mass = 36.97 u     Relative percentage = 24.6

We say that the **natural abundance** of species 1, for example, is 75.4 percent. Both these species behave exactly the same chemically, and so their atomic electron structures must be the same. Therefore their nuclear charges must be the same, equal to the atomic number $Z$ multiplied by the charge quantum $e$. We call nuclei such as this, which have the same charge but different masses, **isotopes** of the element in question.

Isotopic nuclei have the same number of protons but a different number of neutrons.

In order to classify nuclei in terms of their mass, charge, and nucleon number, it is customary to designate an element whose symbol is $X$ as

$$\substack{\text{mass number} \\ \text{charge number}}X \qquad \text{or} \qquad {}^{A}_{Z}X$$

where $A$ is the mass number of the nucleus and $Z$ is the atomic number. For example, the chlorine isotopes we have just discussed would be represented as $^{35}_{17}\text{Cl}$ and $^{37}_{17}\text{Cl}$; both isotopes have the same atomic number, $Z = 17$, but one has a mass number $A = 35$ and the other has $A = 37$. Remember, $Z$ is the number of protons in the nucleus and $A$ is the number of nucleons (protons + neutrons). We refer to these two isotopes as chlorine 35 and chlorine 37. As another example, $^{238}_{92}\text{U}$ is called uranium 238. Its nucleus has a charge $+92e$ and contains 92 protons and $238 - 92 = 146$ neutrons.

In the periodic chart of the elements you are probably familiar with from your chemistry classes, one usually finds listed the *atomic* masses as determined by chemical means. For this reason, the atomic mass given there is an average value of the *isotopic* masses found in nature. For example, the average mass of the two isotopes in chlorine is

$$m_{\text{av}} = 35(0.754) + 37(0.246) = 35.5 \text{ u}$$

which is the value given in the periodic chart in Appendix 1.

The atomic masses* of many isotopes are given in Appendix 2. Notice that these are the masses of the nuclei *plus the atomic electrons*. These masses are given in **atomic mass units** u, and this unit is defined as follows. The atom of the most abundant isotope of carbon (carbon 12) is arbitrarily assigned the value of *exactly* 12 u. All other masses are then measured in this unit by comparison. On this scale, the proton and neutron masses are approximately 1 u. As stated previously, $1 \text{ u} = 1.660566 \times 10^{-27}$ kg.

---

### Example 27.1

The atomic masses given in the periodic table are for the nucleus plus the atom's electrons. What fraction of the atomic mass of $^{235}\text{U}$ (uranium with mass number = 235) is due to its electrons?

*Reasoning*  From Appendix 1, the atomic mass of $^{235}\text{U}$ is 235.04 u. Since the atomic number of uranium is 92 (Appendix 1), it has 92 electrons. Using the fact that the mass of the electron is $9.11 \times 10^{-31}$ kg, or 0.000549 u, we have that the fraction of the mass due to electrons is

$$\frac{92(0.000549) \text{ u}}{235 \text{ u}} = 2.15 \times 10^{-4}$$

Therefore, for many purposes, the mass of the electrons can be ignored.  ■

---

### 27.3 Nuclear size and density

There are several methods by which we can estimate the size of the nucleus of an atom. One is to shoot particles of various types at the nucleus and see how they are deflected or scattered. However, one must use very-high-energy particles to overcome the coulomb repulsion of the nucleus if the bombardment is to be done with

---

* Recall from Chap. 10 that it is customary to use the terms *atomic mass* and *atomic weight* interchangeably.

protons or $\alpha$ particles. Scattering measurements have been made with these particles, as well as with neutrons, electrons, and heavy ions. The results of these measurements show that the nucleus cannot be pictured as a simple hard sphere of uniform constitution.

In spite of the fact that the nucleus has no sharp cutoff radius for its charge or its mass, as would be the case for a hard sphere, its edges are well enough defined for a meaningful approximate radius to be given. As one would expect, bombardment with charged particles measures primarily the charge distribution in the nucleus, whereas bombardment with neutrons measures primarily the mass distribution. Other methods can also be used to measure the nuclear radius. They all agree approximately with each other, and from them it can be inferred that the nuclear radius $R$ is

$$R \approx (1.2 \times 10^{-15} \text{ m})(A^{1/3}) \tag{27.2}$$

where $A$ is the mass number of the atom concerned.

Notice in Eq. 27.2 that the radius of a typical nucleus is of the order of $10^{-15}$ m. For that reason, it is customary in nuclear work to measure lengths in femtometers (fm), where

$$1 \text{ fm} = 10^{-15} \text{ m}$$

Originally this length was designated a *fermi* in honor of the distinguished nuclear physicist Enrico Fermi. It is customary to use the designations fermi and femtometer interchangeably.

The fact that the nuclear radius varies as $A^{1/3}$ gives important information as to how the $A$ nucleons pack together in the nucleus. If we compute the volume of the nucleus, we have

$$V = \tfrac{4}{3}\pi R^3 = \tfrac{4}{3}\pi (1.2 \text{ fm})^3(A)$$

Notice what this says. If the factor $\tfrac{4}{3}\pi(1.2 \text{ fm})^3$ is taken as the volume of a nucleon, then $V$ is simply the sum of the individual volumes of the $A$ nucleons. Apparently the nucleons pack together like hard balls to form the nucleus. As a result, all large nuclei have about the same density, as we shall see in the following example.

---

**Example 27.2**
Find the density $\rho$ of the gold nucleus.

**Reasoning**   If we neglect the mass of the atomic electrons, the mass of a gold nucleus is equal to its atomic mass, given in Appendix 1 as 197 u. The volume of the nucleus is

$$V = \tfrac{4}{3}\pi R^3 = (7.2 \times 10^{-45} \text{ m}^3)(A)$$

Since $A = 197$, and the mass of a gold atom is 197 u,

$$\rho = \frac{\text{mass}}{\text{volume}} = \frac{(197 \text{ u})(1.66 \times 10^{-27} \text{ kg/u})}{(7.2 \times 10^{-45} \text{ m}^3)(197)} \approx 2.3 \times 10^{17} \text{ kg/m}^3$$

Notice that, because the mass number ($A = 197$) is nearly equal to the atomic mass (197 u), the 197s cancel, and so this value is the approximate density within all nuclei. Such extremely high densities are never encountered on a large scale on earth. Only in the interior of certain stars (*neutron* stars) are such high densities found. In these stars, the electron shells of the atoms have been collapsed by the huge gravitational forces at the star's center. ∎

---

## 27.4 Nuclear binding energy

Because like charges repel one another, the electrostatic force between the protons in a nucleus urges it to explode. The gravitational force between nucleons is many orders of magnitude too small to counterbalance this repulsive force. A third force must exist between nucleons to cause them to attract each other and hold the nucleus together. It is the *nuclear attractive force,* often called simply the nuclear force or the strong force.

The nuclear force is unlike the electrostatic and gravitational forces in that it does not obey an inverse-square law. Instead, it has only a limited range. Experiment shows that the nuclear attractive force between two neutrons or two protons is essentially zero for separations larger than about $5 \times 10^{-15}$ m — in other words, a distance equal to about twice the diameter of a nucleon. At separations only slightly smaller than this, the nuclear force overpowers the electrostatic repulsive force between any two protons and binds the protons together. To a first approximation, the nuclear force is the same between two neutrons, between two protons, and between a neutron and a proton.

Consider what happens as a neutron is brought up to a nucleus from an infinite distance. The energy of the neutron as a function of distance $r$ from the center of the nucleus is sketched as the lower curve in Fig. 27.2. At large values of $r$ the neutron experiences no force and its potential energy $E$ is zero. At a distance of about $2 \times 10^{-15}$ m from the nucleus, the nuclear attractive force pulls the neutron into the nucleus. The neutron therefore loses potential energy as it moves to smaller $r$ values. The lower curve in Fig. 27.2 therefore begins to become negative at this distance. Once it is inside the nucleus, the neutron is attracted nearly equally on all sides by its neighboring nucleons. As a result, its potential energy is approximately constant, with value $E = -E_o$ for all positions within the nucleus.

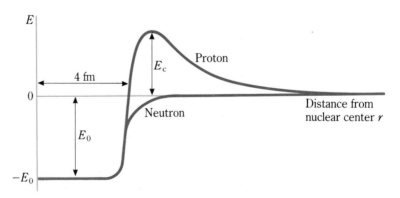

**FIGURE 27.2**

Form of the potential-energy curve of a neutron and proton in a stable nucleus. Typical values might be $E_o = 50$ MeV, $E_c = 8$ MeV.

When a proton is brought close to a nucleus, the situation is somewhat different. Because of the electrostatic repulsion between it and the nucleus, the potential energy of the proton rises as it approaches the nucleus, as shown in the upper curve of Fig. 27.2. At small $r$ values, however, the nuclear force overpowers the electrostatic force. As a result, the curves for the proton and neutron are essentially the same at small values of $r$.

Let us now consider what happens to the energy of a group of widely separated neutrons and protons as they are assembled into a stable nucleus. When they are far separated, their total energy may be taken as zero. However, as they are assembled into a nucleus, each particle will lose energy comparable to $E_o$, the energy shown in Fig. 27.2. We therefore conclude that

The energy of a stable nucleus is less than the original energy of the far-separated nucleons that compose it.

We call this difference in energy between the stable nucleus and the separated nucleons the **binding energy** of the nucleus. Clearly, if a nucleus is to be torn apart into its individual nucleons, work equal to the binding energy must be done. A graph showing the binding energy per nucleon (that is, the binding energy divided by the mass number of the nucleus) for representative elements of the periodic table is given in Fig. 27.3. Notice that the binding energy per nucleon is greatest for elements close to iron (Fe) in the periodic table. These nuclei are therefore very stable, and large energies must be furnished to separate their nucleons. Nuclei with very low and very high $Z$ values have less stability; their binding energies per nucleon are smaller. As you see, Fig. 27.3 can also be interpreted as a rough graph of nuclear stability.

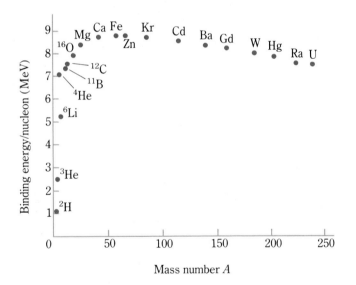

One might well ask what becomes of the energy furnished to a nucleus as it is torn apart into its individual nucleons. According to relativity theory, this energy must appear as an increase in mass of the particles. Therefore the assembled nucleus has a smaller mass than the mass of its separated nucleons. If we call this difference in mass $\Delta M$, then Einstein's relation $\Delta E = (\Delta m)(c^2)$ tells us that

$$\text{Binding energy} = (\Delta M)(c^2)$$

The mass difference $\Delta M$ is called the *mass defect* of the nucleus. It is a measure of the energy required to disband the nucleus; it is also a measure of the energy released when the nucleus is first assembled. We shall discuss this further when we consider practical applications of energy production from nuclei.

---

### Example 27.3
How much energy is required to change the mass of a system by 1 u?

*Reasoning*   We make use of Einstein's mass-energy relation $\Delta E = (\Delta m)(c^2)$. In the present case, $\Delta m = 1 \text{ u} = 1.6606 \times 10^{-27}$ kg. Substituting this value, we find that

$$\Delta E = 1.492 \times 10^{-10} \text{ J} = 931.5 \text{ MeV}$$

This is a convenient fact to remember: *one atomic mass unit of mass is equivalent to 931.5 MeV of energy.* ∎

---

### Example 27.4
Deuterium, $^2_1\text{H}$, is the isotope of hydrogen whose nucleus consists of a proton and a neutron. Its atomic mass is 2.014102 u. Compute its binding energy.

*Reasoning*   The mass of the deuterium nucleus is obtained by subtracting the mass of the atom's single electron from the total atomic mass:

$$2.014102 - 0.000549 = 2.013553 \text{ u}$$

But the sum of the masses of the proton and neutron is

$$1.007276 + 1.008665 = 2.015941 \text{ u}$$

Therefore

$$\text{Mass defect} = \text{mass of nucleons} - \text{mass of nucleus}$$
$$= 2.015941 - 2.013553 = 0.002388 \text{ u}$$

This much mass is lost as the proton and neutron are joined to form the deuterium nucleus. Because 1 u of mass corresponds to 931.5 MeV of energy (Example 27.3), the mass defect corresponds to a binding energy of

$$\text{Binding energy} = (0.002388 \text{ u})\left(\frac{931.5 \text{ MeV}}{1 \text{ u}}\right) = 2.22 \text{ MeV}$$

which agrees with the value (1.11 MeV/nucleon) given in Fig. 27.3. To tear the deuterium nucleus apart requires an energy of 2.22 MeV. ∎

---

*Exercise*   The binding energy of the electron in the hydrogen atom is 13.6 eV. How much mass in atomic mass units, is created as a hydrogen atom is ionized? *Answer: $1.46 \times 10^{-8}$ u* ∎

## 27.5 Radioactivity

As we have seen, the nucleons in a nucleus are subject to two competing forces: the attractive nuclear force between all nucleons and the repulsive coulomb force between the protons. Only certain combinations of neutrons and protons are stable. A combination that contains too many protons relative to the number of neutrons will experience too large an explosive force as a result of the coulomb repulsions. It cannot exist as a stable entity. Other factors also influence the stability of a nucleus, as we shall see later. Only those few combinations of protons and neutrons shown in Fig. 27.4 are relatively stable.

As you can see from Fig. 27.4, large nuclei are stable only if they contain more neutrons than protons. The extra neutrons are needed to "dilute" the positively charged protons and thereby decrease the explosive effect of coulomb forces. Although most of the nuclei indicated in Fig. 27.4 are completely stable, all those with $Z$ larger than 83 are somewhat unstable. You probably know that radium ($Z = 88$) and uranium ($Z = 92$) are unstable, or radioactive; that is, these nuclei undergo change, one by one, by ejecting particles and radiation. Early investigators

**FIGURE 27.4**

Each dot represents a nucleus that is either completely stable or nearly so. The solid line represents the positions of nuclei that have equal numbers of protons and neutrons.

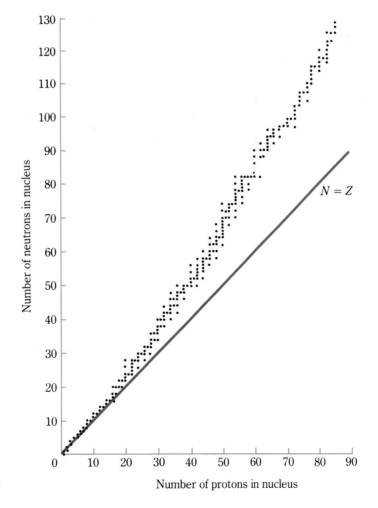

did not know what was being ejected and referred to the emissions as radiation. Hence we have the term *radioactive* to describe these unstable nuclei. Further, the radiations were named $\alpha$-rays or $\alpha$ particles (which we now know to be helium nuclei), $\beta$-rays or $\beta$ particles (electrons), and $\gamma$-rays (photons).

The radioactive nuclei found on the earth today are of two kinds. Some have been synthesized by using "atom smashing" machines and nuclear reactors to bombard stable nuclei; the radioactive nuclei so produced are termed *artificially radioactive.* As we shall see later, these materials are widely used in medicine, industry, and research. Another group of radioactive materials found on earth is the result of nuclear reactions that occurred when the universe was first formed. When the earth was still young, about 4.6 billion years ago, it possessed many radioactive substances.* Although most of these so-called *naturally radioactive* materials have long since disintegrated, some still exist on earth. As pointed out earlier, all elements with $Z$ greater than 83 are radioactive. Let us now discuss how a typical radioactive material disintegrates as a function of time, that is, how it *decays.*

Scientists believe that the particles within a typical nucleus are in continuous motion. As we shall see later, the protons and neutrons in the nucleus are not simple entities. For now we can think of them as being engaged in a continuous attempt to escape from the nucleus. In stable nuclei, they never succeed in this attempt. But an unstable nucleus can reduce its energy and become more stable by ejecting a particle and/or energy. It does so on a purely random basis. We can think of a particle trying to escape from the nucleus, making many attempts each second. Once in a great while, the nucleus is in an internal configuration such that the particle can escape. The nucleus thus undergoes decay.

As you see, this continuous game of chance within all unstable nuclei means that each nucleus has a certain chance for decay during a time interval $\Delta t$. Let us say that the chance, or probability, that a given nucleus will decay in time $\Delta t$ is $\lambda \, \Delta t$, where $\lambda$ is called the **disintegration, or decay, constant.** (Do not confuse this definition of $\lambda$ with that for wavelength.) Then in a sample consisting of $N$ such nuclei, the number that will decay in time $\Delta t$ is $N$ times larger than $\lambda \, \Delta t$, or

Number decaying in time $\Delta t = N\lambda \, \Delta t$

We can therefore write

$$\Delta N = - N\lambda \, \Delta t \tag{27.3}$$

where the negative sign arises because $\Delta N$, the change in $N$, is a negative number, since $N$ is decreasing. We call the quantity $\Delta N / \Delta t$ the *activity* of the sample. It is the number of decays that occur in unit time, and we discuss it further in Sec. 27.12.

Suppose that we have $N_o$ radioactive atoms at a time $t = 0$. We can use Eq. 27.3 to show how the number $N$ of undecayed atoms varies with time. The result is shown in Fig. 27.5. As you see from the figure, $N$ starts at $N_o$ when $t = 0$ and decreases steadily. This type of curve is called an *exponential decay curve,* and we give its equation in the next section.

Each radioactive element has its own characteristic time to decay, even though the general appearance of the decay curve is the same for all elements. We describe the rate at which a substance decays by stating the time that it takes for half the nuclei to decay. In Fig. 27.5 we see that $N$ has decreased to $\frac{1}{2} N_o$ after a time $T_{1/2}$, a time designated as the **half-life.**

---

* The earth solidified about $4 \times 10^9$ years ago, and so this is the age of the oldest rocks found on earth.

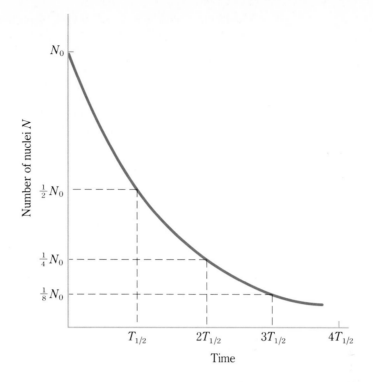

FIGURE 27.5
A radioactive element decays
exponentially.

The half-life $T_{1/2}$ of a substance is the time it takes for half the material to decay.

Half-lives of radioactive substances differ widely. The half-life of uranium 238 is 4.47 billion yr, while that of radium 226 is 1600 yr. Radon gas, the element to which radium decays, has a half-life of only 3.8 days. Many artificially produced radioactive substances have half-lives of only a fraction of a second. Even so, all these elements decay in conformity with the exponential decay law.

An interesting consequence of this mode of decay is the following. Suppose you start with $N_o$ nuclei that have a half-life $T_{1/2}$. After a time $T_{1/2}$, only $\frac{1}{2}N_o$ undecayed nuclei remain. If you wait another time $T_{1/2}$, then half of these nuclei will have decayed, leaving $\frac{1}{2} \cdot \frac{1}{2}N_o$ undecayed. After still another time $T_{1/2}$, half of these remaining nuclei will have decayed, leaving $\frac{1}{2} \cdot \frac{1}{2} \cdot \frac{1}{2}N_o$ remaining. We conclude that after $n$ half-lives, the number of undecayed nuclei remaining will be $(\frac{1}{2})^n N_o$. What fraction of $N_o$ will be left after a time of $4\,T_{1/2}$?* By asking yourself a similar question for various values of $n$, you should be able to construct the decay curve of Fig. 27.5.

We now have two ways to characterize the decay of a substance. Its rate of decay can be described by either $\lambda$ or $T_{1/2}$. Of course, these two quantities must be related. By using calculus, the relation can be shown to be

(Half-life) · (decay constant) = 0.693

or

$$\lambda\,T_{1/2} = 0.693 \tag{27.4}$$

_____

* Answer: $N_o/16$.

We shall have occasion to use this relation often.

---

### Example 27.5

Iodine 131 is a radioactive isotope made in nuclear reactors for use in medicine. When it is taken into the body, it becomes concentrated in the thyroid gland. There it acts as a radiation source in the treatment of hyperthyroidism. Its half-life is 8 days. Suppose a hospital orders 20 mg of $^{131}I$ and stores it for 48 days. How much of the original $^{131}I$ will still be present?

*Reasoning*   Each 8 days, the iodine decays by one-half. We can therefore make the following table:

| Time (days) | 0 | 8 | 16 | 24 | 32 | 40 | 48 |
|---|---|---|---|---|---|---|---|
| Iodine (mg) | 20 | 10 | 5 | 2.5 | 1.25 | 0.625 | 0.313 |

After 48 days, only 0.313 mg of the original 20 mg will remain.  ∎

---

### Example 27.6

A vial holds 1 g of radium. How many radium atoms in the vial undergo decay in 1 s? The half-life of radium is 1600 yr, or $5.1 \times 10^{10}$ s.

*Reasoning*   We use Eq. 27.3 to find $\Delta N$ for a time interval $\Delta t = 1$ s. To use it, we need $\lambda$ and $N$. From Eq. 27.4,

$$\lambda = \frac{0.693}{T_{1/2}} = \frac{0.693}{5.1 \times 10^{10} \text{ s}} = 1.36 \times 10^{-11} \text{ s}^{-1}$$

Also, because the atomic mass of radium is 226 kg/kmol, we know that 226 kg contains $6.02 \times 10^{26}$ atoms. Therefore

$$N = (6.02 \times 10^{26} \text{ atoms/kmol}) \left( \frac{0.001 \text{ kg}}{226 \text{ kg/kmol}} \right) = 2.66 \times 10^{21} \text{ atoms}$$

Now we can use Eq. 27.3 to find the number of disintegrations per second:

$$\Delta N = -\lambda N \Delta t = -(1.36 \times 10^{-11} \text{ s}^{-1})(2.66 \times 10^{21})(1 \text{ s})$$
$$= -3.61 \times 10^{10} \quad ∎$$

---

## 27.6 Exponential decay

The decay curve in Fig. 27.5 is well known in science; it is called an *exponential curve*. As we saw in the last section, it has the property that the curve height decreases by one-half for each half-life along the horizontal axis. The curve can be stated in mathematical terms as

$$N = N_o e^{-\lambda t} \tag{27.5}$$

where $\lambda$ is the decay constant. It is related to the half-life through $\lambda T_{1/2} = 0.693$, as we stated in Eq. 27.4. The function $e^{-\lambda t}$ is called an *exponential function,* and $e$ is the base for natural logarithms, 2.7183.

Use of Eq. 27.5 is facilitated by the fact that many hand calculators have a key for this function. If your calculator does, you may wish to test your use of it by checking the following typical values:

| $x$ | $e^x$ | $e^{-x}$ |
|---|---|---|
| 0 | 1.000 | 1.0000 |
| 2 | 7.389 | 0.1353 |
| 3 | 20.085 | 0.0498 |
| 4 | 54.598 | 0.0183 |
| 10 | 22,026 | $4.54 \times 10^{-5}$ |

If you do not have a calculator with this capability, you can find a table of exponential functions in most handbooks.

### Example 27.7

Uranium 238 has a half-life of $4.5 \times 10^9$ yr. It is believed that the earth solidified about $4.0 \times 10^9$ yr ago. What fraction of the uranium 238 then found on the earth remains undecayed today?

*Reasoning*   We use the decay law, Eq. 27.5, with

$$\lambda = \frac{0.693}{T_{1/2}} = \frac{0.693}{4.5 \times 10^9 \text{ yr}} = 1.54 \times 10^{-10} \text{ yr}^{-1}$$

This gives

$$\frac{N}{N_o} = e^{-\lambda t} = e^{-(1.54 \times 10^{-10} \text{ yr}^{-1})(4 \times 10^9 \text{ yr})}$$

$$= e^{-0.616} = 0.54$$

Today 54 percent of the uranium 238 is still in existence.   ∎

### Example 27.8

Ten percent of a certain radioactive substance decays in 12.0 h. What are the decay constant and the half-life for the substance?

*Reasoning*   We are told that in Eq. 27.5 the value of $N/N_o$ is 0.90 at a time $t = 12$ h. Substitution gives

$$0.90 = e^{-\lambda(12\,\text{h})}$$

To eliminate the necessity of taking logarithms of numbers that are less than unity, let us invert both sides of this equation. Then, because $1/e^{-x} = e^x$, we have

$$1.111 = e^{\lambda(12\,\text{h})}$$

Taking logs of both sides of this equation gives

$$\ln 1.111 = \lambda(12 \text{ h})$$

from which

$$0.1053 = \lambda(12 \text{ h})$$

Solving for $\lambda$ gives

$$\lambda = 0.00877 \text{ h}^{-1}$$

The half-life is found in the following way:

$$T_{1/2} = \frac{0.693}{\lambda} = \frac{0.693}{0.00877 \text{ h}^{-1}} = 79.0 \text{ h} \quad \blacksquare$$

*Exercise* What are the decay constant and the half-life if 20 percent decays in 40 s? *Answer: 0.00558 s$^{-1}$, 124 s* ■

---

## 27.7 Emissions from naturally radioactive nuclei

As stated previously, all nuclei with $Z > 83$ are radioactive. Early workers used the experiment sketched in Fig. 27.6 to examine the radiation from substances such as radium. A small amount of radium is placed in the center of a block of lead. The block has a thin hole drilled in it through which the radiation emitted by the radium can escape. (Lead is very effective in stopping radiation, as we shall see later. Since radioactive materials are serious health hazards, an experiment such as this should be carried out only with appropriate protection.) When the beam of radiation from the radium is allowed to pass into a magnetic field, it splits into three components, as shown. From the directions in which the rays are bent, we conclude that one component has no charge (the $\gamma$-rays), one is positively charged, and the third is negatively charged. As mentioned before, since these radiations were originally unidentified, they were given the names $\alpha$-, $\beta$-, and $\gamma$-rays. Let us now discuss each in turn.

**FIGURE 27.6**

The radiation from radium is separated into three components by a magnetic field. The $\beta$-rays actually follow a more curved path than shown.

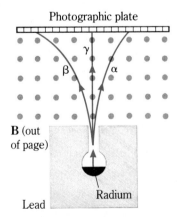

### Gamma radiation

On occasion, a nucleus finds itself in an excited energy state. To reach its ground state, it may emit a high-energy photon called a $\gamma$-ray (gamma ray). If the nucleus makes a transition from a state with energy $E_2$ to a state with energy $E_1$ then it will emit a $\gamma$-ray photon of frequency

$$hf = E_2 - E_1$$

This is completely analogous to the emission of a photon by an atom as its electronic structure adjusts to a lower energy state. Of course, $\gamma$-ray photons are basically the same as light and x-ray photons. However, the term $\gamma$-*ray* is usually given to a photon emitted from the nucleus, while an identical photon emitted during an atomic electron transition is called an x-ray.

The wavelengths of $\gamma$-rays emitted by nuclei can be measured in several ways. They may be reflected off crystals of known structure, as discussed previously for x-ray diffraction, and their wavelength determined by use of the Bragg relation. Another method makes use of *internal conversion electrons*. These are electrons that are thrown out of the atom as photoelectrons when the nucleus interacts with the atomic electrons in such a way as to transfer to the electron the energy normally carried away by the $\gamma$-ray. (Roughly, one can think of the nucleus emitting the $\gamma$-ray, which strikes an electron and loses all its energy to the electron; however, this detailed picture of the process is not verifiable.) The emitted electron will have an energy less than that of the $\gamma$-ray by an amount equal to the binding energy of the electron in the atom. The energy of high-energy electrons can be measured accurately, and so, knowing the binding energy of the electron, we can compute the $\gamma$-ray energy. Other methods for energy measurement based in part upon the interaction of the $\gamma$-rays with matter through the Compton and photoelectric effects also exist.

As in the case of atoms, the photons from a nucleus give us a tool for determining the energy-level structure of nuclei. Even stable nuclei can be investigated by this method. In order to do this for a stable nucleus, unstable nuclei are produced from stable nuclei by bombardment in a nuclear reactor or by some other means. These nuclei eventually decay to some isotope that has a stable nucleus. But since this isotope is usually in an excited state just after its formation, it emits $\gamma$-rays to return to its ground state. The energy-level diagrams of most stable and many unstable nuclei have now been measured in this as well as in other ways.

### Beta particle emission

Many radioactive nuclei emit $\beta$ particles, which are simply electrons. The process that occurs within a nucleus when $\beta$-particle emission occurs is quite complex. There are no electrons in the nucleus, so, in effect, the process transforms a neutron into a proton plus an electron. The new proton is retained by the nucleus while the electron is emitted.

We can represent the emission of a $\beta$ particle from a nucleus whose symbol is $^A_Z X$ in the following way:

$$^A_Z X \rightarrow \, ^{\;\;A}_{z+1} Y + \, ^{\;0}_{-1} e + \, ^0_0 \bar{\nu}$$

where $^{\;0}_{-1}e$ represents the emitted $\beta$ particle, $^{\;\;A}_{z+1}Y$ represents the transformed nucleus, and $^0_0\bar{\nu}$ represents a neutrino, a particle we shall say more about in a moment. The transformed nucleus contains one more proton than the original nucleus did. Its atomic number is therefore $Z + 1$. Its mass number is still $A$ because

## The discovery of radioactivity

Radioactivity was first noticed in 1896 by Henri Becquerel (1852–1908). He, like many physicists of the time, was investigating fluorescence, the fact that many substances glow in the dark after they have been exposed to strong sunlight. One day he postponed his experiments with uranium because the sun did not shine. He wrapped the uranium in black paper and by chance stored it next to an unexposed photographic plate. Later, when he developed the plate, he found it to be completely fogged, as though it had been exposed to sunlight. After a little detective work, he found that all salts of uranium, no matter what their past history, give off invisible radiation that can penetrate paper, cardboard, and many other materials. He concluded that uranium atoms give off a radiation similar to x-rays.

Stimulated by Becquerel's discovery, scientists began searching for other radioactive substances. Among those doing research on this topic were Marie Curie and, later, her husband Pierre. The Curies found that a mineral called pitchblende is slightly radioactive. To isolate the element responsible for the radiation, they started with about a ton of pitchblende and used chemical separation techniques on it. By 1898 they found that two types of atoms carried the radioactivity; one was chemically similar to bismuth and the other to barium. Since neither of these elements is radioactive, the Curies concluded that they had discovered two new elements. The one similar to bismuth they named polonium, after Poland, the native country of Marie Curie. The other, which is chemically similar to barium, was given the name radium. We now recognize polonium and radium as elements 84 and 88 in the table of elements. These were the first of several high-atomic-number elements to be discovered through the use of radioactivity.

there are still the same number of nucleons in the nucleus. Because of its very small mass, the mass number of a $\beta$ particle is considered to be zero.

Unlike the case of $\gamma$-ray emission, where only $\gamma$-rays with definite energies corresponding to differences in energy states of the nucleus are found, $\beta$ particles of widely varying energies are emitted. A typical $\beta$-particle energy spectrum is shown in Fig. 27.7. This is not what one would expect, since, if a $\beta$ particle is emitted, one would think it should carry away a reproducible energy corresponding to the difference in energy between the initial and final states of the nucleus. Another puzzling fact about $\beta$-particle emission is the following. Observation of the recoil of the

**FIGURE 27.7**

The energy distribution for $\beta$ particles emitted from $^{210}_{83}$Bi.

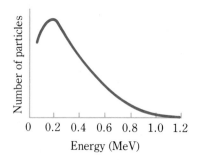

nucleus after $\beta$-particle ejection indicates that the linear momentum of the two objects is not what is expected. In particular, the momentum of the ejected electron is not equal and opposite to the recoil momentum of the nucleus. In order to explain these puzzling facts, it was postulated that a second, undetected particle is emitted with the $\beta$ particle. This particle should have zero rest mass and zero charge; it was given the name *neutrino*. Direct experimental evidence for the existence of this particle was obtained in the mid-1950s. Although the neutrino has zero mass* and zero charge, it must possess angular momentum (or spin) in order to preserve the law of conservation of angular momentum during $\beta$ decay.

### Alpha-particle emission

Some radioactive nuclei emit $\alpha$ particles. These are simply helium nuclei (two protons + two neutrons) and are represented by $^4_2\alpha$ or $^4_2\text{He}$. A typical $\alpha$-particle decay is exhibited by radium nuclei:

$$^{226}_{88}\text{Ra} \rightarrow \,^{222}_{86}\text{Rn} + \,^4_2\text{He}$$

This decay process has a half-life of 1600 yr. We call the original nucleus (radium in this case) the *parent* nucleus and the final nucleus (the unreactive gas radon) the *daughter* nucleus.

Alpha particles emitted by nuclei have characteristic energies that depend on the nucleus involved. Those emitted by radium have energies of 4.79, 4.61, and 4.21 MeV.

---

### Example 27.9

Radon 222 decays to polonium 218 by alpha emission. Find the approximate energy of the emitted alpha particle. Pertinent atomic masses are $^{222}\text{Rn} = 222.01753$ u, $^{218}\text{Po} = 218.00893$ u, $^4\text{He} = 4.00263$ u.

*Reasoning*   The mass loss in the reaction is

$$\text{Mass loss} = 222.01753 - (218.00893 + 4.00263) = 0.00597 \text{ u}$$

Since 1 u is equivalent to 931.5 MeV, the energy released is

$$\text{Energy} = (931.5 \text{ MeV/u})(0.00597 \text{ u}) = 5.56 \text{ MeV}$$

Most of this energy is carried away by the $\alpha$ particle, the observed energy of which is 5.49 MeV. This value differs from the total energy lost because of the recoil energy of the daughter nucleus.   ∎

---

### 27.8 Nuclear reactions

The $\alpha$- and $\beta$-particle decay schemes we have described in the previous section are simple nuclear reactions. Like chemical reaction equations, nuclear reaction equations must be balanced. To maintain balance, nuclear reactions must satisfy the

---

* There is controversy as to whether or not the rest mass of the neutrino is exactly zero. However, its mass, if any, is many orders of magnitude smaller than the electron's mass.

conservation laws of physics. For the present, we shall be concerned only with the conservation of charge and nucleon number.

When we write a nuclear symbol such as $^A_Z X$, the number of nucleons is given by $A$ and the net charge of the nucleus is specified by $Z$. In any nuclear reaction, the sum of all the nucleons (or $A$ values) on one side of the reaction must equal that on the other side. Thus, in the reaction

$$^{226}_{88}\text{Ra} \rightarrow {}^{222}_{86}\text{Rn} + {}^4_2\text{He}$$

we see that these two sums are equal; $226 = 222 + 4$. Moreover, because charge must be conserved, the sums of the $Z$ values must also be equal. In the present reaction, these sums are $86 + 2$ and 88, which, as we see, are equal to each other.

A typical $\beta$ decay reaction is that of thorium decaying to protactinium:

$$^{234}_{90}\text{Th} \rightarrow {}^{234}_{91}\text{Pa} + {}_{-1}^{0}\text{e} + {}_0^0\bar{\nu}$$

The nucleon numbers obviously balance. Since $91 + (-1)$ is 90, it is clear that the $Z$ values are also balanced.

There are other conserved quantities besides nucleon number and charge, and nuclear reactions must also obey these conservation laws. As was pointed out previously, a neutrino is emitted in $\beta$ decay. Without it, the $\beta$ decay reaction would not conserve linear and angular momentum and energy. Energy, including the energy equivalent of mass, must also be conserved in nuclear reactions.

The fact that the total energy before reaction (including the equivalent energy of the rest masses) must equal the total energy after reaction is a useful tool in the study of nuclear reactions. For example, when Rutherford performed one of the very first induced nuclear reactions in 1918, he shot $\alpha$ particles at nitrogen nuclei and observed the reaction

$$^{14}_7\text{N} + {}^4_2\text{He} \rightarrow {}^{17}_8\text{O} + {}^1_1\text{H}$$

In other words, the $\alpha$ particle entered the $^{14}\text{N}$ nucleus, which then disintegrated by ejecting a proton. The original nitrogen nucleus was *transmuted* into oxygen. We now ask whether even very slow $\alpha$ particles could cause this reaction.

To learn more about this reaction, notice the masses of the reactant *nuclei:*

$$\text{Mass of } {}^{14}\text{N} = 14.0031 \text{ u} - 7m_e$$
$$\text{Mass of } {}^4\text{He} = \underline{4.0026 \quad\quad - 2m_e}$$
$$\text{Total mass before reaction} = 18.0057 \text{ u} - 9m_e$$

In the same way, examine the masses after reaction:

$$\text{Mass of } {}^{17}\text{O} = 16.9991 \text{ u} - 8m_e$$
$$\text{Mass of } {}^1\text{H} = \underline{1.0078 \quad - 1m_e}$$
$$\text{Total mass after reaction} = 18.0069 \text{ u} - 9m_e$$

We find that the products have more mass than the original reactants, the difference being 0.0012 u. This mass could be created only if additional energy was added to the reaction. Since 1.0 u is equivalent to 931.5 MeV, as shown in Example 27.3, we see that the increase in mass in this reaction required an external energy of $(\frac{931}{1})(0.0012) = 1.1$ MeV. The incident $\alpha$ particle must have had at least this amount of kinetic energy to make the reaction occur. Actually, since momentum must also be conserved in such a reaction, the end products will not be standing still. As a result, the particle must have more than 1.1 MeV of kinetic energy if the reaction is to be feasible.

Computations like this tell us a great deal about the feasibility of proposed nuclear reactions. Of course, a large variety of reactions are possible. In fact, the field of nuclear reactions in physics is as involved as the subject of organic reactions in chemistry.

## 27.9 Natural radioactive series

You may have been puzzled by the fact that radium 226 is found on earth today. After all, it has a half-life of only 1600 years, while the earth is several billion years old. From the decay law, the ratio of the present-day number of radium nuclei to the number that existed $4 \times 10^9$ years ago should be

$$\frac{N}{N_o} = e^{-0.693t/T_{1/2}} = e^{-0.693(4 \times 10^9)/1600} \approx 10^{-740,000}$$

which is a truly negligible fraction. We must conclude that new radium nuclei are being furnished to the earth as the original nuclei are depleted. Similar calculations show that many other sources of natural radioactivity have half-lives far too short to explain their present-day existence. Let us now see how the existence of these nuclei is explained.

Radium and similar radioactive nuclei are found on earth because they are decay products of extremely long-lived isotopes. For example, uranium 238 has a half-life of $4.47 \times 10^9$ years. Even though it is radioactive, its half-life is so long that large amounts of it still exist on earth. It is the parent nucleus for a whole series of radioactive nuclei. A uranium 238 nucleus decays according to the scheme

$$^{238}_{92}\text{U} \rightarrow {}^{4}_{2}\alpha + {}^{234}_{90}\text{Th}$$

where the daughter nucleus is thorium. The very long half-life of this decay explains why there is still some $^{238}$U left on earth. In addition, we see that $^{238}$U acts as a continuous source for this isotope of thorium.

The thorium formed in this decay is also radioactive. It decays by $\beta$ emission according to the following scheme:

$$^{234}_{90}\text{Th} \rightarrow {}^{0}_{-1}\text{e} + {}^{234}_{91}\text{Pa} + {}^{0}_{0}\bar{\nu}$$

where the daughter nucleus is protactinium. (Why does the atomic number increase in this decay? Why doesn't the mass number change?)

The protactinium in turn decays to $^{234}$U:

$$^{234}_{91}\text{Pa} \rightarrow {}^{0}_{-1}\text{e} + {}^{234}_{92}\text{U} + {}^{0}_{0}\bar{\nu}$$

Several other steps occur in this radioactive series before the final stable element of the series is reached. In this case, it is an isotope of lead, $^{206}$Pb. This series is shown in detail in Fig. 27.8. Notice that in the latter stages of the decay scheme, alternative possibilities for decay exist. As we see, the very long half-life of uranium 238 is responsible for the existence on earth of the members of the series.

Notice in Fig. 27.8 what happens when a nucleus emits an α particle. Because the parent nucleus loses four nucleons in the process, the mass number decreases by 4. Simultaneously, because the alpha particle carries away a charge of +2e, the atomic number decreases by 2. On the other hand, when a β particle is emitted, the mass number does not change, while the atomic number *increases* by 1. Can you explain why?

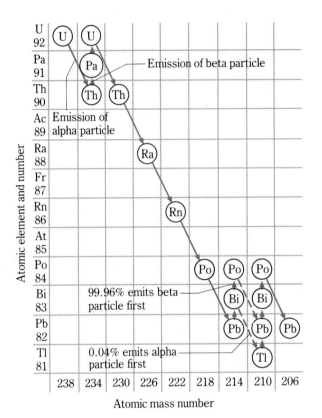

**FIGURE 27.8**

A typical radioactivity series. It is called the uranium series because the parent nucleus is uranium.

One other point should be mentioned in regard to the decay series. As well as α and β particles, γ-rays and neutrinos are emitted. These do not change either $Z$ or $A$ for the nucleus, and so they are not shown in Fig. 27.8. As stated earlier, γ-rays are emitted as the nucleus settles to a lower energy state.

There are two other natural radioactive decay series found on earth. They, together with the one we have been discussing, are summarized in Table 27.1. Notice that they all start with an element that has a very long half-life and eventually decay to a stable isotope of lead. Presumably, other decay series existed on earth at earlier times, but they have decayed too rapidly to be detected at this late date.

■ **TABLE 27.1  The natural radioactive series**

| Series | Starting element | Half-life, years | Stable end product |
|--------|------------------|------------------|--------------------|
| Uranium | $^{238}_{92}U$ | $4.47 \times 10^9$ | $^{206}_{82}Pb$ |
| Thorium | $^{232}_{90}Th$ | $1.41 \times 10^{10}$ | $^{208}_{82}Pb$ |
| Actinium | $^{235}_{92}U$ | $7.04 \times 10^8$ | $^{207}_{82}Pb$ |

**Example 27.10**

If the age of the earth is $5.0 \times 10^9$ years, what fraction of the original amount of $^{232}$Th is still in existence on the earth? (The earth is thought to have been molten prior to about $4 \times 10^9$ yr ago.)

*Reasoning*   The half-life of $^{232}$Th is $1.41 \times 10^{10}$ yr. We know that

$$\frac{N}{N_o} = e^{-\lambda t}$$

However, $\lambda T_{1/2} = 0.693$, and so $\lambda = 4.91 \times 10^{-11}$ per year. Therefore

$$\text{Fraction} = \frac{N}{N_o} = e^{-(4.91 \times 10^{-11}/\text{yr})(5.0 \times 10^9 \, \text{yr})} = e^{-0.246} = 0.782$$

Thus about 78 percent of the $^{232}$Th originally on the earth still exists today.  ■

*Exercise*   How many years will it take for the $^{232}$Th on earth to decrease to one-fourth its present value?   *Answer: $2.82 \times 10^{10}$ yr*  ■

## 27.10 Interactions of radiation with matter

As we use nuclear power and other sources of radiation, the effects of radiation on the human body and on materials become important. When a particle shoots through flesh or other material, it strikes atoms along its path. We use the word *strike* in an imprecise fashion; the particle, if charged, need not actually hit an electron or nucleus to cause damage. The coulomb force exerted on the electrons and nuclei by the charged particle is often strong enough to cause damage even if the particle only passes close to the atom. Even in a near-collision with an atom or molecule, a particle can ionize an atom or cause a molecule to break apart. It is in this way that the major effects of radiation occur.

The effects caused by a high-energy particle depend primarily on three factors: the mass of the particle, its energy, and its charge. An $\alpha$ particle, because it has a mass of 4 u, can cause more damage than an electron (0.00055 u) traveling at the same speed when it collides with an atom, much as a 10-ton truck can cause more damage than a child's wagon. Moreover, the $\alpha$ particle has a charge of $+2e$ compared with the electron's charge of $-e$; it therefore exerts a larger coulomb force on nearby charges than an electron does. For these reasons, an $\alpha$ particle ionizes atoms along its path much more frequently than an electron of the same energy does. However, because both the $\alpha$ particle and the electron continue moving until they have lost all their energy, the electron travels much further before it stops than does an $\alpha$ particle with the same initial energy. In other words, the *range* of an electron is greater than that of an equal-energy $\alpha$ particle.

Typical approximate ranges for a 2-MeV particle in air are 1 cm for an $\alpha$ particle, 10 cm for a proton, and 1000 cm for an electron. The more dense the material through which the particle moves, the shorter its range will be. As a rough approximation, the range varies inversely with density. Therefore, an $\alpha$ particle that has a range of 10 cm in air ($\rho = 1.29$ kg/m$^3$) will have a range of only about 0.005 cm in aluminum ($\rho = 2700$ kg/m$^3$). It should be apparent to you why lead, a material of very high density, is used as a shield against high-energy particles.

Neutrons, which have no charge, are extremely penetrating particles. Coulomb forces do not act on them as they traverse a material. To be stopped or slowed, a neutron must undergo a direct collision with a nucleus or some other particle that has a mass comparable to that of the neutron. Materials such as water and plastic, which contain many low-mass nuclei per unit volume, are used to stop neutrons.

Gamma rays (and x-rays) are not easily stopped because they have neither charge nor rest mass. They lose energy as they penetrate material mainly through the photoelectric and Compton effects,* processes that lead to ion formation. You have seen x-ray photographs of teeth and bones, so you know that x-rays can penetrate flesh and cast shadows of bones. The more electrons an atom of an absorbing material has, and the more dense the absorbing material is, the greater its ability to stop x-rays and $\gamma$-rays. Medical x-rays usually have energies less than 0.1 MeV, but the $\gamma$-rays emitted by nuclei often have energies of several million electronvolts and are very penetrating.

## 27.11 The detection of radiation

Most detectors of high-energy particles and radiation make use of the fact that ions are formed along particle paths. One device that allows us to see the path of an ionizing particle is the *Wilson cloud chamber.* It makes use of the fact that droplets of a supersaturated vapor form preferentially on ions in the vapor. Therefore, if an ionizing particle passes through a region in which cloud droplets are about to form, the droplets will form first along the particle's path, showing the path as a trail of droplets (Fig. 27.9). A somewhat similar device, called a *bubble chamber,* makes use of a superheated liquid, one that is ready to boil. Vapor bubbles form preferentially on ions, and so particle paths are shown as bubble tracks. An example of its use was shown in Fig. 18.13.

**FIGURE 27.9**

Cloud chamber tracks of high-energy particles. A magnetic field perpendicular to the page causes the paths to be curved.

Electronic devices for detecting high-energy particles are convenient to use and are the most common type of particle detector. Typical of these is the *Geiger counter,* illustrated in Fig. 27.10. When no radiation is entering, no charge exists in

---

* A third process, pair production, is important at very high energies. It, too, forms charged particles.

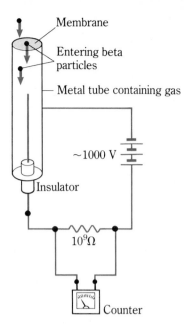

Membrane

Entering beta particles

Metal tube containing gas

~1000 V

Insulator

$10^9\Omega$

Counter

**FIGURE 27.10**
The Geiger counter.

the gas within the metal tube. No current is able to pass from the center wire to the metal tube, and therefore no current flows in the circuit. When an ionizing particle enters the tube, the ions and electrons that it liberates move across the tube under the influence of the electric field between the cylinder and the central wire. The field is made large enough so that the ions and electrons ionize other gas atoms as they move across the tube, causing an avalanche of charge. As a result, the current across the tube is much larger than the current that would result from the original ions alone. Soon after the particle has passed through, all the ions have been collected, and the current stops. Therefore, each ionizing particle gives rise to a current pulse in the resistor. The resulting voltage pulses are applied to a recording electronic system, which then gives a record of the number of ionizing particles that entered the counter.

There are many other radiation detection devices in use today. Which one to use depends on the type of particle (or radiation) being measured, the energy range involved, and the lack of convenience that can be tolerated. Photographic film is acceptable to a dentist, but a hospital carrying out a CAT scan needs an electronic, computerized detection system. Neither of these detectors is acceptable for monitoring the radiation exposure of an x-ray technician as she performs her duties. The detector she carries in her coat pocket is most often no bigger than a pen, yet it gives an accurate record of the radiation she has been exposed to each day.

## 27.12 Radiation units

In our modern world we are more and more concerned about the effects of radiation. Whether it is the result of a medical diagnostic test, of a nuclear accident, or of the radon that seeps into our dwellings from the earth below, radiation has become an important factor in our lives. To describe its effects, we need units in which to measure it. Over the years, many radiation units have been used, and this has led to confusion. Now, however, SI units are becoming predominant, resulting in great simplification. We shall list the most important measured quantities and their units.

### Source activity

The *activity* of a source of radiation is equal to the number of disintegrations that take place in the source in unit time. As an equation,

$$\text{Source activity} = \frac{\Delta N}{\Delta t} \tag{27.6}$$

where $\Delta N$ is the number of nuclei that decay in time $\Delta t$.

The SI unit for activity is the *becquerel* (Bq); a source that has an activity of 1 Bq undergoes one disintegration per second. An older unit that is still in widespread use is the *curie* (Ci), where $1 \text{ Ci} = 3.7 \times 10^{10}$ Bq exactly. To give you an idea of the numbers involved, one gram of radium has an activity of $3.7 \times 10^{10}$ Bq (1 Ci) and is millions of times more radioactive than many medical sources of radiation.

We can use Eqs. 27.3 and 27.4 to obtain the following equation for activity in terms of the decay constant and the half-life:

$$\text{Activity} = \frac{\Delta N}{\Delta t} = \frac{0.693N}{T_{1/2}} \tag{27.7}$$

Use of this equation is illustrated in Example 27.11.

### Absorbed dose

The *absorbed dose* is the energy per unit mass absorbed by a material in the path of the radiation beam. Its SI unit is joules per kilogram, which, in this instance, is given the name *gray* (Gy). Suppose a radiation beam passes through a mass $m$ and deposits therein an energy $E$ as it traverses the mass. Then the absorbed dose given to the material constituting the mass is

$$\text{Absorbed dose (Gy)} = \frac{E}{m} \qquad \text{J/kg}$$

In other words, 1 Gy is equivalent to an absorbed energy of 1 J/kg. Another unit frequently used for absorbed dose is the *rad*, where 1 rad = 0.01 Gy.

### Biologically equivalent dose

The effect of radiation on the human body depends not only on the energy and type of radiation, but also on the region of the body it strikes. To describe the biological effects of radiation, we introduce another measure for radiation dose, the *biologically equivalent dose*. It is simply the absorbed dose multiplied by a factor that compares the effect of the radiation being used with the effect of 200-keV x-rays on flesh. Its unit is the *sievert* (Sv). For example, because an $\alpha$-particle beam is about 15 times more damaging to flesh than are 200-keV x-rays, for a 1.0-Gy dose of $\alpha$ particles the biologically equivalent dose of x-rays would be 15 Sv. When discussing radiation damage to humans and animals, the biologically equivalent dose is the appropriate measure of radiation damage. An older unit that is still used frequently is the *rem*, where 1 rem = 0.010 Sv.

---

### Example 27.11
Strontium 90 has a half-life of 28 yr and is a dangerous product of nuclear explosions. What is the activity of 1 g of $^{90}$Sr?

*Reasoning* From Eq. 27.7,

$$\text{Activity} = \frac{0.693N}{T_{1/2}}$$

In this case, $T_{1/2} = 28$ yr $= 8.8 \times 10^8$ s. To find $N$, the number of atoms in 1 g of $^{90}$Sr, we recall that 1 kmol of $^{90}$Sr (which is 90 kg) contains $6.02 \times 10^{26}$ atoms. Then

$$N = \left(\frac{0.001 \text{ kg}}{90 \text{ kg}}\right)(6.02 \times 10^{26}) = 6.7 \times 10^{21}$$

Using these values, the activity is found to be $5.3 \times 10^{12}$ Bq. ∎

---

*Exercise* How much $^{90}$Sr would produce one disintegration per second? *Answer: $1.89 \times 10^{-16}$ kg* ∎

---

## 27.13 Radiation damage

Since radiation can tear molecules apart, it is capable of damaging any material, including the materials that compose our bodies. One of the most common types of radiation damage to humans is due to the ultraviolet rays in sunlight. These lead to sunburn and tanning of the skin. The high-energy photons disrupt skin molecules upon impact and cause these easily observed effects. In this case, the damage is usually of little importance. Most of the sun's ultraviolet rays are absorbed by the ozone in the upper atmosphere, so normal exposure to the sun's rays need not be avoided. However, in recent years we have become aware that a serious hazard could arise if manufactured chemicals deplete the ozone layer. There is a danger that the increased ultraviolet radiation that would reach us could increase the incidence of skin cancer.

We are continuously exposed to other radiation in addition to sunlight. Nearly all materials contain a slight amount of radioactive substances. As a result, your body is unavoidably exposed to a low level of background radiation. Typically, each person experiences a background radiation dose of about 1 mSv each year. Let us now examine the effects of different levels of radiation dose upon the body.

High levels of radiation covering the whole body disrupt the blood cells so seriously that life cannot be maintained. For whole-body doses in excess of 5.0 Sv, death is likely to occur. Even a whole-body dose of 1.0 Sv can cause radiation sickness of a very serious, although nonfatal, nature. Whole-body doses in the range of 0.30 Sv and above cause blood abnormalities. At still lower whole-body doses, the overall effects on the body are less apparent, but the consequences can nevertheless be serious.

Even very low radiation doses reaching the reproductive regions of the body are potentially dangerous. The DNA molecules in our bodies that carry reproductive information can be disrupted by a single radiation impact. If enough of these molecules are damaged, defective reproductive information will be furnished to a fetus as it develops. As a result, birth abnormalities will occur. Even though there is some evidence that a low level of reproductive abnormalities may be beneficial to the human species, most birth defects are not desirable. For this reason, *no one of childbearing age should be exposed to unnecessary radiation of the reproductive organs.* Of course, a properly given x-ray of an arm, for example, presents no such danger.

In addition to causing birth abnormalities, low levels of radiation present two other hazards. First, there appears to be a delayed cancer effect. Although cancer may not appear at once, low levels of radiation may cause it to develop many years later. Second, a child is particularly vulnerable to radiation. Because the child is growing rapidly, any cell mutations caused by radiation may have serious consequences. For this reason, most doctors are reluctant to prescribe x-ray scans for children unless absolutely necessary.

There is no "safe" limit of body exposure to radiation. We can say only that radiation shoul be kept to the lowest value possible within reason. For example, since we are all subjected to a background radiation of about 1.0 mSv/yr, there is no reason to disrupt our lives to avoid radiation doses less than this. Even though a person who lives in the mountains may experience an annual background dose 0.50 mSv higher than one who lives at sea level, the difference is not large enough to warrant a move. In the last analysis, one must often make a compromise between radiation safety and other considerations. Despite that fact, maximum occupational doses are of value and have been specified. As a rough rule, the maximum yearly dose, except for the eyes and reproductive organs, is about 0.050 Sv.

## 27.14 Medical uses of radioactivity

One of the earliest applications of radioactivity was the use of the radiation from radium and its decay products in the treatment of cancer. Since that time, methods of radiation therapy have greatly advanced because many new radioactive materials have been produced through the use of nuclear reactors and nuclear accelerating machines. One of the most important isotopes available for research and technological use is cobalt 60. It has a half-life of 5.27 years and is an intense source of $\gamma$-rays with energies near 1.2 MeV. Although cobalt 60 is widely used as an x-ray source by industry, perhaps its most important use is in medicine. Its radiation is extremely penetrating and is used to kill cancer cells deep within the body.

The radiation from iodine 131 is used in the treatment of thyroid cancer. This radioactive isotope has a half-life of 8 days. When foods containing iodine are eaten, much of the iodine localizes in the thyroid gland. Therefore, iodine 131 in foods is carried directly to the point in the body where its radiation is needed in the treatment of thyroid cancer. This is but one of many situations in which a radioactive isotope is carried to a specific point within the body for highly efficient localized radiation.

Sometimes radioactive isotopes are used as tracers, to follow the path that various chemicals take in the body. For example, if we did not already know that iodine localizes in the thyroid, we could ascertain that fact by observing the location of radioactivity in the body after iodine 131 had been ingested. Biologists use similar techniques to find out how plants utilize various chemicals.

Another medical use of radioactivity is shown in Fig. 27.11. The patient shown there has had gallium 67 injected into the bloodstream. This isotope lodges preferentially in certain types of tumorous tissue. As you see in the figure, the radioactivity (shown as the darkened regions) has localized in the lymph tissue of the throat and neck. This furnishes a strong clue to the location of cancer in this patient. (See also Color Plate I.)

These are but a few of the many applications of nuclear physics in the field of medicine. There are also many uses in the other biological sciences. Nuclear medicine and biology are rapidly expanding fields that are increasingly important to humanity.

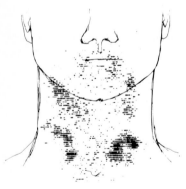

**FIGURE 27.11**
Radioactive gallium 67, traced here in a photoscan, settles preferentially in tumorous tissue. *(Oak Ridge Associated Universities)*

## 27.15 Radioactive dating

One of the most interesting uses of radioactivity is in determining the age of ancient materials. For example, the age of uranium-bearing rocks can be found in the following way. Since uranium 238 decays to lead 206 (see Fig. 27.8), we surmise that the lead 206 that is intimately mixed with uranium 238 in a rock came from the uranium that has decayed over the years. Suppose that analysis of the rock shows that the numbers of uranium and lead atoms per unit volume are $N_U$ and $N_{Pb}$, respectively. Then the ratio of the amount of uranium now present to that present at time $t$ when the rock first solidified is

$$\frac{N_U}{N_U + N_{Pb}} = e^{-\lambda t} = e^{-0.693t/T_{1/2}}$$

where $T_{1/2}$ is the half-life of uranium 238, which is $4.5 \times 10^9$ years. The oldest rocks found on earth have $N_U \approx N_{Pb}$, and so we estimate that the earth solidified about one uranium 238 half-life ago, or about 4 billion years ago. Similar results for the age of the earth are obtained using other isotopes.

To find the age of objects that were once alive, such as wood and bone, scientists employ a technique called *radiocarbon dating* in which a radioactive isotope of carbon, carbon 14, is used. This isotope, which has a half-life of 5730 years, is continuously being produced on earth as cosmic rays from outer space bombard atmospheric nitrogen. Since the radioactive carbon is chemically identical to carbon 12, the common isotope of carbon, all living things have the two isotopes in intimate combination. Over the ages, the ratio of carbon 14 to carbon 12 has averaged about $1.30 \times 10^{-12}$. However, when a tree, for example, dies, the carbon 14 in its wood cannot be replenished; the carbon 14 content decays with a half-life of 5730 years. This fact can be used to determine the wood's age, as we see in the following example.

---

### Example 27.12

An archeologist finds a wooden ax handle whose age she wishes to determine. To that end, she uses special equipment to compare its carbon-caused radioactivity per unit mass with that of a new piece of wood. The ratio of the two radioactivities is found to be 0.034. Estimate the age of the ax handle.

*Reasoning*   We have that

$$\frac{N}{N_o} = e^{-\lambda t} \qquad \text{gives} \qquad 0.034 = e^{-0.693t/T_{1/2}}$$

Using $T_{1/2} = 5730$ years, we solve for $t$ and find the object's age to be 28,000 years. In precise work, corrections must be made for variations in $^{14}C$ concentrations on the earth over these thousands of years.   ∎

---

## 27.16 The fission reaction

After the discovery of the neutron (in 1930), it became apparent that this neutral particle could be used to induce nuclear reactions. Because it has no charge, the neutron can enter a nucleus. Enrico Fermi was the leader in the use of this new projectile and, in the mid-1930s, produced many previously unknown isotopes. One of his major ambitions was to bombard massive nuclei to produce elements with $Z$ larger than any then known. His efforts met with some success; but others have carried on where he left off, and all nuclei up to $Z = 106$ have now been produced.

However, Fermi had the misfortune to misinterpret his perhaps greatest discovery. When uranium was bombarded with very-low-energy neutrons (neutrons with thermal energies, called *thermal neutrons*), a reaction was indeed found to occur. Carrying on where Fermi left off, Otto Hahn and Fritz Strassman (in 1939) carried out a chemical analysis of the reaction products. To their surprise, they found many elements with atomic numbers near $Z = 50$ among the reaction products. Barium, in particular, was one of the reaction products. What could possibly be going on? They had added a neutron to a uranium nucleus ($Z = 92$) and ended up with an element (barium) that has $Z = 56$. Moreover, this nuclide appeared to be highly radioactive, even though ordinary barium is stable.

Seizing upon the work of Hahn and Strassman, Lise Meitner and her nephew, Otto Frisch, found the explanation for these puzzling results. They showed that a uranium nucleus captures a neutron, holds onto it for a fraction of a second, and then explodes into two roughly equal-size nuclei (see Fig. 27.12). The intermediate nucleus is called a *compound nucleus*. Energy and two or three neutrons are also released in the reaction. We call this splitting of a nucleus into a few fragments of roughly equal size the process of *nuclear fission*. The discovery of nuclear fission, a simple scientific curiosity at the time, has greatly altered the future of civilization.

Further analysis of the reaction shows that only one uranium isotope found in quantity in nature undergoes fission in this way. It is uranium 235, which constitutes only about 0.7 percent of the natural mixture of uranium isotopes. The first step in the fission reaction is the capture of a neutron (n) by $^{235}$U to form a compound nucleus:

$$\text{n} + {}^{235}_{92}\text{U} \rightarrow {}^{236}_{92}\text{U}*$$

where we represent the compound nucleus by U*. The compound nucleus then decays by one of many possible reactions. The following is only one possibility:

$$ {}^{236}_{92}\text{U}* \rightarrow {}^{140}_{56}\text{Ba} + {}^{92}_{36}\text{Kr} + 4\text{n} + \text{energy}$$

The reaction products here are not the stable isotopes $^{84}$Kr and $^{138}$Ba found in nature. Hence they decay to other isotopes, and these to still others, until stability is reached. As a result, the products of the fission reaction are highly radioactive, and the reacting material is a strong source of radiation. Even more important, however, the reaction releases large amounts of energy.

We can obtain an understanding of this energy release by referring back to Fig. 27.3, which shows the binding energy per nucleon for various nuclei. Recall that nuclei with high binding energy have less mass per nucleon than do nuclei with lower binding energy. The graph tells us that the mass per nucleon in barium (Ba), for example, is less than that in uranium. Therefore, if a uranium nucleus is split into two nuclei with $Z$ near 50, nucleons lose mass in the process. This lost mass is released as various forms of energy, including radiation as well as the kinetic energy of the neutrons and other reaction products. In the average fission of $^{235}$U, the energy released is about 200 MeV, a tremendous energy indeed.

The mechanism of the fission process is best understood by noticing that a massive nucleus behaves in many ways like a drop of liquid. As shown in Fig. 27.12,

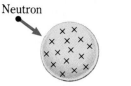

Neutron

(a) Before reaction, n + $^{235}$U

(b) Compound nucleus, $^{236}$U*

(c) Vibrating droplet

(d) Coulomb forces stretch nucleus

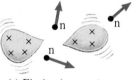

**FIGURE 27.12**
Vibration of the compound nucleus
leads to its eventual fission.

(e) Fission is complete

addition of a neutron to the nucleus sets the nucleus into vibration. Because of the random nature of the vibrations, the situation in Fig. 27.12d may arise. In that case, the effect of the nuclear attractive force is decreased because of the much-increased surface area of the nucleus. Further, the coulomb repulsive force drives the two portions of the nucleus still farther apart, and the nucleus undergoes fission, as shown in Fig. 27.12e. Neutrons are released, and the two fission fragments are highly excited and unstable.

Only a very few large nuclei can be induced to undergo fission when slow neutrons are used. Although all large nuclei will undergo fission if they are bombarded by particles of high enough energy, the high-energy accelerators needed for this purpose consume much more energy than the fission process would produce. However, in the case of uranium 235, only a slow-moving neutron is required. (In fact, the reaction rate is largest for thermal neutrons.) As a result, in this particular case, fission is easily induced, and the process yields much more energy than is required to cause fission.

Since the fission of one $^{235}$U nucleus gives rise, on the average, to about 3 neutrons, and since neutrons induce $^{235}$U nuclei to undergo fission, a self-sustaining reaction is possible. Consider a mass of $^{235}$U so large that the number of neutrons

27.16 The fission reaction      **679**

that escape from its surface is negligible compared with the total number of neutrons. Then, when a single neutron enters a $^{235}$U nucleus, it will give rise to, let us say, 3 neutrons as the nucleus undergoes fission. (The average number found by experiment is 2.47.) These 3 neutrons in turn cause three more nuclei to split, thereby liberating a total of $3^2 = 9$ neutrons. These $3^2$ neutrons cause the fission of other nuclei to produce $3^3$ neutrons, and so on. This process, illustrated in Fig. 27.13, is called a **chain reaction**. After $q$ steps in the chain reaction have occurred, $3^q$ neutrons will be available. If each step of the reaction takes 0.01 s, at the end of 1 s the total number of neutrons will be $3^{100} \approx 10^{48}$. Since 235 kg of $^{235}$U contain only $6 \times 10^{26}$ atoms, it is clear that a reaction such as this could occur with explosive violence.

**FIGURE 27.13**

A chain reaction can be initiated by a single neutron.

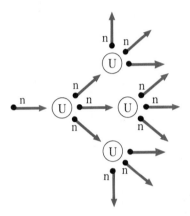

The fission chain reaction forms the basis for the operation of nuclear reactors. In practical applications, several complications arise. In order to maintain a steady, nonexplosive reaction in a reactor, each fission process should cause one additional fission process (not two, because then the reaction would explode, and not fewer than one, because then the reaction would die out). In order to retain enough neutrons in the reaction chamber, the size of the fissionable material must be large enough so that not too many neutrons stray through its surface and become lost to the reaction. There is a **critical mass** (or *critical size*) for the fissionable material. If too little material is available, not enough neutrons can be retained in it to produce a self-sustaining chain reaction.

In addition, the ability of neutrons to be captured by a $^{235}$U nucleus depends upon the speed of the neutrons. Slow neutrons are much more likely to cause fission than are fast neutrons. For this reason, a large part of the total volume of a nuclear reactor consists of a moderator, a nonreactive material used to slow down the neutrons that are emitted in the fission process. Since neutrons have a mass of 1 u, and since, upon collision, a particle is slowed best by particles of nearly the same mass, the moderating material in reactors usually consists of low-atomic-mass substances. Common examples are carbon, water, and hydrocarbon plastics.

## 27.17 Nuclear reactors

The reactor in a nuclear power station serves the same purpose as the furnace in a steam generator. It acts as a source of intense heat, and that heat is used to generate steam. The steam in turn is used to drive the turbines of the electric generator system. A schematic diagram of a typical reactor is shown in Fig. 27.14.

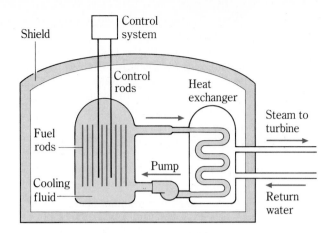

**FIGURE 27.14**

A schematic diagram of a nuclear reactor.

The heart of the reactor consists of the fissionable material (the fuel), sealed in cylindrical tubes. Originally, uranium 238 enriched with $^{235}$U was the principal reactor fuel. Now, however, other fissionable materials are also used in fuel rods. These rods are immersed in a material such as water, carbon, or some similar low-atomic-mass material. This material, the moderator, slows the fission-produced neutrons and reflects them back into the fissionable material. In the design in Fig. 27.14, the moderator also acts as the heat-exchange fluid to carry heat away from the fuel rods.

If a mass much larger than the critical mass is involved, the reaction will build up at a fast rate and an explosion will occur.* This is desirable, of course, if one is making a nuclear weapon. However, in the nuclear reactor, one wants the reaction to proceed smoothly so that a steady but nonexplosive source of energy results. In practice, the number of reacting neutrons in a reactor is controlled by the use of neutron-absorbing rods. For example, cadmium rods readily absorb neutrons, thereby removing them from the reaction. Hence, if such rods are put into the reactor, the nuclear reaction will slow down. The reaction rate can be readily adjusted by positioning control rods of this sort in the reactor.

When a nucleus undergoes fission within the fuel rod, highly unstable intermediate-$Z$ nuclei are formed. These undergo extensive radioactive decay and eject high-energy particles in the process. As these particles are slowed, their energy is changed to heat, which heats the reactor system. This heat is then carried away to a heat exchanger by a circulating fluid.

In the heat exchanger, the heat is transferred to ordinary water in a steam-boiler system. Steam is generated and then used to power electric turbines. As we see, the steam does not come into direct contact with the reactor core. For this reason, its level of radioactivity is low. But the fluid that circulates through the core is bombarded by radiation from the fission products. Like all other portions of the core material, it is often highly radioactive.

When the material in the fuel rods has been used for many months, its original fissionable material is much depleted. The fuel rods are then removed and replaced by new ones. Unfortunately, there is still disagreement as to the best disposal method for the waste material in the old rods. This material consists of highly radioactive, fairly long-lived fission products. It takes centuries for the radioactivity

---

* Since the concentration of fissionable material is much lower in a reactor than that required in a bomb, a reactor malfunction cannot duplicate a bomb explosion.

to decay to harmless levels. Disposal of this waste is one of the major drawbacks of nuclear reactors.

However, reactors also can provide us with radioactive materials for medical, industrial, and other uses. Many of the radiation sources presently used by hospitals, industry, and research laboratories are made by placing suitable materials within the core of the reactor. In addition, research reactors exist in many parts of the world. The intense radiation in their cores can be "piped" outside the reactor to act as powerful beams of radiation. As we see, the fission process possesses vast potential as well as hazards for humankind.

---

### Example 27.13

Suppose a compound uranium nucleus undergoes the following fission reaction:

$$\underset{(236.04564)}{{}^{236}_{92}\text{U}^*} \quad \rightarrow \quad \underset{(141.91635)}{{}^{142}_{56}\text{Ba}} \quad + \quad \underset{(89.91972)}{{}^{90}_{36}\text{Kr}} \quad + \underset{4(1.008665)}{4\text{n}} + \text{energy}$$

The atomic masses of each isotope are given in atomic mass units. Find the energy released in the reaction.

*Reasoning*   The total mass of the products is

$$(141.91635 - 56m_e) + (89.91972 - 36m_e) + 4.03466 = 235.87073 \text{ u} - 92m_e$$

Subtracting this from the mass of ${}^{236}\text{U}$ $(236.04564 - 92m_e)$, we find

Mass loss $= 0.1749$ u

Therefore the energy liberated is

Energy $= (0.1749 \text{ u})(931.5 \text{ MeV/u}) = 163 \text{ MeV}$

Still more energy will be liberated as the very unstable ${}^{142}\text{Ba}$ and ${}^{90}\text{Kr}$ decay further. A major share of the liberated energy appears as the kinetic energy of the fission products, as pictured in Fig. 27.12*d*. ∎

---

## 27.18 Nuclear fusion

The use of the fission reaction as an energy source depends on the fact that nucleons in intermediate-size nuclei have less mass than those in large nuclei. Hence the splitting of a large nucleus destroys mass and releases an equivalent quantity of energy.

If we look back to Fig. 27.3, we see that the small-atomic-number nuclei, such as lithium, have even less binding energy per nucleon than uranium has. This means that the nucleons in low-atomic-number nuclei have more mass per nucleon than do those in nuclei with $Z$ close to 50. We can therefore envision joining small nuclei together to form larger nuclei and, in the process, converting mass into energy. This type of reaction, in which small nuclei are joined together to form larger ones, is called *nuclear fusion*. To illustrate the tremendous energies released in fusion reactions, consider the following set of reactions, which furnishes a large fraction of the sun's energy.

$$^1_1H + {}^1_1H \rightarrow {}^2_1H + {}^0_{+1}e + {}^0_0\nu$$

where ${}^0_{+1}e$ is a positive electron (called a positron) and ${}^0_0\nu$ is a neutrino. The deuterium ${}^2_1H$ reacts further:

$$^2_1H + {}^1_1H \rightarrow {}^3_2He$$

and then

$$^3_2He + {}^3_2He \rightarrow {}^4_2He + 2{}^1_1H$$

As we see, in effect four protons are fused together to form a helium 4 nucleus.

To find the energy liberated in this process, we must find the mass loss. The starting mass is that of four protons,

$$4 \times 1.007276 = 4.029104 \text{ u}$$

while the final mass is that of the helium 4 nucleus, namely,

$$4.002604 - 2m_e = 4.001506 \text{ u}$$

This gives a mass loss of 0.0276 u in the process. It has an energy equivalent of

$$(0.0276 \text{ u})(931 \text{ MeV/u}) = 25.7 \text{ MeV}$$

But in 1 kg of helium, there are $\frac{1}{4}N_A$ atoms. So the energy lost in the formation of 1 kg of helium is

$$\text{Energy} = \tfrac{1}{4}(6 \times 10^{26})(25.7 \text{ MeV}) = 3.86 \times 10^{33} \text{ eV} = 6.2 \times 10^{14} \text{ J}$$

For comparison purposes, the energy liberated when 1 kg of carbon is burned in oxygen is about $8 \times 10^6$ cal, or $3.3 \times 10^7$ J. From this one sees that the energy released in chemical reactions is about $10^{-7}$ as large as that released in nuclear fusion reactions.

Although the source of energy in the sun and stars is a fusion process, the fusion reaction has not yet been made a practical, steady energy source on earth. However, somewhat different fusion reactions are the source of energy for the hydrogen and other types of fusion bombs. The difficulty in obtaining a steady fusion reaction is as follows.

The energy required to shoot two protons close enough together so that they will fuse is of the order of 0.1 to 1 MeV. This energy is easily attainable by means of huge particle accelerators. However, the efficiency of these machines is far too low to make their use in this way practical. To achieve a practical reaction method, one must make use of the thermal energies generated by the reaction itself. Let us see what sort of temperatures are needed to furnish energies of 0.5 MeV, the energy needed to carry out these fusion processes.

Recall from Eq. 10.6 that the average thermal translational kinetic energy of a particle at temperature $T$ is $\frac{3}{2}kT$. Setting this equal to $0.5 \times 10^6$ eV, or $0.8 \times 10^{-13}$ J, we have

$$\tfrac{3}{2}(1.38 \times 10^{-23} \text{ J/K})T = 0.8 \times 10^{-13} \text{ J}$$

from which
$$T = 3.8 \times 10^9 \text{ K}$$

This temperature is far in excess of that produced in any oven on earth.* All materials vaporize at these high temperatures. Work is progressing, however, on achieving very high temperatures in highly ionized gases called *plasmas* that are confined to a region in space by magnetic fields. Although much work is being done in this area, a practical controlled fusion energy system has not yet been achieved. Present results indicate, however, that commercial utilization of the fusion reaction is probably feasible. If this proves correct, early in the next century we may have available a nearly limitless source of energy.

## 27.19 Nuclear models

Research into the structure of the nucleus convinces us that we can describe it as a combination of neutrons and protons held together by the nuclear force. The situation is more complex than that of electrons within the atom because the particles are closely packed within the nucleus. One simple nuclear model, called the *liquid drop model*, pictures the nucleus as resembling a water droplet; the nucleons in the nucleus are analogous to the molecules in the droplet. This representation is useful for large nuclei that contain many nucleons. In particular, the model of nuclear fission that we showed in Fig. 27.12 is based on the similarity between the nucleus and a droplet.

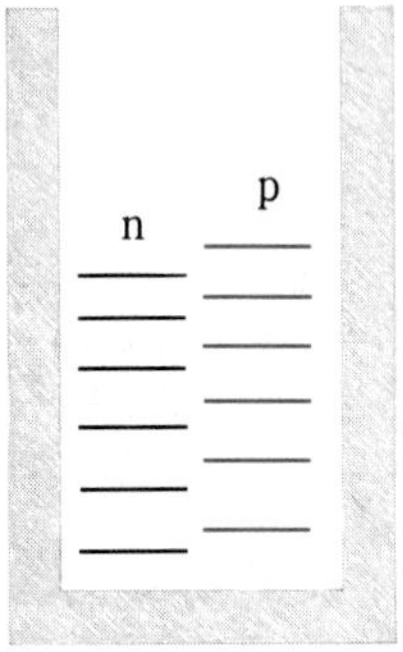

However, to explain the nuclear energy levels and other known properties of nuclei, a more detailed model is required. To begin, let us assume simply that the nucleons occupy certain energy levels within the nucleus. Both neutrons and protons are spin $\frac{1}{2}$ particles and obey the exclusion principle. But because the particles are different, the protons and neutrons will each occupy their own set of energy levels. The situation is sketched schematically in Fig. 27.15. Notice that the proton energy levels (color) are higher in energy than are the neutron levels. This results from the fact that the protons repel each other, and so less energy is required to remove them from the nucleus.

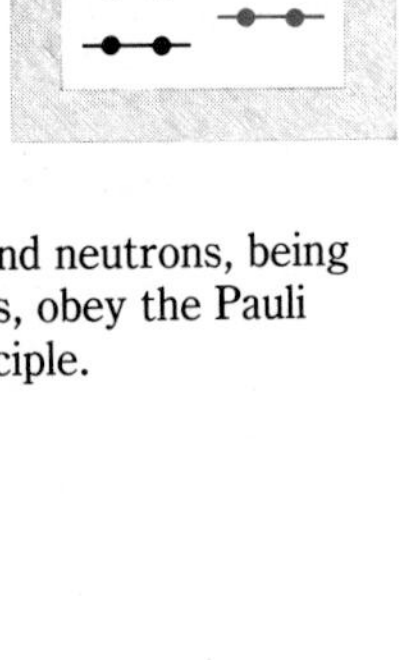

Consider now a nucleus that contains four protons and five neutrons. Its energy levels fill as shown in Fig. 27.16. Because of the exclusion principle, only two particles can exist in a given level. We see at once that a particle that is alone in a level, like the last neutron in Fig. 27.16, is at a comparatively high energy. Just as the single electron in the outer shell of a univalent atom is most easily removed, so too are single nucleons in a level most easily dislodged. Experiment shows that so-called even-even nuclei (those that contain an even number of protons and an even number of neutrons) are most stable; they have no solitary nucleons in high levels. Next in stability are the even-odd nuclei, while odd-odd nuclei are least stable. You can see this from the graph of stable nuclei in Fig. 27.4. Experiment shows that there are about 160 stable even-even nuclei, while there are only four stable odd-odd nuclei.

Figure 27.17 shows two possibilities for the structure of a nucleus that contains many nucleons. Because of the repulsion between protons, the proton energy levels are much higher than those for the neutrons. In nuclei such as this, the situation in *b* is highly unstable because many of the protons are at very high energies and are

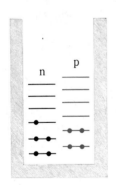

**FIGURE 27.15**
The nuclear energy levels for the protons are at higher energies than those for neutrons.

**FIGURE 27.16**
The protons and neutrons, being spin $\frac{1}{2}$ particles, obey the Pauli exclusion principle.

---

* If, as in the sun, the particles have a wide range of energies, reaction can occur between particles at the high-energy tail of the range at the center of the sun, where the temperature is about $10^7$ K.

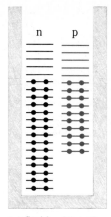
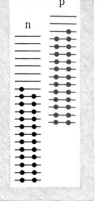

(a) Stable: $N \gg Z$     (b) Unstable: $N = Z$

**FIGURE 27.17**

Although both nuclei have the same number of nucleons,
the one in (a) is much more stable than the other. Why
are all the proton levels higher in (b) than in (a)?

easily dislodged from the nucleus. Thus we see that stable nuclei will take up the
configuration in $a$, a configuration in which the highest-energy neutrons and protons
have the same energy. But this means that, in a large stable nucleus, the number of
neutrons must be much larger than the number of protons. We therefore see more
clearly how the extra neutrons are needed to "dilute" the protons in a nucleus.

To proceed further in our understanding of nuclear structure, we must recognize
the following astonishing fact:

Each nucleon moves rather freely in a static field caused by the average
effects of the other nucleons.

At first, the concept that the nucleons move freely seems absurd; they are packed
quite closely together in the nucleus. But the Pauli exclusion principle must be taken
into account, and it leads to an astonishing conclusion. If we glance back at Fig.
27.17 $a$, we see that nearly all the nucleons are in energy levels that are hundreds or
thousands of electronvolts below any empty energy level. The motion and energy of
such a nucleon cannot change because all the states within its possible energy range
are filled. Hence it cannot have its motions changed by collisions with other nucleons, even though collisions occur frequently. (Two particles, upon collision, can
interchange their states, but the net effect is as though neither had been changed.)
In effect, the nucleons move as though no collision occurred; they behave as though
their movements were collision-free.

With these considerations in mind, it is possible to solve Schrödinger's equation
for a nucleon moving in a spherical potential well, that is, in a spherical box with soft
walls. As with the hydrogen atom, four quantum numbers are found. But the
predicted energy levels do not agree with experiment. Later work has shown that
there are strong magnetic effects within the nucleus as a result of the nucleon spins
and the orbital motions of the particles. When these effects are included, along with
other refinements, reasonable agreement is found between theory and experiment.

Finally, let us use the idea of a nucleon moving in a potential well to find, crudely,
the energy levels of the nucleus. To do this, replace the nucleus by a cubical box with
side length $L$. Then, as was done in Sec. 25.13 for a particle in a tube, we can obtain
the energies of the stationary states for a neutron. If we confine our attention to a

single dimension, we find that the energy levels will be the same as those found in Sec. 25.13, namely,

$$E_n = n^2 \frac{h^2}{8mL^2}$$

where in this case $m$ is the mass of a neutron. Using a typical nuclear diameter for $L$, about 4 fm, we find

$$\frac{E_n}{n^2} \approx 8 \text{ MeV}$$

Although this is at best a very crude estimate, it does point out that nuclear energy levels are in the million-electronvolt range. We conclude from this that the gaps between nuclear energy levels are of the order of millions of electronvolts. Since room-temperature thermal energies are only about $\frac{1}{40}$ eV, it is clear that nuclei cannot be excited by thermal collisions, except, possibly, at million-degree temperatures.

## Minimum learning goals

When you finish this chapter, you should be able to

1 Define (a) nucleon; (b) mass number and atomic number; (c) mass spectrometer; (d) isotope; (e) natural abundance; (f) atomic mass unit; (g) binding energy; (h) radioactive decay; (i) decay constant; (j) half-life; (k) $\alpha$-, $\beta$-, and $\gamma$-rays; (l) becquerel unit; (m) source activity; (n) gray unit; (o) fission reaction; (p) chain reaction; (q) fusion reaction.

2 Estimate the radius of a nucleus when its mass number is given.

3 Interpret symbols such as $^6_3\text{Li}$.

4 Explain why the mass of a nucleus in atomic mass units is nearly equal numerically to the mass number of the nucleus.

5 Sketch the graph of binding energy per nucleon versus $A$.

6 When given the mass of a nucleus, compute the binding energy of the nucleus.

7 Recall that 1 u of mass corresponds to 931.5 MeV of energy.

8 Sketch a graph of $N$ versus $t$ for a radioactive substance. Show the half-life on the graph. Compute the fraction of material that has not decayed after a length of time equal to a given integer number of half-lives.

9 Define each quantity in the equation $\Delta N = -\lambda N \Delta t$ and be able to use the equation in simple situations. Give the relation between $\lambda$ and the half-life.

10 Write the nuclear reaction equation for a given nucleus that emits one of the following: $\alpha$ particle, $\beta$ particle, $\gamma$-ray.

11 Prepare a diagram such as Fig. 27.8 for a series in which the starting nucleus and the emitted particles or rays ($\alpha$, $\beta$, and $\gamma$) are given.

12 Compare the range and ionization effect of $\alpha$, $\beta$, and $\gamma$ radiation passing through matter.

13 Explain why radiation can be harmful to people. In your explanation, point out which regions of the body and which types of persons should be particularly well shielded from radiation.

14 Explain, by reference to the mass-defect and binding-energy graphs, why the nuclear fission reaction should release energy. State what is meant by a fission chain reaction, and relate this to why $^{235}\text{U}$, but not $^{238}\text{U}$, is usable in a nuclear bomb.

15 Sketch a schematic diagram of a nuclear power reactor, showing fuel rods, moderator, control rods, heat exchanger, and output to turbine; and explain the function of each.

16 Explain by reference to the binding-energy graph, why the nuclear fusion reaction should release energy. State why fusion is much more difficult to achieve in a laboratory reactor than fission is.

# Questions and guesstimates

**1** Cobalt 60 is widely used as a source of $\gamma$-rays for radiation therapy for cancer. How many protons, neutrons, and electrons does one $^{60}_{27}$Co atom possess?

**2** Why do chemists consider different isotopes to be the same element even though their nuclei are not the same?

**3** Would the optical spectra of $^{235}$U and $^{238}$U atoms differ in any major way?

**4** Estimate the atomic mass of $^{64}_{30}$Zn from the fact that the binding energy per nucleon for it is about 8.7 MeV.

**5** Tritium is the $^3$H isotope of hydrogen. Its atomic mass is 3.016 u; the atomic mass of $^1$H is 1.0078 u, and that for the neutron is 1.00867 u. What do you predict about the stability of tritium? Repeat for $^2$H, deuterium, which has an atomic mass of 2.0141 u.

**6** A certain metal decays to a stable element by emission of $\alpha$ particles whose energy is about 9 MeV. A tiny sphere of the pure metal is mounted on the end of a pin. Describe how you would find the half-life of the metal if its half-life is about (*a*) five days and (*b*) 2000 years.

**7** A beam of $\alpha$ particles is absorbed in a block of lead.

What happens to the $\alpha$ particles? Rutherford proved the nature of $\alpha$ particles by heating the irradiated lead.

**8** Radon gas, being radioactive, is a dangerous contaminant of air. Because radon seeps into houses from the earth below the house, what factors lead to dangerous radon levels?

**9** What is the source of the earth's helium gas?

**10** A piece of uranium 235 that is smaller than the critical mass may explode if it is placed in a vat of water. Explain. Why may $^{235}$U in the form of a wire not explode, even though the mass of wire exceeds the critical mass?

**11** Most radiologists feel that women beyond childbearing age can safely be exposed to much more x-radiation than can young women. How can they justify such an opinion?

**12** It is possible for a man working with x-rays to burn his hand so seriously that he must have it amputated, and yet suffer no other consequences. However, an x-ray overexposure so slight as to cause no observable damage to his body could cause one of his subsequent offspring to be seriously deformed. Explain why.

# Problems

**1** Evaluate the following quantities for the $^{12}_6$C nucleus: (*a*) nuclear charge, (*b*) number of neutrons, (*c*) approximate radius, (*d*) density.

**2** Find the following quantities for the $^{206}_{82}$Pb nucleus: (*a*) number of nucleons, (*b*) number of protons, (*c*) number of neutrons, (*d*) radius, (*e*) density.

■ **3** The earth's radius is $6.3 \times 10^6$ m, and its density is about 5.5 g/cm$^3$. If the earth were to shrink until its density was about equal to the density within a nucleus ($2 \times 10^{17}$ kg/m$^3$), how large would the earth's radius be?

■ **4** The total mass of the observable universe is thought to be of the order of $10^{54}$ g. How large would the radius of the universe be if it were compressed until its density was $2 \times 10^{17}$ kg/m$^3$, the density within nuclei? Find the ratio of this radius to the radius of the sun, $7 \times 10^{10}$ cm.

**5** In a certain mass spectrometer, a velocity selector is used to obtain a beam of ions with $v = 3.0 \times 10^5$ m/s. Find the radius of the path that a carbon 12 ion (univalent) will

follow when $B$ within the spectrometer is 0.070 T.

**6** For the spectrometer described in Prob. 5, what will be the difference in the radii of the paths followed by a $^{12}$C and a $^{14}$C ion?

■ **7** A certain mass spectrometer accelerates ions through 2000 V and deflects them in a magnetic field of 0.080 T. When a certain univalent ion is examined in the spectrometer, it follows a path with $r = 13.0$ cm. What is the mass of the ion in both kilograms and atomic mass units?

■ **8** In a mass spectrometer like that in Fig. 27.1, it is found that $r$ for $^{12}$C is 9.00 cm. How large would $r$ be for $^{16}$O? Assume identical charges and accelerating potentials.

**9** Potassium found in nature contains essentially only two isotopes. One has an atomic mass of 38.964 u and constitutes 93.3 percent of the whole. The other, constituting 6.7 percent, has a mass of 40.975 u. From these data, compute the atomic mass that the chemists list in the periodic table.

**10** The following isotopes of neon are found in nature:

| Isotope | Abundance, % |
|---------|--------------|
| $^{20}$Ne | 90.9 |
| $^{21}$Ne | 0.3 |
| $^{22}$Ne | 8.8 |

Find the approximate atomic mass of Ne as given in the periodic table.

■ **11** Uranium found on earth has two principal isotopes, $^{238}$U and $^{235}$U. Find the approximate percentages of each from the fact that the masses of the two isotopes are 238.051 and 235.044 u, while the chemical mass is 238.030 u.

■ **12** Carbon found in nature has two principal isotopes; $^{12}$C, which has an atomic mass of 12.00000 u, constitutes 98.892 percent of the whole. The atomic mass determined by chemists is 12.01115 u. What is the atomic mass of the other isotope?

**13** Use Fig. 27.3 to state how much mass is lost as a krypton 84 nucleus is assembled from free protons and neutrons. What is the percentage mass loss?

**14** Use Fig. 27.3 to state how much energy is required to tear a barium 138 nucleus apart into free neutrons and protons. To how much mass (in atomic mass units) is this energy equivalent?

■ **15** Compute the binding energy per nucleon for the $^{12}$C nucleus. *Hint:* Recall that the mass of the *atom* is exactly 12 u.

■ **16** Compute the binding energy per nucleon for the $^{20}$Ne nucleus. Its atomic mass is 19.99244 u.

■ **17** Use Fig. 27.3, together with the proton and neutron masses, to find the mass of an *atom* of oxygen 16.

■ **18** The atomic mass of $^{16}$O is 15.994915 u, while that for $^{17}$O is 16.999133 u. Use these data to find the binding energy of the extra neutron in the $^{17}$O nucleus.

**19** A Geiger counter placed above a radioactive sample registers 628 counts per minute. How many counts per minute will it register after four half-lives have passed?

■ **20** A sample gives 820 counts per minute at one time; 36 h later, it has a count rate of 102 counts/min. What is its half-life?

**21** A sample that contains $4.2 \times 10^{11}$ atoms has a half-life of 0.8 yr. (*a*) What is its decay constant? (*b*) How many nuclei in it undergo decay in 30 s?

**22** A tiny ampoule of radon gas contains $7.0 \times 10^{12}$ atoms of radon. The half-life of radon is 3.8 days. How many disintegrations occur in the ampoule each second?

**23** Watches with numerals that are visible in the dark sometimes have radioactive material in the paint used for the numerals. A student estimates from measurements using a Geiger counter that 700 disintegrations occur on the watch face each second. How many curies of radioactivity exist on the watch if the student's figures are correct?

**24** A tiny piece of rock is radioactive, and a Geiger counter placed above it registers 187 counts in 1 min. Assuming that the counter intercepts radiation from half of the decaying nuclei, what is the activity of the rock?

■ **25** Uranium 238 has a half-life of $4.5 \times 10^9$ yr. What is the activity of a 1-g sample of pure $^{238}$U?

■ **26** The half-life of $^{60}$Co, a radioactive element produced in reactors for use as a medical and commercial radiation source (1.33- and 1.17-MeV $\gamma$-rays), is 5.3 yr. How many atoms of $^{60}$Co are there in 1 mg of the material? What is the decay constant for the material? How many disintegrations occur each second in 1 mg of the material?

■ **27** What fraction of a radioactive sample decays in 80 yr if the half-life of the material is 140 yr?

■ **28** Strontium 90 is a radioactive fission product from nuclear fission reactors and bombs. Since its half-life is quite long (about 28 yr, or $8.8 \times 10^8$ s), it is a persistent contaminant and presents serious disposal problems. What fraction of the original strontium still remains 70 yr after a nuclear bomb explodes?

■ **29** A sample contains $N_1$ nuclei with decay constant $\lambda_1$ and $N_2$ nuclei with decay constant $\lambda_2$. What is the total decay constant of the sample in terms of $\lambda_1$, $\lambda_2$, $N_1$, and $N_2$?

■ **30** Measurements show that only 18 percent of a radioactive material remains after 12.0 h. What is the half-life of this material?

**31** Polonium 210 ($^{210}_{84}$Po) decays by emitting a 5.30-MeV $\alpha$ particle and an 0.80-MeV $\gamma$ ray. What is the resultant isotope? Repeat for $^{209}_{82}$Pb, which emits a $\beta$ particle.

**32** Bismuth 211 ($^{211}_{83}$Bi) decays by the emission of a 6.62-MeV $\alpha$ particle. What is the resultant isotope? Repeat for $^{223}_{86}$Rn, which emits a $\beta$ particle.

■ **33** Radon 220 emits a 0.54-MeV $\gamma$ ray. By what fraction does its nuclear mass change in the process?

■ **34** Thorium 226 emits a 1.11-MeV $\gamma$ ray. By what fraction does its nuclear mass change in the process?

■ **35** The *thorium series* mentioned in Table 27.1 starts with $^{232}_{90}$Th and emits in succession one $\alpha$, two $\beta$, four $\alpha$, one $\beta$, one $\alpha$, and one $\beta$ particle(s). Show that the final product of the series is that in the table.

■ **36** The *actinium series* mentioned in Table 27.1 starts with $^{235}_{92}$U and emits in succession one $\alpha$, one $\beta$, two $\alpha$, one

$\beta$, three $\alpha$, two $\beta$, and one $\alpha$ particle(s). Make a diagram of the series similar to the one in Fig. 27.8.

■ **37** Uranium 238 has a half-life of about $4.5 \times 10^9$ yr and decays according to the reaction

$$^{238}_{92}\text{U} \quad \rightarrow \quad ^{234}_{90}\text{Th} \quad + \quad ^{4}_{2}\text{He} \quad + \text{energy}$$

238.05077     234.04358     4.00260

The isotopic masses (including electrons) are also given. If all the energy becomes kinetic energy of the $\alpha$ particle, what will its energy be in millions of electronvolts? Its actual energy is 4.19 MeV. How can you account for the discrepancy?

■ **38** One step in the radioactive series of Fig. 27.8 was

$$^{222}_{86}\text{Rn} \quad \rightarrow \quad ^{218}_{84}\text{Po} \quad + \quad ^{4}_{2}\text{He} \quad + \text{energy}$$

222.01753     218.00893     4.00260

The isotopic masses (including electrons) are given below the reaction. If all the energy is given to the $\alpha$ particle, what should be its energy in millions of electronvolts? The observed energy is 5.49 MeV for the fastest $\alpha$ particle. Where might the rest of the energy have gone?

■ **39** Consider the following reaction:

$$^{1}_{1}\text{H} \quad + \quad ^{13}_{6}\text{C} \quad \rightarrow \quad ^{13}_{7}\text{N} \quad + \quad ^{1}_{0}\text{n}$$

1.007825     13.00336     13.00574     1.008665

where n represents a neutron and the atomic masses of the particles (including electrons) are given below each reactant. Can this reaction be initiated by a proton ($^{1}_{1}\text{H}$) that has kinetic energy of 1.80 MeV?

■ **40** From an energy standpoint, is the following reaction possible if the incident proton has a kinetic energy of 1.48 MeV?

$$^{1}_{1}\text{H} \quad + \quad ^{7}_{3}\text{Li} \quad \rightarrow \quad ^{7}_{4}\text{Be} \quad + \quad ^{1}_{0}\text{n}$$

1.007825     7.01600     7.01693     1.008665

The masses indicated include the atomic electrons.

■ **41** When polonium 210 decays by $\alpha$-particle emission, it emits an 0.80-MeV $\gamma$-ray along with the 5.30-MeV $\alpha$ particle. The reaction is

$$^{210}_{84}\text{Po} \rightarrow \quad ^{4}_{2}\text{He} \quad + \quad ^{206}_{82}\text{Pb} \quad + \gamma$$

?     4.00260     205.97447

with the atomic masses shown. (*a*) Knowing the kinetic energy of the $\alpha$ particle to be 5.30 MeV, find the approximate recoil energy of the lead atom. (*b*) Calculate the

expected atomic mass for $^{210}\text{Po}$. The measured mass is 209.9829 u.

■ **42** Suppose 1 kg of deuterium (heavy hydrogen, $^2\text{H}$) is combined to form 1 kg of helium according to the reaction

$$^{2}_{1}\text{H} \quad + \quad ^{2}_{1}\text{H} \quad \rightarrow \quad ^{4}_{2}\text{He}$$

2.0141     2.0141     4.0026

where the atomic masses are given. (*a*) How much energy (in joules) is liberated? (*b*) If the confined helium has a specific heat capacity of 0.75 cal/g $\cdot$ C$^\circ$, by how much does its temperature increase as this energy is added to it?

■ **43** Iodine 131 is used to treat thyroid disorders because, when ingested, it localizes in the thyroid gland. Its half-life is 8.1 days. What is the activity of 0.70 $\mu$g of $^{131}\text{I}$?

■ **44** Phosphorus 32 ($^{32}\text{P}$) has a half-life of 14.3 days and is used in medicine because it tends to localize in bone. What is the activity of 0.85 g of $^{32}\text{P}$?

■ **45** How many grams of iron 59 ($^{59}\text{Fe}$) are there in a 1-$\mu$Ci sample of it? Its half-life is 46.3 days.

■ **46** The isotope tritium, $^{3}_{1}\text{H}$, has a half-life of 4600 days. How many grams of tritium are there in a sample that has an activity of 2.0 $\mu$Ci?

■ **47** How much is the temperature of water raised when the water is given a radiation dose of 9.0 mGy?

■ **48** How large a radiation dose must be deposited in lead to raise its temperature 5°C? For lead, $c = 0.031$ cal/g $\cdot$ C$^\circ$.

■ **49** In an effort to date a piece of bone, its count rate due to $^{14}\text{C}$ is computed and found to be only 0.053 that of a similar fresh piece of bone. Estimate the age of the bone. The half-life of $^{14}\text{C}$ is 5700 years.

■ **50** Thorium 232 has a half-life of $1.39 \times 10^{10}$ years and decays through a number of steps to $^{208}\text{Pb}$. The ratio of $^{208}\text{Pb}$ to $^{232}\text{Th}$ in a certain rock is found to have a value of 0.13. Estimate the time since the rock solidified.

■ **51** One isotope of hydrogen, $^{3}_{1}\text{H}$, or tritium, has a half-life of 12.3 years. It is produced in the upper atmosphere by cosmic rays and is intimately mixed with the hydrogen in the air. In order to determine the age of a bottle of wine found in an ancient cave, the tritium in the wine is measured and found to be 7.3 percent that found in new wine. What is the approximate age of the wine?

■ **52** A neutron with a speed of $10^6$ m/s hits a stationary deuterium atom (heavy hydrogen, $^2_1\text{H}$) head on in a perfectly elastic collision. (*a*) Find the speed of the neutron after collision. (*b*) Repeat if the deuterium atom is replaced by an oxygen atom, $^{16}_{8}\text{O}$. Notice that small-mass nuclei are most effective in slowing neutrons.

■ **53** Neutrons are best slowed by collision with particles of equal mass. Suppose a neutron with speed $2 \times 10^7$ m/s strikes a free, stationary proton head on. What will be the final speed of the neutron? Repeat if the neutron collides elastically with a free, stationary gold atom.

■ **54** (*a*) If the $^{235}$U fission process yields an energy of 200 MeV, how much energy will be given off by the fission of 1 g of uranium 235? (*b*) If energy costs 10 cents per kilowatthour, what is the cost of the energy found in *a*?

■ **55** In Sec. 27.16 we stated that one possible fission process for the compound nucleus $^{236}$U * is

$$^{236}\text{U} * \rightarrow \, ^{140}_{56}\text{Ba} + \, ^{92}_{36}\text{Kr} + 4\text{n} + \text{energy}$$

The product isotopes decay in several steps by $\beta$ emission. The decays can be summarized as follows:

$$^{140}_{56}\text{Ba} \rightarrow \, ^{140}_{58}\text{Ce} + 2\beta + \text{energy}$$
$$^{92}_{36}\text{Kr} \rightarrow \, ^{92}_{40}\text{Zr} + 4\beta + \text{energy}$$

Find the total energy released when a uranium 235 nucleus undergoes this form of fission. Pertinent atomic masses are $^{235}$U = 235.043915 u, $^{140}$Ce = 139.90539 u, $^{92}$Zr = 91.90503 u.

■ **56** A nuclear-power-station reactor using $^{235}$U as fuel has an output of $10^8$ W. How much uranium is consumed per hour if the energy released in the fission of one $^{235}$U nucleus is 200 MeV and the reactor has an overall efficiency of 10 percent?

■ **57** In a typical nuclear power plant, the reactor core produces 1000 MW. If the $^{235}$U core material is reduced by 25 percent in 5 years, how much $^{235}$U existed originally in the core? Assume that 200 MeV is released per fission process.

■ **58** (*a*) Use the data on the inside cover of this text to compute how much energy the sun radiates in a year. (*b*) Assume that the sun generates its energy in the way outlined in Sec. 27.18. What fraction of its mass does the sun lose by this process each year?

■ **59** One possible reaction on which a fusion reactor can be based is the following:

$$^2_1\text{H} + \, ^3_1\text{H} \rightarrow \, ^4_2\text{He} + \, ^1_0\text{n}$$

where $^3_1$H is tritium. How many grams of deuterium and tritium must fuse each second to yield 500 MW of power? Pertinent atomic masses are $^2_1$H = 2.014102 u, $^3_1$H = 3.016050 u, $^4_2$He = 4.002604 u.

# 28 The physics of the very large and very small

ATOMS are composed of electrons, protons, and neutrons. Is it possible that still smaller particles exist? Are the proton and the neutron, and perhaps even the electron, divisible? We investigate these questions in this chapter. In addition, we take a look at the early history of the universe. As we shall see, the tiniest particles influence how our universe behaved at its birth; they also may decide its eventual fate.

## 28.1 The ancient history of particle physics

Many, many years ago, in the early 1930s, people spoke of what they called "elementary" or "fundamental" particles. These were the electron, proton, and neutron. Scientists knew that they could describe the smallest pieces of matter then known, atoms, in terms of them. There appeared to be no need to invent or seek other particles because nothing really basic in nature seemed to remain unexplained. Physicists are inquisitive people, however, and refused to believe that there was nothing left to discover. After all, there were a few troubling questions that needed answers.

One question involved the nature of cosmic rays. These are particles with extremely high energy that are hurled toward the earth by unknown forces outside our solar system. Most of these particles are protons. As the cosmic rays hurtle through the atmosphere, however, showers of $\gamma$-ray photons and other particles are formed

691

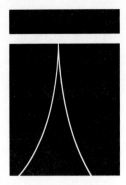

**FIGURE 28.1**

A high-energy γ-ray generates an electron-positron pair in a cloud chamber.

by collision. In 1932, C. D. Anderson observed the strange result shown in Fig. 28.1 as cosmic rays streaked through a lead plate placed in a cloud chamber. A γ-ray entering the chamber at the top underwent collision in the plate and created two particles that left the plate at the bottom. Because of the way the particles were deflected by a magnetic field perpendicular to the page, it was obvious that the two particles had opposite charges. Analysis showed them to be a common electron ($-e$) and what we now call a **positron,** a positive electron ($+e$). Thus a new "fundamental" particle was discovered.*

At about this same time, a need was found for still another particle, the neutrino. As we saw in our discussion of beta decay (Sec. 27.9), the neutrino was "invented" to preserve the laws of conservation of momentum and energy. Because its mass is essentially zero, it travels close to (or at) the speed of light. In 1935, H. Yukawa presented a theory of the nuclear force that required still another particle, a particle with mass intermediate between that of the electron and the proton.

These discoveries and conjectures stimulated many other research ideas. At the time, however, the world was in the midst of the Great Depression; physicists had trouble finding jobs, to say nothing of obtaining research funds. Moreover, in those ancient times physicists were few in number.

Then, in the late 1930s, World War II began. Physicists were diverted from peaceful pursuits to war-related research, research that culminated in all manner of discoveries, including sonar, radar, new materials, nuclear reactors, and nuclear weapons. During the war years, physics changed from a relatively obscure science conducted on shoestring financing in universities (and a handful of industrial labs) to a government-supported effort of huge scale. In the late 1940s, physicists began once again to pursue research topics close to their hearts, but now they had the funding for projects they had previously considered impossible. Thus started the great advances in materials science (lasers, solid-state electronics, and so on) and in nuclear and particle physics that have so greatly expanded our understanding of nature.

## 28.2 Particle accelerators

In order to examine the internal structure of the nucleus, very-high-energy probe particles are required. This is easily understood if we recall that any particle we shoot at a nucleus undergoes diffraction effects because of its wave nature. As we pointed out in Sec. 24.9, details smaller than the wavelength cannot be discerned. Because the diameter of a nucleus is of the order of $10^{-15}$ m, the probe particle must have a de Broglie wavelength much shorter than this if it is to show details within the nucleus. Let us now see what probe-particle energies this requires.

The probe particle will be traveling at relativistic speed, and so its kinetic energy is

$$\text{KE} = mc^2 - m_o c^2$$

For such a very-high-energy particle, $m \ggg m_o$ and $v$ is very close to $c$. Hence the momentum $p$ of the probe particle is essentially $mc$ and the kinetic energy is

$$\text{KE} = pc - m_o c^2 \cong pc \tag{28.1}$$

---

* P. A. M. Dirac had predicted the existence of such a particle in a theory presented in 1930.

Since the de Broglie wavelength is $\lambda = h/p$, we have

$$\lambda \cong \frac{hc}{\text{KE}} = \frac{1.24 \times 10^{-6}}{\text{KE (eV)}} \qquad \text{m} \tag{28.2}$$

To obtain a probe-particle wavelength of $10^{-16}$ m, for example, the particle must have an energy of about $1 \times 10^{10}$ eV. As you see, particles used for probing the nucleus must have energies in the range of 10 billion electronvolts (10 GeV).

We can obtain such high-energy particles by accelerating them through very large potential differences. Although a steady potential difference of a few million volts is relatively easy to obtain and maintain, at much higher potential differences, sparking between the electrodes cannot be prevented. To circumvent this difficulty, more complicated devices for obtaining high-energy particles have been invented. Only two basic designs have survived the quest for higher and higher energies. We shall describe present-day versions of these two types of accelerator.

### The linear accelerator

In the linear accelerator, a long series of metal tubes (in vacuum) is used, as sketched in Fig. 28.2. The particle to be accelerated is injected at the left end with a speed near the speed of light. (An auxiliary accelerator is usually used to provide this initial energy.) Alternating voltage differences between the tubes are adjusted so that the

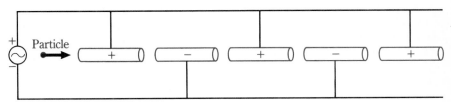

**FIGURE 28.2**

In a linear accelerator, the voltages on the acceleration tubes reverse continually in such a way that the particle is accelerated each time it travels from one tube to the next.

particle is accelerated each time it passes from one tube to the next. The final energy of the particle is determined by the voltage between the tubes and by the number of tubes in the series. Financial considerations place a limit on the ultimate energy of such devices. The Stanford linear accelerator, shown in Fig. 28.3, is capable of accelerating electrons to energies of $30 \times 10^9$ eV (30 GeV). Its cost (in the early 1970s) was about \$100 million. Recently it has been redesigned to produce 50-GeV electron and positron beams that can be brought into collision with each other.

### The circular accelerator

Another way of achieving high energies is to have a particle move in a circular path, falling across the same potential difference each time it goes around the circle. The particle is held in the circle by a magnetic field. (Recall that the field simply furnishes the required centripetal force without changing the energy of the particle.) Two

**FIGURE 28.3**

The Stanford University linear accelerator. By the time an electron has traveled about 3 km through the particle tube, its energy is 30 GeV. An aerial view shows the size of the installation more clearly. (*Stanford Linear Accelerator Center, Stanford University*)

Particle tube

Support

such machines are shown in Fig. 28.4. Both have diameters of about 2 km and accelerate protons to close to 500 GeV. Even though huge superconducting magnets placed around the circle are used to furnish the magnetic field, they are not capable of holding the high-energy protons in a circle smaller than this. At the present time, devices of this type with 500-GeV particles circulating in opposite directions in two adjacent rings are near operation. The colliding beams will, in effect, give collision energies of 1 TeV ($10^{12}$ eV). Machines of still higher energy are in the design and early construction phase.

As you might expect, such machines cost billions of dollars. Because individual nations find it difficult to finance such projects, several have banded together to finance the construction and operation of these expensive tools of pure research. Research in particle physics is an international endeavor that, hopefully, will further promote friendship and cooperation among nations.

## 28.3 Particles within particles: quarks

By the early 1960s, near chaos existed in the field of particle physics. High-energy accelerators made possible experiments in which particle was hurled against particle with ever-increasing energy. The ensuing collisions generated many new parti-

**FIGURE 28.4**

(*a*) An aerial view of the beneath-ground circular accelerator and its associated facilities at Fermilab in Batavia, Ill. (*Courtesy of Fermilab*) (*b*) The interior of the circle at a similar accelerator at CERN in Geneva, Switzerland. (*Courtesy of CERN*)

cles with often exotic properties. Subsequently, by shooting high-energy electrons at protons and neutrons, it was found that these supposedly fundamental particles have structure. Much of the volume of the neutron and the proton appears to be empty space.

Another fascinating discovery concerning particles is that each particle has a corresponding **antiparticle.** For example, the positron is the antiparticle of the electron. Its charge is opposite that of the electron, although its mass is the same. When an electron comes near to a positron, the particle and antiparticle annihilate each other to form $\gamma$-rays. Similarly, an antiproton exists; it can annihilate a proton. Though the neutron has no charge, it, too, has an antiparticle, the antineutron; this particle can be recognized by the fact that a neutron annihilates it. Apparently each particle in nature has a corresponding antiparticle. We represent an antiparticle by placing a bar above its symbol: $\bar{p}$ represents an antiproton, for example, and an antineutrino is represented by $\bar{\nu}$.

An antiparticle has the same mass, spin, and lifetime as its corresponding particle, but they have opposite charge. Each can annihilate the other.

A degree of order was brought to the situation in 1963 when Murray Gell-Mann and George Zweig independently postulated the existence of still other particles, particles that Gell-Mann named "**quarks.**" Although these particles have thus far not been isolated, physicists have become increasingly convinced that they exist. By using them, it is possible to give a unified description of many previously unrelated particles discovered in high-energy collisions. Let us describe what is known about quarks and how they fit into nature's scheme of things.

Although Zweig and Gell-Mann originally thought that there were only three types of quarks, we now believe that there are at least six. Their names, rest masses, and charge are given in Table 28.1. Notice that quarks possess a charge less than $e$. Thus, if these particles truly exist, charge does not always come in integer multiples of $e$. So far, despite an extensive search, no charge smaller than $e$ has been confirmed to exist in isolation. Other evidence, however, indicates that quarks exist and have the following properties.

The radius of a quark appears to be less than $\frac{1}{1000}$ that of the neutron or proton. Each quark has a spin of $\pm\frac{1}{2}$, the same as for the electron, proton, and neutron. Therefore quarks must obey the Pauli exclusion principle. Each of the six quarks has a corresponding antiquark.

■ **TABLE 28.1**
**Quarks***

| Name | Mass | Charge |
|---|---|---|
| Up (u) | $0.005\,m_p$ | $+\frac{2}{3}e$ |
| Down (d) | $0.009\,m_p$ | $-\frac{1}{3}e$ |
| Strange (s) | $0.21\,m_p$ | $-\frac{1}{3}e$ |
| Charm (c) | $1.60\,m_p$ | $+\frac{2}{3}e$ |
| Top (t) | $\sim 43\,m_p$ | $+\frac{2}{3}e$ |
| Bottom (b) | $5.33\,m_p$ | $-\frac{1}{3}e$ |

\* The mass values are still very uncertain.

Although we shall return to this subject later, let us see how the neutron and proton are viewed in terms of quarks. The following compositions are found to explain their observed properties:

$$\text{Proton composition} = u + u + d$$
$$\text{Proton charge} = (\tfrac{2}{3} + \tfrac{2}{3} - \tfrac{1}{3})e = e$$
$$\text{Proton spin} = \pm(\tfrac{1}{2} - \tfrac{1}{2} + \tfrac{1}{2}) = \pm\tfrac{1}{2}$$
$$\text{Neutron composition} = u + d + d$$
$$\text{Neutron charge} = (\tfrac{2}{3} - \tfrac{1}{3} - \tfrac{1}{3})e = 0$$
$$\text{Neutron spin} = \pm(\tfrac{1}{2} + \tfrac{1}{2} - \tfrac{1}{2}) = \pm\tfrac{1}{2}$$

Notice, however, that the quark masses given in Table 28.1 add to give a mass that is less than the observed mass of the proton and neutron. The remaining mass must be locked in the force field that holds the quarks together. We discuss this force field more fully later.

## 28.4 Leptons versus hadrons

As far as we have been able to tell from experiment, electrons are essentially point particles. Their radius, if not zero, is smaller than $10^{-18}$ m. (The de Broglie wavelength of a $1 \times 10^{12}$ eV particle is about $10^{-18}$ m. We cannot see detail smaller than about $10^{-18}$ m with presently existing or proposed accelerators.) The same may be said about the positron. Another particle with a size too small to detect is the neutrino. In all, six pointlike particles (and six antiparticles) appear to exist, and they are called **leptons.** They have diameters less than $10^{-18}$ m. In addition,

Leptons do not respond to the strong nuclear force.

They do respond, however, to the gravitational and electromagnetic forces. In addition, they are subject to another force, mentioned later, called the *weak force*. Table 28.2 lists the presently known leptons. Each particle listed has an antiparticle. The neutrino mass is, at best, very small and is commonly assumed to be zero.

■ **TABLE 28.2 Leptons***

| Name | Mass $\div m_e$ | Charge $\div e$ |
|---|---|---|
| Electron ($e^-$) | 1 | 1 |
| Electron neutrino ($\nu_e$) | 0 | 0 |
| Negative muon ($\mu^-$) | 207 | $-1$ |
| Muon neutrino ($\nu_\mu$) | 0 | 0 |
| Negative tau ($\tau^-$) | 3490 | $-1$ |
| Tau neutrino ($\nu_\tau$) | 0 | 0 |

* Some theories predict that the neutrino has nonzero mass. Present experiments show that the mass of $\nu_e$ must be less than $3 \times 10^{-5}\ m_e$.

A multitude of other particles exist that, unlike leptons, respond to the strong nuclear attractive force. These particles are called **hadrons.** Among them are the proton and the neutron. An abbreviated listing of them is given in Table 28.3. The hadrons are subdivided into two groups: the **baryons** have spins of $\frac{1}{2}$ and $\frac{3}{2}$ and obey the Pauli exclusion principle, and the **mesons** have spins of 0 and 1 and are not subject to the exclusion principle. Notice that the free, isolated neutron is unstable and has a half-life of only about 635 s. However, when it is within the nucleus, the neutron appears to be stable, except for radioactive $\beta$-decay.

All hadrons have a measurable size and are thought to be composed of quarks. We have already seen how combinations of quarks can be used to represent the proton and neutron. In fact, all spin $\frac{1}{2}$ hadrons, the baryons, must contain three quarks; only in that way can the quarks yield the observed spin of the baryon. The mesons listed in Table 28.3, however, contain only two quarks. The spins of the two quarks that compose them are opposite in direction and yield a particle spin of zero. For example, the $\pi^+$ particle is a u d̄ combination.

■ **TABLE 28.3** **Hadrons**

| Family | Name | Symbol | Mass ÷ $m_p$ | Half-life (seconds) | Spin |
|---|---|---|---|---|---|
| Baryons | Proton | p | 1.000 | $\sim\infty$ | $\frac{1}{2}$ |
| | Neutron | n | 1.001 | 635 | $\frac{1}{2}$ |
| | Lambda | $\Lambda^\circ$ | 1.190 | $1.82 \times 10^{-10}$ | $\frac{1}{2}$ |
| | Sigma | $\Sigma^+$ | 1.268 | $5.5 \times 10^{-9}$ | $\frac{1}{2}$ |
| | | $\Sigma^\circ$ | 1.271 | $4 \times 10^{-20}$ | $\frac{1}{2}$ |
| | | $\Sigma^-$ | 1.276 | $1.02 \times 10^{-10}$ | $\frac{1}{2}$ |
| | Plus many particles that are more massive, some with spins of $\frac{3}{2}$. | | | | |
| Mesons | Pion | $\pi^+, \pi^-$ | 0.149 | $1.80 \times 10^{-8}$ | 0 |
| | | $\pi^\circ$ | 0.142 | $5.5 \times 10^{-17}$ | 0 |
| | Kaon | $K^+$ | 0.526 | $8.6 \times 10^{-8}$ | 0 |
| | | $K^\circ_s$ | 0.530 | $6.2 \times 10^{-11}$ | 0 |
| | | $K^\circ_L$ | 0.530 | $3.6 \times 10^{-8}$ | 0 |
| | Eta | $\eta^\circ$ | 0.585 | $5 \times 10^{-19}$ | 0 |
| | Plus many particles that are more massive, some with spins of 1. | | | | |

Many more hadrons than those listed in Table 28.3 have been found. All can be explained as quark combinations in the ground state or excited states. It is important to recognize that in such composite particles the energies between excited states are very large. (As you will see in Prob. 9 at the end of this chapter, a particle such as a d quark confined to a tube $10^{15}$ m long requires an energy of 120 GeV to raise it from the ground state to the first excited state.) Because the energy equivalent of 1 u is only 0.931 GeV, the excited states of the same quark combination must give rise to seemingly different particles of distinctly unlike mass. Thus we see that many types of hadrons are expected as our accelerating machines reach higher energies. It remains to be seen whether the quark model will be able to explain new hadrons found in the future.

## 28.5 Fundamental forces

As far as is known today, there are four fundamental forces, that is, ways in which particles interact. They are as follows:

*The gravitational force:* The attractive force that acts on all forms of mass and energy. (Remember, mass is a form of energy.)

*The electromagnetic force:* It includes both the electrostatic coulomb force and the magnetic-field force. Maxwell was the first to show that the two forces are related. They are, in reality, only a single force.

*The strong nuclear force:* The attractive force that acts between hadrons; it holds the neutrons and protons together in the nucleus. As we shall see, it is actually the result of the attractive force between the quarks that make up the hadrons.

*The weak force:* This force, much weaker than the strong nuclear force, is responsible for beta decay. It has been related to the electromagnetic force in a theory presented independently by Steven Weinberg and Abdus Salam in 1967. This theory is referred to as the *electroweak theory.*

One of the great quests of physicists is to find a way to relate these four forces to each other. A so-called *grand unified theory* (GUT) is under investigation by many people in an effort to unify the nuclear force with the Weinberg-Salam picture of the combined electromagnetic-weak force. Present GUT theories require that the proton be unstable with a lifetime so long that about one proton per year per 1000 kg of matter will decay. Thus far, this instability has not been found by experiment. No success in associating the gravitational force with the others has yet been achieved.

You are familiar with the fact that two of these forces have an associated field, the gravitational field for gravitation and the electromagnetic field for electromagnetism. Moreover, a traveling em field (an em wave) is representable as a stream of photons. Hence the em field has a particle associated with it, the photon. As you know, the photon possesses neither rest mass nor charge. There is no indication that the photon is a composite of other particles. It is postulated that a similar particle, called the *graviton,* is associated with the gravitational field. Because measurable gravitational waves cannot be generated in the laboratory, evidence for the graviton must be acquired through observation of waves sent out from explosions or other motions of stars and similar massive objects. Experiments to detect gravitational waves are now waiting for a wave of detectable amplitude to pass the earth. The graviton should have neither rest mass nor charge. Like the photon, it is probably not a composite of other particles.

It is believed that there is a particle associated with the weak interaction (or force). Indeed, the Weinberg-Salam electroweak theory predicts that three particles, termed *intermediate vector bosons,* are associated with the weak interaction. They are the $W^+$, $W^-$, and $Z^\circ$ particles, which should have charges $+e$, $-e$, and zero, respectively. Unlike the photon and the graviton, these three particles are predicted to have mass. The energy equivalent of their masses, according to theory, is about 80 GeV for the $W^+$ and $W^-$, and about 93 GeV for the $Z^\circ$. (For comparison, the energy equivalent of the neutron mass is about 0.94 GeV.) Collision energies of hundreds of gigaelectronvolts would be necessary to generate particles of this nature. In 1983, an accelerator capable of producing 540 GeV in collisions was constructed at Geneva, Switzerland, and employed to search for the intermediate vector bosons. Using it, the $W^+$, $W^-$, and $Z^\circ$ particles were found. Because of their instability, nonzero mass, and other factors we shall mention soon, there is a general suspicion that these are composites of more basic particles. For now, however, this is only a suspicion.

The fourth force in nature, the strong nuclear interaction, also has a particle associated with it. For many years a particle called the *pion* was believed to be the proper particle for this field. However, just as we suspect may happen for the W and Z particles, the pion has been replaced by a more basic particle in the description of the strong nuclear force field. We discuss this force and its associated particles in more detail in the next section.

## 28.6 The force between quarks: gluons

As we have pointed out before, the attractive force between nucleons is a short-range force; two nucleons must be closer than about $5 \times 10^{-15}$ m for the nuclear force to be felt strongly. In this respect, it differs markedly from the electrostatic and gravitational forces, both of which vary as $1/r^2$ and have essentially infinite range. The discovery that the neutron and proton are composed of quarks provides us with insight into the nature of the strong nuclear force.

Before we discuss the forces between quarks in detail, we present an analogy that will help us understand the nuclear force. The charges on the electrons and the nucleus of an atom give rise to strong electrostatic forces within the atom. Far from the neutral atom, the electric field due to it is zero; the long-range effects of the plus and minus charges cancel. However, very close to the edge of the atom, the electric field is no longer zero because the electron cloud does not have perfect spherical uniformity. Hence, if we had no other evidence for the electrostatic force than this, we would judge it to be a short-range force, rather than the force of infinite extent that it is. In analogy to this, we shall find that the force between nucleons is nothing more than the edge effects of the attractive force between the quarks within the nucleus.

Despite strenuous efforts to knock quarks from the neutron, proton, and other hadrons, a free quark has never been observed. We conclude from this that the attractive force between quarks is exceedingly strong and probably persists to infinite separations. It is speculated that the field due to a quark is independent of distance or decreases perhaps only as $1/r$, where $r$ is the distance from the quark. In either case, infinite work must be done to separate two quarks. Therefore, if the force between quarks varies in either of these ways, an isolated quark can never be found.

Despite that fact, we know that nucleons and other hadrons, all of which are combinations of quarks, exert negligible forces on each other except when they are close together. (We ignore for now the much weaker electromagnetic and gravitational forces that the particles exert on each other.) We are therefore led to believe that the three quarks that compose the proton (or other baryon) combine their fields to cancel each other outside the particle. Similarly, the two quarks in a meson must provide fields that cancel each other far from the meson. Even though the fields of individual quarks extend to infinity, certain combinations of quarks apparently give zero external fields. We now present an outline of the accepted theory of the fields due to quarks, a theory referred to as *quantum chromodynamics* (QCD).

Whereas electric charge comes in only two varieties, plus and minus, quarks appear to be more complicated. To explain the way the quark fields cancel external to the known hadrons, we must assume there are three types of quark "charge." The quark "charge" we are now talking about has nothing to do with electric charge, and so we shall never again call it "charge"; instead, we call it **color** or color charge. Each type of quark comes in three colors and three anticolors. (*Caution:* Color is only a label; it has nothing at all to do with vision.) In analogy with the visible colors, we take the three quark colors to be red, green, and blue (plus antired, antigreen, and antiblue). As with ordinary colors, red plus blue plus green yields white, which we take to be the cancellation of all color. Similarly, the combination of the three anticolors gives rise to cancellation. Moreover, a given color and its anticolor cancel.

With these rules in mind, it is easy to see why individual hadrons should exist. Take the neutron, for example. It consists of three quarks: udd. If each quark has a different color, then the three colors cancel and the neutron has zero color. Therefore the field due to the color charges of its constituent quarks is zero external to the neutron. Even though the individual quarks within the neutron feel strongly the forces due to their color charges, the color field outside the neutron is zero; quarks in other nearby particles experience no long-range force due to the confined quarks in the neutron.

Mesons, on the other hand, consist of only two quarks. Each of the two quarks within a meson must cancel the other's color. This means that a meson is composed of a red quark and an antired quark, or a green-antigreen pair, or a blue-antiblue combination. Although each of the two quarks within the meson is strongly attracted by the other, their canceling color fields exert negligible force on quarks in other particles more than about $10^{-14}$ m away.

The color force between quarks is postulated to have a massless particle associated with it. This particle is called the *gluon*. It plays the same role in color-field interactions that photons do in interactions of the em field. Most people believe that the gluon, as well as the photon and graviton, is a primary particle that cannot be subdivided.

A comparison of the four fundamental forces is shown schematically in Fig. 28.5. We can think of each curve (except for the quark and weak curves) as being the force

**FIGURE 28.5**

The strengths of the fundamental forces between particles are shown as a function of the distance between particles. The graph lines labeled *gravitational, electrostatic,* and *nuclear* show these forces for two protons separated by the distance indicated. Uncertainty exists concerning the force between two quarks, but the *quark* graph line indicates one possibility.

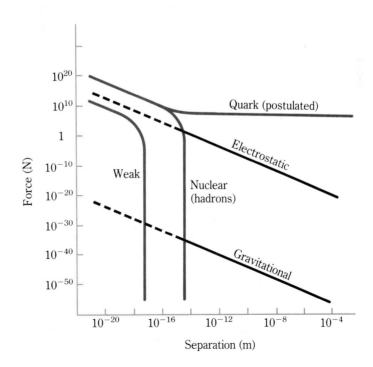

of one proton on another as a function of the particle separation, at least at $r >$ $10^{-16}$ m. Notice the relative magnitudes of the forces; the gravitational force is far weaker than the others. As indicated in the text, the force between quarks is unknown at large distances; at small $r$ values, it is the source of the nuclear force.

Physicists currently speculate that quarks may be composed of still smaller particles. Although at the present time we have no evidence to support such a supposition, the properties of quarks appear sequential enough that an underlying particle structure may give rise to them. As mentioned earlier, the particles of the electroweak interaction ($W^{\pm}$, $Z^{\circ}$) are possibly composites. Even the leptons, such as the electron, may be composed of still tinier entities. We just don't know.

To separate these particles into their (still hypothetical) constituents, collisions at tremendous energies will be required. Our present-day accelerators, even those designed for the future, cannot achieve high enough energies, but these energies were available at the first instant that our universe was formed. And the energies within neutron stars and black holes are also of this magnitude. Particle physicists, in their search for the nature of the tiniest particles in the universe, are looking now to the largest object of all, the universe itself, to enlighten them. We shall explore the current ideas of the origin, and eventual end, of the universe in the sections that follow. This is that portion of astrophysics called *cosmology*.

## 28.7 The primordial fireball

As we search the heavens with our most powerful telescopes, we notice an astonishing fact: the spectral lines from atomic transitions in distant galaxies have different wavelengths from the lines emitted by atoms on earth. For example, the Balmer series of lines from a star in a distant galaxy has longer wavelengths than those we are accustomed to. We say that the light from distant stars exhibits a **red shift** because the spectral lines are shifted to longer wavelengths. The more distant the galaxy, the larger its red shift. As we show in Sec. 28.9, the red shift tells us that the distant objects of the universe are receding from the earth at high speed; the most distant ones are flying away at nearly the speed of light. The Doppler effect (Sec. 14.10) is responsible for the red shift of these receding sources.

Extrapolating backward in time, we conclude that the entire universe must, at one time, have been in a highly compressed state. This is expressed in the **big bang theory** of the universe. According to this theory, about 15 billion years ago the universe existed in a highly compressed state. The mass within it had a density greater than $5 \times 10^{12}$ kg/m³, more than 5 billion times the density of water. This highly compressed universe was at an incredibly high temperature, higher than $10^{12}$ K.

At such extreme temperatures and pressures, atoms and molecules cannot exist. Even the particles we are familiar with as hadrons were fragmented. It is believed that only photons, quarks, electrons, positrons, neutrinos, and perhaps other particles still unknown to us existed at that time. Most of the mass of the universe was in the form of radiant energy. The universe was a fiery cauldron just beginning a violent expansion. No force could successfully resist this expansion, which was driven by the high-energy thermal motion and the pressure of the confined radiation. The present-day recession from us of distant galaxies is the continuation of the expansion of this original fireball.

As the infant universe began its adiabatic expansion, it had to do work against the gravitational forces of attraction among its various parts. This work was done largely at the expense of the thermal energy it possessed, so the temperature of the expanding universe fell rapidly. Within only a few minutes the temperature fell low enough for protons, neutrons, and similar particles to become relatively stable.* The still rapidly cooling fireball was then a very hot gaseous cloud composed of particles and radiation.

So far we have discussed the expanding fireball as though it were a smooth, homogeneous cloud of gas. In all probability, it was not completely uniformly distributed throughout space. Certain regions must have had higher densities than others as a result of factors thus far unknown. Although we have no direct proof of the hypothesis, it seems reasonable that regions of unusually high density acted as focal points for what might best be described as a gravitational condensation. Over huge regions of space, unbalanced gravitational forces were produced that made the local gas start moving toward regions of higher density. Of course, superimposed on this was the continued radial motion of the material of the expanding fireball. These huge, noncompact regions into which matter began to collect are presently huge **galaxies,** aggregate systems containing many stars.

As the now cloudlike cooling fireball began to form regions of higher density, the masses had to obey the usual laws of motion. One, the law of conservation of angular momentum, had a profound effect on the behavior of the condensing cloud. Any small net angular motion $\omega_o$ of the mass within the huge region originally occupied by the galaxy must be multiplied by the moment of inertia of that region $I_o$ to obtain the original angular momentum $I_o\omega_o$. However, as aggregation into a galaxy took place, the radius $r$ of gyration for the galaxy's mass became much smaller, thereby decreasing $I$, since $I \propto r^2$. Since angular momentum must be conserved, $I\omega = I_o\omega_o$; so as $I$ decreases, $\omega$ must increase. Therefore, as condensation took place, the material began to spin about the center of the condensation.

Spinning of the condensing cloud mass would be fastest for regions with the largest rotation at the outset. We can therefore expect some galaxies to be spinning very little, while others have relatively large angular speeds ("large" in this context still means millions of years per rotation). As with any spinning system (a stone on a string or a twirling pizza, for example), the spinning causes the system to flatten into a disk, with the axis of rotation perpendicular to its plane. For the fastest-spinning galaxies, called spiral galaxies, this effect is very pronounced (Fig. 28.6). Our own solar system is part of a somewhat less spiraling galaxy, the Milky Way, and is located about two-thirds of the way out from the galactic center. (There are about 100 billion stars in our galaxy, the diameter of which is of the order of 100,000 light-years. It takes our sun about 200 million years to rotate once around the galactic center with its approximately 250-km/s orbital speed.)

While a galaxy is forming, localized regions in it are condensing much more rapidly into centers of mass, which eventually become the stars in the galaxy. It is easy to see how they become intensely hot. As matter is pulled toward the center by gravitational forces, potential energy is converted to kinetic energy. This means that a tremendous amount of energy is carried to the condensation center, the embryonic star, by the aggregating mass. Hence the temperature of the aggregate can become extremely high; in fact, it exceeds the temperature needed for complete ionization of all the hydrogen atoms in the aggregate. This huge mass, an embryonic star, is now a white-hot, very dense "soup," called a *plasma,* composed of protons, neutrons, electrons, and other basic particles. This is the way a star is born.

---

* For a readable history of the first minutes of the universe, see S. Weinberg, *The First Three Minutes,* Basic Books, New York, 1977, a popularization by a leading theorist in particle physics.

**FIGURE 28.6**
The spiral galaxy M81. (*Lick Observatory*)

## 28.8 Stellar evolution

Pulled radially inward by gravitational forces, the star will continue to contract until its internal pressure balances the gravitational pressure. As in any gas, the internal pressure increases with rising temperature. For small aggregates of matter, the gravitational forces are small, and so equilibrium is reached at rather low temperatures. For the large stars of interest, however, contraction continues until the temperature in the star reaches a few million degrees. At this high temperature, the average thermal energy of a proton is about 1000 eV. In spite of this rather low average energy, enough high-energy protons exist for a fusion reaction to become possible.

At the earliest stages, the fusion reaction of importance is

$$^1_1H + {}^1_1H \rightarrow {}^2_1H + \text{positron} + \text{neutrino}$$

The deuteron, $^2_1H$, reacts again with a proton:

$$^2_1H + {}^1_1H \rightarrow {}^3_2He + \gamma$$

and this isotope of helium reacts as follows:

$$_2^3He + _2^3He \rightarrow _2^4He + 2\ _1^1H$$

In other words, protons are fused in this reaction to form helium; six protons react to form a helium nucleus and two protons. As with any fusion reaction of this type, large amounts of energy are released. The star is now capable of supplying energy to itself without further gravitational collapse.

As soon as the fusion reaction has become powerful enough to increase the thermal pressure within the star to a point where it balances the gravitational pressure, the star is stabilized. The temperature at the very center of the sun, the hottest point, is estimated to be close to 15 million K. Although the proton reaction in the sun has been proceeding for about 4.5 billion years, apparently enough protons remain for the reaction to continue steadily for about that long in the future. In stars that are more massive than our own sun, the interior temperature is higher. (Why?) At these higher temperatures, other nuclear fusion reactions are possible.

Eventually the proton reaction uses most of the available protons, and the reaction slows down. As a result, the thermal pressure decreases and the unbalanced portion of the gravitational pressure causes the star to begin to contract again. When this happens, the outer (formerly cooler) regions of the star are heated enough for the protons there to begin reacting through the fusion process previously outlined. The resulting thermal pressure in this portion of the star causes the outer layers to expand. Hence the outer portion of the star enlarges while simultaneously cooling. At this stage the star changes in appearance from a white-hot star to a considerably larger, redder star, called a **red giant.**

Although the exterior surface of the star is cooled during this transition, the interior is heated as a result of the inner contraction. The core is largely $^4He$, a product of the burned-out proton-fusion reaction. Only after the star has reached a temperature of about 100 million K does the helium begin to fuse. At that stage begins the reaction

$$_2^4He + _2^4He \rightarrow _4^8Be$$

The beryllium then combines with helium:

$$_4^8Be + _2^4He \rightarrow _6^{12}C$$

This reaction is called the *triple alpha* reaction because the stripped helium atom is an $\alpha$ particle.

Later stages of development are uncertain, but laboratory experiments using large accelerators show that reactions between $^{12}C$ and other high-energy particles in the star can lead to the formation of larger nuclei. In fact, it seems likely that the stable elements of the periodic table are formed at this stage of star life. Since our own sun and the planets contain reasonably comparable amounts of the heavy elements, a clue exists as to the evolutionary stage of our own solar system.

Eventually the fusion reactions in the interior of a red giant must die out as the available fuel becomes exhausted. At that stage the gravitational forces are no longer balanced by sufficient thermal pressure, and the star will contract, becoming further heated because of the conversion of potential to thermal energy. After a few tens of millions of years, the star will have shrunk to a very dense, white-hot body called a **white dwarf.**

We have good theoretical reasons to believe that a white dwarf cannot be stable if its mass is greater than about 1.2 times the mass of our sun. Although definitive evidence for this hypothesis is lacking because the masses of only a few white dwarfs have been measured, the existing data do not contradict it. However, since many red giants have masses much in excess of this limiting mass, they must somehow lose mass as they contract to the white-dwarf stage. Exactly how this is accomplished is not known.

One possibility consistent with observation is that the collapsing red giant undergoes explosions as it approaches the white-dwarf stage. These explosions, observed as novas and supernovas, send out into space great masses of gas composed of nuclei. As a result, the galaxy acquires a cloud composed mainly of hydrogen and helium but also containing the nuclei of the heavier elements. This cloud could then undergo gravitational condensation, and new stars would be formed. Consequently the whole process of star evolution outlined above could be repeated. According to this hypothesis, our sun appears to be a star of this type, since it contains the nuclei of heavy elements. Eventually (perhaps in 4 or 5 billion years), our sun should become a red giant; still later, it should contract to become a white dwarf. Of course, during the red-giant stage, the earth would become too hot for human habitation.

## 28.9 The expanding universe

According to the big bang theory, the universe is expanding. If so, we should notice a Doppler effect in the light of receding galaxies. Let us compute the relation between the wavelength $\lambda_o$ of, say, a line in the hydrogen spectrum from a source on earth and the wavelength $\lambda_s$ observed for a source on a star. Suppose the star to be receding from the earth with speed $v$.

Consider a crest of the wave emitted by the star source. It will travel a distance $ct$ toward earth in a time $t$, where $c$ is the velocity of light. However, the stellar source will emit a wave crest once every $\tau_o$ seconds as measured by a timer on the star. According to relativity theory, this time $\tau_o$ will actually be read as $\tau_o/\sqrt{1-(v/c)^2}$ by an earth clock. As a result, the distance the first crest will move during the time between its emission and the emission of the next crest will be (according to an earth observer)

$$\text{Distance} = ct = \frac{c\tau_o}{\sqrt{1-(v/c)^2}}$$

During this time, the stellar source will have moved a distance $vt$, and so the second crest will actually be a distance $ct + vt$ behind the first. This, then, is the wavelength observed on earth for the light from the star:

$$\lambda_s = ct + vt \qquad \text{or} \qquad \lambda_s = \frac{(c+v)(\tau_o)}{\sqrt{1-(v/c)^2}}$$

But since the stellar source is itself in an inertial system, the usual laws must apply there. Hence, $\tau_o = \lambda_o/c$, and so the wavelength $\lambda_s$ of the star's line as observed on earth is related to the wavelength $\lambda_o$ of the line from a stationary source by the equation*

---

\* If you compare this with the Doppler effect equation for sound obtained in Chap. 14, you will see that they are different. At low $v/c$ values, however, where relativistic effects can be ignored, they do coincide.

$$\lambda_s = \lambda_o \frac{1 + v/c}{\sqrt{1 - (v/c)^2}}$$

We see that $\lambda_s$ is larger than $\lambda_o$. In other words, the wavelengths of the spectral lines from a receding star will appear lengthened, or shifted from blue toward the red. This is referred to as the *red shift*.

If we examine the light reaching us from the galaxies, we do find the wavelengths emitted by the atoms to be shifted to the red. We interpret this to mean that all galaxies are moving away from us. By use of the red-shift equation, it is possible to compute their recession speeds. It is found that the more distant a galaxy is from the earth, the faster it is receding from us. For the most distant galaxies, $v$ is quite close to $c$, the speed of light. For them, lines in the blue are shifted into the infrared.

A rather simple experimental relation has been found between the recession speed $v$ of a star and its distance $s$ from earth. It is

$$s = 6.3 \times 10^{17} v \qquad \text{m}$$

where the exact value of the numerical constant is in doubt. Let us now interpret this relation in terms of the big bang theory. We shall see that it allows us to compute the age of the universe.

According to the big bang theory, the earth is part of this now quite cold expanding cloud. It is a property of such an expanding system that everything within it is separating from everything else. The situation is much like that of the blueberries in a blueberry muffin. As the muffin bakes, the dough expands, and each blueberry recedes from its neighbors. We can further illustrate this idea by the diagram of Fig. 28.7. If two objects are moving along the same line but at different speeds $v_1$ and $v_2$, they will be separating at a speed $v_2 - v_1$. If we assume that they started from the

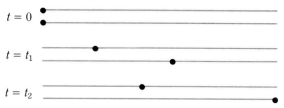

$t = 0$

$t = t_1$

$t = t_2$

**FIGURE 28.7**
Starting from the same place at the same time, the spots recede from each other because their speeds are different.

same point at time $t = 0$, that is, when the fireball first exploded, their present separations should be given by

$$s = (v_2 - v_1)(t)$$

But this is identical in form to the experimental relation found from the red-shift data, namely,

$$s = (v)(6.3 \times 10^{17})$$

From our definition of $v_2 - v_1$, it is the same as $v$. Hence we have

$$t = 6.3 \times 10^{17} \text{ s} = 1.8 \times 10^{10} \text{ yr}$$

as the time from the first explosion of the fireball until now. This gives an age of 18 billion years for the universe.* Although this age has been arrived at from rather tentative assumptions, other data tend to support it.

## 28.10 The future

The universe is still expanding, but its expansion is being slowed by gravitational forces. We are still not sure of the eventual outcome, many billions of years in the future. One possibility is that the universe will expand forever. This will occur if the density of the universe is less than a certain critical density. In that case the gravitational forces will never be able to stop the expansion. If the density exceeds this critical value, however, the expansion will eventually stop. The universe will then begin to contract; after many billions of years, it will contract to the original fireball that existed at the time of the big bang. Currently we do not know which condition prevails, but it appears that the density is close to the critical value.

Much of the density of the universe may be due to neutrinos. We know that the neutrino's mass is extremely small if not exactly zero. Certain theories predict that the neutrino does have finite mass. If the mass of the neutrino is not zero, then, because of their huge number, the density of the universe is increased substantially by them. Thus we see that the fate of the universe may be controlled by the nature of one of the smallest of all particles. Other particles, still speculative, may also contribute nonnegligible mass.

This is but one example of how elementary particle physics is crucial to the study of the universe. An even more intriguing relation between these two areas of study may be found from an examination of the first fraction of a second after the big bang. Energy densities within the fireball were so huge that we cannot hope to duplicate in the laboratory the reactions that then occurred or the forces that then prevailed. Knowing how the universe developed, however, we have clues to what these reactions and forces were. For example, the fact that matter rather than antimatter prevailed is an interesting clue. Moreover, the fact that only a small fraction of the universe's energy is in the form of rest-mass energy is significant. Further, the ratios of the numbers of various particles that formed tells us something about particle behavior at high energies. Why the universe expanded essentially homogeneously in all directions is a question that, when answered, will contribute additional understanding. The fact that the density of the universe is near its critical density is another clue. In addition, because in the first instant of the fireball the fundamental forces of nature were acting most strongly, the nature of these forces in large part directed the development of the universe.

As we see, the study of the very smallest entities of the universe will lead to increased understanding of the universe as a whole and its fate. Additionally, reasoning based on the observed development of the universe leads us to new discoveries concerning the forces and particles of nature. This interplay between the very large and the very small is now a topic of intense research effort that reflects the glorious way in which our universe continues to be revealed to us.

---

* Analysis of these data involves a number of uncertainties. Estimates for the age range from about 10 to 20 billion years.

## Minimum learning goals

When you finish this chapter, you should be able to

**1** Define (*a*) positron, (*b*) antiparticle, (*c*) quark, (*d*) lepton, (*e*) hadron, (*f*) baryons versus mesons, (*g*) quark color, (*h*) gluon, (*i*) big bang theory, (*j*) red shift, (*k*) galaxy, (*l*) red giant, (*m*) white dwarf.

**2** Describe how a typical linear accelerator works. Repeat for a circular accelerator.

**3** Explain why very-high-energy probe particles are needed for research purposes.

**4** Use the quark model to describe the structure of the proton, neutron, and other hadrons. State which particles contain three quarks and which contain two.

**5** Give the basic properties of the following leptons: the electron, positron, and neutrino.

**6** List the four fundamental forces (or interactions), and describe the strength and range of each.

**7** State what is meant by the electroweak theory and the grand unified theory (GUT).

**8** Use the concept of color to explain how the quarks are tightly bound within baryons and mesons by long-range forces, even though baryons and mesons do not exert long-range forces on each other. Define quantum chromodynamics.

**9** State the general features of the big bang theory.

**10** Give the approximate age of the universe, according to the big bang theory.

## Questions and guesstimates

**1** What reasons can governments use to justify the expenditure of billions of dollars to build particle accelerators?

**2** The energy radiated by an antenna increases as the square of the frequency at which charge oscillates on it. What does this imply concerning energy losses of particles orbiting in a circular accelerator?

**3** Suppose the attractive force between quarks does not vary with separation distance. Why, then, could quarks never be separated? If you know calculus, repeat for a force that varies as $1/r$.

**4** In Fig. 28.5, why do the force lines for hadrons and quarks unite for separations less than about $10^{-16}$ m? In what respect are these two lines only schematic?

**5** Is there any evidence that separated neutral atoms exert attractive forces on each other? Estimate the range of the force. How is this analogous to the attractive force between two neutrons?

**6** Perhaps the universe has previously expanded from a fireball and then collapsed once again to a fireball. We would then be in its second pulsation cycle. Can we tell whether previous pulsations of the universe have occurred?

**7** We know that charge is conserved. Does this mean that there are the same number of positive charges in the universe now as there were in the original fireball?

**8** When we look into space, we find that all distant galaxies are flying away from us. Does this mean that we are at the center of the universe?

**9** How is it possible to talk about the density of the universe at times near the big bang if at that time nearly all the energy of the universe was in the form of radiation and there was very little rest mass?

## Problems

▪ **1** What is the de Broglie wavelength of a 200-GeV proton? Of a 200-GeV electron? What is the relativistic mass of each particle?

▪ **2** A proton and an antiproton with negligible energy combine to yield two $\gamma$-rays. (*a*) What is the direction of motion of one $\gamma$-ray relative to the other? (*b*) What must be the relation between the wavelengths of the $\gamma$-rays? (*c*) Find the energy and wavelength of either $\gamma$-ray.

▪ **3** Two protons, each with an energy of 20 GeV, collide head on. If they could fuse together to form a third particle, what would be the following for the particle: (*a*) momentum, (*b*) rest mass?

■ **4** Answer Prob. 3 for a 40-GeV proton fusing with an originally stationary proton.

■ **5** If a photon is to be capable of producing a neutron-antineutron pair, what is the minimum energy the photon must have? Assume that the event occurs close to another particle, so that momentum can be shared with it. What is the photon's wavelength?

■ **6** Experiment indicates that one quark (top) has a mass of about 50 GeV. Consider the head-on collision of two particles, each of mass $m_o$, in an attempt to create such a quark. (*a*) What minimum energy is needed per particle if the particles have equal speeds in opposite directions? (*b*) If one particle is initially at rest, compare the energy then required.

■ **7** What is the frequency of revolution of a proton in a 300-GeV circular accelerator that has a diameter of 2.2 km?

■ **8** A proton with 300-GeV energy follows a circular path of 2.0 km radius. Find the magnetic field required to hold it in this path.

■ **9** Assume that a particle of mass $1.5 \times 10^{-29}$ kg is confined to a narrow tube that is $10^{-15}$ m long. Find the energy levels of the particle according to simple quantum theory. (Neglect relativistic effects.)

**10** What is the average thermal energy (in megaelectronvolts) of a proton in the center of a star where the temperature is $10^9$ K?

**11** Find the wavelength of the first line of the Balmer series received on earth from a galaxy that is receding from earth at a speed of $0.95c$.

■ **12** What should be the present-day diameter of a region that had a diameter of $1 \times 10^{-10}$ m in the original fireball? Assume a steady expansion rate of $3 \times 10^8$ m/s, and assume the age of the universe to be $15 \times 10^9$ years. How long did it take for this region to reach a diameter of 1 cm?

# APPENDIX 1
# The periodic table of the elements

The values listed are based on $^{12}_{6}C$ = 12 u exactly. For radioactive elements, the approximate atomic weight of the most stable isotope is given in brackets.

| Period | $I_A$ | $II_A$ | $III_B$ | $IV_B$ | $V_B$ | $VI_B$ | $VII_B$ | VIII | VIII | VIII | $I_B$ | $II_B$ | $III_A$ | $IV_A$ | $V_A$ | $VI_A$ | $VII_A$ | 0 |
|---|---|---|---|---|---|---|---|---|---|---|---|---|---|---|---|---|---|---|
| 1 | **1 H** 1.00797 | **2 He** 4.003 | | | | | | | | | | | | | | | | **2 He** 4.003 |
| 2 | **3 Li** 6.939 | **4 Be** 9.012 | | | | | | | | | | | **5 B** 10.81 | **6 C** 12.011 | **7 N** 14.007 | **8 O** 15.9994 | **9 F** 19.00 | **10 Ne** 20.183 |
| 3 | **11 Na** 22.990 | **12 Mg** 24.31 | | | | | | | | | | | **13 Al** 26.98 | **14 Si** 28.09 | **15 P** 30.974 | **16 S** 32.064 | **17 Cl** 35.453 | **18 Ar** 39.948 |
| 4 | **19 K** 39.102 | **20 Ca** 40.08 | **21 Sc** 44.96 | **22 Ti** 47.90 | **23 V** 50.94 | **24 Cr** 52.00 | **25 Mn** 54.94 | **26 Fe** 55.85 | **27 Co** 58.93 | **28 Ni** 58.71 | **29 Cu** 63.54 | **30 Zn** 65.37 | **31 Ga** 69.72 | **32 Ge** 72.59 | **33 As** 74.92 | **34 Se** 78.96 | **35 Br** 79.909 | **36 Kr** 83.80 |
| 5 | **37 Rb** 85.47 | **38 Sr** 87.62 | **39 Y** 88.905 | **40 Zr** 91.22 | **41 Nb** 92.91 | **42 Mo** 95.94 | **43 Tc** [99] | **44 Ru** 101.1 | **45 Rh** 102.905 | **46 Pd** 106.4 | **47 Ag** 107.870 | **48 Cd** 112.40 | **49 In** 114.82 | **50 Sn** 118.69 | **51 Sb** 121.75 | **52 Te** 127.60 | **53 I** 126.90 | **54 Xe** 131.30 |
| 6 | **55 Cs** 132.905 | **56 Ba** 137.34 | † | **72 Hf** 178.49 | **73 Ta** 180.95 | **74 W** 183.85 | **75 Re** 186.2 | **76 Os** 190.2 | **77 Ir** 192.2 | **78 Pt** 195.09 | **79 Au** 196.97 | **80 Hg** 200.59 | **81 Tl** 204.37 | **82 Pb** 207.19 | **83 Bi** 208.98 | **84 Po** [210] | **85 At** [210] | **86 Rn** [222] |
| 7 | **87 Fr** [223] | **88 Ra** [226] | ‡ | | | | | | | | | | | | | | | |

*†Lanthanide series*

| | | | | | | | | | | | | | | |
|---|---|---|---|---|---|---|---|---|---|---|---|---|---|---|
| **57 La** 138.91 | **58 Ce** 140.12 | **59 Pr** 140.91 | **60 Nd** 144.24 | **61 Pm** [147] | **62 Sm** 150.35 | **63 Eu** 152.0 | **64 Gd** 157.25 | **65 Tb** 158.92 | **66 Dy** 162.50 | **67 Ho** 164.93 | **68 Er** 167.26 | **69 Tm** 168.93 | **70 Yb** 173.04 | **71 Lu** 174.97 |

*‡Actinide series*

| | | | | | | | | | | | | | | |
|---|---|---|---|---|---|---|---|---|---|---|---|---|---|---|
| **89 Ac** [227] | **90 Th** 232.04 | **91 Pa** [231] | **92 U** 238.03 | **93 Np** [237] | **94 Pu** [242] | **95 Am** [243] | **96 Cm** [247] | **97 Bk** [247] | **98 Cf** [251] | **99 Es** [254] | **100 Fm** [253] | **101 Md** [256] | **102 No** [254] | **103 Lw** [257] |

# An abbreviated table of isotopes

The values listed are based on $^{12}_{6}C = 12$ u exactly. Electron masses are included.

| Atomic number Z | Symbol | Average atomic mass | Element | Mass number A | Relative abundance % | Mass of isotope |
|---|---|---|---|---|---|---|
| 0 | n | 1.008665 | Neutron | 1 | | |
| 1 | H | 1.00797 | Hydrogen | 1 | 99.985 | 1.007825 |
| | | | | 2 | 0.015 | 2.014102 |
| 2 | He | 4.0026 | Helium | 3 | 0.00015 | 3.016030 |
| | | | | 4 | 100− | 4.002604 |
| 3 | Li | 6.939 | Lithium | 6 | 7.52 | 6.015126 |
| | | | | 7 | 92.48 | 7.016005 |
| 4 | Be | 9.0122 | Beryllium | 9 | 100− | 9.012186 |
| 5 | B | 10.811 | Boron | 10 | 19.78 | 10.012939 |
| | | | | 11 | 80.22 | 11.009305 |
| 6 | C | 12.01115 | Carbon | 12 | 98.892 | 12.0000000 |
| | | | | 13 | 1.108 | 13.003354 |
| 7 | N | 14.0067 | Nitrogen | 14 | 99.635 | 14.003074 |
| | | | | 15 | 0.365 | 15.000108 |
| 8 | O | 15.9994 | Oxygen | 16 | 99.759 | 15.994915 |
| | | | | 17 | 0.037 | 16.999133 |
| | | | | 18 | 0.204 | 17.999160 |
| 9 | F | 18.9984 | Fluorine | 19 | 100 | 18.998405 |
| 10 | Ne | 20.183 | Neon | 20 | 90.92 | 19.992440 |
| | | | | 22 | 8.82 | 21.991384 |
| 11 | Na | 22.9898 | Sodium | 23 | 100− | 22.989773 |
| 12 | Mg | 24.312 | Magnesium | 24 | 78.60 | 23.985045 |
| 13 | Al | 26.9815 | Aluminum | 27 | 100 | 26.981535 |
| 14 | Si | 28.086 | Silicon | 28 | 92.27 | 27.976927 |
| | | | | 30 | 3.05 | 29.973761 |
| 15 | P | 30.9738 | Phosphorus | 31 | 100 | 30.973763 |
| 16 | S | 32.064 | Sulfur | 32 | 95.018 | 31.972074 |
| 17 | Cl | 35.453 | Chlorine | 35 | 75.4 | 34.968854 |
| | | | | 37 | 24.6 | 36.965896 |
| 18 | Ar | 39.948 | Argon | 40 | 99.6 | 39.962384 |
| 19 | K | 39.102 | Potassium | 39 | 93.08 | 38.963714 |

| Atomic number Z | Symbol | Average atomic mass | Element | Mass number A | Relative abundance % | Mass of isotope |
|---|---|---|---|---|---|---|
| 20 | Ca | 40.08 | Calcium | 40 | 96.97 | 39.962589 |
| 21 | Sc | 44.956 | Scandium | 45 | 100 | 44.955919 |
| 22 | Ti | 47.90 | Titanium | 48 | 73.45 | 47.947948 |
| 23 | V | 50.942 | Vanadium | 51 | 99.76 | 50.943978 |
| 24 | Cr | 51.996 | Chromium | 52 | 83.76 | 51.940514 |
| 25 | Mn | 54.9380 | Manganese | 55 | 100 | 54.938054 |
| 26 | Fe | 55.847 | Iron | 56 | 91.68 | 55.934932 |
| 27 | Co | 58.9332 | Cobalt | 59 | 100 | 58.93319 |
| 28 | Ni | 58.71 | Nickel | 58 | 67.7 | 57.93534 |
|  |  |  |  | 60 | 26.23 | 59.93032 |
| 29 | Cu | 63.54 | Copper | 63 | 69.1 | 62.92959 |
| 30 | Zn | 65.37 | Zinc | 64 | 48.89 | 63.92914 |
| 31 | Ga | 69.72 | Gallium | 69 | 60.2 | 68.92568 |
| 32 | Ge | 72.59 | Germanium | 74 | 36.74 | 73.92115 |
| 33 | As | 74.9216 | Arsenic | 75 | 100 | 74.92158 |
| 34 | Se | 78.96 | Selenium | 80 | 49.82 | 79.91651 |
| 35 | Br | 79.909 | Bromine | 79 | 50.52 | 78.91835 |
| 36 | Kr | 83.30 | Krypton | 84 | 56.90 | 83.91150 |
| 37 | Rb | 85.47 | Rubidium | 85 | 72.15 | 84.91171 |
| 38 | Sr | 87.62 | Strontium | 88 | 82.56 | 87.90561 |
| 39 | Y | 88.905 | Yttrium | 89 | 100 | 88.90543 |
| 40 | Zr | 91.22 | Zirconium | 90 | 51.46 | 89.90432 |
| 41 | Nb | 92.906 | Niobium | 93 | 100 | 92.90602 |
| 42 | Mo | 95.94 | Molybdenum | 98 | 23.75 | 97.90551 |
| 43 | Tc | * | Technetium | 98 |  | 97.90730 |
| 44 | Ru | 101.07 | Ruthenium | 102 | 31.3 | 101.90372 |
| 45 | Rh | 102.905 | Rhodium | 103 | 100 | 102.90480 |
| 46 | Pd | 106.4 | Palladium | 106 | 27.2 | 105.90320 |
| 47 | Ag | 107.870 | Silver | 107 | 51.35 | 106.90497 |
| 48 | Cd | 112.40 | Cadmium | 114 | 28.8 | 113.90357 |
| 49 | In | 114.82 | Indium | 115 | 95.7 | 114.90407 |
| 50 | Sn | 118.69 | Tin | 120 | 32.97 | 119.90213 |
| 51 | Sb | 121.75 | Antimony | 121 | 57.25 | 120.90375 |
| 52 | Te | 127.60 | Tellurium | 130 | 34.49 | 129.90670 |
| 53 | I | 126.9044 | Iodine | 127 | 100 | 126.90435 |
| 54 | Xe | 131.30 | Xenon | 132 | 26.89 | 131.90416 |
| 55 | Cs | 132.905 | Cesium | 133 | 100 | 132.90509 |
| 56 | Ba | 137.34 | Barium | 138 | 71.66 | 137.90501 |
| 57 | La | 138.91 | Lanthanum | 139 | 99.911 | 138.90606 |
| 58 | Ce | 140.12 | Cerium | 140 | 88.48 | 139.90528 |

| Atomic number Z | Symbol | Average atomic mass | Element | Mass number A | Relative abundance % | Mass of isotope |
|---|---|---|---|---|---|---|
| 59 | Pr | 140.907 | Praseodymium | 141 | 100 | 140.90739 |
| 60 | Nd | 144.24 | Neodymium | 144 | 23.85 | 143.90998 |
| 61 | Pm | * | Promethium | 145 | | 144.91231 |
| 62 | Sm | 150.35 | Samarium | 152 | 26.63 | 151.91949 |
| 63 | Eu | 151.96 | Europium | 153 | 52.23 | 152.92086 |
| 64 | Gd | 157.25 | Gadolinium | 158 | 24.87 | 157.92410 |
| 65 | Tb | 158.924 | Terbium | 159 | 100 | 158.92495 |
| 66 | Dy | 162.50 | Dysprosium | 164 | 28.18 | 163.92883 |
| 67 | Ho | 164.930 | Holmium | 165 | 100 | 164.93030 |
| 68 | Er | 167.26 | Erbium | 166 | 33.41 | 165.93040 |
| 69 | Tm | 168.934 | Thulium | 169 | 100 | 168.93435 |
| 70 | Yb | 173.04 | Ytterbium | 174 | 31.84 | 173.93902 |
| 71 | Lu | 174.97 | Lutetium | 175 | 97.40 | 174.94089 |
| 72 | Hf | 178.49 | Hafnium | 180 | 35.44 | 179.94681 |
| 73 | Ta | 180.948 | Tantalum | 181 | 100 | 180.94798 |
| 74 | W | 183.85 | Tungsten | 184 | 30.6 | 183.95099 |
| 75 | Re | 186.2 | Rhenium | 187 | 62.93 | 186.95596 |
| 76 | Os | 190.2 | Osmium | 192 | 41.0 | 191.96141 |
| 77 | Ir | 192.2 | Iridium | 193 | 61.5 | 192.96328 |
| 78 | Pt | 195.09 | Platinum | 195 | 33.7 | 194.96482 |
| 79 | Au | 196.967 | Gold | 197 | 100 | 196.96655 |
| 80 | Hg | 200.59 | Mercury | 202 | 29.80 | 201.97063 |
| 81 | Tl | 204.37 | Thallium | 205 | 70.50 | 204.97446 |
| 82 | Pb | 207.19 | Lead | 208 | 52.3 | 207.97664 |
| 83 | Bi | 208.980 | Bismuth | 209 | 100 | 208.98042 |
| 84 | Po | 210 | Polonium | 210 | | 209.98287 |
| 85 | At | * | Astatine | 211 | | 210.98750 |
| 86 | Rn | * | Radon | 211 | | 210.99060 |
| 87 | Fr | * | Francium | 221 | | 221.01418 |
| 88 | Ra | 226.054 | Radium | 226 | | 226.02536 |
| 89 | Ac | * | Actinium | 225 | | 225.02314 |
| 90 | Th | 232.038 | Thorium | 232 | 100 | 232.03821 |
| 91 | Pa | 231.036 | Protactinium | 231 | | 231.03594 |
| 92 | U | 238.03 | Uranium | 233 | | 233.03950 |
| | | | | 235 | 0.715 | 235.04393 |
| | | | | 238 | 99.28 | 238.05076 |
| 93 | Np | * | Neptunium | 239 | | 239.05294 |
| 94 | Pu | * | Plutonium | 239 | | 239.05216 |
| 95 | Am | * | Americium | 243 | | 243.06138 |
| 96 | Cm | * | Curium | 245 | | 245.06534 |
| 97 | Bk | * | Berkelium | 248 | | 248.070305 |

| Atomic number Z | Symbol | Average atomic mass | Element | Mass number A | Relative abundance % | Mass of isotope |
|---|---|---|---|---|---|---|
| 98 | Cf | * | Californium | 249 | | 249.07470 |
| 99 | Es | * | Einsteinium | 254 | | 254.08811 |
| 100 | Fm | * | Fermium | 252 | | 252.08265 |
| 101 | Md | * | Mendelevium | 255 | | 255.09057 |
| 102 | No | * | Nobelium | 254 | | 254 |
| 103 | Lw | * | Lawrencium | 257 | | 257 |

* The atomic masses of unstable elements are not listed unless the isotope given constitutes the major isotope.

# A mathematics review

Although this text does not presume prior familiarity with much mathematics, you may find the following review helpful in this and other science courses.

### A. Addition

The order in which quantities are added is unimportant. For example, $8 + 7$ and $7 + 8$ are both 15. If we represent two numbers by the symbols $a$ and $b$, we have $a + b = b + a$.

### B. Subtraction

If $a = 6$ and $b = 4$, then we know that

$$a - b = 6 - 4 = 2 \quad \text{and} \quad b - a = 4 - 6 = -2$$

To subtract a negative number, we proceed as follows: if $a = 7$ and $b = -3$, then

$$a - b = 7 - (-3) = 7 + 3 = 10$$

Or, using two numbers $c$ and $-e$, we have

$$c - (-e) = c + e$$

*To subtract a negative number, we change its sign and add it.*

### C. Multiplication

The order in which common numbers are multiplied is unimportant. For example, $6 \times 3 = 3 \times 6 = 18$, and $a \times b$, which we write as $a \cdot b$ or $ab$, is the same as $ba$. In carrying out multiplications, the following sign rules must be used:

| Rule | Example |
|------|---------|
| $a \times b = ab$ | $5 \times 6 = 30$ |
| $(a) \times (-b) = -ab$ | $(5) \times (-6) = -30$ |
| $(-a) \times (-b) = ab$ | $(-5) \times (-6) = 30$ |

### D. Division

The sign rules for division are as follows:

| Rule | Example |
|------|---------|
| $a \div b = \dfrac{a}{b}$ | $15 \div 3 = \dfrac{15}{3} = 5$ |
| $a \div (-b) = \dfrac{a}{-b} = \dfrac{-a}{b} = -\dfrac{a}{b}$ | $15 \div (-3) = -5$ |

$$(-a) \div (-b) = \frac{-a}{-b} = \frac{a}{b} \qquad\qquad (-15) \div (-3) = 5$$

### E. Parentheses

Parentheses may be manipulated as in the following examples:

$$(a + b) = (b + a) = a + b$$
$$d(a + b + c) = da + db + dc$$
$$(e + d)(a + b + c) = e(a + b + c) + d(a + b + c)$$
$$-(a - b) = -(a) - (-b) = -a + b$$

### F. Fractions

In a fraction $a/b$, $a$ is the numerator and $b$ is the denominator. Notice the following identities:

$$\frac{a + b}{c} = \frac{a}{c} + \frac{b}{c} \qquad \text{and} \qquad \frac{a - b}{c} = \frac{a}{c} - \frac{b}{c}$$

$$\frac{a + b}{c + d} = \frac{a}{c + d} + \frac{b}{c + d} \qquad \left(\text{not } \frac{a}{c} + \frac{b}{d}\right)$$

$$\frac{a}{b} \times \frac{c}{d} = \frac{ac}{bd} \qquad \text{and} \qquad \frac{a}{b} = \frac{a}{b} \times 1 = \frac{a}{b} \times \frac{c}{c} = \frac{ac}{bc}$$

The last example shows that we can multiply both the numerator and the denominator by the same quantity without changing the value of a fraction.

$$a \div \frac{c}{d} = a \times \frac{1}{c/d} = a \times \left(\frac{d}{c}\right) = \frac{ad}{c}$$

This latter identity is often stated in words as follows: *to divide a number by a fraction, invert the fraction and multiply by it.* As a more general example of this,

$$\frac{a}{b} \div \frac{c}{d} = \frac{a}{b} \times \frac{d}{c} = \frac{ad}{bc}$$

Notice what happens when we multiply or divide both the numerator and the denominator of a fraction $a/b$ by the same quantity $c$ to give $ac/bc$. We have

$$\frac{ac}{bc} = \frac{a}{b} \times \frac{c}{c} = \frac{a}{b} \times 1 = \frac{a}{b}$$

Multiplying the numerator and denominator by the same quantity did not change the value of the fraction. In effect, we can cancel like factors from the numerator and the denominator. As examples,

$$\frac{ax}{ay} = \frac{x}{y} \qquad \text{and} \qquad \frac{ab + bc}{bd} = \frac{a + c}{d}$$

Be careful, however;

$$\frac{ab+c}{bd} \quad \text{is not} \quad \frac{a+c}{d}$$

Each term in the numerator and the denominator must possess the same factor, or the factor cannot be canceled.

### G. Exponents

In the expression $a^c$, we call $c$ the exponent of $a$. From the definition of exponent, it follows that

$$a^0 = 1 \qquad a^3 = a \cdot a \cdot a \qquad a^{-1} = \frac{1}{a} \qquad a^{-3} = \frac{1}{a^3} = \frac{1}{a \cdot a \cdot a}$$

The following rules apply:

$$a^n a^m = a^{n+m}$$

$$(ab)^n = a^n b^n$$

$$a^{-n} = \frac{1}{a^n} \qquad \text{and} \qquad \frac{1}{a^{-n}} = \frac{1}{1/a^n} = a^n$$

$$(a^n)^m = a^{nm} \qquad \text{from which} \qquad (a^n)^{1/2} = a^{n/2} = \sqrt{a^n}$$

### H. Equations

Suppose we wish to solve the following equation for $x$:

$$5x + 7x^2 = 3(2 - x)$$

By use of the rule for parentheses, this becomes

$$5x + 7x^2 = 6 - 3x$$

Using the rules given below, we can rewrite this as

$$7x^2 + 8x - 6 = 0$$

This equation is of the form $ax^2 + bx + c = 0$ and can be solved for $x$ using the *quadratic formula:*

$$x = \frac{-b \pm \sqrt{b^2 - 4ac}}{2a}$$

In the present case, this gives $x = -3.32$ and $x = 1.03$. We can use the following rules to simplify equations.

*Rule 1:* Quantities equal to the same quantity are equal to each other. Therefore, if

$$a + b = c \qquad \text{and} \qquad x = c \qquad \text{then} \qquad a + b = x$$

*Rule 2:* Because equal quantities multiplied (or divided) by the same quantity remain equal, we can multiply (or divide) each side of an equation by the same quantity.

$$\frac{a}{b}x = c \qquad \text{becomes} \qquad \frac{\cancel{b}}{\cancel{a}} \cdot \frac{\cancel{a}}{\cancel{b}} x = \frac{b}{a} \cdot c \qquad \text{and so} \qquad x = \frac{bc}{a}$$

*Rule 3:* Cross multiplication can be used as follows. If

$$\frac{a}{b} = \frac{c}{d} \qquad \text{then} \qquad \frac{a}{b}\diagdown\hspace{-1.1em}\diagup\frac{c}{d} \qquad \text{gives} \qquad da = bc$$

This follows directly from Rule 2 if we multiply both sides of the equation by $bd$ and cancel like factors.

*Rule 4:* We can raise both sides of an equation to the same power without invalidating the equality. For example, if $9x^2 = 5$, then we can square each side to give $81x^4 = 25$. Or we can take the square root of each side (that is, raise each to the $\frac{1}{2}$ power) to give $3x = \sqrt{5}$. To prove these facts, use Rules 1 and 2.

*Rule 5:* An isolated quantity can be moved from one side of an equation to the other (transposed), provided we change its sign. Thus

$$x^2 - 5x = 8 \qquad \text{becomes} \qquad x^2 - 5x - 8 = 0$$

This rule simply involves adding or subtracting the same quantity (8 in this case) from both sides of the equation.

## I. Trigonometry

The trigonometric functions are defined in terms of a right triangle such as the one shown. The sine, cosine, and tangent functions are

$$\sin \theta = \frac{\text{opposite side}}{\text{hypotenuse}} = \frac{a}{c} \qquad \cos \theta = \frac{\text{adjacent side}}{\text{hypotenuse}} = \frac{b}{c}$$

$$\tan \theta = \frac{\text{opposite side}}{\text{adjacent side}} = \frac{a}{b} = \frac{\sin \theta}{\cos \theta}$$

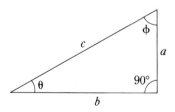

Because the pythagorean theorem states that for such a triangle, $c^2 = a^2 + b^2$, and because $a = c \sin \theta$ and $b = c \cos \theta$, we have that

$$\sin^2 \theta + \cos^2 \theta = 1$$

Two other important relations can be obtained by considering the angle $\phi$ in the figure. Because the sum of the interior angles of a triangle is $180°$, we notice that $\phi = 90° - \theta$. Since

$$\sin \phi = \frac{\text{opposite side}}{\text{hypotenuse}} = \frac{b}{c}$$

and since $b/c$ is $\cos \theta$, we have $\sin \phi = \cos \theta$, or

$$\sin (90° - \theta) = \cos \theta$$

Similarly, we can show that

$$\cos (90° - \theta) = \sin \theta$$

Other trigonometric identities that are sometimes convenient to remember are the following:

$$\sin 2\theta = 2 \sin \theta \cos \theta$$
$$\cos 2\theta = \cos^2 \theta - \sin^2 \theta$$
$$\sin (\theta \pm \alpha) = \sin \theta \cos \alpha \pm \cos \theta \sin \alpha$$
$$\cos (\theta \pm \alpha) = \cos \theta \cos \alpha \pm \sin \theta \sin \alpha$$

Trigonometric addition of vectors is discussed in Sec. 1.5.

### J. Power of ten notation
See Secs. 1.8 and 1.9.

### K. Conversion of units
See Sec. 3.8.

### J. Some useful formulas from geometry
Circumference of a circle $= 2\pi r$
Area of a circle $= \pi r^2$
Surface area of a sphere $= 4\pi r^2$
Volume of a sphere $= \frac{4}{3}\pi r^3$
Area of a rectangle $= ab$
Volume of a right circular cylinder $= (\pi r^2)L$

# Trigonometric functions

| Angle, deg | Sine | Cosine | Tangent | Angle, deg | Sine | Cosine | Tangent | Angle, deg | Sine | Cosine | Tangent |
|---|---|---|---|---|---|---|---|---|---|---|---|
| 0° | 0.000 | 1.000 | 0.000 | 31° | .515 | .857 | .601 | 61° | .875 | .485 | 1.804 |
| 1° | .018 | 1.000 | .018 | 32° | .530 | .848 | .625 | 62° | .883 | .470 | 1.881 |
| 2° | .035 | 0.999 | .035 | 33° | .545 | .839 | .649 | 63° | .891 | .454 | 1.963 |
| 3° | .052 | .999 | .052 | 34° | .559 | .829 | .675 | 64° | .899 | .438 | 2.050 |
| 4° | .070 | .998 | .070 | 35° | .574 | .819 | .700 | 65° | .906 | .423 | 2.145 |
| 5° | .087 | .996 | .088 | | | | | | | | |
| | | | | 36° | .588 | .809 | .727 | 66° | .914 | .407 | 2.246 |
| 6° | .105 | .995 | .105 | 37° | .602 | .799 | .754 | 67° | .921 | .391 | 2.356 |
| 7° | .122 | .993 | .123 | 38° | .616 | .788 | .781 | 68° | .927 | .375 | 2.475 |
| 8° | .139 | .990 | .141 | 39° | .629 | .777 | .810 | 69° | .934 | .358 | 2.605 |
| 9° | .156 | .988 | .158 | 40° | .643 | .766 | .839 | 70° | .940 | .342 | 2.757 |
| 10° | .174 | .985 | .176 | | | | | | | | |
| | | | | 41° | .658 | .755 | .869 | 71° | .946 | .326 | 2.904 |
| 11° | .191 | .982 | .194 | 42° | .669 | .743 | .900 | 72° | .951 | .309 | 3.078 |
| 12° | .208 | .978 | .213 | 43° | .682 | .731 | .933 | 73° | .956 | .292 | 3.271 |
| 13° | .225 | .974 | .231 | 44° | .695 | .719 | .966 | 74° | .961 | .276 | 3.487 |
| 14° | .242 | .970 | .249 | 45° | .707 | .707 | 1.000 | 75° | .966 | .259 | 3.732 |
| 15° | .259 | .966 | .268 | | | | | | | | |
| | | | | 46° | .719 | .695 | 1.036 | 76° | .970 | .242 | 4.011 |
| 16° | .276 | .961 | .287 | 47° | .731 | .682 | 1.072 | 77° | .974 | .225 | 4.331 |
| 17° | .292 | .956 | .306 | 48° | .743 | .669 | 1.111 | 78° | .978 | .208 | 4.705 |
| 18° | .309 | .951 | .325 | 49° | .755 | .656 | 1.150 | 79° | .982 | .191 | 5.145 |
| 19° | .326 | .946 | .344 | 50° | .766 | .643 | 1.192 | 80° | .985 | .174 | 5.671 |
| 20° | .342 | .940 | .364 | | | | | | | | |
| | | | | 51° | .777 | .629 | 1.235 | 81° | .988 | .156 | 6.314 |
| 21° | .358 | .934 | .384 | 52° | .788 | .616 | 1.280 | 82° | .990 | .139 | 7.115 |
| 22° | .375 | .927 | .404 | 53° | .799 | .602 | 1.327 | 83° | .993 | .122 | 8.144 |
| 23° | .391 | .921 | .425 | 54° | .809 | .588 | 1.376 | 84° | .995 | .105 | 9.514 |
| 24° | .407 | .914 | .445 | 55° | .819 | .574 | 1.428 | 85° | .996 | .087 | 11.43 |
| 25° | .423 | .906 | .466 | | | | | | | | |
| | | | | 56° | .829 | .559 | 1.483 | 86° | .998 | .070 | 14.30 |
| 26° | .438 | .899 | .488 | 57° | .839 | .545 | 1.540 | 87° | .999 | .052 | 19.08 |
| 27° | .454 | .891 | .510 | 58° | .848 | .530 | 1.600 | 88° | .999 | .035 | 28.64 |
| 28° | .470 | .883 | .532 | 59° | .857 | .515 | 1.664 | 89° | 1.000 | .018 | 57.29 |
| 29° | .485 | .875 | .554 | 60° | .866 | .500 | 1.732 | 90° | 1.000 | .000 | ∞ |
| 30° | .500 | .866 | .577 | | | | | | | | |

# APPENDIX 5

# Answers to odd-numbered problems

## Chapter 1

**1** 7.20 blocks at 146°
**3** 14.2 cm at 103°
**5** 276 km at 56.4°
**7** 1540 km; 890 km
**9** 38.3 N, 32.1 N; −47.0 N, −17.1 N; 32.1 N, −38.3 N
**11** 367 N at 114° or 367 N at 246°
**13** 43.4 cm at 77.2°
**15** 3.37 N at 331°
**17** 105 paces at 50.3°
**19** 1410 km at 36.4°
**21** 96.6 N at −61.4°
**23** 83.7 m at 44.2°
**25** −7.00 N, +5.00 N
**27** 19.2 cm at −86.8°
**29** (a) $6.04 \times 10^5$, (b) $1.63 \times 10^{-4}$, (c) $4.70 \times 10^{-8}$, (d) $2.95 \times 10^{-9}$, (e) $7.30 \times 10^7$
**31** $4.30 \times 10^{-5}$
**33** $3.21 \times 10^6$
**35** (a) $4.52 \times 10^{-19}$, (b) 246, (c) $3.20 \times 10^{-18}$, (d) 633
**37** $A_x = -1.00$ m, $A_y = +2.00$ m, $A_z = +\frac{4}{3}$
**39** 28.9 N in the $+x, -y, -z$ directions

## Chapter 2

**1** (a) 50.0 N, (b) 20.0 N
**3** 449 N at 249°
**5** 433 N
**7** 728 N, 2130 N
**9** 238 N
**11** 27.5 N, 79.6 N
**13** 1040 N, 671 N
**15** 1410 N, 613 N, 1500 N, 514 N
**17** 261 N; 210 N
**19** 167 N
**21** Zero, −180 N · m, 144 N · m, 364 N · m
**23** $700b/(a + b)$ N
**25** 195 N; 310 N
**27** 351 N
**29** 1240 N, 797 N, 150 N
**31** (a) 2.76 F; (b) −0.444 F, 3.46 F
**33** 56.3°
**35** 61.6 N, 45.7 N
**37** $W(1 + L/2b) \cos \theta$
**39** (a) $\tan \theta = d/h$; (b) $Wgd/2h$

## Chapter 3

**1** $1.59 \times 10^4$ m/s, east
**3** $2.50 \times 10^{-9}$ s
**5** (a) 4.00 cm/s at 37°; (b) 5.60 cm/s
**7** (a) 1.00 cm/s; (b) 0.857 cm/s; (c) −0.400 cm/s; (d) −1.00 cm/s; (e) zero
**9** 40.0 s
**11** Zero; 6.67 m/min, zero, −13.3 m/min, all east
**13** (a) 0.600 cm/s; (b) −0.48 cm/s; (c) 1.30 cm/s
**15** 83.3 m
**17** 0.760 m/s²
**19** −1.79 m/s², 175 m
**21** $-3.21 \times 10^5$ m/s²; $4.67 \times 10^{-4}$ s
**23** 4.62 s, 69.2 m
**25** $-2.28 \times 10^{17}$ m/s², $3.08 \times 10^{-11}$ s
**27** 33.6 s, 4.15 m/s
**29** 80.0 s, 16.0 s
**31** (a) −0.133 m/s²; (b) 1240 m
**33** (a) 15.2 cm; (b) 91.4 cm; (c) 229 cm
**35** (a) 1.39 m/s; (b) $2.61 \times 10^{-4}$ m/s; (c) 11.2 m/s
**37** (a) 8.85 m/s; (b) 0.90 s
**39** 20.4 m; 2.04 s
**41** 11.0 m
**43** 14.1 m
**45** (a) 3.0 m; (b) 2.0 s
**47** 17.2 m/s
**49** 0.988 m; 2.00 m/s, 4.85 m/s
**51** $7.84 \times 10^{-11}$ cm
**53** (a) 28.7 m; (b) 1.66 s
**55** 13.9 m/s
**57** 1.53 m ahead of point
**59** −2.92°

## Chapter 4

**1** 140 N, 24.5 m/s
**3** (a) 2.50 m/s²; (b) 3750 N
**5** 282 N
**7** (a) 490 N; (b) 549 N
**9** 1980 N
**11** 3250 N
**13** (a) 35.0 N; (b) 51.0 N; (c) 28.0 N

## (Chapter 4 continued)

**15** (a) 24.5 N, 11.9 N, 21.2 N; (b) 4.97, 5.05, 0.532 m/s²
**17** 0.869; static
**19** 283 N
**21** 16.4 m
**23** 3.27 m
**25** 112,500 N
**27** (a) 8.70 m/s²; (b) 1.20 m/s²
**29** 36.8 N; 0.542
**31** 29.4 m/s²
**33** (a) 640 N; (b) 340 N
**35** 4.41 m/s²
**37** 0.527
**39** 0.857 m/s²
**41** (a) 3.64 m/s², 10.9 N; (b) 1.68 m/s², 10.9 N
**43** (a) 3.82 m/s², 9.55 N; (b) 1.87 m/s², 12.68 N
**45** 163 N; 0.678 s
**47** $F/2m$; $(F - 2f)/2m$
**49** 0.0526
**51** 0.922 N, 3.00 N, 2.31 m/s², 4.61 m/s²
**53** (a) 102 m; (b) 30.4 m
**55** $(M + m)g/\mu$

## Chapter 5

**1** 20.0 J
**3** 7250 J
**5** 8230 N; $-2.47 \times 10^5$ J
**7** −441 J
**9** 10.0 W
**11** −2.52 J
**13** 42.0 kW; 56.3 hp
**15** 0.443 hp
**17** $2.74 \times 10^{-4}$ hp
**19** 5000 N
**21** 4.55 W
**23** (a) 0.392 MJ; (b) no; (c) 0.0973 hp
**25** $h/3$
**27** (a) $1.84 \times 10^5$ J; (b) 0.0391
**29** 0.995 m
**31** $1.55 \times 10^{-4}$ hp
**33** (a) 2650 N; (b) $2.86 \times 10^{-4}$ s
**35** (a) 7.35 m; (b) 0.440 N; (c) 9.55 m/s
**37** (a) 3.13 m/s; (b) 2.42 m/s
**39** $5.15 \times 10^{-4}$ N

**41** 16.9 km/gal
**43** 5.29 m/s
**45** (*a*) and (*b*) $2\sqrt{Lg/3}$
**47** (*a*) 4.11 N; (*b*) 0.980 m/s
**49** (*a*) 20.0; (*b*) 14.7; (*c*) 73.5 percent
**51** 9.07
**53** 0.664 hp
**55** (*a*) 265 N; (*b*) 0.0265 N
**57** 7.64°
**59** 1.95 m/s

### Chapter 6

**1** (*a*) 20,000 kg · m/s east;
(*b*) 2.70 kg · m/s upward
**3** $m\sqrt{2gh}$ downward
**5** 7500 N
**7** 3250 N
**9** (*a*) 0.300 s; (*b*) 60,000 N
**11** (*a*) $9.09 \times 10^{-11}$ s; (*b*) $4.59 \times 10^{-10}$ N
**13** (*a*) 1.50 ms; (*b*) 0.400 N · s;
(*c*) 267 N
**15** $vM_1/(M_1 + M_2)$
**17** 10.0 m/s to right
**19** 0.100 m/s backward
**21** 9.80 cm
**23** 515 m/s
**25** 0.250 m/s
**27** (*a*) $\sqrt{gh/2}$; (*b*) 0.500
**29** Zero
**31** $v_o/2, 3v_o/2$
**33** $0.0312v_o$
**35** 18.1 m/s
**37** (*a*) $v_x = -30.0$ m/s, $v_y = 20.0$ m/s; (*b*) $-15.0$ m/s, 10.0 m/s
**39** Move in opposite directions with speed $v_o$
**41** $0.800v_o$ and $0.600v_o$
**43** $4.66v_o$ at 185.9°
**45** $3P_a$
**47** Proof
**49** $\sqrt{3gD/8}$
**51** $(VM/L)(V + gt)$

### Chapter 7

**1** (*a*) 0.0750 rev, 0.471 rad;
(*b*) 183°, 0.509 rev; (*c*) 342°, 5.97 rad
**3** 0.0333 rad, 0.00531 rev, 1.91°
**5** 0.105 rad/s, 1.00 rev/min
**7** (*a*) 4.71 rad/s; (*b*) 54.0°
**9** 2200 rev/min²; 3.84 rad/s²
**11** (*a*) 0.00333 rev/s²; (*b*) 0.375 rev
**13** 1.17 rev/s
**15** 236 cm/s
**17** 21.7 rev
**19** 2.00 rad/s²
**21** (*a*) 463 m/s; (*b*) zero

**23** 7.54 m
**25** 6.50 rev/s, 40.8 rad/s, 2340 deg/s
**27** (*a*) 75.0 m; (*b*) 39.8 rev
**29** (*a*) 9.77 rad/s², 175 rev;
(*b*) 55.0 m
**31** 2.25 rad/s²; 71.6 rev
**33** 12,000 N
**35** 1.15
**37** 0.371 rev/s
**39** 15.7 m/s
**41** 176 rev/s
**43** $1.86 \times 10^{-40}$ N; $F/W = 1.14 \times 10^{-14}$
**45** 0.747 as large
**47** About $10^5$ s
**49** $5.97 \times 10^{24}$ kg
**51** 24.8 N
**55** $\sqrt{rg/\tan\theta}$
**57** $0.946\sqrt{gR}$

### Chapter 8

**1** 0.367 J
**3** 3.84 kg · m²; 1.92 kg · m²
**5** $m_ia^2 + m_ob^2$
**7** $I_d + 0.135$ kg · m²
**9** 19.7 J
**11** 0.600 N · m
**13** 0.0105 N · m
**15** (*a*) 15.0 s; (*b*) 4.78 rev
**17** 0.0475 kg · m²
**19** 3.77 rev/s
**21** 0.00688 kg · m²; $T = 0.478$ N
**23** 0.724 rev/s
**25** (*a*) 187 J/s; (*b*) 7.42 N · m
**27** 8.86 rad/s; 6.45 s
**29** (*a*) 1.98 m/s; (*b*) 6.30 rev/s
**31** 1.14 cm
**33** (*a*) 1.50 kg · m²; (*b*) 36.0 rev/s
**35** (*a*) 0.563 rev/s; (*b*) yes
**37** (*a*) $\omega_1 I_1/(I_1 + I_2)$; (*b*) $(\omega_1 I_1 + \omega_2 I_2)/(I_1 + I_2)$; (*c*) $\omega_1(I_1 - I_2)/(I_1 + I_2)$
**39** 0.361 rev/s
**41** (*a*) $\omega_1 - mbv/(I_1 + mb^2)$;
(*b*) $\omega_1 + mbv/(I_1 + mb^2)$

### Chapter 9

**1** 2.42 g/cm³
**3** 161 kg
**5** 1770 kg/m³
**7** $3.69 \times 10^{11}$ N/m²
**9** 4570 N/m²
**11** (*a*) $-2.70 \times 10^{-6}$;
(*b*) $5.60 \times 10^7$ Pa
**13** 107 m/s²
**15** 283 N
**17** 102 kPa; nearly equal
**19** $2.35 \times 10^4$ N
**21** 459 Pa
**23** 9.10 cm

**25** 500,000 N
**27** 5.50 N
**29** (*a*) 0.870 cm³; (*b*) 2.87 g/cm³
**31** 6.47 g/cm³
**33** 965 kg/m³
**35** 0.895
**37** 21.8 g
**39** 121 g
**41** 4
**43** 0.200 cm³/s
**45** 0.0133 N
**47** (*a*) 45.7 kPa; (*b*) 30.6 kPa
**49** 400 m/s
**51** 3.68 cm/s
**53** (*a*) $4.11 \times 10^4 b^3$ N; (*b*) $4.62 \times 10^6\, b^2$ m/s, 0.0443 cm;
(*c*) $9.53 \times 10^{-5}$ m
**57** (*a*) and (*b*) $-2860 - 80(r_1/r_2)^2$ Pa
**59** 339 N toward water source

### Chapter 10

**1** 9.69 m
**3** 6000 N; 306
**5** 77.47 cmHg(103 kPa)
**7** (*a*) 20.0°C, 293 K;
(*b*) $-34.6$°F, 236 K;
(*c*) 1110°F, 597°C
**9** $-40.0$°
**11** $3.27 \times 10^{-25}$ kg
**13** $2.37 \times 10^{23}$
**15** (*a*) $1.66 \times 10^{-23}$ kg;
(*b*) $6.02 \times 10^{19}$; (*c*) $6.62 \times 10^{19}$
**17** $3.01 \times 10^{28}$ kg/kmol
**19** 17.1 g
**21** 279 kg
**23** 590 kPa
**25** $2.71V_o$
**27** 307°C
**29** $-29.0$°C
**31** 21.4 cm; 142 kPa
**33** 0.374 cm
**35** 1.02 MPa
**37** $5.89 \times 10^5$ m/s
**39** (*a*) $1.01 \times 10^4$ K;
(*b*) $1.41 \times 10^5$ K
**41** 0.586 kg/m³
**43** (*a*) $5.47 \times 10^{-13}$ kmol;
(*b*) $3.29 \times 10^8$
**45** $10^{21}\, m_ov$; 0.0729 Pa
**47** $6720/(1 + 12n)$ m/s, where $n = 0, 1, 2, \ldots$
**49** (*a*) 5.42 kg; (*b*) 732 kPa

### Chapter 11

**1** 29.0 kcal; 121 kJ
**3** 2.18 kJ
**5** 45.0 kcal; 188 kJ
**7** 22.0 kcal; 92.0 kJ
**9** 16.3°C
**11** 2.04 g

**13** 1.95 g
**15** 28.2°C
**17** 7.01 g
**19** 0.575 cal/g · C°
**21** $6.74 \times 10^{-20}$ J; 12.4
**23** $2.32 \times 10^4$ s
**25** 43.0 C°
**27** 0.00800 J/s; 9.56 C°
**29** J/kg · C°
**31** (a) 125 C°; (b) 120 C°
**33** $3.75 \times 10^{-4}$
**35** 0.625 cm
**37** $1.25 \times 10^{-3}$
**39** 0.00900 cm$^3$
**41** $3.42 \times 10^5$ N
**47** $4.97 \times 10^6$ J
**49** $1.30 \times 10^5$ J
**51** 1.06 MJ
**53** 0.0267 W/K · m
**55** 7580 W/m
**59** 2.87 m$^2$ · K/W (16.3 practical units)
**61** (a) 61.7 percent; (b) 0.0106 g
**63** 0.0191 C°
**65** 0.0452 C°

### Chapter 12

**1** 6690 J
**3** 3.79 C°
**5** 421 kJ
**7** (a) Zero; (b) −75.0 J
**9** (a) 2350 J, zero; (b) 2350 J, 1570 J
**11** 720 J/kg · K
**13** 250 J
**15** 105 J
**17** (a) 45 J; (b) none
**19** 7730 J
**21** −150°C
**23** 0.0498
**25** (a) 8; (b) 0.125; (c) 0.375
**27** (a) 8; (b) 32; (c) $1.13 \times 10^{15}$
**29** −10.0 J/K
**31** −0.348 J/K · s
**33** 4.01 J/K
**35** (a) Zero; (b) $1.91 \times 10^{-23}$ J/K; (c) $2.47 \times 10^{-23}$ J/K
**37** 0.596
**39** −75.7°C
**41** 39.6 hp
**43** (a) 2.67 GJ; (b) 1.87 GJ; (c) same as (a) and (b)
**45** (a) 2650 J; (b) 3.74
**47** (a) 1395 J; (b) 0.0312 hp
**49** (a) 7.85 J; (b) −7.85 J; (c) 11.1 J; (d) 18.9 J
**51** (a) $gD/c$; (b) $mgD$; (c) $mgD/T_o$; (d) zero

### Chapter 13

**1** (a) 0.181 Hz; (b) 5.52 s; (c) 8.00 cm

**3** 1.46 mJ
**5** (a) 0.495 m/s; (b) 7.00 m/s$^2$
**7** (a) 13.3 m/s$^2$; (b) 0.730 m/s
**9** (a) 0, 75.0 m/s$^2$; (b) 4.74 m/s, 0; (c) 4.11 m/s, 37.5 m/s$^2$
**11** (a) 0.731 m/s, zero; (b) zero, 2.54 m/s$^2$; (c) 0.618 m/s, 1.36 m/s$^2$
**13** (a) 0.0318 Hz; (b) 0.800 m/s$^2$; (c) 4.00 m/s
**15** 1.29 Hz
**17** (a) $2.27 \times 10^4$ N/m; (b) 3.02 cm
**19** (a) 20.0 cm; (b) 0.477 Hz; (c) 2.09 s; (d) 3.60 N/m
**21** (a) 0.248 m; (b) 0.0414 m
**25** (a) 9.80 m/s$^2$; (b) 1.17 Hz
**27** 10.1 cm
**29** Exactly the same
**31** (a) 4.00 cm; (b) 0.500 cm; (c) 5.00 Hz; (d) 0.200 s
**33** 250 m
**35** 27.2 g
**37**

| → | 13 | 15 |
|---|----|----|
| 11 | 12 | ← |

**39** (a) 74.0 cm; (b) 88.8 m/s
**41** (a) 86.7 cm; (b) 433 m/s
**43** 400, 600, 800 Hz
**45** 80.0 Hz
**47** 352, 704, 1056 Hz
**49** (a) 1.25 kHz; (b) 0.625 kHz
**51** 4.80 m/s
**53** 2520 m/s
**57** $\sqrt{g/2\pi^2 L}$

### Chapter 14

**1** 2720 m
**3** 0.850 m; 2.27 cm
**5** 337 m
**7** $2.86 \times 10^{10}$ Pa
**9** 1.07 percent
**11** 7.50 W/m$^2$; 0.200 percent
**13** 64.8 dB
**15** (a) $1.00 \times 10^{-8}$ W/m$^2$; (b) $1.20 \times 10^{-10}$ W
**17** 36
**19** $2.00 \times 10^{-10}$ W/m$^2$
**21** $9.07 \times 10^{-6}$ cal
**23** 48.6, 97.1, 146 Hz
**25** (a) $x = \pm n(8.50$ cm$)$; (b) $x = 4.25$ cm $\pm n(8.50$ cm$)$
**27** $0.150 + 0.300n$  m,  $n = 0$, 1, . . . , 8
**29** $f_o \pm 0.900$ Hz
**31** 104, 311, 518 Hz; 207, 415, 622 Hz
**33** 2130 Hz
**35** $0.0654n$ Hz, $n = 1, 2, . . .$
**37** (a) 75.0 Hz; (b) one end
**39** 16.2 m/s

**41** $0.420 f_a$
**43** $f(v_w + v_s)/(v_w - v_s)$
**45** 3.03 Hz
**47** 2140 Hz

### Chapter 15

**1** 0.211 N, attraction
**3** 5.04 N in $+x$ direction
**5** 4.23 μC and 1.97 μC
**7** 0.0648 m/s$^2$
**9** $8.29 \times 10^{19}$ N
**11** (a) 2.00 N; (b) −0.300 N
**13** $x = 5.97$ m
**15** (a) $x = 0.397$ m; (b) −1.56 μC
**17** 1.05 N at 68.5°
**19** 0.0140 N at 105°
**21** $1.07 \times 10^{-7}$ C
**23** 1.50 μC and 3.00 μC, repulsive
**25** $2.67 \times 10^5$ N/C westward
**27** $7.03 \times 10^{14}$ m/s$^2$ in $-x$ direction
**29** $-4.00 \times 10^{-8}$ C
**31** $7.03 \times 10^4$ N/C inward
**33** (a) Zero; (b) $4.29 \times 10^5$ N/C in $-y$ direction
**35** $x = -83.5$ cm
**37** $1.07 \times 10^5$ N/C in $-x$ direction

### Chapter 16

**1** $6.00 \times 10^{-5}$ J; $-6.00 \times 10^{-5}$ J
**3** (a) 3100 V/m; (b) $4.96 \times 10^{-16}$ N
**5** (a) 233 V/m; (b) −233 V/m
**7** (a) $1.52 \times 10^{-8}$ s; (b) $y = -6.07$ cm
**9** (a) 0.510 g; (b) 0.0100 N
**11** $9.3 \times 10^4$ m/s
**13** 45.5 V; lower
**15** 1.70 V
**17** (a) $9.92 \times 10^{-13}$ J; (b) $1.73 \times 10^7$ m/s; (c) $3.1 \times 10^6$ V
**19** (a) 20.0 eV; (b) 140 eV
**21** $4.0 \times 10^6$ m/s
**23** (a) $3.84 \times 10^6$ V; (b) 3.84 MeV
**25** (a) $-3.21 \times 10^4$ V; (b) $3.28 \times 10^5$ V
**27** $3.39 \times 10^5$ V
**29** $2.40 \times 10^{-5}$ C; 9.6 mJ
**31** (a) 1.41 m$^2$; (b) 0.353 m$^2$
**33** (a) 48.0 nC; (b) 13.0 V; (c) 23.3 nJ; (d) 23.3 nJ
**35** $5.31 \times 10^{-8}$ C
**37** 3.21
**39** $(1/2\pi) \sqrt{(mg + qE)/mL}$

### Chapter 17

**1** (a) 4320 C; (b) $2.70 \times 10^{22}$
**3** 86,400 C
**5** 62.0 Ω

**7** $0.178\ \Omega$

**9** $924$ m

**11** $0.563\ \Omega$

**13** $7.41$ percent

**15** $2.73$ and $27.3\ \Omega$

**17** (*a*) $0.500$ A; (*b*) $240\ \Omega$

**19** (*a*) $18.0$ kJ; (*b*) $0.00500$ kWh

**21** (*a*) $24.0\ \Omega$; (*b*) $1.27\ \Omega$

**23** $6.00\ \Omega$

**25** (*a*) $13.1\ \Omega$; (*b*) $1.14$ A

**27** (*a*) $18.0\ \Omega$; (*b*) $8.40\ \Omega$; (*c*) $0.445$ A

**29** (*a*) $8.20\ \Omega$; (*b*) $1.46$ A; (*c*) zero

**33** $3.36\ \mu$F

**35** $1.11$ A, $0.500$ A, $-1.61$ A

**37** $-4.14$ A, $0.857$ A, $-6.43$ V; $12.4$ V

**39** $I_1 = I = 0$; $42.0\ \mu$C

**41** (*a*) $1.50\ \Omega$; (*b*) $28.0$ V

**43** (*a*) $1.83$ A; (*b*) $8.04$V

**45** $13.5$ A

**47** (*a*) $12.0$ V; (*b*) $36.0\ \mu$C, $48.0\ \mu$C

**49** $3.33\ \Omega$

**51** $7.85$ A

### Chapter 18

**1** $0.128$ N upward

**3** (*a*) Zero; (*b*) $3.90 \times 10^{-4}$ N

**7** $0.136$ N in $-z$ direction; (*b*) $0.0788$ N in $+z$ direction; (*c*) $0.158$ N at $-30°$

**9** (*a*) $3.36 \times 10^{-16}$ N in $-z$; (*b*) $3.36 \times 10^{-16}$ N in $-y$; (*c*) zero

**11** (*a*) $1.23 \times 10^{-16}$ N in $+z$; (*b*) $3.38 \times 10^{-16}$ N in $+z$; (*c*) $3.6 \times 10^{-16}$ N at $110°$

**13** Circle with $r = 14.6$ cm

**15** $14.4$ cm

**17** Negative; $2.43 \times 10^7$ m/s

**19** Helix; $r = 1.42$ cm, pitch $= 15.5$ cm

**21** (*a*) $4.00 \times 10^4$ m/s; (*b*) $4.55 \times 10^{-6}$ m

**23** (*a*) $5.63 \times 10^{-5}$ T; (*b*) $9.38 \times 10^{-5}$ T

**25** $3.14$ mT

**27** $1250$ A

**29** $11.2 \times 10^{-4}$ T

**31** (*a*) $0.788$ N $\cdot$ m; (*b*) zero; (*c*) $0.740$ N $\cdot$ m; (*d*) $11.3$ A $\cdot$ m$^2$

**33** $1.52$ A

**35** Shunt $R = 0.0668\ \Omega$

**37** Series $R = 2920\ \Omega$

**39**

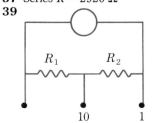

$R_1 = 0.0801\ \Omega$; $R_2 = 0.728\ \Omega$

**41** $qBR/2m$

### Chapter 19

**1** (*a*) $0.600$ mWb; (*b*) zero; (*c*) $0.544$ mWb

**3** (*a*) $2.36A$ mWb; (*b*) zero; (*c*) $1.52A$ mWb ($A$ in square meters)

**5** (*a*) $0.308$ mV; (*b*) zero; (*c*) counterclockwise

**7** (*a*) and (*b*) zero

**9** $3.73 \times 10^{-5}$ V

**11** (*a*) $0.120$ mV; (*b*) clockwise

**13** $0.425$ V

**15** (*a*) $0.094Na^2$; (*b*) $28.3Na^2$ V (*a* in meters)

**17** (*a*) $0.240$ V; (*b*) $6.00$ mH

**19** (*a*) $0.0625$ J; (*b*) $0.280$ T

**23** (*a*) $0.600$ V; (*b*) south; (*c*) not easily

**25** (*a*) $0.799\ Bva$; (*b*) $0.799Bva/R$; (*c*) clockwise; (*d*) $0.638B^2a^2v/R$; (*e*) stop

**27** $0.944\ mgR/B^2a^2$

**29** $2.11 \sin (754t)$

**31** $0.0189$ T

**33** (*a*) $108$ V; (*b*) $30.0$ A

**35** (*a*) $15.0$; (*b*) $1800$ V

**37** (*a*) $0.984$; (*b*) $0.111$

### Chapter 20

**1** $6.00 \times 10^6\ \Omega$

**3** (*a*) $2.40\ \mu$A, $0$ C; (*b*) $0.883\ \mu$A, $22.8$ C

**5** (*a*) $8.00$ s; (*b*) about 8 s; (*c*) $16.0\ \mu$C; (*d*) $2.0\ \mu$A

**7** $0.354$ A, $156$V; $27.5$ W

**9** $7.17$ A; $16.7\ \Omega$; $24.7$ kcal

**11** $375$ W

**13** $0.136$ A; $136$ A

**15** (*a*) Zero; (*b*) $0.317$ A; (*c*) $1.19$ mC

**17** (*a*) $1.13\ \Omega$; (*b*) $11.3 \times 10^3\ \Omega$

**19** (*a*) $63.7$ Hz; (*b*) $0.167$ A

**21** (*a*) $5.00$ A; (*b*) $5.00$ A; (*c*) $500$ W

**23** $1.59$ kHz

**25** $0.592$ A; $8.76$ W

**27** $0.0429$ A; $0.129\ \mu$F

**29** (*a*) $35.2\ \mu$F; (*b*) $2.35$ H

**31** (*a*) $84.8$ mA; (*b*) $7.58$ mA

**33** (*a*) $0.818$ A; (*b*) $20.1$ W, $0.245$

**35** $91.9$ mH; $180$ W

**37** $332$ W

**39** $5.56\ \mu$F

### Chapter 21

**1** $5.00 \times 10^6$ m

**3** (*a*) $5.37 \times 10^{-4}$ s; (*b*) $537$

**5** $8.82$ s

**7** $1.46$ nF

**9** (*a*) $5.20$ V/m; (*b*) yes

**11** (*a*) $2.34 \times 10^{-12}$ T; (*b*) no; (*c*) out of page

**13** $3.33 \times 10^{-14}$ T

**15** $0.240$ mV; along field line; $2.67 \times 10^{-12}$ T

**17** (*a*) $955$ MHz; (*b*) $3 \times 10^{-11}$ T; $9.17 \times 10^{-5}$ V

**19** (*a*) $8.33 \times 10^{-8}$ m; (*b*) ultraviolet

**21** $224$ V/m; $7.48 \times 10^{-7}$ T

**23** $E = 0.150 \sin (7.54 \times 10^6 t)$ mV/m; $B = 5.00 \times 10^{-13} \sin (7.54 \times 10^6\ t)$ T

**25** (*a*) $1.77 \times 10^{-8}$ W/m$^2$; (*b*) $3.65$ mV/m; $1.22 \times 10^{-11}$ T

**27** $0.398$ W/m$^2$

### Chapter 22

**1** $2.56$ s

**3** (*a*) $2.26 \times 10^8$ m/s; (*b*) $443$ nm; (*c*) $5.09 \times 10^{14}$ Hz

**5** $120$ cm

**7** $30.0°$

**11** (*a*) $10.7$ cm, $0.167$ cm, real, inverted; (*b*) $16.0$ cm, $0.500$ cm, real, inverted; (*c*) $24.0$ cm, $1.00$ cm, real, inverted; (*d*) $-24.0$ cm, $2.00$ cm, virtual, upright

**13** (*a*) $p = 300$ cm; (*b*) no; (*c*) same

**15** (*a*) $-12.0$ cm, $0.400$ cm, virtual, erect; (*b*) $-9.38$ cm, $0.750$ cm, virtual, erect; (*c*) $-5.22$ cm, $1.30$ cm, virtual, erect

**17** (*a*) $p = 30.0$ cm; (*b*) no

**19** (*a*) $p = 2f$; (*b*) $\frac{1}{3}$

**21** (*a*) $1.92 \times 10^8$ m/s; (*b*) $406$ nm; (*c*) $15.7°$

**23** (*a*) $30.8°$; (*b*) $0°$

**25** $53.1°$

**27** $48.8°$

**29** $58.5°$

**31** (*a*) $130$ cm; (*b*) $-217$ cm

**33** (*a*) $85.7$ cm, $1.43$ cm, real, inverted; (*b*) $175$ cm, $5.0$ cm, real, inverted; (*c*) $-33.3$ cm, $3.33$ cm, virtual, erect

**35** (*a*) $-27.7$ cm, $0.615$ cm, virtual, erect; (*b*) $-24.0$ cm, $0.800$ cm, virtual, erect; (*c*) $-13.3$ cm, $1.33$ cm, virtual, erect

**37** (*a*) $p = 20.0$ cm; (*b*) $0.500$

**39** $p = |2f|$

**43** (*a*) $-120$ cm from right-hand lens; (*b*) $2.00$; (*c*) virtual, inverted

**45** (*a*) $-16.5$ cm from 2d lens; (*b*) $0.600$

**47** Left from mirror: $20.0$ cm (virtual), $20.0$ cm (virtual), $24.0$ cm (real)

## Chapter 23

**1** 3.21 mm
**3** $-110$ cm
**5** (*a*) Farsighted; (*b*) 66.7 cm converging
**7** f.p. = infinity; n.p. = 87.5 cm
**9** 133 cm
**13** 0.42 cm behind
**15** (*a*) $-3.75$ diopters; (*b*) 0.267 m
**17** (*a*) 3.50; (*b*) 2.50
**19** 4.6
**21** 28 to 33
**23** (*a*) 8.57 cm; (*b*) 11.7
**25** 1.08 min
**27** 200 km
**29** 220 cm in front of eye; virtual, erect; 1.36
**31** $400/\gamma$
**33** $f_e = 1.48$ cm; $f_o = 88.5$ cm
**35** $0.193°$
**37** (*a*) $53.3°$; (*b*) $46.3°$

## Chapter 24

**1** (*a*) $-80.0$, $-160$, $-240$ cm; (*b*) $-40.0$, $-120$, $-200$ cm
**3** (*a*) 18.0 cm; (*b*) 27.0 cm
**5** (*a*) $x = 2.00 \pm 0.250n$ m, $n = 0, 1, \ldots, 8$; (*b*) $x = 2.00 \pm (2n+1)(0.125$ m$)$, $n = 0, 1, \ldots, 7$
**7** (*a*) $7.14 \times 10^{-6}$ m; (*b*) $7.0°$
**9** (*a*) 1.56 m; (*b*) 0.78 m
**11** $4.00 \times 10^{-4}$ m
**13** (*a*) 138 nm; (*b*) 688 nm
**15** (*a*) 2460 nm; (*b*) 1850 nm
**17** 629 nm
**19** $0.116°$ and $0.233°$
**21** 2.38 mm
**23** (*a*) $1.35 \times 10^{-6}$ m; (*b*) $69.9°$
**25** (*a*) $3.26°$; (*b*) $7.26°$
**27** $23.6°$
**29** $n\lambda = d \sin \theta_n$
**31** 6.02 cm
**33** $1.55 \times 10^{-5}$ m
**35** 6150 m
**37** $0.0817n$ nm, $n = 1, 2, \ldots$

## Chapter 25

**1** Displacement = 0
**3** 33 beats/min
**5** (*a*) 820 days; (*b*) $2.40 \times 10^3$ days
**7** $2.85 \times 10^8$ m/s

**9** 15.0 yr; $0.9911c$
**11** (*a*) $0.866c$; (*b*) 36.0 m; (*c*) $3.10 \times 10^{-6}$ s; (*d*) $6.19 \times 10^{-6}$ s
**13** $0.866c$
**15** (*a*) $1 + 5 \times 10^{-7}$; (*b*) 1.005; (*c*) 1.34; (*d*) 3.91
**17** $8.71 \times 10^{-30}$ kg
**19** 1.039
**21** $3.72 \times 10^{-10}$ percent
**23** $0.988c$
**25** (*a*) 0.113 Hz; (*b*) $7.46 \times 10^{-35}$ J, $4.66 \times 10^{-16}$ eV
**27** $3.14 \times 10^{-19}$ J; 1.96 eV
**29** 2.88 eV
**31** (*a*) $4.96 \times 10^{-9}$ eV; (*b*) 41.8 m/s
**33** $6.76 \times 10^5$ m/s
**35** 163 nm; ultraviolet
**37** 91.2 nm
**39** 2.33 eV
**41** (*a*) $1.07 \times 10^{-26}$ kg · m/s; (*b*) $4.43 \times 10^{-3}$ factor smaller
**43** (*a*) $9.70 \times 10^{-6}$; (*b*) $9.70 \times 10^{-3}$
**45** (*a*) $1.60 \times 10^{15}$; (*b*) $1.05 \times 10^{-27}$ kg · m/s; (*c*) $1.68 \times 10^{-12}$ N; (*d*) $3.35 \times 10^{-12}$ N
**47** $3.21 \times 10^{-13}$ m
**49** $(1/n)$ $4.00 \times 10^{-15}$ m; $1.66 \times 10^{-19}$ kg · m/s; 51.4 MeV
**51** $2b/(n+\frac{1}{2})$; $(n+\frac{1}{2})^2 h^2 / 8b^2 m$; $9.4 (n+\frac{1}{2})^2$ eV
**53** $1.06 \times 10^{-20}$ kg · m/s; 417 keV

## Chapter 26

**1** 50.0 m
**3** (*a*) $3.64 \times 10^{-26}$ $r^{-1}$ J; (*b*) $4.73 \times 10^{-14}$ m
**5** $2.18 \times 10^6$ m/s; $0.00729c$
**7** (*b*) $6.46 \times 10^{-13}$ m
**9** (*a*) $-122/n^2$ eV; (*b*) 122 eV
**11** $2.54 \times 10^{74}$
**13** (*a*) 93.8 nm; (*b*) 389 nm
**15** $-122$, $-30.6$, $-13.6$, $-7.65$ eV
**17** 122 nm
**19** 10.2, 12.1, 12.8, 13.1 eV
**21** 4.11 eV
**23** (*a*) 37.4 eV; (*b*) $3.63 \times 10^6$ m/s
**25** $3.14 \times 10^{-14}$ m; 833 keV
**27** (*a*) 0.602 m/s; (*b*) $1.00 \times 10^{-9}$

**29** 18
**31** 10
**33** H, He, C, P
**35** 0.0496 nm
**37** (*a*) 400 J; (*b*) $382$ C°
**39** (*a*) 6000 V; (*b*) 0.27 nm
**41** $1.50 \times 10^5$ W
**43** $600(1 \pm 2.0 \times 10^{-15})$ nm

## Chapter 27

**1** (*a*) 6; (*b*) 6; (*c*) $2.75 \times 10^{-15}$ m; (*d*) $2.29 \times 10^{17}$ kg/m³
**3** 193 m
**5** 0.534 m
**7** $4.33 \times 10^{-27}$ kg = 2.60 u
**9** 39.1 kg/kmol
**11** 99.30 and 0.70
**13** 0.78 u; 0.93 percent
**15** 7.68 MeV
**17** 15.995 u
**19** 39.3 counts/min
**21** (*a*) $2.74 \times 10^{-8}$ s$^{-1}$; (*b*) $3.45 \times 10^5$
**23** $1.89 \times 10^{-8}$ Ci
**25** $1.23 \times 10^4$ Bq
**27** 0.327
**29** $(\lambda_1 N_1 + \lambda_2 N_2)/(N_1 + N_2)$
**31** $^{206}_{82}$Pb; $^{209}_{83}$Bi
**33** $2.64 \times 10^{-6}$
**37** 4.28 MeV
**39** No. Because 3.00 MeV is required
**41** (*a*) 0.103 MeV; (*b*) 209.9837 u
**43** $3.19 \times 10^9$ Bq
**45** $2.09 \times 10^{-11}$ g
**47** $2.15 \times 10^{-6}$ C°
**49** $2.42 \times 10^4$ yr
**51** 46.5 yr
**53** Zero; $1.98 \times 10^7$ m/s
**55** 190 MeV
**57** 7710 kg
**59** $5.93 \times 10^{-4}$ g/s and $8.87 \times 10^{-4}$ g/s

## Chapter 28

**1** $6.20 \times 10^{-18}$ m, $6.20 \times 10^{-18}$ m; $3.57 \times 10^{-25}$ kg, $3.56 \times 10^{-25}$ kg
**3** (*a*) Zero; (*b*) 45.0 u
**5** 1.88 GeV; $6.60 \times 10^{-16}$ m
**7** $4.34 \times 10^4$ Hz
**9** $22.9n^2$ GeV
**11** 4100 nm

# Index

### *Credits for chapter-opening photographs*

**Introduction.** AIP/Niels Bohr Library.  **1.** S. Goldblatt/Photo Researchers.
**2.** E. Erwitt/Magnum.  **3.** Zambelli Internationale.  **4.** S. Caron.  **5.** S. Caron.  **6.** NASA.
**7.** T. Pix/Peter Arnold.  **8.** S. Caron.  **9.** General Motors.  **10.** J. Azel/Contact Press
Images.  **11.** S. Caron.  **12.** Stock Market.  **13.** F. Grunzweig/Photo Researchers.
**14.** T. Orban/Sygma.  **15.** A. Gesar.  **16.** Ransburg Corporation.  **17.** Amtrak.
**18.** D. C. Heath/Education Development Center.  **19.** Studer Revox America.
**20.** ConEdison.  **21.** S. Caron.  **22.** C. Purcell.  **23.** P. Berger/Photo Researchers.
**24.** J. Hogle/Scripps.  **25.** J. Ruskin.  **26.** D. Hailer-Hamann/Peter Arnold.
**27.** P. Koch/Photo Researchers.  **28.** Harvard College Observatory.